CW00799073

Making Noise

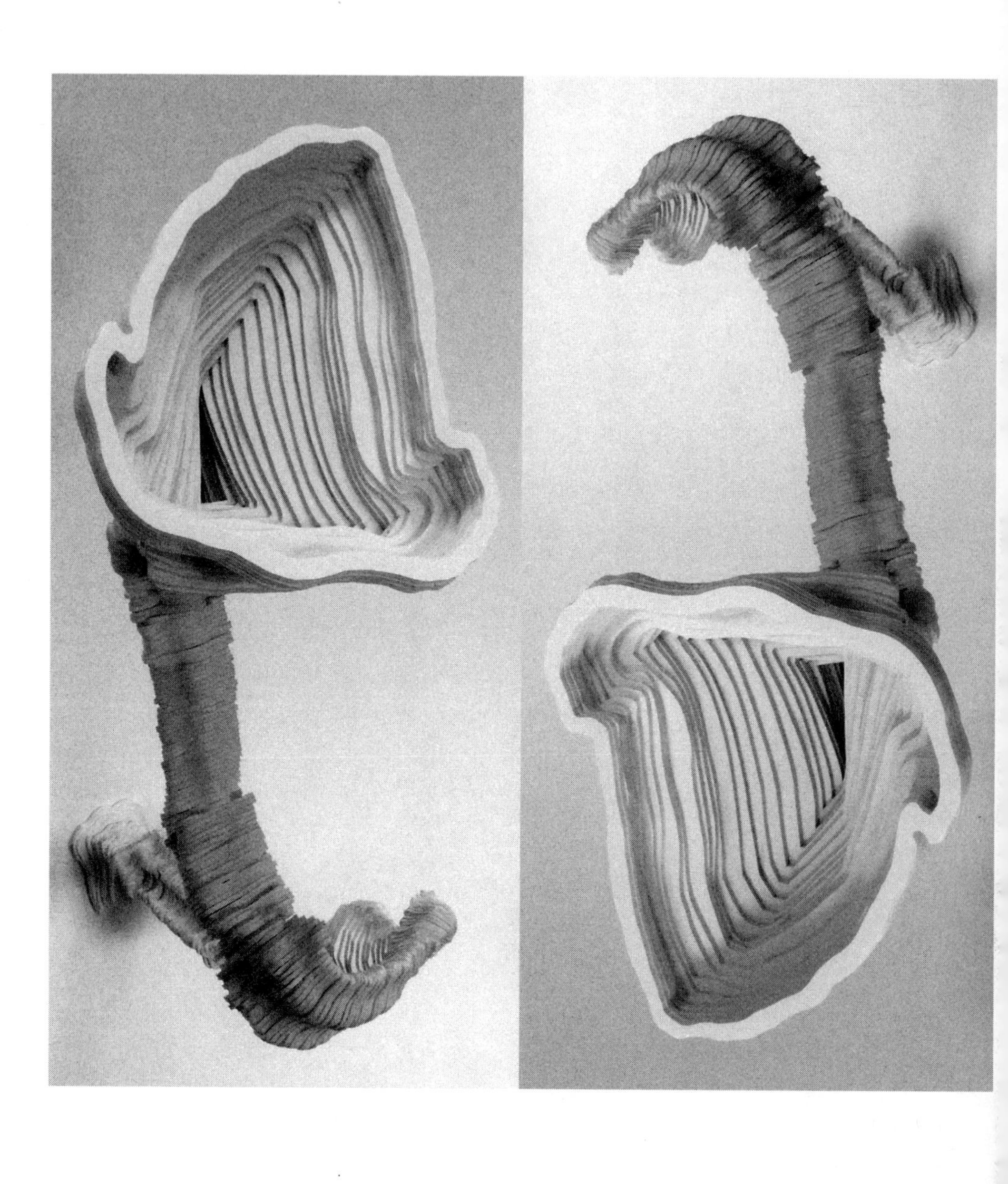

Making Noise

From Babel to the Big Bang & Beyond

Hillel Schwartz

ZONE BOOKS · NEW YORK

2016

ZONE BOOKS
633 Vanderbilt Street
Brooklyn, NY 11218

Second printing 2016.

Printed in the United States of America.

Distributed by The MIT Press,
Cambridge, Massachusetts, and London, England

Library of Congress Cataloging-in-Publication Data

Schwartz, Hillel, 1948–
Making noise : from Babel to the big bang & beyond / Hillel Schwartz.
p. cm.
Includes bibliographical references and index.
ISBN 978-1-935408-12-3 (alk. paper)
1. Noise pollution—History. 2. Noise—Psychological aspects. 3. Noise in literature. 4. Sounds—Psychological aspects. 5. Sound—Social aspects. 6. Sound—History. I. Title.

TD892.S398 2011
155.9'115–DC23

2011014433

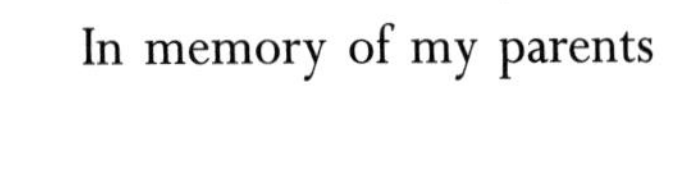
In memory of my parents

Note to Reader

This book is meant to be read aloud.

Now sound in actuality
is always of something
and against something
and in something.

—Aristotle, *De Anima*

The Volume

ROUND THREE: EVERYHOW 551

On hearing what had not been heard, could not be heard, should not be heard. Calibrating and recalibrating noise. Toward what end?

Consonances

To all those within hearing

For transportation and conversation,
Snippets and clippings and aural support,
Theory and music and recalibration,
I thank the following three score and more:

Donna Sasso, Anne Prussing, and June A. Sekera,
Ramona Naddaff and Micki McGee,
Stefan Helmreich and Stefan Tanaka,
John Marino and David G. Roskies.

Joel Best and Kenneth S. Barkin,
Eliza Slavet and Anthony Burr,
Harold Levine and Michel Saint-Germain,
Ben Howell Davis and Michael B. Miller.

Caryn Davidson and Jordan Schwartz,
Mark M. Smith and Robert Piserchio,
Heidi Piltz and Aura Satz,
Daniel and Phyllis Levun-Agostino.

Douglas Kahn and Penelope Gouk,
Ed White, Bill Rable, and Ronald Gervais,
Don Winkler, Alex Magoun, and Carl T. Chew,
Ann Elwood and Aline G. Hornaday.

Alan Hawk, Michael Rhode, and Jorge Machado,
Les Blomberg and Valerie Gibson,
Robert Poe and Harold Boll,
Linda M. Strauss and Michael Schudson.

John E. Talbott and Adrienne Aron,
Karin Bijsterveld and David Kunzle,
Bernard Hibbitts, Greg Leifel, and Henry Allen,
Charles Brashear and Meighan Gale.

Mark Hineline and Ani Patel,
Veit Erlmann, Eda Warren, and Michael Dobry,
Tom Amos, Mark Dresser, and Ellen Gartrell,
and all the Interlibrary Loan Librarians at UCSD.

I am also obliged to all those who consented to interviews/conversations, without rhyme if with some reason, during my research: Rachel Arfa, Leo Beranek, Robert T. Beyer, Ann E. Bowles, Arline Bronzaft, George Bugliarello, Peter Donnelly, Fred Fisher, John Fried, Diamanda Galás, Michael Gerver, Aram Glorig, Earl Gosswiller, Dean Greenberg, Fred Hunt, Bob Kenney, Jack Larkin, The Mudguards, Stephan Munroe, Steve Norton, Bob Ostertag, Allan F. Ryan, Hans Schmid, Fred Spiess, Don Swanson, Billie Thompson, and Christine Vandemoortele.

Librarians, curators, and archivists at 245 institutions and associations in twenty-five states and ten countries have been consistently and ingeniously helpful in my pursuit of the noise rarely indexed but often hidden in glass cases, file cabinets, and boxes of ephemera. Even as research turns digital, I thank them for affording me the chance to lay my hands on unusual materials and to put my ear to nineteenth-century stethoscopes and ear trumpets.

For travel subsidies and occasions for communing with other scholars, I am grateful to the John W. Hartman Center for Sales, Advertising and Marketing History in the Special Collection Library, Duke University; to the National Endowment for the Humanities for participation in a 1994 Summer Seminar led by Joel Best ("Social Problems: The Constructionist Stance"); to the Goethe Institute and De Balie of Amsterdam for a pre-sentation at "Soundscapes (be)for(e) 2000"; to the directors of the 2002 Rotterdam Film Festival; to the Wenner-Gren Foundation for participation in Veit Erlmann's symposium on "Hearing Cultures," April 24–28, 2002, Oaxaca, Mexico; and to the Department of Music of the University of Texas at Austin for participation in Veit Erlmann's conference on "Thinking Hearing," October 2–4, 2009.

Hillel Schwartz
Encinitas, California

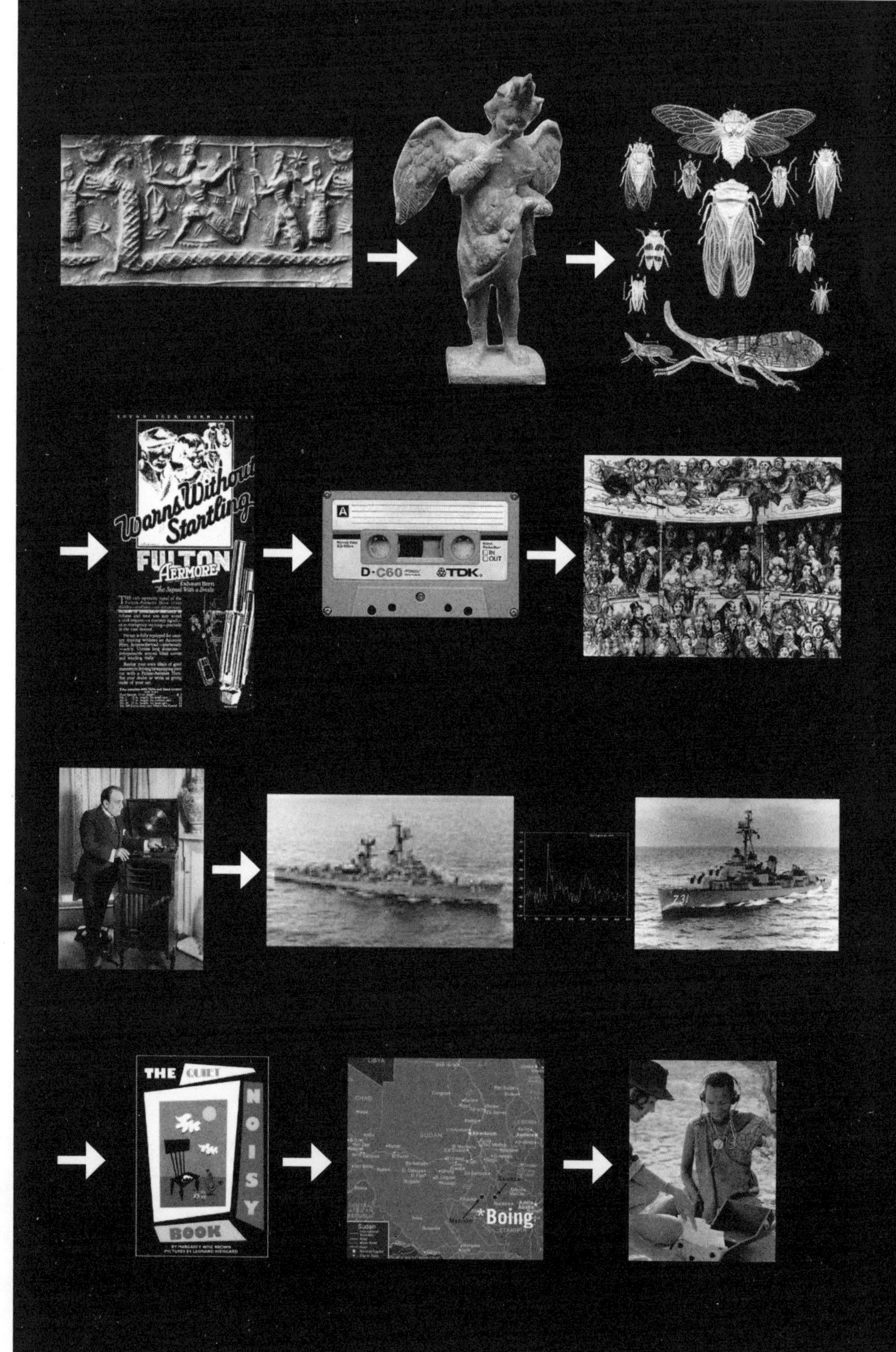

SOUNDPLATE 1

BANG (a beginning)

On hearing out noise. Origins auricular and oracular, mythic and metaphoric. Hardness of hearing.

It is not possible to begin quietly. Ancient Vedic traditions grant us this world through the boom and quake of a Great Breath, a round of vibrations informing all thought, substance, motion, and prayer. Tablet I of the Babylonian Genesis finds Apsû, begetter of the great gods, already enraged by the first generations running riot, their clamor insufferable. Apsû complains to his wife Tiâmat,

> Their way has become painful to me,
> By day I cannot rest, by night I cannot sleep;
> I will destroy them and put an end to their way,
> That silence be established, and then let us sleep!

Tiâmat, she who has given birth to them all, cries out,

> Why should we destroy that which we ourselves have brought forth?
> Their way is indeed very painful, but let us take it goodnaturedly!

The divine vizier Mummu sides rather with stern father than tolerant mother, and things look bleak for the rambunctious younger gods. Getting wind of Apsû's plan, their ringleader Ea devises a subtle incantation to give Apsû the rest he desires. Sound asleep, Apsû is slain by Ea. Forty lines later, Ea's own son starts to make waves; four-eared, four-eyed, fire-breathing Marduk agitates even Tiâmat and drives all the gods into a fury.[1]

Mad dogs, storm demons, and war dragons stalk the world throughout Tablet II. Despite the tumult or dictated by it, humankind is created to dig canals and clear channels, work earlier performed by the gods in their less ebullient or belligerent moments. Or so goes the account in the old Akkadian poem *Atrahasis*, where the Mesopotamian gods themselves, turn and turn about, soon want to call an end to beings as raucous and randy as we are mortal:

> Six hundred years, less than six hundred, had passed
> And the country became too wide, the people too numerous.

The country was as noisy as a bellowing bull.
The gods grew restless at their clamor,
Ellil had to listen to their noise.
He addressed the great gods,
"The noise of mankind has become too much,
I am losing sleep over their racket.
Give the order that *šuruppu*-disease shall break out,
Let Namtar put an end to their noise straight away!
Let sickness: headache, *šuruppu*, *ašakku*,
Blow in to them like a storm."

The gods consent. Humankind suffers pounding headaches, epidemic, years of famine, and a Flood roaring like a bull, the clouds bellowing, the winds howling "like a wild ass screaming."[2]

That flood sweeps over the mountaintops of Genesis 6–8, returning the world to the surge and chaos of *tehom*, "the deeps," of Genesis 1, to whom Tiâmat was cognate. Washed clean, the biblical earth ought henceforth to be serene, a place of warm relations among Noah's offspring and the disembarking animals. But alas! The first lines of Genesis 9 exalt Noah and his sons as "the terror and the dread of all the wild beasts"; the last lines send Noah's son Ham cursed and reeling into slavery. By Genesis 11:9 a ziggurat in the land of Shinar has become a tower of babble on an earth re-peopled but baffled and estranged.[3]

Hermes, said the Greeks, had been the bringer of language and baffler of tongues—Hermes with his winged heels and helm of invisibility, instant messenger and trickster who had given Apollo his lyre, Pan his pipe, and Odysseus the herb to undo Circe's spell. Charmer of sleep and shaper of dreams, the figure of Hermes had been lifted from Egyptian depictions of Thoth, ibis-headed god of wisdom, of night, and of writing whose silent, totemic birds were mummified and stacked by the thousands in the galleries of lunar temples, a counterweight to the solar precincts of hawk-headed Horus who appeared on rings, amulets, and stone reliefs as a naked child with his index finger at (or in) his mouth: the rising sun. (Question: "Why doth a child put his finger into his mouth when he cometh first into the world?" Answer, in *The Problems of Aristotle*: "Because that coming out of the womb, he cometh out of a hot bath; and therefore entering into the cold, putteth his finger to his mouth for want of heat.") Misreading these images, the Greeks took *Har-pa-khrat*, Horus the child, for the god of silence and secrecy, Harpocrates, a finger posted at his lips. This comported well, if incidentally, with the popular *Maxims* of Ani, a twenty-first-dynasty scribe who cautions that "The Sanctuary of God abhorreth noisy demonstrations. Pray thou with a loving heart, and let thy words be hidden." It comported well, and less incidentally, with occult Hellenistic writings ascribed to a magus, Hermes the Thrice-Great, and his Egyptian disciple Asclepius, who advises that the Greeks not be given access to the ancient mysteries lest they botch those too and "the extravagant, flaccid, and (as it were) dandified Greek idiom extinguish something stately and concise," that is, the energetic idiom of Egypt. Enamored of empty talk, Greek philosophy was "an inane foolosophy of speeches. We, by contrast," Asclepius tells Ammon Pharaoh, "use not speeches but sounds that are full of action." Let Harpocrates, bastard of foolosophy, admonish the Greeks toward a silence that a later Harpo, also silent and given to sucking fingers, would expand and explode.[4]

When Phaedrus seduces Socrates into the countryside in order to read aloud to him a lovely speech on love made by Lysias, Socrates wants nothing to do with texts. He wants discourse, the expansion and explosion of meaning. On this hot day, Socrates the inveterate urbanite will indulge for a moment in the shade and fragrance of a plane tree, but what he attends to is "the summery, sweet song of the cicadas' chorus." Such sweetness may seem bizarre to moderns who compare the thoracic contractions of cicadas to the shrill of knife-grinders, but the *Iliad* itself referred to the sweet "lily voice of the cicadas." As the classicist W. B. Stanford has suggested, Homer could count on his auditors appreciating an analogy from the "lustrous brightness" of a single lily, its form distinct, its shimmer consistent, to the clarity, purity, intensity, and constancy of cicada song, unmistakable at long distances. (And capable, entomologists now tell us, of being transmitted at 100 db by virtue of swift muscular compressions of a pair of domed tymbals on either side of the abdomen, buckling the cicada's elastic, expansive ribs with sound-pressures akin to the explosion of a grenade.) From the cicadas, Socrates draws an admonition neither to rapture nor silence nor sleep—but to talk. Rather than nodding off in the sweet warmth of cicada drone, he and Phaedrus should be "steadfastly navigating around them as if they were the Sirens." For, as the story goes, "the cicadas used to be human beings who lived before the birth of the Muses. When the Muses were born and song was created for the first time, some of the people were so overwhelmed with the pleasure of singing that they forgot to eat or drink; so they died without even realizing it." To this day cicadas burst into song at birth and continue singing without food or drink "until it is time for them to die," after which they report to the Muses the (dis)honors done by each human being. In the sustained sonic presence of the cicadas, Phaedrus hence must watch his words, especially words about love. (Among non-social insects, as Darwin would write, the sole rationale for sound-producing organs must be sexual, and he singled out the "marital summons" or siren call of Brazilian cicadas, an abdominal drumming he could hear a quarter-mile offshore.) In an aural cosmos, the alert philosopher learns to put all eloquence to the test: Is it merely euphonious? Is it persuasive because well-spoken? Or is it, however shrill, true? Cicadas, as Asclepius taught, are not just chorus and chitin; they are fully present and sempiternal as the truth, collaborators in discourse. There was once, wrote Asclepius, a god-favored cithara player who had a string break during a contest of music and song; the Mighty One, listening in, caused a cicada "to light upon the cithara and restore the song by keeping the space where the string had been."[5]

Myths of creation and re-creation are sonic, whether cicada and cithara or boom, babble, brouhaha. A Mayan text, the *Popol Vuh*, begins with everything "in suspense, all calm, in silence," but when the Forefathers talk through the making of our world, the first forces to emerge are Caculhá Huracán, Chipi-Caculhá, and Raxa-Caculhá, figures (and word-sounds) of those mightiest surges of thunder and flashes of lightning we call hurricane. The noise of beginnings. Out of primitive stillness, out of darkness, something disrupts the silence—a Great Breath, boisterous youth, thunder and lightning, a sudden "Let there be!"... or laughter. In the Hellenistic *Hidden Sacred Book of Moses Called "Eighth" or "Holy"*, an Egyptian Jew describes the seven laughs of God,

Hha Hha Hha Hha Hha Hha Hha

which engender the seven Fore-Appearers, who get everything moving.[6]

Nor have we, in our newest versions of creation, abandoned sound. Our paleochemists and climatologists return us to thunderstorm as bullroarer, liberating oxygen in proto-atmospheres as bolts of lightning enliven primordial seas, enchain amino acids, twist protein into strands of DNA. Before thunder, before planets and suns, before light itself, we have the astrophysicists' Big Bang. A figure of speech meant sixty years ago in mockery, the Big Bang has become the popular emblem of a prevailing cosmology, putting a loud explosion at the start of time and the heart of matter. And technically, the Big Bang is about sound, arising out of a need to account for the background */@%#! of the universe.[7]

From beginning to end, this is a book about noise, and therefore a book about creation and loss. What disrupts the initial stillness is at first and always noise; having disrupted the stillness, noise is forever conditioned by nostalgic yearnings for a calm that is no more, pleadings for a peace to be restored, oracles of a harmony yet to come. "Our speech," wrote the German poet Novalis two centuries ago, "was at first far more musical, but it has gradually become prosaic and lost its note; it is now noise or a 'loudness'; it must become song again." Note the Romantic speed of his shift from the nostalgic to the oracular; note too his careful separation of "noise or a 'loudness,'" which are not synonymous. Crinkle of candy wrappers or yipping of coyotes, low-frequency hum or high-energy hoopla, noise is never so much a question of the intensity of sound as of the intensity of relationships: between deep past, past, and present, imagined or experienced; between one generation and the next, gods or mortals; between country and city, urb and suburb; between one class and another; between the sexes; between Neanderthals and other humans.[8]

As a register of the intensity of relationships, noise has a fourfold history. First, the chronicle of changing soundscapes: how each era and culture lives within its own ambience of sounds. Next, the annals of sounds earmarked as pleasant or obnoxious: how each era, culture, and rank hears (or does not hear) and welcomes or disdains the sounds around it. Next, the career of noise itself as variously apprehended: how each era, culture, occupation or discipline reconstitutes the notion and nature of noise. Contingent upon these, finally, are narratives of noisemaking and noisebreaking: how noise in each era, culture, and class has been denounced or defended, defiantly produced or determinedly deadened.

Four reciprocal histories follow in the sound-shadow: the history of elected or commanded silence; the history of the deaf and deafness; the history of Arcadian idylls and millennial kingdoms; the history of stillness — of portraiture and death, sedation and paralysis, inner reserve and outward desolation. I cannot do justice to all of these histories, nor to the smoke that hovers as a perpetual nuisance. I shall sound them each for clues to an extraordinary event: how, over the last six-score years, noise in the West has been signally transformed from an exclusively aural experience to a root metaphor about our world, our lives, and the meaning of our lives abroad in the world.

1. Ahem! — Clearing the Throat

Masking the history of noise are ostensibly larger stories of civilization, urbanization, industrialization, mass distribution, and mass communications. With each of these *-tions* come contrary presumptions about the widening presence or progressive absence of noise. Civilization rises above savage drumming and barbarian wailing to produce elegant music, poetry, and discourse. *Or*, civilization confines hearty enthusiasms and restricts the natural body in favor of the artifices of medicine, theater, and etiquette, which make people increasingly sensitive to, and inordinately disturbed by, sounds hitherto negligible. Urbanization promotes concerns with safety and sociability that lead to nuisance laws and campaigns against disorderly conduct. *Or*, urbanization crowds people into densities of thick traffic, thin walls, and relentless pandemonium. The mass production and distribution of goods makes readily available those products and materials that protect shop and home from the hubbub beyond. *Or*, industrialization and mass advertisement shatter the tranquility of the countryside, the silence of the skies, the quiet of kitchen and parlor, and the health and hearing of the worker at the stove, in the office, in the field, or at the factory. Revolutions in communications overcome the aural ravages of urbanization and industrialization by restoring protected lines of sound and by storing the speech and music that would otherwise be lost in the din. *Or*, these revolutions thrust upon us ever-more strident beepers, distressing sirens, potent amplifiers, inescapable background music and intrusive public address systems.

Anticipated by myth and sustained by ideology, such contraries send noise oscillating ever-more feverishly between creation and loss. Yet these contraries have not been irreconcilable; historically, they have often been reciprocal. The first automobiles, for example, were welcomed in the cities as soon as they had rubber tires, with the expectation that they would be quieter than horses clattering and cartwheels rumbling over cobblestone or concrete—a problem since Roman times.[9] In fact, horseless carriages were quieter . . . so quiet that pedestrians, equestrians, coachmen, and aldermen demanded that the newfangled vehicles be furnished with horns to alert the unsuspecting of their unusually swift, silent approach. For the next decades, until engines became far more powerful and the volume of traffic much greater, the loudest part of the automobile, aside from occasional backfires, was its horn, not its motor, gears, brakes, or the thud of its tires. Civility, urban safety measures, and the mass production of a device exemplary of a revolution in signaling—the klaxon or electrical horn—were the forces behind one of the most insistent, abrasive new sounds of the twentieth century. Turn and turn about, that selfsame horn has inspired many an Anti-Noisite.[10]

Studies of anti-noise campaigns are scarce.[11] More common are polemics urging us either to act against a mounting cacophony or to applaud noise as ground and guidepost to political, artistic, or cultural transfiguration. Aroused by the peals of resurgent environmentalism or the quavers of postmodernism, such polemics quickly shift from an ill-defined past to an unrefined present.[12]

An increasing number of studies, increasingly nuanced, have answered the call of a new historical, and ecological, interest in the senses themselves, in the ear and audition.[13] These, like works in biology, neurology, and psycho-acoustics, often posit an invariable physiology: the sounds people hear may change, and their reactions to those sounds do change, but how people hear remains the same.[14] Our hearing, however, is not so historically untroubled. Epidemic diseases with ototoxic side-effects (scarlet fever, measles, typhus) render the people of one era and region more hard-of-hearing than the next, as do the techniques and drugs used to treat diseases (notably quinine, highly ototoxic and widely prescribed for 400 years). Changes in fashions, recreation, housing, manners, and military codes make cultures and orders differently accustomed to modes of hearing (through wigs or helmets), differently sensitive to sound itself (especially at night), and differently aware of certain tones (families with butlers—how often do they hear a knock at the door?). Classes may have physiologically different responses to sounds depending on levels of fatigue, nutrition, and the long-term damage done to hearing by such occupations as artilleryman, blacksmith, boilermaker, carpenter, dentist, drummer, foundryman, goosegirl, hunter, ironworker, millworker, motorcycle cop, quarryman, stock trader.

To be sure, I do not aspire here to a history of sound, which would be a history of all living things and then some: volcano, quake, hail. Neither do I aspire to a history of music with noise as countermeasure, nor to a history of technologies hounded by noise.[15] I shall refer to them, of course, as I shall refer to the history and anthropology of the senses, where noise runs a gamut from the Natural through the Sociable back around to the Unsociable and Ill-Natured.[16] I shall attend to the history of acoustics, particularly of architectural acoustics, where formulae for achieving vividness, depth, clarity, or quiet have run up against one wall after another.[17] I shall trespass on aeronautics, aesthetics, anatomy, astronomy, burglary, dentistry, industrial engineering, marine biology, neurology, nursing, oceanography, penology, psychiatry, warfare, and worship, for "unwanted sound" resonates across fields, subject everywhere and everywhen to debate, contest, reversal, repetition: to history.

Sa'adia ben Joseph, *gaon* or leading sage of the Jews of Babylonia at a time of bitter division among Jews and contentious debate among Muslims, Jews, Manichaeans, and competing sects of Christians, embarked on a history of Jewish philosophy in the hopes that a rational account of "philosophic doctrines and religious beliefs" would help the perplexed of AD 933 (AM 4693–4694) toward a *method* of discriminating true from false. He explained his own method not in terms of syllogisms or split hairs but of sound and hearing. One must treat each set of contrary or confusing statements as "a complex of noises" and one by one proceed to eliminate all sounds that fail to establish the meaning, starting for example with "1) the concussion of bodies, such as the falling of stone on stone, and 2) the cleaving of certain bodies, and 3) sounds like that of thunder and crashing and similar noises, for [the philosopher] knows that from these types of sound he could not derive any proof." Next, strike out "neighing and braying and lowing and the like, since these are no less unintelligible." (Which mulish or sheepish theologians did he have in mind?) Next, cut away all human but inarticulate vowel sounds such as sighing and all unconnected (hummed or hissed) consonants. Next, delete all detached nouns, for as sounds in isolation they are not subject to contradiction or confirmation, until at last one arrives at audible, intelligible claims that are either necessary, possible, improbable, or impossible. So a person who speculates upon the things of Creation, or upon the necessity of the Most Holy as inferred from that

Creation, like a person who hears a world of sounds and hopes to make out a world of meaning, begins with things all mixed up, and sorts them out.[18]

Here, too, listening will be as much a process of reasoning as an act of audience. I myself come to noise by way of the logic of encores. Thirty years ago, I was accustomed to buying blank audiocassettes that had reassurances printed in red, front and back, of LOW NOISE. As an historian of technology, a poet, and a teacher of improvisational dance, I was struck by that LOW NOISE. What did it mean? I wrote out some exploratory thoughts and, in lieu of submitting a paper for publication, recorded a talk entitled "Low Noise" on a LOW NOISE tape to be played in my absence at a seminar. The implication of LOW NOISE, it seemed to me, was not merely that the tapes would be relatively free from hiss and splutter, but that such relative freedom was good and that we had a right to expect just-about-perfect recording and replay. This further suggested that modern audiences must be unique, since only with the advent of phonographs and player pianos could anyone expect to hear a human performance exactly repeated.[19] The musical and theatrical encore, a European phenomenon dating back to the late seventeenth century, must itself have had a vastly different import than it does now, for an audience moved by a musical passage, an aria, a ballet sequence, or a dramatic scene would have had no assurance of witnessing it again, or so well executed.[20]

Encores, then, were far more than tenders of popularity or vehicles for praise. Despite the appearance in certain venues of claques paid to applaud a play or an actress, the request for an encore would not usually have been, as today, *pro forma*, nor was it, as usual today, a request for the performance of *another* piece. An encore was the request for a *reprise in the middle of the action, or* what we call an instant replay. I misspeak: an encore was no "request." Compounded of immediate personal enthusiasm and of cultural despair over impending loss, shouts of *encore! bis!* or *ancora!* were anxious demands. By way of repetition, one might fix in one's mind an entrancing sound, an elemental vision, a tragic moment.

Wouldn't you listen, wouldn't you hear differently as audience to singer or preacher, scientific lecturer or virtuoso pianist, if, even as you arrived in hopes of spiritual exaltation, unbelievable beauty, invaluable information, astonishing invention, or sublime execution, you knew that all this would dissolve on your departure into an incomplete or irretrievable memory? Listening intently, wouldn't you too demand encores while demanding of yourself, your children, your schools, your museums, and your secretaries the practice of those arts of memory that afford or assist comprehensive recall?

Turn and turn about, what would happen as you became accustomed to replay? Would you begin to expect live performances to be equally and exactly repeatable as were phonograph disks, piano rolls, and reels of film? During the first decade after the appearance of phonographs and player pianos, we find audiences at live performances demanding four, nine, twelve successive encores of a favorite aria, their clap-happy applause tyrannical and disruptive, denying all operatic forward motion until they had satisfied a technologically whetted appetite for replay. Were they testing performers scientifically to make sure that the beauty of their tone or precision of their notes was not a freak instance, a singular rise to an occasion, or were they so habituated to replaying their favorite piano rolls and phonograph disks that it seemed only natural to have their favorite singers repeat a favorite aria? At last, after the tenor Enrico Caruso himself had been exhausted by a New York audience's demands for a legendary nineteen encores of the same aria, one right after the other, the Metropolitan Opera Company abolished encores, period.

Or would you begin to troll symphony halls and theaters for precisely those characteristics that distinguish live performance from recording—passion which betrays itself with mistakes in notes or lines or footwork, an inspired improvisation, witty repartee between an actor and hecklers in the second balcony? Within a generation after the appearance of the phonograph and motion pictures, the infinite variations of ragtime and jazz spread across the country as counter-motif to the industrial regularities of records and film. Live music concerts today insist upon (sometimes lip-synched, or preprogrammed) spontaneity, a distinctive wildness that sets them apart from CDs and DVDs.

In *The Culture of the Copy*, I explored the repercussions of our modern hunger for, and our skill at, reproduction and reenactment. Noise is what makes perfect repeatability humanly impossible and, in information theory, inhumanly problematic. But as a "sound in-between," noise also makes significance possible and, like nineteenth-century applause that interrupted a performance in hot pursuit of encores, the role of noise in information theory is as ambivalent as that of redundancy, which at once ensures reception and reduces significance.[21]

A book on noise is thus encore and twin (sometimes evil twin, sometimes companionable, sometimes vanishing) to a book on likenesses and facsimiles. As encore, a history of noise must resume our infatuation with instant replay and our lust for copies-more-perfect-than-originals in order to account for our attitudes toward interruption, incompletion, and imperfection. As evil twin, noise in this book will sound inchoate, obtrusive, indelicate, obstreperous, nerve-wracking. As companion, noise will holler in the vanguard of protest, progress, advertisement, celebration, discovery. As vanishing twin, noise will be the statistical limit on the intelligibility of messages and the no-longer-so-theoretical force behind fluctuating values in the futures market.

I've just slipped out from the binary shackles of noise as good or bad, and away from the all-too-common belittling of noise as mere epiphenomenon, a fleeting byproduct. Evil, companionate, or vanishing, noise inhabits an historical space that is at least triune, and in any guise may be appreciated as an active historical agent, be it provocateur, saboteur, ally, or poltergeist.

2. "Noise Spokes"—The Tonkin Gulf

Consider: In April 1962, the U.S. Seventh Fleet initiated destroyer patrols off the Chinese mainland, surveillance which was extended in 1963 to the coast of North Vietnam. In January 1964, President Lyndon B. Johnson approved intensified surveillance and closer approaches to the coast, and on July 30–31 the Navy began the bombardment of North Vietnamese shore installations. On August 2, 1964, North Vietnamese patrol boats in the Gulf of Tonkin shot at an American destroyer, the *Maddox*, which had been mauled by kamikaze pilots in 1944, reported sunk during the Korean War, and was now monitoring naval traffic and communications. The United States issued a stiff warning to Hanoi about the rounds of fire. On the night of August 4, sonarman Third Class David Mallow, two years in the navy, one year aboard the destroyer *Maddox*, heard something disturbing in the Gulf. He alerted the bridge, where Captain Herrick, listening to Mallow over the intercom, could hear in the background the sounds which the young sonarman called "noise spokes" but which Herrick took to be an incoming torpedo. (Herrick, and his second in command, Lieutenant Commander Buehler, and the Junior Officer on Deck were all having problems with their hearing: their eardrums had been ruptured by the loud noise of the five-inch guns the *Maddox* had used during the attack two days previous.) Over the next two hours, Mallow reported "noise spokes" on average of two or three a minute while the *Maddox* and its consort, the *Turner Joy*, veered to elude one torpedo after another. Radarmen on both destroyers heard the distinct pips of enemy PT boats and initiated the firing of 300 rounds against the "skunks," two of which were sunk by automatic gunfire keyed to radar lock-ins. The Pentagon was notified, and then the President, who ordered retaliatory air strikes on PT boat bases in an operation code-named Pierce Arrow. On August 7, by Congressional resolution, the Secretary of Defense and President were broadly empowered to use further force against the Hanoi regime. Ever since, critics of American military escalation in Vietnam have accused the Secretary of Defense, Robert McNamara, and the Johnson administration of fabricating the August 4th battle out of thin air as a pretext for expanding the war—given that, aside from recordings of "noise spokes" and radar blips, there turned out to be no evidence that PT boats had ever attacked or twenty-six torpedoes ever been launched, and no one ever found signs of wreckage from two sunk "skunks."[22]

The battle in the Tonkin Gulf was fabricated out of thick air, not thin, and was less the spawn of militarism or imperialism than, at first, of noise.

The night of August 4th was stormy and moonless; the incident took place in pitch dark except for distant flashes of lightning, and the ships themselves were totally darkened. In calmer seas and clearer skies, the Gulf was already well-known to radarmen for its "Tonkin spooks," blips that were ghosts and pips that were echoes of ghosts, furiously appearing and disappearing, signifying nothing. In the sonar room of the *Maddox*, David Mallow had apparently not been told about something his more experienced shipmate, Patrick Park, had recently determined: that one of the rudder motors of the *Maddox* made an underwater noise audible on sonar each time the ship turned. This noise, and the noise of the ship's propellers, were the "noise spokes" Mallow reported 178 times, originally as the ship pursued its sinuous course around the Gulf, then each time the speeding *Maddox* swerved to avoid the putative torpedoes. In effect, the ship with more than eighty full-rudder course changes was chasing itself.

Aboard the *Turner Joy*, sonarmen that night picked up no suspicious sounds or "noise spokes," in part because the peculiar (rudder) noise was localized beneath the *Maddox* and obscured by the beat of both ships' propellers, in part because the two ships were trying hard to keep their distance from each other so as to avoid collision in the "freakish" aural and visual conditions of the Gulf. Churning through the waters at 32 knots, the *Turner Joy* was producing turbulence that quenched most returning echoes and outside noises. At this high speed, almost double that to which they were accustomed and 5 knots faster than the maximum speed at which they had been trained, the sonar operators could not differentiate self-noise from external sounds and did not realize that both ships were creating sonic dead zones. They did know that PT boats, capable of 50 knots, could hide in the wakes of destroyers, and they were on edge after alerts from the *Maddox* about incoming torpedoes, as were radarmen who reinterpreted the recurring "Tonkin spooks" as enemy craft. In the clouds and fog above the Gulf, jets sent to scout the waters failed to locate any North Vietnamese boats, but this the pilots attributed to the bad weather and low ceilings. Hearing panic and confusion on the destroyers' radio frequencies, they were convinced that a battle was underway, a battle that lasted about 150 minutes across 85 miles of water.

Drawn up six months earlier, the Tonkin Gulf Resolution took neither its motive nor its language but its public cue from five kinds of sounds commonly described as noise: atmospheric noise in the Gulf, due to local cloud layers and topography; mechanical noise from a ship's rudder and propellers; electrical noise, or static, on the intercom over which Captain Herrick heard the "noise spokes" in the background; visual noise or "spooks" on radar, due to weather conditions and the ducting of electromagnetic waves; the vocal noise of panicky, shouting, cursing men as heard over the radio by the flyers above. (Not to speak of the otological noises of tinnitus commonly associated with the ruptured eardrums of the three men in command of the *Maddox*, and likely some of the gunners.) Separately and together, these noises were mistaken for the sound of battle and the passage of the Resolution taken for the equivalent of a declaration of war.

Just as noise is what we make of certain sounds, the meanings we assign to noise are no less consequential than the meanings we assign to other sounds. Noise may be unwanted or incomprehensible sound; it is never insignificant sound. Noise impinges on our ambitions and our crimes, our weddings and divorce courts; it impinges on our customs of birth and burial, our senses of publicity and retirement, our practices of peace and war, our allegories of apocalypse and paradise. In each of its forms, be it biological, vocal, mechanical, electrical, electromagnetic, geological, atmospheric, cosmic, literary, or spiritual, noise is a player on the historical scene.

If from time to time I ascribe independent agency to noise, I hope to catch myself at paragraph's end and return us to the people, hard-pressed, middling or wealthy, young or middling or centenarian, urban or rural or suburban, military or civilian, who give meaning to noise.

3. SHHH . . . bang! From the Age of Four

Noise personified is a trickster. We cannot fix noise to a single meaning, nor can we completely rein it in. By definition noise is what breaks through and becomes obvious, sometimes painfully obvious. Which reminds me: I have said that I come to noise by way of encore. That was a scholarly evasion. The home truth is that I come to noise by way of a children's book by Margaret Wise Brown. Not her classic *Goodnight, Moon*, which ends with "Goodnight noises everywhere," and not her almost-unknown *SHHH . . . bang! A Whispering Book* (written, she said, to get even with all prim librarians); rather, *The Quiet Noisy Book*. Four years old, I took that book to bed and to heart, listening intently as my parents read and reread it, night after night. It was a story about what breaks through. As usual in her series of seven Noisy books, a little dog much like her own becomes aware of a sound he cannot quite make sense of. In the case of *The Quiet Noisy Book*, it was "A very quiet noise." What was it? "Was it a bee wondering?" No. "Was it butter melting?" . . . No. Muffin, "the little dog who heard everything," had been asleep all night. What had awakened him? "Was it a skyscraper scraping the sky?" No. Then what was it? "It was a very quiet noise. Such a quiet noise. . . . Quiet as someone whispering a secret to a baby." What was it? . . . "Muffin knew what it was! It was"—turn the page—"the sun coming up." It was "the milkman whispering to his horse" and "the flies opening their million-cornered eyes," and . . . and. It was, of course it was, and each night falling asleep I knew on waking that it would be, the new day.[23]

Such wonderful reassurance: each evening silence and loss; each dawn noise and creation. First the noise, then the light. When a tree falls in the forest and no one hears, does it make a sound? Yes. Far away, someone catches wind of something unidentifiable, a very quiet noise, or feels a tremor underfoot, like "an elephant tiptoeing down the stairs," and the fall of the tree resounds. Am I alluding to the devastation of tropical rainforests and other outrage thousands of kilometers away? Am I thinking of the long-term effects of the tremendous crack of an ice shelf as it heaves away from Antarctica? Or am I working toward a parable for the doing of history, which is about listening over one's shoulder, and for the making of a history of noise, which is all about the attenuation of sound over time?

Listening to books read aloud by my parents and to the deep booming voice of the Great Gildersleeve on the radio, I have had since childhood a readiness to credit the slightest sound and least noise with powers of imminent domain, of making things likely, taking things over. This readiness, personal and mayhap generational, shapes a history of noise begun upon the premise that noise may be unwanted or unintelligible but never insignificant. This same readiness leaves me open to the metaphorical bearings of noise as something other than acoustic. Each Round in this book bears reference to the annual chanted perambulation of a community's borderstones (the cultural geography of noise), to the daily tour of patients in hospital (the psychophysiology of noise), to the circling of the seasons, festivals, and analog hours (the historical temporalities of noise), and to the *caccia* or rounds in which musical voices, ostensibly discordant, chase each other toward a final breathless unison.

4. Boing

We are all hard of hearing. Unassisted, we cannot hear the high frequencies employed by bottlenose dolphins or the low frequencies enjoyed by manatees and blue whales. Unassisted, we cannot hear most of the internal sounds our own bodies make, like the flux of lymph or the flow of cerebrospinal fluid; rarely do we realize that our ears themselves may broadcast. Men are in/sensitive to different frequencies than women, especially in older age, and each of us, female or male, young or old, is psychologically idiosyncratic with regard to what we hear as noise, physiologically idiosyncratic with regard to the volume and duration of sound our ears can take. We do not attend equally well to all that we are each capable of hearing; daily we dispute among ourselves over what we do manage to hear, arguing about loudness, softness, nearness, tunefulness. Some of us are deaf to the intricacies of birdsong, some to the pitches of a hot-rod's engine. Without training and sound recorders, few of us would detect the subtleties in the susurrus of crickets and katydids (or the sweetness in a chorus of cicadas), but every parent knows every innuendo of an infant's cries, the weakest of which can penetrate the parent's deepest sleep.[24]

There are few cultural constants in the realm of sound. The screech of chalk or fingernails on a slate blackboard may be one. Just thinking of that screech raises my hackles, and the neurological reaction to this sound, or gride, could be universal. Why, no one rightly knows.[25] Was that the noise of a nightmare beast, a prehistoric predator? Around the world, onomatopoetics for animal noises are diverse, related more to the syllabic fund and tonal qualities of local speech than to the anatomy of dog, monkey, eagle, or bee. "Was it a fish breathing?"

Breath and heartbeat produce sounds everywhere familiar, but some cultures have conflated breath with soul, others have made a science of breaths classified by affliction or sentiment. The heart itself has been only irregularly bound up with love or life-force. We may agree about generic sounds denoting in-breath, out-breath, and heartbeat; we will not agree about the connotations for inspiration, compulsion, heartiness, compassion. Such disagreement has by no means been resolved within the medical world, where chest sounds heard through stethoscopes are to this day debatable. "Was it a mouse sighing?"[26]

Bells might be another constant. Their ringing seems to have functioned in much the same way since the Bronze Age: shamanic or priestly contact with spirits, weapon against demons, warning of earthly dangers, announcement of strangers or friends, signal of festivity or fire. The correlation of a specific bell sound with a specific meaning or function has however been local, not global. As objects, bells (or anciently, rattles and drums) may be broadly recognizable, but the sounds they make, pure tone or subharmonic, share no consistent meanings. "It was an alarm clock about to ring." Or was it?

For half a millennium in the West, longer in India and China, philosophers and linguists have sought in vain for sounds or syllables whose meanings have been so clear and constant across the continents that one might locate at least the remnants of an *Ur*-language, root of all other tongues spoken by humanity. The search has been half-mythic, half-mystical, for the ultimate mission has been the recovery of a language so integral to the human psyche and the world itself that each word would vibrate in essential sympathy with its referent: to speak such words would be instantly to enact their meaning. No confusion, no obfuscation, no circumlocution. Speaking that *Ur*-language, people before Babel had to have been of one mind about each sound. No longer.[27]

Well, if no fully and mutually comprehensible *Ur*-language lies within our grasp, what of *Ur*-ears? Unable to fix universal baselines for the significance of the sounds we hear, surely we can fix baselines for how much we ought ideally to be hearing. In December 1960, the New York physician Samuel Rosen traveled with four colleagues, Moe Bergman, Dietrich Plester, Aly El-Mofty, and Mohammed Hamid Satti, to a remote part of the Sudan that had just recently been opened to foreigners. Some 650 miles southeast of Khartoum, on the Ethiopian border, in the southeastern region of the Upper Nile, lived twenty thousand Mabaan, a "pagan, primitive, tribal people whose state of cultural development is the late Stone Age" and whose straw-thatch villages were situated in a territory as "noise-free" as any New Yorker could imagine. Aside from the bleating of lambs, mooing of cows, and crowing of roosters, little would break the Stone Age quiet except for their five-stringed lyre and some shouting, hooting, and choral singing during their spring harvest rituals.

Without drums or guns, bell-metal or Oldsmobiles, the Mabaan should have had *Ur*-ears, exemplary of the hearing that humans must have begun with before Babel and blacksmiths. Rosen and his colleagues traveled to the village of Boing, where they tested 543 of these "Stone Age" people. Two were hard of hearing from head injuries. Three were deaf from bouts with meningitis. The rest could hear across a wider range of softer sounds than 98 percent of modern Americans. What's more, unlike modern Americans, their hearing did not deteriorate as they grew older; Mabaan of every age could hear as well or better than the healthiest young people who had been tested at the 1954 Wisconsin State Fair. In a follow-up report published in 1964, the researchers observed that 10 percent of Mabaan youth could distinguish sounds from 500 Hz to 24,000 Hz, an unusual range upward across normal human limits into the ultrasound that delights bats and dolphins. There was individual variation in hearing acuity among the Mabaan and evidence of mild decline with age, but well into their seventies most could hear high-frequency sounds inaudible to urban Americans. Rosen et al. concluded that at least part of hearing loss through life is noise-induced, and that the Mabaan "for the most part do not die until they slowly wear out completely, whereas in modern civilizations we die before the process of slowly wearing out can be completed."[28]

Implication: Had we remained in Eden, or close thereunto, as hunter-gatherers and pastoralists with a chiefly carbohydrate diet of unadulterated grains,[29] we would inhabit a richer soundscape, for it is our post-Stone Age hustle and muscle, our pounding of iron and hammering of steel, our thumping of breastplates and trumpeting of war, our meat-eating and sugar-licking, our Manhattan traffic and Manchester factories, that put a damper on our acoustic universe and constrain our powers as audience. Ideally, we should share with manatee and dolphin, throughout our lives, a world of sounds twice or thrice as expansive as the one with which we now make do.

One might dismiss this ear-fantasy as a scientific romance, sharing the environmentalist impulses of Rachel Carson's *Silent Spring*, that "explosive bestseller" of 1962 which took acoustic evidence in the opposite direction (pollution → deathly quiet) but was also rooted in the audiovisuals of an earlier, healthier, ecology: "There was once a town in the heart of America where all life seemed to live in harmony with its surroundings," she began—she who had been born in Springdale Township, Pennsylvania, three years after the Harwick Mine Disaster of 1904, and had graduated from Parnassus High School four years before the nearby explosion in the Kinloch workings that killed forty-six. Nonetheless, in this "town in the heart of America," grain filled the fields, fruit the orchards, birds abundant and variegated sang through the pines, and oak, maple, and birch blazed with color in the autumn, when "foxes barked in the hills and deer silently crossed the fields, half hidden in the mists of the fall mornings." Then came the blight, which was us, fallen out of Eden and never so much at a loss as when facing the unintended consequences of our own industry and insecticides, so that now (at work in her cottage on Southport Island, Maine, fifty nautical miles south of Margaret Wise Brown's cottage on Vinalhaven), Carson chose lines from the consumptive Keats for her epigraph:

> The sedge is wither'd from the lake,
> And no birds sing.[30]

One might also dismiss the comedy, to English ears, of seeking out a quiet, isolated territory only to perform audiometric tests in a village called Boing, although the Western Nilotic language spoken by the Mabaan has contextual pitch variations whose subtle tonalities must have played a part in what the Mabaan claimed to be able to hear.[31] One cannot dismiss the tragedy of southern Sudan, where civil war (1983–2005) and recurrent famine left 2,000,000 dead, with the 20,000 Mabaan (or Meban) on the borderlands between North and South at risk of attack from both the Sudan People's Liberation Army and government troops. The noise of modern life (and death), which in 1960 had already begun to affect the hearing of young Mabaan men drawn to the excitements of Khartoum, shattered any "Stone Age" peace they had back home. By 1998, most of the Mabaan had fled to Ethiopia, Kenya, and Egypt, some to Canada, others to Rosen's New York. But years before, in 1936–1937, the Mabaan had already met up with the guns, bombs, and strafing planes of the industrial West, as Italian forces occupied Ethiopia and environs.[32]

Is there no escaping noise? Rosen et al. were eager to find a place that had dodged noise, but the only way they knew to do this was to find a dodgy group who were, so to speak, out of time, atavisms belonging to another Age. This led to a serious research flaw, according to Aram Glorig, the contemporary American otologist most concerned with research into presbycusis (loss of hearing with age) and coiner of the term "sociocusis" (the psychoacoustic impact of environmental noise): the Mabaan kept no track of birthdays, and Rosen et al. were unsure of their ages. Contrasting by age-groups the "Stone Age" Mabaan with Industrial Americans, the researchers overconfidently correlated hearing loss and other Western signs of aging—high blood pressure, arteriosclerosis, heart disease—with "the stress and strain which afflicts modern civilized man."[33] To hear exceedingly well throughout life was part of lifelong well-being, creditable to the *Ur*-status of the Mabaan. The scientific descent into the Sudan was thus a Sixties return to Camelot; it had echoes in back-to-the-land movements, treks to Nepal, quests for fountains of youth in apricot groves, searches for clusters of centenarians who might divulge the dietary or spiritual secrets to longevity. It was also, for Rosen himself, a mission driven by impulses redemptive if not messianic. In 1952 Rosen had begun a battle "against human stubbornness and professional conservatism" to introduce a new operation to mobilize the stapes of the ear and so restore hearing to the congenitally deaf; he would preface his later autobiography with an English proverb, "None so deaf as those who will not hear," and a quotation from *King Lear* (act IV, scene 6): "A man may see how this world goes with no eyes. Look with thine ears: see how yon justice rails upon yon simple thief. Hark, in thine ear: change place; and handy-dandy, which is the justice, which is thief?" He had headed for the Sudan in 1961 after a talk with a German otolaryngologist, Dietrich Plester, who described a safari in Ethiopia that had gone awry. Mauled by a lioness he had shot while she was protecting her cubs, Plester had been taken to a hospital just across the border in a Sudanese village inhabited by "Marvelous people, Sam, absolutely marvelous ... tall, mostly naked, erect, sinewy, very peaceful and friendly." What's more, said Plester, "Their remoteness from everything civilized seems to have made them in many ways more genuinely civilized than the men we think of as 'civilized.'" Rosen at sixty-four immediately made plans to go to Boing, and when he got there, "it proved to be, if not a novelist's idea of Shangri-La, then an otologist's dream of one," with background noise at a tenth of that of an electric refrigerator. What's more, the Mabaan were a heart-healthy people, much like the residents of the Shangri-La of *Lost Horizon* (1937, and a television musical in 1960), so their intact hearing could be correlated with cardiovascular criteria for longevity and a healthy diet of whole grains, fish, and wild dates. Rosen described the Mabaan as "lean," "dignified," "upright," "energetic," the very kind of folk that the fourth of five children of a Jewish peddler who had sold crockery in a "desultory" sort of way in the ghetto of Syracuse, New York, would hope to find in a Pure Land.[34]

Inspired, Drs. Y. P. Kapur and A. J. Patt climbed 7,000 feet into the Nilgiri Mountains of South India in September 1964 to study the hearing of a tribe of buffalo herders who lived a quieter life than even the Mabaan. The Todas, all 706 of them, were vegetarians without weapons, musical instruments, or roosters. Their tools were few: a split bamboo stick for churning milk and a club-shaped spatula for stirring food. From their rolling hills and secluded valleys came nothing other than the sound of "gurgling brooks, singing birds, human voices, lowing cattle, rustling leaves, and whistling winds." Sadly, a quarter of those tested were hard of hearing, probably from otosclerosis, but as "a pure race who have lived in the same area for the last 800 years in a noise-free environment" and whose ages were reliably recorded by priests at an annual temple ceremony, three-quarters had excellent hearing—especially the women—and, like the Mabaan, demonstrated only a slight deterioration of hearing with age. Like the Mabaan, too, the Todas were admirable, impressive folk, wrote Kapur and Patt, with "European features, Roman noses, bright hazel eyes, and glossy hair . . . friendly to all, having a grave and dignified bearing."[35]

What of those whose features were less European, less "upright," or less welcoming, such as the Bushmen of the Kalahari Desert? They were somewhat less willing to be studied, and they had a cultural notion, elaborated in the acoustics of their rock art, of the spiritual aptness of veiled sound. They also had a propensity for paying more attention to the sound of the instructions (mimicking voices saying "This is the Pitch Test. Ready for Column A") than following the instructions per se, and it was hard to find a quiet test site in the desert, "with the hordes of flies, Bushmen chattering, babies crying, wind, etc." And their ages, after puberty, were almost anyone's guess. Even so, inspired by accounts that they could hear a single-engine aircraft seventy miles off and the approach of another party of Bushmen at inconceivable distances, a series of studies between 1962 and 1973 appeared to show, once again, that a relatively low level of ambient noise (in comparison with Danish and American environs) might best explain their unusually good hearing at higher frequencies and their retention of good hearing throughout their lives. When they went to work for local farmers and were exposed to the noise of farm machinery and the lowing of cattle, their blood pressure might rise and hearing decline, but this was speculative. In the meantime, as one pair of South African anthropologists found, they who sang and danced to the stamping of a drummer beside a hollow in the earth covered with animal hide became quickly bored with symphonic music, preferring Viennese waltzes and "Red Sails in the Sunset."[36]

Around the same time, two investigators went to the Mexican barrio of San Enrique, outside the village of Ticul, to test the hearing of seventy-one Indians in a quiet "primitive suburb" on the Yucatan Peninsula. These people did not have *Ur*-ears, and their hearing, as they aged, followed the same downward curve as that for industrialized Americans farther north. Perhaps the people of the barrio did not inhabit such a Pure Land, and measurements of background noise in San Enrique failed to register deafening sounds that occurred at irregular intervals or different seasons. Perhaps the Indians had been making frequent trips into noisier parts of the Yucatan for commerce or ceremony. Perhaps they were being dodgy, unwilling to reveal their aural talents to outsiders who might inadvertently strip them of a valuable survival tactic—claiming not to hear while actually and easily overhearing. Perhaps, since hearing tests in the field rely upon the subjects themselves to indicate which sounds can be heard, they were not listening the way a person in Los Angeles or Manhattan listens.[37]

Marcos V. Goycoolea of the Minnesota Ear, Head and Neck Clinic went therefore to Rapa Nui, Easter Island, that legendarily mysterious and "most isolated inhabited place in the world," 2,010 nautical miles west of Chile in the middle of the Pacific, to resolve the issue. "The island of the great silence" and enormous stone faces (with elongated ears) had recently undergone a census, so there was demographic and medical data for each of its 1,828 inhabitants when Goycoolea and his team of physicians arrived on the island in 1984. Although the population was genetically homogenous, many had been born and/or lived abroad before returning to the island, so Goycoolea could also test the hearing of those who had rarely left home against those who had endured years of more crowded, industrial, or martial soundscapes. The results were definitive: the more years spent on the continent (primarily in Santiago), the greater the decline in both men's and women's hearing at all frequencies and the greater the deficit in the hearing of higher frequencies. And yet... the median threshold levels of both males and females who had never lived abroad were similar to those of women in the United States. This would seem to vitiate any conclusion about the frightful effect of industrial civilization on long-term noise-induced hearing loss, unless men in industrialized countries are exposed more regularly to the loud sounds of armies and factories, and these are louder than electric blenders, televisions, cap guns, and screaming babies. Then again, the men and women of Rapa Nui had been punishing their ears with deep-sea diving, and many had upper respiratory tract allergies that would temporarily limit their auditory acuity. In addition, a fifth of Rapa Nui women had been beaten by their husbands, in a societally sanctioned violence called Púa Púa severe enough to cause head trauma and fractures. So much for an acoustic Eden. To their surprise, Goycoolea's team found little evidence of middle-ear infection or bone-conduction loss due to such beatings (did the Rapa Nui men, they wondered, take care not to strike women about the ears?), but they made no comment on the surprising equivalence of auditory acuity among stay-at-home Rapa Nui and American females, almost half of whom in 1984 were actively part of the (out-of-home) workforce.[38]

Whatever evidence we have of superb hearing among ancient or isolated communities, that evidence is tainted by historically variable notions of what it means to hear well and tempered by fables of perfect hearing. For example: a current etiology for autism is that the autistic have heightened senses, especially of hearing, so that the squeak of a swivel chair or the distant growls of a lawnmower are to them noises so horrendously loud that they must hide away or scream to drown out the world beyond, for they have none of the filters necessary to scale down the soundscape to endurable levels; they hear too well for their own good.[39] This is a modern fable of perfect hearing; the moral is that it behooves no one to hear so well that the inevitable clangor of society becomes a cruel debility. If the postmodern sensitivity to sound is conditioned by postmodern expectations of being able to fine-tune it, the most desperate of humans must be those born without the proper tuning circuits. The circuitry, or circularity, of this cultural logic approaches fable when we seek out the Mabaan or Rapa Nui in order to prove that it must be the loudness of modernity that makes us hard of hearing, while our deafness requires, for our own security, an ever louder world—a world in which the fate of the autistic child is at once revealed and sealed.

Standing in a room sealed off from external sound and deadened to internal sound, John Cage found himself agitated by the beating of his heart and the rush of his blood. This too is a fable of perfect hearing. A modernist composer who attended to all sounds with wry equanimity but insisted on a monastic intensity of listening, Cage was as vociferous about silence as he was vengeful about sound. He arranged concerts in which the only sounds were those of the audience waiting for music to begin, their uneasy chatter fed back through auditorium speakers. He used random number tables to generate pauses, longer pauses, and apothegms. He realized earlier Futurist and Dadaist programs for a new music whose outer limits were defined rather by patience or fortitude than by noise or quiet. Stepping into the anechoic chamber, Cage was entering a test site for microphones, amplifiers, and machines of every kind, jackhammers to jet engines; he was also entering a place baffled against reverberation, against any hint of the few seconds it takes for sound to be reflected back. In a place out of time and without society, Cage became his own timekeeper and his own audience, his body itself still reverberant, his blood rushing and nerves throbbing for him alone.[40]

At least, so he claimed. As with all hearing, the fable is labyrinthine: John Cage, mycologist and musician with a portfolio for unearthing all fungi and all frequencies, claims suddenly to become aware of the cage of his noisy body inside a cage of baffles used to "sound condition" those devices that make us most aware of the cage of sound through which we all, as living beings, must make our way.

Abandoning therefore any claim to imperturbable sanctuary or impeccable hearing, we are free to move on to what is left: the history of noise.

ROUND ONE

Everywhere

On apprehensions of noise on all sides. How this comes to be, and from which directions.

Begin high and wide with waterfall. Of ceaseless sounds, the roar of a cataract was the loudest in this world until the advent of enormous waterwheels. Of steady but not unceasing sounds, loudest was the storm-tide pounding of surf, loud as fulling hammers. Unsteady and inconstant were the sounds of volcano, tornado, rolling thunder, and the crack of arctic ice at spring thaw, loud as jackhammers or cathedral organs; sometimes, like volcano, louder. Momentary were avalanche, earthquake, hail, and claps of thunder directly overhead—these claps the loudest transient sounds familiar anywhere until men rang great bells and fired great cannon. Across the last two thousand years, humankind on occasion has matched the volume of the rest of the natural world.[1] Across the last two hundred we have regularly surpassed it at every point along the spectrum: sirens and steam whistles on the high end, more piercing than screech of parrots or shriek of sandstorm; klaxons and loudspeakers at midpoint, howling above wolves and high winds; supersonic jets and nuclear explosions on the low end, booming above calving glaciers.

Yet the cry of a peacock in an abandoned courtyard, the whimper of a baby in an empty stadium, or the groan of oaks under the winds of dark woods may feel louder, noisier, than any siren. Not everything about loudness or noisiness is covered by the science of acoustics or the anatomy of the ear. Engineers have refined devices for measuring levels of sound, but their meters have not been up to the task of measuring *sentiments* of loudness and noisiness, which have less to do with volume, pitch, timbre, attack, and decay than with the contexts in which sounds appear, endure, vanish, reappear. Pigeons on the grass.

True, sentiments of noise are often bound up with sensations of loudness, and above 130 db all sound, alas, is fatal to the cilia of our inner ears, but what one anywhere encounters as noise are sounds that must be listened *through*, regardless of decibels: hard candies slowly unwrapped at harpsichord concerts, a chainsaw upriver of a wilderness camp.

SOUNDPLATE 2

NOISES FOR ALL POCKETS · from teeth-cleaning to spring-cleaning
our house really safe from Burglars? Can you really enjoy your "evening
knowing that your home is dark and silent, an open invitation to the criminal
? Why not set your mind at rest? There is something to suit YOU in our
Comprehensive Range of Noises from
"Bijou" Domestic Bump and Tinkle Set......£4 10s. 0d.
TO
"Kountrihouse" Ball Set with orgy attachment ..£27
WE BEG TO INFORM OUR REGULAR
Clients of the following NEW LINES
NEW!
MISCELLANEOUS Gurglers
OTHER DETAILS ON APPLICATION TO:
HOUSEHOLD NOISES INC. Dept. O. Prancing Lane. London. E.C.4
TAB. VII.

Elsewhere, metrically loud sounds may be welcome cover for lovers whispering sweet nothings.

"Nothing is noisy," declared Gertrude Stein, liberating sound from any categorical judgment and, in the same breath, registering the obtrusiveness of silence.[2] Where one expects, relishes, and enforces quiet, any sound may be loud and all sound may be noise. The greater the anticipation of silence, the greater the ruckus of sound.

1. Tschitschatschots, or Soundbites

Take, for example, a reading room. Library etiquette would seem always to have required a private reading so quiet that the softest voicing of words might grate upon sensibilities. To this day, the most animated and vociferous of librarians must wrestle with ancestral audiotypes of a hushed, pursed figure shushing the multitudes.[3] Twenty-one hundred years ago, however, in the great library of Alexandria, all but a stentor's proclamations might have gone unremarked, because much reading was done by larynx, throat, tongue, teeth, and lips, in soundbites.

"Soundbite" is not an anachronism. Greek and hieroglyphic texts were ordinarily incised or inscribed with LETTRSALLINCAPSRUNNIN GONEWORDINTOTHENEXTSANSPUNCTUATIONORSPACESABB REVS&NUMERALSCXVIMIXEDLXVIN. Ancient Latin texts initially had spaces between words; these were soon dropped. Subvocally or in full voice, readers had therefore to sound out the syllables to arrive at the words or numbers and then, decoding syntax and declensions, the aggregate meaning. Taking in each passage by chunks of sound, they had also to articulate to make sense of the text; mumbling, A SENTENCE'S WORDSTREAM could become A SENTENCE SWORDS DREAM.

Whether such tumbling wordstreams could be read without making a noise is no longer in question. Recent scholarship has proven that the bookrolls and correspondence of classical antiquity could be read to oneself, silently, and—prose or poetry, dispatch or dictate, commerce or philosophy—they often were. But in many situations, among Hellenistic elites as among congregations of Jews and early Christians, reading was a seriously shared performance, an acoustic means for building or affirming community. At once a communal event and a personal interpretation, reading-out relied upon the oral skills and wisdom of readers who understood the text well enough to insert revealing pauses that made each sentence intelligible and each (unmarked) paragraph persuasive. This demanded a mastery of the "eye-voice span," a twentieth-century term for what Quintilian had already explained in the first century as the challenge of reading-out, "since it is necessary to keep the eyes on what follows while reading out what precedes, with the resulting difficulty that the attention of the mind must be divided, the eyes and voice being differently engaged." In practice, then, readers-out had to develop intellectual and rhetorical tactics for selecting, recalling, and repeating soundbites while reading ahead for context. Since anyone who was literate necessarily became, on occasion, a reader-out, it would follow (I suggest) that most readers, reading silently or aloud, would both hear and see the passages before them.[4]

Medieval monks, living under a rule of silence and the spiritual exercise of copying scripture, began to indite their Latin with words•regularly• separated•by•dots•or spaces, phrases indicated by commas, sentences set apart by dashes—so that they could move more confidently through sacred texts without disturbing fellow monks. Their silence was relative: close readers and assiduous copyists of scripture were expected to be "ruminators," chewing the cud of each word, sounding out the meaning through meditation and mutter.[5]

Paul Saenger, who has sounded out the meaning of punctuation marks in medieval texts, suggests that the Christian stutter-step from reading aloud to reading silently, advocated by St. Isaac of Syria in the sixth century, advanced by Irish monks of the ninth century, and adopted across Western Europe by the eleventh, was more than a shift from mutter to mute. The quiet of late medieval readers facilitated new forms of scrutiny, closer comparison among discrepant versions of hand-copied texts whose parallel passages could be more quickly located once words were spaced, phrases segregated, Proper Names capitalized, indexed, alphabetized. Silent reading allowed critiques to be shaped without being overheard and preemptively quashed, thus (overreaching the evidence, wherefore the rise of parentheses) the appearance of Humanism and literalizing heresies. Generations of proliferating dashes, commas, colons, semi-colons, periods and question marks (?) underlay the conception and technical perfection of printing by moveable metal type in neatly paragraphed blocks. Such modularity enabled a change in ideas of personhood: modeled by the silent and solitary reader, a private "self" appeared at a punctuated, upper-cased distance from every other body. With the headlong rush of letters now leashed, one could set the type for an ethos in which each *italicized* body was to be quietly respectful of the other.[6]

This supposed shift from reading aloud to reading silently has been interpreted as one element of the larger transition from an oral, communal, agrarian society to a visual, individualist, industrial society. Elites of the early modern West themselves argued loudly about the preeminence of sight over sound among the senses, of painting over music among the arts, of geometry over harmony among the stars, of vision over voice among sources of truth, of eyewitness over earwitness among the means of proof. Scholars still rattle each other's cages concerning the relative cultural power and historical presence of the oral and the visual.[7]

Day to day, one cannot so readily parse audition and vision. What can be heard of the early modern world tells us that much reading remained vocal, voluble, and social. Noblewomen read to each other or had ladies-in-waiting read to them, daily. Courtesans learned to speak well so as to converse with and read aloud to their clients, nightly. Each morning, magistrates, merchants, bankers, barristers, and bishops had their correspondence read to them by secretaries. Afternoons, university faculty lectured from prepared papers or well-worn books; in anatomy theaters, surgeon-demonstrators wielded the knives while professors, aloof from flayed corpses, pontificated from ancient pages by Galen or a recent rehash by Mondino. Priests read canonical passages in Latin to uncomprehending peasants; Cistercian and Benedictine monks heard the words of Church Fathers read to them while eating silently in the refectory; Protestant fathers read vernacular scripture to their families at table (mothers and other women, by an Act for the Advancement of True Religion under England's Henry VIII, were forbidden to read the Bible aloud even in private, a provision soon repealed by Edward VI). Town criers cried the news, advertisements, and official proclamations at each crossing, preceded by wooden clappers, drums, trumpets, horns, bells, fifes. The illiterate dictated letters to street scribes, and public notaries reread contracts back to wary customers. Executors read last testaments to survivors and heirs, aware that statements of faith had been written into wills for the audible consolation or spiritual coaxing of a captive audience awaiting word of an inheritance.[8]

Nor did "moderns" abandon the experiences of reading aloud and being read to. Consider the totems of Western modernity. Thrice daily, Charles Darwin had family correspondence and the novels of Sir Walter Scott or Jane Austen read aloud to him while he lay on a sofa pondering the evolutionary value of birdsong and laughter. Home from his labors under the echoing blue dome of the British Library, Karl Marx, "an unrivalled storyteller" whose tales were told as he strolled and so "measured by miles not chapters," read his children the whole of Homer, the *Nibelungenlied*, and most of Shakespeare. On vacation after months of quietly listening to neuroses, Sigmund Freud elocuted from tomes of poetry and took it as "a familiar fact that in reading aloud the attention of the reader often wanders from the text toward his own thoughts"—an occasion for further reflection on the psychopathologies of everyday life. Gertrude Stein, touring in a robe of Chinese blue, read aloud from anything she could get her mouth around, with "metallic resonant glamor."[9]

Postmodern or retro, we still read aloud each evening to our young, though rarely with Stein's resonance or from her *The World Is Round*, designed in blue print on pink pages by Clement Hurd, who illustrated eight books by Margaret Wise Brown (*Goodnight, Moon*) and here had to manage the contours of Rose (is a Rose, is almost ten), who wonders "would she have been she used to cry about it would she have been Rose if her name had not been Rose," and of her cousin Willie, who likes to sing to owls (*Reader, please read aloud*):

Once upon a time the world was round
The moon was round
The lake was round
And I I was almost drowned.[10]

Heaving themselves out of narrow cots and slumping into class on their daily round, college students long discouraged from voicing printed words or reading aloud in the library stacks still hear professors reading from scribbled outlines or rapid-fire from textbooks (sometimes alas their own), still hear scientific lecturers repeating what they have scrawled on the glass screens of overheads. "Pigeons large pigeons on the shorter longer yellow grass alas pigeons on the grass. If they were not pigeons what were they."[11]

And we still have hawkers barkers fast-talkers con men luring pigeons with words loud and repetitious or soft and seductive, ranks of pigeon-toed lawyers objecting, rows of stool pigeons testifying, oral arguments before appeals courts supreme courts. Middle managers speak truth to Power Points, electronically projected colorized thematized schematized words nonetheless read aloud word-for-word as if to the red bouncing ball of a matinee movie sing-along. Congregants respond aloud from sacred texts, shout amens to passages in sermons, flock to mass chorales for Messiahs and Hallelujahs.

Praise be! No mere artefacts of an outmoded oral culture, such oratorical, jurisprudent, pedagogical, managerial, and liturgical acts reflect how people live today, at heart, environed by talk shows, books on tape, televised preaching, cell phones, public address systems, elevator music, and traveling albums on CD, MP3, and iPod. In the sixteenth and seventeenth centuries, when men and women were learning to their imperial delight or indigenous devastation that "the world was round and you could go on it around and around," such media were wide-flung fantasies of sounds quick-frozen in air, then thawed by the chunk: soundbites. Punctuation had frozen blocks of sound in place; typesetting had frozen sentences on the page; now sound itself could more easily be imagined to linger in time and space, frozen.[12]

Almost literally. On the second of four evenings of conversation among the elegant guests of the Duke of Urbino in 1507, the Magnifico Giuliano added to a round of "lies that are so big and splendid that they make us laugh." It seems that a merchant of Lucca, on business in the north one winter, arranged with some Poles to buy furs from some Russians. Due to hostilities between Poland and Muscovy, neither party dared cross the border, the Dnieper. The Russians shouted across the frozen river to begin the bargaining. In the bitter cold their shouts froze halfway across. The Poles rushed out to the middle of the ice and built a fire. Soon, the suspended words "began to melt and descend with a murmur like snow falling from the mountains in May; and so they were immediately heard very clearly." But the Russians, cold and frustrated, had gone off, and in any case the price they had been asking was too high. The merchant returned furless to Italy.

Coming on the heels of an anecdote about a woman shuddering at the shame of having to appear naked with everyone else at Judgment Day, and just before the tale of a chess-playing Brazilian monkey, the yarn of the sable-trader was looped between the World-to-Come and a World-just-Come, that "country or world just discovered by the Portuguese mariners." A story of suspended words suspended between worlds was the story of the storyteller himself. Youngest son of Clarice Orsini, who died when he was nine, and of Lorenzo the Magnificent, who died when he was thirteen, Giuliano de' Medici had been in exile since 1494 from his native Florence and stood at twenty-eight on the verge of formal adulthood, a well-spoken young nobleman who had yet to accomplish much of anything. If, like a courtier's best efforts, Giuliano's story was a splendid lie, it was also his life—a fact understood by his friend Baldesar Castiglione. At work from 1508 through 1516 shaping the Urbino conversations for his *Book of the Courtier*, Castiglione would have seen Giuliano return briefly and ineptly in 1512 to govern Florence, would have watched Giuliano disappoint his brother Giovanni, now Pope Leo X, as a commander of papal legions, and would have lamented Giuliano's death at the age of thirty-seven. *Il Magnifico* had had a career of unresolved promise, an irresolution which was in musical terms the theme of his life, as we hear on the first evening when Giuliano asserts that "it is very wrong to have two perfect consonances one after the other; for our sense of hearing abhors this, whereas it often likes a second or a seventh, which in itself is a harsh and unbearable discord." Why? Because "to continue in perfect consonances produces satiety and offers a harmony which is too affected; but this disappears when imperfect consonances are introduced to establish the contrast which keeps the listener in a state of expectancy"—like that state of dissonant expectancy in which lived a fellow exile and Florentine, Niccolò Machiavelli, forced upon Giuliano's death in 1516 to dedicate *The Prince* to another, Giuliano's cousin Lorenzo, Duke of Urbino, who would also die young, in 1519, at the age of twenty-seven.[13]

Laboring under intermittent papal (i.e., Medici) commissions from 1520 to 1534 on a memorial chapel in Florence for four of the family's early- and dear-departed, Michelangelo sculpted a muscular Giuliano sitting in *contrapposto*, ready to rise from his niche above a turbulent figure of Day and a restless Night, his head and long neck turned strongly to the left, intent on . . . what? On the sudden eruption of the Reformation and the

Peasant Wars? On the floodwaters supposed to arrive with the conjunction of Saturn, Jupiter, and Mars in the sign of Pisces in 1524? On the French sack of Rome in 1527 and the contortions of his family's political fortunes? Improbable. Rather a patron of the arts than a defender of the faith or a power broker, Giuliano would more likely have been intent on Leonardo painting *John the Baptist* under his patronage, intent on that celestial voice toward which da Vinci's virile, adolescent John is pointing.[14]

Intent, in any case, on the audible: Giuliano and his cousin Lorenzo, situated directly opposite, are stone-blind. The two figures, unlike all of Michelangelo's other finished work, have no pupils, cannot see. Their heads turned toward what was originally the main door to the chapel, Giuliano of the *Book of the Courtier* and Lorenzo of *The Prince* must be *listening* for our entrance, listening for a posterity that affirms all they might have done had they lived to maturity. Vice versa, for contemporary women in wooden platform shoes and men in wooden clogs, as for modern visitors in high heels and hiking boots, the experience of the Medici Chapel of the Church of San Lorenzo had also to be impressively acoustic: each click and scuff echoing off marble floors and walls under a small vaulted dome.[15]

Similarly fine senses of resonance, of affect and after-effect, characterized the courtier and, said *Il Magnifico* on the third evening, the perfect court lady. A lady must be candid but circumspect, resourceful but unobtrusive, continually monitoring her own presence and impact. For all of her accomplishments, she must not draw disproportionate attention. "For example, when she is dancing I should not wish to see her use movements that are too forceful and energetic, nor, when she is singing or playing a musical instrument, to use those abrupt and frequent *diminuendos* that are ingenious but not beautiful." Such drawn-out dwindlings of sound, not (yet) marked on sheets of music, were individual ostentations, overtly and overly dramatic.[16]

Courtliness was theatrical, but it was a theater of graceful virtuosity and sonic restraint in which conversation was more than amusement, music more than blandishment, and consonance more than incidental metaphor. To whatever extent Europeans of the sixteenth century felt themselves to have *inner* lives (a question with profound confessional consequences), they were urged by Castiglione and other expositors of elegance to act in public in a manner befitting a thoroughgoing nobility. What was voiced had to be consonant with one's finest, deepest character. Conversation literally sounded one out, and one's voice was oneself. At century's end, when *As You Like It* opened in London and William Shakespeare resumed the commonplace that "All the world's a stage," each of his seven stages of man had a defining voice: the *mewling* infant; the *whining* schoolboy; the lover *sighing like furnace, with woeful ballad*; the soldier *full of strange oaths*; the justice seasoned with *wise saws*; the old slippered man, *his big manly voice, / Turning again toward childish treble, pipes / And whistles in his sound*; second childhood, *Sans teeth, sans eyes, sans taste, sans everything* but hearing, which sense physicians and clergy at deathbeds knew to be the last to go. Hearing was as well the most anchored of the senses, insisted Helkiah Crooke in his *Microcosmographia* of 1616, for "the Act of Seeing is sooner ended and passeth more lightly by the Sense than the Act of Hearing. Whence it follows necessarily that things seen do not stick so fast unto us."[17]

Diminuendo then was no small matter. At stake was oblivion: how people, their words and voices, stick fast or melt away. *Diminuendo* touched Giuliano de' Medici to the quick; in the form of *fama* it touched families far less comfortable and courts far stricter. Men and women of all estates were players when it came to *fama*, a word of enormous capacity encompassing news, information, rumor, reputation, credibility, public opinion, glory, and infamy. *Fama* embodied the power of words to linger and the lingering power of public talk. In medieval law courts, where eyewitnesses played second fiddle to ears'-full of public opinion, each person who testified might be asked, "*Quid est fama?* What is fame?" or "How many people constitute fame?" Knowing the boundaries of *fama* was critical in societies where rights and truths were fashioned in public and out loud. Week to week, *fama* was the word on the street; over the years, it was the "noise" a person made in the world, for better or worse. For worse, *fama* grew from public scoldings and insult, the barratry of backbiting and "jangling" of quarrel. In addition to these "Sins of the Tongue," *fama* accrued from furtive note-passing, lettered slander, the gossip of alphabets. In early modern law, as personal credit was increasingly allocated by financiers and allocuted by lawyers, *fama* was determined by the degree to which gossip might be binding.[18]

Castiglione's courtiers calibrated *fama* by the quality of the company one kept, the quality of the words that passed between, and how those resounded abroad; Machiavelli's *fama* referred back to republican Rome and forward to a free Italy established by "embarking on great enterprises," unfazed by flattery or scorn. Planning the Medici Chapel, Michelangelo intentionally left uninscribed the tombs of Giuliano and Lorenzo, who had far less claim to fame through what they had made of themselves than through what others made of them. Instead, Michelangelo (into his own fifties) anchored *fama* to the chapel's underlying form, noting that "Fame holds the epitaphs *a giacere*," that is, "recumbent" or "in abeyance." Giuliano and Lorenzo turn to hear us coming not because we disturb their mortuary peace but because our noiseful entrances extend the resonance of their lives, otherwise cut short. Our noise shapes their triumph, like the noisy triumphal entries of conquerors and patron dukes into the walled cities of the Renaissance, heralded by peals from every church, trumpeters in turrets, and scores of musicians at the gates. Frozen in stone, Giuliano and Lorenzo are fixed on our presence, for *fama* is a collective enterprise of audience and encore.[19]

The rightness and repercussions of a life, like the rightness and repercussions of music, could not be accomplished with ingenuity alone; proper *diminuendo* was driven by an intrinsic beauty and led toward a fortunate memorial. Often read as imperatives toward artifice, the *Book of the Courtier*, *The Prince*, and the Medici Chapel may also be heard as recapitulations of bespoken principles and principals, on behalf of that which needs must last. Sites of active recall, they shade into those Renaissance arts of memory through which a universe of cues established a parallel universe of discourse, that no bit worth remembering be lost. Though court ladies might get swept up in exorbitant *diminuendo* and pedants pursue memory schemes so intricate as to require clues to the cues, Giuliano's story of frozen words was as much a consolation as a joke, assuring recuperability.[20]

Apart from court or cloister, the Renaissance soundbite had a more prosaic purchase. In *The Shortest Way to Reading* (1527), Valentin Ickelsamer advocated a classically Roman (and rabbinic) strategy, taking youngsters through the alphabet sound by sound, then orally through each syllable and word, so that the skill of reading would be first a skill in listening and enunciation, next a skill in repetition and the analysis of constituent sounds. As with elaborate mnemonics, this popular pedagogy could lead to extravagant demonstrations of the ability to handle any concoction of syllables, any *tschitschatshots*, in verbal displays that one critic called a "babble factory." The Dutch had an adage, Better to pass by a slew of smiths at their anvils than a roomful of kids at their abc's.[21] Implicit in Giuliano's critique of *diminuendo* had been a caution against the thoughtless embrace of indiscriminate distinctions, holding on to the last syllables of words for no other purpose than the holding on. Renown follows the upswing of an extensible intentional soundfulness, expressed politically through *fama*, musically through *diminuendo*; on the downswing comes Babel.

An allegory of idolatry and arrogance, the Tower of Babel was taken up by medieval Jewish kabbalists curious about the bricks themselves and how the spiritually adept might use the material world to access the divine. An allegory of godless empires risen and fallen, Babel was constructed by medieval Christian kabbalists as a puzzle locating the one true tongue spoken by Adam and his descendants—before the Lord Almighty, offended by the hubris of the Tower's builders, made each nation unintelligible to the other. An early modern Catholic allegory of intellect without spirit, the Tower was painted three times by Pieter Bruegel the Elder, in his last and largest version (1567) an architectural wonder still a-building but already in ruins, its outer skin of arched doorways rising like the shell of an ancient coliseum around a rough tongue of masonry that licks at the clouds while Nimrod and his fellow architects study their plans, oblivious to the heavens. A Lutheran allegory of works without faith, of a ruined Rome and vainglorious papacy, Babel had a matching goddess of Confusion, wrote the Anglican playwright Thomas Dekker in 1608, "and when she spake, her *Voice* sounded like the roaring of many *Torrents*, boisterously struggling together, for between her Jaws she did carry 1000000. Tongues." When all of the tongues jabbered at the same time, "It was a Noise of a thousand sounds, and yet the sound of the noise was nothing."[22]

Babel anchored this noise in sacred time and place, for it marked a second, *super*-sonic, Fall. The first Fall (Genesis 3) toppled upon two sound/bites: Eve hearkening to the serpent and eating of the fruit of the Tree of the Knowledge of Good and Evil; Adam hearkening to Eve and eating of that selfsame fruit, after which *too much* became comprehensible and everything, in reprisal, laborious or precarious, setting man against woman and removing both from prospects of immortality. The second Fall (Genesis 11) toppled upon the power of words themselves and culminated in a wide-ranging incomprehensibility. "And the Lord said, 'Behold, the people is one, and they have all one language, and this they begin to do: and now nothing will be restrained from them, which they have imagined to do.'" What unrestrained thing had the people, confabulating among themselves, imagined? They had imagined a version of immortality that the translators of the King James Bible in 1610 understood as *fama*: "let us build a tower, whose top may reach unto heaven, and let us make us a name, lest we be scattered abroad upon the face of the whole earth." Confounding their language, "that they may not understand one another's speech," the Lord halted the Tower in mid-spiral, dividing nation from nation and mortals from immortality by splitting, as it were, their tongues. First Fall and Second Fall both came of speech, its reach and its bite; both were accelerated by acts of listening and anticipation; both ended in penalties of dispersal and travail. But only after Babel did humanity become noisy unto itself, the syllables of one tribe mere soundbitten babble to others, "a Noise of a thousand sounds."

Speaking in English, Italian, French, or German, "Babel" was homophonous with "babble," "*balbettio*," "*babil*" and "*babillage*," "*babbeln*" and "*plappern*," as in biblical Hebrew the Tower of Babel punned upon the verb "to confuse" and a noun, "gate of God." Plain as it had been that this pun was not encysted in all languages (in Greek, *lalagein*; in Spanish, *farfullar*), it was also absent from the bewildering number of tongues encountered by Europeans now scouring the coasts of Africa, India, Indonesia, and the Americas for precious metals, slaves, and semi-precious souls. Native North Americans alone had over two hundred mutually unintelligible languages, multiplied into the thousands by mutually unintelligible dialects, scarce a one with a homophonic "babble."[23] That very absence was fresh evidence of the extent of the primal confounding at Babel. Equally plain, even "blatant" (coined in 1596 from the Latin *blattire*, to babble), was that the world, like the poet Edmund Spenser's "blatant beast," was roaring in a thousand tongues.

Newly troubled by this old trauma, philosophers and churchmen would formulate ambitious responses that will come into play later in this Round. Here I pause to make blatant the mode and scope of these pages. Weaving between figural and material environments, I have begun reconstructing the cultural space in which, gradually, noise would come to be heard as ubiquitous. For all of the contentious hurly-burly that was the urbanizing Europe of mid-millennium, the cultural space of noise was imaginal: a question of how women and men registered the sounds they heard. Babel had been loud with the industry of thousands of workers, but it became a primal site of noise only when riven into a hubbub of tongues, mutually and distressingly unintelligible. While sound in 1507 or 1608 had to do with anvils and cobblestones, noise had to do with the habit of mind of those to whom the run-of-the-mill racket of life could seem disruptive once they bent to reading silently. While sound had to do with street cries and mewling infants, noise had to do with an awareness—antic or pedantic—that words could melt from thin air and syllables stray from their moorings. While sound had to do with rebecs and hunting horns, noise had to do with personal and collective fears of malicious sputter and whispered conspiracy, *fama* gone awry. While sound had to do with funeral bells and the keening of women hired to mourn, noise had to do with individual demeanor and collective disposition, with what could be countenanced and what (ghoulish wailing? roustabout spirits?) could not. While sound had to do with public debate and private wooing, noise had to do with styles of deportment intolerant of "rude" interruption, "roughness" of voice, the "stinking breath of hisses."[24]

Was noise simply sound gone wild? Better to reverse the blade of this old saw and set the cutting edge where Renaissance poets might have whetted it to epigram: Sound comes to us; noise we invent. "Invention" in the 1500s was rarely the act of creating something out of nothing; it was the act or rhetorical process of *coming upon* something. One "invented" by astute recognition, not improvisation. Invention demanded preparation, attention, and the recalibration of background against foreground. To draw in perspective was a form, and soon the model, of invention; to reveal founding documents as forgeries was a form of invention; to lay bare the principles of politics was a form of invention; to restore the genera of ancient Greek music was a controversial invention; to print texts using moveable type cast in metallic fonts was a notable invention, taking the idea of the frozen word another step beyond that ancient invention, writing, which one Spanish author defined as "a portrait that captures the moment of speech or the glorious rumor of words that is left after you have spoken; a likeness or reanimated life."[25]

Best, then: sound comes to us; noise we *come upon*. To *come upon* noise on all sides would be an invention of signal import, a cultural, social, and ecological invention for which the two Falls had been but starting bells. I am not claiming that modernity or capitalism invented noisiness, nor that Europe in the 1500s was newly noisy. Millennia before, the gods had been upset by a clamjamfry down on earth substantial enough to impress the authors and audiences of Mesopotamian epics. Rather, I am piecing together the process by which Europeans and their American outliers *came upon* so much noise that they began to feel that there could be no escaping it, that noise was everywhere and (next Round) all the time. Such a potent invention cannot be explained through accountancies that mount sound upon sound; one must circumambulate and then stand within specific cultural spaces, imaginal yet historical, listening.[26]

2. All Over (Echoes) A Lover

Listening, just now, for echoes. For echoes, and for the echo-effects that Renaissance poets and musicians came most happily upon. Echoes are a genus of soundbite related to *diminuendo*, where some syllables fall away and others assume a surpassing resilience. Echo-effects exaggerated the temperamental timeliness of echoes, repeating and recombining end-syllables into remonstrance, irony, or affirmation while prodding monologue into dialogue. Through echoes and literary echo-effects, early moderns registered the displacement of sound and found themselves coming upon noise at every turn.

"At every turn": that was the key. Castiglione's four days of conversation in *The Book of the Courtier* relied upon a respectful taking of turns; it was no accident that the dénouement of Giuliano's yarn implied that the frozen syllables of the Russians melted in sequence and drifted down in an intelligible flurry, nor that Giuliano should be upset by *diminuendo*, which indecently perpetuated the ends of things. Echoes, though they too relied upon a pattern of call and response, were hardly respectful of person, sequence, or conclusion, arriving from unknown quarters, in overlapping snatches and tailing refrains. They did not share the full simultaneity of *fama*, her gallimaufry of words "on every mouth," but moved to a rhythm peculiarly their own, interjecting, interrupting. How peculiar we will--

"Once a noisy Nymph," wrote Ovid, re-sounding Greek shape-shifting into Latin *Metamorphoses* amid the social, legal, and political changes effected by Augustus Caesar during the first years CE,

> Once a noisy Nymph
> (who never held her tongue when others spoke,
> who never spoke till others had begun),

and who by her chatter delays Hera from catching her husband Zeus *in flagrante delicto* with any number of nymphs, Echo has been cursed by the goddess so that she can only repeat *repeat* . . . *repeat* . . . *reap* words spoken by others: woman as antiphon, a respondent chorus and despondent recording. Falling eyes over ears for a gorgeous youth, she must wait to make her passion known to him through his own words as he loses track of his hunting companions and halloos into the wild: "Anyone here?" Close by but hidden in the woods, Echo can at last declare herself: *Here?* "Come on over here!" shouts the youth. Echo: *Over here!* No, he insists: "Join *me!*" At this mis-taken invitation Echo runs to throw her arms around him, *Join me!*, but he is after all Narcissus and denies her: "Hands off! I'd rather die than have you hold me!" *Hold me!* Echo must call out as she is spurned, as she turns away *hold me!* to burn with love in lonely caves *old me*, languishing, her body kernelling to the sternness of bone and finally to voice alone . . . *old me*, untouched and intangible. Narcissus himself will languish and perish for love of an inaccessible twin, watery vision and mirror self, optical inverse of the audible fable of Echo—Echo, whom Ovid places by the woodland pool, looking on invisibly as Narcissus, fatally attracted to so-lovely a reflection and anorexic with frustration, breathes a barely audible "Farewell." *Farewell*, murmurs Echo, moving on, restless and fully vocal, unlike silent immobile Narcissus, and my obligation *ligation* here *ear* is to her, to her shards of sound *hardſound*, making half- and therefore new sense of what reaches her, asking us to listen again to words no longer our own.[27]

Poets of the fifteenth, sixteenth, and seventeenth centuries had Echo waiting on masquers, melancholics, warriors, and lovers to get her words in edgeways. No: edge*wise*.

> When all the tale is tolde and sentence said,
> Then I recite the latter words afreshe
> In mocking sort and counterfayting wies:[28]

Mocking and counterfeiting, Echo shifts the rounds of discourse, co-responding from unseen places even as she holds to nubs of sound, her antiphonies the ultimate précis, hearts of our natter. Renaissance writers reveled in the classical echo-effect through which a word or phrase becomes other by loss, confusion, or elision of letters, syllables, or sounds. Echo-effects had been used by Aristophanes in his satire of Euripides, *Thesmophoriazusae* (412–411 BCE), where a bumpkin Archer, chasing Echo, shouts, Οὐ καιρήσεις, "You no get free!" and hears his κ tauntingly slipped back to Οὐκ αιρήσεις, "You no get me!" Echo continues to reply in Greek to the sixteenth-century Latin of an earnest youth (tinged with Narcissus) wondering if he should "prefer a monastic life to wedlock." "Lock yourself in?" asks Echo (singed by Narcissus). The Dutch humanist Erasmus shunts this dialogue abruptly from marriage to melancholy to mockery in puns best read aloud:

Youth: But it's wretched for a lad to live without a lass.
Echo: Alas!
Youth: What do you think of most modern monks, eh?
Echo: Monkeys!

Impatient with circumlocutions, with words that do not touch her, Echo repeats only what bears repeating, askew yet pointed and personal. In the Italian of Battista Guarini's influential play, *Il pastor fido* (The Faithful Shepherd, 1582), *dolore* (pain) slips into *l'hore* (time), *amore* (love) into *more* (I die). Addressed directly, Echo locates identities half-known, feelings half-grown:

Silvio: Who is it answers? Tell me, are you Echo?
Or are you Love that imitate the same?
Echo: The same...[29]

Stepped up a notch during the 1550s from name and noun to English verb, Echo actively intervenes. More than sound (the Greek *eche*), in her boldness and bodilessness Echo comes to bespeak a cultural physics of noise that will redound upon these pages as interference, feedback, compression, stochastic resonance.[30] Forward and afterword, she muffles, intensifies, truncates, and distends, each process leaning toward a different sensibility:

Echo muffling: nostalgia; uprooted sound; ironic loss.
Tumult all around, will you be able?
Babel.
Echo intensifying: exaggeration; multiplicative sound; focused repetition.
Are holy leaves the Echo then of blisse?
Yes. yes. y e s. sssss.
Echo truncating: immediacy; abbreviated sound; sudden relevance.
Echo! What shall I do to my Nymph when I go to behold her?
Hold her.
Echo distending: anticipation; translocated sound; wry recapitulation.
Flint all over.
A lover.[31]

Even tardy or utterly unresponsive, as to the importunate songful Lady in John Milton's *Masque of Comus* (1634), Echo is a power to be reckoned with, the "Sweet Queen of Parley, Daughter of the Sphere." Should she not respond, her silence too may be too telling: listen to your own lay, Lady. For Renaissance Neoplatonists drawing upon the rabbinic tradition of Echo as "Daughter of the Voice," *bat kol*, and upon the kabbalistic magic of words anagrammatically rearranged, Echo late or silent still cannot abide the nominative, must work in the manner of the Lord's "still small voice," intervening by her very stillness or retard.[32]

Muffling, intensifying, truncating, distending, delaying, Echo stands in perpetual indenture to our soundfulness; without us, she's mute. That of course was her curse. But the curse reverses: the more dynamic we imagine her in response to us—whether ironic, prophetic, obstreperous, intimate, or admonitory—the more decisive is her redirection:

O who will show me those delights on high?
I.

So Echo in her various forms will lead us on through this book, much as she has led one modern poet-scholar, John Hollander, to claim without qualm that "a certain amount of echoing overhang is necessary in every actual human acoustical situation in order to avoid a sense of sound death." That would seem to be an eloquent recasting of recent textbooks on speech science; the impulse toward such a claim, framed so squarely in mortal and spatial terms, is much older, stemming from early modern notions of the liveliness and ubiquity of Echo.[33]

Neither Ovid nor his first Renaissance interpreters focused on Echo's ubiquity. If Ovid was the first to twine Echo with Narcissus, Ovid's references were bound to Boeotian woodland, to the legendary home and resting spot of Narcissus northwest of Athens. Almost as afterthought did Ovid's phantom sound and nodding daffodil, *Narcissus Pseudo-Narcissus*, spread beyond this pale. There was however another competing story that drew its pathos from Echo's everywhereness, one that joined Echo not with coldly beautiful Narcissus but with hot hornèd Pan. Pan, animal-man. Pan the player of pipes. Pan the "lover of merry noise." Pan the pal of Dionysus. Pan the ever-lecherous, ever-confident in his lechery, whose advances are rejected by Echo, a virtuous nymph. Humiliated, the goat-footed demigod of peak and pasture sends a madness, a panic, "among the shepherds and goatherds, and they in a desperate fury, like so many dogs and wolves, tore [Echo] all to pieces and flung about them all over the world her yet singing limbs."[34] So wrote the Greek novelist Longus in *Daphnis and Chloë*, where Echo in her remnancy becomes a haunting presence the world over. While Ovid early in the first century gave us a dwindling temporal Echo who takes up our words to return them surprisingly reconsidered, Longus in the second century bequeathed a ubiquitous Echo who takes up our worlds, ready to surprise us at every turn.

Really, the two poetic strands are complementary. Echo carries with her the prospect of sound displaced—uprooted, multiple, abbreviated, translocated—*and* the displacement of sound at every prospect. Pining for Narcissus, Ovid's Echo withers to a *genius loci*, the spirit of a desolate place who, melancholy or bitter, distorts; fleeing Pan, Longus' defiant Echo is scattered so widely that she expands into the spirit of every place, pure and singing. Plutarch, struck like Ovid by the sadness of sound defaced, heard Echo keening over the death of Pan, whom she had secretly loved; the mythographer Macrobius, four centuries after, heard Echo's song as a pan-celestial harmony. Later poets and philosophers caught Echo's song in registers fluctuating between epithalamion and epitaph. Through the efforts of Renaissance Neoplatonists, Echo with "her yet singing limbs" became grace itself, in the form of song—which was, wrote Marsilio Ficino in 1489, "a most powerful imitator of all things." And "when it imitates the celestials, it also wonderfully arouses our spirit upwards to the celestial influence and the celestial influence downwards to our spirit." Echo thus transfigured was residual everywhere, as was the offering of grace beneath harmonic heavens defined note by note, chord by chord, planet by planet, by Ficino and then, with *glissandi* between aphelion and perihelion, by Johannes Kepler. Swept up in one more Renaissance retrieval of the ancients, Ficino and Kepler listened for a music of the heavenly spheres that had much earlier been scaled by the Pythagoreans.[35]

An illustrious contemporary of Kepler's was skeptical: "The Heavens turn about in a most rapid motion, without noise to us perceived, though in some dreams they have been said to make an excellent Musick." Not that Francis Bacon, the statesman, philosopher, and lawyer, decried astronomy or music; rather, as a detail-man, he was dubious of celestial harmonies, which rhetoric flew too close to the Idol of the Marketplace. Yet he too exalted Echo. Like other land-wealthy but debt-ridden Englishmen embroiled in a politically treacherous, religiously divided island culture, Bacon was preoccupied with access, mutability, and catholicism; his Echo was reassuring, a figure of attentiveness, permanence, and universality. The power of Echo's song lay in the purity and patience of her society: how she stuck by men, confirming and correcting. With enough distance from the human to be admirably neutral, Echo was counterpoise to the "crookedly-set predisposition" of the human mind, which Bacon described in visual, narcissistic terms as "so far from being like a smooth, equal, and clear glass, which might sincerely take and reflect the beams of things, according to their true incidence; that it is rather like an enchanted glass, full of superstitions, apparitions, and impostures." The true philosophy he described in sonic terms, as that "which echoes most faithfully the voices of the world itself, and is written as it were at the world's own dictation; being nothing else than the image and reflexion thereof, to which it adds nothing of its own, but only iterates and gives it back." Bacon extolled Echo for her honesty, retentiveness, and responsiveness; all else was forgiven. Indeed, for two younger contemporaries, for the French friar, musician, and mathematician Marin Mersenne, who by 1636 had calculated Echo's speed, and for the German Jesuit and polymath Athanasius Kircher, who in 1650 showed Echo fleeing Pan over mountaintops in plate XI of his *Musurgia Universalis*, *echometry* and *echosophia*—the measurement and understanding of "reflected, repercussive or reciprocal voice"—were key to advances in the playing and composing of music, the acting and design of drama, the monitoring of public life, the proceedings of government and self-government. To take the full measure of Echo would be to take the measure of the world.[36]

Executive and commanding, this notion of Echo agreed with the original Latin of Echo's penultimate exchange with Ovid's Narcissus, who says, *emoriar, quam sit tibi copia nostri*, "I'll die before I give you all [power] over me," to which Echo responds, *sit tibi copia nostri*, "I give you all [power] over me." Disregarding the slipperiness of each local echo, listening instead for Echo's pan-aurality, Bacon, Kircher, and Mersenne took the mastery of Echo to presage a mastery of Nature, Echo's rapt and neutral observation a model of method. Perpetual critic of cliché, solipsism, and arrogance, she heralds a way beyond the Idol of the Cave (the self conspiring with its own thoughts), the Idol of the Theater (cultural scripts), and the Idol of the Tribe (parochial logic). Quite the opposite of siren songs that swamp reason and swipe memory, Echo's song repeats in the manner of a physical law: impassive, regular, recurrent, ubiquitous.

Aristotle (or rather, pseudo-Aristotle, perhaps Strato) had written that all sound was accompanied by echoes, just as all light was accompanied by reflection, and the Latin poet Ausonius had allowed Echo an independent voice through which she announced herself essential for audition. To enlist her in maintaining the lawfulness of sound was no great stretch, except for the problematic ear, a most recalcitrant site of functional anatomy. For Bacon in 1626, the ear was a "sinuous cave, with a hard bone to stop and reverberate the sound; which is like to the places that report echoes." During the prior century, that cave had yielded some of its mysteries to European spelunkers who described and named the hammer, stirrup, and anvil, the three smallest bones in the body; the round and oval "windows"; the semicircular canals; the cochlea; the path of the pharyngo-tympanic (Eustachian) tube; the stapedius and tensor tympani muscles and their role in protecting the ear from damage by loud sounds. The labyrinth of the inner ear, however, remained mysterious well into the 1800s. Meanwhile, the chthonic quality of the living ear as a dark echo-chamber with its own atmosphere, a drumful of air preserved from the womb, would keep the antiphonal art of hearing/responding at a distance from the science of sound in the colder air beyond.[37]

"Acoustica," as Bacon called it in 1605 before "acoustics" was pluralized into a science at century's end, had long been governed by a pair of analogies, one visual, one tactile. The visual analogy was driven by the ghost of Narcissus: sound is a disturbance moving in widening gyres, like the ripples formed by a pebble tossed in a pond, each surrounding circle larger and fainter than the one before. The tactile analogy was pandemoniac: sound is concussive, particle colliding with particle until fatigued and inaudible.[38]

Occasional traveling companions since Hellenistic times, the two analogies had been deftly mitred by Chaucer, who built his *House of Fame* (1374–1385) upon a passage in the *Metamorphoses* where the villa of *Fama* (Fame and Rumor) thrums with "confused noises," doubling all sounds off walls of "echoing brass." Chaucer, himself shaken by the blaring blatting horns of the Great Peasant Uprising of 1381, subdivided Ovid's villa and carted in new materials: for Fame, Infamy, and Oblivion, a castle of beryl staked to a cliff of ice in high heaven; for Rumor, a roughly-latticed cage of twigs in a cloud-valley just below. Both are open day and night and noisy with the shouts and pouts, lullabies and war cries that rise from earthbound voices, but the pinnacled castle is devoted to steady glory or notoriety, the sixty-mile cage to huggermugger. Fame and Infamy spread in concentric waves, rounded with the sound of trumpets but impossible at last to grasp, like Narcissus and his reflection; Rumor churns like Pan chasing Echo, avid for the tumble of particulate bodies, chaotic with burble and brief collision.[39]

How could this be? How could every sound on earth, reverent or scurrilous, secretly communicated or publicly drummed, fly up to the cage and castle? Chaucer happily explained, merging the two analogies. First, the tactile:

> Sound is nought but air that's broken,
> And every speech that is spoken,
> Loud or private, foul or fair,
> In its substance is but air

twisted by pipers piping sharply or tanged by harpists intently plucking, so that, soundbitten and broken, particles could independently move off and up into higher reaches. But would they? Aye, considering that if one throws a stone into the water, it forms a roundel or circle, and

> That wheel will cause another wheel,
> And that the third, and so forth, brother,
> Every circle causing another,
> Wider than itself.

Wider, wider... and here Chaucer worked a marvel. The visual analogy, like Narcissus himself, is superficial, keeping sound to the surface of things along an unbroken plane; its virtue is expansiveness, not profundity. The tactile analogy, more volatile, bounces sound around; liveliness—manic, Pan-ic—is its virtue, not broad coverage. In order to translate soundbites through leagues of sky into utmost heavens, Chaucer had to give depth to the visual and expansiveness to the tactile. He did this by insisting that the roundels multiply, sight unseen, throughout the body of water, below the surface:

Above, it goes yet always under,
Though you may think this a great wonder.

Deepening the third dimension of the familiar ripples, Chaucer argued that as a stone sinks through the water, it causes layer after layer of unseen repercussions. Back now into the air, rambunctious, percussed:

And right thus every word, be it known,
That aloud or private may be spoken,
First moves air about,
And from this moving, without a doubt,
Another [part of] air anon is moved,
As I have of the water proved,
That every circle causes another.
Right so of air, my dear brother;

and right so a sonic ascent makes sense, for speech and/

Or voice, or noise, or word, or sound,
Rises through multiplication,
'Til it be at the House of Fame...[40]

Himself a creature of Fame and casualty of Rumor, Francis Bacon further merged the visual and tactile analogies of sound as he composed the volumes of his *Instauratio Magna* during forced retirement from 1621 to his death in 1626. Sound was the environs of the Great Instauration through which Bacon intended to restore his reputation as he restored the world to its proper senses. If he overlooked the recent Italian rescue of ancient Greek music; if he neglected to meld music, mind, and inner character as would Thomas Hobbes and Robert Hooke; if he relegated to spirits animal or angelic the mystifying operation of the acoustic nerves: even so, Bacon's had been a thoroughly echoic universe where "the eye is the gate of the affections, but the ear of the understanding." Sound for him was a material presence, whether in the schoolyard, where Bacon abruptly left off his play to investigate an echo; or in Paris under the arches of the Charenton bridge, where Bacon heard his voice echo thirteen times; or in London in December 1600, where the playwright Ben Jonson, crafting *Cynthia's Revels* for Queen Elizabeth's court, restored Echo's body, that she be "enricht with vocal, and articulate power."[41]

Bacon was likely absent from those revels; his friend and patron, the Earl of Essex, was by then in the Queen's worst graces, and Bacon too, until he guided the quick prosecution that led to his friend's beheading. Later, as Learned Counsel to Elizabeth's successor, Bacon was more probably in the audience for the 1609 *Masque of Queenes*, where Jonson elevated James I's queen Anne to a House of Fame blissfully free of Rumor. This was courtly antithesis to another house in another Jonson play of 1609, *Epicoene, or the Silent Woman*, where Morose, a detestor of noise who has built for himself a room "with double walls and treble ceilings, the windows close shut and caulk'd," and who goes about in public with a turban of nightcaps over his ears, is tricked into wedding a person so shy as to be speechless and so female as to be recessive, only to find her a spitting, coughing, sneezing, cavorting, snorting fury and a cross-dressed boy to boot.[42] Certainly, after Bacon's downfall as Chancellor of England amid parliamentary tricks and mean rumors, he was persuaded as much from political experience as from experiment that "All *Sounds* (whatsoever) move round, that is to say, On all sides, Upwards, Downwards, Forwards, and Backwards." Visual analogy was inadequate to such relentless dimensionality: "*Sounds* do not require to be conveyed to the *Sense* in a right Line, as *Visibles* do, but may be arched, though it be true they move strongest in a right Line; which nevertheless is not caused by the rightness of the Line, but by the shortness of the distance." Oh? Take a wall: "you speak on the one side, you hear it on the other; which is not because the sound passeth thorow the Wall, but arched over the Wall." What's more, "If the Sound be stopped and repercussed, it cometh about on the other side, in an oblique Line: So, if in a Coach, one side of the Boot [luggage carrier] be down, and the other up, and a Beggar beg on the close side, you would think that he were on the open side."[43]

Strange that a beggar should pop up out of nowhere. Is sound a class act? To be sure: wherever sound comes to *resound* in the early modern world, one must reflect upon the nature of the second (be it the aye-saying courtier, the maid-in-waiting, the dutiful apprentice, the wife). I'll take up anon the echoes of this "seconding" class. Here, the lowest ranks must be attended to. How better in 1626 to evoke sound everywhere and out of place, visibly expanding wave and invisible pummeling particle, than to call into play a beggar from among the "Infinite numbers of the Idle wandering people" whom the ruling and seconding classes were increasingly likely to stumble over, and who in turn were being subjected to stiffer countermeasures? Spreading across England and Wales in the wake of agricultural dislocation, spreading across the Continent in the wake of religious expulsions and the rapine of the Thirty Years' War (1618–1648), migrant laborers, evicted widows, second sons of second marriages (like Bacon), and maimed or footloose soldiers were a threatening noise, resonant of bread riot, tavern brawl, stray matchlock shot, arson and fire-bell. From within the handsomely appointed coach of one who had advised England's constables concerning "sturdy beggars, vagabonds, rogues, and other idle persons," who had asked the Lord God to "pierce the tongues, that seek to counterfeit / The confidence of truth, by lying loud," and who had been disgraced for accepting but *not* acting on bribes, the ubiquity and displacement of noise could be no more starkly personified than through the unseen scrofulous figure of a sudden pleading voice.[44]

Participant in an inventively acoustic culture of "earwitnesses" (1594) and "quire-birds" (choir-birds, ex-cons who had recently "sung" in prison: 1562), of "street cry" (1596) and market "squabble" (1602), of door "creak" and "night-shrieks" (1605), of "night alarm" (1606) and "hubbub" (riot: 1619), the English beggar was a notably "sonorous" (1611) figure whose "snuffling" (1600) and "canting" (begging: 1566) were within continual "earshot" (1607) of "eardrums" (1615). It was a time when convicted felons like Ben Jonson might have their ears clipped and disturbers of the peace their ears nailed to a pillory. It was a time when a Puritan minister could use the image from Exodus 21:6 of a servant's earlobe bored through with an awl in perpetual service to his master as a figure for listening obediently to the preaching of God's Word, in contrast to the "dull Ear" of the drowsy, the "itching Ear" of the fashionably dainty, and the "adulterous Ear" of those who think to find loving words elsewhere than at church. It was a time when "artillery" (a snuff box: 1608) came easily to hand and musicians in small bands called "Noises" barged into houses at night, demanding, "Will you have any music?" When a "ragged regiment" of rogues heaped curses upon constables, "grinding their teeth so hard together for anger, that the grating of a saw in a stone-cutters yard . . . makes not a more horrible noise." A beggar, then, could be no silent petitioner. Like sound itself, beggars were "local motions" meeting resistance, and though shut up temporarily in a workhouse, inevitably "infestious" (1593). "It is evident," wrote Bacon, "and it is one of the strangest secrets in Sounds; that the whole Sound is not in the whole Air only, but the whole Sound is also in every small part of the Air," as it was on Prospero's island in Shakespeare's *The Tempest* (1612), with its music "I' the air or th' earth?" And back to Bacon: "So that all the curious diversity of Articulate sounds of the voice of Man or birds, will enter into a small cranny, inconfused." *Un*confused, that is: clear and penetrating, capable of occupying every nook and cranny, an Echo in every cave.[45]

Scarcely strange for Bacon then to make his way in four short sentences from an out-of-sight beggar to a bell echoing in a windowed chamber to sound's spherical motion to the question of whether "Sounds do move better downwards, than upwards." Questions of class—of instruction, command, and authority—were critical to an analysis of the clarity, power, and ambit of sound. "*Pulpits* are placed high above the people," observed Bacon, who had repeatedly defended royal prerogative. "And when the Ancient Generals spake to their Armies, they had ever a Mount of Turf cast up, whereupon they stood." Were they merely standing above obstacles? Well, no, "there seemeth to be more in it; for it may be, that Spiritual [i.e., extremely rarefied] Species, both of things visible, and Sounds, do move better downwards than upwards." From above, ruling or seconding, visual patterns come clear, "as Knots [designs or mazes] in Gardens show best from an upper Window"; from below, looking up as beggar, digger, or day laborer, the heights are vague. Did the same directionality hold for sound? Are not edifying sermons, galvanizing speeches, and proclamations from on high heard clearly by parishioners, infantry, and commoners below, while from above one hears the rabble as riot or babble?[46]

Chaucer, drawn into the skies as a clerk-dreamer, had to explain how words could be whisked up to the abodes of Fame and Rumor; Bacon, the Baron Verulam, accustomed to the heights of command, was obliged to explain how sound might travel more expeditiously downward. This was personal: under impeachment, Bacon had written directly to the King, protesting, "I have been no avaricious oppressor of the people. I have been no haughty or intolerable or hateful man in my conversation or carriage." Grandson of a sheep-reeve tending a minor estate, son of a Lord Keeper of the Great Seal, Bacon meant to keep things in perspective. In the first of his *Essays*, "Of Truth," he had para-quoted the philosopher Lucretius: "'no pleasure is comparable to the standing upon the vantage ground of truth' (a hill not to be commanded and where the air is always clear and serene), 'and to see the errors, and wanderings, and mists, and tempests, in the vale below;' so always that this prospect be with pity, and not with swelling or pride." With specific regard to sound, he proposed experiments to verify its directionality, about which, at the end of a career beset by enemies who "think to reign, and work their will / By subtle speech, which enters every where," he was less than certain. In his *New Atlantis* (ca. 1623), he lit upon a distant island utopia whose Solomonic guardians had erected perspective-houses to study light, perfume-houses to study smell and taste, engine-houses to manufacture motions, and "sound-houses, where we practice and demonstrate all sounds, and their generation." They had devices to imitate the voices and notes of beasts and birds, to help humans hear, and to propagate "strange and artificial echoes, reflecting the voice many times, and as it were tossing it: and some that give back the voice louder than it came; some shriller, and some deeper; yea, some rendering the voice differing in the letters or articulate sound from that they receive."[47]

Hadn't such renderings all along been the problem? Hosts of "strange and artificial echoes" that startle and betray, that make it impossible for

> The man of life upright, whose guiltless heart is free
> From all dishonest deeds and thoughts of vanity

to live a clarified life, with

> Good thoughts his only friends, his wealth a well-spent age,
> The earth his sober inn and quiet pilgrimage.

For Bacon as Great Instaurator, Echo unhampered was a model of the plain upright speech he aimed for but regularly failed at, in notes to his mother-in-law (of a wife thirty-one years younger than himself) as in official correspondence ending with propositions. Guiltless heart to one side, Bacon was also a dandy, vanity lapping at the heels of a man who appeared in public with full retinue and ruffle. The couplet of well-spent age and quiet pilgrimage was a poetic shell game: Bacon lacked the temperament to play the Christian Stoic,

> The man whose silent days in harmless joys are spent,
> Whom hopes cannot delude, nor fortune discontent...[48]

Echo herself would not rest content with idyll. The *old me* of ancient Echo had been a nymph out-of-doors, mountainous, sylvan; the *hold me* of early modern Echo was a figure of interiors. Naturally, echoes are as ancient as the geophysical world they take in. Neolithic and Bronze Age peoples assimilated them, painting on cliffs or cave walls at points of greatest acoustic reflectivity. Priestly Mayan architects placed stepped pyramids at the narrow ends of stone ball-courts whose sonic delays might reproduce the voices of divinities and plumed serpents. Greek amphitheaters had been hewn out of rock depressions that amplified most sound, imitated in concrete by Romans who controlled for echo. Early modern Europeans learned from the resurrected works of Vitruvius (first century BCE) about his acoustical plans for amphitheatres, whose ruins often lay nearby, plundered for their stone, now revisited for what they said about a golden past.[49]

Experiences of Echo began to take on a sharper, more personal character, in keeping with a European "self" emerging from printed texts, cityscapes, theological fractures, and a new golden age under conscious construction.[50] Leon Battista Alberti, whose 1452 book *On the Art of Building* refitted Vitruvius for the Italian, French, English, and Polish Renaissances, proposed that each house have "public, semi-private, and private zones," with separate though conjoined chambers for husband and wife and a bedroom of her own for grandmother, who, "being weary with old age and in need of rest and quiet, should have a bedroom that is warm, sheltered, and well away from all the din coming from the family or outside." Alberti's suggestion that a Christian house should have private parts was almost indecent. Although the rooms of country villas had been represented by Roman writers as refuges from urban crunch and collision, and although Mozarabic elites on the Iberian peninsula enjoyed a domestic architecture that accommodated seclusion by gender and task, medieval Christian homes had almost never been designed for quiet, let alone for privacy. The poor ate, slept, and worked in windy single-room shelters; the middling shared larger, higher, more insulated houses and beds with apprentices and servant-relatives; the rich made love within sight and hearing of their servants, ate with retainers at their table, and slept with handmaidens at their feet, their sleeping quarters at first indistinct from other spaces, then separated by a two-sided hearth, then by walls pocked with doors, since bedrooms remained passageways as well as centers for the conduct of daily business. Early Christian traditions of hermit caves had been reconfigured by monastic houses, whose rules by the eleventh century discouraged absolute solitude especially at night, lest rampant succubi seduce dreaming monks or incubi lie with dreaming nuns. Lay or religious, beggared or bejewelled, Europeans had slept in common, risen in common, labored in common, eaten in common, worshipped and confessed in common, died in company. If there was privacy, it was in the closet—the prayer closet, for the wealthiest; a "close," or walled yard or garden, for the well-to-do; a "privy" or "close stool" for the middling; alleyways for the poor.[51]

Dormitory environments, with their nighttime snuffling snoring rustling and moaning, their daytime clomping of wooden shoes, their shouting and scraping, were sonically and psychoacoustically rich, but rarely echoic. Domestic walls were thin and uneven, or hung with tapestries to keep in the warmth; floors were dankly earthen, or stone strewn with rushes, or roughly planked; ceilings were low and composite, or higher and timbered; windows, if any, were narrow openings covered with waxed cloth or shutters. Echo in these circumstances remained a wilderness nymph, as she had to be, given that echoes require distance between sound and rebound and are hampered by absorbent materials and sound-scattering surfaces. Technically, an echo is a reverberation that arrives at the ear at least one-fiftieth of a second after its prompt; absent such a delay, one hears sounds as augmented or overlaid, not repeated. Bacon in 1626 could not effectively measure fiftieths of a second, and his own hearing was limited by the opiates he took for infirmities plaguing him lifelong, but he did distinguish between two kinds of sonic reflection, "the one at Distance, which is the Echo, wherein the original is heard distinctly," and the other "in Concurrence; when the Sound reflecting (the Reflexion being near at hand) returneth immediately upon the original, and so iterateth it not, but amplifieth it." His physics may have been inexact, but his sociology was telling: "concurrent" reflection was the sound of crowds and close quarters; the clarity of Echo implied a private hollow between oneself and the world where one might sustend the gist of things. Granted her honest company, Echo was kissing cousin to the modern "self" who would seek out solitude, attend to "inner musics," venture into the mountains to be struck "with the hoarse echoings of every sound within the spacious caverns of the wood," where space might astonish but did not terrify.[52]

Space had been opening up since Alberti's time and the completion in 1436 of his friend Filippo Brunelleschi's dome for Santa Maria del Fiore in Florence, the largest dome ever built with bricks and mortar.

This cathedral did not reach the heights of Old St. Paul's in London (the world's highest manmade structure until its 527-foot spire was destroyed by lightning in the year of Bacon's birth, 1561). Nor did it top Notre Dame in Strasbourg (finished in 1439 at 472 feet). But its double-shelled dome was 176 feet across, wider than the Roman Pantheon, wider than would be the dome of St. Peter's (1564) or, afield, of the Taj Mahal (1648). Almost as capacious as the Hagia Sophia in Constantinople, Santa Maria del Fiore shared with that formidable Byzantine church (a mosque, as of 1453) the long reverberation times that come of arched volume and hard surfaces—marble floors, smooth stone walls and, for Brunelleschi's dome, 4,000,000 bricks.[53]

Had this experience of spaciousness, with its attendant resonance and echo, so possessed Alberti, a superb organist and canon of the Florentine cathedral, that he must insist that areas of a house meant to be "shut off or free from noise should be vaulted"? Was Echo becoming vital to privacy, confirming a necessary distance between the individual and the world? Though a shallowly vaulted ceiling in a villa would yield less delayed reverberation than the domes of churches and mosques in Florence, Rome, London, Constantinople, or Agra, it would still bespeak those who spoke beneath it, one at a time. Invested with the air of a contemplative rather than a melancholic, Echo enabled intimate conversations, antidote to the uproar of public office, where "You hear constant complaints, innumerable accusations, great disturbances, and you are personally beset by litigants, avaricious creatures, and men of the grossest injustice who fill your ears with suspicions, your soul with greed, your mind with fear and trouble."[54]

Avarice and acoustical terror appeared in the third of Alberti's four books *On the Family*, written as Brunelleschi's dome was nearing completion. For a bastard who was born to a father in exile, a boy who lost his mother at two and father at seventeen, a youth whose fortunes were constrained by executor-cousins, a man whose murder was plotted by relatives, and a respected elder who never married or sired children, "family" was a fiction of security, long life, economy, loyalty, and lineage. More famous among Florentines for his use of linear perspective, his art, his mapmaking and architecture, Alberti limned the ideal family during the same decade that he was obsessively drawing, painting, and writing up his own profile. Family occupied him not because he meant to marry or to withdraw from public life but because, turning twenty-eight in 1432, he had reached the conventional age of adulthood, a time to map his biography and philosophy onto the life-course of his father and ancestors, in their Tuscan vernacular. Presented in a series of dialogues among men, *I Libri della Famiglia* was more concerned with fraternity and solidarity than with sex or childrearing. Women appear obliquely, "almost all timid by nature, soft, slow, and therefore more useful when they sit still and watch over our things"; gossips, their "peak of dignity" is silence. Husband and wife should enjoy "conjugal friendship and solidarity," which "preserves the home, maintains the family, rules and governs the whole economy." Solidarity was the objective; great passion could be unwieldy. The architectonics of the family, like that of a well-designed house, depended upon right proportions of closed and open spaces, physical, emotional. Under one roof, family and house were to be surveyable and serene, but privacy for Alberti was less an ambition than a contingency, as Echo herself was contingent. A life lived too much in private was a life, like Echo's, straitened to fragments, so Alberti's domestic floorplans kept bedchambers as crossroads, their several doors linking different rooms, on the premise that, as one historian reminds us, "company was the ordinary condition and solitude the exceptional state."[55]

Nearly two centuries later, architectural renderings of large houses new or renewed began to show isolated bedrooms with doors bordering on "one common way in the middle through the whole length of the house," commodious enough that servants could pass unremarked through the long hallway without disturbing secretaries or sleepers. Back stairs were constructed so that cooks, scullery boys, and laundry maids need never appear in public view of masters or their guests. "That which is chief," wrote Sir Roger Pratt in 1660, "the dwelling house [must be] kept sweet, neat, and free from all noise and such like other inconveniences."[56]

Neatness was possible, with supervision. Sweetness was implausible, given the odors spreading from kitchens hung with game, from unwashed bodies overlaid with weeks of perfumed powders, from stewing chamber pots and unventilated commodes. Quiet was next to impossible. The long hallways echoed, stairways (grand and back) echoed, and behind their hinged doors the wealthy family walked among paintings and mirrors whose illusions of spaciousness colluded with large glass windows but made for altogether more reverberation. Unless... unless echoes were requisite to a sense of privacy, heard not as noise but as the sounding of a proper distance, a distance that made for domestic and other harmonies.

Harmony and distance: Baroque composers across Europe were enamored of Echo, from the echo *fantasias* of Jan Sweelinck in Amsterdam and Adriano Banchieri in Bologna to the *Sonata in Echo for Three Violins* of Biagio Marini in Germany; from echo-effects alternating *forte* and *piano*, used repeatedly by Giuseppe Tartini in Padua, to the echo-lines of the *Christmas Symphony* of Gaetano Schiassi in Lisbon; from the final "Echo" movement of the *Partita in B minor* by J. S. Bach in Leipzig to the overlapping, contesting fronts of sound from separated choirs as exploited by Giovanni Gabrielli in Venice. Audiences to Georg Philipp Telemann's *Overture in F (The Alster Echo)* heard frogs and gulls echoing off Lake Alster in Hamburg and a pandemoniac "Village Music of the Alster Shepherds," full of missing notes, bleating, and blare. Chambers for chamber music blared in sympathy, often miserably, from too-lively plastered walls and ceilings echoing the implacable beat of a conductor's long wooden staff on hardwood floors.[57]

Hardwood from Brazil, from the Americas.... Where Columbus mistakenly heard the song of the European nightingale upon Caribbean shores while receiving raucous, word-echoing red macaws presented to him by welcoming "Indians."... Where Jesuit, Gallican, and Puritan missionaries listened hard for Hebrew syllables among the orations of Inca and Iroquois chiefs and the gabble of (the Ten Lost?) tribes while foisting Roman alphabets on new, tongue-twisting languages, and mistaking repetition for faith.... Where the spaciousness of the new hemisphere was more often audible than visible to explorers and hunters struggling through jungle or thick forest, dense underbrush, high prairie grasses, roaring rivers.... Where the noise these men made in their violent passage was as disconcerting to the indigenes as the polished armor they wore and the pale skins beneath. Europeans came with steel blades, hammers, and iron anvils that parted the air with a brightness unlike anything of bone or stone; with muskets and artillery percussing louder than all but hurricane (< *Huracan*, a Mayan creator god); with two-hundred-pound growling mastiffs; with neighing, snorting horses shod in metal; with trumpets and hawkbells, then with hurdy-gurdies and churchbells.[58]

There, in the New World "wilderness," Echo kept for a time to her ancient settings; in the Old World she was becoming a fixture of the built environment. She was active in the stone corridors of narrow city streets, in the hallways of country houses, in the lyrics and staging of songs and operas, in artificial grottoes hollowed out for aristocratic gardens and public amusement parks, in the echo-organs of cathedrals whose vaulted domes sometimes (as at St. Paul's) had whispering galleries. The urban merchant had to be wary of her in the grand edifices of commerce, like London's Royal Exchange, where Marforius cautioned Pasquill in 1589 to "Speak softly," for "This Exchange is vaulted and hollow, and hath such an Echo, as multiplies every word that is spoken by Arithmetic, it makes a thousand of one." The urban poor encountered Echo in local churches and in cellars or garrets across from the shops of blacksmiths, tinsmiths, coppersmiths, coopers. Everyone, urban or rural, heard her during the Thirty Years' War (1618–1648), under the impress of heavier field cannon whose pounding echoed across battlefields to settlements miles away...six, ten, twenty miles away.[59]

Or more. At sea, nothing on the horizon for a hundred miles, the giant Pantagruel hears voices in the springtime air. His shipmates fear for their lives, but the skipper calms them: they are on the edge of the Frozen Sea, where a battle took place last winter, and the "shouts of the men, the cries of the women, the slashing of the battle-axes, the clashing of the armor and harnesses, the neighing of the horses, and all the other frightful noises of battle became frozen on the air." Now thawing, they can be heard again. Pantagruel grabs handfuls of the more colorfully cool-whipped sounds out of the upper air. Thrown on deck, they melt into gibberish, except for one large crystal that startles the company, exploding "like an uncut chestnut tossed on an open fire"... in its day, a cannonshot. Other sounds go off: drums, fifes, trumpets. The company hear sharp words, terrifying words, ugly words, and a final

> hin, hin, hin, hin, his, tick, tock, crack, brededin, brededac, frr, frrr, frrrr, bou, bou, bou, bou, bou, bou, bou, bou, tracc, tracc, trr, trrr, trrrr, trrrrr, trrrrrr, on, on, on, on, on, ououououon, Gog, Magog, and goodness knows what other barbarous sounds

which François Rabelais's rock-solid Captain interprets as "the noises made by the charging squadrons, the shock and neighing of horses." Where the trope of the frozen word augments the echo of clashing sabers and cannonade, the noise of war becomes at once babble and Babel, intimate and inescapable.[60]

On the battlefield itself, reverberations were being felt more intimately, more inescapably. What would much later be called "shellshock" was first identified on the Dutch front of the Thirty Years' War among Spanish troops fatigued by long sieges, wracked by the whinge of bullets from "dragons" (carbines) and concussion from loudly inaccurate cannon. Soldiers who became despondent, who lost their appetite, and who displayed a "disordered imagination" were diagnosed *estar roto*, "broken" as on a torturer's wheel. They were in effect soundbitten: they had lost their resolve amid life-threatening sounds that could *ricochet* (a new usage) from any direction. Lucan had described the Roman troops of Caesar and Pompey at the Battle of Pharsalia (48 BCE) hearing "utterances / of their own madness repeated by the entire earth," and medieval romances had mélées ringing with blows of metal upon metal, but by the seventeenth century, revised tactics and weapons deadlier from greater distances had led to a newly inchoate war at ground-level, deafening vulnerable infantrymen from any sense of their position. In the fog of battle, the sounds of trumpets, muskets, artillery, shouts and convulsive screaming might be all a man knew before his life was cut short. There was so much "noise and terror, that you could not [but] have thought that now at last we had moved to Hell's gates," wrote a captain after the battle at Marston Moor in 1644 during the English Civil War. "The two main bodies joining made such a noise with shot and clamour of shouts that we lost our ears, and the smoke of powder was so thick that we saw no light, but what proceeded from the mouths of guns." By 1678, soldiers who could no longer stand such noise were coming down with "nostalgia," which could be fatal unless they were sent home or dosed with opiates. Copywriters and interior decorators may today associate nostalgia with a rewarding timesickness and profitable homesickness; at the onset it was a military and acoustic affliction, a trauma of separation, meaninglessness, and echoic delay.[61]

Perhaps because the displacements of echo had become widespread in the course of daily life and nearly yearly war; perhaps because they had a subtle bearing on senses of privacy and selfhood; perhaps because they were too obvious and ordinary an artifice in a world of baroque contrivance (the reasons must be multiple), echo-effects fell for a time into disrepute as "false wit." The phrase was that of the essayist Joseph Addison in 1711, but ridicule had begun as early as 1662 with Samuel Butler's mock-epic, *Hudibras*, where Echo's dialogue with a discombobulated mock-knight shrinks his simplemindedness to silliness: grudge it/*Mum budget*; dish/*Pish*. Addison himself, contributing to a joint translation of the *Metamophoses* by the finest English poets of his era, evaded Ovid's echoic wordplay in the chase scene between Echo and Narcissus. Though cursed "To sport with ev'ry sentence in the close," Echo does no sporting whatsoever; rebuffed by Narcissus, the "love-sick Virgin" wanly weeps herself away to a skeletal voice, until

> Her bones are petrify'd, her voice is found
> In vaults, where still it doubles every sound.

A *she* reduced to an *it*, Echo decays from babbling nymph into prosaic architectural element, uniform and unchallenging.[62]

Taken as evidence for the ongoing suppression of woman's voice, reducing it (and women) to a sheer seconding, Addison's translation was of a piece with a passage further along in the same volume. John Dryden's version of Ovid's tale of the maidens Iphis and Ianthe emphasized that their love for each other dare not be consummated until the goddess Isis transformed Iphis into the man she should always have been, and

> Her features, and her strength together grew,
> And her long hair to curling locks withdrew.
> Her sparkling eyes with manly vigour shone,
> Big was her voice, audacious was her tone....

A man's voice was by nature spacious and audacious; smaller was woman's, and meeker. But the audacity of Echo, persisting in baroque music, would return soon in Gothic and Romantic poetry. Addison's maneuvers were rather a reaction to the urban omnipresence of Echo, familiarity breeding a mild contempt. As a weekly columnist on the alert for gossip, fiddlefaddle, and the flack of language, Addison remained fascinated with discrete, displaced soundbites: scraps of (ghostly) echoes among the burial vaults of a ruined abbey; the (anonymous) "twanking of a brass kettle" on the street; the (baffling) calls of scissors-grinders and bellows-menders. Half a year before he scowled upon "the conceit of making an echo talk," Addison had written delightedly of "Frozen Voices," framing his *Tatler* essay as a "splendid lie" from the lips of that prime counterfeit, the medieval knight, Sir John Mandeville. Butler in *Hudibras* had alluded to "words congealed in northern air"; Addison filled in the backstory, couching it among Mandeville's fabulous accounts of his world tour. Reaching Nova Zembla in the frigid north, Sir John, his English crew, and the crews of a Dutch and a French ship, put in to shore to refit their vessels. Building cabins against severe weather, the men cannot hear each other speak, their halloos and curses congealing instantly in the cold air. After three weeks of arctic silence, Nova Zembla is relieved by warm winds, and Sir John's cabin fills "with a dry clattering sound, which I afterwards found to be the crackling of consonants that broke above our heads, and were often mixed with a gentle hissing, which I imputed to the letter S, that occurs so frequently in the English tongue." At length, as if riffling the pages of a primer, he can discriminate syllables, then short words, then entire sentences, as well as the "posthumous snarls and barkings of a fox." Next cabin over, the "harsh and obdurate sounds" of Dutch require more reheating before much can be heard other than brandied sighs and belches. The French, *par contre*, are at sixes and sevens; in the first heat of the moment, their passionate yet flimsy words have fallen apart and utterly dissolved. Above, however, playing for hours on end, Sir John hears the defrosting minuets of a French pocket fiddle.[63]

Why does music played indoors sound better in frosty weather? asked Francis Bacon, who was intrigued with processes of freezing and would catch his death of cold, so it was said, after a wintry day out of doors experimenting with the properties of snow as a preservative. Cold, reasoned Bacon, could not aid the progress of music through the air, but it could make instruments of wood and gut more hollow, thus more resonant. In general he dismissed temperature as a factor in the transmission of sound, which was a mistake, though some of his critics were no less mistaken to believe that sound travels more easily in winter because cold air, being thinner, presents less obstacles. (The colder or thinner the air, the less elastic, so the *more slowly* sound travels through it; sound velocity is directly related to the elasticity of the medium through which it passes, inversely related to density.) Bacon had also been mistaken to conflate echo with all sonic reflection from a distance, though neither he nor his critics understood that atmospheric temperature inversions can refract sound, especially low-frequency sound, downward across miles of flat terrain. This phenomenon, more common in winter, and likely to have been more frequent during the longer, colder winters of the Little Ice Age that benumbed early modern

Europe, lay behind both the expansion of the ambit of echoes and extended play with the classical trope of the frozen word.[64]

That said, the cultural climate in which noise could spread in all directions was shaped by more than the temperature of the outside air. Only by tracking the historical confluences of material experiences, equally material poetics, and always-material habits of perception can we make sense, for example, of the *London Spy*'s 1698 punchline to one more tale of frozen words, which, when thawed at a warmer latitude, broke out together "like a Clap of Thunder, that there was never such a Confusion of Tongues ever heard at Babel."[65] Thunder we have yet to take up, but it was the expanding notion of sounds bitten off one from another or gone astray—uprooted, wildly multiplied, truncated, distended, isolated, and unintelligible—that contributed to the likelihood of coming everywhere upon noise, linking as it did the trope of the frozen word (sounds translocated) to Babel (unintelligible sounds, from every mouth), to Echo (truncated sounds, out of nowhere), to *fama* (multiplied sounds, public or loudly private), to the mouth of hell.

3. Roaring Shrieking Bleating: Aftersounds

Hell had not always been a locus of noise. Mesopotamian, Jewish, Greek, and Roman underworlds from time to time launched vengeful spirits on missions into the land of the living, like fumes hissing from those sulfurous steaming clefts in volcanic rock that pantheistic Romans and medieval Christians heard as portals to the nether regions, but for the dead the underworld was deathly quiet. That is, until Aeneas, in Virgil's Latin epic (28–19 BCE), caught wind of a frightening din from below, the scrape of iron and drag of chains. Competing against cults of Isis, Dionysus, and Mithra with their subterranean rituals, early Christians quitting their own catacombs had motive and opportunity to reconceive the underworld as more than merely lugubrious. Elaborating upon Jewish folk motifs about Gehenna, they went well beyond the Gospel's "outer darkness" and "gnashing of teeth" (Matthew 8:12, 22:13) to picture an actively vindictive Hell that taught fiery lessons to noisy schismatics and backsliding believers. The farther off the Second Coming and more obdurate the "pagans" in their resistance to an evangelical, imperial Christianity, the more harshly instructive Hell had to be. *Harsher,* and more alert to crimes of language, since it was through the Lord's Word that the Good News must come unadulterated to each ear. In the second-century Apocalypse of Peter, blasphemers hang by their tongues and roast over open fires, slanderers and false witnesses gnaw their own lips as flames scorch their offending mouths. *Harsher, and louder,* for if the agony of the sinful were soundless, it would mock the saintly endurance shown by Christian martyrs under Roman torture. In the fourth-century Apocalypse of Paul, Hell is heard before it is seen, a dark realm of grief where the voices of the damned, crying out for a still-plausible mercy, so move the Archangel Gabriel that he pleads successfully with the Lord to grant one day's respite. *Harsher, louder, and unremitting,* for as Heaven's rewards were painted in ever-more sensual detail, the contrast with Hell had to be sharpened. By the eighth century, Hell was spewing up globes of black flame that fell back into its great pit, and rose, and fell back, each globe hot with condemned souls, while the volcanic underworld echoed with demonic laughter and horrible screams. So noted the Venerable Bede, trusting to reports from a Northumbrian named Drythelm, who had by chance risen from the dead well before Judgment Day, and who while dead had received a tour of Hell, where he heard "the noise of a most hideous and wretched lamentation, and at the same time a loud laughing, like that of a rude multitude insulting captured enemies. When that noise grew plainer and came close to me, I observed a gang of evil spirits dragging the howling and lamenting souls of people into the middle of the darkness, while the devils themselves laughed and rejoiced...and as they went down deeper, I could no longer distinguish between the lamentation of the people and the laughing of the devils, yet I still had a confused sound in my ears." Laughter and lamentation meet in roars or shrieks; roar or shriek, whichever Drythelm heard, evinced a thoroughgoing agony, for Hell was acutely corporeal and the damned had to suffer from more than a tormented conscience. Medieval Christians who knew not the Venerable Bede heard the roars of the damned issuing from terrorized mouths on frescoes and tapestries of Hell, or through images of the Day of Judgment woven into chancel screens, sculpted on altarpieces, carved over church portals. The mouth of Hell may have gaped wider during apocalyptic eras when Doomsday loomed, but its jaws opened everywhere, and loudly.[66]

It was a loudness of sound audibly deranged, shredded, turned upside down. The hornblowing and potbanging of *charivari* that hounded weddings of codgers to teenagers, as well as the "rough music" of drums, pans, pot spoons, lids, and cauldrons that hounded husbands or wives who abandoned their spouses—these were replayed as sound-effects for Hell during performances of medieval miracle, morality, and mystery plays at the same time that artists were painting musical instruments into the hands of angels. Latin, the language of the liturgy and psalters, was garbled by playwrights and ritual clowns into half-assed fragments and fractious grammar, *ffragmina verborum / tutiuillus colligit horum / Belzabub algorum / belial belium doliorum.* The Divine Kiss of Christian eschatology was reduced to scatology—*Osculare fundamentum,* Kiss My Ass. God's Word and Breath were everted into the breaking of wind, thunderous farts. On squares in front of churches, on carnival stages, in sermons and visions, in music and manuscript, Hell exploded the aural consequences of the First and Second Falls: the curse of the two sound/bites of Eden—a mortal, anxious, painfully physical life at odds with itself—became an everlasting, bellowing bodily terror; the curse of soundbitten Babel—of people driven apart and made mutually incomprehensible—became an irremediable exile in a barbarous place where any petition for mercy would be spat back in flames.[67]

As Hell went, so went the Devil. Angels, not demons, had overseen Hell in early apocrypha attributed to Peter and Paul, and St. Augustine had clipped Satan's wings, fearing that an independent fallen angel could seduce the innocent, as he himself had once been seduced by the comforts of Manichaeanism with its utterly binary forces of good and evil. However, as the punishments of Hell became more inevitable and unabating, Satan by popular referendum was sworn in as governor and his various faces (serpent, rebel, tempter, dragon-tyrant, prince of darkness) were consolidated into one diabolic countenance. His minions, who had begun their careers as close-mouthed receiving clerks or mute guides and whose numbers grew from 30,000 in the second century to 7,405,926 in the sixteenth, danced through late medieval Hell barking and taunting and scraping music out of new instruments of torture—violins, bagpipes, and portable organs that would take on, in the art of Hieronymous Bosch, a violence practiced upon human organs. As hideous in their sounding as in their wounding, Hell's workforce wielded tongs and pincers that tore at the tongues of sinners, producing the unholiest of howls and stigmata, so reproducing in their subjects what they were themselves: sound fiends.[68]

Sound fiends, though at home in Hell, were by no means bound to it. They surfaced, figuratively and materially, in the accelerating number of cases of possession reported from the twelfth century on. In fact, given the perplexingly similar gestures, postures, rhythms, and rhetoric of divine and demonic possession, it was sometimes only through the quality of the sounds produced that communities could distinguish those inspired by God from those deceived by Satan or gone mad. Since the Holy Spirit swept into the body through the heart, its presence should prompt ecstatic sighs, emotional prayers, fits of weeping; fiends, tunneling in through the viscera, should pump out throaty screams from abdominal depths, or a diaphragmatic growling and rumbling. Whether the possessed were out of their minds or in the hands of evil spirits was a judgment call that was shifting toward evil spirits; either way, the wildly gesticulating woman in a twelfth-century church whose screaming disrupted sermons and worship was audibly not on a path to sainthood. True, medieval physiology had it that women's bodies were hollower and more vulnerably porous than men's, so their weaker hearts could overflow with love of God, expressed in "groaning, sobbing, lamentation, song, and clamors," but such heartbursting clamor, noted around 1420 by a German churchman, Johannes Nider, could be told easily enough from the shrieks of Hell.[69]

Easily? Satan was nothing if not a master ventriloquist. If he could put obscenities in the mouths of innocents, he could as easily disguise foul motives and fouler sentiments with pretty words and lilting tones. At full strength, his zealous lieutenants could take control of the senses of those they possessed, producing impressions of sanctity. Under such devious circumstances, the discernment and ejection of evil spirits was an arduous, often sonic, art. Exorcism manuals, a genre new to the fifteenth century, posted comprehensive lists of the ways that one could tell a bad spirit from a good, lists that relied heavily upon a good ear. Of the twenty-two signs listed by the exorcist Zaccaria Visconti, the first four (and nine all told) were acoustic:

1. their tongues are black and swollen, their throats inflated or constricted, as if strangling;
2. demoniacs "weep aloud and do not know why they are weeping";
3. "They answer questions angrily in a loud, quarrelsome voice";
4. vice versa, they do not speak when pressed to do so;
7. "They say many things whose meaning it is impossible to decipher";
13. they make animal sounds, "the roaring of lions, the bleating of sheep, the lowing of oxen, the barking of dogs, the grunting of pigs, and so on";
14. "They hiss through their teeth, froth at the mouth, and show other signs like rabid dogs";
20. "They cry out when one places any saints' relics on their head";
22. last but hardly least, they avoid speaking sacred words, "and if they do pronounce them, they try to stutter, or corrupt the words, and demonstrate extreme boredom. At length they cannot say them at all."

Although the weeping and unusual silences could characterize the divinely inspired, *in toto* Visconti's signs constituted the portrait of a sound fiend who makes a hash of language, uprooting, truncating, exaggerating, distending. It was not for lack of other tools, therefore, that exorcists themselves relied heavily upon the power of the Word and their own resolute voices to battle for the souls of the possessed, much like the popular preacher Bernardino of Siena, who was not afraid to use his "clear, sonorous, distinct, explicit, solid, penetrating, full, rounded, elevated and efficacious" voice in the manner of *Isaiah* 58, shouting like a trumpet, high and low, to return his flocks to the Lord amid loud weeping and shrieking. At Bernardino's bonfire of vanities in Florence, wrote one observer in 1424, "The shouting, I can't begin to describe, it seemed like thunder." Exorcists too would turn up the volume, believing that pious shouting, like churchbells, annoyed the Devil no end.[70]

Oral *and* aural processes, exorcisms were audiopathic cures, provoking by force of the Word audible crises of expulsion, *and* they were intent conversations with fiends. Though the Devil could never "so perfectly imitate the human voice that the falsity and pretence of it is not easily perceived by their hearers," and accordingly demons spoke "as if their mouths were in a jar or cracked pitcher," exorcists listened patiently for the tale behind the possession, listened through soundbitten sentences "confused, ambiguous, muffled, and feeble"; listened through the "horrible roaring voice" of young Alexander Nyndge; listened through the fits of Thomas Harrison, who would "howl like a dog, mew like a cat, roar like a bear, froth like a bear." Exorcists had begun listening, like physicians, for etiology; by the 1470s they were listening, like lawyers, for conspiracy. "I have listened, most malignant demon," they would say, following a script. "I have listened, damned creature, reprobate and cursed by God." Then, "I exorcise and declare you anathema.... I command that you tell me by what means, whether by means of food or drink or by any maleficent superstitition or trickery somehow wrought upon this servant of God, you have rashly presumed to enter this servant of God." They would neither relent in their anathemas nor rest easy in their listening until the invasive demon had given up the names of conspirators.[71]

Thirty editions of the *Malleus Maleficarum* from 1487 through 1669 defined for priests, lawyers, and inquisitors across Europe the nature of this maleficence, or witchcraft, and detailed dozens of methods of eliciting confessions that incriminated others, for witchcraft had been reconfigured by the Church from a folk mischief (acts of sorcery) to a contract with Satan (diabolism). No longer a heathen remnant or creedal ignorance but a full-blown heresy that appeared to be gaining ground on all sides, its extent had to be exposed, its operatives stamped out. In this long campaign, the *Hammer of Witchcraft* was a blunt instrument, its rules of evidence unruly, its arguments arguably incomplete, its conclusions inconclusive. Its author, Heinrich Krämer, a Dominican Master of Theology and Inquisitor for Germany, belabored the stealth, ruthlessness, avidity, and hence—it was his job to make this connection, and the connection made his career—the pervasiveness of the Devil and the Devil's consorts. Knowing how craftily a cabal of evildoers could lurk behind facades of piety, naïveté, or senility, Krämer sanctioned deceit in the cornering of the deceitful, bluster in cases for which nothing turned up but a case of bad *fama*. What exercised him most? Silence.[72]

Out and about, across empty fields or on moonlit missions, women witches were supposed to be noisy, their cackling and cursing part of a set of behaviors akin to the blatancies of the possessed but more reprehensible because witches were in bed with Satan, whose lovebites made them more sensual, more quickly aroused by touch and smell, far-seeing and far-hearing, and more likely to have second sight than a third nipple, yet insensible to pain or cold. Copulating, witches made noise not love: kissing Satan's ass, performing fellatio, screeching at rough penetrations. As wild night riders in the late medieval folk imagination or as orgiasts in black Sabbaths of the baroque theological imagination, as village vipers or town shrews in local courts, witches cried out, spoke out, broke out into unseemly laughter. Neighbors found them out because, disrupting daily life with freak disasters, they were too loud not to betray the source of their charms, potions, curses, and worse. Once found out, witches in prison characteristically could not keep quiet, though they might speak the devil's tongue: "Mr. Brown a Master of Arts of Trinity College Oxford went to the prison being about two a clock in the morning," where nine women of Northampton in 1612 awaited arraignment, "and heard a confused noise of much chattering and chiding but could not discern a ready word," which noise was cited as probative that the women *were* witches.[73]

Ergo, only the power of the Devil himself could account for the unnatural silence of some women (the garrulous gender, that will "keep up a clamor all day like a barking dog") subjected to interrogation for witchcraft. Read with an ear to resistances, the *Hammer* had been damped three years before it was written, when Walpurgis of Hagenau, "notorious for her power of preserving silence," was found to have taught "other women how to achieve a like quality of silence by cooking their first-born sons in an oven." Outrageous, such collective defiance; inquisitors relied upon chains of confessions to dramatize the peril of this heresy and prove that witches were conspirators. Having shown that witchcraft could be found in every unlit corner and that most witches were women, Krämer noted "the great trouble caused by the stubborn silence of witches," a silence sometimes unto the death, when accused women preferred to hang themselves rather than come clean. "In conclusion we may say that it is as difficult, or more difficult, to compel a witch to tell the truth as it is to exorcise a person possessed of the devil." This would justify the severity of means used to get at the truth, but the results could be circular: the greater the torture, the more frustrating the silence. So perturbing was the silence of these women that the Dominican took cold comfort in his conclusion, and a few pages later was at it again, worrying over Walpurgis and her infernal "gift of silence" (those powdered grounds of roasted first-borns), deciding that "a hundred thousand children so used could not of their own virtue endow a person with such a power of keeping silence" unless God willed it, which had to mean that the making of the powder was but another of the Devil's ploys to get Christians to commit heinous acts that would forever bind their souls to Hell.[74]

Should these taciturn women break their silence, their voices could bewitch "by the mere sound of the words they utter, especially when they are exposed to torture." Such words did not depend upon clarity for their force; issuing from the lips of women "put to the question," they would have been heard as the distended soundbites of hours stretched on the rack. Nor would these sounds have been a forgiveable weeping; their hearts turned icy by Satan, witches were incapable of tears or repentance. Echoing off the stone walls of the Inquisition's subterranean chambers, their syllables would have drawn their power not from definition or despair but from distortion, as in Hell itself, where the excruciation of the damned merged with the ecstasy of demons. In several senses, the European "witch craze" heaved up the underworld into the land of the living and spread it abroad, sometimes by its flames or fumes, always by its hissing, roaring, and shrieking.[75]

Acoustics was the anvil, the shaping base, to the *Hammer*. Acoustical analogies underlay Krämer's explanation of how the earthly and unearthly could intersect to Satan's advantage, how an ethereal demon could be felt as a material actor, how witchcraft was no delusion. Could witches truly fly? Was sex with Satan for real? Yes, using a certain unguent and following the Devil's lift-off procedures for broomsticks or chairs, witches could fly through the air (see the Book of Daniel and Aristotle's *Physics*). As for sex, one had to appreciate the process by which the Devil assumed and consumed bodies, which entailed an analysis of the turbidity of air and the physics of sound. Air was "in every way a most changeable and fluid matter: and a sign of this is the fact that when any have tried to cut or pierce with a sword the body assumed by a devil, they have not been able to; for the divided parts of the air at once join together again." Air could not hold a shape unless imbued with other matter, for example, the condensations that give body to fog; in other words, air needed a co-conspirator, and the Devil likewise. Fully "inspissated" with an agreeable body, the Devil made himself at home like an occupying army. Housed within, he could have intercourse with a witch in every meaning of that word: he could speak with her, hear her out, eat with her, lie with her. Moving from dialogue and commensality toward embrace and coitus, the narrative logic of bedevilment had always to begin with true conversation, to which four things were requisite: a percussable atmosphere; lungs to pump air; an inner voice; a means to shape and loft sound. Stumbling over his own criteria, Krämer, known for his pulpit eloquence, jumbled mouth, meaning, and mind so fully that the mouth appeared to be the place where mind made itself known to the speaker as well as the auditor, by way of teeth, tongue, lips and surrounding air. Devils had no lungs or tongues of their own, but "by some disturbance of the air included in their assumed body, not of air breathed in and out as in the case of men, they produce, not voices, but sounds which have some likeness to voices, and send them articulately through the outside air to the ears of the hearer." In cause and in effect, the Devil's speech was a fine semblance, his hearing finer, "far finer than that of the body; for it can know the concept of the mind and the conversation of the soul more subtly than can a man by hearing the mental concept through the medium of spoken words." Sex with the Devil was just like that, a fineness of communion, though complicated by the messiness of the semen that had to be stolen from oblivious young couples in the midst of innocent copulations, for the Devil had none of his own. Witches were demoniacs in more demeaning intercourse with the Devil than were the simple possessed, so they were more culpable in the sharing of voices and the befouling of the Lord's Word that had, in the beginning, made the world what it was.[76]

In sum: the more cases of possession, the more sorcery to be uncovered; the more demoniacs treated and witches found out, the more expansive the scouring of towns and villages and the greater the anticipation of noise on every side, from neighbors and strangers, young and old, male and, especially, female. Distressed, distressing fragments of sound might come from sickened animals, colicky infants, stricken adults, or from those who had bewitched them; from the afflicted and their healers, from witch-heretics and witch-hunters. What's more, demoniacs and witches were frequently discovered and diagnosed by the qualities of their soundmaking, noisiness a trait so distinctive that it could be complemented only by a preternatural silence. That silence broken, the twisted bits of sound that might come from the mouth of a person shattered on the rack could themselves be deadlier than contagion. Finally, the central process by which the Devil took hold of a demoniac or witch was explained by analogy to principles professedly acoustic.

Most of this sound system held as well for Protestants as for Catholics. In order to understand how this could be, we must return to the afterworld, this time to Purgatory, which was nowhere to be found in primitive Christianity except as a "sleeping in the bosom of Abraham" (or later, Jesus), the dead faithful slumbering toward an ever-delayed Second Coming. As the thermostat was turned up on Hell and as martyrs, holy men, and a few holy women were formalized into Christian saints who breezed up to Heaven at the instant of death, theologians came upon scriptural hints of a middle place for those believers who had committed venial sins but had been on the whole decent. The Last Judgment receding before the assured continuities of the Church, the dormitory of the faithful-in-waiting became crowded with half-awake souls increasingly in need of reassurance that repentance in life, deathbed remorse, and intercessory prayers from the living would suffice to hoist them out of their graves into the good graces of Heaven. Lay discussions of the afterlife made it clear that the living expected the "not-very-good" to be rewarded for enduring a season or two of Hell; encounters with ghosts made it noisily clear that absolution had to be on the horizon, no matter how many hundreds of years it took. In one of her visions, the twelfth-century saint, Elisabeth of Schönau, saw her deceased uncle stepping out of a fiery cave. "Can he never be freed?" she asks. The Lord tells her he can, "if in his memory and that of all the faithful departed, thirty masses and as many vigils are celebrated, and alms are given as many times." Elisabeth persists, demanding exactness: Would her uncle "ever be wholly delivered"? To her relief, the Lord nods. With relatives lobbying on behalf of the recent dead, with a host of angry wraiths returning to voice their impatience, and with theologians themselves recalculating the potency of contrition, the Church deeded Purgatory in 1274 as a place of penitence and refining fires.[77]

Escorted by the poet Virgil through Hell to the peak of Purgatory, Dante described in breathtaking Italian the torments of those he fixed in each circle of Hell, then struggled to give a comparable presence to Purgatory, a place of special moment to the *Divine Comedy*, situated as it was in the Jubilee Year of 1300, during which Christian souls not condemned eternally to Hell received passes into Purgatory as the prayers of Holy Year pilgrims released them from centuries of chastening. Even so, the slow improvement of remorseful souls was less dramatic than the remorselessly inventive torture of sinners in an *Inferno* of open wounds and complex passions. The redemptive theater of *Purgatorio* was bound to be much less captivating, its rounds of expiation deliberate, its joys preliminary.

Dante's solution was partly topographic: he turned Hell on its head and mapped Purgatory as a hard climb on a precarious path spiraling up a precipitous island-mountain. The climb was analogous to the purgatorial task of sloughing off old sins and letting earthly desires drop away. By its very charge, however, and in its mediate status, Purgatory could not tend to the delights of touch, taste, or smell; could not counter Hell's pitchforks, hot coals in the mouth, and sulfurous fumes with the swansdown, spoonsful of nectar, and celestial perfumes proper only to Heaven. Although heavenly visitors and visions could vouchsafe to the penitent an eventual, glorious welcome, these could only be (as on earth) guest appearances, inadequate to sustain a drama. What to do? Compose a stunning soundtrack, the souls of Purgatory avid for the Word, whose harmonies resound from above. On his descent, Dante had heard Hell before he saw it, heard the "cries and shrieks of lamentation" echoing through "starless air," but the *Inferno* engages and afflicts every sense; on his ascent through Purgatory, Dante's cantos are driven almost entirely by the ears and the voice.[78] From lowest to highest circle, crowds of indistinct souls rush past the poet, weeping in efficient sorrow, mourning their sins and praying for mercy, extolling God's forgiveness and singing hosannas to the Most High. Climbing, Dante needs visual clues to mark each moral hairturn up the mountain, but the altitudes are essentially aural, the vistas choral.

Iconic of Purgatory, Siren herself appears in the fourth circle, just above the prideful, envious, and angry, at an elevation where the wailing of Hell-bent souls has become inaudible. She who (involuntarily?) drew Greek mariners to shipwreck with her seductive voice is now (in Canto XIX) a dream-figure, pale, stammering, cross-eyed, maimed. On seeing

Dante, she feeds on the heat of his gaze; blood returns to her face, desire to her throat,

> And with her speech set free
> she started singing in a way that would
> have made it hard for me to turn aside.

Virgil calls Dante away three times; rapt, he hears Virgil not. It takes a holy woman, interrupting the dreamsong, and Virgil himself, tearing open the Siren's robe and exposing her putrid belly, to estrange the poet from the song and wake him from the lyric. It takes another, a winged angel, to draw him ahead, upward:

> "Come, here is the passage,"
> spoken in such gentle, gracious tones
> as are not heard within these earthly confines.

The Siren's call is the lure of life on mortal strand, the lure of sensual embodiment, inevitably if opaquely corrupt; in sonic terms, hers is the noise of mortality overlaid with a superficial tunefulness. Those ascending Purgatory must deafen themselves to profane song and attend to the sacred, exercising their virtues to a heavenly music circulating from the clear air above.[79]

Again, the physics of air and sound come to the fore, made explicit in Canto XXVIII on Dante's approach to Eden, that earthly paradise regained at the peak of Purgatory. There, Dante, who has been quizzing Virgil as to how disembodied souls still feel the pangs of embodiment, wonders about the rustling boughs he hears in the garden. Matilda, his guide now that pagan Virgil can go no farther, explains:

> "Now, since all the air revolves in a circuit
> with the first circling, unless
> its revolution is at some point blocked,
>
> that movement strikes upon this summit,
> standing free in the living air, and makes
> the forest, because it is so dense, resound."

No such natural history had been given of the sounds of Hell; its acoustics of imprisonment and incessant repetition (and so, of echo) could be inferred from the voluble suffering of a millennium of sinners and centuries of narrative. Purgatory, whose logic demanded an upward circulation, air and sound swirling around its peak with a freedom suggestive of the radiance just above, lacked such precedents. More precisely, it had risen beyond them.

Deep beneath the caldera of a volcano and distant from the province of the dead lay the forge of Hephaestus/Vulcan, the rough but steady god of metalwork, skilled enough to craft a golden bevy of mechanical maidens and rude enough to be the Olympian foil to Aphrodite/Venus. Molten streams had flowed through his Greek foundry long before they fed Hell; pounding had echoed from his Roman anvil long before metallic hammering rang off Hell's cavernous walls. The more ethnocultural territory commandeered by the Christian Hell, the more it encroached upon other less vicious underworlds, and by the time of the Visions of St. Brendan (900s) and Tondal (1100s), Hell was as much a foundry as a burning sea or a torture chamber. Brendan heard "the noise of bellows blowing like thunder, and the beating of sledge hammers on the anvils and iron"; Tondal saw desolate souls melted in a furnace, "cooked and recooked to the point where they were reduced to nothing. Then the devils would take them with their iron forks and place them on burning anvils to be forged together with big hammers, so that twenty or thirty or fifty or a hundred of them would become one mass. Tormenting them, the devils would say, one to the other, 'Are they forged enough?'" A horrible joke: all this smelting and smiting could result, one would suppose, in nothing more than infernal filigree. Yet, before the founding of Purgatory, the recasting of imperfect souls had been done by demons at forges, instruments of a God who extended forgiveness to much of Hell. Once Purgatory was declared open and demoniac possessions were heard to be multiplying right and left, the job of purification was withdrawn from the claws of demons and put in more therapeutic hands. Hell's thunderous bellows and anvil choruses remained in business below, counterfeit of the less strident if utterly audible refining that had to happen in Purgatory. Like the workshops of smiths in the growing cities of Europe, Hell's forges were removed from centers of (spiritual) commerce, their noise no longer a tolerable prelude to redemption.[80]

Silence also no longer. Early Christians, trying to resolve the problem of the fate of the pious who died too early to be redeemed by the Transfiguration, and medieval Christians, trying to resolve the problem of the fate of newborns who died before they could be baptized, had conceived a quiet place in the afterworld, a "Limbo" for those souls whose only stain was Original Sin. Literally a borderworld, Limbo hemmed the edges of Hell, waiting to be harrowed by Christ. Its deferral of redemption and its geopsychical position set precedents for Purgatory; its silence did not. Purgatory had need of sound to be demonstrably effective; a silent atonement could seem morose, evidence rather of fear than joy. Even Dante, revering Virgil and other virtuous ancients, could not allow the good residents of Limbo a restful silence. Crossing into the otherwise classical underworld that he finds in the topmost circle of his *Inferno*, Dante feels the air of this Limbo a-tremble with sighs in a collective sorrowing for a mis-timed life. Farther down, Hell is noisy with shrieking and bellowing—an "anti-musical" world that parodies sacred hymns. By contrast, his *Paradiso* is full of lyres and choirs. Aurally intermediate, *Purgatorio* makes up for its constricted range of emotions by swinging between sigh and song: between sighs of regret that filter up from Limbo and songs whirling from the muse Calliope, whom Dante invokes in the first lines of his climb out of Hell; between sighs of contrition and songs of praise; between sighs of relief and songs of transport; between sighs of anticipation and songs of celebration—each variety of sigh a demonstration of deepening repentance, each kind of song a different part of the program of expansive grace.[81]

Grace too expansive would raise the hackles of later Reformers as pulpits became raucous market stalls for the sale of papal indulgences that offered to reduce terms in Purgatory on a sliding cash scale. Such offers of pardon seemed to arrogate to the Pope a munificence reserved to the Lord. The sale of indulgences also marginalized those fervent acts of contrition that Dante had put at the center of his *Purgatorio*. Finally, indulgences minimized the fear of Judgment and horror of Hell, which should be sufficient "to constitute the penalty of Purgatory, since it is very near the horror of despair." Thus Martin Luther in 1517 in the fifteenth of his *Ninety-Five Theses* on "the Power and Efficacy of Indulgences," denying that, of themselves, the Church and the Pope could remit any guilt whatsoever.

Numbers of Luther's single-sentence theses were theologically fine critiques; others were, in the most modern of respects, soundbites: clauses or phrases meant to be easily detachable, so as to catch the ears of folk far beyond the university town of Wittenberg. This was more than a matter of style or of the conventions of debate; it was a riposte to the soundbattles in a Europe now flooding with printed broadsides, ballads, jokebooks, almanacs, chronicles, and polemics read silently *and* aloud, then digested, abbreviated, and recirculated in oral, written, caricatured, and pirated print versions. Johannes Tetzel, whose aggressive hawking of indulgences with fill-in-the-blank printed forms drew Luther's venom, condensed his own salespitch to a jingle:

> As soon as the coin in the coffer rings,
> The soul from Purgatory springs.

Luther retorted:

> 28. It is certain that when the penny jingles into the money-box, gain and avarice can be increased, but the result of the intercession of the Church is in the power of God alone.

A penny jingling into a money-box, *in cistam tinniente*, sounded tinny, false, a counterfeit of true charity, and

> 45. Christians are to be taught that he who sees a man in need, and passes him by, and [instead] gives [money] for pardons, purchases not the indulgences of the Pope, but the indignation of God.

Non indulgentias pape sed indignationem Dei—alliteration and balance were there in the original, a phrase waiting to be circulated like superior coinage. In church Latin and in the muscular German to which he resorted for his hymns and his translations of Old and New Testaments, Luther maintained at first that nothing should stand in the way of the experience of Holy Writ. Taken at his word by men and women who began to hear God speaking directly to them with prophetic or radical readings of scripture, Luther retreated, insisting instead on the guidance of a trained ministry—not "inspired," not female—that was expert in exegesis and pastoral counseling. Taken at his word by men and women eager to debate the Word, who organized theological spinning bees or tavern talk of salvation, who sang anti-clerical songs in town squares, who heckled in church and interrupted sermons to dispute a preacher's gloss on scripture, Luther and other elders of Reform were soon on the defensive, upset by a personal faith whose enthusiasm had become noise and whose exuberance had become belligerence, contributing to the Peasants' War of 1524 and turmoil across Switzerland, Germany, and the Low Countries. In 1528, a council in Basel forbade the public contradiction of ministers; in 1531 Luther spoke to the topic of "Much babbling not desirable for real prayer." Still, though suffering from tinnitus (a painful "ringing and buzzing in my ears"), Luther never retreated from his emphasis on the hearing of the Word; though upset with vaulted churches more suitable for the "bellowing and bawling of choristers" than for preaching, he never gave up on the delivery of the gospel through word and song; and though willing to consign lightning and thunder to the work of devils, he remained skeptical of Purgatory as a safe house for those snookered into the belief that snoring through the Lord's Word could always be remedied with silver. A symbol of spiritual deafness, of venality in the guise of compassion, and of a corrupt Church interposing itself between believers and God, Purgatory had to go.[82]

Gone, and the City of God rid of an abomination, the results were as predictable as with any drastic act of urban renewal. The dilemma of what to do with decent hardworking dead returned to haunt the Reformers, who had to find a place for the dislocated or risk a horde of disturbed and disturbing spirits. There could be no outright dismissal of beliefs in an active, voluble spirit world: its roots reached much farther back than did Purgatory, and its vines were woven more finely into the fretwork of early modern communities. Having vacated Purgatory, Reformers had to revamp the rhetoric of dying and review the ambit of the supernatural. In the process, they enlarged the operations and range of spirits, strengthened convictions in/of/for witchcraft, and firmed up a soundsystem that led people to anticipate, from all directions, strange noises.[83]

First, the art of dying. Medieval Christians had approached the deathbed as the site of a no-holds-barred wrestling match between good and evil in which family and friends by their prayers, and a priest by his ministrations, battled like a tag-team to keep a departing soul from the Devil's hammer-lock. Woodcuts showed demons fighting dirty, grabbing at the head and foot of deathbeds unless a saint herself was about to go, in which case celestial figures hovered overhead like referees; ordinary folks heard in sermons or from manuals of *ars moriendi* how to prepare for death and how to comport themselves on their deathbeds. Where purgatorial ascent lay just beyond, however, dying would seem less conclusive, less like a miniature Judgment Day. With Purgatory gone, and intercession by saints and the Virgin inadmissible, Protestants faced a nervewracking death, fearing that they might lose all hope of Heaven in a single hour near the end when, taxed with excruciating pain, they might scream delirious curses at God. Sensitive to this widespread anxiety about a soundbitten and suddenly hell-bent death, the Swiss Reformer Ambrosius Blarer assured his congregants that faith, not reason or the noise of its loss, was determinative *in extremis*. Further, as Reform theologians implied, the cleansing of Original and venial sins that Catholics relegated to Purgatory actually took place in the throes of death itself, clearing the way for the souls of the Elect to sleep deeply, dreamlessly, until the Second Coming (Luther), or go directly to Heaven (Heinrich Bullinger), or take up quarters at a halfway inn of happiness, awaiting the joy of Judgment Day (Calvinists, Anglicans). Whichever aftermath Protestants favored, it allowed for no moral improvement and no further commerce with the living. Dead was dead, burial basic, mourning moderate. The souls of the damned, punished at death and ever after, had no chance to revisit old haunts; the souls of the saved, who could count on an eventual bodily resurrection, had no cause to return. And rest assured, Blarer comforted the most anxious: hosts of angels are at hand to help you from the time when body and soul are joined at conception to that fateful instant when they are (temporarily) detached.[84]

Angels took over many of the offices of the ghosts that Protestants were being asked to leave behind. Providing a still-desirable communion with the world beyond, angels outshone ghosts as helpmates, messengers, intelligence operatives. Because they were troubled neither by remnants of physical desire nor by regrets for past wrongs, angels could be trusted in ways that ghosts could not. "In the Old Testament, the Lord forbade the seeking of the truth from the dead, and any sort of commerce with spirits," declared the Second Helvetic Confession (1566), citing Deuteronomy 18—but those were understood to be evil spirits; sanctified spiritual beings were fine, active throughout the Bible, attendant upon man, woman, and the Son of God alike. Trusting to the presence and powers of angels, Reformers could hardly be cynics of the supernatural. If there could be no theological truck with ghosts as souls on the loose or commuting from some Middle Place, belief in ghosts was sufficiently strong that in Lutheran and Calvinist campaigns for their dismissal, angels had to take up some of the slack. The remainder was taken up by demons. This world is chock full of spirits, many truly evil, advised Swiss Reform theologian Ludwig Lavater, in a tract translated in 1572 as *Of Ghostes and Spirites Walking by Nyght, and Of Strange Noyses, Crackes, and Sundry Forewarninges*. Lavater listed the criteria for telling an evil spirit (a demon) from a good spirit (an angel), then wrote up a protocol for how people "ought to behave" should they meet up with either kind. Since spirits of both sorts could assume the visage or voice of a long-lost loved one, and since a demon just might assume the raiment of an angel, the safest bet was silence: Do not engage a malign spirit in conversation; remain worshipfully quiet before apparitions apparently angelic.[85]

Under such clouds of unknowing, angels had to be monitory, warning against devilish imposture and urging repentance so as to certify their own credentials. When the Lutheran vintner Hans Keil met an angel on a hill near Stuttgart one day in February, 1648, after decades of devastation during the Thirty Years' War and after praying that very dawn for God to relieve the distress of the people, the angel told him flat out, "God will punish the land and its people because of their sin if they do not repent. For the Lord has visited all of Christianity for the last thirty years with war and bloodshed, hunger, pestilence, and all kinds of punishment, but no one pays any attention, and every day brings new provocation. Therefore He cries, 'Oh woe, oh woe Württemberg, oh woe, oh woe Germany, oh woe, oh woe all Christianity, oh woe, oh woe the great sinfulness.'" In lieu of the full-throated sighing of Catholic Purgatory, the Protestant angel echoes God's own sighing; in lieu of purgatorial afflictions, the angel predicts more of the same on earth, "fire from heaven, the Turkish sword, and plenty of hunger," should great sinfulness continue. What great sinfulness? First (and sixth, with regard to gambling on the Sabbath, and seventh, with regard to priestly hypocrisy, and eighth, with regard to rowdy hunters), cursing, "which is common among all people, so that in one hour often over a thousand cursings occur. Whoever curses the smallest curse drives a nail into the limbs of my Lord Jesus.... The Lord will give power over these people to accursed Satan and to a thousand wicked spirits, who will fly in the air and drag the accursed in their misery away, tear at them in anger, and with body and soul take them to hell." Among abandoned fields and villages on the tooth-edge of famine, why should cursing be the first of the sins to set Keil's vines bleeding like veins? Because it was so direct and thoughtless a challenge to the Lord? Because it encompassed all classes? Yes, and more: because it poisoned the Word. Using such soundbitten syllables, people opened themselves up to other blasphemies and became speaking puppets for Satan. In contrast, Keil, the village reader who read aloud for the entertainment and education of his neighbors, had opened himself up to prophecy and taken "a great pleasure in broadsides and songs about all kinds of wonders, signs, and apparitions"; in his Bible he kept several printed sheets, one describing "the appearance of an angel to a poor gardener who sighed deeply from his heart over the wretchedness and hardship in this life and over the misery of the times." His torturers, trying to get to the bottom of Keil's visions, and historian David Sabean, trying to get to the bottom of the communities through which Keil's visions spread swiftly by woodcut and word of mouth, found that the vintner had collaged his visions from a passage in *Acts*, assorted broadsheets, his pastor's sermons, and encounters with thoughtless blasphemers, "so he said that anyone who swore three times could not be saved." That was to say, as Sabean has observed, no one who curses God has any reason to *expect* to be saved, and humankind's good works do not change God's mind, but in truly asking for mercy, a penitent Christian may look forward to the central miracle of Christ, God's grace. Prophesying against cursing, Keil was making a counterclaim for the power of God's Word and the good spirits that carry it abroad and aloud.[86]

Keil had been arrested as much for challenging local taxes as for prophetic imposture; he was tortured on suspicion of sorcery. Counterweight to an enlarged and expressive angelology was a Protestant demonology so delicately balanced that inspiration could torque of a sudden into possession, possession into witchcraft. Things might start simply with poltergeists ("noisy spirits" < German), but they led quickly—how quickly was spelled out by the French Protestant minister François Perrault in his *Démonologie* after his family was assaulted by poltergeists in 1612—to dead-serious affairs, for now that mysterious daytime bangings and nighttime crashings could no longer be explained as attempts at communication by relatives in the afterworld, such unmannerly noises had to be attributed to Satan himself, his devils, or those in league with devils. Absent the moderating rationale of ghosts as discomfited dead, suspicions of witchcraft could arise more quickly in a Protestant's mind than in a Catholic's, no matter how boldly the Reformers painted the Church of Rome with the black and scarlet of cruelty and superstition.[87]

Perrault's *Démonologie* was not published until 1653, by which date the rate of prosecution for witchcraft was slowing in all but a few places around the Atlantic ecumene (Hungary, Estonia, Sweden, Iceland, British North America, parts of the Caribbean). Estimates for the total number of formal accusations of witchcraft between 1430 and 1760 have settled around 150,000, the number of executions and associated deaths around 60,000. Given that more than ninety-five percent of these occurred between 1500 and 1700, the toll from the witchcraze would have amounted on average to 15 formal accusations and 6 executions *each week* for two hundred years. Adding in the capital judgments rendered on (other) heretics and on necromancers, traitors, rebels, spies, murderers, pirates, brigands, kidnappers, infanticiding mothers, bigamists, counterfeiters, and high-stakes swindlers, as well as the intermittently severe sentencing of rapists, adulterers, housebreakers, and shoplifters, it would have been rare before the late eighteenth century for a city to skip any day but the Sabbath in its round of public whippings, torture, and executions.[88]

Had I chosen to dismiss demonology as the scholarly excitements of a few sexed-up men and possession as the febrile excitements of a few sexed-up women; had I dismissed witchhunts as offshoots of social dislocation; had I dismissed debates about the afterlife as theological horseplay—nonetheless, in listening for the cultural positioning of noise, there could be no escaping the cries on the scaffold. If Hell tormented the ear "with the yellings and hideous outcries of the damned in flames," demons themselves bickering, dickering, drumming, and yowling, and everyone "howling and roaring, screeching and yelling in such a hideous manner, that thou wilt be even at thy wits end, and be ready to run stark mad again for anguish and torment," executions of those on their way to Hell were expected to sound the opening notes of that unholy ruckus. Executions were held in public because authorities civil and ecclesiastic meant them less as free entertainment than as repeated lessons in power and rectitude—lessons to be learned as much from sound as from sight. Explained the English civil servant Lodowick Bryskett in 1582: "And whereas the things which we learn by the eyes, are but dumb words: so do the ears hear the lively voices, by which we learn good disciplines, and the true manner of well living. And therefore Xerxes said, that the mind had his dwelling in the ears, which were delighted with the hearing of good words, and grieved at the hearing of unseemly." Equally disappointing to local authorities and to the raucous crowds around scaffolds were those hangings, burnings, breakings on the wheel that proceeded in silence, where the unseemly did not sound its moral point, where the condemned did not shriek.[89]

4. Squeaking, Whistling, Listening Through

Then something happened. Something that led to a deeper "grieving" at the "hearing of the unseemly." Something that made shrieks from the damned and the condemned more distressing. Something that (in Romantic diction) shook Hell to its foundations and moved Heaven within hearing of earth. Something that made diabolical witchcraft sound unlikely and restored some of those suspected as witches to their herbs and midwifery, retiring many more, male and female, to workhouses, poorhouses, madhouses. As with the disappearance of plague from Western Europe around the same time (1680s–1760s), no single explanation can account for it.

My set of explanations for the something-that-was-happening will be broadly acoustic, not in order to account for a concomitant disappearance of noise, but very much the opposite. As Hell contracted and Purgatory lost its pull, as demons had their passports revoked, as possession and witchcraft gave way to differently constituted interior voices, as torture was withdrawn from view and prison replaced public executions, noise was installed to run rather in the background of all parts of life. To be sure, eerie or tempestuous fragments of sound would take a leading role in Romantic scenes of the horrible and sublime, and the "rough music" of *charivari* would muss up the brocades of the Rococo. However, as noise diffused from the dramatic into the daily, people might come upon it not only under conditions of terror or terrible cold, of riot or revel, of sorcery or torture, but in the most banal of places and benign of circumstances, where muffled, multiplicative, truncated, and distended sounds would come to seem elemental. That's where these pages are heading: toward noise as elemental.

When Samuel Willard, a thirty-one-year-old minister of Massachusetts, found his new servant carrying on "in a strange & unwonted manner," convulsed by "sudden shrieks" and "extravagant laughter," he inquired after her health, which was fine, or so she said. Then, sitting by the fire on October 30, 1671, the sixteen-year-old cried out, "Oh my legs! & clapped her hand on them, immediately oh my breast! & removed her hands thither; & forthwith, oh I am strangled, & put her hands on her throat," gasping. For months thereafter, Willard and others of the town of Groton were audience to a dis/concert of "leapings, strainings, & strange agitations, scarce to be held in bounds by the strength of 3 or 4: violent also in roarings & screamings, representing a dark resemblance of hellish torments," mixed with choruses of "money, money" and "sin & misery" to which the young minister listened with Solomonic patience. His *Briefe Account of a Strange & Unusuall Providence of God befallen to Elizabeth Knap[p] of Groton* was all about listening, just as Knapp's agitated maneuvers were all about the sound and sense of interruption as the only surviving child of James and Elizabeth Knapp, who sent her (as was the custom, oft ill-rewarded) to serve for a time in another's house. The theatrics of this possession were the theatrics of unaccustomed locution and uncustomary recusal, her "fits" alternating between cursing, barking, laughing, sobbing, "yelling extremely, & fetching deadly sighs, as if her heartstrings would have broken," and an acrobatic speechlessness where "her tongue was for many hours together drawn into a semicircle up to the roof of her mouth, & not to be removed." She pulsed also between tearful confessions of the Devil's connivances and a swollen-throated voice through which the Devil spoke his own mind. The demonic ventriloquism impressed Willard, who noted that "the labial letters, diverse of which were used by her, viz. B. M. P. which cannot be naturally expressed without motion of the lips, which must needs come within our ken, if observed, were uttered without any such motion." Yet he waited her out, disregarding her accusations of witchcraft against several neighbors, refusing to give the Devil so much due that Elizabeth would have no other recourse for companionship. That winter, at last, she settled down, and soon made her peace with Groton and the Word of God as fully as she might: in 1674 she married Samuel Scripture of Cambridge and proceeded to have a child for each of the commandments.[90]

Mrs. Scripture outlasted Willard, who left to become minister of South Church in Boston, where he also served as acting president of Harvard College until his death in 1707. She lived into the 1720s, long enough to have held in her hands the thousand-odd pages of Willard's posthumous magnum opus, *A Compleat Body of Divinity* (1726), the largest single volume printed in English North America to that date. This corpus of two hundred fifty of Willard's sermons was an intently listening body, arranged according to the oral practices of the Westminster Shorter Catechism and organized as a point-by-point defense of God's Word against those who heard it amiss or spoke of it dismissively. (This was in accord with Harvard's own rules that students avoid "all swearing, lying, cursing, needless asseverations, foolish talking, scurrility, babbling, filthy speaking, chiding, strife, railing, reproaching, abusive jesting, uncomely noise"; and it was also in accord with Harvard's physics textbook, which defined "Unpleasant or Unelegant Sound" as "that which causeth a Motion of Air unsuitable to the motion of our Spirits; Mathematically, Such as discords in Music, Strings out [of tune]; or Physically, Such as Shrieks of filings, or Scraping of trenchers, etc: which do [as we express it] Set our Teeth on Edge.") In a sermon of 1702 in the *Compleat Body*'s mid-riff on the Third Commandment, Willard attacked wickedly witty Deists who set *his* teeth on edge by making "facetious and drolling Discourses from the Word of God," a topic he had addressed before in *The Danger of Taking God's Name in Vain* (1691) and *The Fear of an Oath, Or, Some Cautions to be Used about Swearing* (1700). For his sermon of 1702, he jumped directly from a condemnation of the jeering use of scripture to a condemnation of its opposite, the over-zealous straining of the Word "to a sense contrary to the analogy of Faith, and contradictory to itself." Then he drew a startling analogy: "When Men go about to torture the Scriptures, to make them to confess things which they abhor, is an Affront offered that carries the highest indignity in it." Such torture was more reprehensible than "careless neglect of Reading & Hearing the Word of God," which was "the great *Medium* of our acquainting ourselves with him"; it was more reprehensible than the "*horrible abounding*" of cursing and swearing, where "even Children in the Streets, have gotten their Mouths blistered with such Profaneness." Most reprehensibly, it made the Lord himself "to be rash and ignorant, not to know his own Mind, and to contradict himself, than which what greater indignity can be offered to the Alwise God?"[91]

SOUNDPLATE 3

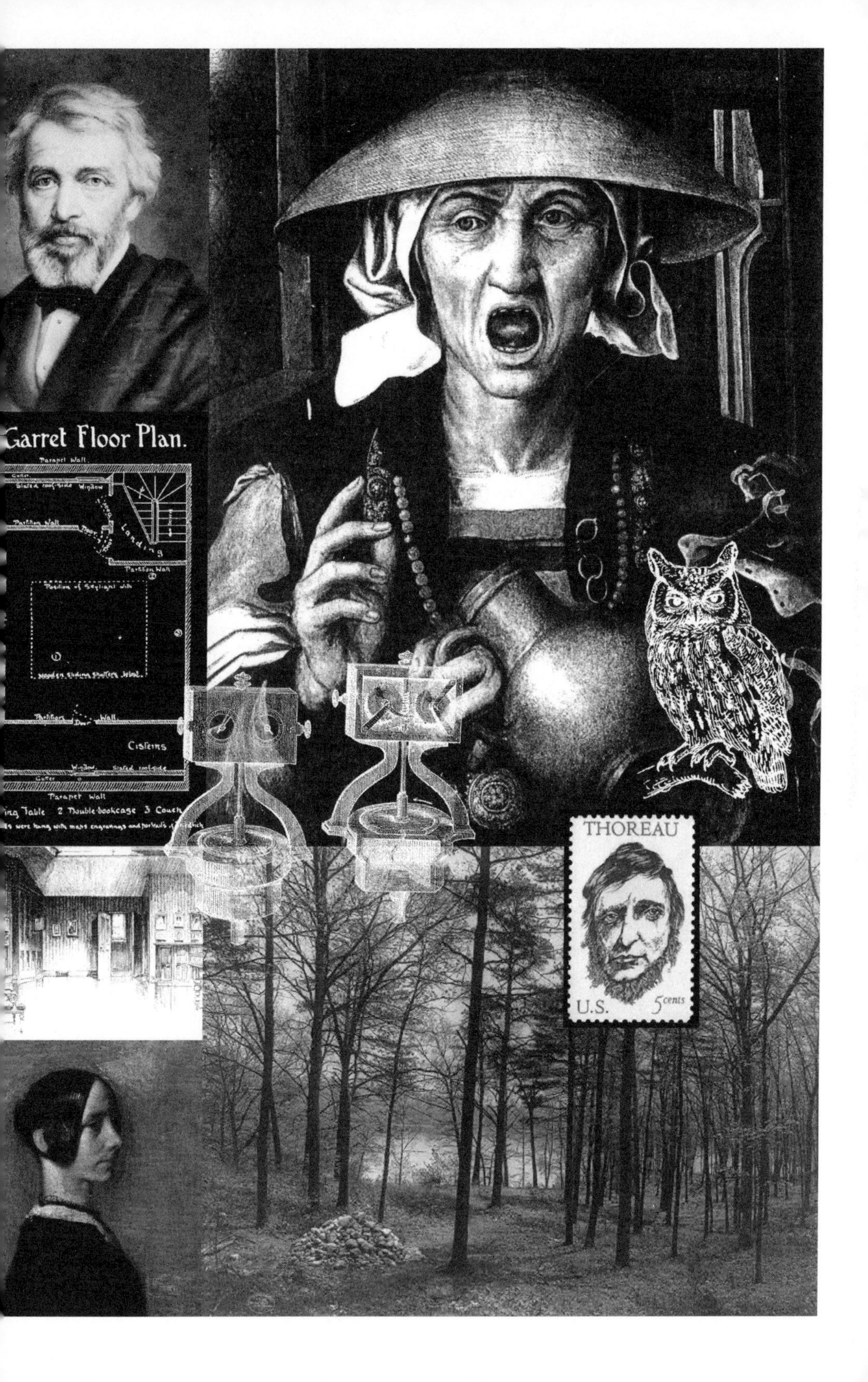
Garret Floor Plan.
Parapet Wall.
Gutter.
Slated roof-side
Window
Partition Wall
Landing
Partition Wall
Position of Skylight with
Cisterns
Window
Slated roof-side
Gutter
Parapet Wall
2 Double-bookcase 3 Couch
THOREAU
U.S.
5cents

None. Wherefore Willard in 1692 had anonymously composed a dialogue between Boston and Salem, blasting the logic of the ongoing witchcraft trials, their rash confessions and strained convictions. Wherefore he would promote a fast day of public atonement in 1696 for the wrongs committed in Salem. As he had been listening *through* Elizabeth's fits for the maidservant he knew, he had been listening *through* the voices of a ventriloqual Satan for the good people of Salem. The hard facts of Original Sin and predestination notwithstanding, Willard must listen with "the Rule of Charity, which is to think no evil." That rule attached also to forms of interrogation: though physical torture was not used at Salem, "let me tell you, there are other ways of undue force and fright, besides Racks, Strappadoes and such like things as Spanish Inquisitors use," such as combustions of public opinion and the scorching of good names. "This is no light matter to have men's names forever Stigmatized, their Families ruined, and their Lives hazarded," wrote Willard, whether proved by "the Devil, turned informer," or by courtroom ordeals in which eager witnesses underwrote their testimony by being rendered speechless and spastic in the presence of those at whom they were about to point an accusing finger. In the latter case, wrote Willard, "The effect is preternatural, and the thing unaccountable: and men's wild guesses in such an affair, ought not to pass for Maxims, where life is concerned." As for himself, "I profess I should be loath to die upon a mere point of Philosophy that is at most disputable."[92]

Willard was no Modern and no moderate. He was a strict Calvinist partial to jeremiads, taking both modernity and the Devil as formidable adversaries in *The Christians Exercise by Satans Temptations* (1701) and death as an instructive result of sin in *The Mourners Cordial against Excessive Sorrow* (1691, "Very suitable to be given at funerals"). Yet even he, socially conservative, theologically orthodox, was a player in what was happening, which in his case could be described as an ear cleaning. Willard's response to "the Deism which so horribly prevails in this perilous Age" and to jeering rationalists who made scripture out to be noise, "senseless, unintelligible, and self-contradicting," was to urge the faithful to listen more carefully to the inner workings of the Spirit, the "Voice of God speaking to the Soul of a Sinner, inviting and alluring him to come over to the Lord Jesus Christ." That *Effectual Calling* was what made of a Christian a new creature, and it was what the Elect strained to hear amid a noise that Western mystics had earlier ascribed to distractions of the mind but would now be ascribed to the tussles of the world. In Puritan and later Pietist and Methodist diaries, convictions of sinfulness were underscored by an anxious regret that, caught up in the daily tattle and rattle, one might have missed the Call, for "how many are there who regard not this Call in the least?" asked Willard. "Who stop their ears, and turn their backs upon it. It is Christ's complaint." One had to listen *through* the noise.[93]

Close, persistent listening *through* was at the core of the broader phenomena of private journaling and autobiography that arose in the sixteenth and seventeenth centuries, engaging a newly literate middle class of Catholics, Protestants, and Jews, male and female. Expanding upon a medieval genre of monastic reflections and tales of miraculous healing, Catholic women wrote out their lives in order to live out their faith, sometimes in obedience to visions, as with Catherine Burton (1668–1714), an English Catholic "much disturbed and almost half turned [crazy]" in a thwarted desire to become a nun. At the command of the sainted voice of Francis Xavier (†1552), an early Jesuit missionary to India and Japan (painted by Rubens in the act of restoring life to a dead Indian child), Burton wrote and rewrote accounts of her tribulations, among these an illness of seven years punctuated by convulsions so violent that they rattled the windows. "At other times my joints would turn out of their places, and snap so loud, that the noise was heard all over the room and in my father's closet at the further end of it," and while her father stuck to his prayers "till my bones had done snapping," she was brought to mind of God and "what he suffered when stretched upon the cross," prompting further visions of angels and of Jesus dragging his cross at the foot of her bed until, at length, after palsy and paralysis, she had listened through her own disjointedness and the disapproving clucks of her family to a less conventional, finally conventual life.[94]

Glikl of Hameln (1645–1724), a Jewish widow of similar fortitude, wrote up her own life in Denmark and northern Germany as a series of misgivings, of signals misheard and signs garbled, through which she listened for the Lord's guidance in matters more material than mystical. Wedded at fourteen and for two decades perpetually pregnant in what appears to have been a happy match, she was desolated at her first husband's death, "and my woes came 'new every morning.'" Still, she recovered to deal profitably in seed pearls and, at fifty-four, to remarry with some forebodings a rich, honorable widower who "groaned in his sleep." She asked him many times what troubled him, "but he always answered, nothing; it was, he said, his nature and habit to groan in his sleep." Though relatives assured her that he had been a night-groaner from early in his first marriage, the groaning presaged disaster, and by a strange chain of events he went bankrupt and died, leaving Glikl once again on her own and unable, on the Sabbath of Shavuot of 5475 AM (1715), to make sense of what was a most tragic noise. Just as the cantor was leading morning prayers, chanting "Thou art God in Thy power and might" in honor of the Almighty who had given the people of Israel the tablets of the Law, the congregation heard a rumbling overhead, like "stones grinding loose." Fearing that the temple roof was about to cave in and hearing their menfolk below shouting at them to flee, women in the balcony trampled over each other to get down the narrow staircase, until more than fifty lay "knitted and writhing together on the steps, glued to one another as with pitch, the living with the dead in one mass." The roof, it turned out, was solid, but six women had died on the stairs, and "to this day we do not know the reason for the terrible rumbling noise," wrote Glikl, taking this as an epitome of the disaster she heard at every next corner.[95]

Another widow, the English Protestant Anne Halkett, began her memoirs in 1677 at the age of fifty-five, centering the pages upon an act of disobedience committed at twenty-two, when she encouraged the attentions of a young man her mother thought unsuitable. Forbidden to see him, "it presently came in my mind that if I blindfolded my eyes, that would secure me from seeing him and so I did not transgress against my mother. And he might that way satisfy himself by speaking with me." Her joy at this clever solution was short-lived, as the charade proved too much for her suitor, who sought and found a more accessible woman; meanwhile, for fourteen months her mother spoke to her only in reproach, a silence so dark that Anne considered running off to Holland. She weathered this love-storm, but for a woman to put a ruse of listening-through at the turning-point of a tale of love's labors lost thirty-three years before, this deserves notice.[96]

For that most observant and appetitive diarist Samuel Pepys, scurrying around on Admiralty and amorous business during the London day, poring over paperwork in the evening, going home to sing and play the pipes with his wife at night, his skill at listening *through* was how he made his way maritally, politically, and aesthetically. A lover of good music and a critic of the mediocre, he made his listening *through* most explicit after an evening of song, when he confided to his diary, "I am more and more confirmed that singing with many voices is not singing, but a sort of Instrumental music, the sense of the words being lost by not being heard . . . whereas singing properly, I think, should be but with one or two voices at most, and that counterpoint," so as to hear the lyrics through the harmonies. Completing this entry for the Lord's Day, September 15, 1667, he wrote: "my wife and I to my chamber, where through the badness of my eyes she was forced to read to me, which she doth very well; [it] was Mr. Boyle's discourse upon the Style of the Scripture, which is a very fine piece. And so to bed."[97]

Some Considerations touching the Style of the Holy Scripture (1661) was a fitting conclusion to the Sabbath, for Robert Boyle had there undertaken to defend the literal truth of the Bible as he could best hear it through centuries of mangled editions, mistranslations, mistaken figures of speech, bad punctuation. He meant to help Christians hear the Word, which by its very nature had to be plain. How else could the Word be effectual with people of all callings or capacities? To do its work, the Word had to be as clear and powerful as the thunder that the twelve-year-old Boyle had heard while visiting Geneva in 1639, waking "in a Fright with such loud Claps of Thunder" and dazzles of lightning that it seemed to herald Judgment Day. It was first this sound, breaking into his sleep, and then the light, vivid in the summer night, that had confirmed him in his youthful religious conviction.[98]

Returning to Geneva in 1642 after further travels, Boyle met François Perrault, who gave him a manuscript version of his *Démonologie*, which had also been precipitated, as we have heard, by fearsome noises, and which, after its publication in French, Boyle sought to publish in English, convinced by Perrault's account of strange noises and stranger spirits, despite "my settled indisposedness to believe strange things." The translation appeared in 1658 with a prefatory letter from Boyle at the same time that he was arranging experiments using the newly invented vacuum pump. These experiments were not unrelated to "strange things," for the question of a vacuum had import for both physics and metaphysics. The German military engineer Otto Gericke, as devout a Copernican as a Lutheran, appreciated that import; reversing the airflow of a foundry bellows to create suction, Gericke meant his pumps to reverse the ancient dictum that nature/God abhorred a vacuum, so clearing the way for the void posited by Copernicus in the space between planets. Proving emptiness in an atmosphere of poltergeists and other acceptable invisibles was hardly simple, but ever since an ingenious little test performed in 1615 by Gianfrancesco Sagredo, philosophers had resorted to their sense of hearing to corroborate the im/possibility of a vacuum. Granted, as most did, that sound required a physical medium in which to travel, the ringing of hawk bells and the ticking of spring-driven watches came to occupy a signal position in the debate, for both sides presumed that in the absence of *any* medium, i.e., in a (visually indeterminable) vacuum, no ringing or ticking would be heard. If, as Sagredo did, you hung a small falconry bell from a thread inside a glass ball presumably evacuated of air, then gently shook the ball so that the bell moved without hitting the glass, you should hear no ring. The results of a succession of such experiments had been notably ambiguous. Sagredo, hearing less of a ring than he would in the open air, concluded that the chamber was pretty much empty and that sound was damped because of the near-absence of air. Athanasius Kircher, using a magnet on the outside holding an iron hammer above a bell on the inside of a sealed vial, heard "a very clear sound from the bell" once the magnet was pulled away and the hammer released, whence, he reported in 1650, "certain more obstinate-minded persons who were present at once inferred that sound could be made in a vacuum." To this third side of the debate he gave short shrift, for "the more sensible deduced at once from this very clear experiment that it was quite impossible for there to be a vacuum in a place where such manifest signs of air were displayed in the sound." Shortly thereafter, using "a mechanical timepiece with a high-pitched ring," a hammer, and a bell, Gericke heard the sounds gasp into inaudibility as he slowly pumped the air out of a glass chamber, but putting his ear next to the glass, he could still hear "a dull noise" though "by no means a ringing," concluding thereupon that he had achieved a vacuum and that bells, cymbals, and strings make their sounds by beating the air, while "the noise or din which is aroused by mere friction or rubbing together of things is aroused, not through the medium of air, but by the Sounding Virtue itself," which led him away from the study of sound-conducting materials toward a theory of intrinsic noise.[99]

All of these results were severely affected by another kind of intrinsic noise, or what is now called experimental noise: imperfections in devices, materials, or modes of observation. Here, in perhaps the first series of experiments whose designers seriously pondered the problem, the noise was audible, not metaphorical. What these men heard or could not hear was at the crux of discussion, and in order to conclude as they did, they had to listen *through* noise—through the hissing of air out of imperfectly sealed chambers, through the heave and grind of air pumps, through the tramp of the boots of "strangers that chanced to be then in the room," through the ring of handbells from peddlers on the street. Otto Gericke, too much the mechanic, was distracted by a "noise or din" he could not listen through, those dull sounds that should not have made it past his vacuum. Boyle, who stuttered severely "as if he were constrained by an internal force to swallow his words again and with the words also his breath, so that he seems so near to bursting that it excites compassion in the hearer," was himself a walking experiment in speech, breath, and the pumping of air, and was accustomed to having others listen patiently through his stammer. Collaborating with Robert Hooke, a younger and more mechanically adept investigator who refined Gericke's pump and reconfigured the glass chamber, Boyle was able to listen through the "squeaking, whistling" air of a still-imperfect seal to not-hear a ticking watch suspended by a pack-thread within the bell jar, its hands moving in almost-perfect silence.[100]

Contemporaries struggled to replicate that air pump and its experimental results. In fact, when Boyle and Hooke performed the classic bell-in-a-jar test, they and their auditors failed to discern "any considerable change (for some said they observed a small one) in the loudness of the sound," which led the pair to consider that the residual sound might be due to small amounts of residual air or to a "more subtle matter" acting as a sonic medium. A "more subtle matter"? Was there, as Descartes and the Greeks before him had supposed, a *pneuma* or other invisible, rarefied, infinitely permeating substrate called "aether" that enabled all interactions among physical bodies, including sound and, maybe, the brush of an angel's wing? A "Christian Virtuoso" *and* a "Sceptical Chymist," Boyle set no impenetrable border between the material and the spiritual, but his philosophy of corpuscular action had little use for a neutral, aetherial medium that did not need to be taken into account, for example, in his law of the reciprocal relationship of pressure and expansion in air and other gases, or in his study of the loud crack and flash of gold fulminate, or in his inquiries into how well noises or words could be heard through a flexible pipe by a man in an air-filled leather diving sac seven yards underwater. Placing a lit candle inside their evacuated bell jar and watching it gutter out, Boyle and Hooke speculated about the upward spread of its smoke (borne by aether, or by residual wisps of air, or by the natural tendency of products of flame to flee the earth?) and about the basic nature of fire (an element or a chemical process?), but both paid more attention to a parade of tiny bubbles than to an imponderable aether, and the squeak and whistle of the bell jar reminded them that they had achieved a vacuum nowhere tight enough to allow them to specify the precise role of air in combustion, certain though they were of its importance to the communication of sound.[101]

Absent a theory of chemical elements and molecular oxygen, the details of respiration as well as combustion eluded them, albeit not without a few animals suffocating under glass, and not without the suffering of Hooke himself, who in 1671 tried out a man-sized bell jar from which one-fourth of the air was removed. He emerged with the first artificially-induced case of altitude sickness and a ringing in the ears that plagued him thereafter. Having seen through the world of the obvious, bending over a much-improved microscope that gave him access to realms previously invisible, for the latter half of his life he had to listen through a "great noise in my head" or "a strange noise in right ear upon waking like a horn or bell" to continue his studies of music, a first love. When Pepys, who had been amazed by Hooke's *Micrographia* of 1665, met up with him the next year, Hooke was excited to have added to his fine observations of flies and bees (whose "glassy" wings made for noisy flight) and butterflies and moths (whose "feathered" wings made for quiet flight) a chart of the number of vibrations needed to produce each musical tone, so that now he could tell how many wingstrokes a buzzing fly made "by the note that it answers to in Music." Then he became the victim of a buzzing far less musical, afflicted by a tinnitus that, aggravated by the vacuum chamber, was probably another symptom of a genetic bone disease that had made him an ailing fragile child, had bent his back with scoliosis during adolescence, and further deformed him during adulthood, accompanied by severe pain and insomnia that he treated with marijuana and opium. Coupled to an apparent adult-onset diabetes and its related hearing problems in the upper registers, a diagnosis of *osteogenesis imperfecta* would help account for Hooke's 1668 construction of "otoacousticons" or amplifying hearing aids and for his reaction to high frequencies, where, "if the string be very short and be stretched very stiff, it makes so shrill a noise that it becomes very offensive."[102]

Attempting to explain the troublesomeness of this "troublesome noise" as something less personal, Hooke wrote in 1676 that fast vibrations induce such quick continual changes in the air and thence on the "drum and organ of the ear" that the ear has no time "to tune itself all the while," like visual confusion caused by a too-quickly vibrating object. This was not the first time Hooke had drawn an analogy between sound and light. As author of a theory of light in which color was "nothing but the Disturbance of that light," and as Curator of Experiments for the Royal Society, Hooke in 1672 had been asked to critique a paper on the separation of colors, presented by a young Cambridge man who claimed that white light was a collation of all colors. Contemporaries would have as much difficulty replicating Isaac Newton's results with prisms as they did the results of the Boyle-Hooke air pump, but Hooke found Newton's experiments beautiful. The stumbling block was Newton's conclusion: it was no more reasonable to assume that all individual colors were commingling in white light than that "all those sounds must be in the air of the bellows which are afterwards heard to issue from the organ-pipes."[103]

Hooke himself just missed making the direct connection between colors and wave-lengths that Newton proceeded to make in response to the Society's critiques, in part because Hooke stuck to a mechanical view of light, in part because he stuck with the word "pulse" to describe the vibrations associated with light. His sense of vibrations was rather acoustic than optical, and we know from the work of Penelope Gouk that the theory and playing of music conditioned Hooke's analogizing in many fields, as had been the case earlier with Marin Mersenne, who had gone so far as to claim that "all movements that occur in the air, in water, or elsewhere, can be called sounds, inasmuch as they lack only a sufficiently delicate and subtle ear to hear them." Listening *through* was an especially apt method for Hooke, with his tinnitus and hearing loss, and it was fitting that he should extend Mersenne's claim by postulating regions of sound beyond human hearing that were occupied by the songs and rubbings of insects and small animals. On one page of his "Method of Improving Natural Philosophy," Hooke laid out a vast soundscape of noises as yet inaudible or barely audible, noises he expected one day would disclose secrets that vision could never illuminate. Although he prefaced this page with a formal disclaimer that hearing was less instructive than sight, and although he professed to be skeptical that even a much-improved otoacousticon could pick up "Noises made as far off as the Planets," he imagined devices akin to telescopes and microscopes for equally valuable sonic revelations. Far away, distant thunder and acoustic signs of bad weather; close up, in plants, "one might discover by the Noise the Pumps for raising the Juice, the Valves for stopping it, and the rushing of it out of one Passage into another." Closer in still, as one detects the proper running of a watch by listening through the casing to "the beating of the Balance, and the running of the Wheels, and the striking of the Hammers, and the grating of the Teeth, and Multitudes of other Noises; who knows, I say, but that it may be possible to discover the Motions of the Internal Parts of Bodies" by listening through the skin to more than was already audible—the beating of a heart, the rumbling of the guts, the wheezing of the lungs, "the humming and whistling noises, the slipping to and fro of the joints in many cases, by crackling, and the like." With an artificial tympanum, one might tell the motions of internal parts through sounds that would be as meaningful to the physiologist as they are to the al/chemist: "the hissing noise made by a corrosive Menstruum in its Operation, the Noise of Fire in dissolving, of Water in boiling..." Having explored worlds that became more populated under lenses microscopic and telescopic, Hooke was prepared for a world filled with noises at every scale, noises that people were already, unawares, listening through, noises that would eventually deserve to be listened *to*.[104]

5. *Earwork and the Buzz*

"Ear-dropper" was a new English word in 1670 for one who "eavesdropped" (1606), that is, stood outside under the overhang of a house and listened through walls or windows to conversations. "Auricle," its reference expanding in 1653 from the human ear lobe to the external ears of all animals, became more intimately figurative when applied in 1664 to the two ear-shaped upper lobes of the heart. As physicians after William Harvey's mapping of the circulation of blood (1628) had greater reason than ever to listen in on the beating of the heart, the heart in its own way listened back, auricles to auricle. The medical profession, however, was socially and ideologically unprepared for the bent-back game of auscultation, which would not commence for another century. What followed more immediately from Hooke's general scheme was a sense of noise generally everywhere.[105]

Thomas Willis, the Oxford physician who had first recommended Hooke to Boyle, differentiated light from sound by the capacity of "sonorific particles... more subtle than the little Bodies of the Air" to "insinuate themselves within... bending pores and blind holes, like the flowing of Waters." Insinuation was a strong word, but sinuosity and surreptitiousness were crucial to sound's infiltration. Once gathered in by an auricle (or an ear trumpet, for the hard of hearing), the sonorific particles had to make their way through "turning and winding passages" toward the eardrum, where their vibrations would be magnified and sent toward the auditory nerve. The notion that sonorific particles could wend their way into lightless mazy places contributed to convictions of sound's ubiquity, which was old news to Bacon, older news to Echo. New, noteworthy, and correspondent to a posture of listening *through*, was a passage a couple of paragraphs further on in the *Two Discourses concerning the Soul of Brutes* (1672), where Willis wrote that hearing is much diminished by loose or damaged eardrums and in such cases a "vehement" sound that increases the tension of the drumhead may temporarily restore hearing. This oft-cited, oft-contested phenomenon of paracusis (or "hearing through") Willis himself had heard at two removes, "from a Credible Person, that he once knew a Woman, tho' she were Deaf, yet so long as a Drum was beaten within her Chamber, she heard every word perfectly; wherefore her Husband kept a Drummer on purpose for his Servant, that by that means he might have some converse with his Wife." That Willis—the Sedleian Professor of Natural Philosophy, a famous anatomist of the nervous system and the first to identify typhus and sugar diabetes—could entertain the prospect that noise might improve hearing and that there are advantages to listening *through* noise, this would seem to merit an exclamation point.[106]

Except that this exclamation might better be awarded to a discovery by the Parisian physician G.-J. Duverney, who in his 1683 treatise on the ear endorsed paracusis but went a step further, a surprising step for one who preferred the anatomy table to the practice of medicine and mute corpses to mewling patients. Anatomy had been a wise choice, given that his *pharmacopoeia* harkened back to Middle Kingdom Egypt and his prescriptions for ear ache included a woman's breast-milk beaten into egg-white and water drawn from the boiling of a blessed thistle in a pot of earthworms, ant eggs, and wood lice. When it came to anatomy, on the other hand, Duverney was in top form, relieving millennia of maternal anxiety and medical ambiguity about the leakage of pus from a diseased ear into the brain (impossible), explaining how the deaf might hear through their teeth (by bone conduction), and differentiating real from unreal noises in the ear. This last required the analytic skill of an anatomist, because earlier theories of tinnitus had presumed the shaking-up of "implanted air in the Ear" or a miniature aural storm of internal air and fluids. Neither theory held up to the rigor of an anatomist's scalpel, and Duverney was among the first to propose that the ear's disturbing "Whistling and Tinklings" were "false Sounds," neural ghosts in the same way that the "stars" we see when knocked out come not from meteors but from internal shocks. Ear noises could be vibrations in the *lamina spiralis* of the labyrinth or through the semicircular canals, or more seriously could proceed from the brain's agitation of the auditory nerve during delirium or epilepsy. There could, however, be "a true noise" in the ears, a real internal buzzing, a pulsation that came from a dilated artery beating in time with the heart:

There was once a fine lady of Picardy
whose ear pulsed so loudly and heartily

it seemed to her that she had
a pendulum stuck in her head
and indeed it struck others most smartly!

The limerick is mine, but the paraphrase is accurate. I mean by this raw poesy to emphasize how striking such an anecdote would have been and how the ear itself was becoming participant in the creation of a world of background noise. It would be two full centuries before "otoacoustic emissions" made neurological sense or had diagnostic value; in 1683 Duverney's tale of the loud ear of the lady of Picardy was worth no less than an exclamation point.[107]

Duverney hunted with a pack of Parisian anatomists led by Claude Perrault, another physician who left off doctoring for dissecting and later became renowned for his neoclassical architecture. From his dissections of humans and other animals (often in collaboration with Duverney), and from an amateur's love of new music, came the four volumes of his *Essais de physique*, the second of which was devoted to sound. More exactly, to noise: *Du Bruit*. Perrault had precedents for taking *bruït*, a medieval French synonym for *fama* and rumor, as a covering term for sound, but he contentiously elevated *bruit* to a higher order, as the category of which all sounds are subsets. This reversed a long tradition of definitions that held *bruit* at bay as an unwonted interval or horrid dissonance in music, a collision of "two hoarse sounds in opposite motion" in the air (Thomas Hobbes). Rather than confining *bruit* to the inharmonious and clamorous, wrote Perrault in a defiant preface, he was going to use *bruit* to make sense of all sounds, establishing a typology of *bruits* simple or composite, continuous or successive, clear or closed, sharp or dull, "verberant" or "excessive" as thunder and artillery, and inventing names for kinds of *bruit* as yet unnamed, such as *le bruit rompu* for the "broken" noise of fat flies buzzing and *le bruit rude et continu* for the scrape of a hoe across smooth hard rock. Rejecting any wavelike motion for the "corpuscles" of *bruit*, barely Cartesian and basically nonmathematical in his physics of *bruit*, and more descriptive than analytic in his typology of *bruit*, Perrault had none of the technical facility of a younger contemporary, Joseph Sauveur, who made major contributions to—and in 1701 named—the science of sounds, "*l'acoustique*." But my narrative hews to the side of the apparently antiquated Perrault, eldest brother of that Charles Perrault whose fairy tales were as seminal to modern literature as they were, in 1697, deft arguments against the Ancients. If Claude Perrault did not publish a detailed frequency table of musical pitches, as would Sauveur, he did publish 185 pages devoted to listening *through* noise for all species of sounds; if he did not map octave intervals using logarithms, as would Sauveur, he did map the labyrinth of the ear in exquisite cross-section; if he did not establish the frequency contours of human hearing, as would Sauveur, he did couch sound within a larger universe of noise, so contributing to the broad cultural shift toward a notion and experience of noise as omnipresent.[108]

Several of Claude's brother's fairy tales are so monstrous that publishers routinely abridge them for today's children. Few parents realize that the spellbound silence around *La Belle au bois dormant* protects only the first half of a tale that becomes louder and uglier when the mother-in-law, an ogre-queen, hungers for the freshly slaughtered flesh of Sleeping Beauty and her children Dawn and Day. Set opposite the archaic roughness of his brother Charles's unedited tales, Claude Perrault's *Du Bruit* appears quite modern, at least with regard to the distinction that he drew between noise and sound, which according to recent enginears is the difference between "radiant mechanical energy in air" and "our perception of that energy." With no prince in the vicinity, no griffin or ogre around to respond to its crash, a tree falling in a modern forest *does* make a noise, cannot make a sound. Claude Perrault never heard of "radiant mechanical energy" and the philosopher George Berkeley (all rumors to the contrary) never contemplated a tree falling in a forest, but at the core of the something-that-was-happening around 1700 was a concern with a corresponding difference between event and perception.[109]

"Sitting in my study I hear a coach drive along the street; I look through the casement and see it; I walk out and enter into it; thus, common speech would incline one to think I heard, saw, and touched the same thing, to wit, the coach." Not so, wrote the twenty-four-year-old Berkeley in 1709: from hearing he had inferred one object and its distance from his window; from seeing, a second object and a second estimate of distance; from touching, a third. The all-in-one coach rocked and rolled only in his head, like sounds themselves, that "have no real being without the mind." The sound of the coach inhered neither in the coach nor in the air around it. If one cited the bell-in-a-jar experiments as evidence that sound was air in motion, this did not prove that sound inhered in air unless one could prove that air itself was sentient, for sound was no more than a sensation, which could not exist without mind. While such philosophical idealism (viz., the world consists only of perceptions) was too extreme to be widely embraced, and while Berkeley as an Anglican preacher meant by his logic to drive readers from skepticism into the arms of God (whose thought-full omnipresence assures the substance of human perceptions and ideas), his work put a permanent pause between effect and cause.[110]

Cause has always had a churlish relationship with effect, a churlishness that has provoked much of the human enterprise—our history, philosophy, theology, and science. In comparison with earlier Hindu, Buddhist, Neo-Confucian, Taoist, Jewish, and Islamic thinkers on cause and effect, George Berkeley was a greenhorn; in his own small segment of the globe,

Berkeley's pause was a milestone along the path of a change in how Europeans would draw arrows of causation. We have heard something of that pause in Samuel Willard's patience with Elizabeth Knapp. We might go farther back and rejoin Francis Bacon, who advised a pause in the too-eager acceptance of the confessions of "witches." Neither the first nor fiercest critic of witchhunters, Bacon worried most about the mistaking of accident or coincidence for causation. Earlier critics, like the Lutheran physician Johann Weyer in 1563, the English Protestant squire Reginald Scot in 1584, and the French Catholic lawyer Michel de Montaigne in 1588, had warned against the mistaking of a "diseased imagination," tortured confessions, or melancholic delusion for evidence of actual witchcraft, but they did not bar recourse to supernatural causes when ordinary laws of the physical world seemed to have been violated. Like the Italian physician Girolamo Cardano, who prided himself on his reasoning, his astrological charts, and his mathematics of chance, they could at best advise hesitation upon encountering the inexplicable. "I do not ascribe this incident to a miracle," wrote Cardano in 1575, describing one morning in Pavia when, not quite awake, he heard a knock on the wall of his bedroom, though the adjoining house was empty, then another blow, "as of a hammer," at the very hour that a friend had died. "First, the whole matter, owing to the way one sound followed the other, may be ascribed to a dream. Again, it can easily have proceeded from some natural cause, as from a puff of steam." Scrupling over this poltergeist, he would still include in his autobiography a chapter on "Things Absolutely Supernatural," among them the portent of swine grunting in the street the livelong night and the omen of a mysterious thumping "as of a wagon-load of boards being unloaded" above his bedroom just before his mother died. He devoted another chapter to his own guardian angels and repeated without proviso his father's story of a goblin also knocking on a bedroom wall, "as if (my father used to say) someone were tapping the wall with one of those metal hoops which were usually used as a restraint to chain people up." For Bacon fifty years later, the chains of causation had to be tighter, causal arrows more precisely targeted, and simile had to yield to series before speculation could be confirmed. Faith did not instantly abandon the field to reason after Bacon or Galileo, nor alchemy to chemistry, metaphysics to physics, astrology to astronomy, as the early publicists of a "Scientific Revolution" would once have had us believe. Rather, there was a gradual change in the kinds of evidence that would be given a serious hearing, and a co-variable extension of patience in scrutiny.[111]

Patient scholars like Barbara Shapiro have charted this change in scientific and legal thought, where facts and conclusions began to be framed in terms of degrees of probability that entailed a prolonged listening to more witnesses and a listening *through* more testimony, with cross-examinations and tests for stability of information. Berkeley's emphasis upon an absolute space (though, by God's grace, not an inviolable vacuum) between events and perceptions had already become audible in judicial proceedings where judges listened through *fama* and hearsay. In English, Dutch, French, and Italian cultures of coffeehouses that "buzzed" (a new usage) with voices commercial and political, judgments had to be rendered amid a public noise that was hardly figurative, a noise that on occasion had literally overrun the courtroom. Consider the trial of a spokesman for freedom of speech and conscience, the Puritan and Leveller John Lilburne, accused in 1649, as twice earlier, of sedition. From a shorthand account as exact "as it was possible to be done in such a crowd and noise," we learn that Lilburne in the thronged courtroom had appealed beyond the justices "to all the people that hear me this day." Justice Keble assured him that he would be heard, "But I must tell you that Words are but words, and it were well that you do as well and as Rationally as becomes a Rational man." Unintimidated, Lilburne rose to "extravagancies and heats" that got the crowd into an uproar. The court crier had to shout, "The Lords, the Justices, do straitly charge and command all manner of persons to keep silence!" but when the twenty-page indictment was read out, the noise was so great that Lilburne, himself the author of such pamphlets as *The Apprentice's Out-cry*, demurred that he had been unable to hear the charges. Pages were reread. Lilburne appealed beyond the Justices to God Himself, an appeal followed by a collapse of the scaffolded bleachers where stood squalls of spectators, "which occasioned a great noise and some confusion by reason of the people's tumbling." Silence at length restored, Lilburne asked for a chamberpot, made water, and then began his defense after a warning from Keble to "quietly express yourself." Lilburne did so but finished by praying of the jury that "the presence of the Lord God omnipotent, the Governor of Heaven and Earth, and all things therein contained, go along with you, give counsel, and direct you, to do that which is just and for his glory," at which "The People with a loud voice, cried Amen, Amen, and gave an extraordinary great hum, which made the Judges look something untowardly about them, and caused Major General Skippon to send for three more fresh companies of foot soldiers." As custom dictated, the Attorney General spoke last, but Lilburne interrupted, "I pray, sir, do not now go about to tire the Jury with tedious repetitions, nor to sophisticate or adulterate their understandings with your falsehoods and untruths." About to retire, the jury wanted its customary quart of sack to ease deliberations, but strong wine was forfended in cases of treason. The disappointed men filed out, discussed the case drily for less than an hour, and returned a verdict of Not Guilty, "Which *Not* being pronounced with a loud voice, immediately the whole multitude of People in the Hall, for joy of the Prisoner's acquittal, gave such a loud and unanimous shout, as is believed, was never heard in Guild-Hall, which lasted for about half an hour without intermission." The Justices turned pale. Lilburne stood silent. Court was adjourned. A mob ran through the streets cheering and setting bonfires. A week later Lilburne was released, only to be hauled up for trial in 1652 on criminal libel, found guilty, banished eternally from England, arrested again for returning in 1653, whereupon another trial ensued at Guildhall, with another noisy acquittal after Lilburne showed that the State could not prove he was the same John Lilburne who had been eternally banished. The Protector, Oliver Cromwell, recognized his old if outrageous friend well enough to hold Lilburne 2 in the Tower and then in Dover Castle. There, "giving fullness of scope to that divine and heavenly voice of God speaking plainly in my heart" and "giving up my own reason, understanding, will, wisdom, and affections, to be crossed and crucified by the Will and Wisdom of JESUS" after brooding over seventeen hundred pages of theology, Lilburne 3 was "knocked down off, or from my former legs, or standing." With "the light of God speaking in my soul," he gave himself up "wholly to be guided by his Divine teachings shining within me, before which I now stand ready to give ear to what by it shall be told me," which was to join the Quakers at a time when they physically quaked with the ferocity of God's Word and shook with prophetic indignation.[112]

Earwork, all of this: the "full hearing," the calls to quiet, the reading aloud of the long indictment, the oral arguments, spectators hanging on each spoken word, the collapse of the bleachers, the sound of pissing in the pot, the speechifying, the prayers, the "extraordinary great hum," the hectoring of the prosecution, the verdict delivered, the cheers, the shouting in the streets. Remarkable in all this earwork was the tolerance of Lilburne and the justices for the noise, as if a given. Even more remarkable was how completely the process of discovery-through-noise in Guildhall was echoed by Lilburne's conversion while confined at Dover Castle: the full hearing of the heavenly voice, the focus, the need to move beyond personal rationale to get at the truth, the long reading, his own physical collapse, the speaking in and from the soul, persuasion, prayer, awaiting a verdict, freed speech, shouts of joy—as if everything in his life had been a noise he had at last been able to listen through. And most remarkable was Lilburne's attentiveness, amid the noise, to the rules of evidence and witnessing, legal and evangelical, which saved him more than once and *saved* him forever, before he died of fever at forty-two.

Portraits of Lilburne rarely show him with the short hair and "huge long ears" of a Roundhead who had fought alongside Cromwell and Parliament against the King and his Cavaliers with their gorgeous ear-concealing locks, but his ears were clearly open, and "if we look into Primitive Records," wrote Jonathan Swift in 1704, "we shall find, that no Revolutions have been so great, or so frequent, as those of human *Ears*." Swift's

"Primitive Records" went back only a century, to the lopped, and mutilated ears of those punished for challenging royal or ecclesiastic authority in the years before the English Civil War, and to the years after, under Cromwell's Puritan Protectorate, when "many Endeavours were made to improve the Growth of *Ears* once more among us. The Proportion of Largeness was not only looked upon as an Ornament of the *Outward* Man, but as a Type of Grace in the *Inward*." Swift pressed on toward the purely anatomical: "Besides, it is held by Naturalists, that if there be a Protuberancy of Parts in the *Superior* Region of the Body, as in the *Ears* and the *Nose*, there must be a Parity in the *Inferior*: And therefore in that truly pious Age, the *Males* in every Assembly, according as they were gifted, appeared very forward in exposing their *Ears* to view." (Was this why Lilburne had been so public in his pissing?) Then came the Restoration of Charles II in 1660, "who raised a bloody Persecution against all *Ears*, above a certain Standard: Upon which, some were glad to hide their flourishing Sprouts in a black Border, others crept wholly under a Periwig: some were slit, others cropped, and a great Number sliced off to the Stumps." But, were one to continue in like vein, ears scientific, philosophical, legal, astrological, and medical would shortly acquire an inveterate forwardness, listening in the pause between event and cause as much for the bell-jar of theory as the jubal of grace, a chapter absent from Swift's facetious promise of a "general History of Ears."[113]

6. Borborygmus—Audio Out

Pendent to such a history of ears would be the history of madness and melancholia, whose treatment was gradually shifting from the pastoral, curatorial, or punitive toward the medical. Madness and melancholia were known to afflict one's ears and one's sense of cause and effect in proportionate measure, for when the mad were beset by voices or their own vociferating, the melancholic by migraines and noises in their ears, how could they track external sequences or make sense of befores and afters? Dizzied by the churn of crowds and the least tumult in the air, the "light-headed," the "distracted," and the "mopish" had since ancient times been treated with everything from music to vomits to isolation. Soranus, a widely-cited Roman physician, had prescribed for them "a moderately light room with regulated temperature, the quiet of which no noise can disturb," which was a teasing luxury, especially for the off-kilter poor who were abandoned to wastelands or shuffled in among beggars at shrines. There was some communal caretaking of village "fools," but medieval records show that the urban mad were hustled into chapels, workhouses, or jails in their worse moments, less to shield them from the noise of the world than to put some distance between their own roaring and a neighborhood's thin walls. Some of the mad were subjected to exorcism, as if possessed; some to pastoral counseling, as if dis-spirited; others to interrogation, as if heretics or criminals in disguise; many to purges, blisterings, and bleedings, as if out of their humours; a few to admiration, as if inspired.[114]

During the mid-seventeenth and eighteenth centuries, the mad and sad were increasingly sent to asylums dedicated to the long-term care of lunatics, where their prognoses hardly improved. Robert Hooke, as an architect and surveyor of London after the great fire of 1666 had destroyed 13,000 houses and left 70,000 homeless, designed a newly hopeful hospital for St. Mary of Bethlehem, a small institution that had been receiving the "curable" mad since 1402 and putting them on display, for a fee, since the 1500s. Hooke's five-hundred-foot-long facade, its entry framed by lifesize sculptures of melancholic and raving madness, attracted even more visitors; his plan for ranks of individual cells, clean and quiet, allowing for meditation and the conversation of physicians and patients, was reduced to "bedlam" (<Bethle'm) by crowding and by paying audiences who came to gawk at loud entertainment provided in/advertently by inmates who, on the outside, rarely achieved such theatrics. Hallucination of voices, weeping, raging, inapt laughter, screaming, and cursing were, it is true, prominent among the symptoms of insanity documented in the daybooks of such as the astrological physician Richard Napier, who treated over two thousand women and men for mental disorders between 1597 and 1634, but the major change in early modern Europe came neither with confinement to contain such noises nor with the heavy use of opiates to deaden the noises. It came with medicalization, in order to listen through the noises for underlying circumstances, earthly or sidereal—as did Napier, who heard out his patients at length in order to construct their charts, and whose own portrait in full ruff and conical cap featured an unusually long right ear.[115]

Many of those listed by Napier as "troubled in mind," "mopish," "frantic," grieving, or melancholic were also suicidal. The pause between event and cause, which allowed physicians to retrieve reasonableness from behind the embroiled or self-inflicted noises of the insane, was manifested in a movement for quiet asylums and in a shift of legal sympathies with respect to suicide. Quieting took a while. In 1632, a visitor had observed that "the cryings, screechings, roarings, brawlings, shaking of chains" at St. Mary's would drive a sane man mad; a century later, the novelist Eliza Haywood had the orphan Annilia bound, gagged, and conveyed into what by then was a stereotypical asylum, where "The rattling of Chains, the Shrieks of those severely treated by their barbarous keepers, mingled with Curses, Oaths, and the most blasphemous imprecations, did from one quarter of the House shock her tormented Ears, while from another, Howlings like that of Dogs, Shoutings, Roarings, Prayers, Preaching, Curses, Singing, Crying, promiscuously joined to make a Chaos of the most horrible Confusion." A poem of 1733 described Bedlam's

dreadful Din of heterogeneous Sounds;
From this, from that, from ev'ry Quarter rise
Loud Shouts, and sullen Groans, and doleful Cries...

A visitor to the actual Bethlehem twenty years later found it still very much bedlam, with scores of spectators running through the wards, making such "sport and diversion of the miserable inhabitants" that they became furious with rage. Although Bethlehem's chief physician in 1765 opposed the construction of a watchhouse next door, lest his patients be disturbed by rambunctious shifts of watchmen with bells or rattles, and five years later forbade putting the mad on public display, it would be another half century before most European asylums managed a modicum of silence, regardless of the medical dictum that it was "Absolutely Necessary...in all Cases of Lunacy and Acute Diseases that the Patients should be kept Extremely Quiet."[116]

Legal sympathies took effect more quickly, as eighteenth-century courts in northwestern Europe grew reluctant to punish those who had violated God's law and society's codes by attempting suicide. In a sense that was newly moving to juries and judges, the insane and insanely suicidal embodied the problem of cause and effect, since they acted rashly on—or were paralyzed by—misapprehended events and hasty, overdrawn, or obsessively frozen conclusions. Understood no longer as men and women who had been led so far afield by Satan that they were willing to deny God by ending their own God-given lives, those who attempted suicide were seen rather as caught up in spirals of despondency. The first steps in any remedy had therefore to be to declare them *non compos mentis* and hand them over to "mad-doctors" who could institute a pause (or strait jacket) in which they might learn to listen through the noises that sounded everywhere, for everyone, in the ordinary background of life.[117]

Silent systems for prisons and silence for funerals would come next, but those silences must keep, for they were responses to a deepening sense of the ubiquity of noise, a sense whose contours I have yet to square up. On paper, the cells Hooke outlined for St. Mary of Bethlehem had been the antithesis of bedlam; they followed the lines of early modern cartography, astronomy, and solid geometry that projected round worlds and complex volumes onto explicitly gridded planar surfaces, enabling more precise topographical surveys and more imperial commerce. The hybrids and ogres doodled on the margins of medieval illuminated manuscripts and the bellowing giants who had given weight to unknown lands on the edges of late medieval maps were becoming generic "chimeras," no longer the fire-breathing, lion-headed, goat-bodied, snake-tailed Chimera of Greek mythology, but mere daymares. Though people of all orders would remain fascinated with oddities fantastic (mermaids) and actual (conjoined twins, hermaphrodites), Europe's intellectual elites were in the process of demoting wonders to curiosities, monsters to miscues, grotesques to scalar extremes. In 1668 the members of the Academy of Painting in Paris were shown wonder, astonishment, and veneration according to a grid laid upon a standardized face by Charles Le Brun, First Painter to the King, who merged the old habit of physiognomy with new habits of analytical geometry and estate mapping.[118]

Mapping impulses reached to the farthest of the antipodes: Hell. Where was Hell, exactly, and how large? Was it at the earth's core, or did comets ferry the damned to the surface of the burning sun? Conducting *An Enquiry into the Nature and Place of Hell* in 1714, Tobias Swinden opted for the spaciousness of the sun, in contrast to the cartography of a Bavarian preacher who had cornered Hell into one square mile containing 100 billion suffering souls, an overcrowding that put Bedlam to shame and was "a poor, mean, and narrow Conception both of the Numbers of the Damned, and of the Dimensions of Hell." The Jesuit Leonardo Lessio figured on four square miles and 80 million damned, which was something of a relief, though the fantastic exactness of these numbers corresponded little to Church doctrine, much to a newly statistical approach to populations. In general, Catholic surveyors found that Hell, instead of expanding like London or Paris to contend with crowds of the damned, had by intrinsic perversity contracted to squeeze them into a roiling "sordid mustard of excrement." Baroque Hell was less capacious than Dante's with its thirty-eight square miles; if Nature or God no longer abhorred a vacuum, Satan did. The depth of the abhorrence led to such compression that a forced promiscuity was inevitable, the interpenetration of bodies rampant, the ears themselves "martyred" by dissonant voices and harsh forked tongues, violated by sharp fingers and demonic cocks.[119]

Used to be, that last sentence would have shocked readers, offending a sense of propriety nurtured by books of courtesy or civility that transposed models of deportment and discourse from the Renaissance court to the seconding classes of early modern Europe and its colonies. Erasmus, who had had his fun with Echo, and in another colloquy had scolded schoolboys for digging wax out of their ears in the middle of reciting lessons, was responsible for this transposition, having tossed off a short book in 1530 that within eighty years enjoyed eighty editions in fourteen languages—in English, *A Little Book of Good Manners for Children*. The novelty and power of this work lay in its plainspoken address to children *as* children, regardless of rank, and its hearty adoption by teachers, tutors, and parents of modest means. The *Little Book*'s reach was so wide because it steered clear of the contentious language of sin and punishment, relying instead (notes Jacques Revel) upon the Humanist program of fluent social intercourse through clear codes of gesture and behavior. Whatever punctured or muddied communication or bent it toward uncivil ends was inexcusable. Noisy noseblowing, spitting after every third word, and whispering during Mass were thus deserving of as much attention as the need to speak distinctly, listen religiously, respond respectfully, and eat with clean hands. Most problematic (by page count) was laughter, for the belly laugh, the rolling-in-the-aisles laugh, the fall-down-dead laugh, the bray, the neigh, and the laugh-for-no-good-reason were audible signs of an animality, obtuseness, or insanity that crippled discourse. The history of painting and portraiture, noted Erasmus, proved the tight association of propriety with a pursed mouth. Only a ventriloquist, of course, could talk with lips pressed together, but any excessive public opening of the lips or show of teeth in loud talk, unseemly laughter, sloppy eating, or sudden sneezing belied an underlying brutishness, while the deformation of the face during such episodes belied a deformed character.[120]

Sarah, astonished by her pregnancy at the age of ninety, had laughed her newborn into his name Isaac (*Yitzhak*, "he'll laugh"); Aristotle had classified man as the "laughing animal"; troubadors had put bawdy laughter on the lips of dainty women; Dante had thrilled at Beatrice's laughter in Paradise; Ficino had applauded the "gracious laughter" of graceful people. Yet none of this stood up to the Gospel fact that the Son of Man had never laughed. Whereupon medieval churchmen imposed penances upon monks who smiled during services and a "special fast" on those who broke out "into the noise of laughter." Worship, not laughter, wrote theologians, made Man distinctive; laughter rang with the libertine echoes of paganism, the oafishness of boors, the cruelty of the arrogant. "Inasmuch as laughter is caused by something ugly," wrote the French physician Laurent Joubert in 1560, agreeing both with the Church and with a classical sense of laughter as a reaction to incongruity, "it does not proceed from pure joy, but has some small part of sadness"—a shallow sadness that inclined toward irritation, not pity. For him, laughter could be little more than a noisy spasm of guts and lungs, disturbing and demeaning: "Some men, when they laugh, sound like geese hissing, others like grumbling goslings; some recall the sigh of woodland pigeons, or doves in their widowhood; others the hoot-owl; one an Indian rooster, another a peacock; others give out a peep-peep, like chicks; for others, it is like a horse neighing, or an ass heehawing, or a dog that yaps or is choking; some people call to mind the sound of a dry-axled cart, others, gravel in a pail, others yet a boiling pot of cabbage." Whatever the sound, laughter was unhealthy, particularly after long cacchinations, when the lungs heat up and melt their resident mucus, "which then tickles, stings, and irritates them, and forces coughing," producing disorders of the throat. Laughter was also uncomfortably ambiguous (was a laugh empathetic or sarcastic? shocked or relieved?), a sound as loud and inarticulate as the Barbary apes who were its totems. In fact, laughter might be so animalistic as to make you shit in your pants. As for Aristotle, well, no animal on earth "picks wax from his ears."[121]

I am not sure what to make of recurrent references to ear wax, except to note that Europeans lagged behind East Asia in yet another technology (the ear pick) and another field of medicine (aural hygiene), and that the figurative "ear cleaning" I attributed to Samuel Willard was of literal concern to contemporaries. They spoke of *sordum coitus* with reference to the "dirty coming together" of wax in the ear and called devious flatterers "earwigs" (1633), after a flat, pincered, half-inch-long insect, *Forficula auricularia*, which dispenses a foul yellowish-brown stream similar to the worst ear wax but which has never crept into a sleeping ear or (as was thought) bored its way through the brain.[122] I am sure, however, that the history of early modern laughter was exemplary of a prescriptive literature overwhelmed by practice. "Polite" laughter, managed by the few at court, was unsatisfying to the many, and abundant sources show that laughter in most places was scarcely restrained, even in church, where a Lancashire pastor in 1647 had to rebuke the flightiness of "smiling and laughing in time of God's worship." All mirth was not sinful, allowed a Massachusetts Puritan, John Barnard, eighty years later, for there was "a sort of Mirth, which is an *innocent property of human Nature . . . salutary* or *wholesome*," and if the Jesus of the Gospels was a Man of Sorrows for whom we have no record of laughter, he was an exceptional Man whom we cannot imitate in all things, which Jesus himself acknowledged during the festive nuptials at Cana. Yet, outlining *The Nature and Danger of Sinful Mirth*, Barnard was at pains to keep mirth in its place and out of the meetinghouse and away from the Sabbath and not just before or just after worship and not on days of public fasting or humiliation and not for heads of households and not for elders and not for youth prone to vanity or profanity and not in night revels or drunken frolic and not at the calamities of others or at the infirmities of the aged or at those of weak brain or twisted limb and not as a custom or business and never no never in the House of God.[123]

Oh so much mirth, so hard to contain! The genial proverb, "laugh, and be fat," kept company everywhere with cheap jestbooks, roving comedians, and tankards of ale. In Baroque and Mannerist art, mouths that had been closed except to taunt Christ or shriek in Hell began to open wide in pain, shock, amazement, scolding, joking, gluttony, lovemaking, and highjinks. Depictions of gameplaying and carnival, of witchcraft and possession, of singers and sots began to show mouths indecorously open in boisterous, musical, jesting, or drunken laughter. In early modern novels, plays, public ritual, and politics, laughter (especially a newly remarked species of "explosive" laughter) was a pervasive prelude to resistance, revolt, and reversal: the world turned upside down. Refiners of laughter, like Joubert at the French court and Sir Philip Sydney at the English, were right to fear it as something difficult to regulate, something that might slip from a delight in the disparate to mockery of all that divided rich from poor.[124]

To the degree that etiquette operated by way of ridicule, laughter was both object and instrument of repression, and the more forcibly laughter was repressed, the more it bubbled below the surface, like the "snicker" (1694), the "snigger" (1706), and the titter, "a laugh smothered in its birth" but meant, in the very act of repression, as contempt. Snickers and titters breathed just below the surface of much of Morvan de Bellegarde's *Reflexions upon Ridicule; or, What it is that makes a Man ridiculous; and the Means to avoid it.* "'Tis a thing that provokes Laughter," wrote the misogynist *abbé* in 1696, "to see toothless, decrepit Women equally fond of all sorts of Pleasures, as when they were but twenty; to see them mark all the Cadences with their Hands and Head, when the Young ones Dance; to beat time in Consort, and make passionate Exclamations when an Air affects them." By 1753, *An Essay On Ridicule* would have nothing to do with such scornful tittering, for ridicule was itself indecorous and inefficient. Good manners had instead to be secured by a habit of self-monitoring already in place among religious diarists and the characters of epistolary novels; modern individuals had to develop such an inner sense of proper behavior as would be subverted by neither oddity nor foible. Thus would the titter be hushed to a tsk-tsk.[125]

Still and all, etiquette could do little to squelch laughter, not even the horse laugh, "a most formidable imitation of the true and natural laugh . . . often practiced in public assemblies, to the great terror of speakers." Etiquette had greater success squelching those noises speeding at ten feet per second from the other end. Farting and laughter followed one upon the other in early modern tales. Londoners in 1650 still delighted in the stories of the rude giantess Mother Bunch, twenty thousand and one-half inches tall, whose laughter made bulls and bears roar with terror and who was once "wrung with wind in her belly, and with one blast of her tail, she blew down Charing-Cross, with Paul's aspiring steeple." Panurge, Rabelaisian tutor to the giant Pantagruel, "would fart like a horse, and they would say with a laugh! 'How you *do* fart, Panurge.'" "'Oh dear no, Madam,' he would reply. 'I'm merely tuning myself to the counterpoint of the music you are making with your nose.'" As comic or satirical entertainment, well-timed and well-tuned farting had an enviable antiquity and continuity from Greek and Roman theater to Italian *commedia dell'arte*, English pantomimes, and later German and French cabarets, but books of etiquette clamped down on its moral expressiveness and ritual ambit. Stained glass no longer told the lesson, as it did in a medieval French church, of the adulterous wife of Gengoult, who, recently widowed and loudly skeptical of the miracles reputed to be flowering near the tomb of her late husband, swore that "Gengoult works as many miracles as my ass!" and from that moment had to fart whenever she spoke. The emblematic bear of St. Blaise would soon cease breaking wind as it descended from its cave to announce the rebirth of the year in earliest spring, as men in ritual bear-drag had done for centuries, like one in the French town of Romans in 1579, where peasants and the poor resumed that farting with wooden trumpets and drums that sounded, briefly, a rebellion. French Jesuits in North America found natives everywhere flatulent and prone to laughing at their own after-dinner "music," a laughter and music in which the Jesuits themselves, sharing Indian meals, may have joined, but the comic, colonic relief of farting had eventually to be tamed along with all else savage.[126]

Each of us, whatever our tribe, expels 400 to 650 cc of flatus each day, more when consuming the indigestible oligosaccharides in beans, peas, oats, potatoes, cooked cabbage, raw apples, chestnuts, onions, cucumbers, whole wheat, barley, and weak beer, which approximates the diet of most Europeans before 1900. For the meat-eating minority (and for colonial North Americans) who could afford sugar and tobacco, their smoking and decayed teeth compensated in this respect for a less leguminous diet: sucking on pipes and gulping down rather than chewing food increases aerophagia, the swallowing of excess air that adds to intestinal gas, as does diverticulosis from insufficient dietary fiber. Newly sensitized to foul odors and newly disturbed by the frequency of rearward "booms" and "cracks" (fifteen to twenty times a day by men, less by women; more by smokers, gluttons, and the elderly), polite company by the 1770s required that farting be suppressed. This went against longstanding medical opinion that "a fart ought to be as free as a belch," and that

> Great harms have grown, and maladies exceeding,
> By keeping in a little blast of wind:
> So Cramps and Dropsies, Colics have their breeding,
> and Mazèd Brains for want of vent behind.

A work devoted to this subject by the sixteenth-century French physician Jean Feyens and translated as *A New and Needful Treatise* in 1668, had detailed the dire consequences of flatus that "cannot get out." Trapped but on the loose, intestinal winds whip through the body, chilling the blood, stirring up phlegm, twitching at the meninges of the skull to cause headaches or vertigo, stretching the membranes of the ears to cause tinnitus. Flatus also produced borborygmus, sound from the stomach and bowels, for which Feyens had a scale indexing the kind and direction of these winds. If the sounds were sharp and shrill, air was moving up through "the straight gut"; if humming like an organ pipe, air was moving through empty "thick guts"; if a *bombus* or rumbling, then moving down past a moist stool. But etiquette drew an ever-sterner line between the airs of borborygmus and the "trumps" of flatus, the former involuntary and unexpected, the latter voluntary and predictable, therefore avoidable. In 1640 the courtier-poet Sir John Suckling, preying upon this sociomedical predicament, drew the same line for Love:

Love is the fart of every heart;
It pains a man when 'tis kept close,
And others doth offend when 'tis let loose.

By 1778, *Narcissus; or the Young Man's Entertaining Mirror* accepted no waffling: "Belch not, nor break wind in any manner, so loud as to make others notice of it; much less be proud of it or attempt to defend the necessity of it, under pretence of it's being good for your health; for it's no commendation, for one to live in health and behave like a swine." So would the fart be hushed to a practiced fizzle.[127]

Farting was one of those rare categories in which rules of etiquette overrode medical wisdom. Etiquette books also prescribed degrees of cleanliness and aromatic relief to which medical practice did not then aspire, but only with farting and belching did etiquette, alive as it was to sonic disruption, run directly counter to medical premises, in part because etiquette favored disguise while medicine favored discovery, in part because medical notions of the body were increasingly out of joint with courtesy's ideals of circulation. The late Renaissance medical body had become increasingly a zone of exchange, of exhalations and perspirations, inhalations and miasmas, while etiquette's body was a drill field of measured movements, like the forty-two steps for firing a musket in the newly-regimented quick-firing battalions that moved swiftly across early modern battlefields. As well-drilled infantry learned to ignore their bodies and listen through canonades for their commands, so rules of etiquette compelled the suppression of bodily urges and a listening *through* the fart, for if a person should fail to control those nether winds, neighbors were now obliged to carry on the conversation as if nothing had occurred, and perpetrators to cover up any fart with a cough. From an entry in Jacob Eliot's Connecticut diary of marital battles with his wife Ann around 1762, we can hear the contest between an older acknowledgment of breaking wind and the modern etiquette of obliviousness: "[On] my singing to the child to get him to sleep (being mad before), [Ann] cried out with great vehemence and spite, O don't don't don't don't Mr. Eliot make that noise! I am almost killed with it already &c.—Soon after, she letting a rousing fart, I pleasingly and jocosely said that was as bad a noise I thought as my singing, at which she flew into a prodigious Rage." Seeming to defy the requisite obliviousness, a 1798 French *Eloge du pet* (Elegy for the Fart) turned out to be rather a tongue-in-cheek obsequy, recalling the medical and musical virtues of the fart, citing Aristophanes to prove that farting preceded all other languages, quoting Cardano on the sixty-two different sounds, and defending *le pet* in the name of a republican freedom of expression. Etiquette by then had so demoted and suppressed the more obvious expostulations of flatulence that it could be associated only with barbarians, peasants, the medieval, and the uncouth. When the novelist Emile Zola brought Jesus down to earth by making him a prodigious *péteur*, the offended outcry of readers in 1887 was louder than the explosions of this bum-kin, who "despised timid little squeaks, smothered between the cheeks, squirting out uneasily and ashamedly." Jesus was, in this sense, less a man of the people than a man of another time.[128]

Erasmus had promoted manners as a corollary to Christian tolerance and to honest diplomacy, an oxymoron in which he had high hopes. Quieting was at the heart of the peaceful, comprehensible conversation he urged upon religions, nations, and families. Contrariwise, manners were more often taken as ladders for personal advancement and marks of distinction between the honorable and the rabble at a time when middling city folk and ambitious merchants, neither paupers nor princes, had need of social definition. Publishers of books of manners turned the rules of medieval monastic houses and the courtesies of Renaissance courts into travel guides for youth making their way into adulthood, for apprentices making (or marrying) their way into mastery, for the bourgeoisie making (or marrying, or buying) their way into nobility. Such readers would learn how best to "pass" in a strange land.[129] In this context, quietness had less to do with an engaged moral discourse than with self-consciousness, noise with obnoxiousness and foolishness. According to Samuel Smithson's *The Figures of Nine*, nine sorts of foolish things were unseemly: "to eat pottage with a ladle, to blow one's nose at supper, to fart before a justice, to slabber in eating, to laugh at prayer, to kiss like a clown, to stand like a fool, to jeer continually, and to laugh at a dog's tail wagging." Most of these foolish things of 1662 were unseemly because noisy; in Smithson's day even the unseemliest of places, the land of Cockaigne (German *Schlaraffenland*) where fools earn money by sleeping, belching, and letting off loud farts, had quieted to a stuporous snoring. Offend no sense, advised Counsellor Manners in 1672, foremost the ear: "don't talk loud like a Clown . . . forbear also singing, especially if thy voice be harsh and untunable." Yawning, don't "yawl, and roar, as some do," and don't "sneeze, or cough too loud, and violently if thou canst help it, but (if possible) repress it." Keep a proper distance from others, "interrupt no one whilst he is talking, either by making of a noise, nor by speaking out of thy turn," and speak to the point, for "the weakest wheel in the cart screeks the loudest, and the emptiest Hoghead gives the greatest sound." So, advised the author of *Youth's Behavior*,

5. Sing not within thy mouth, humming to thyself, unless thou be alone in such sort as thou canst not be heard by others. Strike not up a Drum with

the fingers, or thy feet.

6. Rub not thy teeth, nor crash them, nor make any thing crack in such manner that thou disquiet any body. . . .

10. When thou blowest thy nose, make not thy nose sound like a trumpet.

At table suck no bones to get at the marrow and "make not much noise with thy teeth, neither in supping, nor in grinding too hard." Further, advised the *Narcissus*, make no "gnashing or creaking noise, through friction, or rubbing of iron, brass, stones, or any other hard substances together, which is what is vulgarly called, setting the teeth on edge, though it is the ear, in fact, that is annoyed by it; and this is the most poignant offence that can be offered."[130]

Intensifying aural sensitivities to (in)voluntary physiological processes, speech levels, nervous habits, and grides "setting the teeth on edge," etiquette multiplied the number of sonic events heard as noise and the kinds of noises felt as poignant. Sound, after all, was the central avenue of etiquette, argued a physician of Rouen in an award-winning treatise on the sensations and passions, published in 1767–1768 with a frontispiece embellished with cupids holding ear trumpets. "Through the ear alone, or almost alone," wrote Claude-Nicolas Le Cat, "is the soul beautified by that science of manners that drags man up out of the savage and barbaric state into which he is born." More than mere catalogs of the social universe of sound, books of manners were machines for excavating noises—dredging up and defining obnoxious sounds. The aural specificity of etiquette manuals was unflinching, as in caveats against interrupting with an "uncouth sound, not much unlike a dumb man trying to speak" and against the loud hawking up of phlegm in order to spit, which should be done into a handkerchief, folded instantly without inspection, for "good breeding consists in not bringing to people's attention anything that might offend or confuse them." A Russian directive to servants delivering messages assumed an exquisite sense of levels and qualities of sounds: "At the door of a storeroom, a cottage, or a monk's cell, wipe your muddy feet, blow your nose, and cough. Say a prayer as a test. If no one replies with an 'Amen,' say another prayer, even a third, louder than the first. If you still get no answer, knock lightly."[131]

7. Tears ♪ ♪ ♪ Music

Doubtless these sonic gradations were as polite a fiction as the prettiest of manners, but aural sensitivities did make their influence felt. Counter-Reformation Catholics, taking the Fifth of Loyola's *Spiritual Exercises* (meant to be read aloud), were to meditate on Hell and "hear with the ears [the] wailings, howlings, cries" of sinners, so that they might acquire an "interior sense of the pain which the damned suffer" and, remembering that pain in body and soul, keep to the straight and narrow. Lutheran and Calvinist positions on predestination (of both the elect and the damned) put a theoretical cramp on the power of Hell's torments as a deterrent, but in practice the faithful heard sermon after sermon that urged them to heed the cries of the damned, lest they backslide or take too quick comfort in a baptism among the elect. During the late seventeenth century, empathies began to flow in the opposite direction, first among freewheeling sectaries, Christian kabbalists, Quakers, and chiliasts; next century among deists, pietists, mystics, Enlightenment encyclopedists, novelists, Masons, political radicals, and working-class agnostics. How could those in Heaven be happy, ran the new emotional logic, knowing that others were convulsed under infinite tortures? Hearing "wailings, howlings, cries" from below, how could the redeemed not be saddened by such poignant sounds venting from Hell? Indeed, how could Heaven be a place of contentment when Hell was shrill with pain? John Milton, departing from medieval tradition, made Hell into a land of hard silences and made Pandaemonium, the capital at its center, a walled and dour parliament—perhaps in response to his own blindness, which required sound and spoken words to assure him of society and poetry. After years spent dictating *Paradise Lost* (1667), he conceived a Hell whose silence was archetypal of the lost and incommunicado. A few years later, the philosopher Anne Conway denied outright the eternity of Hell and decried the "horrible idea" of God as "a cruel tyrant." Herself afflicted since adolescence by headaches so cruel that they could be relieved neither by drugs nor by a temporary opening of her jugular veins to bleed away the pressure in her skull, Conway protested the glorification of suffering: "Since the grace of God stretches over all his work, why do we think that God is more severe and more rigorous a punisher of his creatures than he truly is?" Adapting Platonic, kabbalist, and Quaker positions concerning the processes of creation and redemption, she resolved the dilemma of empathy by proposing a universal salvation and a gradual emptying of Hell. Given that a creature made in God's image "cannot proceed infinitely toward Evil, nor fall into inactivity or silence or utter eternal suffering, it irrefutably follows that it must return toward the good." The visionary Jane Lead, taking her *Enochian Walks with God* in 1694, found Adam and Eve jubilant as the dead left Hell for Heaven after a period in those "dark Centres" from which Satan himself would eventually be released.[132]

Poetry, philosophy, and prophecy of themselves were insufficient to arouse that sense of poignancy accompanying what D. P. Walker has called the "decline of Hell." Etiquette's heightening of aural sensitivities and the consequent sharpness of the cries of the damned occurred within a new weather of tears. Notwithstanding Calvinist constraints on grief, medical accounts of melancholy, and Nicolaus Steno's dry anatomy of the ducts from which drip a salty liquid to keep eyes moist, weeping came to be expressive of a personal sensibility, an inner virtue, and a shared compassion that was favored in public (at the theater, at funerals, soon at executions) and savored in private (while reading, courting, and consoling). Margaret Cavendish, Duchess of Newcastle, a scientific woman, described herself in 1656 as inclined less to merriment than melancholy, "not crabbed or peevish melancholy, but soft melting solitary, and contemplating melancholy; and I am apt to weep rather than laugh, not that I do often either of them; also I am tender natured, for it troubles my conscience to kill a fly, and the groans of a dying beast strike my soul." The irrepressible tears of tender natures with such tender ears would damp the flames of eternal punishment even as they sustained the faithful. With tears, wrote the Puritan Anne Bradstreet, "I sought him whom my Soul did Love," and "He bow'd his ear down from Above." Then,

> My hungry Soul he fill'd with Good;
> He in his Bottle put my tears.
> My smarting wounds washt in his blood,
> And banisht thence my Doubts and fears.

Though neuroscientists now claim that the limbic systems of women are activated eight times more frequently than those of men, the eighteenth century welcomed both the weeping woman and the "man of feeling," whose tears could be mindful or passionate, ennobling or repentant, nostalgic or anticipatory, aesthetic or sensual. Fathers exchanged tears with children, proving familial bonds; suitors exchanged tearful confidences with their beloveds, proving an attachment beyond friendship; explorers proved the savagery of savages by noting how unmoved they were by tears. Under this new economy, as European men and women were more readily and publicly moved to tears, tears were understood to move them toward ethical and spiritual insights; rather than a veil, tears availed, making sense of sensibility. Settled in salons to hear chapters of a novel read aloud, sitting in pietist circles hearing the testimonies of the Spirit, on the edge of their pews or standing in fields to hear revivalist preachers of a Great Awakening, they learned to listen through their tears for the passions of the world and the needful compassion of its Lord.[133]

Sounds, wrote Jean-Jacques Rousseau, elicit tears more effectively than images. Gestures and attitudes may suffice in law and commerce; depth of passion comes only with sound, with *accens* or tones that make us thrill-and-shudder, *tressaillir*, for in times of passion those personal *accens* cannot be disengaged or disputed and, true as they are to one heart, so swiftly they penetrate another's. In this respect, the new weather of tears was primarily aural; to be seen weeping was less moving than to be heard weeping, especially through the upper-class walls of newly private bedrooms and sitting rooms from which, appearances left behind, true feelings could be inferred. Rousseau, who throughout his *Confessions* weeps often and well, began an essay in 1755 that addressed the relative effects of word and gesture, an issue raised by the success three years previous of his word-rich operetta, *The Village Fortune-teller*, which opens with a lovelorn girl sighing for a country boy and wiping tears from her eyes. Listening through the tearful *accens* of performers and the tears of audiences, Rousseau pushed on to the larger issue of the origins of language, an issue implicit in his famous *First Discourse* (1750) in praise of the primitive, "natural" life. He put final touches to his essay on language in 1762, about the time that the public got its first look at *Émile*, his pedagogical novel in which tears mark the first stage of social integration and a first step in the making of comprehensible signs. If tears did not render speech superfluous, surely they preceded language as one of humanity's essential means of expression and a key, like music, to communication across cultures.[134]

Like music? Wherever Europeans had gone, they found people making music

> |: African slaves, who were "great Lovers of Music, and much pleas'd with such Instruments as make a certain delightful noise, and a kind of harmony" :| *a Frenchman in the Caribbean*

or what was clearly meant to be music.

> |: Persians, whose "Music consisted of Lutes and Viols, very poorly played on; as also of Tabors and Voices, which made a wretched kind of Harmony" :| *a German in Isfahan*

Knowing it was meant to be music, they listened through what seemed to be noise

> |: Cuna Indians, who made "most hideous Yellings and Shrieks; imitating the Voices of all their kind of Birds and Beasts. With their own Noise, they join'd that of several Stones struck together, and of Conch-shells, and of a sorry sort of Drums made of hollow Bamboo" :| *an Irishman on the Isthmus of Darien*

for the musical impulse, imperial,

> |: Aztecs, whose chieftains were accompanied "with a great deal of music on flutes, bells, shell trumpets, bone scrapers, drums and other instruments, of little pleasure to Spanish ears" :| *a Spaniard in Mexico*

shamanic,

> |: Lapps, whose painted, illustrated drum "being beaten, and some Songs sung, they bring the designed Sacrifice to Thor, who if he signifies by a ring in the Drum, that the Sacrifice is pleasing to him, they fall presently to work" :| *a German/Swede in Lapland*

diplomatic,

|: Siamese, who play a March at the entrance of the King and his ambassadors, "a confused noise" of violins, oboes, gongs, drums, and bells, "and this noise, as fantastical and odd as it is, has nothing unpleasant, especially on the river" :| *a Frenchman in Thailand*

or just to pass the time before breakfast.

|: Floridians (Apalachites), who "In the morning, as soon as they are up, they commonly play on the Flute or Pipe; of which Instrument they have several sorts, as well polish'd and as handome as ours, and some of those made of the bones of their Enemies: and many among them can play with as much grace as can well be imagin'd for Savages" :| *a Frenchman in the Caribbean*

They made analogies to their own folk, court, playhouse, or church music,

|: Tupinamba Indians, who began a ritual with "a very soft sound, as you might say the murmuring of those who mumble their hours-devotions" :| *a Swiss Huguenot minister in Brazil*

describing the exotic instruments, how they were made and played,

|: Tartars, who play Kerrenai, somewhat like oboes, "save that they are of Brass, being above eight foot in length, and at the extremity, above two foot Diameter," played with their ends pointing up into the sky "and making a noise, which hath not only nothing of harmony in it, but is more like a dreadful howling than any thing of Music" :| *a German in Central Asia*

and their kinship by shape or shade of sound with more familiar instruments,

|: The Abundi, with their Marimba, "sixteen Calabashes orderly plac'd along the middle between two side-boards join'd together, on a long frame, hanging about a Man's Neck with a Thong. Over the Mouths of the Calabashes there are thin sounding slips of red Wood called Tanilla . . . which being beaten with two little wooden sticks, returns a sound from the Calabashes of several sizes not unlike an organ" :| *an Italian in the Angolan Congo*

all a far cry from the outright condemnations of "the music of the people" by early Church Fathers such as Cyril of Alexandria, "That where there is the Sound of the Harp, the Beating of Cymbals, the Consort of Fiddlers . . . there is also all kinds of Filthiness." A far cry too from Puritan condemnations of the misuse of music, echoed as late as 1711 by a sour Anglican vicar who had no use for notes "newly invented" (quavers, semiquavers) that unseasonably quickened the tempi of his favorite slow organ pieces, their "solid *grave Harmony*, fit for a Martyr to delight in, and an Angel to hear," now so transformed by virtuosi that "I know not any sober

Person, who can understand any thing in it, except a *Jargon* of *Confusion*, without *Head* or *Tail*," hearing "nothing but *Noise*, *Rattle*, and *Hurry*." Such intramural controversies over forms and theories of music (in which Rousseau himself took up a cudgel) had long been part of European religious, political, mathematical, and artistic milieux in which styles and philosophies of composition did change (as Rousseau advocated), in which music notation did vary (like Rousseau's own scheme of 1741), in which churches and nations did differ (as Rousseau thought they must), so music on distant shores could be heard as naturally human even if unappealing. The Chinese court, listening politely to the music of their French guests, "like the *European* Music well enough, provided there be only one Voice to accompany the Instruments. But as for the most curious Part of [our] Music, I mean the Contrast of different Voices, of grave and acute Sounds . . . , they are not at all agreeable to their Taste, appearing to them a confused Discord." Since musical traditions were found everywhere, early modern travellers recognized the diversity of musics as an intriguing aspect of cultural difference meriting detailed description, especially in those cases of clear divergence from European modes, as in the rituals of Abyssinian Christians, where (according to Portuguese Jesuits in Ethiopia), "the whole service is partly in the Chaldean and partly in the Ethiopic language. Whatever they say, they accompany with suitable movements of face and body. Thus they speak weeping the words of tears; laughing those of laughter; leaping those of leaping; loudly those of clamour; softly those of silence; in short, whatever they utter, they show with such bodily gesture as to be understood no less through the speaker's signs than through his voice."[135]

Which returns us to Rousseau, who believed that language was originally figurative rather than mimetic—and sung rather than spoken. Passion made speech necessary, music made it mutual. Voice had been "wrested" from the body neither by hunger nor thirst but by love, hatred, anger, and above all pity, those moral feelings that demand duels and duets, the sparring/sharing of selves and sounds. Those sounds, like the score for his popular operetta (performed over four hundred times in fifty years) would have been melodic, matching syllables to notes in a "natural" syntax—as opposed to putatively artificial systems of harmony with their simultaneities of notes that were subject to shifting, arbitrary conventions about "good" and "bad" chords. Melody was to harmony as Adam-in-Eden was to Nimrod-at-Babel, the former calling things by their right names, the latter calling down confusion. Rousseau, who had become hard of hearing when he was twenty-four and suffered ever since from a "grave hollow buzzing" in his ears, harped on the inadequacy of harmony for the most elemental communication. "By itself harmony is even inadequate for the expressions that appear to depend uniquely upon it," he wrote, referring to natural sounds with no intrinsic melody. "Thunder, the murmuring of waters, winds, and storms are poorly rendered by simple chords. Whatever one may do, noise alone says nothing to the mind, objects have to speak in order to make themselves heard, in every imitation a type of discourse always has to supplement the voice of nature." Nature does not speak to us, it speaks through us; Nature is essentially noise until we listen through it with our sound-selves. "The musician who wants to render noise with noise is mistaken; he knows neither the weakness nor the strength of his art." Music's strength: passionate human translation. Music's weakness: locality, for as passions are rooted in person and circumstance, so music is rooted in culture, place, and change. "Why is our most touching music but an empty noise to the ear of a Carib?" Asked, and answered: Because the more touching, the more particular.[136]

Having particularized music, Rousseau could claim that music-making was universal but could point to no universally pleasing melody; some music might be dramatically healing, as with infamous outbreaks of tarantism in southern Europe, but those bitten by a tarantula and sent into muscle spasms were healed of their dancing mania only by familiar, local, whirling tunes. Having localized music, Rousseau could claim that music was a language but could point to no universally intelligible language, unless it was that language of numerical ratios by which he hoped to reform musical notation. Having exalted melody over harmony, Rousseau could claim that melody was original to music and music original to speech, but could point to no "first" melody. And having drawn noise itself into the orbit of music ("teach [a composer] that he should render noise with song, that if he would make frogs croak, he has to make them sing"), Rousseau could claim that music has the capacity to lift all sound into intelligibility, but he could not claim that a Russian song of thunder would make a Carib think of rain.[137]

Tears might be a better candidate for the comprehensibly pan-human. Tears, after all, were the seeds of salvation: after the Deluge, Prometheus had refloated humankind from soil mixed with his own tears; the Great Mother of Egypt's Middle Kingdom had wept man into being; Christ's tears had redeemed; the Virgin's tears still restored. No two peoples made the same music; people everywhere wept. Yet the meaning of tears was no less slippery than that of laughter. One could weep for joy or relief or shame, in physical pain or emotional agony, or in sympathy for the joy, relief, shame, pain, or agony of others, or in a play for the sympathy of others, or as an actress in a sad play, or as one of *les grandes pleureuses*—great weepers—in eighteenth-century audiences, or at the startling sight of a live baby in swaddling clothes on stage during a play at Fontainebleau in 1769. The title of that one-act, *Le Cri de la nature*, echoed Rousseau's analysis of the origins of language in his Second Discourse, "On the Origins of Inequality" (1755): "Man's first language, the most universal, most energetic, and only language he needed before it was necessary to persuade assembled men, is the cry of Nature." Most authentic of human sounds because least tempered by society, the cry of Nature got society moving by inspiring pity, which confirmed the sharing of a world. A quarter-step beyond noise (meaningless sounds to which we are therefore indifferent), the instinctual cry made itself heard as the sound of a person *in extremis*, a sound that others could not ignore. Unlike the roaring of the 1755 Lisbon earthquake, heard and felt from North Africa to Switzerland and beyond, with many repercussions for Euro-Atlantic philosophy, Rousseau's cry of Nature was a cry of passion, and it was this near-noise of passion that had detonated history.[138]

From the terrible rumbling in Glikl of Hameln's synagogue to the rumbling of the Lisbon quake, from the strange musics of Asia, Africa, and the Americas to screams in European asylums, from the night-groans of Glikl's anxious second husband to the jarring laughter and open farts of boors, from the squeak and whistle of a bell jar to the whistle of tinnitus and the squeak of the ear itself, one had to listen *through* apparent noise. At every scale, and confined neither to the figure of Echo nor to the fulminations of Hell, there was noise. Thanks to a more finicky etiquette (and to manners in greater demand), to a more fashionable compassion (and the tearful sensibility by which it became evident), to more precise sciences (and their intrigue with ever-smaller phenomena), to heated debates over music (and the aesthetics of dissonance), to concerns about privacy and a shifting economy of sex (of which more anon), noise was increasingly heard and felt as part of the matrix of life and was explicitly theorized as such by men with clout—Mersenne, Hooke, Perrault, and Rousseau—who heard Nature itself in terms of noise.

Could there be no escape from this Babel? Not for Dr. Alexander Hamilton, a Maryland doctor and amateur musician on a summer's jaunt around New England in 1744, who in Philadelphia went into the street to hear war being publicly declared on France among "a rabble of about 4,000 people," after a procession of "roaring sailors" and "8 or 10 drums that made a confounded martial noise, but all the instrumental music they had was a pitiful scraping negro fiddle, which followed the drums, and could not be heard for the noise and clamour of the people and the rattle of the drums." Who visited Cohoes Falls, where "the noise is so great that you cannot discern a man's voice unless he hollers pretty loud," a noise "easily heard at four miles' distance" and twelve miles in spring when the ice breaking was "heard like great guns all the way at Albany." Whose ears, throughout his tour of New York, were "perpetually invaded and molested with volleys of rough sounding Dutch," but whose patience was rewarded in Manhattan by the glorious violin and voice of Mr. Wendall, who for an encore "imitated several beasts, as cats, dogs, horses, and cows, with the cackling of poultry, and all to such perfection that nothing but nature could match it." Whose chambermates in a tavern included two "greasy" weavers and "a raw boned boy" who "went reeling downstairs making as much noise as three horses." Who in Boston "had a deal of discourse in the disputatory way" with the loud talking "Mr. Clackenbridge (very properly so named upon account of the volubility of his tongue)." Who, on his way back, stayed at an inn where he was disturbed "by the noise of a heavy tread of a foot in the room above," because his own room was "so large and lofty that any noise echoed as if it had been in a church." Who, nearing home, had another chambermate, an Irishman who "made a hideous noise in coming to bed, and as he tossed and turned, kept still ejaculating either an *ohon* or *sweet Jesus*," and who woke him early in the morning with his "groanings, *ohons,* and yawnings," now and then "bawling out, 'O sweet Jesus!' in a mournful, melodious accent; in short he made as much noise between sleeping and waking as half a dozen hogs in a little pen." And who, on his roundtrip of 1,624 miles, twice made the acquaintance of Captain Noise himself, "a dealer in cattle, whose name and character seemed pretty well to agree, for he talked very loud, and joked and laughed heartily at nothing." A tour de farce, Hamilton's *Itinerarium* was also a tour de force of backgrounded noise in all of its eighteenth-century expansiveness, as met on the street among roaring sailors and the rattle of drums, at coffeehouse table or private dinner amid strepitant debate, through wilderness to waterfall, upstairs in a tavern. Among his own, at the "ancient and Honorable Tuesday Club" founded in Annapolis in 1745, Hamilton himself was no sonic wallflower, for he enthusiastically joined in when

> Each member in chorus did roar and did bellow
> With the horn and the fiddlers and violoncello.
> You'd have thought the Grand Harmony of the high Spheres
> Came down in a Sudden, ding dong in your ears.

Noise in Hamilton's earthy world swelled across categories of the natural and the man-made, the animal and the human, the raw and the refined, informing his prejudices of class, race, ethnicity, and space. Measured in decibels, that noise may have been no greater than a century before, but it was felt to have spread in every direction—as if, upon the cultural contraction of Hell, the noise of the underworld had been ejected to the world above. Which is exactly what Anne Bradstreet had found a century before, one "silent night" in 1666, when suddenly she heard a "thund'ring noise / And piteous shrieks of dreadful voice," that came not from Hell, as one might expect from such language, but from "That fearful sound of 'fire' and 'fire'" as her own house burned to the ground.[139]

Across the Atlantic in that same apocalyptic year had been heard the "Noise and cracking and thunder of the impetuous flames, the shrieking of women and children, the hurry of people, the fall of Towers, Houses and Churches," as first plague and then fire tore through London, after which Robert Hooke and his friend Christopher Wren began laying out a grander, stone-faced city. Backtracking, *reculer pour mieux sauter*, we are back around again to Hooke, though less for the Neoclassical syntax of his architecture than for his concern with syntax in general and the need for clarity amid abounding noise, which led him to decipher messages in the arcane writings of a sixteenth-century magus and to read a "curious Discourse" about the Tower of Babel to his fellows at the Royal Society.[140] Pages ago, on the topic of Babel, I deferred discussion of the early modern response to the trauma of the thousand tongues and the dilemma of "communication"; now, with music and tears just behind us and noise everywhere in the background, the time has come.

8. Moans Groans Sighs Sex

Conjuring spirits in 1583, the medium Edward Kelley, sitting with John Dee, heard a voice and then spied a husbandman appareled in red, kneeling on the table and praying in a strange language, "*Oh Gahire Rudna gephna oh Gahire*, &c." The man brought a message about the end of Time and the sorrow of the Earth, whose "Waters pour forth weepings," but the words of the prayer were a mystery, as would be many of the words dictated by a shining figure named Galvah, who commanded Dee to start in on a book entitled "*Logah*: which in your Language [Latin: *Logos*] signifieth *Speech from GOD*," to be written back to front, as in Hebrew, with a first leaf that would be a "*hotchpot without order*, So it signifieth a disorder of the World, and is the speech of that Disorder." Dee dutifully transcribed her words:

> Loagaeth seg lovi brtne
> Larzed dox ner habzilb adnor — Now seas appear
> doncha Larb vors hirobra
> exi vr zednip taiip chimvane
> chermach lendix nor zandox.

And much more of the same: forty-nine "calls" or prayers and ninety-five gridded tables inscribed with letters, part of a series of angelic books intended to facilitate a re-reading and renewal of the world. He was eager to learn from the angels how to pronounce the words of this heavenly language and how it related to Adam's primordial Hebrew, so that he too might speak sentences that told the world its truths. An avid student of kabbalistic methods for rooting out the core of scripture and of alchemical processes for getting to the heart of matter, Dee was happy to subscribe to a celestial secret. Kelley in a fit of *pîque* upbraided the spirit-form Madimi for speaking in Syriac, "Unless you speak some Language which I understand, I will express no more of this Ghybbrish," but Dee was not dismayed at having to write up volumes of gibberish spelled out by a colloquium of angels.[141]

Robert Hooke was. Dismayed. Apologizing to the Royal Society in 1691 for calling its attention to a book published decades earlier, he admitted that *A True & Faithful Relation of what passed for many Years Between Dr. John Dee . . . and Some Spirits* seemed to be "a Rhapsody of incoherent and unintelligible Whimsies of Prayers and Praises, Invocations and Apparitions of Spirits, strange Characters, uncouth and unintelligible Names, Words, and Sentences." He had thumbed its five hundred pages several times since the book's posthumous printing in 1659, and "wherever you open and begin to read, you may find cause enough in a very little time to throw it aside and neglect it," as so many other readers had done. For a man of Dee's caliber—physicist and horologist, linguist and geometer, conjuror to Queen Elizabeth and guest of the Holy Roman Emperor—to peddle such twaddle did not add up, until at last Hooke did the arithmetic: Dee must have been a spy or courier in the Queen's service, and "I do conceive that the greatest part of the said Book, especially all that which relates to the Spirits and Apparitions, together with their Names, Speeches, Shows, Noises, Clothing, Actions and the Prayers and Doxologies, &c. are all *Cryptography*." In specific, the cryptography of a German abbot, Johannes Trithemius, whose *Steganographia* of 1499 had proposed using angels as invisible conduits for secret correspondence, with keys for encoding that Dee had patently borrowed. "Besides"—and here Hooke knew that he was clinching his argument—"the Words that he sets down, as delivered by his Spirits, are many of them inarticulate, according to the commonly accepted Sounds or Pronunciation of those Characters they are written with, and therefore were not put to signify those Letters." That is to say, the phonics of Dee's angelic language ("brtne," "vr") were too impossible to pronounce, too much like noise, to be other than code.[142]

Dee, it so happens, had copied out and studied the *Steganographia*. While Hooke's hunch about the spying remains speculative (Dee *was* a cryptographer for the Queen's secret service), Hooke's familiarity with Trithemius was indicative of the close garden of texts and theory in which early moderns labored to exhume or prune a universal language. Ideally, one should be able to undo the effects of Babel—confusion, isolation, frustration, suspicion, fear, war, noise—by finding a means for delivering the truth that would be harmonious, unintimidating, and broadly intelligible. Suffering through decades of religious warfare and theo-political battles over speech and images, with growing recourse to casuistry (telling half the truth to others so as to be wholly true to oneself), philosophers and proto-scientists of the late sixteenth to eighteenth centuries felt a need for, and often a mission toward, clear communication. John Durham Peters has shown that "communication" in the sense I have just used it was constructed during the 1600s while its earlier meaning, "making common," was nudged toward the cosier notion of the sharing of thoughts. Where before communication had denoted the act of transferral, by 1700 it referred to the success of the transfer: a means had become an end. The test of a universal language, for which a new proposal was floated every other year between 1600 and 1800, had once been the completeness and efficiency of its lexicon; now the test would be the ability of two people, despite inevitable differences, to exchange ideas *through* that language. A year before Hooke uncovered the meaning and purpose of Dee's balderdash, but a decade after Hooke had been in earnest discussion about the prospects for a universal language, John Locke clarified what must be meant by communication: "To make words serviceable to the end of Communication, it is necessary...that they excite, in the Hearer, exactly the same Idea, they stand for in the Mind of the Speaker. Without this, Men fill one another's Heads with noise and sounds; but convey not thereby their Thoughts, and lay not before one another their Ideas, which is the end of Discourse and Language."[143]

Spirit-beings could do the trick, seeing as how they had no bodies, baggage, or other personal impediments to flawless intercourse, and were graced besides with extrasensory speech. Cleverly apt had been Dee's angelic cryptography: he used a transparent medium to communicate through code the backstage intrigues at the court of the Holy Roman Emperor, Rudolf II, who was himself a mopish enthusiast of the occult. In the absence of angelic messengers, however, what assured reliable communication? The "natural language" of the hands as developed by the deaf, by monks, by mimes and international merchants, illustrated by John Bulwer in his *Chirologia* of 1644? The Lord's Prayer translated into and transliterated from 150 "languages, dialects, and character systems," including Chinese, Yiddish, and Choctaw, as compiled in 1715 by a polyglot Englishman? European-language dictionaries for Bantu, Japanese, Malaysian, Nahuatl, and the ancient languages of Aramaic and Sanskrit, prepared by Continental lexicographers in Hooke's day? Grammars (Ethiopic) and dictionaries (French) organized by root words, revealing common meanings within the language groups known to seventeenth-century linguists? Restoration of the root language of humankind, attempted among others by Athanasius Kircher, who kept an Ecstatic Heavenly Journal of his travels among the stars with a guide called Cosmiel but who, when back home, used cryptography and comparative linguistics to penetrate the chaos of post-Babel languages? Mathematics, which was the language of Nature and perhaps of human nature, as conceived by Galileo and pursued a century later by Leibniz, who developed the calculus while seeking an infinite means of expression that did not give rise to anomaly or superfluity? Glyphs or written characters, because less ambiguous and less corruptible than spoken words, as proposed by an Anglican bishop, John Wilkins, who lost a first draft to the Great Fire but recalled the text well enough to have his *Essay Towards a Real Character and a Philosophical Language* published in 1668? A "language absolutely new, absolutely easy, absolutely rational, in brief a Pansophic language, the universal carrier of Light," imagined that same year by Jan Comenius, the Great Didact of Moravia?[144]

Well, and what about sex? It was practiced everywhere, had common signs and sounds of arousal, begot a host of recognizable positions, and enabled the communion of utter(ing) strangers. What else could anyone ask of a universal language? The second Fall, at Babel, undid none of the physical arts of lovemaking that women and men had refined as consolation for lives of hard labor to which they had been sentenced after the first Fall, when nudity of a sudden turned to nakedness and companionship to sex. After Babel, sex survived as a last(ing) remnant of mutual intercourse. In the form of the erotic, sex was lifted above animality to an aesthetics of anticipation and caress; in the form of the bawdy, sex became a battle axe for attacks on inequities of law, power, and wealth; in the form of the obscene, sex abraded the line between the licit and illicit; in the form of the pornographic—a blend of the erotic, bawdy, obscene, and freethinking that emerged during the 1600s—sex was the prime mover toward a combinatorial utopia where bodies, minds, and passions meshed in endless permutations. Whatever its form, sex served as a *lingua franca*, making sense where little else did. In encounters across Asia, Africa, Oceania, and the Americas, the erotic became so entangled with the exotic (Turkish harems, Tahitian maidens, South Indian dancers, North African water girls and boys, West African princes and princesses) that no further evidence was needed to show that sex could be a *clavis universalis*, a universal key.[145]

John Cleland's exotic *curriculum vitae*, from soldier to agent to whistleblower to novelist, from London to Smyrna and Bombay and back to London, was itself key to the fact that this author of the notorious *Fanny Hill, or Memoirs of a Woman of Pleasure* (1748/1749) was also the author of *The Way to Things by Words and to Words by Things* (1766), and of the *Specimen of an Etymological Vocabulary, or Essay, by Means of the Analitic Method, to retrieve the Antient Celtic* (1768). In the *Specimen* Cleland showed that Celtic was the original language of Europe and, courtesy of the Israelite tribe of Gomer that anciently set sail for Wales, heir to primordial Hebrew. An intimate, sometimes dangerous liaison bound eighteenth-century studies of language to the tumescent literary genre soon to be called "pornography." John Locke, his own library well-stocked with classical Latin and Renaissance erotica, had maintained that language represented not Nature transposed to a different register but the operations of the mind itself, so it was fruitless to snuffle through the copse of languages for root syllables that typified, once and for all, a dildo, the act of fellatio, or sexual ecstasy. Instead of fixing a universal language upon an absolute set of true names for things, one had rather to trace the flexible spine of language, the syntax by which were threaded speech and thought. In grammatical terms, philosophers of language had to redirect their energies from nouns to verbs, from sign-lists to particles and conjunctions. Distinguishing ideas from the mental operations by which a person arrives at ideas, Locke located the power of language in copulative and connective "particles," copulatives joining words into propositions, connectives joining propositions into arguments. Through joinings that were brief propositions of inextinguishable lust and longer arguments for infinitely supple subordinating and coordinating conjunctions, the copula of grammar would be engaged with the myriad copulations of a woman of pleasure. "Grammar," explained grammarian Richard Johnson in 1706, "is the Art of Expressing the Relations of Things in Construction," and it was just this relation of things *in media res* that grabbed pornographers. As philosophers of language turned away from archaic symbols toward a practical anatomy of the ligaments of language by which propositions were fashioned and arguments held together, composers of sexplay turned away from taxonomies of fixed positions toward couplings in which the libidinous body became intrepidly pliable and through which every sort of relationship—emotional, social, economic, political—was reconstrued.[146]

Tempting as it would be to diagram the erotic as transitive (pleasuring another) and the pornographic as intransitive (self-pleasuring), the ribald as subjunctive and the obscene as ablative, such parsings distract from the crucial question with regard to sex as a universal language: How could one be sure that sex communicated truthfully? Was not the brevity of sex itself a default on eternal spiritual love and a sop to enduring passion? In late medieval romances, truth had been assured as often through longings denied as through desires consummated; either way, truth lay with the heart, and bodies when they clung to each other clung to the spirit beneath the flesh. Radically materialist and experimentalist, the new pornography drew upon the truth-telling capacity of the senses themselves: hearts palpate, lips tremble, torsos sweat, flesh tumesces, liquids flow, all in service to a lustfulness that cannot be denied. "Truth! stark naked truth, is the word," Cleland's Fanny vows in the preface to her narrative, "and I will not so much as take the pains to bestow the strip of a gauze-wrapper on it, but paint situations such as they actually rose to me in nature." Scientific societies had begun to publish transactions that adopted a plain style for accounts of a natural world laid bare under telescope and microscope; it would hardly do to have descriptions of the most basic of behaviors veiled in the circumlocutions of prudery or "false delicacy, that takes umbrage at everything, and gives a criminal sense to the most innocent actions and words." The candidacy of sex as a universal language was advanced as much through its nakedness as through its flexibility and ubiquity. This nakedness, which led to Cleland's prosecution for obscenity in 1749, had earlier gotten him in hot water as an agent for the East India Company, whence he had been fired for trying to expose the Company's darker doings in Bombay. From debtor's prison in London, where under a twenty-guinea contract he completed *Fanny Hill*, Cleland entered the company of more secure British novelists who were exposing the limited options of young unmarried women of small means and weak connections when subjected to the sexual depredations of rampant capitalism. To the degree that Fanny's directness was in line with Cleland's own life, it was also a function of her class and gender, in that her progress from innocence to experience could not proceed under cover of coyness if she were to profit from the profession of sex, unlike the shower of euphemisms used by Cleland in his less successful sequel, *Memoirs of a Coxcomb* (1751), whose hero learns little because, as a man and a wealthy libertine of leisure, he is doubly defended in his desires.[147]

Man's desire is visibly consummated; a woman's orgasm can be inscrutable. Pornographers often insisted upon suppositious female ejaculations in tandem with the male, but even before the eighteenth-century reversal of European images of women's bodies from lesser versions of men's bodies to physiological opposites, there were clear differences in depictions of male and female orgasm. Aural differences. Men snorted and panted on their way to coming while women flushed and fluttered; at climax, while men were triumphantly quiet or monosyllabic, women explicitly confirmed orgasm, sighing long and loud, weeping with joy, or making ostensibly involuntary cries heard as evidence of a mutually successful fuck. In the older literary-legal traditions of male lovesickness and of Courts of Love that articulated the rules of lovemaking, sex was as frequently a last resort and remedy for a man's unwonted attentions as it was a means for sustaining a relationship; Courts of Love, with their women barristers and justices, gave women the chance to tell sexual truths and to affirm sexual pleasures because empowered by fictions of a public realm that promised both intimacy and honesty. In the newer fictions of European pornography, which early on had a considerable female audience and took female pleasure seriously, the sexual act was itself the truth-teller, stripping away social conventions, abiding no hypocrisies, daring the senses to revel in unspeakable conjugations. And for all of the anxiously listening partners, their ears "erect," the truth of sex lay in its noises, its unpreventable noises.[148]

Authors of Renaissance erotica had been alert to those noises, as with Pietro Aretino's description of a father confessor's sodomy of an abbess, his cock in and out of her ass making sounds "like the slurping [that] pilgrims' feet make when they walk on a road of sticky clay." Aretino's *Dialogues* (1534–1536), illustrated often with his friend Marcantonio Raimondi's etchings of sixteen sexual positions, for each of which Aretino had composed a sonnet, featured scenes in which males and females were aurally aroused, orally expressive, and equally noisy, as in a convent orgy where "all agreed to do it together as choristers sing in unison, or more to the point"—and more noisily—"as blacksmiths hammer in time." With an "Oh my God, oh my Christ!" and "Push out that sweet tongue!" and "Wait, I'm coming!" and "Oh Christ, drive it into me!" this anvil chorus mounted to crescendo. "Some were whispering, others were moaning loudly—and listening to them you would have thought that they were running the scales, *sol, fa, me, re, do*—their eyes popping out of their heads, their gasps and groans, their twistings and turnings making the chests, wooden beds, chairs, and chamber pots shake and rattle as if the house had been hit by an earthquake," until at last "eight great sighs rose all at once." But these dialogues, or *Reasonings* (Ragionamenti), were social and political in thrust and gender-balanced in parry, while the readership of the eighteenth century was more fixated upon orgasm itself, reports of which were now coming in from all parts of the sexual compass. Pornography was fixed especially upon the orgasms and eargasms of women. Not only were the gentler sex's least gentle noises needed as proof of female climax, they were vocal recalls of what was being expunged from medical treatises on the nether regions: references to female sexual pleasure and to female orgasm as prerequisite for successful conception.[149]

Spying on her first sex scene in the London brothel to which she has been taken after both parents have died of smallpox, Cleland's virginal fifteen-year-old becomes excited as much by sound as by sight: "now the bed shook, the curtains rattled so, that I could scarce hear the sighs and murmurs, the heaves and pantings that accompanied the action, from the beginning to the end; the sound and sight of which thrill'd to the very soul of me, and made every vein of my body circulate liquid fires," and Fanny is moved to pleasure herself. The next night, peeking through "the long crevice of a partition" in a closet, she observes her friend Polly with a young Italian, both of them headed toward climax: "'Oh! Oh!—I can't bear it—It is too much.—I die.—I am a going—' were Polly's expressions of extasy: his joys were more silent; but soon broken murmurs, sighs heart-fetch'd, and at length a dispatching thrust, as if he would have forced himself up her body, and then motionless languor of all his limbs, all shewed that the die-away moment was come upon him, which she gave signs of joining with, by the wild throwing of her hands about, closing her eyes, and giving a deep sob, in which she seem'd to expire in an agony of bliss." Many scenes and auditions later, no longer a virgin, and "Lifted then to the utmost pitch of joy that human life can bear," Fanny herself "touch'd that sweetly critical point, when scarce prevented by the spermatic injection from my partner spurting liquid fire up to my vitals, I dissolv'd, and breaking out into a deep drawn sigh, sent my whole sensitive soul down to that passage where escape was denied it, by its being so deliciously plugged and choak'd up." Women's sobs and deep drawn sighs were no mere *organum* to the sex act; they were necessary "signs of joining in," testimony to a climax otherwise indeterminate, which explains Fanny's (Cleland's) rhythmic conflusion of touch and sound: "Then began the driving tumult on his side, and the responsive heaves on mine, which kept me up to him; whilst, as our joys grew too great for utterance, the organs of our voices, voluptuously intermixing, became organs of the touch... and oh, that touch! how delicious!... how poignantly luscious!" Novice whores learned these vocal techniques as swiftly as Fanny learns the cries essential to the lucrative pretense of being a virgin, giving "such a timely jerk, as seem'd to proceed, not from the evasion of his entry, but from the pain his efforts at it put me to: a circumstance too that I did not fail to accompany with proper gestures, sighs and cries of complaint, of which, 'that he had hurt me—he kill'd me—I should die—' were the most frequent interjections."[150]

London in 1725 had 107 brothels around Drury Lane alone; in 1758 the city was serviced by at least three thousand full-time prostitutes and more part-timers, most of whom practiced to perfection "the quick-drawn breath," "the quivering sigh," "the dying gasp." From midcentury on, as estimates of their numbers took flagrant flight toward fifty thousand, the better-settled professionals were listed in lucrative directories such as the *List of the Sporting Ladies* and Thomas Harris's *List of Covent-Garden Ladies*, which detailed the technical prowess of each and often noted the quality of their soundmaking: Miss D. G... y in Fisher Row had good eyes "and a *melodious* voice"; Miss Jen-s at a shoemaker's in Tower Street had "a very good voice"; Miss Ur–art, at the jelly-shops, talked "with all the cant of a methodist preacher"; Miss Cl–l––d (no kin of John Cleland's?) at the barber's on Newport Street took "sometimes a cup too much, but is not very noisy"; Miss F-wl-r of Carter's Bagnio, she of the serpent tongue, swore "like a trooper"; Miss C–p–t of Wild Street had, "to be sure, black eyes and hair; but her blasting method of speaking, were she the greatest beauty in London, would destroy the whole"; Sally P-w-l, fervently religious, was a plaintive "wheedler." By 1795 an editorial in the *London Times* complained that city streets were nightly "infested by a number of impudent though unfortunate Women, who not only assail the ears of [passersby] with the most blasphemous and obscene language, but even go to the length of assaulting their persons."[151]

Generations earlier, sexual slander hissed out on the street had been taken as a personal assault and was subject to suits for defamation of character, but the hubbub of a distended urban environment in a London growing furiously past its first half-million inhabitants rendered such suits inaudible. In a place so large, it was hard to prove that any local, fleeting utterances did lasting violence to *fama*, reputation. Scolding, which had long been a public means of keeping a village in good order, lost its force in the city, where Justice Bulstrode expressed a metropolitan impatience with litigants who had been upset by a public scold: "[T]his silly woman only makes a noise amongst her neighbours," declared Bulstrode, addressing the Middlesex Grand Jury in 1718, and her offence "only grates upon the organ of hearing in the tender and distinguishing ears of a wise person, and sinks no deeper." The scattered noise of slander, sexual or otherwise, was still ungenerous, but it no longer had much standing at the bar; it was merging into the background noise of the city, which was always a little at war with itself and whose authorities worried more about mob and riot, collective noises that formed out of the swirl of hundreds of indistinct voices. After 1727 London courts no longer heard suits filed by women who had been called "whores" in public, and since public whoring would not be a statutory crime in England for another century, prosecutions relied upon indictments for disorderly conduct or disturbing the peace—indictments pursued by self-certified moral reformers or newly constituted constables itching to prove their superiority to older systems of communal watchmen.[152]

Soliciting the noises of climax from women a-swiving, attentive to the tenor of the voices of Covent Garden ladies, alert to the "drunken screams" and "loud ribald talk" of "rude whores" whose very vulgarity attracted many clients, eighteenth-century society never narrowed sexual desire to the eyes. Although voyeurism was as basic to porno*graph*y as it was to fashion and to the newly modish habit of museumgoing, sonic elements were no more incidental to sex than to the whirr of newly modish hoop skirts in 1710—"What strange noises are those that meet our ears? What is this creaking, and crumbling and crushing? Wherever is this whirr, as of some ten thousand wind-mill sails in full career?" Layered with petticoats of silk or crackling taffeta, hoop skirts were at once bastions of modesty and floating invitations to a rollicking nakedness beneath, the tilting up of such hoops to become a convention of the bawdy. For those less well girded, there was the street ballad of a damsel accepting a ride from a jolly wagoner, who

> whip'd on his Fore Horse, he jingl'd his Bells,
> Such Music ye Eunuchs, your Music excels,
> She kept Time to his Tuning, and sigh'd at each Sound,
> O dear says she, *Robin*, the Waggon goes round,
> *Geehup* Robin, *geehup geeho*.

"Licentiousness is not bred in a moment," observed Hannah Woolley. She approved of romances that treated of generosity and constancy, since such texts furnished material for conversation and rescued women from standing "like so many Mutes or Statues when they have happened into the company of the ingenious," but she was skeptical of "loose songs," cautioning the readers of her best-selling *Gentlewoman's Companion* of 1675 that "songs of wantonness will breed in you a more than fitting boldness, which will put you in the confidence of practising what you read or sing." Young men, too, learned from the *School of Good Manners* that "Lascivious Songs and Ballads" were treacherous, "for the Practice of them will greatly Debauch you." Nevertheless, apprentices, servants, ladies' maids, clerks, scriveners, delivery boys, tradesmen, stock jobbers, libertines, and gallants rarely preferred their Bibles to such songs, jests, and riddles, which figured prominently in early modern street-life and circulated widely through fairs, cafes, taverns, and coffee houses. As invitation, as encouragement, as penultimate incitement, or as sure conclusion, the rapture of words and rhythm of sighs was integral to the truth of this universal language, especially as pornography was drawn into the embrace of Enlightenment concepts of the free and the natural, concepts that made much of women's sexual responsiveness. With its multiple partners and positions, pornography shared eighteenth-century society's fascination with carnival and masquerade, where gender was put into play in sexualized spaces that deceived the eye and hand much sooner than the ear. In *Nunnery Amusements* (1786), the anonymous poet-pornographer began his story with layers of sound:

> Oft has been thought within a Convent's walls
> No voice but pray'r and sorrow ever calls;
> Be mine to shew the sounds of amorous love
> Are not unknown—and every veil remove...

until the saintly orgiasts scream in ravishment and a "deep-fetch'd sigh th'approaching bliss" declares, with at long last "soft sighs and murmurs."[153]

With his four aunts nuns, his father a bisexual *roué*, his mother *in absentia* (having decamped after his birth in 1740), and his uncle (who raised him) an *abbé* with a private seraglio, Donatien Alphonse François was overdetermined to be the Marquis de Sade, that free-spending, freethinking, ritualist composer of drab plays and dark pornography about which so much criticism has been written that it surpasses the volume of his own corpus. De Sade urges himself on this chapter not by virtue of the tedious lectures and overhearings that were embedded in his novels, nor by vice of the screams of painful pleasure that erupt from rounds of his real and fictive penetrations and flagellations, nor even by dint of his dramatic sketch of an orgy *interruptus*, "when a dreadful shrill screech was heard, all the candles snuffing out that very instant" as a startled owl takes wing from one corner of a subterranean chamber, terrifying every other tender body with the blur and whirr of its feathers. No, de Sade urges himself on us through an analogy that appears in his *History of Juliette*, fifth of the ten volumes of

La nouvelle Justine, completed around 1797. There, the philosopher-queen Mme. Delbène disabuses young Juliette of the need for a conscience, which is little more than an acoustical shadow, an "inner voice which cries out when we do something—it makes no difference what—we are forbidden to do," a voice activated by alarms or hurts suffered in childhood. "Guilt, thus, is merely an unpleasant reminiscence." The inner voice of conscience can be muted as easily and fully as Madame can demonstrate the dullness of rectitude, which relies upon tattered echoes of habit and memory rather than upon any lively sensibility to what is happening *now*. Her demonstration invokes the acoustics of percussion: "All moral effects," declares Mme. Delbène, "are to be related to physical causes, unto which they are linked most absolutely: the drumstick strikes the taut-drawn skin and the sound answers the blow: no physical cause, that is, no collision, and of necessity there's no moral effect, that is, no noise." Doubtless de Sade savored the "taut-drawn skin" of that drumhead, with its triple entendre to the ear's tympanum and the uncircumcised tip of an aroused prick, but what he meant philosophically to do with that tautness was to leave no room for regret in the rebound of effect from cause. Mme Delbène continues: "There's the drumhead struck, the cause of a vicious or a virtuous act; one hundred *louis* stolen out of my neighbor's pocket or transferred as a gift from mine to someone in need, there's the effect of the blow, the resultant sound. Are we answerable for these subsequent effects when the initial causes necessitate them? May the drum be beaten without there being a sound emitted?" Of course not, booms the answer, so long as the skin is taut, for those truly alive quiver to the instant. "And can we avoid these reverberations when they and the blow are themselves the consequences of things so beyond our control, so exterior to ourselves, and so dependent upon how we are each constituted?" No, we cannot. Our passions compel us even as the world swirls around and shunts us from Parisian bordello to Provençal castle to provincial court to a cell in the Bastille. "And so 'tis madness, 'tis true extravagance to refrain from doing whatever we please, and, having done it, to repent thereof." Insofar as we can be the causes of our own effects, we should exult in those moments of freedom and the noise we make; insofar as we are at the mercy of other causes, we have no reason to blame ourselves. Sex in all its utterable forms and fleet sensations was, in de Sade's lasting analysis, an implicitly perfect candidate for universal language because it freed speech of particular pasts, laid the field of invention and neologism open equally to all classes, and enabled fluidly fraternal, sororal, and intramural intercourse—just as the sounding of the "taut-drawn skin" of the drumhead could beat in all breasts.[154]

Off the printed sheet and onto stiff new calico-print cotton bedsheets, sex was a notably audible experience, not only because sexual pain-pleasure released all manner of noises, some of which were taken to be necessary and irrefutable statements of orgasm, but also because life was so spatially constituted that few could avoid overhearing the thrashing of active intercourse. Most eighteenth-century families still slept together in a single room, often in a single bed, when fortunate enough to own one; in better-off families, where the parental bed was separate from others, it was defended by thin curtains; in the houses of the wealthiest, where couples enjoyed separate bedrooms, their wooden walls were scarcely thicker than drapery and remarkably resonant at lower frequencies, carrying for some distance a couple's "tumbling and tossing in a good Feather-bed" and the wife's "sighing and groaning, as if her very twatling-strings would break, making her moan to the curtains, fumbling, and biting, and tearing the sheets." If indoor privacy was now more sought after than it had been, it remained in most houses acoustically unachieved, so that domestic sexual activity was more often out of sight than out of earshot. Evolving desires for privacy in lovemaking went acoustically unmet, as readers could hear in *A Dialogue between a Married Lady and a Maid*, a 1740 English adaptation of an earlier influential piece of French pornography, Nicholas Chorier's *Satyra Sotadica*. Communing with her daughter Octavia about her upcoming wedding night, Tullia the mother confides that "This Silence and Quiet in the Family, Tomorrow Night, would be to thee and thy Philander much more welcome; but, instead of that there will be nothing but Noise about you, till you are both abed, and People placed in every Corner, to hear what you do and say, and if the Bed shakes, as certainly it will, upon the first Eagerness of enjoying thee, thou wilt hear them . . . Laugh or Giggle, to the no small Disappointment of thy Pleasure at first." Promising Octavia a less disconcerting first night, Tullia arranges for a prenuptial tryst, but first she instructs her daughter in the ways of the clitoris and the telling sounds of a man's climax, his "thousand Murmurs, and kind dying Words." Her own deflowering, Tullia confides, had taken place with *her* mother and a friend hidden behind the hangings of the bridal bedroom as *her* groom, gallant Horatio, had plowed steadfastly ahead despite the shrieks of his virginal bride, who upon the third turn was roaring and weeping until she began moving together with him, crying out, "Oh! Oh! — My Horatio! my Soul! Soul! Melt! Oh, what it is to feel my Love! My Horatio!" whereupon Pomponia the friend had leapt out and come "running to me, and catching me naked, as I was, in her Arms, began to kiss me with great Joy, telling me how she pitied me, when she heard me cry out; but that since a kinder Sound, and softer Expressions of my Joy, had assured her of my Felicity; she was now at Ease." Then Tullia's mother scampered out from behind the draperies and gave Horatio a thousand thanks. "Thy victory, said she, was made known to me by Tullia's Cries; but I make no Question they will hence forward be turned into kind Expressions between you for ever." Tullia, embarrassed, had been "so ashamed to hear my Mother's Voice so near, that I endeavoured to cover Horatio (who was almost naked) with the Bed-Clothes, and turn myself under 'em, not being able to look my Mother in the Face," yet after partaking of light refreshment Horatio had jumped his bride once more, calling for his mother-in-law and maid not only to listen but to watch. As for the new generation, who welcome the daylight prelude that Tullia has so thoughtfully arranged, Philander is just about to enter Octavia when Tullia bursts in upon them to rescue (she says) her daughter's petticoat from the "turmoil" of sex; noticing the young man's huge erection, she pleads with him to be merciful, then departs. Just as Philander has made his manful way through the hymen, Tullia bursts in again, having heard her daughter shriek at that crucial accomplishment. She starts to reproach Philander for his insensitivity only to hear Octavia cry in delayed orgasm, whereupon she pulls her daughter roughly out from under the young man lest the wedding guests below realize that their anticipations of standing audience to an audible deflowering have been short-circuited. When bride and groom late that night at last retire to a properly married connubial bliss, can anyone blame Philander for leaping out of bed to check the locks on all doors and peer intently into each corner of the candle-lit room?[155]

Excusing himself under pretext of a bad catarrh, the Margrave of Baden Dourlach retired to his chamber in Florence and lay abed alone with his copy of the *Memoirs of a Woman of Pleasure*, to hear and enjoy the book for himself. To "hear" the book? Yes: whether between or under the covers, pornographic texts were probably sounded out, given that entirely silent reading was not yet the norm, that the pulsations vital to pornography were most provocative when lips were in motion and ears pricked up, that pornography fêted language until it made a fetish of sexual sounds, and that the noises of masturbation ran in hot dialogue with words rubbing off the page. While other literary genres of the eighteenth century may have refocused newly bespectacled readers upon motions of the psyche, the test of pornography lay in its corporeal effects, from the *frisson* of holding a forbidden text in one's hand to the swoon after one's other hand had finished elsewhere. Jacques Ferrand, a French physician who drew the contours of "erotomania" in a seventeenth-century treatise on erotic melancholy, had cautioned that sexual titillation might arise through any of the senses, but especially through the sense of hearing—under which he included "the reading [aloud] of lascivious and dishonest books," even of medical texts "which discourse of seed, Generation, and many secret diseases." Painters during the eighteenth century developed a corresponding motif: a young woman, sprawled in an armchair in a dressing gown, head back, mouth slightly open, one tremulous hand letting fall a novel, the other hand drooping in the post-coital languor of an erotic fantasy privately fulfilled.[156]

"Merry question" from a popular riddle-book of 1782:

My lady has a thing most rare,
Round about it grows much hair,
She takes delight with it in bed,
And often strokes its hairy head.

Answer: *A lap-Dog*. Or not.... Was it coincidence that soon after pornography emerged as a genre, masturbation emerged as a moral-medical problem and the generic erotomania of unrequited love became the gendered nymphomania of insatiable sexual desire? The Catholic Church, unalarmed except where succubi and witchcraft might be implicated, had treated occasional spending sprees of male or female bodily fluids as indiscretions meriting mild penance. In the later seventeenth century, however, Continental Protestant theologians reasserted an early Christian theory that the soul was transmitted through conception rather than infused directly by God, which made sperm spiritually seminal and its scattering more reprehensible. Braced by the recent microscopic identification of spermatazoa, this reassertion was linked scripturally to the fate of Onan in Genesis 38 and medically to the awful *sequelae* of venereal disease as described in *Onania: or, The Heinous Sin of Self Pollution and all its Frightful Consequences, in both SEXES Considered*, anonymously published in 1712, with revisions, supplements, and printings well into the 1800s. Complemented by more than sixty editions and translations of a 1760 French thesis on onanism by Samuel Tissot, a respected Swiss physician in correspondence with Voltaire and Rousseau, self-pleasuring was recast by Enlightenment writers as moral self-pollution and physical self-abuse.[157]

What was it about masturbation that drew the sudden wrath of philosophers and physicians? The historian Thomas Laqueur suggests that, quite aside from theological disputes over the origin of souls, masturbation was of burning concern because it violated three canons of the Enlightenment: rational self-government, a workable balance between public and private life, and politically responsive sociability. "The dangerous supplement," as Rousseau called it, inclined both men and women toward an unmanageable sensuality, solipsism, and selfishness, in addition to a material depletion of bodily fluids disposing the body to ailments that mirrored societal dysfunction: the unpredictability of heart murmurs, the inanition of nervous exhaustion, the solipsism of epilepsy and consumption, the selfishness of hysteric fits and a diseased imagination, the isolation of barrenness or blindness. Yet these putative risks of "solitary sex" could not have been dreadfully worrisome in comparison to the probabilities of contracting the veritably social diseases of gonorrhea and syphilis, had not sex been assigned a major role in communication between genders and among peoples, either explicitly as a universal language or implicitly as a prime vehicle of human relations. Absent such a strongly positive program for intercourse, there would have been little logic to exaggerating the dangers of "onanism" during a time of declining plague and expanding populations, a century of increasing numbers of pregnant brides and illegitimate children, a century when one-third to one-half of all first children were born out of wedlock in both Protestant England and Catholic France, a century when the age of female consent in Great Britain was effectively reduced from twelve to ten in an effort to provide more virgins for men seeking sex free of the taint of venereal disease.[158]

Censoriousness about solitary sex recalled Renaissance suspicions of solitude, which Stefano Guazzo had deplored in each of its species, whether of time ("the stillness of the night, or the instant wherein one speaketh alone in the presence of many"), of place ("the chamber or private dwelling which every one chooseth of purpose to sequester himself"), or of mind ("when one is present in person amongst many others, and yet is absent in mind and thought"), lest solitude of any kind become a habitual silent resort, sealing one off from the benefits of good society. It recalled as well Renaissance satires of horny monks cavorting with horny nuns and the Reformation's contempt for celibacy as a disaster waiting to happen. But the eighteenth-century attack on solitary sex also built upon Enlightenment predilections for a scientific economy of bodily quantities and, as Laqueur argues, cultural uneasiness with new mass market economies, such that profligate sexual spending was thought to be equivalent to moral bankruptcy and onanists who secluded themselves from all forms of sexual exchange were thought to live in a world of phantasms as perilous as the new financial world of unstable credit and stock market bubbles. Solitude helped one catch one's (newly tobaccoed, ginned, and chocolated) breath, check one's (newly in-sewn) pockets, and steady one's (newly individuated) resolve; but over-indulged, solitude led to "Th'ungenerous, selfish, solitary Joy" with its limitless demand and limitless if imaginary supply. Like etiquette's campaign against farting, which overcame medical opinion favoring the release of internal winds no matter how loud or pungent, the Enlightenment campaign against masturbation overcame medical opinion favoring the release of semen through nocturnal emissions or other more conscious means, lest excess seed, too long retained, turn poisonous. In both cases, one imposed one's will upon otherwise innocuous, involuntary bodily effluvia so as to promote smoother social intercourse; reckless fingering, though, with its sighs and moans, unlike reckless farting with its pops and sputters, had to be understood to result in much worse than a noisy breach of etiquette. In 1707, for example, J. F. Ostervald, a Swiss Calvinist minister, had published a treatise on *The Nature of Uncleanness Consider'd*, defending chastity of every sort, but the author of *Onania* found Ostervald's opus "wanting that Horrour, with which the Reader ought to be fill'd against Self-Pollution," since "Self-Pollution is a Sin, not only against Nature, but a Sin, that perverts and extinguishes Nature, and he who is guilty of it, is labouring at the Destruction of his Kind, and in a manner strikes at the Creation itself."[159]

Back again to Babel. Farting may have been a reminder of the audibly animal aspects of being human, but it carried little mythic force. Secretive and compulsive masturbation, on the other hand, now represented a loss of potential souls and, worse, a loss of actual communion, its hidden sounds echoing the aftermath of Babel when people became estranged from each other as different languages bore them apart. Within Enlightenment schemas of sociability that demanded intelligible, undisturbed intercourse, masturbation or "self-conversation" was a withholding of meaning, a refusal of audible relationship. As Onan (in the King James account) had withheld his seed from his slain brother's wife, spilling it on the ground, so onanism reflected a hardness of heart. "I entreat you to assist me, and will pay whatever the Charge is, for my Husband wonders what is the matter with me, for I have not the least desire he should lie with me, nor the least pleasure in the Act when he does," wrote a wedded gentlewoman to the author of *Onania* in 1723, after a decade of self-pollution from the age of eleven through and beyond her marriage at sixteen, taking more delight in masturbation "than when my Husband lay with me, altho' he is a young brisk Man." Worst, masturbation was madness; it did not just lead to insanity ("my Brain is as stupefied, and I have not a clear Thought, my Memory is extraordinary bad," wrote a young man), it was insanity itself, for it made one thrall to one's senses and inaccessible to reason. Immured in immoderation, the onanist "humiliates human nature," wrote Tissot; onanism was as criminal as suicide, since from such venery came "lassitude, weakness, numbness, a feeble gait, headaches, convulsions of all the senses, but especially of the sight, and dullness of hearing, an idiot look, a feverish circulation, exsiccation, leanness, a consumption of the lungs and back, and effeminacy."[160]

Nymphomania, which D. T. de Bienville defined in 1771 as an extreme form of sexual incontinence that arose from the ungratified desires of curious virgins, young widows, and wives of cold husbands, also turned the tables on gender distinctions that were just then being set in stone. While excessive masturbation made men impotent and effeminate before it knocked them off, it made women barren and masculine. Fired up by the "burning-glass" of romance novels, bawdy songs, or voluptuous etchings, and in the throes of a *furor uterinus* with its inflamed imagination, irregular menses, and a clitoris swollen to the size of a penis, women became not only infertile but bold as rakes, pressing themselves on the objects of their overheated affections with the most brazen declarations and lewd proposals: "the excess of their lust having exhausted all their power of contending against it, they throw off the restraining, honorable yoke of delicacy, and, without a blush, openly solicit in the most criminal, and abandoned language, the first-comers to gratify their insatiable desires." Apart from their physical demonstrativeness (in an era whose men's fashions were shedding the last vestiges of the Renaissance codpiece), nymphomaniacs manifested an acoustic distemper in which the "fair sex" assumed the acoustic deportment of the grosser: "they hiss, applaud, deny, affirm, assume ridiculous gestures, throw their bodies into strange contortions, attempt to stimulate the passions of the men by the loosest language." Public in their rampant use of obscenities and, close to the end, in their loud self-pleasuring, de Bienville's patients required isolation, swathing, cold baths, vaginal injections, bleeding, oaten-chaff ticking for their mattresses, opiate sedatives for their insomnia, and frugal broths to return them to civility, "and if I despaired of being able to cure them," wrote the determined doctor, "it was rather on account of their blasphemies and continual imprecations, than because their disorder was of such a particular nature." The uncontrolled orality of nymphomania was reverberant of two sets of classical "monsters in human shape," the Sirens with their haunting, deadly singing and the Furies with their howling, deadly rage. De Bienville therefore defended his own aural tactics, his use of plain language and "expressions, less eloquent, than natural, and alarming, like so many thunderbolts, by which the most obstinate, and infatuated minds must be stricken with conviction and remorse." Distressed by the impudence of an epidemic nymphomania that was sounding ever louder, he prayed, "May my pencil be sufficiently expressive, may my colours be sufficiently natural to inspire all that horror, with which so detestable a vice should be surveyed!"[161]

Vicious as could be the effects of masturbation, even unto heart failure and nymphomania, such natural colors tended to blend into a vicious circle. By dint of their detail, tirades against lewdness and onanism themselves constituted a library of sexual excitements, beginning with the fifth edition of *Onania* (1719) to which a self-certified surgeon, John Marten, appended prurient letters from (spurious?) readers unashamed to reveal their need for his Strengthening Tincture and Prolific Powder. By 1726, an additional dangerous supplement to *Onania* was promoted as containing "many very remarkable, and some of them even astonishing letters from persons of both sexes, young and old, single and married, concerning their self abuses, &c. Also letters from clergymen, physicians, schoolmasters, &c. some of them casuistical, of cases of conscience, &c. with answers to them, and one from a lady very curious concerning the lawful use and sinful abuse of the marriage bed, with histories of cases and cures, &c." If the original author of *Onania*—Marten himself?—purposed to write "with the utmost Circumspection and Caution," sexual explicitness was nonetheless vital to the project(ion) of horror upon which he was engaged, for "Bashfulness and Ignorance are very often the Companions of the Sin I treat of." None of his letter-writers confessed to enjoying the salaciousness of *Onania* itself, but some did confess to having been first aroused by the reading of gynecological pages in medical texts, and there was a regular conflation of education and ejaculation by secret readers of the many editions of *Aristotle's Masterpiece*, ostensibly a scholastic compilation of texts on midwifery and female complaints. "Solitary indulgence," admitted Dr. John Ware in his *Hints to Young Men on the True Relation of the Sexes* (1850) was a subject upon which it was still "hard to think, to speak, or to write, without seeming to partake in some measure of its pollution." Strong testimony, that, to the power of the language surrounding sex and to the purchase of sex upon language, even in the grip of horrible simile,

> Our legs in Cupid's Fetters lock'd,
> Our lips, like Birdlime, sticking fast. . . .[162]

Sex and language were also found *in flagrante delicto* when it came to religious enthusiasm. Enlightenment attacks upon masturbation appeared side by side with reports of cases of abnormal yearning, exhaustion, insanity, and suicide due to incitements of the Word spoken too warmly into ears too sensitive. "His *Gestures* in preaching are *theatrical*, his Voice *tumultuous*," wrote a critic of the revivalist George Whitefield, "his whole Speech and Behavior discovering the *Freaks of Madness*, and *wilds of Enthusiasm*." This same critic, the Congregationalist Samuel Hopkins, had heard the Truth in more dulcet tones at another revival, and had studied for the ministry under the formidable Connecticut revivalist Jonathan Edwards, whom he defended for his quieter demeanor: "His words often discovered a great degree of inward fervor, without much Noise or external Emotion, and fell with great weight on the Minds of his Hearers. He made but little Motion of his Head or Hands on the Desk, but spake so as to discover the Motion of his own Heart." Not only skeptics and orthodox critics but partisans like Hopkins described the more unsavory excitements of the Great Awakening of the 1730s and 1740s as if they were hand jobs and sessions of voluble masturbation rather than "seasons of grace." How else make sense of the unseemly ejaculations, weeping fits, raptures, and faintings-away of Methodists in open-air meetings in England and Ireland, of Scots in the natural amphitheatre at Cambuslang, of inspired Pietists prophesying in Dutch cities and German villages, of New Light Puritans and Baptists groaning with remorse and relief in New England meeting houses, and of aroused Anglicans in Virginia and South Carolina, stirred up by ordeals of introspection that seemed to awaken desires more earthly than spiritual? And how not, with such as John Wesley preaching God's free grace, teaching how to make one's way to heaven through love, and reaching the hearts and infirmities of the faithful through the electric tones of his voice, the electricity of his touch, and sparks of a portable electrical apparatus, "a species of fire, infinitely finer than any other yet known"?[163]

Electricity in the eighteenth century was a fluid—a fluid in excess, an energetic fluid in excess. As a medium of communication, it mostly communicated excitement, so it had physical affinities with unbridled sex and the possibilities of sexual renewal. Until Volta, Galvani, Coulomb, Ohm, and Ampère put their minds and names to the relationship between current and resistance, electricity was as vivid as riot, as stunning as lightning, and as invisibly ubiquitous as desire. All animal bodies, reported the *London Magazine* in 1748, were "being constantly electrized by the earth." Once it could be visibly and audibly generated in some kind of globe, run through some kind of thread, and stored in some kind of jar, electricity was some kind of thrilling: If Benjamin Franklin in Philadelphia in 1747 could "electrize a person twenty or more times running, with a touch of the finger on the wire," his nemesis Jean Antoine Nollet in Paris in 1752 could enchain 180 royal guards, command the men at each end to touch the poles of a Leyden jar, and so launch a company into the air of Versailles in one electrifying leap. ("Electrifying" was already figurative, part and participle of an agreeable shock, a shock that Nollet extended to a 900-foot-long line of monks.) Physicians got the general idea and used electricity to relieve blockages: cramps, gout, kidney stones, arthritis, rheumatism, paralysis, deafness. "Being electrified morning and evening," wrote John Wesley in his journal at the age of sixty-two, "my lameness mended." Sexual inertias—frigidity, impotence, infertility, autoerotic fatigue—might also be unblocked by electricity, a possibility exploited in 1779 by a Scot late of Franklin's Philadelphia, James Graham, eye and ear doctor, who recrossed the Atlantic to open a Temple of Health in London, its rooms beaming with the "Celestial Brilliance of Medico-Electrical Apparatus." Uterine and genital refreshment were the promise of his magnetic thrones, electrical spas, and a huge mirror-domed Celestial Bed resting on twenty-eight glass pillars, serenaded by musical automata, crowned by sculptures of Cupid, Psyche, and Hymen, grounded by fifteen hundredweight of magnets, and "sparkling with electrical fire" from an Influence Machine.[164]

9. *Cries . . . Creaks . . . Shudders . . . Sensibilities*

Ah, the Influence Machine. Francis Hauksbee, who had revisited the bell jar experiments of Hooke and Boyle and been much engaged with sound, had built the first Influence Machine at the start of the century. Hauksbee's globe generators of 1706–1708 used a crank to spin a glass sphere whose contents, air under low pressure, glowed with electrical discharge, the light issuing (as he noted in italics) "with *a considerable Noise*, much like that of Wheezing, tho' something smarter; and it was easy enough to distinguish it from the Noise made by the working of the Engine, which notwithstanding was not a small one." Since Graham's version of this noisy device was hidden beneath the Celestial Bed, and since his other friction devices could have made electricity visible only sporadically (unlike the much later, much larger Tesla coils of mad-scientist motion pictures), it was the sound of electrical influence—the hum, the wheeze, the crackle—that must have persuaded toward rejuvenation, along with the faintest of electrical vibrations, the thrum of organ music, the lilt of woodwind melodies piped in through secret tubes, and the huffing duets of bed partners encouraged to sing their way toward bliss. The celestial fire of electricity in the 1700s had more sizzle than flame.[165]

Down the line apiece, electricity will amplify forms of communication other than sex, but electrical forces will be recurrently confused with sexual stamina and strong nerves. And down the line, as here, electricity will be instrumented by men intent on sound: Here, Hauksbee (who redefined "acoustics" for the Royal Society), Franklin (who investigated thunder and echoes, invented the glass harmonica, and relied upon his ears as his eyesight deteriorated), and Nollet (who sat underwater listening for pistol shots, whistles, bells, and human voices, to determine whether it was worth their while for fish to have a sense of hearing); and down the line, Thomas Edison (who, nearly deaf, strained for clarity in sound recordings), Alexander Graham Bell (whose mother had become deaf, whose father taught deaf mutes, and who was himself a professor of vocal physiology), and Nikola Tesla (whose hearing was hyperacute, never more so than during a nervous breakdown in his youth, when a fly alighting on a table caused "a dull thud" in his ear).[166] Here in the 1700s, electricity was aural-tactile evidence of universal surplus and a panacea for depletion. While the members of the Beggar's Benison, one of several male sex clubs of eighteenth-century Scotland and England, called themselves to order with the blowing of a horn (phallically embossed), read aloud from pornographic texts (an early draft of *Fanny Hill*), made crude toasts (to "the mouth that never had the toothache"), ritually masturbated before one another (each to his own), and loudly enacted scenes of pleasurable spending (in keeping with their motto, "May Prick nor Purse never fail you"), electrotherapy crackled with a similar ethos of freely available, morally unproblematic abundance. Matching the rising fertility of an England whose population would nearly triple between 1681 and 1831 as the rate of illegitimate births quadrupled and the age at first marriage declined for both men and women, electricity heaved with potential.[167]

Evidentiary of orgasm and almost constitutent of organism, inexpungable by etiquette and almost inextricable from experiment, noise had become as elemental as the rush of charged fluids. Amid stock bubbles and bankruptcies, bread riots and panics of famine, smallpox epidemics and scarlet fever, high infant mortality and world war between the French and the British, noise was a figure of surplus. Cry of nature, sigh of love, rumble of bowels, tumble of mechanism, noise turned up everywhere, under guises philosophical, pornographic, medical, and industrial. Early modern and Enlightened responses to the impasse of Babel led not to the exclusion of noise but again and again to a listening *through* noise, whether in exotic music, erotic arts, the erratics of chance as subsumed by mathematicians under a law of large numbers,[168] or the elasticities of population and the quixotics of dearth. It would be a while before noise would be heard to be literally universal, but in the 1700s there were intimations of an earthy and unearthly omnipresence to noise nearly as extensive, from country crossroads and city streets out to the drear abbeys, dark castles, and craggy cliffs of gothic romance.

Earthy: criers of artichokes, asparagus, baskets, beans, beer, bells, biscuits, brooms, buttermilk, candles, six-pence-a-pound fair cherries, chickens, clothesline, cockles, combs, coal, crabs, cucumbers, death lists, door mats, eels, fresh eggs, firewood, flowers, garlic, hake, herring, ink, ivy, jokebooks, lace, lanterns, lemons, lettuce, mackerel, matches, mussels, nosegays, onions, oranges, oysters, pastry, pears, pins, purses, rabbits, radishes, socks, songbirds, toasting forks, toy horses, turnips, lily-white vinegar, and whetstones had to contend with the cries of an equally importunate service corps of chairmenders, chimney sweeps, dog skinners, glassmen, grease collectors, kennel rakers, ratcatchers, scissors grinders, sow gelders, tinkers, and traders of old clothes. Desperate to make themselves heard as well above ballad singers on street corners, above the shouts of sedan-chair carriers threading their way through traffic, above the snort and whinny of horses, the rumble of wagons, the hammering of smiths, and the rattles and whistling of cowherds, gooseherds, and pigherds driving their squawking and grunting animals to market, street vendors stretched their calls into loud, long, and progressively unintelligible wails. Rather pitch-pockets than determinable phrases, these calls in all their profusion were a sonic surplus, an audible sign of availability. Was it for this reason that Marcellus Laroon enlarged the ears of the figures in his widely imitated *Cryes of the City of London Drawn after the Life* in 1687? Street cries had long been grist for composers; now the criers themselves were a profitable subject for artists and their publishers, whose sale of portraits of mop-sellers, knick-knack peddlers, and fishmongers furnished a steadier income than that taken in by the folk who cried mops or apples from dawn to dusk along streets festering with pigshit. Laroon's path-setting prints of sixty individual criers quarantined them from the emptied chamberpots, potholes, unholy ruffians, and rough weather that were their lot. Framed in isolation from dross and traffic, and captioned with an intelligible spiel, each crier, though ragged or stooped, was quietly composed. Half peepshow, half puppet theater, the images were as attractive to early modern audiences as their associated soundbites were familiar: "Rushes green rushes! Wellfleet oysters! Crab, crab, any crabs?" On the streets of Paris, similar prints and a similar welter of cries: "*Ça, tôt le pot, nourrices!*" for morning milk, "*Échaudés, gâteaux, pâtés chauds!*" for warm pastries, "*Tinettes, tinettes, tinettes!*" for firkins of butter. On the streets of Boston, faithfully transcribed: "'Ar-co,' 'ar-co,' 'ar-co'!" for charcoal. Not until the mid-nineteenth century would the presence of these criers be generally protested, though a milk vendor might intrude upon a neighborhood like the "scream of a person being stabbed in the bottom by a cobbler's awl" and a custard-monger's bawl be so loud that "he puts his hand behind his ear to mitigate the sensation which he inflicts upon his own tympanum." If Jonathan Swift in 1709 complained of being plagued at home by a man "crying cabbages and Savoys," Addison two years later merely joshed London's criers with a proposal for a new post of Comptroller-General in charge of tuning the sow-gelders' horns, sweetening the voices of brick-dust sellers, and improving the enunciation of bellows-menders. In 1787, abroad in Venice among its street harlequins and canal musicians, Goethe enjoyed the crowds in the piazza of San Marco, where "All merge into one great stream, yet each manages to find his way to his own goal. In the midst of so many people and all their commotion, I feel peaceful and alone for the first time. The louder the uproar of the streets, the quieter I become."[169]

Goethe, who rejected Newtonian optics and considered colors the property of healthy eyes, would halt most oddly in the midst of outlining his theory of sight to make a plea for hearing. He knew of no principle by which colors could be properly correlated with sounds, and he left unfinished an inquiry into audibility, but he was sure that the "world of tone brings about a *sensory-moral inspiration of the inner and outer senses.*" Redeploying italics in the next paragraph, Goethe disputed the passivity of the ear, for it had "an advanced organic nature and we must credit it with *counteracting and asserting its own demands.*" Let others speculate about the coincidence of the seven notes of an octave with the seven colors of a prism; let Locke's "studious blind man" associate the sound of a trumpet with the color scarlet; let Castel in France and Krüger in Germany contrive "ocular harpsichords" to play music with matching colors. The real work was to listen: "Let us shut our eyes, let us open our ears and sharpen our sense of hearing. From the softest breath to the most savage noise, from the simplest tone to the most sublime harmony, from the fiercest cry of passion to the gentlest word of reason, it is nature alone that speaks, revealing its existence, energy, life, and circumstances, so that a blind man to whom the vast world of the visible is denied may seize hold of an infinite living realm through what he can hear." Goethe went boldly on: "Thus Nature also speaks to other senses which lie even deeper, to known, misunderstood and unknown senses. Thus it converses with itself and with us through a thousand phenomena.

No one who is observant will ever find Nature dead or silent."[170]

Nature morte as a still life on canvas, an aesthetic attitude, or a metaphysical position, drifted out of favor in Goethe's time.[171] Author of the intensely popular *Sorrows of Young Werther* (1774), Goethe himself was responsible in no small measure for that drift. Unmoved by treatises that reduced nature to tables and aligned the organism with clockwork, the reading public responded more warmly to the vague vitalism of "an infinite living realm" and young Werther's infinitely precious sensitivity to it: "when I listened to the birds that bring the forest to life, while millions of midges danced in the red light of a setting sun whose last flare roused the buzzing beetle from the grass; and all the whirring and weaving around me drew my attention to the ground underfoot where the moss, which wrests its nourishment from hard rock, and the broom plant, which grows on the slope of the arid sand hill, revealed to me the inner, glowing, sacred life of Nature—how fervently did I take all this into my warm heart, feeling like a god in that overflowing abundance, while the beautiful forms of the infinite universe stirred and inspired my soul." What then were those other senses, inner and outer, known or unknown, to which Goethe alluded and to which a buzzing, whirring Nature spoke? Robert Hooke would have had some interesting answers, but a century later it was the readers of fiction, rather than microscopists, physicists, or anatomists, who had the readiest response. From sentimental novels, and then from Gothic tales, readers would have deduced the existence of synaesthetic senses that accommodated opposites and managed the simultaneous concourse of disparate impressions: a sense of time and loss, a sense of proximity and absence, a sense of wonder and repulsion, a sense of beauty and deformity, all folded within a disposition called "sensibility." It was sensibility that produced decades of public weeping by women and, in 1771, Henry Mackenzie's capstone novel, *The Man of Feeling*. It was sensibility that made eighteenth-century juries unwilling to punish those who had tried to kill themselves and eighteenth-century readers willing to be moved by Werther's reasons for—and fatal accomplishment of—a lovelorn suicide. It was sensibility that turned churchgoers away from an eternal hell toward a universal love. It was sensibility, as shared by the milkmaid Ann Yearsley in her plea,

> But come, ye souls who feel for human woe,
> Tho' drest in savage guise! Approach, thou son,
> Whose heart would shudder at a father's chains,
> And melt o'er thy lov'd brother as he lies
> Gasping in torment undeserv'd...

that led to the abolition of slavery in Europe.[172]

Sentimental literature exaggerated the sway of pity, love, and anguish to the same degree as I have exaggerated the historical sway of "sensibility," but it would be no exaggeration to claim that the authors of sentimental plays and novels cued their characters to a system of vocables nearly identical to that of pornography and expected of their auditors and readers a like response: tremors of voice, moannns and groannns, breath*less*ness, ejaculations of surprise!! of revelation!! sighs o yes sighs and here and there a swoooooon . . . . . . whether reading aloud in company or alone, or in congregations singing the hymns of Charles Wesley and appealing to "Jesus, Lover of My Soul" as a deserted wife or workhouse orphan might tug at the heartstrings:

> Other refuge have I none;
> Hangs my helpless soul on thee;
> Leave, ah! leave me not alone,
> Still support and comfort me.

Persons of taste, when absorbed in a landscape, play, painting, poem, or book, experience an "instantaneous Glow of Pleasure which thrills through our whole Frame, and seizes upon the Applause of the Heart, before the Intellectual Power, Reason, can descend from the Throne of the Mind to ratify its Approbation." This, from the 1755 *Letters concerning Taste* by the sentimental poet John Gilbert Cooper, who was sure that the Almighty "attuned our minds to Truth, that all Beauty from without should make a responsive Harmony vibrate within." At the core of both the pornographic and the sentimental was the capacity not to think but to feel; depth and subtlety were nowhere near as valued as the volcanic eruption of feeling. In pornography, the result was the noise of orgasm, on and beyond the page. In sentimental texts, the result was a surplus of exclamation points, dashes, asterisks, ellipsis marks, and broken syntax at moments—frequent moments—of inexpressible, overwhelming emotion - - - - and beyond the text, a susceptibility to the startle, the catch in the throat, the gush of tears, the sense of alarm or foreboding.[173]

With foreboding, with anxious foreboding, noise assumed also an unearthly bearing. As sentimental literature was wounded to the heart by a generation of critics who found it florid and naive, cultural education in synaesthesia was taken up by the Gothic, a genre that further heightened those senses under cover of a psychological realism *in extremis*. The ordeals of the Gothic novel, often erotic or sexually violent, infused the sense of time with a feeling for the uncanny, the sense of proximity with a feeling for the eerie, the sense of wonder with terror, the sense of beauty with sublimity. "When archetypal phenomena stand unveiled before our senses," wrote Goethe, "we become nervous, even anxious. Sensory man seeks salvation in astonishment, but soon that busy matchmaker, Understanding, arrives with her efforts to marry the highest to the lowliest." Which marriage was exactly what a countryman of Goethe's had in mind in his *Observations on the Feeling of the Beautiful and Sublime.* A man of "sensitive nerves," attuned to the sentimental novels of Richardson and the romantic pedagogy of Rousseau, in 1762 Immanuel Kant was a thirty-eight-year-old university lecturer who had completed two dozen works in philosophy, anthropology, physics, education, and astronomy, but was not yet the standard-bearer of German critical philosophy. When the University of Berlin two years later tried in vain to lure him away from Königsberg, it was with the offer of a professorship in poetics. Drafting his *Observations*, Kant waxed by turns poetic and anthropological, arguing that beauty is appreciable because harmonious and complete, while the sublime is at best approachable, defying—and defining—one's limits as a rational being. One can wrap oneself around the beautiful as the sexes wrap themselves around each other for pleasure or comfort; the sublime draws one beyond the mortal physical self. We are tempted, Kant would write later in his *Critique of the Aesthetic Power of Judgment*, to regard as sublime "thunder clouds towering up into the heavens, bringing with them flashes of lightning and crashes of thunder, volcanoes with their all-destroying violence, hurricanes with the devastation they leave behind, the boundless ocean set into a rage, a lofty waterfall on a mighty river, etc.," but these heavily acoustic dynamics of Nature are sublime only to the degree that, by making us fearful, they arouse in us the courage to acknowledge our own powers "even over Nature." Beauty shows what we can achieve, alone and together; the sublime reveals all that overpasses us, confirming the distinctively human potential to be humbled, astonished, mystified. "Tall oaks and lonely shadows in a sacred grove are sublime; flower beds, low hedges and trees trimmed in figures are beautiful." In short, "The sublime *moves*, the beautiful *charms*."[174]

Being *moved* demanded more in Kant's world than was demanded of the characters in and readers of sentimental tales, where pathos and sympathy, joy and grief appeared at the least hint of an indrawn breath. That kind of subtlety trifled with the emotions, making them seem frivolous or forced; the subtlety of Kant's sublime was lofty, upholding the dignity of mankind: "Temperaments that possess a feeling for the sublime are drawn gradually, by the quiet stillness of a summer evening as the shimmering light of the stars breaks through the brown shadows of night and the lonely moon rises into view, into high feelings of friendship, of disdain for the world, of eternity." Beauty was merely convivial, a social reservoir of the pleasant; hence the province of women. The sublime could be splendid, noble, or terrifying, but it was always epic, always penetrating, almost always disconcerting; hence the province of tougher, more resilient men. These gendered definitions were proved by typically feminine responses to beauty, with "laughing delights" and "audible mirth"; responses to the sublime were more masculine: "quiet wonder," solitary meditation, deep conversation, or the internal quiver of adventure.[175]

Transported to the realm of the Gothic, beauty turned darker and the sublime noisier. A young Edmund Burke had anticipated this twist when, a decade before Kant, he conducted his own inquiry into the sublime and found that "Excessive loudness alone is sufficient to overpower the soul, to suspend its action, and to fill it with terror." Terror instigated the sublime: "Whatever is fitted in any sort to excite the ideas of pain and danger, that is to say, whatever is in any sort terrible, or is conversant about terrible objects, or operates in a manner analogous to terror, is a source of the *sublime*." With an Irish Catholic mother and a solicitor father who had converted to the Church of England, Burke pursued truth unto its innermost parts and predictably arrived at that which was insusceptible to reason, a sensorial and imaginal surplus that he depicted in sonic terms: "The noise of vast cataracts, raging storms, thunder, or artillery, awakes a great and aweful sensation in the mind, though we can observe no nicety or artifice in those sorts of music." The incommensurability of such noise might lead to mob scenes, where the shouting of multitudes "by the sole strength of the sound, so amazes and confounds the imagination, that in this staggering, and hurry of the mind, the best established tempers can scarcely forebear being borne down, and joining in the common cry," but Burke's larger point, as a passionate orator and dispassionate philosopher, concerned the powers of persuasion. Reason could convince, but amazement and anxiety *moved*; whatever was indeterminable yet strikingly apprehended was more thoroughly persuasive than what was clear and perfectly comprehended, for it was human nature to fear the worst of things unknown, ambiguous, sudden, or out of (one's) control, wherefore "Few things are more aweful than the striking of a great clock" in the silence of the night. Heard at an indefinable but threatening distance, the "angry tones of wild beasts" could also cause "a great and awful sensation," as could a "low, tremulous, intermitting sound" whose source was frightfully uncertain. Kant, the Enlightenment son of stern Pietists and a lifelong bachelor accustomed to regular constitutionals around his provincial Königsberg, fixed the sublime upon a sense of astonishment and inadequacy in the presence of the ungraspable or morally majestic; the experience of sublimity was consolation for the finitude of being human. Burke, a conservative critic of the Commonwealth and a politico haunting the halls of Parliament, fixed the sublime upon a steep precipice of fears—"astonishment is that state of the soul, in which all its motions are suspended, with some degree of horror"—and, ultimately, upon a mortal terror of absences: "All *general* privations are great, because they are all terrible; *Vacuity*, *Darkness*, *Solitude*, and *Silence*."[176]

Graveyards lay beneath much of this sublimity, dark and solitary if scarcely empty or quiet. In the wildly successful mid-eighteenth-century school of "graveyard" poetry familiar to Burke, general privations produce specific tremors, as in "The Grave" by the Scottish cleric and botanist Robert Blair:

> The Wind is up; Hark! How it howls! Methinks
> Till now, I never heard a Sound so dreary:
> Doors creak, and Windows clap, and Night's foul Bird
> Rook'd in the Spire screams loud:

while Echo sends back the sound "Laden with heavier Airs" from the vaults of the dead, and

> Again! The Screech-Owl shrieks: Ungracious Sound!
> I'll hear no more, it makes one's Blood run chill.

Though Hell, too cruel and unrelenting for the sentimental eighteenth century, may have sunk out of hearing, the cemetery with its "wild shrieks" served audibly in its stead to enliven faith by putting readers in mind of a tenuous mortality and the ghastliness of an unhappy or unbelieving death.[177]

Gothic novelists, preponderantly women who were more attentive to spirit than to creed, thrust their characters scene after scene into emptiness, darkness, solitude, and silence, demonstrating (*contra* Kant) that women enjoyed equal access to the sublime, as at a ruined abbey where "The sky was clear and serene, and nothing but the light trembling of the leaves, heard at intervals in the breeze, disturbed the silence of the place. It was a moment sacred to meditation, and wrapped in sublime contemplation, she beheld the deepening veil of the twilight, which had just shaded the meek blue of the heavens, stealing upon the surrounding scenery." But Gothic novels were not the night thoughts of men *or* women looking toward a pious Christian reckoning, nor the postulates of men *or* women philosophizing about the rightness of the world. Gothic plots turned upon something being *ever wrong* with the scene—the emptiness gaping, the darkness bleak, the solitude barren, the silence ominous, with "trees that waved mournfully over the place, and replied to the moaning of the rising blast." Oscillating between Burke's dreadful sublime, which "comes upon us in the gloomy forest, and in the howling wilderness," and Kant's ennobling sublime, which amid tragedy stirs us to self-sacrifice, Gothic fictions used sounds apocalyptically to move readers and listeners from fear through agony to revelation. Sound of all sorts. Unrecognizable sounds, or sounds a sane person would do better not to recognize. Sounds almost inaudible or inchoate. Sounds suspiciously delayed or muffled or screened and uncoupled from their source. Sounds cut short or distended. Sounds outrageously intensified by echo and expectation. Sounds abominable, ear-curdling. In other words, noise. Noise was the aural correlate of every other kind of distortion upon which the Gothic played—the gnarled, the hideous, the tortured, the grotesque (Kant's most frequent pejorative).

And since these visual distortions were played out by guttering candlelight, in fog or mist, under the pall of dusk, or amid a dark and stormy night, it was primarily the succession of noises that heightened the tension and pushed the horror toward a climax. In 1773, John Aikin helped fashion an exemplar for his sister Anna Laetitia's essay "On the Pleasure Derived from Objects of Terror," which followed Burke's maxim that "To make anything very terrible, obscurity seems in general to be necessary." In their story, one Sir Bertrand, lost in a desolate waste, throws himself on the ground in despair. Drawn by "the sullen toll of a distant bell" to a "ruinous mansion," he spots a light in one of the turrets, but otherwise "All was silent.... All was still as death." He pauses. "After a short parley with himself, he entered the porch, and seizing a massy iron knocker at the gate, lifted it up, and hesitating, at length struck a loud stroke. The noise resounded through the whole mansion with hollow echoes. All was still again—He repeated the strokes more boldly and louder—another interval ensued—A third time he knocked, and a third time all was still." Another pause, another "deep sullen toll" from the turret, and Sir Bertrand's heart makes "a fearful stop." Yet another pause, and then resolve: "The heavy door, creaking upon its hinges, reluctantly yielded to his hand—he applied his shoulder to it and forced it open—he quitted it and stept forward—the door instantly shut with a thundering clap," chilling Sir Bertrand's blood. He feels a dead cold hand take hold of him momentarily, hears a "deep hollow groan" and...and you tell me.[178]

Dreadfully familiar by now is this sonic universe of horror that was contrived in the later eighteenth century and amplified by a Reign of Terror in revolutionary France from June 1793 through July 1794. Shortly before, in 1791, Ann Radcliffe had published her landmark Gothic novel, *The Romance of the Forest*, in whose first pages readers could hear a grind of tears from the tenderly sentimental to the terribly sublime. Fleeing his creditors, the dissipated Pierre de la Motte, a man "infirm in purpose and visionary in virtue" whose conduct was "suggested by feeling, rather than principle," dashes out of Paris with his wife and servants on a night "dark and tempestuous." Lost on a wild heath three leagues beyond the capital, he seeks aid from an "ancient house" where, like Sir Bertrand, he has spotted a light in the window. Like Bertrand, "Having reached the door, he stopped for some moments, listening in apprehensive anxiety—no sound was heard but that of the wind which swept in hollow gusts over the waste." Like Bertrand, once inside, he "heard the door locked upon him." Voices from the other side of a wall. Stillness. Wind. The "sobs and moaning of a female." More apprehensiveness. "A noise in the passage." A snarling blackguard drags in a girl, unkempt, "aghast with terror." At pistol-point, La Motte the Infirm must choose: take this girl away with him or be shot. La Motte takes Adeline, who heaves "deep convulsive sighs." Back in the carriage with his wife and servants, they dash madly through the dismal night. At dawn, "the terrors of La Motte began to subside, and the griefs of Adeline to soften." Sweet Adeline is "delicately sensible to the beauties of nature," to "the flowery luxuriance of the turf, and the tender green of the trees," and to "the magnificence of a wide horizon" she has rarely seen, but these optical delights will be undone by new acoustic terrors on the first pages of Round Two. There they are, La Motte & Cie., caught up in a "romantic gloom" as they approach the Gothic remains of an abbey in a twilight diffusing "a solemnity that vibrated in thrilling sensations upon the hearts of the travellers." High grass waves in the breeze. Moss somehow whistles in the wind. Above, from decrepit battlements, birds of prey circle and swoop. And there is La Motte again, like Bertrand, advancing to the gate, lifting a heavy knocker. "The hollow sounds rung through the emptiness of the place." He waits, then he forces open the gate, "which was heavy with iron work, and creaked harshly on its hinges." Time for indentation:

> He entered what appeared to have been the chapel of the abbey, where the hymn of devotion had once been raised, and the tear of penitence had once been shed; sounds, which could now only be recalled by imagination—tears of penitence, which had been long since fixed in fate. La Motte paused a moment, for he felt a sensation of sublimity rising into terror—a suspension of mingled astonishment and awe!

The tremor passes, but the gloom deepens as he continues to explore the desolate chapel where "the sound of his steps ran in echoes through the place, and seemed like the mysterious accents of the dead, reproving the sacrilegious mortal who thus dared to disturb their precincts." Gathering in the abbey's great hall, the entire company hears "an uncommon noise, which passed along the hall. They were all silent—it was the silence of terror." They hasten to depart but suffer further dark passages, further odd noises, further echoing footsteps, until relieved to find that the noises were nothing but owls and rooks with their "confounded clapping" of wings. We have made it through the first 18¾ pages.[179]

Maria Edgeworth, three years younger than Radcliffe and destined to rank among the best-selling Irish novelists of the early nineteenth century, read *The Romance of the Forest* start to finish a few months after it came out, judging "the horrible parts" to be "well worked up, but it is very difficult to keep Horror breathless with his mouth wide open through three volumes." How could a novelist keep Horror, "loud and keen, and chill," breathless? Stephen Hole in his 1792 "Ode to Terror," kept ravens croaking and distress shrieking. By 1797, a contributor to London's *Monthly Magazine* had the formula down: "construct an *old* castle... built in the Gothic manner, with a great number of hanging towers, turrets, and pinnacles. One half, at least, of it must be in ruins... the doors must be so old, and so little used to open, as to grate tremendously on the hinges; and there must be in every passage an echo, and as many reverberations as there are partitions." The battlements must be "remarkably *populous* in *owls* and *bats*," for "The *hooting* of the one, and the *flitting* of the other, are excellent engines in the system of terror, particularly if the candle goes out." There must also of course be a heroine, "with all the weakness of body and mind that appertains to her sex; but, endowed with all the curiosity of a spy, and all the courage of a troop of horse. Whatever she hears, sees, or thinks of, that is horrible and terrible, she must enquire into it again and again." Whatever she hears, above all whatever she hears, for the synaesthetic senses as taught in Gothic novels were each of them coupled to the ear. As Matthew Lewis rhymed in the preface to his darkly Germanic *Tales of Terror*, all senses together would be enwrap't by

> The night-shriek loud, wan ghost, and dungeon damp,
> The midnight cloister, and the glimm'ring lamp,
> The pale procession fading on the sight,
> The flaming tapers, and the chaunted rite...

In Gothic fictions the inauspicious, the terrifying, the dangerously hidden, and the deviously base were most often first detected by the auricle, tympanum, labyrinth, and basilar membrane of the ear. If Anne Radcliffe's emphases upon

> Frantic Fury's dying groan!
> Virtue's sigh, and Sorrow's moan!

were nothing new to theatrical tragedy, in Gothic fictions the liminal verges of life and death most often first manifested themselves through the "listening ear"—the ear of such as the poet Hannah Cowley, who

> 'midst the shrieking wild wind's roar,
> Thy influence, HORROR, I'll adore.[180]

So the synaesthetic algebra of the Gothic invoked equations of sound + *x*.

Equation 1. The uncanny = sound + sight. Sounds so much out of time as to be half-noise (bells tolling the wrong hours long past midnight, voices calling from beyond the grave, improbable echoes, *déjà entendu*) + vision perplexed (dreams coming alive, curses taking their toll, doppelgängers appearing, *déjà vu*).

Equation 2. The eerie = sound + touch. Sounds so problematic as to be half-noise and too close for comfort (stifled moans, muffled wailing, ghoulish *diminuendos*, cries piercing the silence, wolves howling, branches cracking) + a fitful tactility (the brush of a wing, the touch of a cold hand, the drift and cling of a spider's web).

Equation 3. Terror = sound + touch + sight. Sounds so painful as to be half-noise (the hiss, the shriek, the *decrescendo* of a scream) + feelings of alarm (dry mouth, hair rising at the nape of the neck, goosebumps, shivers, temporary paralysis) + loss of sight (narrowed vision, temporary blindness or blinding tears, fainting).

Equation 4. Sublimity = sound + sight + touch + smell. Sounds so overwhelming as to be nearly indescribable (glaciers cracking, thunderbolts, earthquakes, angelic choirs, a sudden silence) + sights so fine or vast as to be inexpressible + a touch of the untouchable + a whiff of implausible sweetness or unimaginable carnality (be it in ecstasy, cruelty, or decay).

QED: A Gothic heroine's curiosity and courage were most bound up with a desire to hear and a grand capacity for listening through the frightening half-noise of what could be heard. In *The Italian: or, The Confessional of the Black Penitents* (1797), Radcliffe's Marchesa reads a black-letter inscription over a confessional booth, which bears "these awful words, '*God hears thee!*' It appeared an awful warning"—and an awful responsibility, for she herself (like other Gothic heroines) is exceedingly alert to strange sounds, waiting patiently (like the readers and auditors of Gothic novels) for the darkest secrets to emerge from behind the gasp, the moan, the sigh, the wail, and the groan.[181]

10. The Gasp The Hiss The Listening For

"Whillaluh" or "ullaloo," according to the glossary that Maria Edgeworth appended to her first novel, *Castle Rackrent* (1800), was an Irish lamentation and funeral song that, in its classic form, included a first and second semi-chorus and "a full chorus of sighs and groans." Maria, who never had a room of her own and wrote fourteen volumes of novels amid the traffic and chatter of a common sitting room, observed that the song had declined from the lyrical into the brutish, for "The present Irish cry or howl cannot boast of much melody, nor is the funeral procession conducted with much dignity." Led by keening women who were hired for the occasion, a thousand mourners might assemble around the hearse, "and when they pass through any village, or when they come near any houses, they begin to cry—Oh! Oh! Oh! Oh! Oh! Agh! Agh! raising their notes from the first *Oh!* to the last *Agh!* in a kind of mournful howl." This gloss was no more than an afterthought to the main ambitions of *Castle Rackrent*, a moralizing work of social criticism that was sympathetic to the griefs of Irish peasant farmers; but the Gothic novels and Romantic poetry that Edgeworth herself read took death as a central motif and headed it in the direction of chain-clanking ghosts, just as Irish literature was taking more formal note of the banshee. While James Beattie's *The Minstrel* (1770) had its young Edwin dreaming of

grave, and corses pale;
And ghosts, that to the charnel-dungeon throng,
And drag a length of clanking chain, and wail,
Till silenced by the owl's terrific song,

the Lady of Death and Spirit of the Air, *Beansighe*, was screaming, clawing at the windows, scraping at the walls of those of ancient Irish stock, wailing the imminent death of someone languishing inside. A woman heard but not seen, rising against all convention, the banshee was heard with greater frequency and at greater volume as late-eighteenth-century memoirists and novelists began to listen for her.[182]

Sirens, of course, had been calling for ages, and a quieter figure, the *femme fatale*, would leave her calling card on Late Victorian nights. A singular bird, however, calls just now. Not the wooden cuckoos popping out of Black Forest clocks since 1730, nor the mechanical warblers singing their way out of Switzerland in the 1780s, but a living, breathing cockatoo. A cockatoo in danger, sealed inside a glass sphere perched on a slender hardwood stand running upward past a decomposing skull sunk in the preservative liquid of a transparent bowl. A white cockatoo, its voice stifled by air withdrawn. A *struggling* white cockatoo, its one spread wing pointing toward the mantic eyes of a long-haired scientist, his right hand extended in explanation near the crank of an air pump, his left poised at a stopcock above the glass sphere. A rare white cockatoo, the *absence* of whose song is central to a circle of onlisteners who are bent around the sphere: one man gripping a watch; a young couple bearing witness (the woman to the man, the man to science); two fearful girls (one who cannot look, one who wishes she could not), and an avuncular man who reassures both; a philosophical middle-aged sort, eyeglasses off, reflecting as much on the skull as on the suction of the pump; and two boys, one at each end of the room, unevenly illumined—as are all of these faces—by the flame of a candle.[183]

Shortly after he finished *An Experiment on a Bird in the Air Pump* in 1768, Joseph Wright of Derby sold it to Dr. Benjamin Bates, a member of the sexually riotous Hell-Fire Club and an art lover who already owned Wright's *Three Persons Viewing "The Gladiator" by Candlelight*, in which three men, including the painter himself, glory in a small copy of a famous Hellenistic sculpture of a naked soldier stretching to parry a blow, his musculature prominent, his penis bravely unshielded. Wright does not figure in *An Experiment*, and among those gathered around the air pump, only the couple has been identified: Thomas Coltman and his fiancée Mary Barlow, whom Wright would paint again upon their marriage the following year. However, the full moon drifting behind dark clouds outside the room's sole window alludes to the Lunar Circle of Birmingham, a group of instrument makers, naturalists, physicians, physicists, and engineers who met monthly on the Sunday nearest each full moon. The Circle included

Richard Lovell Edgeworth (Maria Edgeworth's father, a gentleman-conjuror and indefatigable inventor of such devices as a turnip-cutting machine and a signal telegraph), Josiah Wedgwood (the ceramics manufacturer, in business with Ann Radcliffe's uncle), and Erasmus Darwin (botanist and physician, grandfather of Charles and, according to family tradition, the man holding the watch). A Scottish scientific lecturer who had toured the midlands in 1762 may have shown the Lunar men his air pump, but public sensibilities dictated that no live specimen be sealed up in the glass sphere or "receiver." As James Ferguson had explained in his 1760 *Lectures on Select Subjects*, "If a fowl, a cat, rat, mouse or bird be put under the receiver, and the air be exhausted, the animal is at first oppressed as with a great weight, then grows convulsed, and at last expires in all the agonies of a most bitter and cruel death. But as this experiment is too shocking to every spectator who has the least degree of humanity, we substitute a machine called the "lung-glass" in place of the animal; which, by a bladder within it, shows how the lungs of animals are contracted into a small compass when the air is taken out of them." Yet Wright placed a live bird in jeopardy, a bird so rare in England that he may never have seen a breathing example: an umbrella cockatoo, *cacatua alba*, as delicately white as the dove of the Holy Spirit that fluttered at the ear of Mary in medieval pictures of the Annunciation and flew heavenward with the redeemed in Renaissance pictures of the Last Judgment. A cockatoo from the East Indies that could also, perhaps, speak.[184]

Psittacidae, the parrot family, are arguably the noisiest of birds. Cockatoos, even those that learn forty or fifty words, are arguably the noisiest of parrots, screaming for sheer pleasure mornings and evenings. And white cockatoos are among the loudest and noisiest of cockatoos. In vogue as exotic pets for the households of prosperous artisans and merchants as well as gentry and nobility, parrots in eighteenth-century Europe were tailed by longstanding, nearly contradictory associations with the haptic, the optic, and the acoustic. Doting, affectionate, sharply beaked, they were emblems of unrestrained eroticism and jealousy; bonding lifelong to their mistresses and pining for them at death, they were emblems of unswerving loyalty. Voracious, messy, prone to screeching obscenities at indiscreet moments, they were emblems of wildness; strutting with clipped plumage and trimmed talons, they were emblems of a dandified civility. Shrill and shrieking, they were noise; mimics of inflected human speech, they were the closest thing to a neutral carrier of language, a universal medium. The organic embodiment of the soundbite, parrots were as disputatious as echo, as retentive as frozen words.[185]

Ordinarily, had a live bird been subjected to the cold rigors of a vacuum, it would have been (as with earlier decades of experiments) a cheap, restless bird, a chirping sparrow or thinly warbling wren—small, plain, unspeaking. Wright's choice of a cockatoo lifted his *Experiment* out of the ranks of the many "paintings of light" for which he became so famous. His nocturnes of technical subjects lit solely by candles or flamepits—paintings of alchemists and blacksmiths, of glass furnaces and iron forges, of boys blowing up a bladder or girls amazed by an orrery—all had at their visual center a crucible of light; in the *Experiment*, the glass sphere is also highlighted, but as a crucible of *sound*. The effect of sealing a cockatoo within a gradual vacuum would have been first and most apparent to the ear, the cockatoo's constricted silence more telling than its slower, softer, subtler ceasing to breathe. Five of the nine spectators are not even looking in the direction of the cockatoo at this critical moment, and if not for an acute awareness of an overlong silence that must be the cue for the reopening of the stopcock, the cockatoo will die, since the demonstrator, the mage, is also looking away, far away, toward some empyrean (if empirical) universe. Of the four other spectators, Thomas Coltman is staring not at the cockatoo but at the stopcock, and Erasmus Darwin (?) is gazing at a point of equilibrium between the sphere above and the round watch in his hand below, as if listening simultaneously for the last sounds of the bird and the ticks of his timepiece. Only the two youngest witnesses, a rapt boy and the girl-who-wishes-she-could-not-look, are facing the (absent) music, their necks twisted and strained, their mouths open just enough to help the bird suck in its next breath. From the shape of the girl's mouth and the tenseness of her cheeks, it appears that when we can no longer hear the cockatoo, we must hear her.

Asked by Wright to listen through the silence for a redeeming noise, what are we to make of Mary Barlow's white bonnet that repeats the pattern of the cockatoo's feathers, as do the ribbons of white lace glowing in the hair of the girl-who-wishes-she-could-not-look, as does the brocade of the mage's dark-red gown and, most patently, the white ruffled cuff of his shirt hovering above the stopcock? What, or who, is in jeopardy? What and who will be redeemed? Isn't this the question posed by the open mouth of an antic mask carved into the edge of a mantelpiece in the brown darkness behind the mage? Isn't this the question posed by the second, close-mouthed, boy, at his station near the moonlit window, waiting to lower the cage back down around a revived cockatoo at what has to be an audible sign, a word from the mage, a squawk from the bird, or an obscenity?

Noise, then, as it came to seem ubiquitous, would come to be a newly acute index to sentience, liveliness, responsiveness, and power—sounds not only to be listened *through*, like the squeal of a bell jar imperfectly sealed, but listened *for*. In the context of street cries, air pumps, and Gothic novels, this listening for noise was much more than the soldier's or hunter's; it was bound up with a sense of human relationship, notions of the irrepressible, intimations of the tremendous, and strong expectations of something unexpected. Noise was what happened when one was drawn beyond oneself, or off exploring.

"Their features were far from disagreeable, the Voices were soft and tunable, and they could easily repeat many words after us," noted James Cook of a group of aborigines he encountered in July 1770 as his ships teased their way through the Great Barrier Reef off what he called New South Wales, "but neither us nor Tupaia," the Tahitian chief who acted as translator, "could understand one word they said." Cook, who had set off for the South Seas and East Indies on the *Endeavour* while Wright put the finishing touches to *An Experiment*, was much impressed by these "tranquil" people who "set no Value upon anything we gave them, nor would they ever part with anything of their own for any one article we could offer them." To his mind, these parroting but otherwise unintelligible people lived admirably within a sphere of their own making, with "no superfluities," in contrast to the sly Easter Islanders he would encounter in 1774 upon his second voyage, "as trickish in their exchanges as any people we had yet met with." Why so trickish? Because these islanders, who knew their island as Te Pito te Henua, navel and uterus of the world, had met up with Europeans fifty years before, on the day after the Christian day of Resurrection, and knew now what to expect: muskets and massacre. Because, to Captain Cook and his second lieutenant Charles Clerke, these almost fair-skinned men and women with their huge "Auricular Ornaments" were the obverse of the darker coastal natives who had listened and parroted effortlessly without meaning. With ears so stretched that the lobes hung "fairly down upon their Shoulders," the islanders listened *for* something, as did their *moai*, those hundreds of great stone faces listening out toward the sea with unusually long ears carved down each side. On Easter Island, Cook and his English crew were caught up in the suspense of "the sound in-between": soundings (as Paul Carter calls them) on both sides of a cultural divide during first contacts, each side badly parroting the other and listening *for* the half-noises that become, in succeeding generations, pidgin.[186]

Upon the return of Cook's three expeditions during the 1770s, the English became acquainted with thousands of new species of plants, fish, and birds, including new cockatoos, and Erasmus Darwin had a wealth of material to inform three book-length reflections upon the origin and variety of life. In *The Botanic Garden* (1789–1791) he explained in verse how life (and vegetable "sensation," vegetable "love") evolved from the ocean. In *Zoonomia* (1794–1796), he explained in prose how organisms changed and different species came into existence, beginning "from a single living filament; and that the difference of their forms and qualities has arisen only from the different irritabilities and sensibilities, or voluntarities, or associabilities, of this original living filament." In *The Temple of Nature* (1803), he explained in verse and prose how society evolved from organic life and, in a long note on the "Analysis of Articulate Sounds," how language evolved from society. Gesture and cry were inadequate even in small island societies as people developed more complex needs for communication and

> Love, pity, war, the shout, the song, the prayer
> Form quick concussions of elastic air.

Each society responded to a unique set of environmental and historical conditions, so each expressed itself differently. Any search for a primeval universal language was therefore futile, but one could discern a common sequence in the construction of language, with nouns and verbs coming first, then adjectives, "hard, odorous, tuneful, sweet, or white." Parts of speech flourished in almost organic profusion, syntax became the sex of talk, words commimingled. This commimingling most intrigued Darwin, himself a stutterer: how do humans shape and articulate their words? By 1791 the English actor and elocutionist John Walker had compiled *A Critical Pronouncing Dictionary* to compensate for the unsystematic spelling of English words (which irked Darwin) and to counteract the slovenly articulation of English speakers, against which Samuel Johnson the lexicographer and Thomas Sheridan the playwright had also railed. Tired of listening politely *through* bad pronunciation and worse enunciation, Sheridan in 1762 had directly addressed the problem, lecturing about "stammerers, lispers, a mumbling indistinct utterance; ill-management of the voice, by pitching it in too high, or too low a key; speaking too loud, or so softly as not to be heard, and using discordant tones, and false cadences." Poets too listened *for* the noise of improper elocution, the better to instruct in skilllful cadences. Robert Lloyd of London's Nonsense Club that year advised a fellow club member, a thespian, that

> 'Tis not enough the Voice be sound and clear,
> 'Tis modulation that must charm the ear.
> When desperate heroines grieve with tedious moan,
> And whine their sorrows in a see-saw tone,
> The same soft sounds of unimpassioned woes
> Can only make the yawning hearers doze.

Similar concerns were expressed in France, where the parliamentarian Charles de Brosses, completing in 1765 a long treatise on the formation of language, declared that the nasal tones affected by nobility (and newcomers thereto) sounded false; a trustworthy voice was harmonious and resonant, shaped by lips, teeth, tongue, palate, and throat in the manner he laid out on a series of anatomical plates. Vocal tone, to be sure, was not the theme of his two volumes. De Brosses was arguing that it was no use trying to locate a so-called Adamic tongue by listening through the *mélange* of words in a language, "no matter how impoverished or barbarous," for the number of words borrowed from other languages was always too great and the distance from ancient origins too remote; instead, one must examine the physical structures by which all humans are daily enabled to speak. Darwin would have agreed, and around the time of Wright's *Experiment*, he had already become so intrigued with the mechanics of speech that he had "contrived a wooden mouth with lips of soft leather, and with a valve over the back part of it for nostrils." Inside, between two hollowed-out pieces of wood, he stretched a silk ribbon. Using a blacksmith's bellows in lieu of diaphragm and lungs, Darwin could vibrate the ribbon and modulate the mouth to produce the English sounds *p*, *b*, *m*, and *a*, arriving at such combinations as *mama* and *papa*, "and a most plaintive tone"—like the cockatoo?—"when the lips were gradually closed."[187]

Speaking machines, whether costumed figures for lucrative public exhibition or skeletal contraptions illustrating the anatomy of talk before scientific audiences, were much celebrated in the eighteenth century, capable as some were of managing longer sentences than any parrot could, though breathier and more wooden in their inflection. Similar machines would lead several lives in the nineteenth century, as demonstration models for professors of speech and therapists of stammering, as fronts for stage magicians and mindreaders, or as prompts for fictions of more responsive automata, usually in tandem with forces electrical, as would be the case on the shore of Lake Leman in the stormy June of 1816. But we are not quite prepared for Mary Shelley and must spin back into the orbit of the Lunar Circle toward another kind of speaking machine.[188]

Matthew Boulton, master of a foundry and founding member of the Circle, could have been in attendance upon Wright's white cockatoo; James Watt, who moved down from Glasgow to work with Boulton in 1774, joined just as the Circle became a Society. 'Tis a pity that neither Watt nor a later Lunar man, Joseph Priestley, can be placed on the scene of *An Experiment*, for both had the strongest of reasons to be painted in proximity to an air pump, Priestley for his isolation of "dephlogisticated air," or oxygen, that would help explain how the cockatoo was being asphyxiated, and Watt for his improvement of steam engines, which drew their operating principles from observations of air pressure with regard to the vacuums created by air pumps. When James Ferguson was on his scientific tours, he would demonstrate how a piston-rod with a vacuum created beneath it might be pushed downwards by atmospheric pressure, and Watt patented a new air pump that drew off the vapors from his improved engine, which used steam instead of atmospheric pressure to force a piston down.[189]

Steam engines laid in a new noisetrack for the workaday world, strikingly parallel to the noisetrack of Gothic novels: the screams and shrieks of a high frequency hissing and metallic clanging; the moans, groans, and gasps of a lower frequency wheezing, shuddering, and heaving. Heard early in the eighteenth century echoing down the tunnels of tin mines and coal mines where engines pumped water out of the depths in which caged canaries sang until their suffocated silence sounded an alarm, and heard later in that century shaping metals in foundries and running the looms and stocking frames of textile mills, by the nineteenth century the noisetrack was put on rails and sent out across the land. All along the way, as engines grew more horsepowerful and pressures mounted within their iron boilers, they waxed louder. Watt and the managers maintaining these engines listened diagnostically *for* their noises; prospective investors also listened *for* the noise, the louder the better, for it was in their noise that these machines spoke to them of industry and power: "The velocity, violence, magnitude, and horrible noise of the engine give universal satisfaction to all beholders, believers or not," wrote Watt. "I have once or twice trimmed the engine to end its stroke gently, and to make less noise; but Mr. Wilson [the manager] cannot sleep unless it seems quite furious, so I have left it to the enginemen; and, by the by, the noise seems to convey great ideas of its power to the ignorant, who seem to be no more taken with modest merit in an engine than in a man." In his honor or to the shame of such a quiet and cautious man, the energy emitted by a sound source per unit time, its sound power, would come to be expressed in terms of watts, though audio purists might more fittingly speak of "trevithicks." When the Cornish engineer Richard Trevithick the Younger, at home with the "puffers" of the Ding Dong mines, built a new engine that heated the water not from below but from within a thicker, more cylindrical boiler that could contain steam pressures twenty times as great as Watt had ever dared, the noise was greater still, and his engines' exhausts were heard booming five miles off. The first to build a commercially viable steam vehicle, something that Watt, Darwin, and Richard Edgeworth had each contemplated and that another Lunar man, William Murdock, almost achieved in the 1780s, Trevithick set his "puffing devils" of 1801–1804 to work with as much huffing as puffery. His "Catch-Me-Who-Can" locomotive of 1808, which sped twelve miles per hour around a circular track in the middle of London, could equally well have been called "Shut-Your-Ears-Though-You-May," and the successors to the steam-powered rock-boring machines and threshers that he patented in 1812 would set the British countryside screeching. Two years later, when the former coal shoveler George Stephenson built his first locomotive, capable of pulling eight wagons at four miles per hour on two cylinders with an eight-foot-long boiler, he called it—and with the steam escaping loudly from its exhaust, it must have sounded exactly like—*Blücher.*[190]

Already in 1785 the sonic impact of steam engines on the countryside had been registered by Anna Seward, a tangent of the Lunar Circle whose adopted sister married Edgeworth and who, a neighbor of Darwin, tended his invalid wife. Widely known for her *Elegy on Captain Cook* (1780), Seward was ever regardful of her own genius and would claim that Darwin had stolen posies from her own work for his *Botanic Garden*, but listening rather to her environmentalism than her envy, we can find her protesting the effects of the early industrial revolution, first in a letter of 1765 on Eyam Dale, whose "craggy steeps" were "continually growing less and less distinct, picturesque and noble; broken and ravaged" for the construction of houses and roads, "and this by the force of gunpowder, and by the perpetual consumption of the ever-burning lime-kilns. They, and the smelting-houses in this dale, sully the summer skies with their convolving smoke," though at night "they are very fine, emitting their lurid flames, which seem so many small volcanoes." Twenty years later, nothing so luminous redeemed another dale, Colebrook Dale outside Birmingham, its choir of naiads silenced by "rattling forges, and their hammer's din," and its wildwood glens overcome by

> the vast engine, whose extended arms,
> Heavy and huge, on the soft-seeming breath
> Of the hot steam, rise slowly;—till, by cold
> Condens'd, it leaves them soon, with clanging roar,
> Down, down, to fall precipitant.

Contrast these with the uplifting lines of *The Task*, also published in 1785, by a poet who had left London to take up residence in Olney, a quietly bedraggled town where lace was still manufactured by hand:

> Nor rural sights alone, but rural sounds
> Exhilarate the spirit, and restore
> The tone of languid Nature.

To William Cowper and others recalling the classical pastoral tradition, it was the city with its crowds, traffic, street criers, and roaring commerce, that represented the "noisy world," and the countryside the lyrical, or at least bucolic, for "Sounds inharmonious in themselves and harsh," like cawing rooks, clamorous infants, hissing teapots, thumping flails, or the twanging of the post horn,

> Yet heard in scenes where peace for ever reigns,
> And only there, please highly for their sake.

Were not village bells, even those pealing

> loud again and louder still,
> Clear and sonorous as the gale comes on

in homeopathic efforts to fend off dangerous thunderstorms, much preferable

to smoke, to the eclipse
That Metropolitan volcano's make,
Whose Stygian throats breathe darkness all day long,
And to the stir of commerce, driving slow,
And thund'ring loud, with his ten thousand wheels?

Certainly the lyricism of the country was preferred by Cowper himself, who had had several breakdowns before settling in Buckinghamshire, where he became absorbed in the voyages of Captain Cook while composing evangelical hymns from a garden shed, his "summer parlor" where "the sound of the wind in the trees, and the singing of birds, are much more agreeable to our Ears, than the incessant barking of dogs and screaming of children." Cowper's lyrical lines rang bells well beyond Olney: *The Task* was the most widely read, and heard, poem of its era.[191]

Henceforward == as the number of iron furnaces in Great Britain quadrupled between 1788 and 1825 and production of iron multiplied tenfold (steam engines assuring hotter fires, more tremendous blast, and an endless roar) === and as the miles of rolled iron track expanded twentyfold from 500 in 1838 to 10,000 by 1860, nearly matched by France and Germany, and exceeded by the United States (everywhere screeching under the weight of ever heavier locomotives) ==== and as the number of rail passengers in Britain alone rose twentyfold to 110 million by 1855 and the tonnage carried by railroads throughout Europe and North America rose with similarly exponential bravado ("A nobleman would really not like to be drawn at the tail of a train of waggons in which some hundreds of bars of iron were jingling with a noise that would drown all the bells of the district," wrote a critic of the 112-mile London–Birmingham line that opened in 1838) ===== and as the coal they hauled was trundled into longer and longer freight trains as the number of British coal mines quadrupled (1842–1856), while coal production multiplied sevenfold in Germany (1800–1850) and sixty-five-fold in the United States (1829–1859) ====== and as coal-driven rock-boring machines tunneled through mountainsides while mechanical threshers and reapers cut swathes through the countryside ======= it would become less and less possible to keep the rural sonically separate from the urban. In 1815, thirty years after Seward's lament but twenty years before the London–Birmingham railway was opened, the novelist Mary Brunton passed through the Midlands and found them more disturbing than ever. This author of *Self Control* (1810) and *Discipline* (1814) enjoyed viewing one of the graceful iron bridges over the Severn, but then she came to (Coal)Brook Dale: "The steep and lofty banks of the Severn have been torn and disfigured in search of materials for manufacture, till they exhibit such appearances as might be supposed to follow an earthquake—fissures, cavities, mounds, heaps of broken stones, and hills of ashes and scoriae. The dell, which seems intended by nature for a quiet solitude, soothed by the hush of waters and the wooings of the cushet, resounds with the din of hammers, the crackling of flames, and the groanings of engines and bellows."[192]

***11. Cacophony*]*"I will not hear you"*[**
April 1815: two months before the Battle of Waterloo, a million and a half tons of volcanic material exploded with a noise louder than any human ear had heard in sixty thousand years. Though some in Britain claimed to have caught snatches of the cannon of Waterloo two hundred miles distant, the "super-colossal" eruption of Mt. Tambora in the Dutch East Indies was heard sixteen hundred miles off as it sent as much as twelve cubic miles of magma into the sky. For a year and more its sulfuric ash rode the winds into the stratosphere, joining ash from recent eruptions in the Celebes and Philippines. By the spring of 1816, pandemics of typhus and cholera were reported in China and India, and aerosol clouds were shuttering the sunlight over the Northern Hemisphere. Crops came up short, damaged by summer frosts. Thousands of New England farmers left their lands, heading west during "the most extraordinary year of drought and cold ever known in the history of America" (wrote Thomas Jefferson). Soup kitchens opened in New York City; the poor rioted for bread in Wales and the Midlands; a record number of Parisians starved to death. Cantonal officials declared an emergency to handle famine in parts of Switzerland, toward which were headed Percy Bysshe Shelley and his wife Mary, with their infant son William and Mary's stepsister Claire Clairmont. Claire hoped to join Lord Byron, whom she had hustled into a one-night stand the month before and who himself was headed toward Switzerland via Waterloo, over whose fields he rode singing Turkish songs and rehearsing the battle—the bagpipes "savage and shrill," the thundering cannon, the fifty to sixty thousand dead, "of whom each / And one as all a ghastly gap did make / In his own kind and kindred"—for canto III of *Childe Harold's Pilgrimage*. Leaving cold Albion during this "year without a summer," the Shelleys made their way down through France and the Alps amid "a violent storm of wind and rain," through unseasonable snow, past avalanches heard "as of the burst of smothered thunder," and alongside advancing waves of glacial ice that were sensitive to the slightest of sounds. Arriving in Geneva a fortnight ahead of Byron and his physician-companion John William Polidori, they had time to sail on Lake Leman, but after Byron showed up and was cornered by Claire, with Percy and Mary in tow, a "perpetual" rain began to fall. "The thunder storms that visit us are grander and more terrific than I have ever seen before," Mary noted in her journal, taking the same pleasure in these storms as did Byron in his canto, where echo amplifies all:

From peak to peak, the rattling crags among,
Leaps the live thunder! Not from one lone cloud,
But every mountain now hath found a tongue,
And Jura answers thro' her misty shroud,
Back to the joyous Alps who call to her aloud![193]

Quartered within five minutes' walk of each other on the quieter side of the lake and confined indoors together during the June squalls, the quintet amused themselves by reading aloud from a French edition of a German collection of *Fantasmagoriana*. Tales of "Apparitions, Spectres, and Ghosts"—of a midnight's entertainment using a skull that harbors the spirit of the ventriloquist's dead father, or of a young woman embraced unto death by the spirit of her dead sister—had dark subtones for these twenty-somethings. Byron, twenty-eight, was just quit of a hard year-long marriage and would never again see England or his daughter Ada, to whom that summer he sent a crystal necklace on her first birthday. His companion John William, aged twenty, had just finished a medical dissertation on somnambulism and was obsessed with philosophies of suicide and the pharmacology of poison; he would die in five years' time at his own hands. Percy, twenty-three, burdened by money problems and bad reviews of his long poem, *Alastor, or The Spirit of Solitude*, had crossed the Channel newly relieved of fears that months of lung spasms signalled a fatal consumption; he would drown six years later off the coast of Italy. Mary, eighteen, had suffered through the premature birth and death of her first child a year before ("Dream that my little baby came to life again—that it had only been cold, & that we rubbed it by the fire & it lived") and was still shaken; a second child, her infant son William, who was with her in Geneva, would die in Rome three years later of malaria. Claire (a.k.a. Jane, Clara, Clare), eighteen and pregnant by the unloving Byron ("Is the brat *mine*?" he asked), suffered from night terrors; by year's end she would give up her newborn to Byron, who sent the baby girl to an Italian convent school where she died at the age of five.[194]

Challenging each other to write ghost stories of their own, Byron, Percy, and Claire broke off as soon as the weather broke, but Mary and John William followed through. Polidori took up a thread of Byron's and wrote the first vampire tale to ennoble one of the undead. In *The Vampyre*, an honest orphan, Aubrey, drawn to the charismatic Lord Ruthven, follows him through Europe much as Polidori had followed Byron; in Greece, riding out on a dark and stormy night, Aubrey meets up with the monstrous side of vampirism. His is a violently sonic encounter, where "echoing thunders had scarcely an interval of rest." In one rare interval, the quiet is broken by the "dreadful shrieks of a woman mingling with the stifled, exultant mockery of a laugh, continued in one almost unbroken sound;—he was startled: but, roused by the thunder which again rolled over his head, he, with a sudden effort, forced open the door of the hut. He found himself in utter darkness: the sound, however, guided him." A thud in the night and "He found himself in contact with some one, whom he immediately seized; when a voice cried, 'Again baffled!' to which a loud laugh succeeded," he is held in a superhuman grip until the glare of torches drives his enemy off. *The Vampyre* contributed to a definition of noise that was being refined in the darker recesses of the Gothic aesthetic: noise as sounds you do not *want* to identify, unforgettable sounds you try desperately to forget, sounds whose implications so repulse as to demand for sanity's sake a generic masking, sounds that should be so unfamiliar as to be incomprehensible—sounds of cannibalism and necrophilia, of bloodsucking, of sex-unto-death. Such noise was the acoustic underside of the sublime: if the sublime takes your words and breath away, at its most horrifying the sublime becomes noise, sounds you cannot dismiss but will not define, sounds you wish you had never heard.[195]

Frankenstein is the tale of a speaking machine asking to be heard, a creature who horrifies his creator but demands of him a hearing. Locating its genesis in a spirited conversation between her husband and Byron (actually, Polidori) about the "principles of life," with reference to Johann Ritter's recent electrification of corpses in Germany and Erasmus Darwin's studies of revivification, Mary Shelley had been anything but the "devout but nearly silent listener" she later claimed to be. Familiar with the electrochemistry of Humphrey Davy, a friend of her father's, she rested her tale on Davy's premise that "all the products of living beings may be easily conceived to be elicited from known combinations of matter," and she studied Davy's *Elements of Chemical Philosophy* while she finished her tale back in England. As a girl she had also met Erasmus Darwin and knew his theories about biological origins. No mere auditor, she was more critical of her protagonist, Victor Frankenstein, than readers could have understood from the text published in 1818 after heavy editing by Percy Shelley, who intruded a materialist scientism on what began as a critique of science and scientists for *not listening* to the person imbedded in the flesh.[196]

"Begone! I will not hear you. There can be no community between you and me; we are enemies," cries Victor when the creature catches up with him, long after Victor has fled, trembling less at what he has attempted to do than at what his creation looks to be. "Listen to my tale," the creature pleads: "when you have heard that, abandon or commiserate me, as you shall judge that I deserve. But hear me. The guilty are allowed, by human laws, bloody as they may be, to speak in their own defence before they are condemned." Victor had run off at first sight of his waking creation; having "lost all soul or sensation but for this one pursuit," his eyes grown "insensible to the charms of nature," his compexion "shriveled," Victor saw the worst of himself mirrored in a handmade creature with yellow eyes, ill-fitting skin, straight black lips. "Listen to me, Frankenstein," the manufactured creature pleads. "You accuse me of murder; and yet you would, with a satisfied conscience, destroy your own creature. Oh, praise the eternal justice of man! Yet I ask you not to spare me: listen to me; and then, if you can, and if you will, destroy the work of your hands." Though large, the creature was not disproportionate, not incomplete, and not inherently evil. What Victor ran from, at one in the morning, with "breathless horror and disgust," was rather the obligation than the outrage of his act: the obligation, as with any bloody wrinkled squalling newborn, to respond to its noises and nurture it to meet an imperfect world. When the creature rose of its own and went to find its creator, Victor was hiding in bed, nightmarish with dreams of holding his dead mother in his arms, graveworms crawling through her shroud. The creature parted the bed curtains. Victor, half-awake, half-saw the "miserable monster" open his jaws, grin, and mutter "some inarticulate sounds. He might have spoken, but I did not hear; one hand was stretched out, seemingly to detain me," but the young doctor would have nothing of this gesture, nothing of the sounds, and rushed down into the courtyard. "Begone! relieve me from the sight of your detested form," Victor later repeats on the slopes of an ice field across from Mont Blanc, confronted by a creature now more eloquent than his maker. Should we believe what Victor has told us, that the creature had always been ugly but once animated became dreadful, though he had chosen its features for their beauty and shaped them to be pleasing? "Thus I relieve thee, my creator," says the creature, placing his hands in front of Victor's eyes; "thus I take from thee a sight which you abhor. Still thou canst listen to me, and grant me thy compassion." Pacing in the courtyard that night, Victor had listened "attentively, catching and fearing each sound as if it were to announce the approach of the demoniacal corpse to which I had so miserably given life." He did not want to hear it out; he wanted to be always out of range of its hearing, out of range of hearing it. "By the virtues that I once possessed, I demand this from you," says the creature: "Hear my tale."[197]

Hear it Victor must, this tale of a child spurned by his sole parent and left to fend for himself in an untrustworthy world. Delivered unto the world as something of an automaton, mobile but unprogrammed, and abandoned without guidance, as something of a wild child, Frankenstein's creature becomes a viable individual by reason of his nature and a distinct person by nature of his reason. He begins, as did a memorable figure who came to light and sound in 1754, by sorting out sensations. Étienne Bonnot de Condillac, a man of the cloth and friend of the young Rousseau, was encouraged by his mentor, the brilliant invalid Elisabeth Ferrand, to postulate a statue on which he could operate like a surgeon, implanting one sense at a time within an immobilized body. At each step in his *Treatise on the Sensations*, Condillac posed a basic Enlightenment question: When in the exercise of the senses might a creature—a conscious statue—become aware of itself as part of a society of like-reasoning others? The outcome of his thought-experiment was an empirically isolated being whose ideas stemmed entirely from lingering sense-impressions, the organization of which was by itself no mean achievement, especially with regard to hearing. The world is full of sounds, which may arrive one after another or, like chords, all together, wrote Condillac, who acknowledged that it would take the statue a while to hear such chords as music. The world is also full of simultaneous sounds that remain irreducible to music or meaning; these Condillac defined as noise.[198]

Dissonance so multitudinous was just beginning to be synonymous with cacophony—a classical medical term for harshness of voice (symptomatic of lung disease), an early-modern term for "vicious utterances or pronunciation" (symptomatic of the lower classes), and, by the mid-eighteenth century, a term for the discord of contesting voices and boisterous confusion (symptomatic of mob rule). Given, as Rousseau maintained, that the chief pleasures of the ear consist in the hearing of melody rather than harmony; and given that language becomes possible only through the ordering of syllables beyond the metaphors of gesture (as another of Rousseau's friends, Denis Diderot, had written); the task of a suddenly hearing statue must be to filter and sequence sounds into appreciable series—to detach the *kakos*, the bad, from the *phonē*, the sound. But though sounds vibrate through a body, they are weak proof of an actively impinging exterior world. "*The statue limited to hearing is what it hears*. . . . When its ear is struck, the statue becomes the sensation that it experiences. It is like the echo of which Ovid says: *sonus est qui vivit in illa*; it is the sound that lives in it." Only after Condillac invested his statue with touch, through which it feels the gap between itself and other bodies and bangs those bodies until they rattle, does it realize that sounds may arrive from a distance, from a world of noises for which it must listen.[199]

Letter of L. at Nottingham to the Prince Regent, June 4, 1812: "The cry of your sins, is gone into the Ears of the Lord of hosts and the day of repentance will shortly be gone by - - - The cry of your hard and immoveable heart to the sufferings of your *Poor Starving* Subjects is gone into the Ears of General Lud." The sounds arriving from a distance as Mary Shelley struggled to give voice to the creature were the cries of food rioters shouting "Bread or Blood!" and of men being hanged in November 1816 and April 1817 for having wounded a factory watchman and wrecked fifty lace-making frames during an action in Leicestershire on June 28, 1816. General Ludd's men, organized bands of once-privileged woolcombers and less skilled artisans who invoked an eponymous Ned Ludd, were asking to be heard above the sound of the power-loom, which made "such a clattering and noise it would almost make some men mad." Their quarrel was less with steam-powered machinery than with a deafened economic and political system that reduced wages below subsistence and denied their children a future. Address from the Framework-Knitters to the Gentlemen Hosiers of the Town of Nottingham: "Our Children, instead of being trained up by a regular course of Education, for social life, virtuous employments, and all the reciprocal advantages of mutual enjoyment, are scarce one remove from the Brute . . ."[200]

Daughter of the philosophical anarchist William Godwin and the feminist radical Mary Wollstonecraft, friend now of Byron who had spoken out in Parliament against a bill to make frame-breaking a capital offense, Mary Shelley heard that summer from her half-sister Fanny Imlay about miners demonstrating and thousands unemployed. Back in England by September and laboring over her *Frankenstein*, Mary could not but have heard the noises of protest even as news came to her of Fanny's suicide in October, could not but have heard (of) the March of the Blanketeers in March 1817 when dragoons turned back seven hundred Manchester weavers and spinners headed toward London with blankets on their backs and petitions in their arms, could not but have heard (of) the hanging of forty-five stockingers and ironworkers who, out of food and work, had risen against the government near Nottingham in June 1817. It was an audibly unsettled time that the reformer Jeremy Bentham heard as verging on "cacotopia," and so "It is with considerable difficulty that I remember the original era of my being," says the creature as he begins his story; "all the events of that period appear confused and indistinct." Sparked into life and then spurned, Frankenstein's "hideous" brute is born into a Condillac noise: "A strange multiplicity of sensations seized me, and I saw, felt, heard, and smelt at the same time; and it was, indeed, a long time before I learned to distinguish between the operations of my various senses." Feeling cold and pain, squinting at the moon, alert to the harsh notes of sparrows and sweet songs of thrushes, it learns to sort out its sensations. Trying its tongue at birdsong, it fails; its own expressions come to naught but noise, and "the uncouth and inarticulate sounds which broke from me frightened me into silence again."[201]

Now we are running with wolves and wild children, in particular a wild boy found running naked on all fours in a Lithuanian forest in 1694, his countenance "hideous," his mouth uttering "sounds that in no way resembled human sounds." Once caught and taught to stand and speak, the boy could not recall his prior life, "could not give much better account of it, than we can do of our Actions in the Cradle," though he continued to make "a strange sort of Noise not unlike Howling." His partial amnesia was solid evidence for Condillac that consciousness has no time for memory when in survival mode and confronting, during infancy or a state of nature, unmanageable cacophony, "a large number of perceptions that act on us with roughly equal force" at the same time. Other *enfants sauvages*, girls and boys, would also be taken as case studies for the Enlightenment research program into what comes innately to a human being. Only one of these children was subjected to a determinedly scientific protocol: a boy first glimpsed on the slopes of the Tarn mountains in southern France in 1797, the same year that frames the tale of Victor Frankenstein, the same year that Mary Wollstonecraft bore her namesake daughter and died of puerperal fever in an exchange of lives that lay at the heart of Mary Shelley's own ambivalence about acts of creation.[202]

Victor (that's right, the boy would be called Victor) was twice "rescued" and twice escaped his rescuers before turning up at farms in Aveyron, silently begging for food and a few minutes' warmth beside a hearth during the severe winter of 1799. In January 1800 a local commissioner took him in hand and sent him to an orphanage to be cleaned up, then in February to Rodez, the departmental capital, to be looked over and after by a professor of natural history, *abbé* Pierre Joseph Bonnaterre, his gardener Clair, and Clair's wife. In November he was taken to Paris to be studied by a Society of Observers of Mankind. That Society had been made up out of the blue by L. F. Jauffret, a man known for his children's stories, *Charmes de l'enfance* (1791) and ever since a protégé of the *abbé* Roch Sicard, who as newly reinstalled director of the National Institute of the Deaf and Mute was intrigued by an article in the *Journal des débats* about a young deaf-mute found living alone and naked in the mountains. On behalf of Sicard and the Observers of Mankind, Jauffret had written instantly to authorities in Aveyron, asking that the boy be sent forthwith to Paris under the Institute's auspices—forthwith, before the savage was civilized, so the Observers could probe the mind of a child who had been "abandoned to himself." Departmental processes moved glacially through a fruitless search for kin; the *sauvage* meanwhile was housebroken and learned to roast chestnuts and potatoes, peel string beans, sit in a chair. Bonnaterre, himself an encyclopedic observer of plants, reptiles, and fish, protested that the child in his charge was not deaf and not exactly mute—so not within Sicard's purview. But there was political clout in the group of savants and clerics who swiftly subscribed to Jauffret's Society, attracted by a boy on the edge of puberty who had been living under circumstances just a step beyond Condillac's statue.[203]

Citizen Bonnaterre would have to surrender the child, though not before he published a memoir on this "wild boy of Aveyron." Accustomed in his own work to reviewing one by one the senses of each realm of animal life, and having gone into particular detail in his review of the old controversy as to whether fish can hear (settled in the affirmative by eighteenth-century anatomists and by Jean-Antoine Nollet two fathoms underwater),

Bonnaterre carefully assessed each sense of this "new member of the social body." Dismissing stories about his swimming like a duck and climbing like a squirrel, the professor took the boy to be more than animal if less than articulate; he went to thoughtful lengths to explain how this round-faced, dark-eyed, chestnut-haired child with the gracious smile could be so much more attentive to smells than sounds, so ticklish and prone to outbursts of laughter yet so immune to cold, so little adept at touch, and so unwilling to imitate speech. Maybe that classical index of active reason, the need to speak, had been inhibited by years of solitude aggravated by an early childhood of physical abuse: if one brought a rope into his presence, the boy held out his hands, waiting to be tied up and led on a leash; he was covered with scars that resembled burns rather than vestiges of claw, tooth, or bramble; the 41 mm-long gash on his throat bespoke a violence suffered at human hands, as if kin had meant to silence him forever. Had he gnashed his teeth at the throngs that followed him to Rodez because he was a beast fit only for the wild, or because some years ago he had been treated bestially by humans and left to die in the mountains? What meant the murmurs he made while eating, the low sounds he made while rocking back and forth on his haunches? Was he incapable of speech as a result of that gash, or after privation and isolation had he forgotten how to use his healthy tongue? Without partners to please or passions to share, were not smell and taste most crucial for survival? As for his hearing, he was understandably a specialist; insensible to the sharpest of cries, the sweetest of musics, the boldest of noises, he reacted instantly to the sound of a nut cracked behind his back and the click of a latch to a cupboard of potatoes. His mind, too—hadn't it been narrowed in focus by overweening needs for food, sleep, safety? Once safe in Clair's garden, assured of plentiful food and a protected sleep, would his mind open, his eyes widen, his ears clear? Would a voice emerge?[204]

Delivered to Paris by a reluctant Bonnaterre at the command of Napoleon's brother Lucien, Minister of the Interior, the still-speechless boy was examined by a most prestigious Observer, a leading clinician and specialist in *aliénation mentale*, Philippe Pinel, chief physician of the two major hospital-asylums of Paris. Not only was the boy no deaf-mute, pronounced Pinel: he was no wild child. The boy lacked the skills and acuities that would have been crucial to surviving in the Midi-Pyrénées, near where Pinel himself had been raised. The boy had unsteady eyes, poor coordination of touch and sight, a weak sense of smell, and "His organ of hearing was equally insensible to the loudest noises and to the most touching music": he could not possibly have fed or protected himself in the wild. With no active memory, judgment, or language—"only a uniform guttural sound," occasional cries, and laughter that arose from a "meaningless stupefaction"—he was technically an idiot, and a hopeless case at that, deserving of care but not of celebrity. Dismissive of Bonnaterre's account of what the boy had accomplished in six months at Rodez, Pinel declared that the boy was beyond help. What could one do with such noise?[205]

Years alone in the woods, huddling in huts he built from branches and leaves, scratching for roots and acorns with dozens of scars to show for his troubles: how could the boy be an idiot? Ignoring (as had Pinel) Bonnaterre's suspicions about abuse, Dr. Jean Marc Gaspard Itard took on the boy as a test of the power of education and socialization to bring the rawest of humans within the province of equality and fraternity. Perhaps because Itard had also been a scrambler, a bank clerk in Marseille who escaped the draft (and infantry service) by pulling strings to get hired on at a military hospital, then studying his way into medicine and surgery; perhaps because he too came from wooded elevations, from the Provençal town of Oraison, "Prayer"; perhaps because he was a generation younger than Pinel, so less respectful of Enlightenment strictures about reason, more open to Romantic notions of selfhood; and perhaps because, as Sicard's newly-appointed chief physician, he was more at ease with the voices of the deaf and mute, "or rather the gutteral cry which continually escapes them in their games," Itard listened *for* the boy's noises and heard them as promises of something more. "During the night by the beautiful light of the moon," he wrote in his initial report of 1801, "when the rays of this heavenly body penetrated into his room, he rarely failed to waken and place himself before the window much of the night—motionless, his neck bent, his eyes fixed upon the moon-lit fields, giving himself up to a sort of contemplative ecstasy, the silence and immobility of which were only interrupted at long intervals by deep inspirations nearly always accompanied by a plaintive little sound." If one listened for subtle tones, the boy did express himself, did have an "exquisite sensibility" for sounds that mattered, like the delectable cracking of a chestnut. Spoken at, he demonstrated a preference for the sound of a French *O*, so Itard baptized him in the name of water, *l'eau*, and called him vict*O*r.[206]

Everything about Victor, as it turned out, had to do with Condillac noise: in a wilderness of simultaneous sounds, sights, smells, and textures, he had needed to make only a few basic sensory discriminations in order to survive. Testing Victor's hearing by firing pistols near his ears, Itard was stunned at the boy's indifference to the sound, but instead of classifying him as deaf or a dullard, he considered the life of a child who could not risk being overwhelmed by the sonicity of forests, where his "ear was not an organ for the appreciation of sounds, their articulations and their combinations; it was nothing but a simple means of self-preservation which warned of the approach of dangerous animals or the fall of wild fruit." To speak to him, and to teach him to speak, one had to get him to hear speech as something other than another undifferentiable element of Condillac noise. "One might say that speech is a species of music to which certain ears, although perfectly constituted otherwise, may be insensible," or one might blindfold an adolescent and train him for months to distinguish the sound of a bell from the sound of a drum, the striking of a shovel from the word for milk. This education of the senses—which extended to exercises in touch, sight, and smell—is exactly what the creature in *Frankenstein* deserves of *his* Victor; without it, the storm of sensations builds to a fury and the creature becomes a monster. In Shelley's novel, names and roles were reversed, for the Victor taught by Itard is "a creature who could develop only by carefully graduated steps and who advanced toward civilization only because I led him on imperceptibly," while the Victor of *Frankenstein* retreats into the mountains and ends up in the polar north where he has a final shout-out with his creature, rejecting its plea for a partner in life, a second laboratory creation. In that expanse of frozen words, Victor dies. The creature disappears among floes of ice, "lost in darkness and distance," heading toward what he vows will be a self-ignited pyre; on the last page of the original manuscript, he is simply "lost sight of."[207]

Itard's Victor remained in Paris, out of sight on a dead-end street near the Institute of the Deaf and Mute. After five methodical years of ingenious pedagogy, Itard had abandoned Victor, on the verge of a torturous puberty, to his devoted caretaker, Mme. Guérin. At his peak Victor had been able to recognize the referents of a few dozen written words, had invented a chalk-holder, and had learned to copy out simple sentences; but he would never read a Gothic tale, never recount his life, never confide in a friend, never make love. "How many times did I regret ever having known this child, and freely condemn the sterile and inhuman curiosity of the men who first tore him from his innocent and happy life!" exclaimed Itard in 1806 in a self-serving retrospect, for he too had embraced Victor as "an ideal problem of metaphysics," of what it takes to become fully human. Enlightenment thinkers like Pinel, who believed that humanity becomes itself through intelligent discourse, had listened for audible, sociable words; in their hearing, Victor was a failed specimen, an imbecile whose noises were less telling than those of a dog. Sicard, inheriting from Charles de l'Épée, founder of the Institute, an elegant language of the hands, argued on behalf of the deaf and dumb that intelligible sociable discourse could be silent, that fluency with signs and the written word was enough to form a good citizen; but he too, with Itard, felt impelled to drill the deaf in speech and music so that they might vocally affirm, and claim a proper share in, the Declaration of the Rights of Man.[208]

Revolutionary politics had engaged so many competing voices that a satirist called his report on debates in the "High Chamber of Parnassus" a "cacophoniana." The acoustic liberties of the revolution had early been heard in the provinces as lists of grievances read aloud before provincial *parlements*; Parisians would have heard these liberties most dramatically as outbursts from the stage. For too long, three royal theaters had held exclusive rights to mount drama and classical comedy with speaking actors. All other performers, whether strolling players at fairs or troupes housed in shabby halls on the *Boulevard du Crime*, were supposedly confined to wordless pantomime, puppetry, acrobatic buffoonery, or comic poses and grimaces, punctuated by narrative scrolls and sound-effects, animal imitations, whistling, humming, farting. By 1780, the players on the Boulevard, where courtiers and bluestockings rubbed perfumed shoulders with apprentices and streetwalkers, were breaking into song and dialogue, vocal and articulate. These performances were prelude to the sonic exuberance of the Revolution itself, whose deputies orated in highly theatrical spaces, took acting lessons, and were cheered by audiences in boxes while philosophers debated the wisest or widest avenues for the expression of public opinion. And while the public heard subversive noises rising from Parisian sewers, the Revolution's dramaturges used forms of melodrama and vaudeville as well as massive choirs at civic festivals. Among the first melodramas was one based upon a novel of 1796 so popular that the play took the same title: *Victor, or the Child of the Forest.*[209]

Since the deaf often make involuntary throat sounds when they gesture or sign, and since the sounds they make voluntarily when distraught or delighted can be extreme, it seemed rational and revolutionary to expect of the deaf a modicum of speech, however uninflected or abnormally loud, now that the raucous voices of others previously gagged or muffled could be heard unintimidated out in the open. Since "deaf" was itself a catch-all term for mild to moderate to major hearing loss, the diagnosis of which depended on the ability of a patient to listen for the whisper of a voice or tick of a clock in a busy room, those identified as "deaf" usually did have some residual hearing, so it could be doubly reasonable to expect them to want to invoke a newly available citizenship through newly rewarded efforts toward speech. At the same millennial time, it did not seem fantastic to anticipate total cures for total deafness, whence a flood of nostrums that were essentially cold remedies or fluids for ear cleaning and a parade of itinerant healers whose ear spoons and drops may have done more hearing damage than did Revolutionary cannon. In these contexts arose a heated dispute between "oralists," who insisted upon active speaking parts for the deaf in a speaking world, and "signers," who wanted to nourish a genre of expression richer than any seriously deaf person would achieve by audible speech alone. The dispute, which began in the 1780s and would continue into the 1980s, began as a medical debate about the permanence of hearing loss, a political debate about the grounds of citizenship, and a philosophical debate about language and how a person becomes fully human.[210]

Siding at first with the oralists, Itard never thought to teach Victor to sign, never built upon the gestures that Victor had invented on his own. Although he later promoted signing in the education of the deaf and as an indispensable adjunct in teaching them to speak, Itard's successes even then would come with those who retained a modicum of hearing and for whom speech was rather a restoration than an innovation. In the absence of any memory of speech, in the absence of babbling as run-up to speech, in the absence of a habit of imitation to perfect the sounds that become speech, after years of patient experiment Victor yielded little more than long silences and restive noises. Itard in 1825 would recall privately how excruciating had become his daily work with the boy; in 1828 he published a paper on mutism in which he noted a "discordance"—and supposed a lesion—between the sensory and intellectual functions of persons "mentally debilitated," who could hear but not listen, shriek but not speak. Victor died that year, maybe that spring when Mary Shelley was also in Paris, confined to her room with smallpox. Recuperating for months, she wrote a story, "The Mourner," which begins on the Virginia Water below Windsor Castle, the lake "an open mirror to the sky" under a setting sun, where "the eye is dazzled, the soul oppressed, by the excess of beauty." But it is the ear that fixes the scene, with strains of sound "more solemn than the chantings of St. Cecilia," strains that "float along the waves and mingle with the lagging breeze." Then comes a sentence that could stand as Victor's epitaph: "Strange, that a few dark scores should be the key to this fountain of sound; the unconscious link between unregarded noise, and harmonies which unclose paradise to our entranced senses!" Like the Mourner, the Savage was (so far as anyone knows) buried in an unmarked grave with "no stone, no name."[211]

12. Twang → Movements of the Inner Ear

Pinel had been sufficiently impressed by Itard's early successes with Victor to change his mind and allow for hope, but Victor's final state seemed to confirm his initial judgment that the boy was a hopeless case. That judgment had relied on oralist premises and had been swayed by aural evidences, but it signaled rather a philosopher's disappointment than a physician's discharge. Like Itard in his later career with the mentally disabled, Pinel would be remembered for his creative and humane treatment of the insane at the women's hospital-asylum of Salpêtrière, doing away with ice-cold baths and beatings, improving their rations, removing their chains, and encouraging them to exercise, all of which conduced toward quieter wards and lower institutional mortality. Pinel was also celebrated for his practice of moral therapy, adapted from English Quakers and evangelicals, who alternated calm conversation and gentle correction with stern allusions to a Supreme Judge who knew and saw all. *In loco*

Dei, Pinel commanded an extensive system of surveillance to isolate his many charges from family and friends who carried with them the miasma of the outside world, a Condillac noise that provoked attacks of mania in those patients not yet steeled to the simultaneities of daily life. According to Pinel's medical anthropology, mania itself, unlike idiocy with its weak mentation and obtuse hearing, did not render a person less human. In fact, given that only humans go mad, Pinel viewed mania as a stripping agent exposing the otherwise invisible, inaudible, but vibrantly human connections between reason and the senses, connections which among the mad seemed to have lost their capacity for mutual self-correction. Culturally and physiologically complex, insanity was bound up with a newly defined interior sense of *caenesthésie*, *Gemeingefühl*, or synaesthesia, but it was not inherent, not "constitutional," not even in the case of a young chemist who had closeted himself in his laboratory, taking stimulants instead of naps, focusing his energies upon a discovery that would have made his fortune had not he been overtaken by mania. Between the first (1801) and second (1809) editions of his treatise on the care and treatment of the insane, Pinel even dropped "mania" from the title, hoping thereby to disaggregate *la folie* (constitutional madness), *la manie* (episodic derangement), and *aliénation mentale* (becoming a stranger to oneself). More than a restoration of reason, healing entailed speaking in one voice, coming publicly to terms with *one*self.[212]

Against the odds, the furious child that is Frankenstein's creature gains a voice, but it is a voice in five registers: dislocated laborer, social critic, hunted murderer, lonely bachelor, and self-monitoring machine with superhuman strength, size, speed, endurance—and a trail of unintended consequences. When *they* speak, the five voices speak what he is, a commotion of selves. He oscillates among figures of wildness, madness, and mechanism—of moving mechanisms, figured automata—for unlike Condillac's statue, this multiplex creature has been capable of movement from the start. Its first movements are the necessary proof of Victor's success and sufficient proof of his awful mistake: "I saw the dull yellow eye of the creature open; it breathed hard, and a convulsive motion agitated its limbs." A glimmer, a gasp, a shiver, nothing more; but with that seismic instant came a slippage of epochs from the Enlightened to the Romantic. Enlightenment concepts of being-in-motion were incremental and cumulative; power moved in straight lines or along tempered curves. Romantic self-propulsion, whether realized as will, as love, or as super-heated steam, was transfigurative, and it was this transfigurative, transgressive energy, as revealed in the creature's moving countenance, that after one glimpse (and surely a gasp or shiver) had sent Victor rushing from the room. To her third edition Shelley added contortions "that ever and anon convulsed & deformed his un-human features." Analogous to the most significant of innovations in steam technology—the redirection of forces from reciprocal to rotary, the application of feedback devices like Watt's centrifugal flyball governor, the encapsulation of superheated systems—the manufactured creature, once roused, looked to be metamorphic and self-monitoring, self-contained yet explosive.[213]

Metamorphic and self-monitoring, self-contained yet (potentially) explosive: this was the pattern on which carceral forms would now operate. As the bell jar and glass sphere with their imperfectly held vacuums had been transformed into steady generators and collectors of (still-fractious) electricity, and as air pumps had been subcontracted to steam engines with (dangerously) powerful boilers that restrained and released ever more productive energy along iron rails, so asylums that had been open to visitors and spectators were closed in upon themselves to shut out the Condillac noise of daily life and better effect the improvement of (easily disturbed and disturbing) patients through a system of surveillance, therapy, and moral suasion. Bodies too, as a carceral form, were understood as no wardrobes of skin and bone but as enclosed, interfused systems of circulation and sensation whose complexities (and eruptions) intrigued "observers of mankind" in every guise—popular audiences fascinated with showpiece automata; medical students studying exquisitely layered anatomical waxworks; bourgeois readers engaged with the contending forces of love and destructiveness detonated by a self-conscious, organically reconstituted creature in the fiction of Mary Shelley or by a sedately unselfconscious, mechanically lovely Olympia in the "Sand-Man," a story begun (like *Frankenstein*) in 1816, by the German writer E. T. A. Hoffmann.[214]

Paradigmatic of this new pattern was a new physiology of the inner ear. When Itard in 1821 summed up his experience with deafness and diseases of the ear, he described the interior of the ear as *terra incognita*, an anfractuous part of the body "deeply hidden" within hard bone. Searching the medical corpus, he found "a vast lacuna" with regard to the inner ear, "an empty literature." Yet he owned up to debts to the work of Duverney in the seventeenth century and of Valsalva, Morgagni, Meckel, Cotugno, and Scarpa in the eighteenth, all of whom had contributed amply to excavations of the ear, so what was he going on about? Upset that anatomists and physicians clung to the old idea that the inner ear was "empty," he projected that emptiness onto the field itself. Plato and Aristotle had put in play the mythic notion of an ear whose labyrinthine hollow retains air from the fetal sac, as if each person hears the world from birth to death through the pure air of the womb, and as if the inner ear were a capsule preserving a primal innocence, protected by a maze of bone and flesh from the suddenness of the present. These as-ifs are mine, but they were implicit in the language used by such as Jean Fernel, in whose 1567 *Physiologia* Itard would have read that "The ear is always open to hear, by an entrance that is rigid and very tortuous, to prevent a sound from suddenly entering far enough to damage the sense," that it has "many windings that resist injury from outside, so that nothing could unexpectedly force its way in," and that the "particular instrument of hearing is a very rarefied air inherent in the ears, enclosed in a membrane and situated at the inmost cavity of the ears, at the point reached by the end of the auditory nerve running out from the brain." It was correct to say that of all sense organs the ear was most highly developed in utero, but it was wrong, warned Itard, to believe that our ears emerge from the womb with prizes of embryonic air, safeguarded unto death. Already in 1684 a German physician, G. C. Schelhammer, had shown "victoriously" that the labyrinth was not filled with "innate" air but with a liquid—albeit an elusive liquid that drained away an hour after death, which explained why other anatomists, exploring corpses, found no sign of it.[215]

Old bodies and poor technique were no excuse. The inner ear now ought to be pictured rather as an inland sea than a subterranean chamber of inviolate air, and if Itard could "not yet say" whether the labyrinthine fluid was crucial to hearing, he penned page after page to disabuse readers of the ear's hollowness. He had neither discovered the fluid nor "victoriously" confirmed its existence, so why this fierce concern? Because its presence made of the ear a more environmentally responsive and temporally sensitive organ. No treasure cave, no inert vacuum chamber or "receiver," the ear with its inland sea obtained a resilient fluidity and an active role in maintaining lymphatic flow and balance. It obtained, too, a self-contained fullness that was acoustically reasonable given the greater speed at which sounds travel through liquid than through air, as had just been calculated for fresh water by Ernst Chladni in Wittenberg and Pierre Simon de Laplace in Paris and as would be demonstrated experimentally in 1826 in Geneva when a Swiss physicist on one side of Lake Leman and a French mathematician on the other timed the flash of gunpowder against the underwater ringing of a 65-kg. bell and its reception 13,487 meters distant by a long tube (resembling an ear trumpet) four-fifths under water. By way of the inland sea the ear also obtained the dynamism of fluids in motion, which could account for the oceanic hum and the crash of waves that often tormented sufferers from tinnitus.[216]

'Twould remain for the Scottish naturalist Robert Brown—who was taught the word for "ear," *twang*, by Western Australian aborigines while Itard was shooting off a pistol near the ear of a wild boy—to describe the microscopic motions of granules of pollen suspended in water, and then coal dust, soot, bits of a meteorite, all moving in irregular flurries and all pointing by 1828 toward "the general existence of active molecules" everywhere. But Itard in 1821 had already anticipated the nonlinearity of aural functions in consequence of the viscous fluid in the ear. Brown would not be able to account for the "rapid oscillatory motion" of each kind of particle he placed in colloidal suspension, nor could he explain why scrapings from the Sphinx or from his own teeth should go into unceasingly irregular motions in water. Such strange randomness awaited the equations of Einstein, and Itard was no physicist, but problems of flow and fluid viscosity were in the scientific air, and there was something to the motions of lymph within the ear that appealed to him, something more otologically active than a series of concussions reaching a dark sanctuary that harbored a pristine remnant of uterine air.[217] So too the mathematics of elasticity were not yet up to the task of explaining how the designs demonstrated in Paris in 1808 by Ernst Chladni had arisen from the acoustical vibration, at different pitches, of metallic plates powdered with sand. Even by 1821 Chladni's vibratory sand-stars, sand-circles, and other sounded geometries were not fully predictable, but Itard was keen to approach hearing as a process of listening through different more or less elastic media. Granted the fact of a labyrinthine fluid, one had to come to terms with the implication that each act of human hearing ordinarily engages with sound simultaneously through three media: air, liquid, and bone.[218]

Bone? Itard knew from the work of Chladni and from years at the Institute that bone conduction could be the primary vehicle for some individuals otherwise deaf, and that metallic ear trumpets might be effective less as expansions of the collecting area of the pinna than as resonance devices, translating vibrations to the inner ear through the bones of the skull, circumventing other damage and tight compactions of ear wax, another medium he studied. There was no truth to the conjecture that slack-jawed bumpkins, gaping idiots, and the hard-of-hearing kept their mouths open in order to collect sound via the Eustachian canal that leads from the pharynx to the middle ear: Itard had disproven this by inserting a pocket watch into a subject's mouth so tenderly that it rested solely on the tongue, distant as possible from teeth and hard palate; mouth closed and the hardy watch bathed in saliva, but touching no bone, the subject heard no ticking. An experiment peculiarly echoic of the putative fetal air suspended in time in the recesses of the ear, this suspension of time in the recesses of the mouth was necessary to show that the act of hearing was not merely the result of newly shaken air from outside impinging (through the Eustachian tube) on old air trapped inside the ear. Hearing operated through several media at once, and those physically or mentally unable to manage such multiple channels suffered the Condillac consequences, a world of unintelligible sounds, worldwinds of noise.[219]

Beethoven, for example, whose deafness, stemming perhaps from a bout of typhus in 1797, and exacerbated by lead poisoning, made itself known through an obnoxious buzzing in his ears, the gradual loss of high notes and soft tones, then a severe distress in company, as much psychological as otological, since he, a lover of discourse, became disoriented and angry when unable to track separate voices in a crowd of talk, the aching loudness and confusion of words driving him to the use of "conversation books," hundreds of them, in which he silently, furiously, wrote down, and traded, repartee. The choppiness of the ninety-eight words with which I have opened this paragraph hints at the exasperating choppiness of sound that Beethoven, a musician tremendously sensitive to the dynamics of *piano* and *forte*, of rhythm and flow, would have experienced for the last thirty years of his life, helped a little by a quartet of ear trumpets made for him by the inventor of the metronome, and a little more by a resonating plate made for him by a company that manufactured louder and louder pianofortes for his failing ears. The more that Beethoven felt confined by an increasing deafness and oppressed by storms of noises in and at his ears, the more he sought as a composer to break from the confining tonal and rhythmic systems of Western music, as in his last two piano sonatas (1821–1822), and to test the boundaries of contrapuntal forms with music that was, as musicologist Anthony Burr suggests, "a crowd of competing voices." The disjunction between what Beethoven's mind could administer and what his ears could physically manage yielded neither a brooding silence nor mechanical imitation but aggressive invention among the sonic media of his time—horns, reeds, strings, kettle drums, cymbals, cannon—and a mastery of dynamics and timbre.[220]

Timbre puzzled Chladni, an amateur musician who invented a euphonium, tested tuning forks, and studied the acoustics of concert halls. For the same note in the same key, different instruments effected different sand patterns. This occurred, Chladni guessed, because timbre was a mix of musical tones with incidental though unavoidable noises, just as one can hear behind the melody of a violin the scrape of the bow against the strings. Acquainted with Leonhard Euler's mathematical analyses of soundwaves and the layered harmonics produced by bells, Chladni speculated that the musical side of timbre arrives in the form of coordinated waves while the noisy side of timbre arrives as conflicting rays or particles. This had a kind of visual logic to it, noise appearing as that which pierces and distorts the waveform orderliness of consolidated sound. Dissonance had long been been part of music theory and composition, and the intervals or chords judged forbiddingly dissonant had been challenged by each generation from the *Ars nova* of the fourteenth century through William Croft's 1713 *Ode with Noise of Cannon* to Beethoven's new music. At each challenge, conservatives dismissed the new music as *Katzenmusik*, the caterwauling of alleycats, and by the 1580s they had the first schematics for an iconic *Katzenklavier*, which had the tail of a screaming cat hitched to each key or chord. But the noises that Chladni brought hesitantly within the fold of timbre were of a different breed, neither discordant intervals nor clumsy partials nor awful catgut but accidents of performance, artifacts of the physicality of musical instruments meeting up with human bodies out of range of standard notation. Some of these noises matched in complexity the forms of Brownian motion and would demand a similar sophistication of mathematical analysis. More immediately to the point, if Chladni's speculation held good, noise was always to be found in the thick of music. As it had been in the thick of the "rough music" of the pans, ladles, and pot lids of *charivari* and the drums, trumpets, and bagpipes of battle, now noise was to be in the thick of the chamber music of the court and the symphonic music of public concert halls.[221]

Wherever one turned, noise. If incumbent in the sweetest consorts of flute and pianoforte, where could it not turn up? Present in the Condillac noise of an infant's experience of the world and in the death rattle of those near their end, noise was not simply ubiquitous but elemental—as elemental at least as air and laughter, steam and tears. Noise told the truth of sex and of the sublime; it bespoke the highest feelings and the lowest digestions, the politics of the gathering crowd and the protests of the laboring poor, the extremes of spiritual desolation and revival, the power of engines and the pit of horror. At the center of Gothic fictions and of Enlightenment philosophies of language, it was slowly emerging from the wings of mathematics, physics, and physiology.[222] Noise was narrowing the circumference of the countryside and redefining the urbanity of the city. As vampirical sounds one hoped never to hear, it elicited fear. As tortured sounds one hoped never to hear again—public execution by burning or breaking on the wheel, the flogging of slaves, the baiting of bears and monkeys, the chaining of madmen—it elicited compassion and reform. As sounds heard commonly but unintelligibly—the drawn-out call of a street vendor, the endless barking of a cur, the guttural shouts of a deaf-mute, the "little triumphing noises" that a six-month-old makes "when she thinks she is going to be picked up"[223]—it elicited suspicions of a meaning just out of reach. Noise was making more and more sense, and as the roil of Itard's inland sea, one came upon it in the very act of hearing, so that from within one listened *through* noise, roundabout one listened *for* it, and *almost* everywhere it was there to be heard.

13. [***silence***]

Where not? In unseen, unheard Antarctica, on any windless day beyond the scrawk of emperor penguins, or in Alaska in July of 1786, during the long pauses between the crash of enormous masses of ice into Glacier Bay, where "The air is so calm, and the silence so profound," wrote a French explorer, "that the single voice of a man may be heard half a league." At Ottoman courts and seraglios, whose "silentiaries" or chiefs of protocol assured a silence that continued to astonish European diplomats. With a Meadow Brown butterfly atop Mont Blanc, in whose thin air in 1787, just a year after the peak had first been scaled, Horace Bénédict de Saussure's pistol shot had no more oomph than the thinnest of Chinese firecrackers (though, thanks to Saussure, his valet, eighteen Chamonix peasant-guides, and the eight volumes of his *Voyages dans les Alpes*, these mountains would soon be loud with other climbers). Lower down and closer in, through the mute halls of Cistercian monasteries whose system of signs and gestures was as extensive as that of the Turkish harem. Lower and closer still, in penitentiaries. When William Godwin in 1794 described the "massy doors, the resounding locks, the gloomy passages" of London's Newgate Prison, he found some prisoners "noisy and obstreperous, endeavouring by a false bravery to keep at bay the remembrance of their condition; while others, incapable even of this effort, had the torment of their thoughts aggravated by the perpetual noise and confusion that prevailed around them." This would change, as it was changing at Pinel's quietened Salpêtrière—part hospital, part insane asylum, part poorhouse, and part prison where sixty-seven were serving life sentences in 1790. Long sentences were new, and pivotal. Previously, those found guilty would face quick punishment, whether the lopping off of an ear or a head; jails had been detention halls for those awaiting trial or transportation to a colony, prisons mere holding cells for debtors and their families, with a special area for felons awaiting execution. The noisiest of limbos, jails and prisons might have more visitors than Bedlam, more commerce than a market square. In contrast stood the workhouse or house of "correction," less porous institutions for the restraining and retraining of vagrants, petty thieves, prostitutes, beggars, and disobedient youth. Spreading from Amsterdam in 1596 across Europe to the Americas, workhouses were meant to redeem the reprobate while quarantining them from honest society as well as from infection by "hardened" villains, so inmates were set to spinning wool or rasping hardwood logs, packed off to separate cells each night, and instructed in the ways of the Lord each Sabbath for the length of their sentence, seven days to fourteen years. A mix of the monastic, medical, and military, the workhouse was clumsy and eventually unworkable, operating as did Amsterdam's *Rasphuis* and *Spinhuis* under divergent mottos, "'Tis valor to subdue what all men fear," and "My hand is stern, but my heart is kind." Even so, unlike branding or a debilitating stint at the oars of a galley, houses of correction did not maim or permanently mark a person as a criminal (though a visitor in 1668 saw the men in the *Rasphuis*, "being naked, and in a sweat, and the dust of the Brazil wood flying upon them . . . all over painted of a beautiful red color"). With its orderliness, individual cells, and relative quiet, the house of correction anticipated the silence of the penitentiary.[224]

Penitentiaries arose from a confluence of legal, philosophical, economic, demographic, and religious forces that had little to do with sound, yet silence would be as overriding a concern for prison architects, wardens, guards, inspectors, and reformers as sanitation or security. Nothing sonic other perhaps than screams from the scaffold and a heightened propensity for tears prompted a judicial disinclination toward capital punishment, such that the number of executions in England dwindled from eight hundred in 1603 to sixty-eight in 1805, from five executions annually during the 1740s in Amsterdam to one or none by 1800, and their total abolition in Tuscany in 1786. Nothing sonic other perhaps than an association between the hue and cry of hot pursuit, the whimpers of pickpocketing boys and shoplifting girls waiting months in jail for their arraignments, and the cry of *oyer* (*o yes!* or *hear ye hear ye!*) of curt trials prompted a shift in the theory of deterrence from severity to certainty of punishment. Nothing sonic other perhaps than the creak of wooden hulls and the groans of chained men under the lash of galley masters or wheezing in the tumid holds of vessels bound for Surinam, the Carolinas, or Australia prompted the on-again off-again policies of transporting criminals aged nine and older to colonial outposts as unfree labor. Nothing sonic other perhaps than the street cries of growing cities with their unsettling poor and the gulping breaths of typhus ("jail fever") prompted anxieties about the unhealthy crowding of prisons and the indiscriminate mingling of old hands and young minds. Nothing sonic other perhaps than the miserable clanking of the chains of convicts doing street repairs prompted a reconsideration of public hard labor (in Pisa in 1818, Mary Shelley "could never walk in the streets except in misery" because everywhere she saw criminals "heavily ironed in pairs . . . and you could get into no street but you heard the clanking of their chains"). And nothing sonic other perhaps than the gnashing of a soul toiling with sin and the swelling sighs of religious conviction prompted English Quakers and evangelicals, Gallican republicans, and Italian Catholics to campaign for a new carceral form that promised a lasting repentance.[225]

Nonetheless, the soul of the penitentiary was soundproofing, for without it there could be no solitude, and it was solitude that conduced toward repentance, "SOLITUDE" that was incised above the entrances to five new houses of correction in Gloucestershire. Serving time, which by 1800 had become standard punishment in Western Europe and North America, meant time served quietly, alone with one's thoughts. "This proposal stands upon a principle, that man is a *reasoning, as well as a passionate creature*," wrote Jonas Hanway in *Solitude in Imprisonment*, a tract grounded upon the eighteenth-century paradox that made of human beings insistently social beings while relying upon innate intellect and inward communion to choose a moral course. Solitude, as in Gothic novels, could then be represented as punitive *and* illuminating: "Reflection cannot lose all its power; nor can the heart of man be so petrified, but the consideration of his *immortal* part, under the terrors of solitude, will *open his mind*." An orphan who became a wealthy merchant, a loquacious Anglican who took up quill and philanthropical umbrella on behalf of the infant poor and seamen adrift, Hanway had directed his proposal of 1776 at those convicted of serious crimes and awaiting execution or transportation, but he expected that first offenders would also benefit from a forced "retirement," which was "the best friend to virtue." Under impositions of solitude and freed of leg irons, every prisoner could be led to the light, since "The idea of being excluded from all human society, to converse with a man's own heart, will operate potently on the minds and manners of the people of every class," as such interior conversations did in Quaker meeting and Anglican prayer closet.[226]

Disgusted with the state of the nation's prisons in the 1770s, where bandits, idiots, bankrupts, lunatics, and beggars of both sexes and all ages shivered and shat in cramped and noisy quarters, sometimes without clean water or fresh air, Hanway's compatriot John Howard toured the Continent looking for healthier models of incarceration. He found them in Belgium, the Netherlands, Switzerland, and Italy: clean, orderly, well-ventilated, and above all quiet places, unlike English prisons where there was "such *confusion* and *distress*, and such shrieks and outcries, as can be better conceived than described." Indeed, Howard could not believe that the Dutch houses, "so quiet, and most of them so clean," were houses of correction. He returned to England to lobby on behalf of hygienic prisons with separate wards for each kind of offender, all to sleep alone at night in their own cells. Separation made the plotting of escape more difficult, slowed the spread of lice (and typhus), removed the naive from the clutches of the corrupt, protected those held simply as witnesses, and gave inmates the space in which to hear themselves think: "Solitude and silence are favourable to reflection; and may possibly lead to repentance." Toward this end there had to be strict rules against quarreling and cursing; bells to signal hours of rising, of scripture reading, of meals, work, and sleep; and inspectors to inquire after each prisoner: a regimen akin to forms of work-discipline being implemented in new factories run by the likes of Boulton and Watt, whose superintendents had the approval of local magistrates to whip thieving or singing workers. "In English factories," wrote a French visitor, the radical Flora Tristan, "you never hear snatches of song, conversation, and laughter. The master does not like his workers to be distracted from their toil . . . he insists on silence, and a deathly silence reigns, so much does the worker's hunger reinforce the employer's command!"[227]

"Inspectable," insisted Jeremy Bentham; the rooms "must all be inspectable yet perfectly air tight." Writing in August 1816, just as Mary Shelley was getting up to speed with *Frankenstein*, Jeremy Bentham had hopes at last that one version of his Panopticon would make it beyond his drafting desk—in the form of a prison underwritten by the Juvenile Delinquency Preventing Society. "A magnificent instrument" with which he dreamed of "revolutionizing the world," Panopticons in splendid variety had taken flight from Bentham's pen for thirty years, during which time he had lived alone in Westminster "like an anchorite in his cell, reducing law to a system, and the mind of man to a machine," his chief exercise "post-prandial vibrations" inside a private garden, brisk walks in the company of such as his neighbor William Hazlitt, who gently mocked the short man's "grand theme of UTILITY." Arising in 1785–1787 out of a joint mission with his brother Samuel to start a shipbuilding and textile manufacturing enterprise on the Russian estate of Prince Potemkin, the idea of the Panopticon was initially a response to the difficulties of training serfs in industrial processes and of managing dozens of unruly British artisans. Back home, Jeremy elaborated on Samuel's plans for a two-story circular mill in which workers and foremen alike were fixed within spheres of surveillance. He applied the Panopticon idea to the design for a school where "All play, all chattering—in short, all distraction of every kind, is effectually banished by the central and covered situation of the master, seconded by partitions or screens between the scholars," and then to an "all-efficient" circular prison, each prisoner in a separately visible grated cell on the circumference, overseen by vigilant officers in a concealed center, "hence the sentiment of a sort of omnipresence" and "*Solitude*, or *limited seclusion, ad libitum*." The central station, having "the most perfect view of every cell," would itself be reviewable by master eyes. A geometer's solution to problems of containment that were being elsewhere solved by sturdy cylindrical boilers and clever feedback mechanisms, Bentham's Panopticon partook of an Enlightenment admiration for circularity, but like the architecture of steam engines, what remained was the problem of noise, "the only offence by which a man [in a Panopticon cell] could render himself troublesome." To address the noise, Bentham proposed to thread his prison with a maze of tubes linking each cell to the central station and exploiting "the power possessed by metallic tubes of conveying the slightest whispers to an almost indefinite distance," so that officers would be instantly alerted to noisemakers. Obversely, using the labyrinth as a system of speaking tubes, officers could reprimand one prisoner without disturbing the solitude of others. Finally, "If found otherwise incorrigible," a noisy prisoner could be "subdued by *gagging*—a most natural and efficacious mode of prevention, as well as punishment," but Bentham did not try to architect an inhumanly perfect silence. Instead, he used noise as insurance of security, his Panopticon also a Panoticon. There would be a warning bell atop each door through which prisoners passed, another bell to notify the warden of visitors, another in the belfry in case of fire. Most inventively, Bentham would have prisoners shod with wooden shoes, a symbol of servitude in England, cheaper and louder than leather: "By the noise they make on the iron bars, of which the floors of the cell-galleries are composed, they give notice whenever a prisoner is on the march." As much an ear trumpet as a spyglass, the Pano(p)ticon's acoustic functions were reflected in its annular, auricular design, about which he was adamant. A machine for corrective living, what made his Pano(p)ticon metamorphic and self-monitoring was the premise that the more clearly prisoners could listen to their inner rational selves, the greater the marginal utility of conscience; what made this machine self-contained yet potentially explosive was the essential noisiness of living beings held within tight walls. Bentham's schematics for the Pano(p)ticon revealed a canular interiority that resembled less the inversions of an oblong eye than the windings of an ear.[228]

Alack, no Panopticon was ever built, not as a workhouse for the poor or as a prison for juvenile delinquents, not as a *paedotrophium* for rearing infants or as a *sotimion* for restoring fallen women, not even as a *ptenotrophium* for raising hens in multi-storied coops. During the 1820s, Hazelwood School near Birmingham came close to matching the discipline of Bentham's *paedotrophium* under the supervision of another Utilitarian, Rowland Hill, who had "The boys form in classes, march to their places, to their meals, to bed, down in the morning, &c., to [military band] music; the consequence of which is that every movement is made in one-half of the time it formerly occupied, and with one-tenth part of the noise." The music was cued by regular bells, part of a sonic regime that remains familiar: "The bells ring, the classes assemble, break up, take their meals &c with such clock-like regularity that it has the appearance almost of magic." Several new penitentiaries also came close to matching Bentham's model, namely London's Pentonville, and across the Atlantic, Pittsburgh's First Western Penitentiary and Richmond's Virginia Penitentiary; but First Western and Virginia failed on counts of sanitation, Pentonville and Virginia on ease of surveillance. Bentham had managed, however, to make materially evident what had been spiritually central to the work of Hanway, who realized that prison solitude had to be an "imprisonment within imprisonment," walls around walls, chambers within a larger chamber, like the ear itself with its middle and inner ears, and as consecrated to overhearing as to overseeing. This structural principle would guide the design of many later prisons, where for the sake of ventilation Bentham's "air-tightness" was interpreted instead, and logically, as sound-tightness.[229]

Three years after his death in 1832, after his body by direction of his will was dissected for the benefit of medical students and his head mummified under the bell jar of an air pump then bundled for safekeeping inside the thoracic cavity of his skeleton, and after a wax replica of his skull and face, inset with glass eyes, was skewered on an iron spike to sit atop those eighty-four-year-old bones outfitted in one of his usual black suits, and after this mannequin was folded into a chair and propped up by the steadfast companion of his last years, a walking stick named Dapple, and after the ensemble in its entirety was enclosed in a nearly air-tight mahogany case with folding glass doors—three years after, in other words, Bentham had sentenced himself for the rest of his useful days to a personal Pano(p)ticon,[230] open to endless inspection and an intently acoustic introspection, his brain and ears now literally in place of his heart... (wait for the beat)... two British committees set out to anatomize the acoustics of rooms. The first committee was charged by the House of Commons to determine the best shape of a hall for clear exchanges during debate—a pressing political issue for the raucous if Loyal Opposition, which seemed always to be seated in Parliament at muffled lengths from, or at inaudible angles to, the bench of the Prime Minister and his cabinet. The second committee was charged by the Inspectors of Prisons (among them Rowland Hill's brother Frederic) to determine how best to blunt the spoken word—a pressing carceral issue, given that "evil communication" among prisoners vitiated the moral force of solitude.

Led by Sir Robert Smirke, designer of an expanding British Museum and brother to Sydney, architect of its later domed and echoic reading room, the prison committee studied the transmission of speech and sounds through a dozen kind of walls. Their hope was not to make prisons soundless, for total silence was as undesirable as it was economically prohibitive; the solitude of prisoners in separate cells was reenforced by the metallic noises effecting their isolation: a regime of harsh bells from dawn to dusk, the rumble of the wheels of food carts down long corridors, the bray of trays, the "crash of levers and grind of bolts" as cell doors were opened or closed, the pound of hammers at hard labor, the "clang of sabres" against the leggings of the guards, the jangle of keys. The committee's hope was rather to ascertain the most practical form of soundproofing, such that the muttered messages and night whispers of prisoners would be muddied into unfathomable noise, unequal to conspiracy or criminal apprenticeship. Erecting experimental cells at London's Millbank Penitentiary, itself environed by factory smoke and "dreadful groans" from coal-fed steam engines, Smirke & Friends tested different materials, thicknesses, textures. Try 2¼ inches of sand between two 9-inch brick walls: a prisoner on one side can be understood by a prisoner on the other if he talks loud and slow. At Michael Faraday's suggestion, make the inner surfaces of the brick walls jagged, so as to break up sound the way that zigzag surfaces break up light, and keep 5 inches of air between them: a man can still make out speech in adjoining cells, though monosyllables only at sporadic intervals. Add a cloth curtain to the void between the walls: words disappear, but a man can hear the "mere sound of the voice" as well as knocking and whistling, and besides, such soundproofing would be too costly. Try something cheaper, like foot-thick stone walls: talk is surprisingly easy. Try 9 inches of stone, 6 inches of dead space, and 9 more inches of stone, or compacted earth, or or or: what was really needed, the committee decided, was "efficient supervision" and super-audition, since a prison telegraph of taps, coughs, and cleared throats would arise wherever there were the narrowest of conduits for ventilation or plumbing. Smirke's committee concluded that "the Silent System, originating in a deep conviction of the great and manifold evils of gaol association, and designed to guard against those mischievous consequences which attach to it, is cumbrous and intricate in its construction, inadequate to the purpose which it contemplates, and dependent for its successful working upon circumstances which can neither be universally secured nor relied upon." Preference should be given in future to the Separate System.[231]

Separate System and Silent System were American shorthands, for it was in Pennsylvania and New York State that prisons became full-fledged penitentiaries, driven by reasoning that was one-third evangelical, one-third enlightened, one-third republican and wholly hopeful that prison sentences (understood as reprieves from death or maiming) could produce a truly penitent person—so long as there was a "perfect quietude, in which it is intended the prisoner should be kept, and converted, as it were, into a new being," as one of the earliest foreign visitors noted. But it was an imperfect quietude, actually, for some harsh mechanical sounds were as valuable as they were unavoidable, like the grating of prison doors that Benjamin Rush, dean of American physicians, proposed in 1787 to amplify "by an echo from a neighboring mountain, that shall extend and continue a sound that shall deeply pierce the soul." Iconic sounds, then, but no talk. Neither Philadelphia's Separate System nor Auburn's Silent System tolerated speech or its proxies, whistling, humming, or "laughing, singing, and bawling" from inmates, but the Separate System isolated prisoners at all times while the Silent System convened them for mute meals and labor. Justifying the risks of such togetherness by economies of scale, wardens of the Silent System were more physically brutal, resorting to cats-of-six-tails at the least blink of communication between prisoners (a mouthed syllable, a wink, a nod, a moue, a tic) and would immure the most obdurately sociable in solitary. "Let not the voice or face of a friend ever cheer them," pronounced Auburn's Board of Inspectors in 1821; "let them walk their gloomy abodes, and commune with their corrupt hearts and guilty consciences in silence and brood over the horrors of their solitude, and the enormity of their crimes, without the hope of executive pardon." The ultimate punishment for those violating the Silent System was, ironically, the Separate System, whose relentless solitude (wrote critics and physicians) led to masturbation, self-mutilation, and madness more often than mildness, for which reason Bentham and other English prison reformers opted for cells shared by two, three, or four. A defender of the Separate System objected that madness resulted only when administrators denied prisoners the requisites of a healthy solitude—work, light, decent food, exercise, a Bible, and religious instruction; in turn he attacked the Silent System's "atrocious cruelty," by which he meant the frequent floggings and hard labor but which for others meant also the demeaning regime of masked or hooded prisoners marched in lockstep from cell to meals to workbench and back. Properly run, he insisted, the Separate System could be humanely silent, alleviating the "Miseries of Public Prisons" that had deafened such as Philadelphia's Walnut Street jail, with its "frantic yells of bacchanalian revelry; the horrid execrations and disgusting obscenities from the lips of profligacy; the frequent infliction of the lash; the clanking of fetters; the wild exclamation of the wretch driven frantic by desperation; the ferocious cries of combatants; the groans of those wounded in the frequent frays... mingled with the unpitied moans of the sick... and the continued although unheeded complaints of the miserable and destitute." Led blindfolded to cells whose walls were built "so thick as to render the loudest voice perfectly unintelligible," kept in the dark about the outside world, and stalked by turnkeys trained to report every murmur, Separate convicts had nowhere to turn but to the still small voice of the Lord.[232]

After the roaring falls of that divinely natural wonder at Niagara, no site was more visited by European tourists than the penitentiary at nearby Auburn and its Sing Sing spawn to the southeast (built in 1825–1828 by convicts setting "stone upon stone" or, in mangled Indian syllables, "sint sinks"). Foreigners exploring the new nation had often been all ears: the French naturalist Jacques Milbert, who tramped the northeast for a decade, noted in his journals the pound of the caulker's mallet near the docks, the "hoarse clamor" of black crickets, the "wearisome din" of grasshoppers, the thrum of sky-darkening flocks of passenger pigeons overhead, the whirr of a rattlesnake "like that of a watchspring being released," hymns sung with open lungs by Methodists at camp meetings in the woods, their "cries of inspiration" piercing the trees and their morning call to prayer blown through "a long tin horn shaped like a speaking trumpet," all culminating in a trip to Niagara, which from afar sounded like "the far-away rumble of a carriage at night" but up close was deafeningly indescribable, louder than "the roar of an angry sea." When such tourists visited Silent or Separate institutions and came upon countenances "cold, respectful, sorrowful, and calm," the silence cut them to the quick. Basil Hall in 1828 had huffed at the fuss of American firefighting—the "deep rumbling sound of the engines" at two in the morning "mingled in a most alarming way with the cheers of the firemen, the loud rapping of the watchmen at the doors and window-shutters of the sleeping citizens," the violent clanging of church-bells, "a whole legion of boys all bawling as loud as they could"—and had noted disparagingly that "A small amount of discipline, of which, by the way, there was not a particle, might have corrected the noise." At Niagara, the "unusual rumbling noise" alarmed his infant but never this captain of the British Royal Navy, who stolidly differentiated "the loud splashing noise" of the rapids above the falls from the deeper, louder cascade below, and concluded that "the sound of the falls most nearly resembles that of a grist mill of large dimensions." Yet when Hall entered the gates of Sing Sing, he stood dumbfounded: "There was something extremely imposing in the profound silence with which every part of the work of [prisoners] was performed. During several hours that we continued amongst them, we did not hear even a single whisper." Silence—not security, not rectitude—was the engine of the penitentiary: "Silence in fact is the essential, or I may call it, the vital principle of this singular discipline." Hall was cowed by the completeness of the quiet as enforced by guards in moccasins, whose tread could not be heard and whose ears were tuned to "the faintest attempt at communication," abetted by an architectural happenstance that "the space in front of the cells seems to be a sort of whispering or sounding gallery" that amplified all sounds. (Boston's Prison Discipline Society discovered the same at Auburn, where "The space in front of the cells is a perfect sounding gallery; so that a sentinel in the open area, on the ground, can hear a whisper from a distant cell, in the upper story.") Wrote Hall: "I do not remember in my life to have met before with anything so peculiarly solemn as the death-like silence which reigned, even at noon-day, in one of these prisons, though I knew that many hundreds of people were close to me." And at night? Two well-born Frenchmen, sent to study American penitentiaries, felt a silence like death when Sing Sing's nine hundred men retired each evening to their cells. Preachers came to preach in the evenings, for "The truth strikes upon the ear, when the men are sobered by the labors of the day, when no mortal eye sees them, and when the twilight, and the silence, and the loneliness combine in causing it to make a deep impression," said the Boston Society. What made the deepest impression on a prisoner like Horace Lane were the mortal noises of night: spoons rapping on gratings to get keepers to remove the body of a convict who had just died; a coffin hammered shut.[233]

Pittsburgh's new penitentiary, opened in 1826, would be demolished seven years later because it was too noisy, its walls "so built as to admit free conversation between prisoners." Pittsburgh itself, growing by industrial leaps and sounds, was famous even then for its dense smoke from foundries and wood-fired glasshouses, rather than for any residual hush as a frontier town—despite the "deep-mouthed tones" of its watchmen and an ordinance promulgated in 1816 that "If any person or persons shall, within the said city, beat a drum, or without lawful authority, ring any public bell, after sunset, or at any time except in lawful defence of person or property, discharge any gun or fire arms, or play at or throw any metal or stone bullet, or make a bon-fire, or raise or create any false alarm of fire, he, she, or they so offending, shall for every such offence, on conviction thereof, forfeit and pay the sum of four dollars." Such civilian ordinances everywhere went unenforced, but when it came to prisons the enforcement of silence was critical, in part because penology now insisted on solitude, in part because "free conversation" bore old allusions to "a crime which is not fit to be named among Christians," as fitfully ambiguous as "unrestrained intercourse." Regimens of moral discipline through silence and solitude had strongly homophobic overtones, separate cells acting as prophylactics against buggery. These sexual fears were doubled by undertones of racism and a dread of black-on-white sodomy; by 1830, penitentiaries from Maryland to Massachusetts were already seen to be disproportionately populated by blacks.[234]

Seen, and *heard*: South and North, "whitemen" heard "coloreds" along the same compressed acoustical spectrum as they had from the start heard "red" Indians, who were stealthily silent, suspiciously taciturn, outrageously oratorical, or savagely loud. "When we came first a land they made a doleful noise," reported George Percy upon his arrival in Virginia in 1606, "laying their faces to the ground, scratching the earth with their nails," and when they gathered to dance, they began "shouting, howling, and stamping against the ground, with many antic tricks and faces, making a noise like so many wolves or devils." These Powhatans, wrote John Smith in 1612, welcomed prominent visitors "with a tunable voice of shouting" followed by orations made "with such vehemency and so great passions, that they sweat till they drop, and are so out of breath they can scarce speak," but Europeans also noted the Indian courtesy of uninterrupted speeches and long silence afterward, allowing time for each speaker to add whatever had been forgotten in the flourish of the rhetorical moment. Other ornamental flourishes confirmed Indians as sonic contradictions: James Whitelaw in 1773 described a Mohawk brave sporting a long fur down to his waist "at the end of which hung about 20 or 30 women's thimbles, [and] you may easily conjecture what a noise these trinkets made as he walked along," regardless of his moccasins. Well-decorated shamans, or *Pawwaw* (*powah*, Algonquian for "he dreams"), were conflated by whites with the councils in which they spoke and danced, thus called "powwows." And while the whooping of native warriors was so terrifying as to be accounted responsible for the rout of General Braddock's forces at the Monongahela River in 1755 ("the yell of the Indians," confessed one British infantryman, "is fresh on my ear, and the terrific sound will haunt me until the hour of my dissolution"), no less threatening was their "skulking way of war," weaving through woods in quiet moccasins, paddling through rivers in quiet canoes, lofting arrows from quiet bows that could kill from the quiet shadows of trees five hundred yards distant. Taken prisoner, white women and men who survived to tell tales of captivity and "redemption" came closest to depicting the quiet daily routine of Native Americans, all the while tautening the extremes of the acoustic spectrum; in her canonical account of 1682, Mary Rowlandson shuddered at "the roaring, and singing and dancing, and yelling of those black creatures in the night, which made the place a lively resemblance of hell." A century later, as colonists reread new editions of her story, a countervailing European literature of the Noble Savage was driven by the premise that Native American languages lacked the lexicon and syntax to be fully expressive, so noble savages had also to be nobly quiet. Americans could hear both extremes, native silence and native noise, in the popular work of a French emigrant and New York potato farmer, Michel Guillaume Jean de Crèvecoeur, whose *Letters from an American Farmer* (1782) reveal a man on the frontier on-the-listen for war whoops even while admiring the happy society of wigwams, where "Instead of the perpetual discordant noise of disputes so common among us, instead of those scolding scenes, frequent in every house, [visitors] will observe nothing but silence at home and abroad." Indians may have passed their time at a gambling game called *hubbub*, but "Nothing can be more pleasing, nothing surprises an European so much as the silence and harmony which prevails among them," except when compromised by the white man's liquor. Then, abrasively drunk, they might find themselves in jail or prison.[235]

Blacks, enslaved or free, faced white audiences that often as not heard their silences as surreptitious, their celebrations as raucous, their dance and song as animalistic, their antiphonal worship as anything but angelic. Under slavery, silence was a discipline too ostensible. Was it proof of obedience or a subtle reproof, a sign of respect or a suspicious restraint on the part of beings otherwise too childlike to keep such quiet? Though it might come of terror or fatigue, silence burned in the air after the final lashes of a flogging, after the squeal of the last hog slaughtered, after the conch sounded curfew. Night and day, owners and overseers listened through the silences of slaves for signs not of religious meditation but of criminal premeditation, the plotting of an uprising or escape from a Georgia plantation, a colonial New York farm, or a Maryland ironworks: "Run away from the Subscriber, at the Gunpowder Ironworks, Baltimore County, Maryland, on the 28th of August last [1745], a Negro Man, named Squallo, about Thirty Years of Age, of middle Stature, born near Burlington, in the Jerseys, and formerly belonged to Dr. Redman." The subscriber, Stephen Onion, noted Squallo's pea jacket and jockey cap, but the "impudent fellow" could be most readily identified by the sounds he made, for he was "often singing and making a Noise like a slow Trumpet with his Mouth." Even on the run, Squallo was not thought to be able to keep silent; wasn't it the nature of these Africans to make barbarous sounds wherever they went?[236]

Slavery and escape from slavery were both played out along this tensed spectrum of silence and noise. To the significant degree that slavemasters' control of the acoustic environment was key to their sense of security and dominion, to that degree or more were silence and "shouting" keys to redemption for those enslaved. Silence: to avoid the blows of headmen and the beatings of mistresses such as "the wife of master George," who would sit motionless for hours or move "her small fingers over the strings of her guitar, to some soft and languishing air," but at any misstep of her house slaves became a virago, her very voice, "soft and sweet when speaking to her husband, and exquisitely fine and melodious when accompanying her guitar," becoming "shrill, keen, and loud." Shouting: what whites heard as "a hideous Noise" of screams accompanied by pots, gourds, or (forbidden) drums, and what whites saw and felt as a wild stamping, whirling, clapping, and thigh-slapping, was to African-Americans a way out and a way back—affirmation, remembrance, pledge, and prayer. The chanted sermon, congregational call-and-response, and the "ring shout," central features of slave religion by the time of the first penitentiaries, established communities of aural expectation distinct in dynamics and timbre from the most vociferous forms of white religion. Oh, the occasional white preacher, like Charles Giles, a Methodist circuit rider in the area around Niagara Falls, might offer "a Plain Vindication of Sonorous Adoration," conceding that "no loud empty noise ever did, or will do, any good," but protesting still that "when the heart is filled with the power and love of God, and the mouth speaketh out of the abundance of the heart, [the voice] will come like mighty showers of rain from loaded clouds, and will raise a noise like torrents of water falling on dried ground," so that backsliders hearing the word of the Lord as "a voice within a voice" would find that "a dreadful sound is in their ears!—the power of God arrests them in their flight!—cries exhorting!—Christians praying!—praises ringing!" Oh, and the occasional black minister, like "Uncle Jack" the Presbyterian, would oppose "both in public and private, every thing like noise and disorder in the house of God," upbraiding parishioners when they commenced clapping, groaning, and shouting in church, saying, "You noisy Christians remind me of the little branches (streams) after a heavy rain. They are soon full, then noisy, and as soon empty. I would much rather see you like the broad, deep river, which is quiet, because it is broad and deep." These were exceptions common enough to be recognizable on each side of the racial divide, but in African-American settings the position of Old Elizabeth was the rule when she told off a watchman sent to silence her prayer meeting, saying, "A good racket is better than a bad racket" of drunken sinners, and what better racket than to hound noisy sinners into the joyful noise of righteousness? Did not Saint John, in Revelation 19, "hear a great noise of much people in heaven, saying allalujah, salvation, and glory, and honor"?[237]

Escape from slavery reversed the entailments of noise. In pursuit, white hunters with whips, pistols, rifles, horns, snarling dogs, and criers of runaways became the primary noisemakers. Slipping off, slaves had to rely upon those skills at elusive silence by which they had survived servitude and that skill at listening by which they had anticipated vicious moods, skirted capricious commands. James Williams, fleeing an Alabama plantation, counted on his friendship with his master's dogs, who were chasing him down; once safe in their company, he had to muffle the bells on their necks and keep them from baying at every wild animal, then to listen for wolves and other strangers in the night and identify by sound the dash of a deer, the bustle of a bear, and "the chirp of black squirrels in the trees." Louis Hughes, born of a white man but born into slavery and sold from master to master, failed four times to escape from the cracking of bull whips laid upon men so fiercely that "one would have thought that an ox team had gotten into the mire, and was being whipped out, so loud and sharp was the noise!" On one of his attempts, with a handful of other slaves, he thought he was safe until he heard "the yelp of blood hounds in the distance," for which a fellow runaway, Uncle Alfred, had come prepared with a foot ointment of turpentine and onions to throw the hounds off the trail. So anointed, each slave ran in a different direction, Alfred warning, "Don't let the bushes touch you," but he himself "ran through the bushes with such a rattling noise one could have heard him a great distance. He wore one of those old fashioned oil cloth coats made in Virginia; and, as he ran, the bushes, striking against the coat, made a noise like the beating of a tin board with sticks." They were all caught. On his fifth try, after thirty years in captivity, Hughes escaped.[238]

Freedmen and freedwomen in the North, celebrating Independence Day, Pinkster (Pentecost, with the freedom of its apostolic speaking in tongues), or the First of August (the end of slavery in the British West Indies in 1834) with brass horns, tambourines, drums, clarinets, banjos, and "sonorous metals," were heard as essentially noisier than exhilarated whites. Describing a Pinkster gathering where "everything wore the aspect of good-humor," James Fenimore Cooper, creator of the nobly silent "last of the Mohicans," fixed noise deep within the bodies of New York's blacks, thousands of whom had collected in the fields and were "singing African songs, drinking, and worst of all, laughing in a way that seemed to set their very hearts rattling within their ribs." Journeyman whites in the galleries of New York's black theaters made the rude noises they expected blacks would make, to the discomfiture of genteel black audiences center front;

white actors in blackface exaggerated the thick lips, wide-open mouths, drawn-out syllables, and exultant volume of what they heard to be African-American song and dance.[239]

White abolitionists, turning the tables, excoriated Northern whites for the "noise, nonsense, and shams" of their horned Fourths of July—"Horns musical, horns unmusical—horns to march by, and horns to fall by—gunpowder and glory—drumming and drinking—lying in the gutter and lying on the platform"—while escaped slaves were hunted down on New England streets. William Lloyd Garrison condemned "the noise and confusion, the turmoil and uproar, the splendor and squalid poverty of the great American Babylon!" that was New York. "Accursed city! Abandoned of God, and utterly controlled by a satanic spirit!—the central point of all that is lawless, mobocratic, licentious and demoniacal in the land! . . . pandemonium of demagogues, swindlers, ruffians, and rogues of every description! . . . a city crowded with churches, dedicated in solemn mockery to the worship of God, and with public halls which may be readily obtained for the vilest purposes, yet not one of them all can be hired for the use of the American Anti-Slavery Society, lest it should be torn to the ground, and a bloody riot be the consequence." Some of this language was evangelical, some splenetic, and some astutely political, tapping into a deep strain of American anxiety about the path of democracy. "We may please ourselves with the prospect of free and popular Governments," John Adams had written in 1776. "But there is great Danger, that these Governments will not make us happy. God grant they may. But I fear, that in every assembly, Members will obtain an Influence, by Noise not sense—By Meanness, not Greatness—By Ignorance not Learning—By contracted Hearts not large souls." Such "noise" was more than the commonplace "noise of politics" with its inevitable tussles; Adams feared that the voices of democracy could turn too easily (in)to the noise of demagoguery. "There must be a Decency and Respect, and Veneration introduced for Persons in Authority, of every Rank, or We are undone." David Howell, a delegate to Congress in 1782, believed that "all my Constituents have the same right to act for themselves as I have. I wish them to take the same pains to be informed & then to act according to their own minds without being duped, or cajoled by a noise & rattle of words, or overborn by the frightful authority of numbers, or majorities." That noise, however, was it not the *demos* making itself heard? Congress, meeting in New York City in 1785, had asked that chains be placed across nearby streets to reduce the "almost unceasing noise" of passing carriages, a rattle and clatter that disrupted its deliberations, but Congress could not so chain speech as to eliminate the political "rattle of words" that unnerved proceedings well into the nineteenth century, inside stately halls or along urban parade routes. The journeymen who protested a market economy that was breaking up old forms of apprenticeship and craftsmanship, these too were noisemakers, their marches and strikes a white noise sharped by declining real wages and flatted by straitened circumstances, three or four families forced together into a tenement apartment where "pots break, chairs crack, pans ring, and jarring notes of harshest discord rise on every side" in audible proximity to bawdy houses, brawlers, and cockfighting pits.[240]

Unruly, rash, irreverent, indecent, chaotic, rapacious, demagogic—this was how Southerners heard the noise of the North with its traffic and crowding, its loud industry and louder machinery, its labor unrest and thieving gangs, and its anti-Catholic, anti-immigrant, anti-black rioters. As Mark M. Smith has shown in *Listening to Nineteenth-Century America*, antebellum Northerners in turn heard the noise of the South as cruel, arbitrary, unchristian, atavistic, and despotic, and the Northern God, "having His ears continually open to the tears and groans of His oppressed people," He heard it too—young and old chained and flogged as beasts of burden, snarling dogs, hotblooded dueling. Quietude was no less partisan: the South heard the North's silences as the taciturnity of the inhospitable, the secrecy of scheming financiers, the slyness of lawyers, the contraband of the Underground Railway; the North heard the silence of the South as economic sluggishness, rural dullness, physical exhaustion, fear, and loss. When Southern whites sang slow hymns in church, Northern whites heard a duplicitous serenity; when Southern blacks sang with "tones, loud, long and deep," ex-slave Frederick Douglass in the North heard their songs as "a testimony against slavery, and a prayer to God for deliverance from chains." Northern white audiences, who enjoyed the minstrelsy of "Zip Coon," were also taken with the *coloratura* calls of slaves in work gangs and the plaintiveness of the African-American spiritual, while Southern white plantation families preferred "cheerful music and senseless words" from their blacks, whose supposedly irrepressible joyfulness as they weeded fields or shucked corn ("Pull de husk, break de ear") was taken as evidence of a contented workforce—a jollity of song and dance upon which masters insisted when chaining women, men, and children into coffles on their way to slave auctions, jokey affairs full of "hollering and bawling." Such differences in reception and perception of sound on each side of the Mason-Dixon line, argues Smith, helped polarize sectional identity in the decades leading up to the Civil War.[241]

Audited from prisoners' cells, South and North were not so different, and in one regard penitentiaries resembled plantations: Silent or Separate, prisons South and North hoped to recover their expenses through in-house industries run by wardens or private contractors. Although income from visitors' fees at Auburn Penitentiary amounted to $1,942.75 in 1845 and at Ohio's Penitentiary to $1,038.78 in 1844, the real money was expected to come from the forced labor of inmates. Georgia convicts built 371 railcars, the cotton processing plant in Mississippi's state penitentiary turned an annual profit of $20,000, and the nailmaking in Auburn and shoemaking of prisoners in Ohio helped keep those penitentiaries afloat. Elsewhere, when such industries went into the red, as often happened, wardens kept them going. Kept them going despite opposition from churchmen, who demurred that work of any kind led inmates astray from their one true job, penitence; opposition from artisans and tradesmen, who objected that cheap prison labor was unfair competition; from political radicals, who suspected that demands for a stable prison workforce would lead to longer sentences and fewer pardons. Still, convicts almost everywhere were put to work, separately in their own cells, or in workshops, or in chain gangs on outside projects. Penologically debatable and commercially prejudicial, compulsory labor was maintained because it was integral to what had become a primary mission of the prison, the silencing of inmates, less likely to rumble or riot while concentrated on jobs calibrated less by quality than by time.[242]

Serving time, as a silent service in the name of the greater goods of factory capitalism, had a visible impact on the prisoners, whose faces seemed all to wear the same expression. So remarked the English radical Richard Cobden in 1835, who had turned east from the "stunning noise" of Niagara Falls to inspect the penitentiary at Auburn with its new island of cells five tiers high, "the most interesting sight after Niagara that I have been shown in America" and to his mind a "salutary system of moral discipline and reformation." Yet the prisoners' faces bothered him; he wondered whether the sameness of expression arose "from the identity of their condition, as we see a peculiar resemblance of feature or expression in the faces of almost all *deaf* men?" What Cobden was remarking was not some inevitable countenance of deafness (in Hartford's American School for the Deaf, another tourist destination, students signed to each other with vivid motions of hand and face); it was the mute despondency of a tyrannous silence, the hangdog look. As a report of 1842 had it, the prisoner "rises and lies down, goes in and goes out, moves and stands still, begins work and quits work, always under authority, never from free choice, except a choice of evils." That was the Silent System, but the Separate System of the Eastern Pennsylvania Penitentiary sounded much the same to the ears of another foreign visitor: "The dull repose and quiet that prevails is awful," wrote Charles Dickens. "Occasionally there is a drowsy sound from some lone weaver's shuttle, or shoemaker's last, but it is stifled by the thick walls and heavy dungeon doors, and only seems to make the general stillness more profound." Held to such slavish monotony, prisoners at Wethersfield penitentiary in Connecticut passed their days singing to themselves "without audible sound," noted their chaplain. "They can judge of the measure, harmony, and melody of a piece of music, and highly relish its performance, by merely causing imaginary sounds to pass through their minds." So prisoners kept "quiet" while they were kept at work. And the work—could it also keep them sane?[243]

Insane asylums, a carceral form shaped by many of the same constituencies that shaped the penitentiary, were not yet the quiet havens for which the Society of Friends and other reformers had been campaigning since the late 1700s. In the time of John Lilburne, Quakers had quaked with the still-large-voice of the Lord and had endured charges of religious mania. As they settled down over the course of the eighteenth century, they developed practices of personal and mutual "discernment" by which they listened together in silence through the daily clangor of their own lives for spiritual guidance. They applied this model of self-restraint and communal esteem to the running of mills, foundries, and trading companies that eventually strayed from quietist principles, but they retained strong sympathies for the insane and their treatment. With the substitution of penal servitude for execution, public torture, and criminal transport, asylums became in effect the penal system for women, who were sentenced to prison only if "the vilest of the vile," whose brazen vulgarity and lewdness jangled male prisoners, exciting them to masturbation or worse. Unhappy with the disruptive presence of such women but lacking separate cells or matrons to keep them in check, wardens hid them away in jerryrigged quarters without much work, food, air, sanitation, or supervision. At Auburn's crowded women's rooms in a third-floor attic, the English traveler Harriet Martineau found that "the gabble of tongues among the few who were there was enough to paralyze any matron." This was saying something, since Martineau was quite hard of hearing and used ear trumpets for conversation. Back in England, the Quaker Elizabeth Fry had already begun to publicize the plight of women prisoners, founding associations to improve prison conditions and turn women convicts "from clamour to quietness, from obscenity to decency" through education and better sanitation. In 1843, across from Sing Sing in Mount Pleasant, the first American prison built specifically for women, the inmates rioted; the next year a twenty-eight-year-old Quaker, Eliza Farnham, took on the job of matron and restored an overall quiet by introducing the relief of song and prayer, instruction in reading and arithmetic, and half-hours of quiet prisoner conversation. But her reforms on the basis of an "Artistic Maternity" and a belief that "Music, the audible expression of Melody and Harmony, belongs to Woman, by her interior nature," sat uneasily with male overseers who felt that hers was a too lenient, too melodic penal code. This, despite a set of draconian acoustic rules she willingly enforced (given that "the highest external expression which spiritual harmony has yet reached, is the harmony of sounds"): for "Noise," gagging; for "Noise and violence in her room," buckets of cold water dumped heavily on offenders from several feet above; for "Disobedience and noise in her room," twelve days in solitary confinement; for "Noise in her room at night," a nocturnal straitjacket and bread-and-water for a week. Dismissed in 1847, Farnham went to California, where for a year she directed the Stockton State Insane Asylum.[244]

Scouting most carceral institutions between Maine and Virginia in the 1840s, Dorothea Dix counted less than two hundred women in prison. Many more, with bastard child, consumptive, lunatic, epileptic, "feeble-minded," decrepit, or simply vagrant, were in almshouses, workhouses, jails, private asylums, cellars, or the rare mental hospital, put there by husbands or families unable to cope with chronic physical or mental illness and by courts reluctant to send wayward women into dens of thieves, whores, and manslaughterers. When distraught, distempered, disabled, or insomniac, the female sex, being of a "purer spirit" though a weaker constitution (as men averred and women writers oft agreed) would best benefit from asylums redrawn along lines suggested by the English Quaker Samuel Tuke, where "Quiet, silence, regular routine, would take the place of restlessness, noise, and fitful activity." Dix in her thirties saw herself for a while as an invalid, victim of "an *embodied cough*" and headaches, and she would invoke her gender as qualification for her new-found mission at thirty-eight, the "peculiar sensitiveness" of her "woman's nature" allowing her to gauge the humaneness of conditions in prisons, hospitals, and insane asylums, then to press relentlessly for their reform, in North and South. "There is a physiology of the mind, as well as a physiology of the body," she wrote in one of her commonplace books, and it was this physiology, as well as her own abusive upbringing in a hell-fire-and-damnation household, that made her peculiarly sensitive to noise. Visiting Boston's almshouse, where the poor and the insane were herded into barred cages, "all loathsome and utterly offensive," she followed Unitarian and Gothic directives to expose secret evils, most of which were revealed in the darkness, like Edgar Allan Poe's tell-tale heart of 1843, by their sound: "they howled and gibbered, and shrieked day and night, like wild beasts raving in their dens. They knew neither decency nor quiet, nor uttered anything but blasphemous imprecations, foul language, and heart-piercing groans." Unshackled, bathed, clothed, and treated as redeemable, "In a few months, behold the result! recovering health, order, general quiet, and measured employment." Although impressed by the bell-rung orderliness and kindness of New York's Bloomingdale Asylum in 1851, Dix still felt compelled to submit a Supplementary Report urging enlargement of the building "occupied by the more excited, noisy and violent class of females, known as the 'Women's Lodge.'" She could accept their violence and excitement; she could not accept their noise: "Perhaps the most crying defect in the architecture of the present building is the entire absence of any arrangements whereby the noise of excited and vociferous patients can be shut off from those who are quiet, or would be, but for the noise of others & [who] are greatly molested & injured by loss of sleep by night & unquiet by day. Vociferation or other great noise in any one room in the building can readily be heard in all the other rooms," and "I do not a whit exaggerate," she continued, "when I say that during the warm season when patients are most excited & noisy, no individual can sleep therefore on an average of five nights out of seven." Yes, a few "demented oblivious persons" could sleep through "as perfect a pandemonium as was ever witnessed on the earth," but "As I now write a patient leaves us who has prevailed upon her husband to remove her chiefly on account of the suffering she has experienced from loss of sleep & hearing almost constantly by day & by night blasphemous vulgar & invective declamation."[245]

Bloomingdale's directors responded within sixty days by recommending as "urgently necessary" the construction of new rooms "suitable for the confinement of noisy patients by themselves in apartments connected with the main edifice yet at one of its extremities and facing outward. These rooms being separated from the main building by a hall and double doors, would not permit the outcries of the excited to be heard within the other wards, because the sound would be reflected into the open space beyond while the noise made in the lodges is now wafted toward the rear of the asylum and therefore penetrates into every window along the reflecting surface or wall." The alacrity of their response was a credit to the power of the words of a woman who during her career would prompt the construction of thirty-two new mental hospitals in fifteen states, but the detail of the directors' plans for noise control reflected a larger cultural engagement with the nexus between character and soundscape as managed by architecture. Thomas Kirkbride, a medical man who had been educated in Quaker schools and who took his first residency at a Friends' Asylum, had become by 1854 one of the major American advocates for a new architectonics by which to change habits of force into the force of quiet habit. The thick outer walls of his ideal asylum would be ten feet high, enclosing rustic acres whose interior security features would be masked by evergreen trees. The asylum buildings would be divided into five, seven, or eight distinct classes of patient of each sex, with the better(-off) classes in wards on upper floors near the center of the complex, which in its grandness was intended to assure a favorable prognosis for patients and to express the "generous confidence" of a new profession, psychiatry. "Everything prison-like" must be avoided. Bolts and locks would be inconspicous, forged to close noiselessly. Kirkbride would have ticking clocks everywhere but no gaping, guffawing, fee-paying public. Noisy, violent, or suicidal patients would have individual rooms, since "a single incurable can turn an entire ward crazy" and the impact of excited patients upon "the comfort, quiet, and usefulness of the establishment, cannot be too highly estimated." The far side of one wing would house this "most noisy class," and somehow the speaking tubes in each ward, the hard plaster on the walls, and the hidden canals of sewage pipes and ventilating flues would not amplify the more egregious sounds of patients, due perhaps to the calming presence of smiling attendants or the cheerful domesticity of the interiors.[246]

Far cries from prison hulks—decaying warships that held overflows of convicts in reverberant quarters below deck—the new, comparatively dry and quiet asylums and penitentiaries failed less on the grounds of architecture than of anticipation. Hulks had been a vicious tease hinting at transportation and release into the open air of, perhaps, Tasmania. Penitentiaries and reformed asylums had begun more virtuously with promises of reclamation that appeared reasonable so long as the numbers of convicts and mentally ill did not grow at a rate comparable to the general populations of a more productive, less plague-ridden Europe. But Euro-American nations were incessantly at war with each other or with indigenes resisting empire, and at the end of each war between 1760 and 1816, prison hulks were inundated with demobilized soldiers whose skills at rapine on the march and mawling on the battlefield were useful back home only (it was said) for brigandry and brawling. In the wake of further demobilizations, of industrial displacements of farm laborers, and of the scattering of refugees from the suppressed revolutions of 1848, prisons and asylums on both sides of the Atlantic were flooded. Kirkbride had been confident enough of the effectiveness of his asylum architecture to sketch out a General Collection Room in which a maximum of 250 patients would calmly assemble for church services, concerts, and gymnastics, but over the years most asylums and penitentiaries became lower-case "general collection rooms" for large numbers of adults and youth who could no longer be kept Separate or Silent.[247]

Europeans, who considered Americans on the whole to be a loud, boisterous people and were all the more moved by the quietness of American penitentiaries of either system, cobbled the two systems together for their own prisons. Daniel Nihill, chaplain to London's hybrid Millbank Penitentiary, acknowledged that the Silent System had the laudable goal of preventing moral contamination among prisoners, but in practice, "In spite of the most vigilant superintendence, [prisoners] will sip enough of communication to defeat discipline and to subserve bad purposes, while the hindrance of harmless conversation excites a sense of injustice, or confounds their moral notions, taking off the conscience from real guilt." In other words (and other words there would always be: that was his point), the rule of silence made silence too ostensible. Inflamed by such disproportionate punishments as six lashes for laughing or a week on bread-and-water at Canada's Kingston Penitentiary for "making a great noise in a cell by imitating the barking of a dog," prisoners would leave off meditating upon their own crimes to focus upon more immediate hurt. Of course, if the wills of prisoners were always in unison with those of wardens, "a prison would form a beautiful specimen of moral order, and would indeed be a school of virtue. But," sighed Nihill, "it is otherwise," and noise would out—sobbing, coded tapping, nightmare cries, masturbatory groans. Better some noise than "gall and bitterness" sapping all inclinations toward penitence. So Millbank wardens imposed stiff silences during the day while convicts were assembled for work, then allowed soft conversation during the evening, when prisoners in twos or threes were back in their cells.[248]

Crowded with four or five, as prison cells often came to be, Separate was less than separate, Silent less than silent, and neither of the twain were mete. Nihill argued that the social, spatial, and emotional privations of the penitentiary were hardly severe enough to impress penitence upon men of the laboring classes, from which most inmates came. Being sent to a penitentiary might be "cheerless and depressing to an individual habituated to the comforts of a peaceful home," that is, to an individual of the comfortable classes. "But the majority of offenders have no peaceful home to contrast with it." In architectural terms, he was right: the average size of an Irish cottage or a sod hut on the American prairie, like the floor space occupied by a family in a shared tenement flat in London or New York, was about the same as that of a prison cell, except that prisoners were provided with steadier food, cleaner water, better air, and did not sleep three or four to a cot. The starkest difference between the digs of the poor and the digs of prisoners, aside from the absence of a few furnishings (a mirror, a ticking clock) was solitude, with its accomplice, silence. Gothic novelists and evangelical reformers had made solitude hauntingly horrifying, and Silent System penitentiaries thrust new prisoners into solitary confinement for the first months of their sentences so as to rasp away rough edges before being brought into the mute society of other inmates. For men and women of the "dangerous classes," such solitude must have been shocking, experienced by some as a monstrous isolation, by others as a welcome luxury. Once Gothic tales and evangelical tracts lost their grip on the carceral imagination, it became unclear whether the "self-adjusting principle" of solitude or the complex machinery of silence could spur a remorse deep enough to return prisoners to their better selves. But even as many American penitentiaries became "reformatories," abandoning solitary confinement except as a substitute for the lash, and even as European penitentiaries became so congested that inmates could not be rigorously audited, silence was still invoked as the prime directive. And even as the guards of crowded prisons admitted that silence could no longer be effectively maintained, silence remained the first of the rules posted in each cell, and prisoners themselves felt the impress of that rule, however weakly enforced: "There is a strange humming in my head," said one inmate of the 1920s, "caused by this eternal quietness." Shorn of its distinctive American resonances with slavery and racial segregation, silence in German, Swiss, French, Italian, British and Irish prisons remained a potent signifier of what it meant to be unfree.[249] In a modernizing Western world that was prone to hear noise as ubiquitous and to appreciate it as elemental not only to urban trade and heavy industry but to the climax of sex and the physics of music, an imposed silence was finely emblematic of exclusion and weakness.

14. Stethoscopies of Stertor and Râle

Bodies too quiet, as cellmates knew too well, were all too weak. In the fall of 1816, while Shelley was taking Frankenstein's monster through the education of his senses and the "year without a summer" was taking its toll in cases of phthisis and catarrh, Dr. René Théophile Hyacinthe Laennec was listening to the noises of a body through a tightly rolled quire of note-paper that he had pressed to the breasts of a heart-stricken woman. Close listening-in upon prison cells through pano(p)tic tubes, as proposed by Bentham, had been a solution to penitentiary problems of conspiracy and noise; close listening-in upon ailing bodies, as proposed by a Viennese physician in 1761, was a solution to the medical problem of grave illness arising from unseen causes. Expanding upon the long-neglected practice of auscultation, Leopold Auenbrugger had thumped patients while pressing an ear to the back or the chest, then drawn diagnostic inferences from the strength of the returning sounds. Not until 1808, however, when Laennec's mentor and Napoleon's doctor, J.-N. Corvisart, turned the ninety-five Latin pages of Auenbrugger's *Inventum Novum* into over four hundred pages of commented French, did physicians across Europe begin to practice "percussion" and listen for its repercussions.[250]

Thumping a patient had been known to Hippocrates and to Roman experts on heart disease and bronchitis, but European physicians had been reluctant thumpers for reasons scabrous, social, and semiological. Scabrous, because there was something indecent in bending close enough to a (female) patient to hear the returning sounds. Social, because in their long campaign for professional respect, physicians had held themselves apart from barber-surgeons who plunged elbow-deep into patients' bowels while they, educated gentlemen, stood above the messy particulars, watching from a distance that allowed for generalization. Semiological, because the signs that doctors had learned to remark—color and moistness of skin, texture of tongue, sharpness of breath, smoothness of gait, clarity of sputum and urine, odor and consistency of feces—could be consolidated into diagnoses only through listening to the stories of their patients' symptoms, which demanded a different set of skills from those of listening-in on the body itself. Rather annalists than analysts, European physicians had listened for narrative and character; they gained the trust of patients by attending to the tales of their ailments and, with a few deft questions, arriving at a meaning. Some might take the pulse, feel for fever, and palpate a tumorous mass or distended abdomen; most rarely touched adult patients, drawing instead upon personal archives of cases and their skill at inquiry from a respectful, respectable distance. From a respectful distance, they assessed the gravity of a cough; from a respectable distance (sometimes by post, as did Erasmus Darwin) they made out prescriptions and dispensed prognoses.[251]

Consider what it would mean, then, to lay one's ear upon the body of a man or woman. Auscultation narrowed the distance between patient and doctor and changed their aspects: the closer the doctor approached, the more he had to look away from the patient and the more the patient confronted the unmoving pinna of a doctor's ear rather than a pair of observant, responsive eyes. Momentarily, the ear became the physical matrix of an exploratory communion, and momentarily, patient and doctor became an oddly conjoint twin, doctor's head pressed to patient's breast. Although this "immediate" listening was often decorously muffled by layers of clothing that rustled between doctor's auricle and patient's skin, and although auscultation was therefore practiced with the professional ear at some remove from the pained body, it was kinaesthetically, often erotically, charged. Novices might mistake the swoosh of a silk garment for pneumonia and veterans might argue over vagaries of tone, but their bending toward a patient would be felt and understood as a serious undertaking. The only other time a physician ordinarily approached so closely to a patient was at the deathbed.

Relying upon the audibility of internal resonances beyond conscious control, doctors could hear sounds their patients did not know they were making. Like autopsy, auscultation was a form of mining, but while autopsy scraped at lesions, auscultation unearthed hidden processes, yielding a series of sounds that could be construed into a counter-narrative or taken as proxy for the voices of comatose, mute, or delirious patients and speechless infants. The percussing doctor was tapping into an involuntary sound system that ran on a separate track from patients' volunteered narratives, confirming or belying tales twice-told—or revealing untold jeopardies of lung and heart. "I have learned from much experience," wrote Auenbrugger, "that diseases of the worst description may exist within the chest, unmarked by any symptoms, and undiscoverable by any other means than percussion alone." This was a radical claim, defying traditional protocols that hinged upon a patient's recital of symptoms. The poles of discourses were reversed: a physician would trust first to an acoustic realm he alone could summon, secondarily to the words of a confiding patient. It was the difference between listening-in to a body and hearing a person out. Auenbrugger's First Observation ignored testimony, sped straight to the body: "The thorax of a healthy person sounds, when struck." The ensuing reverberation resembled "the stifled sound of a drum covered with a thick woollen cloth" (as with the muffled drums of funeral processions), but it was distinct enough to be diagnostic. Admittedly "more distinct in the lean, and proportionately duller in the robust," these reverberations were almost inaudible in the obese. Method therefore mattered: the patient's blouse must be drawn tightly over the chest so that the ausculting ear be not deceived by ripples of cloth, or the physician must use a glove, for "If the naked chest is struck by the naked hand, the contact of the smooth surfaces produces a kind of noise which obscures the natural character of the sound."[252]

Clearly an ear of no mean sophistication was needed to tell such noise from natural bodily sounds and to catch fateful inspirations, expirations, and retentions of breath. Librettist in 1780 for a comic singspiel composed by his friend Antonio Salieri, Auenbrugger had an ear sufficiently sophisticated to write a virtuoso aria about the limitations of operatic virtuosos, who should not be expected to be equally proficient at *cantabile*, *falsetto*, *crescendo*, *tenuti*, *staccato*, *tempo rubato*, etc. His libretto was panned by critics as "wretched from beginning to end, miserable, and nonsensical," but his aural talents were more happily applied to identifying "languishing legatos" and "yearning diminuendos" in patients with consumption, pericarditis, scirrhus of the lungs, and nostalgia, that wasting disease of young men forcibly impressed into a lifetime of military service and grieving so deeply for home that eventually their moaning lungs clung to their pleura, which Auenbrugger could hear as a "preternatural sound on one side of the chest." The duller the sound, the more deeply imbedded the disease; the weaker the percussed sound, the thicker the obstruction from morbid fluids or inflamed tissue. Total destitution of natural sound was a fatal sign, implying a body whose organs were infested or enlarged and had no resonant spaces.[253]

Aside from the beating of the heart, the rumble of an empty stomach, the belch of a full stomach, or the squelch of a fart, the healthy body had long been presumed to be a quiet body. Not deathly silent, but quiet. Noises had traditionally been signs of disorder, whether the flatus and abdominal murmuring of "dropsy of the chest" as heard by Auenbrugger *before* percussing, or "a hissing and crackling noise in the middle of the chest around the joints where the ribs meet the sternum," as heard at arm's length from a blacksmith in the care of Dr. William Brownrigg in 1737, whose bread-milk-and-butter poultices failed to prevent the smith's death two months later from "asphyxiation by purulent material." Sick bodies gave off other noises loud enough to be heard without the provocation of a thump, such as the aortic regurgitation that drew the attention of Dr. James Douglas to a patient at St. Bartholomew's Hospital in London, "And which is almost incredible, sometimes the trembling and throbbing make such a noise in his breast, as plainly could be heard at some distance from his bedside." In 1809, as Auenbrugger lay dying of pneumonia, Allan Burns of Glasgow reported hearing "a hissing noise" at some distance from a patient with mitral regurgitations. Doctors also incorporated into their diagnoses the noises they had to listen through as patients recounted their ailments in clauses punctuated by stertorous or whistling breaths, raucous coughs, repeated wheezes or short gasps. Given the obviousness of these noises as contrasted with the finesse of auscultation (for which, like Auenbrugger, one must have learned "the modifications of sound, general or partial, produced by the habit of body, natural conformation as to the scapulae, mammae, the heart, the capacity of the thorax, the degree of fleshiness, fatness &c. &c."), and given the increasingly refined noses of physicians reluctant to come too near unwashed patients with putrid sores or rotten teeth, Auenbrugger's work went unappreciated for half a century. Even Rozière de la Chassagne, a Montpellier physician who in 1770 turned the *Inventum Novum* into abridged French, swore neutrality as to the value of this new method, which he himself never put to the test. Despite the rise of clinical medicine with its avowed interest in physical examination, few patients before 1800 had ever felt the brush of a physician's ear, and for those many who had been cupped and bled, their skins would have been too painfully tender to suffer the gentlest tap or softest pinna. Corvisart himself, dedicated as he was to "the medical education of the senses," only bent near his patients, preferring to place an open palm on their breasts to feel the quality of vibrations (or *frémissements*, "thrills") while percussing with his other hand. François Double, in his *Séméiologie générale* or *Treatise on Signs and Their Value in Illnesses* (1811–1817), did urge physicians to shake off the ridicule associated with the gyrations requisite to putting an ear squarely on a patient's chest, for there was much to learn from unmediated listening to the palpitations of the heart, the vagaries of the voice, and such breath noises as that plaintive respiration whose sighing resembled the sounds made by bakers kneading dough. But when G.-L. Bayle, another of Corvisart's students, did go all the way, like his younger friend Laennec, both found that immediate auscultation had serious limits even for recognizing the stages of consumption upon which Auenbrugger himself had focused and from which, in May 1816, Bayle died.[254]

Five or six months later, Laennec took up what was essentially a tube that acted as a wave guide and physicians thenceforth would hear how noisy were the interiors of every living body, healthy or not. Here was a clinching moment in the history that I have been tracking, the history of the notion and experience of the ubiquity of noise. With the arrival of the stethoscope, noise that had been impinging from every other direction was found to be domiciled in, and revelatory of, man's innermost parts. Or rather, to begin with, woman's innermost parts, as Laennec explained in the first edition of his treatise on "mediate" auscultation. Consulted during the fall of 1816 concerning a young woman presenting with symptoms of heart disease, he had been stymied by her innocence, gender, and plumpness, which made palpation and percussion inexcusable or unintelligible. It occurred to him that "if the ear is placed at one end of a log, the tap of a pin can be heard very distinctly at the other end," a "well-known acoustic phenomenon" that had been exploited earlier by William Wollaston, who had measured the vibration of his own muscles by sounds (as of a carriage rumbling over cobblestones) translated from his leg through a notched stick to his cushioned ear. Semi-spontaneously, Laennec "took a paper notebook, made it into a tight roll, one end of which I applied to the precordial region, and, putting my ear to the other end, I was just as surprised as I was satisfied to hear the beating of the heart in a manner that was clearer and more distinct than I had ever heard it by the direct application of the ear." He did not say what treatment the woman received as a result of this historic examination, of which no clinical record has ever been found, and he would be accused of devoting more time to detecting noises than effecting cures, but Laennec's stethoscopy spread quickly among physicians, for it—and he—promised to resolve the problem of correlation between the diseases of a malfunctioning body and the imbedded pathologies of a cadaver.[255]

Anatomy and "active hearing" were key to demonstrating these correlations. Like Auenbrugger, Laennec had a passion for music, lyric poetry, and anatomy; he had taken up the flute before entering his uncle's medical school in Nantes at the age of fourteen, where he began anatomical studies in a dissecting room directly and audibly beneath the sick ward and across from chicken coops and a pig sty. Like Auenbrugger, he made his surgical way through military hospitals, in his case noisy field hospitals. Like Auenbrugger, he sought to predict, from the sonic course of an illness, the damage likely to be found at autopsy. Like Auenbrugger, who had heard his innkeeper father thumping at casks to inventory beer and wine, Laennec would opt for acoustic clues to what was going on inside live bodies, listening-in according to the mode of "active hearing" that had been defined by Régis Buisson, a devoutly Catholic and brilliantly catholic twenty-five-year-old finishing up at the École de Santé in Paris when Laennec arrived there at age twenty to complete his own studies. Jacalyn Duffin, Laennec's recent biographer, has shown how basic was this concept of "active hearing" to the young and pious Laennec, who in 1802 published a detailed review of the doctoral thesis in which Buisson revamped the system of physiological classification so that inspirited aspects of being were given their due. Assigning a passive and an active component to each organic function and sense, Buisson defined passive hearing as mere *audition*; active hearing, which engaged both mind and ear, was *auscultation*.[256]

Auenbrugger's auscultation had been keyed to the percussions of the new pianoforte, at which two of his daughters would become so adept that Haydn in 1780 dedicated a set of keyboard sonatas to them, "for their way of playing and genuine insight into music equal those of the greatest masters." Composed several years earlier but now marked up with dynamics that could be achieved only at the piano, these sonatas marked Haydn's own turn away from the harpsichord and clavichord. The pianoforte, that "soft-loud" instrument invented in 1700 but deemed unseemly in public until the 1760s (a "cauldron-maker's instrument," wrote Voltaire), began to attract composers as well as players during the next decades as spring-loaded levers were added to each key so that hammers would not rebound onto the strings after a keyboard barrage. With harpsichord and virginal, there would have been no point to vehemence since these instruments, whose strings were plucked not struck, had little capacity for changes in tonality or loudness. The strings of a clavichord were struck with limited and almost invariable force; apt for accompanying singers, it could be coaxed "from barely audible to extremely soft," so it was useless for instrumental ensembles or concerts in large halls. The piano, on the other hand, responded strikingly to *fortissimo*, *crescendo*, and *diminuendo*, soon augmented by pedals that allowed for notes to be cut short, sustained, or softened at the nudge of a toe. Adaptable in volume and timbre to halls great or small, the piano also and easily accommodated other concert instruments and solo voices. More sensitive to the pressures of a fingertip and the torque of a wrist, for that very reason the piano responded with a greater musical inconsistency. Under amateurish fingers, its sound became cloudy or dynamically listless, losing the crispness that even a novice could muster at a harpsichord. And the more awkward or agitated the pianist, the more quickly the piano went out of tune. These vices and virtues of piano percussion bore analogy to medical percussion, which took the body as a sounding case that was passively responsive to the knowing thumps of a physician. Laennec acknowledged that immediate percussion, as taught to him by Corvisart and Bayle, was useful as a first pass at detecting lung and abdominal conditions and discriminating between fluid and solid blockages. However, like performances on a banged-up piano, percussion could be confusing or misleading where the chest was deformed by rickets or obscured by layers of fat. Worse, percussion was least informative where the interior of the body was most naturally dense, around the heart, and its repertoire of elicited sounds was too coarse to tell apart any number of related conditions. And worst, slight differences in the angle, pressure, or timing by which one tapped at the thorax could produce differences in resonance that were artefacts of style, like the idiosyncracies and inconsistencies of a pianist at her keyboard. Such artefactual resonances were exacerbated, like piano concerts in halls with low ceilings, by the acoustics of the rooms in which percussion took place: ausculting a patient in a large open room yielded sounds quite unlike those from the same patient when enveloped by bedcurtains or leaning forward in an armchair in the middle of a small room full of overstuffed furniture (the better to damp the sounds of a piano?)—or within the bare walls of a hospital ward, even those now distanced from operating theaters so that patients would not hear the cries of those under the knife.[257]

Laennec's "mediate auscultation" operated under the sign of the flute, the instrument most closely associated with the human voice. His first "cylinders" were ten to thirteen inches long and looked and felt like half-size baroque flutes. His first title for the cylinders, *le pectoriloque* or chest-speaker, referred back to a 1752 text by J. J. Quantz, a virtuoso German flautist whose famous *Essay of a Method for Playing the Transverse Flute* compared its ideal sound to the tones of a contralto or what singers called the "chest voice." By the 1790s, when Laennec picked up the instrument, French flutes were being tuned toward the soprano and French flute teachers were tut-tutting the hissing noises of the Quantz school of embouchure, but flautists continued to strive for a calm chest action and to model their sound upon vocal tones in song and speech. In practice, Laennec's *pectoriloque* was neither an end-blown nor a transverse flute; it was a flute in reverse, in the same way that an ear trumpet was the reverse of a speaking trumpet. This reversal was vital to "active hearing," for the noises Laennec heard through *le pectoriloque* had to be understood as autonomous, issuing unprovoked from within living bodies. Only then could ear and mind catch hold of them as the sounds of internal organs in natural or hampered motion. As a flautist and asthmatic, Laennec was most attuned to passages—that is, to the melodic flow and harmonic release of air and other fluids. Auenbrugger and Corvisart, percussionists, had concentrated on volume, treating the living body as if it were a maze of communicating caverns carved out around a rhythmic heart. Their auscultation was but a thump beyond the "succussions" practiced by Roman doctors who had patients lifted up and shaken in order to assess the quantity, viscosity, and asymmetry of obnoxious fluids. Laennec listened-in upon the body sounding off on its own, speaking out.[258]

Literally. Laennec asked his patients to speak while he listened through his *pectoriloque* for changes in the passage of sounds through chest and lungs, then made assessments in terms of *pectoriloquie*, the brightness of a voice that "seemed to come directly from the chest and pass intact through the central canal of the stethoscope," or *egophonie*, the "natural resonance of the voice in the bronchial branches, transmitted by the intermediary of a thin, trembling substrate of effused liquid, and become more sensitive due to the compression of pulmonary tissue." The former could be heard clearly with *le pectoriloque* poised over a specific square inch of the chest;

the latter, audible across the thorax, resembled the bleating of a goat, or the nasal ventriloquy of a carnival puppeteer's Punch and Judy show, or "as if a kind of silvery voice, of a sharper and shriller tone than that of the patient, was vibrating on the surface of the lungs, sounding more like the echo of the voice than the voice itself." These curious inner voices were not the voices of conscience but of prescience; they were prior and preeminent carriers of meaning because they sounded the body from the diaphragm through lungs to throat and mouth before a patient ever recited a symptom. They were related to another newfound voice, that of the siren—not the call of a seductive, devouring, aquatic half-woman but something "analogous to a human voice" that issued from an invention by Charles Cagniard de la Tour, an engineer who had already built a device to calculate the number of vibrations corresponding to any pitch. In 1819, the same year in which Laennec published the first edition of his treatise on mediate auscultation, Cagniard de la Tour announced his *sirène*, so-called because it could produce as much sound underwater as above, using either currents of air or water to rotate a perforated disk; the faster the rotation, the higher the pitch. Just as Laennec's reverse flute was not for music-making but medical detection, so the siren, shrill though it could be, was not yet for noise-making but for taking acoustic measurements.[259]

Post-horns ("heart-shaking, when heard screaming on the wind"), pianos (heavier, with iron frames and sometimes a "Turkish music" pedal that unleashed triple brass bells), flutes (longer), orchestras (larger), artillery (more powerful), steam engines (heavier, longer, larger, and more powerful)—so much in Laennec's world was becoming louder that the voice was in danger of being overwhelmed. Whereupon people raised their voices in giant chorales during and after the French Revolution, in North American camp meetings and the shape-note singing of Protestant congregations, in street cry and hortatory above the grumble and thunk of much heavier drays, above the clatter and ringing of new horse-cabs with leather belts of bells, and the snap of coachmen's whips over thicker traffic—the "most inexcusable and disgraceful of all noises," wrote the German philosopher Arthur Schopenhauer, "this sudden, sharp crack, which paralyzes the brain, rends the thread of reflection, and murders thought... much as the executioner's axe severs the head from the body." Recording *The Miseries of Human Life, or the Groans of Samuel Sensitive and Timothy Testy* in 1806, Rev. James Beresford had Sensitive change his London lodgings "to escape from a brazier at the next door, who counted his profits so very distinctly upon the drums of my ears" deploring the "Bombalio, clangor, stridor, tarantantara" of the city, with its "savage jargon of yells, brays, and screams, familiarly, but feebly, termed 'the Cries of London.'" Despite a French translation of 1809, Laennec never mentioned Beresford's satire, but in the new silence required of examining rooms for the undisturbed exercise of *le pectoriloque*, Laennec found that the breathing of young children was normally and normatively louder than that of adults, a result (he wrote) of their need for more air to more fully oxygenate their blood. It was consistent with sensations of a louder world that a stethoscopic loudness should characterize the voices of the vital young well before they could speak.[260]

Asthmatic breathing could be far louder, audible at twenty paces. When, however, Laennec put one ear to his instrument and stuck a finger in his other ear, he heard less noise from the lungs of asthmatics than from the healthy or from those who had not known how ill they were. The power of the "stethoscope" or "chest-explorer" was, so to speak, its short-hearing, its intentness upon subtle internal noises that could be posthumously correlated with visible pathologies during autopsy. In order to hear these noises, Laennec had to play a nice duet with his subjects, getting them to talk clearly, breathe regularly, cough on command, or hold their breaths—much like the breath control exercised by a good flautist—while he held the cylinder of the stethoscope in place, one end in perfect contact with the patient's skin, the other end pressed securely within the tragus of his ear. His eyes, directed away from the patient, and surmounted always by horn-rimmed glasses, had to be a flautist's eyes, glancing obliquely toward his fingering as he laid his instrument on one site and then another, listening for a *râle crépitant*, a *râle muqueux* or *gargouillement*, a *râle sonore sec* or *ronflement*, a *râle sibilant sec* or *sifflement*, a *bruit de râpe*, a *bruit de scie*, or a *bruit de soufflet*.[261]

Naming such noises engaged Laennec in extraordinary acts of simile meant to transform each stethoscopic *bruit* into a cognizable sound. The wheezing crackle or *râle crépitant* of the first stage of peripneumonia was like the noise made by raw salt when heated in an evaporating pan. "It would be difficult to describe it any better," he apologized, "but after having heard it just once there can be no mistaking it." (His English promoter, John Forbes, did mistake it, translating *râle* as a rattle rather than a wheeze.) The mucosal gurgling of pulmonary catarrh, which was audible to the naked ear as a "death rattle" but could be earlier and more accurately pinpointed by stethoscopy, required no elaboration, but a dry booming *râle* corresponded to the sound made by air blown into animal bladders, or to a man snoring, or so exactly to the cooing of a wood pigeon that you would think a bird had nested under the patient's bed. Sibilant whistling from obstructed bronchi reminded him of the chirp of a small bird and the slick squeal made when layers of oiled marble slabs were pulled brusquely apart. The ringing quality of the cough of a patient with lung disease sounded like "a fly buzzing in a porcelain vase." As for the heart, wrote Laennec, the ear of a musician would trump any hand or eye at discerning cardiac rhythms, and with a wooden cylinder at his ear a flautist physician could hear the *choc* of the heart even when a palm laid on the chest felt no quiver.

Although the healthiest were the quietest, all beating hearts made two sounds, the first clear and swift like the click of a valve on a bellows, the second slower and duller, like the lapping of a dog. Noisier, more troubled hearts produced a sound like a rasp or a saw, or "like the sighing of the wind through a keyhole." Moving his instrument up to the carotid artery of a consumptive woman, Laennec was surprised to hear a melody, which he annotated with notes rising and falling on a stave, and which, "at the risk of employing a bizarre comparison," was "like the sound of military music, on a march, every now and then interrupted by the hoarse roll of the drum."[262]

Some noises he did not know what to do with: the pericardial friction rub, aortic regurgitations. Some he misunderstood: a hum like that of a seashell at the ear—was it arterial or venous? a leathery creaking like that of a new saddle—did it come from the lungs or the heart? Some he got all wrong: the beats of the heart, which he reversed, and heart murmurs, on which he reversed himself. And some he missed: fetal heartbeats, reported without fanfare in 1818 by a Genevan doctor who had unabashedly put his ear to the bellies of women in childbed; fetal heart rhythms and the placental *souffle*, or murmur, reported in 1821 by Laennec's friend and eulogist J.-A. Le Jumeau de Kergaradec, who on one knee during a proper obstetric examination had been startled to hear through his stethoscope the *tic-tac* of a watch coming from the abdomen of a gravid, erect, and fully dressed Mme. L., soon understood as the combined beats of fetal heart and maternal pulse; and the whizzing sound of the fetus during labor, which the New York physician Isaac E. Taylor would compare to the sound of ice expanding on a cold day.[263]

Over the next fifty years, as the handiness of the stethoscope was improved by, among others, Taylor's friend George Cammann (who in 1852 designed one of the first binaural instruments), clinicians piled metaphor upon simile to make sense of all the new noises they heard and to complete the catalog of noises begun by Auenbrugger, Corvisart, and Laennec. The greater the apparent reach of noise inside and out, the greater the drive toward a full accountancy of sound. This was one of the grand cultural enterprises of the nineteenth century, the scientific isolation and poetic triangulation of noise, sound, and music. Chladni had taken their measure as much optically and haptically as arithmetically, running a violin bow across the edges of glass and brass plates to produce an intriguing geometry of vibratory figures that Sophie Germain and then Charles Wheatstone formalized in equations of planar and curvilinear elasticity. Cagniard de la Tour took their measure aurally and mathematically with the siren, whose fundamentals and overtones would be differently heard and differentially analyzed in the 1840s by August Seebeck and Georg Ohm, then by Hermann von Helmholtz and William Wundt. Charles Bell and Johannes

Müller took their measure neurologically, locating acoustic sensations in a specific set of nerves. Laennec and his successors took the measure of sound physiologically and poetically, their reports splendid with metaphor through which the body's secret noises were magnified and prognosed into intelligible sounds.[264]

Reporting on four cases of heart murmurs, one "exactly resembling the cooing of a dove," Prof. John Elliotson reminded his London audience in 1830 of an earlier virtuoso of sound, Robert Hooke, who had heard the beating of a heart and dreamed of "otoacousticons" through which one might listen-in on the universe at every scale. Stethoscopists, applying their otoacousticons to head, neck, heart, lungs, abdomen, and intestines, arrived at congeries of sound that were still dreamlike, as perplexing as they were minutely specific. John Forbes, examining a blacksmith in 1822, heard under the right clavicle "the grating noise of a saw with fine teeth, or that produced by the forcible action of the piston in a large, empty syringe," two decidedly different noises; Thomas Hodgkin in 1828 described what would become a key murmur for cardiologists as "a purring, thrilling, or sawing kind of noise," three different noises; and William Stokes, paraphrasing Laennec, described the vaguely metallic sound of pneumo-thorax (air or gas in the pleural cavity) as a "drop of water falling into a deep vessel, a grain of sand into a glass, or a pin striking a porcelain cup." In 1835 J.-B. Bouillaud, imitating Laennec, used musical notation to capture the rhythmic tones of a heart "gallop"; he also found valvular fibrosis to have "a timbre so dry, so snapping and so hard, that one would think he was hearing the sound which two sheets of parchment produce in striking each other abruptly and forcefully," and he called the noise of the jugular vein a *bruit de diable*, alluding to a children's top that spun as if possessed. German physicians heard this same *bruit de diable* as the "white nun's murmur," *Nonnensausen*, a faintly onomatopoetic term that invoked the underlying anemia of which this murmur was thought to speak.[265]

James Hope, who did much in England and Scotland to promote mediate auscultation by publicly demonstrating the ease and accuracy of the stethoscope, dismissed nuns and demons from his lexicon. Having discovered, by pressing his instrument over large veins, that "the venous murmur may often be raised, by a gradual swelling, into a more or less musical hum, such as is yielded by a child's humming top," he proposed to call this the "*Venous Hum*, for without being unnecessarily squeamish, I think that this is not only a rather more euphonious epithet, but more intelligible than 'noise of the devil.'" Arguments over the validity and euphony of names, driven though they were by medical one-upmanship, arose in the context of a growing anxiety that the numbers of stethoscopic sounds, with their trailing vines of similes and extended metaphors, were getting out of hand. One Brooklyn aurist, Joseph Colgan, breathed an audible sigh of relief in

1858 when he realized that cardiac noises fell into one of two categories, blowing or rubbing. He announced, under the heading of "Progress in Auscultation," that a conscientious doctor need no longer "trouble himself about the cooing, the bellows, the whistling sounds, the bleating, the rasping, the sawing, the to-and-fro sounds"; he had simply "to bear in mind that blowing or puffing, or the like, no matter the key or the variety, so as it is blowing or puffing, is pathognomic of endocardial trouble; and that rubbing or rasping, or to-and-fro sound, is an indication of pericardial trouble." A similar bent toward simplification characterized the German school of stethoscopists, following the lead of Joseph Škoda, a specialist in diseases of the chest and lungs. During the 1830s, after ausculting hundreds of patients at Vienna's General Hospital, Škoda had concluded that not every blurp or sough had diagnostic import. He even advanced a catch-all category of "indeterminate noise" (*unbestimmte Geräusche*) for stethoscopic sounds that had no convincing medico-physical correlate. What was key, to this son of a Bohemian locksmith, was the acoustical physics of anatomical pathology, and how exactly a voice or cough would be affected on its way through the labyrinthine densities of tissue and bone. Why strain at simile? What help was it that Dr. Samuel Gee could liken the sound he often heard when percussing the chests of healthy screaming babies to "the sound produced by clasping the hands loosely together, and striking the back of one of them upon the knee"? Nonetheless, medical dictionaries in French, English, and German had already begun codifying dozens of terms for the various rubbings, crackles, pings, and musical murmurs enlivening the suddenly noisy interiors of bodies.[266]

Stethoscopies, whether acoustic manuals or campaign brochures for a medical innovation, belonged to a new genre: the sonic catalog. Drawing its first breaths in Gothic novels and Romantic verse, the sonic catalog emerged full-blown with Charles Nodier's dictionary of onomatopoetic words. Nodier, a librarian, philologist, novelist, and entomologist of insect antennae, compiled a first (1808) and second (1828) edition of French words that imitated "the natural sound of the thing signified." He opened his lexicon with *aarbrer*, an old word for the neighing of horses when one pulls too tightly on the bridle and they rise on their hind legs. Eight pages later came *ahan*, the inbreath of a woodcutter as he readies himself for the next axeblow. *Âme* and *asthme* were swept into the realm of onomatopoeia by way of the anatomy of speech, for as they were spoken they released or abruptly, aptly, withheld the (soul) force of the respired breath. No less a meta-physiological inquiry than a set of metaphorical correlations, Nodier's lexicography yielded four hundred pages of sound-words, including the *cliquetis* of metal shoe buckles, the *upupa* of a hooting owl, and *cahot* (jolt) from "the shock you experience in a badly-sprung coach travelling over a rough, stony road, and from the effort you make to regain your breath so harshly interrupted"—a sound as peculiar to bodily circumstance as the stethoscopic *bruit de rappel* of mitral stenosis identified by Bouillaud in 1835, who compared this cardiac snap (at a narrowed atrioventricular opening) to a hammer, "which after striking the iron, falls on the anvil, rebounds, and falls again." Guided by a larger principle that the soundsystem of a language is shaped by its physical environment, Nodier believed that language arose from the imitation of natural noises; but he was under no illusion that onomatopoeia was the key to a universal language, since each language had words for sounds that other cultures never heard, or could not discriminate. For instance, *tintamarre* or "uproar" came from the noise made by French peasants striking their *marres*, iron hoes, to alert friends farther afield of the lunchbreak; *bruit* was onomatopoetic for "noise" in French because it was a confused sound (*b-rrr-oöo-whee*) "like that which arises from within a forest shaken by violent winds or from the *fracas* of torrents and the rush of great waters."[267]

Surveying *The Music of Nature* in 1832 in an *Attempt to Prove That What Is Passionate and Pleasing in the Art of Singing, Speaking, and Performing upon Musical Instruments Is Derived from the Sounds of the Animated World*, William Gardiner claimed that "There is nothing in nature that arouses our attention, or impresses our feelings more quickly, than a sound." So Nodier's text, distributed to school libraries by France's Committee for Public Instruction, complemented the five hundred English pages of Gardiner's sonic catalog, which was meant for a readership that had experienced a century's worth of controversy over song and music in theater, at work and, most heatedly, in church.[268] Owner of a hosiery factory whose passion was music and who advised on the acoustics of concert halls, Gardiner had an ear that took in everything from the clacking of a flea leaping over a lady's nightcap to the rattle of a pie-man's metal plates. "Noise," he wrote, was "a confused mixture of sounds produced by the concussion of non-elastic bodies; whereas musical sound is a pure harmonious effect emanating from a simple elastic body, as the tone of a bell." This was hardly a workable definition, since two pages later Gardiner observed that "Probably the most appalling sound in nature is that of the falls of Niagara* (*An Indian term for the voice of thunder)," where elastic bodies of water were in fine fettle. And if it was a "curious fact," it was not indisputable "that musical sounds fly farther, and are heard at a greater distance, than those which are more loud and noisy," nor that "Women talk better than men, from the superior shape of their tongues," nor that "There is a natural tendency in all languages to throw out the rugged parts which improper consonants produce, and to preserve those which are melodious and agreeable to the ear." Gardiner's ear was an imperial ear, indiscrimate in its collecting of the sounds of hurricanes, house flies, and hunting horns, of oxen, avalanches, and opera soloists, of street cries, tower bells, and growling dogs—yet absolute and British in its judgments. The languages of frigid climes were nasal and guttural because natives dared not open their mouths; the thick lips of Africans made them mumble; the thin lips of Hindus made them lisp. European lips were "best shaped for agreeable speech," but French actors had adopted a stage voice "disagreeably high and chanting," while the voice of William Pitt the Elder "was both full and clear; his lowest whisper was distinctly heard, his middle tones were sweet, rich, and beautifully varied," and as the formidable prime minister primed the machinery of empire in Parliament, elevating "his voice to its highest pitch, the House was completely filled with the volume of sound."[269]

Imperial too were the stethoscopists in their collecting and tagging of noises. "Never had the physician more reason to be proud of his mastery over the secrets of nature," wrote Robert Spittal of the Royal Infirmary of Edinburgh, "than of the power which auscultation has given him"—especially as physicians began to use a flexible *gutta percha* tube with bell-shaped endpieces. Early stethoscopes, short and rigid, had entailed the same awkwardness as ear trumpets and ear-to-chest auscultation, with the doctor facing away from the patient and, holding his instrument like a pen, struggling to direct it exactly where it was needed while placing an ear squarely over the listening end. "It is almost needless to add," added the thorough authors of *A Practical Treatise on Auscultation*, "that the physician should have everything quiet around him; that he ought to listen for a sufficiently long time, and to abstract his attention," all of which implied a newly secured acoustic space around the ausculting doctor, who "acquires the habit, not only of being insensible to the different sounds made in the neighborhood, but also of disengaging from several morbid phenomena the one on which he ought chiefly to fix his attention." Listening-in *past* transient heart, lung, or tracheal irregularities from patients unnerved by the new device, stethoscopists had also to take care that their own fingernails did not click against the cylinder and ignore the "peculiar buzzing sound produced in the ear of the listener after the first application of the instrument." That buzzing was one of many odd noises that came of pressing one's ear to a wooden tube planted on another's body: linking a patient's thorax to doctor's cranium, stethoscopes of cedar, ebony, sycamore or rattan acted like the dowels that the deaf used for conducting sonic vibrations through teeth and skull. P.-C. Potain, listening-in on the heart's diastolic thuds, found that many of these "gallops" vanished when he switched to a flexible stethoscope: they had been artefacts of bone conduction. Longer, stronger rubber tubes and then binaural earpieces put the doctor once again face to face with his patients, allowing a simultaneously acoustic, visual, and tactile exam that extended his imperium. Under the shared acoustics of percussion, patients had been able to hear some of what the physician heard, more so when the percussion was done with a small hammer mediated by an ivory disk (invented in 1826 by P.-A. Piorry), allowing for a more precise thumping and yielding louder "pleximetric" sounds that obviated the need for an ear to the skin. In stethoscopy, neither interior noises nor interior silences were shared; a patient might be oblivious to mortal afflictions heard by a physician and, vice versa, malingering soldiers or illicitly pregnant women could be found out. "The manifestation of these physical phenomena is independent of the caprice or ignorance of the patient," wrote J.-B.-P. Barth; "he carries them about him, without being able to simulate or to conceal, to exaggerate or to lessen them." As the chief instrument of what amounted to a blatantly secret interrogation at a safe distance from sweaty, pustulous, or scrofulous bodies, the stethoscope was on its way to becoming the insignia of diagnostic prowess and medical authority.[270]

General practitioners, cardiologists, pulmonologists, internists, and physicians' assistants to this day wear that insignia with professional nonchalance, yet they do not always hear the same noises in the same patient; should they agree on a noise, they may disagree on the inner workings that give rise to it. Heir to what Jonathan Crary and Jonathan Sterne have termed "the separation of the senses," as hearing, vision, touch, taste, and smell were picked apart from what had previously been a common "sensorium" and neurophysiologically segregated from each other during a long nineteenth century, they have not quite managed to reassemble the whole.[271] The auscultatory empire could not be consolidated into an indisputable regime of sounds because physicians and nurses (like other people) vary in the acuity of their hearing by age and avocation, because noises (like pain) can be referred from one site to another, because identical noises (like other portents) can stem from contrary phenomena, and because simile (like metaphor) is ripe with ambiguity. Intimate but obstreperous, stethoscopic sounds have hovered uneasily close to noise as defined by Neil Arnott, a widely travelled surgeon to the East India Company and author of popular volumes on physics: "Where a continued sound is produced by impulses which do not, like those of an elastic body, follow in regular succession, the effect ceases to be a clear uniform sound or tone, and is called a *noise*.—Such is the sound of a saw or grindstone—or the roar of waves breaking on a rocky shore, or of a violent wind in a forest." The similes chosen by Laennec and his successors—the grating of a saw, the rasp of a file, a man snoring, air venting from a bladder—scarcely escaped the perimeter of Arnott's noises. The sicker the body and the more it murmured and crackled, or resounded with martial music and the hoarse roll of drums, or sang like a top with a fiendish hum, the more it reproduced the sonic conditions of a crowded metropolis, "including the rattling of wheels, the clanking of hammers, the voices of street-criers, the noises of manufactories, &c.: which rough elements, however, at last mingle with such uniformity, that the combined result is often called the hum of men."

Listening inside that hum, stethoscopists heard the interior of the living body as an urban streetscene where carriages rumbled over cobblestones, new saddles creaked, wind squealed through keyholes. At its quietest, the interior of the body resembled "the stillness of night in a great city, where the astronomer [stethoscopist] contemplating the wondrous spheres above [below], hears only the tongues of passing time in the [beating heart of] church-towers, or the call of watchmen [bronchi expanding and contracting], faintly sounding in the distance."[272]

Simile (analogy), protested the more prosaic, was not explanation. Isidore Bricheteau's "mill-wheel" sound—how did it help one understand the physiology of hydropneumopericardium? Robert J. Graves's "wet finger rubbing on glass"—how did it further the analysis of the mechanics of pericardial friction? F. J. Bigelow's "rapid tinkling sound in the throat" that very much resembled "the clicking of the electro-magnet attached to the telegraph"—how did it lead to a better idea of the operations of the thyroid? Wasn't it unavailing and unscientific to rely on fanciful associations for noises welling up through a stethoscope? "Who ever thought of describing the sound of the wind or the rain except for poetical purposes?" asked Peter Mere Latham, a celebrated lecturer at St. Bartholomew's Hospital in London who had bravely declared his faith in the experimental truth of "auscultating signs above all others, and oftentimes before all others, and oftentimes in the place of all others" with regard, at least, to cardiac inflammations. But, he warned, "Men may *hear* ghosts as well as see them," so a "well-disciplined ear" must listen long and hard enough so as never to mistake heart sounds for lung sounds and to prick up at each murmur of disease. The fault lay less with a compulsive listening-in than with a poetic overreaching, perilous especially for beginners.[273]

Twenty and just graduated from medical school in Edinburgh, where students were known less for their disciplined ears than for noisy disrespect—hissing their duller professors, shouting obscenities, and shooting off pistols—Charles J. B. Williams suffered from tinnitus and deafness in his left ear. Nonetheless, while a sixteen-year-old Charles Darwin headed reluctantly up to Edinburgh to study medicine, Williams left excitedly for Paris to study with Laennec, smitten with this "new idea of bringing another sense—the sense of hearing—to aid us in investigating the organs in health and in disease, and, through studying its indications, learning as it were a new language." On returning to England, Williams taught what he had learned and also improved upon it. Laennec had used a concave wooden stopper at the bell-end of his stethoscope to hear the lower tones of the heart and the unstoppered bell-end to hear the higher tones of the lungs; Williams stretched a band of rubber across a trumpet-shaped end to create a vibratory diaphragm that enhanced the stethoscope's sensitivity to higher tones. In 1828 appeared his *Rational Exposition of the Physical Signs of the Diseases of the Lungs and Pleura*, after which he experimented further, finding that the metallic buzzing Laennec had associated with "pneumothorax with liquid effusion from perforation of the pleura" was "nothing more than a high-pitched echo, produced by the rapid reflections of sound from side to side within a cavity, as in a bottle, or an india-rubber ball, containing air." Along the same demystifying lines, Williams reduced the sounds of the healthy heart to a minimalist *lubb-tup* or *lubb-dupp*. German physicians followed suit with *tik-tak*, *tom-tum*, *dohm-lopp*, *ohm-ik*. Still, what caused hearts abnormal or normal to produce these syllables—or P.-L. Duroziez's *fout-ta-ta-rou* of mitral stenosis, or Hyde Salter's *rub-a-dub dubb* of pericarditis—was subject to dispute, as in Williams's short collaboration and long feud with James Hope over the causes of cardiac sounds. Williams explained these by reference to a Newtonian mechanics of mass-and-collision; Hope worked from more recent theories of hydraulics.[274]

Tubes were the riddle: What happened to sound inside a wooden cylinder or *gutta percha* "tubing" (a word coined in 1823), between a stethoscope's bell-end and its earpiece seven to sixteen inches away? What happened to sound inside bronchial tubes during bronchitis? What were the acoustics of bloodflow through arteries and veins under duress? What caused the regular and irregular flapping of valves on the larger of these dynamic inner tubes? What did it mean to hear a sound "like a leak"? And how did a physician register such noises in that tube on the other end of stethoscopy, the fluid-filled (as Itard insisted) coils of the inner ear, invisible even by the mirror-directed light of a new device of 1825, the "Inspector Auris"?[275] As had occurred with the experiments of Hooke and Boyle at their air pump, where noise was simultaneously evidence and error, so with the stethoscope noise was simultaneously what physicians listened for, what they analogized and catalogued, and what they strove to forestall.[276] Noise *through tubes*, for stethoscopy did its cultural work by double homology: homology between the tubular windings of the cochlea and its unwound extension, the listening tube of the stethoscope; homology between those listening tubes and the inner tubing of the rest of the body, the acoustical physics of which were as yet unplotted.

Attending Laennec's clinic and lectures, Williams had been surprised by the master's frail grasp of physics, finding that "his knowledge of acoustics was by no means profound; and, clever as he had been in tracing the signs empirically, he was not equally successful in explaining them rationally." But this was Williams writing from sixty years' retrospect about the Paris of 1825–1826 (a city of the "greatest splendor, the greatest filthiness," wrote Chopin, finding there "nothing but cries, noise, din and mud, past anything you can imagine"), the last years of Laennec's life. Always fearful of infection, superintending dissections with a long pair of forceps that kept him as safely distant from the dead as his stethoscope kept him from the living, Laennec yet died of consumption, though four physicians who ausculted him in his final months heard little to confirm the diagnosis. He was buried without autopsy—would the linings of his lungs have "cried out" when scraped with a scalpel, as did those of a tubercular corpse Laennec himself had dissected? He died in August 1826, well after Siméon Denis Poisson had begun work on a mathematical theory of soundwaves (1808) but well before anyone could solve a difficult problem of which Poisson had made note and which bore on stethoscopy: "the propagation of sound in a tube whose diameter is not constant." He died, vocal to the end, shortly after the publication of Félix Savart's book on the acoustics of the human voice (1825), but just as Savart was measuring the speed of sound through fluid-filled elastic tubes like those of the cardiovascular system. He died, disabused of dreams of profits from canal stock, two years before J.-L.-M. Poiseuille defended a medical thesis on blood pressure and a dozen years before Poiseuille found the law governing the laminar flow of viscous liquids in straight tubes, a law still applied (with some jockeying for non-Newtonian liquids like blood and curving tubes like arteries) to calculate the uneven flow of blood through capillaries, airflow through alveoli, and the effect of narrowed valvular openings on blood supply to the heart. He died at the age of forty-five, three years before a Dublin physician named Corrigan showed that the simple passage of water through any rubber tube could give rise to the stethoscopic *bruit de soufflet*, and six years before Joseph Rouanet in Paris showed that "all membranes, in passing suddenly from flaccidity to tension, always produce a sound." He died a decade before Williams showed that by attaching a rubber tube to a stopcock and reservoir, one could reproduce "every variety of blowing, whizzing, or musical murmurs, by varying the pressure on, or the obstruction in, the tube, or by altering the force of the current" (1837). As for more variable musical sounds "like the cooing of a dove, the humming of a fly, or the whistling of wind," these could be produced with "weak currents passing through a tube much obstructed." By 1845, H. M. Hughes in London would compare the heart's second (ventricular) sound to "the recoil noise which is heard when the closure of a stop-cock suddenly arrests the flow of water through a pipe," or what plumbers call "water hammer." So much for Romantic simile. Sonic cataloguing, however, proceeded apace.[277]

Until recently, wrote the French clinician J.-H.-S. Beau in 1856, there had been no "general history of the noises that occur within the human body." The stertor of apoplexy, the tracheal whistling of asthma, the hissing wheeze of *la coqueluche* (whooping cough)—each had been heard and understood in isolation. Now, with the deft use of stethoscope and pleximeter, a general history of bodily noises was within reach, including "abnormal" sounds: laughter, hiccuping, sighing, yawning. Beau neglected to mention stutters and cries, which had already been scrutinized by a younger contemporary of Laennec's, Marc Colombat. A poet and playwright, a political prisoner in Switzerland, and later a physician and the founder of an Orthophonic Institute in Paris, Colombat died in 1851 after a long illness that had deprived him of speech, an ironic end given his dedication to restoring the fluency of stammerers. In a short, rich essay on the physiomechanics and meaning of cries, Colombat had argued that the voice, though reliant upon vocal "cords," acts less like a string instrument than a wind instrument, and less like a reed than a flute or trombone, drawing its sound from a vibrating column of air moving through the cartilaginous tube of the larynx. Stutterers, having been embarrassed into laconic responses or brooding silence, needed exercises to loosen up a lower vocal tract that, unexercised, had dried out and was constricting the flow of that column of air. Like a flautist or trombonist, a stutterer had to wet his whistle in order to resume the rhythms of spoken language. With regard to extra-linguistic noises, each had a characteristic intonation, from which Colombat hoped to construct a diatonic scale of cries. To each cry he attached both a prophylactic and an expressive value, for momentary as a cry might be, it is "nonetheless the most energetic and immediate expression of the great motions of the soul." Citing Montaigne, who recommended a healthy cry before meals so as to dispel sadness and improve digestion, Colombat prescribed cries as antiphlogistics to reduce fever and counter-stimulants to release tension. He then catalogued a host of cries according to situation: the cry of getting burnt (*e-ah!*, from low *c* to *e*), the cry of getting cut (*e*, *ah! e*; *ah! la*, *la*, from *g* to high *a*), the cry of having a boil lanced (*a-on!*), the cry of giving birth (*e-ah!*), and more generally, cries of distress, of joy, and of disgust (*pou-ah* or *fé-hi*). There were also clamorous, disdainful, and hounding cries, or *huailles*, added the French composer Georges Kastner, who took Colombat's work two steps further in his own study of the cries of Paris. Combining an interest in the iconography of the (classical) siren and aeolian (wind) harp with a love of military bands and orchestral tympany, Kastner mixed the concert sounds of the elite with the popular sounds of the street. The cries of Parisians, he wrote, were the voice of the people, elemental to the language of the masses, uttered with a "naive energy" that should not be dismissed. In the *océan sonore* of the city, cries were "the secret link" between nature and art. Had not the abbé Sicard defined interjections as a separate part of speech in order to underscore the latencies for language in the throaty cries of the deaf? Like that heart-rending outburst of Christ nailed to the cross, the cry was suspended between the naturally rough noises of storm and the civilized blends of spoken language. What composer could fail then to be inspired by the wail of mourners, the crying of royal deaths and processions, loud advertisements for public baths, mirrors, wine, mushrooms? So, although Kastner had been the first orchestral composer to use that brassy new wailing instrument called the saxophone, he was hardly the first to put the cries of the masses to music; his unique contribution was to compose in 1857 a *Grande Symphonie humoristique* that followed the cries of Paris hour by hour, just as George Sala was doing in prose for London, twice 'round the clock.[278]

Cries caught up in this quotidian round were correlative to the noises heard by stethoscopists on their medical rounds: natural, involuntary, revelatory, consistent with an inner state. Music had long been considered to have healing powers by dint of vibrations that resonated with inner states, and physicians had prescribed music of specific keys as therapy in particular for the ailments of women, whose sensitive nerves were held to be more responsive to such vibrations. Music was also known to restore internal rhythms, such that stutterers set to singing were instantly at ease with every syllable and choirs of the mad calmed down, brightened up. Though "the Voice be harsh and inharmonious," the lively use of the lungs and the diaphragmatic vibration of internal air dissolved purulent deposits and stabilized the pulse. The Romantic poet Novalis declared that "every illness is a musical problem—its cure a musical solution," and nineteenth-century German psychiatrists took this to heart, their asylums organized around choirs, marching bands, chamber orchestras, and rhythmic gymnastics. Statistics from French asylums in 1860 showed that instrumental music had to be tenderly fitted to the moods of the demented, while pure song, which was to Romantic ears the true language of the heart and "the most natural of human expressions," was more broadly therapeutic. But the cry, as heard by nineteenth-century physicians and musicians, patients and paying audiences, arose from somewhere deeper in *soma* and *psyche*, from a place before and beyond language, such that Samuel Mathews of Virginia knew of "many cases of persons resuscitated" sheerly by the shrieks of neighbors stumbling upon the newly dead. At once desperate and redemptive, primitive and exalted, the cry was a noise so intense that it could transfigure opera with a single sob, as at the climax of Wagner's *Tristan und Isolde*.[279]

Sung to its innermost, the cry had been at the acid-edge of opera from its birth, with Euridice's tragic "Ahi!" as Orpheus looks back at her and so condemns his beloved forever to the Underworld in Claudio Monteverdi's first opera (1607). The eighteenth-century cries of Euridice and of Iphigenia in the operas of Glück, and of Don Juan in Mozart's *Don Giovanni* (redoubled cries that critics termed *la criaillerie* or screaming), had cut away the contours of Italian *bel canto* and *falsetto*, straining toward expressions of absolute loss which could only result in loss of speech and the assertion of wordless voice, "trans-sensical," piercing. (Thus an eloquent modern critic, Michel Poizat, whose analysis rises to nearly the same trans-historical pitch as that of Hegel.) A contemporary of Laennec's, G. W. F.

Hegel maintained in his idealist *Lectures on Aesthetics* (delivered in Berlin between 1823 and 1829) the supremacy of the ear as pick-up for correspondences between inner tremblings of the soul and the vital trembling of all things embodied. The human capacity to appreciate the world and the self in their co-presence was grounded in those shared vibrations called sound. It was the shiver at the heart of being that revealed the heart of meaning—an invisible shiver felt most deeply in music, whose chief task "consists in making resound, not the objective world itself, but, on the contrary, the manner in which the inmost self is moved to the depths of its personality and conscious soul," thus shaking off the bindings of language and soaring wordless toward the cry, the "'och' and 'oh' of the heart." But to return to earth and the premiere of *Tristan und Isolde* in 1865, Isolde's cry could most sharply pierce the air, the ear, and the heart because Wagner could invoke a new sort of theatrical listening that was confluent with the listening-in of stethoscopists: private, focal, rapt.[280]

15. Clap + Rap = Audience

Traditionally, theater audiences had been as noisily inattentive as crowds milling around the street booths of puppeteers. Easily distracted by scuffles in the pit, all the while cracking nuts, crunching apples, guzzling drinks, gambling, and smoking cigars or spitting tobacco, audiences anticipated flashes of wit, anger, and bawdry from the floor, the boxes, and the gallery that could be every bit as entertaining as the snatches overheard from professionals (and here I mean the performers, not the applauding claques under the thumb of gangleaders hired by the performers, nor the experienced hecklers hired by rivals). Competing with spectators whose seats did not always face the stage and who might be scribbling business dispatches or mash notes during shows where the lights never dimmed and refreshments were hawked non-stop, conductors pounded out the time with thick batons, stage managers blew whistles to cue stage hands for scene changes, and actors and singers strained to project their voices above an uproar that was sometimes genial, sometimes ruthless. If by 1800 young strutting noblemen no longer sat in loges on stage, rapiers at hand, redirecting actors or crashing scenes, and if brawling apprentices rarely hurled each other up onto the boards, and if there were actual silences for a favorite scene or aria, audience accessories still included horns, rattles, squeakers, seven-toned whistles, and a small device that made a noise like a delirious manx, whence the term "catcall." Hector Berlioz in 1858 attributed the increasing shrillness of opera divas and the preference of composers for the highest soprano registers to the extreme measures needed to keep the attention (and esteem) of such audiences, especially Italian audiences who "made it their habit to talk as loudly during performances as they would at the stock exchange." Should their attention be drawn to something on stage, audiences openly expressed their opinions by "shrieking, stamping and battering with their canes against the backs of the seats in front," or by pelting performers with nuts, fruit, pits, and rinds... or by hissing. "Damn 'em, how they hissed!" wrote Charles Lamb, an English essayist and dramatist "constitutionally susceptible of noises," recalling the hostile reception of his farce, *Mr. H.* "It was not a hiss neither but a sort of frantic yell, like a congregation of mad geese, with roaring sometimes, like bears, mows and mups like apes, sometimes snakes, that hiss'd me into madness." It was not much better when audiences liked what they heard, for they would call so loudly and persistently for encores that most productions could not escape the immediate repeating of soliloquies, arias, duets, sometimes entire scenes or movements. Wrote a later critic, "The right of the *encore* is one which the sovereign public, in the uncertainableness of its enthusiasm, is continually abusing." And vice versa. Playing to such audiences in 1818, the popular tenor John Braham so anticipated the call-backs that he preemptively repeated a solo before anyone had a chance to cry *encore!* When he returned of his own accord to repeat it yet again, the indignant audience shouted "Off, off!" and "No, no!" and drowned out his too-accommodating voice.[281]

However, just as medical spaces around stethoscopists were being quieted under the pressures of auscultation and its intent listening-in, so musical and dramatic spaces were being quieted under the pressure of an aesthetic that demanded a more intent theatrical listening. As the art historian Michael Fried, cultural historian James H. Johnson, and music historian Michael P. Steinberg have shown, ideals of painting and musical performance arising in Paris and Vienna during the late eighteenth century had demanded a "collective act of heightened attention"—figures in paintings so completely focused upon an event or object that spectators found themselves drawn into the scene, actors and singers so fully embracing their roles that audiences became totally absorbed in the drama. If actors should not go slack between speeches, if singers on stage awaiting their next musical cue should not groom themselves or wink at their claques, neither should audiences disrupt the emotional cogency of a play or an opera by stomping, hooting, forced coughing, derisive farting, or untimely applause, whether a *bombi* like humming bees, *imbrices* like rain plummeting on tiled roofs, or *testae*, like clattering castanets. A work of art should be so compelling, its sentiments so exquisite, as to create a community of interests where performers and auditors put the momentum of the piece before their personal glory or rapture, respecting the aesthetic integrity of the work and deferring all honors to the end. "No Audience," claimed a 1763 treatise on the corruption of poetry and music, "demands the Repetition of a pathetic Speech in Tragedy, though performed in the finest Manner, because their Attention is turned on the Subject of the Drama: Thus if the Audience were warmed by the Subject of an Opera, and took part in the main Action of the Poem, the Encore, instead of being desirable, would generally disgust."[282]

Decades would pass before silence gripped the larger audiences of theater, opera, and symphonic music as deeply as it gripped readers of Gothic tales, museum visitors struck by sublime paintings, housebound invalids (like Charles Darwin in 1839) who required perfect quiet to stifle palpitations of the heart or blunt the hammered anvils and demonic drills of migraines (depicted by Honoré Daumier in 1833, George Cruikshank in 1835), or worshippers in New England pews so hushed that one young woman would be indelibly impressed by the preliminaries for a final hymn, "the rustling of singing-book leaves, the sliding of the short screen-curtains before the singers along by their clinking rings, and now and then a premonitory groan or squeak from the bass-viol or violin, as if the instruments were clearing their throats; and finally the sudden uprising of that long row of heads in the 'singing-seats'" of the choir. A millworker from the age of eleven to twenty-one amid "the whir and dust" of Lowell, Massachusetts, "City of Spindles," Lucy Larcom had one day a week off from "the buzzing and hissing and whizzing of pulleys and rollers and spindles and flyers." In old age she would call this industrial noise "tiresome," an understatement given that the one hall whose spinning machinery has been restored for Lowell's National Historic Park is so loud as to pound past today's foam, form-fitting earplugs unknown to Larcom's co-workers as they tended to ten thousand power looms. Another Lowell worker likened the sound to the thunder of Niagara Falls, "and a cotton mill is no worse, though you wonder that we do not have to hold our breath in such a noise." Mechanical pickers that separated and cleaned the fibers had a hum that "would a deaf and dumb man stun," wrote a professor of eloquence in an anti-poetic "Picture of a Factory Village"; Larcom herself worked from five a.m. to seven p.m. at a dressing-frame, a "half-live creature, with its great groaning joints and whizzing fan." At church on the Sabbath, then, she was soothed by the quiet and stirred by the hymns but did not think her voice equal to the music, speculating instead that "Our natural desire for musical utterance is perhaps a prophecy that in a perfect world we shall all know how to sing." 'Til that millennial time, there would be singers and there would be audiences, as at the St. Charles Theater in New Orleans, the largest opera venue in America when its doors were thrown open in 1835 by James Caldwell, whose program notes declared that he was "determined to keep strict order in the establishment; to put down, at every risk, every attempt to disturb the quiet and attention which ought always to be ascendant in a public assembly, but which is too often violated by ignorant, at other times by disorderly persons, who think because they pay their money and because it is a theatre, they may make as much noise as they please."[283]

Long before they attended an opera, if ever, urbanites would have heard its melodies sung by street singers peddling sheet music, performed by students at recitals, played in piano transcriptions for the parlor or saloon, edited for barrel organs or roving brass bands, recast for ballrooms or dancehalls, hummed in snatches on the avenue, or whistled in the kitchen. During the 1820s, annoyed by the popularity of Carl Maria von Weber's *Der Freischütz*, a German gentleman advertised for "a servant who was *not* able to whistle its songs." Whether ticketed for a spacious box or waiting for cheap gallery seats, people regarded opera as a hardy, open form that could weather incidental outrage and survive juxtaposition in the same evening with high-wire artists and comedians acting such farces as *Presumption; or, The Fate of Frankenstein*. If bourgeois concertgoers listened to orchestral concerts as practiced amateurs who might know the program intimately from sheet music that they themselves had played, operagoers of lesser means who had never handled a concert instrument or properly hit a sung note had yet enough familiarity with opera to rise "hissing, howling, whistling, kicking and screaming" when a troupe at the St. Charles in 1837 omitted the last scene of Rossini's *Semiramide*. Seat cushions were ripped open, chairs went flying, canes were thrust at Caldwell's grand chandelier.[284]

Regardless of this and many another outburst, the momentum by 1865 was toward a quieter auditorium with less give-and-take between players and audiences—always excepting Italian opera houses, French cabarets, German *tingel-tangel* saloons, British music halls, North American vaudeville, and burlesque everywhere. After sixty-six nights of "hissing, hooting, groaning, growling, and vociferation" at Covent Garden in 1809, numbers were arrested, and the Court of King's Bench sat for the case of a barrister in that audience who was suing for false imprisonment. The Court decided that audiences have the right to hiss or applaud at any time, but should individuals plot together to make a noise so great as to render actors inaudible, though they commit no other violence, they are guilty of riot. Chief Justice Lord Mansfield added, "I cannot tell on what grounds many people think they have a right, at a theatre, to make such a prodigious noise as to prevent others hearing what is going forward on the stage." Mayhap, as the French novelist Stendhal proposed in 1823 with regard to opera, the noise and ill humor arose from packing audiences into stifling auditoriums. "It may be surmised that the human ear is surrounded by a kind of musical ether . . . ," wrote Stendhal in his *Life of Rossini*. "Thus, to obtain perfect pleasure it is essential to be situated in a kind of *isolated field*, as we observe in certain experiments involving electrical magnetism, which in practice means that it is necessary that the space of at least one foot should intervene between oneself and the nearest adjacent human body. The animal heat which emanates from a foreign body appears to me to have a most pernicious effect upon the enjoyment of music." Was it to counteract this effect and overcome the restless noise of audiences, driving them back to their proper places, that Grand Opera was contrived, with its crowd scenes and overpowering choruses (singing through megaphones in Meyerbeer's *Robert le Diable*)? And did not theater also require that each member of the audience, surrounded by a kind of dramatic ether, be situated in an isolated field? Richard Wagner thought so, and fought to have the audiences for his "musical dramas" swathed in darkness, each person to her own field of vision and audition; noise should come only from the stage and the orchestra, as with *Das Rheingold*'s infamous anvil chorus, which at its premiere in 1869, and decades after, overtook audiences with a stridency hitherto unknown among the wildest evenings at Covent Garden or at the most passionate of labor union meetings where solidarity was reported in one insistent parenthesis after another as "(Cheers)...(Loud cheering)...(Shame!)...(Cheers)...(Hear, hear)." Once dimmable gas lights were fixed in auditoriums and carbon arc lamps and recessed footlights directed toward the stage, audiences elsewhere than at Wagner's Bayreuth would find themselves in the dark—and more focused on the performance. Until the advent of cinema, it is true, Americans in particular objected to darkness as unsociable and total darkness as unsafe in an era of deadly theater fires and deadlier panics, but they did become increasingly pliable to instruction in proper audienceship by masters of etiquette such as James Smiley and maestros such as Theodore Thomas. In *Modern Manners and Social Forms* (1889), Smiley made silence imperative: "Perfect quiet should be maintained during the performance, and attention should be fixed on the stage. To whisper or do anything during the entertainment to disturb or distract the attention of others, is rude in the extreme," and those disrupting a performance were no better than "thieves and robbers." Thomas, an imperious conductor who expected musicians to be riveted to his baton ("a delicate little white wand, hardly larger than a pencil"), expected of symphonic audiences an equally riveted listening: no latecomers, no shuffling or moving about, no gabbing, no fussy unwrapping of candy or perpetual lighting of cigars, no huzzahs for a solo, no clapping during a movement, and no cries for encores (such greed!) until the very end. "When we are most deeply moved and interested by works which speak to the intellect and to the soul," wrote a music critic, "we are inclined to the most silent form of approbation." At least such was the case for persons of a meditative and harmonious temperament, but the younger and more thoughtless rose to encore the most hackneyed of pieces by any Signorina Screecherina, and a good conductor's job was to protect "the silent, music-loving many" against the encore fiends.[285]

Flocks of children, women, and men since the 1790s had willingly paid to feel their way through subterranean corridors and climb spiral staircases into twilit or skylit rotundas to be awed by suddenly brilliant panoramic paintings or by the slow dawns and glowing dusks of dioramas. Encompassed by an architecture that imposed upon them an aesthetics of absorption they may never before have heeded, audiences to such spectacles felt almost alone amid alpine meadows or a naval battle. "As soon as you enter," wrote a visitor to a London reconstruction of the Battle of the Nile, "a shiver runs down your spine. The darkness of night is all around, illuminated only by burning ships and cannon fire." The highlight of this battle between the fleets of Napoleon and Nelson was the explosion of *L'Orient*, Napoleon's flagship, from which rained down "heads, limbs, cannon mounts, yards, masts, muskets, crates, shreds of rope," but the triumph of this circumambient horror required surround-sound, which our German visitor of 1799 thought fit to provide: "The sea is churning furiously, stirred up by the hundreds of cannonballs that have missed their mark, the timbers crashing down from above, and (one imagines) the shock waves rending the air." Visitors to other panoramas also insisted upon hearing things in these circular hushed rooms of optical illusion and auditory hallucination where evocation became sonification.[286]

Acoustically focused audiences were explicitly coordinate with the listening-in of telegraphers, who by the 1850s were transcribing messages across thousands of miles of wire by the *dit-da* of Morse Code as voiced by a mechanical "sounder" and amplified by an empty tin of Prince Albert tobacco wedged next to the brass plate on which a small pen-lever clicked and tapped. Acoustically focused audiences were also coordinate with the listening-in of mid-century spirit mediums who dematerialized the telegraph only to rematerialize its audible pulse. Originating in mystical channels mapped a century earlier by Emanuel Swedenborg, nineteenth-century Spiritualism followed this Swedish physicist, naturalist, engineer, and theologian in his emphasis upon hearing as the connector of matters of the spirit with the spirit of matter. Abetted by strong coffee, Swedenborg had entered visionary states where gossipy, belching spirits tugged at his ears or, to share his precision, at his eardrums, cochlea, and the elevator muscles of his ear lobe. From 1744 to his death in 1772, he learned to ignore such imps and listen with an "inner ear" for spirits more metaphysical, investigating these with no less intensity than he had investigated crystals, the spinal cord, the cerebral cortex, phosphorescence, and ear trumpets. Spiritualists supplemented Swedenborg's arcana with mesmeric healing, electrotherapy, and clairvoyant trance, but they too relied heavily upon a good ear to receive tapped, knocked, and rapped messages from the Other Side. In 1848, when the family of John and Margaret Fox began to encounter mysterious noises from the floors and walls of their wood-frame house in upstate New York, they thought it must be rats, or a nearby cobbler working late, or loose boards shaken by the wind—that is, until the raps responded to Mrs. Fox's claps and questions. With none for no and two for yes, the family and excited friends heard from the ghost of a murdered peddlar (or a shifting bedstead), then more formally, through seances, from Benjamin Franklin and other, less electric dead (or the loudly cracking knee joints of the Fox daughters Margaretta and Catherine). A physician, asked by sceptics to apply his stethoscope to Maggie and Katy during one of their spirit-rapping sessions, found no sign of those ventriloquisms Laennec called pectoriloquy, egophony, and bronchophony, or anything phony at all. Whence began another sonic catalog, as earwitnesses compared the intriguing noises to a falling pail of bonnyclabber, a hand slipping on a newly varnished post, or the sawing and planing of boards and the hammering of nails into a coffin. Such noises, akin to the knocking and rapping of a 1789 ghost-event from downstate New York, were occasionally mimetic, rarely diagnostic. Their variety and volume were instead declarative of earnest, voluble, otherworldly presences, for Spiritualism aimed at a more direct response to Samuel F. B. Morse's first telegraphed question of May 1844, "What hath God wrought?" In lieu of the results of a political convention, which had been the first telegraphed answer, they received consolation and encouragement from the Departed. If a Philadelphia newspaper editor in 1845 predicted that "The wires of the Telegraph will be the nerves of the press, vibrating with every impression received at the remotest extremities of the country," the *Spiritual Telegraph*, like other 1850s Spiritualist newspapers, vibrated with impressions received at the remotest extremities of the soul, here and There, with a democratic *élan* that was strongly feminist, stirringly personal, and potently acoustic. So potent that Elizabeth Doten's 1863 *Poems from the Inner Life* included new stanzas from the late lamented Edgar Allan Poe and two poems from Shakespeare, who had had some years to rethink a certain soliloquy:

> "To be, or not to be" is not "the question;"
> There is no choice of life. Ay, mark it well!
> For Death is but another name for Change.[287]

Sadly, Doten by her forties had withdrawn from successful tours as a spirit medium, improvisational poet, and inspirational public speaker, perhaps because she had found that she could not tell where her own mental operations stopped and the voices started ("It is often as difficult to decide what is the action of one's own intellect and what is spirit-influence, as it is in our ordinary associations to determine what is original with ourselves and what we have received from circumstances or contact with the minds of others"), perhaps because of "declining health," perhaps because she soon married into Boston society and had to contend with other, more demanding voices. Physically and intellectually active for another forty years, she became intrigued with California, chemistry, and electricity, and looked ahead to a true science of psychic phenomena, confident that

"The Intelligence of the Universe exists in us, and operates through us. As individual entities and conservators of that great force, we stand co-related to it and to each other, and it is both a logical and legitimate conclusion, that there should be a direct communication along the whole line, to the uttermost parts of the universe. We only lack the ability to perceive and understand it." Scientific spiritualism would untangle the origins of spirit communications, make them clear and indisputable. ("You can fill any house with noises if you will," wrote a sceptical Charles Dickens in *The Haunted House* of 1862, "until you have a noise for every nerve in your nervous system.") Spirit mediums, of course, like all seers and sibyls, must present themselves as clear conduits for otherworldly messages, and it was believed in Doten's era that women were best suited to such clarity, since they had a natural disposition for pure receptivity, the temperament requisite to becoming a "sensitive," and the maternal habit of yielding their bodies selflessly to others (children, husbands, spirits). But women were also believed to be more suggestible than men, with less mental rigor and stamina, so there was ever the danger that the voices they fronted would take over, especially as rapping was augmented by automatic writing and a tranced bespeaking during which mediums, sometimes enclosed in dark cabinets for isolation from the noises of the world, were possessed by the hands, minds, and voices of the Departed.[288]

Hearing voices had been a problem, as Julian Jaynes has speculated, ever since the resolution of an archaic cognitive split between percepts understood as human and imbedded voices understood as gods. In the generations after *The Iliad*, voices formerly deific were consolidated into the "voice of conscience," and the hearing of inner but clearly other and commanding voices was reserved for the demonstrably "inspired": oracles, prophets, saints, mystics. All others' experience of interior audition would be consigned to childhood fantasies, demonic wiles, senility, dreams, or *dementia praecox*. With the startling emergence of stethoscopy, of telegraphy, of mediumship that sometimes used brass ear trumpets as otherworldly speakers, and of mind-reading acts that used automaton heads as centerpieces, it became plausible to imagine this process in reverse, not as inner hearing but as a deliberate listening-in. Quoth one Rod'rick, as early as 1829:

> "I'll a place contrive
> So dark and safe, no man alive
> Shall to our private meetings grope."
> "Egad," cries Johnny, "that won't do,
> If there's no crack to listen through,
> They'll take 'reports' by Stethoscope!"

Troubled souls picked up on this new trope, convinced that wherever they went, their thoughts were being monitored. A symptom of "paranoia"

(1857), it was also a canny extrapolation of the potential of these new tools for listening-in on hidden, invisible, or distant realms. Weren't there already professional "detectives" (1856) who, like prison guards, could be recognized by their quiet soles ("sneaks," 1862; in 1906, "gumshoes")? A shady occupation fringed by snitches and "snoops" (1864), "private detection" with its wardrobe of disguises and its eavesdropping devices glided perceptibly toward the underworld. "The pastor is being stethoscoped by a gang of thieves who are around 351 Rodney street, Brooklyn," T. H. Elderd warned Rev. Behrends in a pencilled note after services at the Central Congregational Church in 1889. "*I live there and they are stethoscoping me also.*" Elderd then called upon the pastor to alert him to the fact that "It is not God that you think is inspiring you; it is a lot of thieves with a physician's stethoscope." The gang would listen at one end of the stethoscope and whisper to you through the other. "They are preventing you from thinking and talking naturally. They are listening to the private secrets of families and interfering, by sensational tricks, with the relations between man and wife." Using their stethoscope, the gang could hear the quietest word at a hundred yards and listen-in from alleyways. If you blew on their stethoscope you'd go blind. Foulness was afoot, and Elderd and his mother were being driven into a poverty worse than that depicted by a reporter who found him living in the front room of a wooden "shanty" with broken windows. Neighbors regarded Elderd "as a victim to some nervous disorder," as did a local judge whom Elderd called upon for a warrant to arrest the gang, who had cornered him at the docks and tried to assassinate him with an electrical stethoscope.[289]

16. Beaters, Grinders, Bangers, Clashers, Worriers

Gramophones, telephones, radios, radar—these too would eventually figure in the paranoias of "schizophrenics" (a word coined in 1911) as devices by which dark intelligences could eavesdrop on and commandeer a mind, but I am getting ahead of myself. The brief to be made in respect of Mr. T. H. Elderd concerns less a shift in psychiatric categories than the cultural installation of a new mode of listening that entailed a heightened sensitivity to the ubiquity of noise. In the hearing of his contemporaries, Elderd had taken to a recognizable extreme that listening-in first modelled by stethoscopists, then by telegraphers and spirit mediums at mid-century, and then amplified in the 1880s as Americans began to bend their ears to the horns of phonographs and the receivers of telephones. It was a mode of listening that crossed genders, ranks, and faiths, extending far beyond silenced audiences in darkened theaters or at seance tables.[290] It was the mode of the Romantic poet and sentimental versifier, listening-in to the workings of Nature and the human heart; of Transcendentalists, listening-in to the Brahma of the universe or Thoreau's "sphere music";[291] of machinists, listening-in for loosened bolts or worn cogs; of tenders of power looms, listening-in to pulleys and rods for the rough edges of mechanical fatigue; above all, of writers and scholars, who had to hear themselves think and were liable to blankness or "breakdowns" (1832) under siege of noise.

Carlyle, say. Thomas Carlyle: the Scottish essayist, historian, philosopher, poet, and biographer. Biographer in 1831 of Diogenes Teufelsdrökh, a philosopher of raiment who urged the exchange of old customs/costumes for new after the July Revolution (of 1830) in France and who expounded upon "the benignant efficacies of Concealment," by which he meant the virtues of silence (said Carlyle, who had fabricated Herr DevilTurd out of whole cloth): "Silence is the element in which great things fashion themselves together; that at length they may emerge, full-formed and majestic, into the daylight of Life, which they are thenceforth to rule." Hold thy tongue but for a day, and "on the morrow, how much clearer are thy purposes and duties; what wreck and rubbish have those mute workmen within thee swept away, when intrusive noises were shut out!" Biographer in 1838 of the likes of Dante and Oliver Cromwell and pretenders to the heroic like Rousseau, who "had not 'the talent of Silence'" and did not "*hold his peace*, till the time come for speaking and acting." Cromwell the Protector had the talent and timing, as did Samuel Johnson the Grand Definer, and Robert Burns the Poet. "Ah yes, I will say again: The great *silent* men! Looking round on the noisy inanity of the world, words with little meaning, actions with little worth, one loves to reflect on the great Empire of *Silence*. The noble silent men, scattered here and there, each in his department; silently thinking, silently working; whom no Morning Newspaper makes mention of! They are the salt of the Earth." Born in a cottage in Ecclefechan, eight miles north of the English border, Carlyle kept returning to that moorland of "greenness, and the deepest quiet": Craigenputtoch Hill where, in 1833, he and a young Ralph Waldo Emerson walked and talked through the immortality of the soul; the farms of Hoddam and Scotsbrig where on sweet days he found "the air noiseless, the very barndoor fowls fallen silent; only the voice of some husbandman peaceably directing his thatchers and potato-diggers." These farmers, thatchers, and diggers, salts of the Scottish earth, toiled on despite an encroaching "Mechanical Age" against which Carlyle had written his "Signs of the Times": "It is the Age of Machinery, in every outward and inward sense of that word; the age which, with its whole undivided might, forwards, teaches and practises the great art of adapting means to ends." A smash-and-grab era, this Age, and morally dubious. "For all earthly, and for some unearthly purposes, we have machines and mechanic furtherances; for mincing our cabbages; for casting us into magnetic sleep. We remove mountains, and make seas our smooth highway; nothing can resist us. We war with rude Nature; and, by our resistless engines, come off always victorious, and loaded with spoils"—or ourselves spoiled. Yet in 1834 he and his wife Jane Welsh left rural Scotland to make their home in "that monstrous tuberosity of Civilised Life, the Capital of England," with the "stifled jarring hubbub" of its old-clothes markets and "all the wagons, hackney and glass coaches, dog-carts, night-men, daymen, noblemen, gentlemen, gigmen, grinding along."[292]

Jane had no qualms about renting a house in Chelsea around the block from their sweet lilting friend Leigh Hunt, for she had found Craigenputtoch "the dreariest place on the face of the earth ... the most dreadful, lonesome, barren of places," while Thomas dismissed then-suburban Chelsea as "defaced with green-paint, cockneyism, dust and din,—an abominable *aping* of country." Metropolitan London, no matter, was the place to be for someone as avid as he was to be heard. "One might erect statues to Silence" and remind oneself that a loquacious Scot would do well to set aside time for not speaking, but one must also speak out, heroically, when the time was ripe. Perhaps not so loudly as Jane's old tutor/suitor and Thomas's close friend Edward Irving, who had come "screaching" down to Regent Square in 1831 to turn the Church of Scotland toward primitive Christianity, then turned London on its head with a millennial ministry of "tongue-work." Women in his congregation began "hooting and hooing," prophesying from the pews; men sang, then spoke the syllables of God's language. "It is a mere abandonment of all truth," Irving insisted when called upon the carpet, "to call it screaming or crying; it is the most majestic and divine utterance which I have ever heard," whereupon he was expelled from his pulpit, sent to the Highlands and, in 1834, to a premature death that grieved both Carlyles. "A man of antique heroic nature, in questionable modern garniture, which he could not wear!" Thomas mourned. "The voice of our 'son of thunder,'—with its deep tone of wisdom that belonged to all articulate-speaking ages, never inaudible amid wildest dissonances that belong to this inarticulate age ... which cannot speak, but only screech and gibber,—has gone silent so soon." Still, thou should'st not closet thyself from the world, as Thomas had written in the garb of Teufelsdrökh. Wert thou, say, "suddenly covered-up within the largest imaginable Glass-bell—what a thing it were, not for thyself only, but for the world!" Thy daily mail would flutter up against the glass unread, thy thoughts fall short of any ear, and thou'dst be caught in the vacuum of a bell-jar, "no longer a circulating venous-arterial Heart, that, taking and giving, circulatest through all Space and all Time: there has a Hole fallen-out in the immeasurable, universal World-tissue, which must be darned up again!"[293]

Darning here and darting there, Jane did her best with a dyspeptic husband who went into funks about morning scales scraped along a piano by the daughters of a "wall-neighbor," afternoon screeches of macaws, supper-to-breakfast crowing of cocks, the voices that were visited upon him morning, noon, and night by street singers "howling under my windows, uncertain with what object, for indeed I never heard such screeching and braying combined in one till now—like the melody of jackasses concerting with gigantic he-cats!" Anxious nonetheless to cut a swath through the "Bedlam of Creation," through "its wail and jubilee, mad loves and mad hatreds, church-bells and gallows-ropes," Thomas in society was a conversational terror. Carlyle the Sensitive and monumentally Insensitive, he who would persuade the English of the Swiss proverb that "Silence Is Golden," showed himself in most blatant contradiction one evening in 1838 at table with three eminent Victorians named Charles. Across from him were Charles Lyell, author of *Principles of Geology*, a gregarious man of science who doubted that the "feverish spasmodic energy" of quakes and eruptions of biblical proportion could be the sole force driving geological change, whose own speech was as slow and crystalline as schist, and who praised a dinner table "so arranged as that everyone could hear and be heard by others" without the barricade of centerpieces; Charles Darwin, who, having abandoned medicine and then theology to take up geology, botany, and zoology, was back from five years aboard the *H.M.S. Beagle* and for the moment relishing company; and Charles Babbage, of whom more in a flash. Carlyle "silenced everyone by haranguing during the whole dinner on the advantages of silence," wrote Darwin. "After dinner Babbage, in his grimmest manner, thanked Carlyle for his very interesting lecture on silence."[294]

Everyone at table would have best appreciated the grimness of that gratitude a quarter-century later, by which time not only Carlyle's but Babbage's campaign against street music had become legendary. A mathematician and political economist, Babbage met Carlyle when both were in their forties and near top form, Babbage known for his lustrous Saturday soirées, his prowess with numbers, and ambitious projects for constructing "difference" and "analytical" engines that could be programmed to produce complex tables, Carlyle the translator and expositor of Goethe and a dashing lecturer on Heroes. Shortly after the Three-Charles dinner, Carlyle sketched Babbage most unheroically as "A mixture of craven terror and venomous-looking vehemence; with no chin too: 'cross between a frog and a viper,' as somebody called him." Portraits and photographs confirm Babbage's intensity of glare; venom and terror were artefacts of Carlyle's own crabbiness, and he had conveniently overlooked the ears, where he and Babbage were alike, sharing an irascible sensitivity to sound. Babbage's sound-mindedness had led him to applaud the "simple contrivance of tin tubes" for communicating between factory halls, one of the Advantages of Machinery listed in his 1832 book, *On the Economy of Machinery and Manufactures*. And he had magnified the sense of hearing itself in one of the more daring sections of a treatise on natural religion drafted in 1837, refined in 1838. Discussing "the Permanent Impression of Our Words and Actions on the Globe We Inhabit," Babbage supposed a vast republic of impressions based upon the principle of "equality of action and reaction" as applied *ad infinitum*, and acoustically: "The pulsations of the air, once set in motion by the human voice, cease not to exist with the sounds to which they gave rise." Though inaudible to most ears, those pulsating waves persist across the globe and "in less than twenty hours every atom of its atmosphere takes up the altered movement due to that infinitesimal portion of the primitive motion which has been conveyed to it through countless channels, and which must continue to influence its path throughout its future existence." Future existence? "If we imagine the soul in an after stage of our existence, to be connected with a bodily organ of hearing so sensitive, as to vibrate with the motions of the air, even of infinitesimal force, and if it be still within the precincts of its ancient abode, all the accumulated words pronounced from the creation of mankind, will fall at once upon that ear." Echoing Chaucer's House of Fame and the Renaissance trope of frozen words, this biophysics gave permanence and ubiquity to noise: "Thus considered, what a strange chaos is this wide atmosphere we breathe! Every atom, impressed with good and with ill, retains at once the motions which philosophers and sages have imparted to it, mixed and combined in ten thousand ways with all that is worthless and base." If it were the case that "The air itself is one vast library, on whose pages are forever written all that man has said or woman whispered," it was the noisiest of libraries, bearing forever the vibrating particles of mirth and murder, capable of testifying—he concluded—to the atrocity of each chained African who had ever been thrown overboard in mid-ocean by the captain of a slave ship needing to lighten his load: "Interrogate every wave which breaks unimpeded on ten thousand desolate shores, and it will give evidence of the last gurgle of the waters which closed over the head of his dying victim."[295]

Such vividly abolitionist language would seem to make of Babbage's running battle against organ grinders a trifling matter, but Babbage was a man of detail. His idea of genius was not "the observation of quantities inappreciable to any but the acutest senses," as one might suppose; genius attached rather to the finding of means to force Nature to reveal and record "her minutest variations" so that they were sensible to all. His calculating engines demanded the tightest of tolerances for their rotating plates and metal columns, the building of which (in the case of his difference engine, the only one almost completed) took a decade because he and master mechanics had to devise new tools and methods for precision machining. Similarly, fancying himself a sort of intellectual engine, he had little tolerance for interruption while listening-in upon his own tight reasoning. Accosted on the street by beggars, he spent days investigating and disconfirming their pleas; when Labor turned from grievance to strike, he rushed to the side of Capital and continuity; in the 1850s, when his once-bourgeois neighborhood acquired a more plebeian tenor, with lodging houses, beer shops, and a horse-cab stand, he became greatly agitated by migratory concerts of "organ grinders, brass bands, fiddlers, harps, harpsichords, hurdy-gurdies, flageolets, drums, bagpipes, accordions, half-penny whistles, tom-toms, trumpets," etc., whereof the "lower classes are the chief encouragers." He made charts: in one period of 80 days he had remonstrated with street musicians 165 times; in one year he had spent £104, a skilled worker's annual wages, in legal proceedings against them; "one-quarter of his working power" over a dozen years had been lost to them. More than his own powers were at issue; estimating that on any given day 4.72% of Londoners were at home in sickbed, Babbage accused street musicians of sapping the recuperative powers of all respectable Londoners, whether or not they worked at home as he did.[296]

Working at home, in this Age of Machinery, was newly gendered. Men went out to work while women remained behind to manage children, guests, coal fires, meals, cleaning, deliveries, repairs and, in better-off households, servants who had to be hired, trained, and supervised. Men occupied the public realms of trade, finance, politics, government, and science, where they could change the universe; women occupied the private realm, charged with domestic upkeep, childrearing, minor illnesses, interior decoration, aesthetics and, more generally, emotional expression. Or so the tale has been told. In practice, these "separate spheres" were never totally disengaged, especially for the hundred thousand men, women, and children who could be found in garrets and basements painting buttons or artificial flowers, or on the street crying the cheapest of wares, lugging sacks of coal, grinding organs. When Babbage scurried out his door to find a bobby to dislodge a hurdy-gurdy player, it was these same people, scratching a living that flew in the face of any doctrine of separate spheres, who ran after him, throwing stones. It was Capital's reserve army of the unemployed that cried "Babbage!" to his face, his name itself an epithet. Born in comfort (his father a merchant banker) and made more comfortable by marriage (his in-laws well-seated in a Shropshire manor), Babbage sired eight children across thirteen years and endured a domestic scene until five died young, along with his wife Georgiana, after which Babbage the widower manned both spheres, public and private. Unlike the "vagabonds" who found both their work and entertainment out of doors, Babbage demanded an acoustic separation of spheres, interior from exterior, house from avenue, the lathes in his laboratory from the iron hoops trundled by children on the street. "Every kind of noisy instrument, whether organ or harp, or trumpet or penny whistle, if sounded, should be seized by the police and taken to the station, also all hoops and instruments for playing games." His was a perverse demand given that the laboratory in his coach house must have been as obnoxiously noisy to his neighbors as any machine shop, unless one realizes how much Babbage at home honored the need to listen-in to himself and how much he understood intellectual work as a matter of *focus*. He who had a box at the German Opera House and thought to redesign its stage lighting with new electric-arc lights and mirrors to focus them, advised those who gave coins to street musicians to buy tickets to music halls instead, where they would hear better music in more focused environs. "No man having a brain," Babbage was sure, "ever *listened* to street musicians."[297]

Brainwork, whether of the emerging professions of statistics, engineering, chemistry, and physics, or of the established professions of finance, law, civil service, and (now) medicine, was hard on the nerves, almost by anatomical definition. Harder on the nerves, because questionable to a society utilitarian in its economics if not its decor, was the brainwork of dawbers and scribblers, men who meant to earn their livings or reputations by painting in their studios and writing in their studies. Uneasily at home with projects of incalculable value, such men could appear to be, or feel themselves becoming, ornamental, feminine, invalid. Carlyle, appreciating the jeopardies, exploited the rhetoric of invalidism in his petition to a neighbor: "We have the misfortune to be people of weak health in this house; bad sleepers in particular; and exceedingly sensible, in the night hours, to disturbances from sound. On your premises, for some time past, there is a Cock,—by no means particularly loud or discordant,—whose crowing, would of course be indifferent or insignificant to persons of sound health and nerves; but alas, it often enough keeps us unwillingly awake here, and on the whole gives a degree of annoyance which, except to the unhealthy, is not easily conceivable." Jane herself suffered from excruciating headaches and shared murderous fantasies about cocks and macaws, so Thomas's "we" was a true alliance against loud fowl; but, as biographers have observed, it was Jane who endured one noisy renovation project after another to fortify their leased house against the outside world, while Thomas stomped around upstairs in huffs, struggling to make good on the "strenuous idleness" of a man in the arms of the Muses. He also struggled to make himself publicly useful by giving lectures to paying audiences. "I am again at my old trade of basket-making, or lecturing," he wrote to a friend in 1838. "It goes along very tolerably; though at a frightful expense to my nerves." Speaking extempore, he had woven each basket in his head and at his desk, so when he felt depleted by the "whirligig (of a London City, with its million-and-half all scrambling together)," how disastrous it was to find no calm at home. "I go home from the City, the brain overwrought, feverish, and fatigued," complained another brainworker, Victor Baune, on returning from the office, "and I require rest and change of occupation—reading, writing, music—and these are impossible with the horrible street music from all sides."[298]

Baune's was one of numerous letters that appeared in *Street Music in the Metropolis,* published in 1864 by the brewer Michael Bass, a liberal M.P. who was campaigning for a legal remedy against these nuisances, whom the police could not remove except by cause of a sick or dying person in the vicinity. But what was it about street music, among all the enemies of brainwork at home—the bawling of infants, the pounding of washboards, the hammering of repairmen—that persuaded Babbage, Carlyle, Dickens, Tennyson, and eighteen other writers and artists to sign on to Bass's Bill for the Suppression of Street Music? Of all noises, why did these men of stature find themselves most "worried, wearied, driven nearly mad" by "beaters of drums, grinders of organs, bangers of banjos, clashers of cymbals, worriers of fiddles, and bellowers of ballads"? Was it the brazenness of brass bands that, unsolicited, played loudly and at length and then had the gall to knock on each door for payment? "And if any person with the slightest musical taste could tolerate a street band at all, it was perfectly clear they could not tolerate four bands at once, shrieking, blasting, counter-blasting, and creating the most horrible discord"—so was it the horrible simultaneity of musicians clamoring for business? Or was it the chaos of boy bands, "every member of which attempts a different tune at the same time on a damaged wind instrument"? Was it the "nigger" melodies, or the fact that sweet voices and nimble coordinations of fingers and tongue were being shoved aside by "music by handle," rusted metal reeds, and monstrous mechanisms that reproduced the "piercing notes of a score of shrill fifes, the squall of as many clarions, the hoarse bray of a legion of tin trumpets, the angry and fitful snort of a brigade of ragged bassoons, the unintermitting rattle of a dozen or more deafening drums, the clang of bells firing in peals, the boom of songs"? Or was it the literal foreignness of these sounds, arising often from Germans who had fled after the failed revolution of 1848 or Italians running sweet youth through the mill of an ugly poverty? Was it the itineracy itself, like that of "the notorious widow, whose infant phenomena perform irritating sonatas on a jingling pianoforte placed on a costermonger's vegetable truck, drawn by a small donkey"? Or was it, as John Picker suggests, a boundary dispute in cities whose populations were doubling every twenty years (charted by Babbage), a dispute between two differently precarious groups—the poor, who were being dislocated by the levelling of shanties for train tracks or gas works, and ill-at-ease cadres of brainworkers, who were caught between definitions of gender and of power, both groups disputing what was basic to urban life, the aplomb of sociability or assurances of privacy?[299]

Answer: all, and more. Street music vexed many of the publicly articulate because it was heard&seen to be inescapable, heard&thought to be noise, and heard&felt to be a distinctly *penetrating* kind of noise. Why was it heard&felt to be so penetrating? After all, until purpose-built rail stockcars and viable mechanical refrigeration became available in the 1880s, urban meat markets depended upon supplies of thousands of cattle, sheep, and pigs driven in the early morning through city streets, bleating and squealing down ramps into abbatoirs whose walls scarcely damped the sounds of slaughter, sounds that should have been psychoacoustically as penetrating as most organ grinders—to make an historically justifiable pun, given the new sentiments that led to the founding of popular societies for the prevention of cruelty to animals (1824 in London, 1845 in Paris, 1866 in New York). Given also the frequency and ferocity of fires, with all the cries of alarm, the grating of watchmen's rattles, the sharped whistles of police, the clanging of brass bells off metal fire engines, the hallooing of firemen and their apprentices, the whinnying of frightened horses in nearby stables, the cursing and fisticuffs between competing fire companies rushing to the scene, the screams of people caught in the blaze, the harsh commands of brigade chiefs through speaking trumpets, firesounds should have been as evocative and penetrating as any motley of trombonists—here was how one Londoner heard a fire being fought in 1854: "shouts of the crowd as it opens to let the engines rush through it; the foaming head of water springing out of the ground...; the applause that rings out clear above the roaring flame as the adventurous band throw the first hissing jet; cheer following cheer as stream after stream shoots against the burning mass, now flying into the socket-holes of fire set in the black face of the house front, now dashing with a loud skirr against the window frame and wall, and falling off in broken showers." And let us not forget, in the quietest of nights, that dogs whining or howling in their pens, roosters crowing, and alleycats snarling and wooing could reach more penetrating tones than any street operator taking on *Il Trovatore*. Trying to make the best of "this 'very *penetrating* world'—as a maid of my Mother's used to call it, in vapourish moods," Jane Carlyle lost sleep and patience when a neighbor had not one but two cocks crowing so near "that it goes through one's head every time like a sword," waking her every quarter hour, Thomas at three a.m., again at five. "If this goes on, he will soon be in Bedlam; and I too."[300]

Bass, however, pushed for no bill silencing slaughterhouses, fire engines, or roosters, let alone his own firm's drays of beer barrels tun-rumbling down narrow streets at dawn, drivers counting on that very noisiness to warn off pedestrians. Babbage himself was unaffected by the noise of trains he rode while measuring and recording the vibration of railway carriages; at home he relished the companionship of squawking parrots. Charles Dickens, who was of several minds about "street minstrelsy" (happy when strains of opera reached the masses, unhappy with "barbarous jangles" and raunchy ballads, fond of "Ethiopian Serenaders" knocking tambourines on their heads and "rattling their bones with fifty-horse power"), felt "no nervous discomfort" with the bawling of fishsellers and claimed that "the two merriest and most consoling sounds of common life are the squeak of [the puppet] Punch and the clink of a blacksmith working at his forge." The clink?! Sometimes indeed Dickens missed London's crowds when writing in quieter towns, and his reactions to Italian organ grinders or German bands varied, he knew, with the state of his liver. What sounds always drove him mad? "I shudder at the shrill, screaming, ceaseless whirr" of power saws, he wrote in his magazine *All the Year Round*, and at the "diabolical screech of the knife-grinder's wheel," the scrape of pencil on slate, the repulsive rip of calico.[301]

Street music, then, required parliamentary action not because it was louder, shriller, more repulsive, or more ubiquitous than dogs, horses, cocks, parrots, beer wagons, firebells, police whistles, power saws, or screaming children, but because it was heard&felt to penetrate the ear and brain in ways more devious and hurtful. By mid-century, music was understood to command attention, and the public had more enclosed venues in which to pay music its proper respects than it did fifty years before, when there was but a single hall for London concerts and (reminisced one William Harvey in 1863) the finest of singers took their places in the open air, where a man strolling down the street could hear the marvelous Peter Links singing,

> When the wolf in midnight prowl
> Bays the moon with hideous howl,
> Howl, howl— —

and other "tender, natural-toned airs." Now meant to be listened to, not through, music when on the avenue was more distracting than cats or traffic or begging "grizzlers" deft at making miserable noises that drew passersby to commiserate—and all the more disturbing when played badly, loudly, unrelentingly within the hearing of artists, intellectuals, and paying audiences trained to pay discriminating attention to it. The cultural practice of listening to professionals with well-tuned instruments in gilded rooms seemed to be mocked by itinerants on the road who commanded attention through loud repetitions and industrial rhythms generated by music machines like barrel organs, or who, in "any alley, mews, cul-de-sac, fore-court, or garden near any house" produced almost-random chords from frayed strings and raggletaggle horns. As the English physicist Thomas Young had stated in 1807 and as the German physiologist von Helmholtz confirmed in 1862, the very definition of noise was singularity, a rapid alternation of sounds insusceptible of being organized into anything regular, while "a musical tone strikes the ear as a perfectly undisturbed, uniform sound which remains unaltered as long as it exists." Interruptedness was the essence of noise, an interruptedness that, as Helmholtz would show, was as much psychological as physical. Played haphazardly, street music could be a peculiarly insinuating torture to a mind seeking periodicities, a torture no less painful than that aimless humming and tuneless whistling against which etiquette books had begun to rail. Heard in snatches between cries for "chairs to mend" or the dustmen's bells, "the low and feeble strain / Of some sweet, simple, sadd'ning childlike song" could be as maddening as it was saddening for both the passerby and the Italian girl who strove in vain "to pierce surrounding din / With feeble voice and shattered violin." And in an era when both "folk" and "classical" musics were being aestheticized and detached from daily labor, street music could also appear to be a class tantrum, breaching the cultural line between hard menial tasks and intellectually rewarding, spiritually fulfilling lifework just as it breached the physical lines between tenement and townhouse.[302]

Wherefore Thomas Carlyle acted upon a dream he had entertained for a dozen years, namely "an apartment free to air and absolutely inaccessible to sound,—at once perfectly ventilated and *deaf*," which became in 1853 "a top story put upon the House, one big apartment 20 feet square, with thin *double* walls, light from the top &c. and artfully ventilated,—into which no sound *can* come, and all the cocks in Nature may crow round it, without my hearing a whisper of them!" This would be one of the first rooms ever built specifically to be impenetrable by external noise ("airy, and yet utterly *deaf*, inaccessible to sound"), where Thomas hoped "to be independent of all men and all dogs, cocks and household or street noises,—which are waxing every year in this Chelsea with the furious building &c that now goes on in the once quiet suburb!" His dream of "reading, thinking, living in an absolute *divine silence*" was, he confessed, the demand of "a wounded and humiliated mind," the mind of an "unvictorious man" struggling to complete a twenty-one volume biography of the Prussian monarch Frederick the Great, all of whose eighteenth-century victories were undone by the tumult of the French Revolution. The Carlyles's own tumult began in August as their contractor set to work, and after a month's "crisis of discomfort," Thomas fled to the countryside, leaving Jane as always to deal with falling plaster, clumsy whistling workmen, and a hell of hammering. In "perfect solitude" south of London, Thomas awaited "a perfectly *sound proof* apartment," by which he meant "*deaf* utterly, did you even fire cannon beside it, and perfect in ventilation; such is the program,—calculated to be the envy of surrounding 'enraged musicians,' and an invaluable conquest to me henceforth, if it prosper!" Prosper it did not. He returned in October to find the room almost done, almost done again in November, by which time he feared that "my room is irremediably somewhat of a failure," though each window shutter be stuffed with cotton wool. In December the "Demon Fowls" next door were still a-crowing, and his "sound-proof" room, a phrase of his own coining, seemed a "totally futile issue." Jane wrote, "Alas! And the silent apartment had turned out the noisiest apartment in the House! And the *cocks* still crowed and the *Macaw* still shrieked! And Mr C still stormed." In January his mother died, and it was May of 1854 before Mr C could spend a full day in what he called his "garret," which turned out to be spacious and bright, but "As a room that was to be *silent*, inaccessible to sound, it is a most perfect *failure*, one of the undeniablest *misses* ever made!" Despite its "silent remoteness" atop the house, behind double doors, "I do hear all manner of sharp noises,—much reduced in intensity, but still perfectly audible; it is only the *dull* noises that are quite annihilated." Yet he had to admit, in this glowering spring between England's "mad" declaration of war on Russia and the Turks and the first disastrous battles in the Crimea that fall, "for the rest, it *can* be defined as an eminently *quiet* room, far *less* noise in it than any other . . . and perfectly isolated from the rest of the house."[303]

Distance and space can "purify" sound so that the inharmonious becomes harmonious, declared a Boston physician. On that point the Carlyles would have read J. Baxter Upham's *Acoustic Architecture* (1853) with scepticism, but as residents of a populous Chelsea permeated with sounds viciously inharmonious, they would have agreed with a quote he lifted from a fellow Bostonian, Dr. Luther V. Bell of the McLean Asylum, who in 1848 had lectured the Massachusetts Medical Society on "true ventilation." Thomas, who had specified good ventilation for his New Room because he liked to recall the air of country walks, and Jane, who knew that otherwise his studio would hang heavy with cigar smoke, would both have been partial to Bell's analogy (the earliest such?) between air pollution and noise: "At first thought it may appear somewhat fastidious and even unphilosophical, to regard merely unwelcome sounds, recognized by the ear alone, in the light of offensive impurities, or rather as disagreeable additions to the medium in which we live and intercommunicate, yet no explanation will be needed by a body of medical practitioners, residing, in part, in large towns and cities, for considering the means of obviating, or alleviating, the annoyances and injurious irritations of painful sounds, in the same category with the disagreeably offensive or actually malarious contaminations." If love of alliteration had led Carlyle to write of London's "grinding dust and discord," it was also the case that dust and discord had become twinned, physiologically and psychologically; where noise in waves failed to penetrate his garret by direct assault upon its double doors, it could filter in like particles of dirt through every crevice, any window cracked open. How craft a habitable sanctuary from noise that had to be as air-tight as the bell jar disparaged by Teufelsdrökh?[304]

President of the Boston Music Hall Association, Upham in forty pithy pages surveyed what was known about architectural acoustics after a generation of efforts to redesign all kinds of listening chambers—choral and symphonic, parliamentary and congressional, operatic and theatrical, jurisprudent and ecclesiastic—for new audiences being directed to listen-in. Stringing wires overhead to absorb stray or tinny vibrations was a common if admittedly stopgap measure and worked nowhere near as well as the thick draperies, carpets, and plush seats of Victorian interiors or the mid-century's wholesale rebuilding of churches and concert halls with lower, less vaulted ceilings and horseshoe seating around a raised stage. To tell the truth, overhead wires did nothing whatsoever to "purify" or amplify sound, but the premises for their use were as analogically clear as they were mistaken: like acoustic lightning rods, they would draw off the excess; like telegraph wires, they would conduct unseen impulses, with the same selectivity that enabled mediums, shivering like highly-strung wires, to repel irrelevant or irreverent spirits from the Other Side. Although Upham found that "The subject of Catacoustics, or the doctrine of reflected sound, is, perhaps, the most unsatisfactory in its results of any branch of physical science," he believed he knew enough to take into account the number of right angles in a room, the hardness of its plaster and panelling, the depth of its upholstery, the variety of its recesses, and the "constant discord" of life just outside. He also knew what sounded sublime—"the mingling of a thousand voices and instruments in an open field," or concerts at the Boston Music Hall itself—but he had no equations to offer and was stumped by the multifocal: he found the problem of designing a legislative chamber where each speaker could be heard from his own place "incapable of being solved by any of the known principles of science." It had not been solved, for example, by Benjamin Latrobe, the architect who completed the United States Capitol. Latrobe's 1803 "Remarks on the Best Form of a Room for Hearing and Speaking" began by noting how little had been done on acoustics since Vitruvius, proceeded to correlate shape with echo, and concluded by endorsing humbler versions of neoclassical domes, despite their noisy "buzz," because therein a speaker would find himself at ease, "as it were, floating upon the sound of his own voice." Orotund congressmen doubtless enjoyed that buoyancy, but for debate and discussion (where parliamentary rules directed that "no member is to disturb another in his speech by hissing, coughing, spitting, by speaking or whispering") the House of Representatives was a notable failure. Fifty years later the American physicist Joseph Henry was consulted on plans for a new, acoustically improved chamber for the House; unimpressed with the Capitol Building from the day he first visited Washington, Henry now listened for resonating keynotes and clocked the reverberations in good halls from Philadelphia to Boston, but he did not have a handle on the multifariousness of sound that, by definition, made noise what it was. As a Scottish contributor had written in the prestigious *Journal of the Franklin Institute*, even buildings that reproduced the dimensions of halls known for their clarity could end up in acoustic shambles, and those inventors who had shunted a mobile pulpit around a Liverpool church had never found the perfect spot from which to preach. For a lecture hall at the Smithsonian, Henry proposed a fan-shaped room with the podium "placed in the mouth as it were of an immense trumpet" and the audience seated along a "panoptic curve," a layout that presumed a habit of listening-in. Where the trumpet was reversed—where it seemed as if all the noises of the world were funneling in toward a single person or house, as it had seemed to the Carlyles and (with more statistics) to the hounded Babbage—what could anyone do?[305]

Parliament passed Bass's bill in July 1864. By August, Thomas had informed Jane that the "yellow demons," his ostentatious smear for Italian street musicians, were "rifer than ever." As Babbage lay dying in 1871, his son Henry, who shared a suspicion of beggars from years of military service at outposts of the East India Company, found "a man inciting boys to make a row with an old tin pail" while organ grinders converged on Dorset Street day and night to triumph in his father's passing. Later campaigns against street musicians in London, Paris, and New York would have greater success when drawn away from the music itself—which, said many, was a rare source of beauty in the lives of the poor, protested only by "frumpish old maids and bitter bachelors, who have no music in their souls." It was more effective to direct social animus against the cruel treatment of street-musician children as "slaves of the harp." Thomas Carlyle meanwhile abandoned his own campaign against organ grinders and even against roosters, as Jane reported in a letter from Chelsea on Christmas Day, 1865: "For years back there has reigned over all these gardens a heavenly quiet—thanks to my heroic efforts in exterminating nuisances of every description," but that December the cocks had returned to wake her—"crows at 2, 3, 4, 5, 6, at seven! Oh, terribly at seven!" without the slightest protest from her husband, for "his whole attentions having been since his visit to Mrs. Aitken morbidly devoted to—railway whistles."[306]

Pharmacologists might explain Jane Carlyle's "terribly at seven!" by the quinine powders she had been taking steadily since the 1820s to deaden the pain of her debilitating headaches; quinine produces tinnitus and hypersensitivity to noise, side-effects reported since the 1820s but generally ignored until the 1870s, given the obvious effectiveness of quinine against high fever, pain, ague, and malaria.[307] Ophthalmologists might consider Babbage's acoustic sensitivity a compensatory response to his longstanding bilateral monocular diplopia, or double vision in each eye, which he could work around only by squinting through a small hole in a card or by use of a concave lens (and which diplopia may have been the motor in his drive for a reliable calculating-and-printing machine, since the most difficult of formats to create and to correct by a man with double vision would be columns of numbers). Cardiologists and audiologists might attribute

Babbage's outrage at street musicians to restricted cochlear blood flow caused by the chronic peripheral arterial disease that eventually killed him; with cochlear degeneration would have come loss of hair cells, a reduced dynamic hearing range, and "loudness recruitment" in which the ear, straining to amplify softer sounds, is stressed by louder sounds, especially at higher frequencies which are typically the first to become inaudible and to which newly recruited bundles of cilia have been sensitized.[308]

Seductive though it is, such physiological determinism would be as hazardous as accounting for the exaltation of the ear over the eye in "On the Power of Sound" by recourse to the problems that William Wordsworth was having with his eyesight in 1834 while revising the poem for publication. His blurred vision and his recoil from "dazzling lights" may have conditioned the final stanza's apocalypse in which the visible world vanishes, outlasted by "audible harmony, and its support in the Divine Nature," but this was the same poet who fifty years earlier

> would walk alone,
> Under the quiet stars, and at that time
> Have felt whate'er there is of power in sound
> To breathe an elevated mood, by form
> Or Image unprofaned:

And it was the same poet who some forty years earlier had come upon "a perfect quietness" in the air;

> The ear hears not; and yet, I know not how,
> More than the other senses does it hold
> A manifest communion with my heart.

The lyrical Wordsworth of Book Two of *The Prelude* had not suddenly embraced audible harmonies in fear of incipient cataracts at the age of sixty-four. Within a culture whose readers regularly read aloud to each other and patiently read aloud to those who could not read, and where companionship meant the singing of songs and telling of tales well into the night, he trusted to the power of sound. Within a culture whose well-off households of the 1830s had begun to gather, servants and all, for family prayer, Wordsworth could conclude,

> No! though earth be dust
> And vanish, though the heavens dissolve, her stay
> Is in the WORD, that shall not pass away.[309]

So too, when Thomas Carlyle went to visit his sister, Jean Carlyle Aitken, he was upset by the train whistles not so much because he was "shivery in the nerves" and once again overcome with "a great feeling of *thin-skinnedness,*" nor so much because there was little love lost between him and the

Age of Machinery with its "horrible squeaking choking railway trains," but because those steam whistles had the most powerfully penetrating and intentionally abrasive new sound of the nineteenth century, piercing every ear in the market town of Dumfries and across the countryside from Ecclefechan and the Scottish moors to the mines of Wales.[310]

17. Shrill

Adrian Stephens, chief engineer since 1827 at the Dowlais Iron Works in Wales, had been researching an automatic alarm system to alert workers to boilers about to explode. High pressure steam boilers, now powering ships, locomotives, mills, foundries, mine-pumping stations, and factories, too often did explode, due either to a shortage or blockage of water or to structural weakness, about which there was, in 1829, "no concurrence of opinion, even among scientific men." Among the non-scientific, "Even in cases where the danger arising from an accident is trifling," wrote one correspondent to an investigating committee in the United States, where boilers were kept at higher pressures than in the United Kingdom, "the noise, and the vapour that issues from a very small opening in the boiler, or other parts of a high pressure engine, are so much more alarming to persons ignorant of the cause, than when a similar accident happens in an engine of low pressure." The first locomotive boiler explosion in America occurred in June 1831 aboard the *Best Friend of Charleston*, the first locomotive built entirely in America, when the fireman, disturbed by the perpetually shrill whistle of steam escaping through the release valve, tied the valve down; pressure built until the boiler burst and killed him. Later accidents could be even more sudden, loud, and violent, as in 1833 with the *New England*, a steamboat whose water levels had been good just three minutes before its copper boilers "were torn asunder, and folded in massy doublings, like a garment; and they were so crushed, flattened, and distorted, that as they lay upon the wharf, after they were raised from the bed of the river, it was difficult for a common observer to discover how the mutilated parts were ever connected." Low-pressure boilers for the heating of brewing vats and tobacco sheds, though less likely to explode, already had built-in alarms of some sort; it was imperative to find a system for the more volatile and dangerous high-pressure boilers. Neither ordinary whistles nor organ pipes had proved loud enough to be heard above the fierce pounding at the Iron Works, so around 1832 Stephens crimped one end of a copper tube tight as the tip of a boatswain's whistle and secured the open end to a safety valve on a boiler. Experimenting, he found that the longer the tube, the louder the sound, until "I had got the harsh and unpleasant, though useful, sound at last." The physician and physicist William Wollaston had noticed in 1820 how useful a harsh unpleasant sound could be: the hard-of-hearing registered sharp or shrill sounds much better than low dull sounds, and people with "exhausted ears" continued to hear the rattle of a chain or the "scraping of catgut unskilfully touched" long after the rumble of a carriage had become inaudible; now, the fife-like shrillness of escape-valve steam channeled at high pressure through a narrow metal mouth would warn iron workers, despite exhausted or half-deaf ears, of impending explosion. Within three years, the "Eldritch shriek" of Stephens's steam whistle alerted teamsters, shepherds, equestrians, and pedestrians to locomotives miles off. Within a decade, this jet-powered noise became a fixture on trains rattling at fifty miles per hour across Great Britain, taking the explosive soundmark of modern industry from Midlands mills and Welsh mines directly into the cities then out again into the countryside far past Ecclefechan,

Roaring o'er the trembling land,
Mountains piercing, vallies crossing . . . ,
Shouting aye, to wood and fell,
My infernal roundelay
Over moor and pasture screaming

and on to the United States, where larger and more various steam whistles—piston, compound, banshee—would be run through their gamut than in any other nation.[311]

Thoreau at Walden Pond in 1845–1847 could hear the whistle of locomotives on the Fitchburg line three miles distant. It "penetrates my woods summer and winter, sounding like the scream of a hawk sailing over some farmer's yard." In his rambles, Thoreau met trains as if fiery comets, but what struck him most (he who claimed that the "five senses are but so many modified ears") was their constant audibility, morning, noon, and midnight: "All day the fire-steed flies over the country, stopping only that his master may rest, and I am awakened by his tramp and defiant snort at midnight, when in some remote glen in the woods he fronts the elements incased in ice and snow." Even when he could not map their rails or track their smoke, the engines made their presence known by whistling steam, heard almost everywhere, through rain and snow, almost all the time: "The startings and arrivals of the cars are now the epochs in the village day. They go and come with such regularity and precision, and their whistle can be heard so far, that the farmers set their clocks by them, and thus one well-conducted institution regulates a whole country." Back in town, he undertook a systematic study of botany and zoology, conducted escaped slaves on Concord's Underground Railroad, made a living as a surveyor, and revised *Walden, or Life in the Woods* for publication while walking to lectures and lecturing on "Walking." In his journal for 12 June 1852, he would write, "The steam whistle at a distance sounds even like the hum of a bee in a flower. So man's works fall into nature."[312]

More must be said about that "fall into nature," and the fall of noise itself "into nature." Next chapter. For, as Thoreau understood shortly before factories began to whistle the change of shifts, before tugboat captains, loggers, and rail cabmen began to shrill new languages of steam, the screams of steam whistles completed the ambitions of noise, Round One: Everywhere. Penetrating Thoreau's "Dismal Swamp" and "unfrequented woods," the steam whistle took noise into the last refuges of dwindling Echo, into those places where had been heard only the creaking of trees, the plunge of water, the wail of singing dunes. Offblow of industrial explosions too loud, too deadly, and too common, the steam whistle signalled the start of Round Two: Everywhen, Everyone.

ROUND TWO

Everywhen, Everyone

On ears of all sorts. On who is hearing noise, under what conditions and at what time of day or year of life.

> ... as if an enchanter or a stage manager,
> at the first peal of the whistle from the first locomotive,
> gave a signal to all things to awake and take flight.[1]

So recalled the photographer, cartoonist, and balloonist Nadar, who was twelve when a locomotive first huffed along French rails, thirty-two when a steam-driven dirigible luffed through the Paris air, eighty-six when winged aircraft sputtered up from the gardens of the Bagatelle. Himself the first to shoot aerial photographs, Nadar was also the first to take studio

SOUNDPLATE 4

portraits by electric light and the first to take lights and camera down into the sewers of Paris. His descent in 1861 came at the invitation of the chief engineer, eager to document the metamorphosis from what Victor Hugo knew as a "fetid, savage" underworld. Reconstituted after epidemics of cholera, typhus, typhoid fever, and revolutionary fervor, the sewer system was vital to Baron Haussmann's plans for the metropolis at mid-century: wider avenues aboveground for the unimpeded march of the forces of public order, wider avenues below to carry off the putrid masses that threatened the public health. While some 350,000 of the "dangerous classes" (and hundreds of street criers) were raked from the central *quartiers* by Haussmann's "strategic beautification," sixty miles of sewer were resurfaced and another three hundred miles added with such alacrity that the Paris Exposition of 1867 could offer subterranean tours aboard canal boats ushered along by trumpeters all in white. Nadar also shot the catacombs with their walls of skulls, his arc lights hissing off bone and socket with a mortuary closeness; the sewers were of a different order, curving vistas of sculpted hollows waiting to be filled with well-dressed tourists in the colors of the 1860s—Perkin mauve, Verguin fuchsine, Hofmann violet, Martius yellow—all recently synthesized from coal tar. The new hues were so vivid, even at night under thousands of newly installed gas lamps, that they seemed to glare at the older generations still partial to their "muddied, quenched colors (the so-called fashionable colors)" prepared from mollusk, insect, or vegetable dyes. The fading generations glared back, already appalled by the "loud" dress of dandies and shirts "too 'loud' in pattern," decrying now the loudness of synthetic colors in word and thought, crossing over, in this description, from sight to sound as the Paris sewers had crossed under from organic stink to artificial light.[2]

Alerted by the steam whistle, this chapter listens for those crossings over and under that transported noise from a set of singular if pervasive sounds into a series of analogs of sound, persistent and prevailing. As noise in the later 1800s came to be experienced as ubiquitous *and* interminable, it came to be understood not only as inescapable but intrinsic. Through the enchantments of science and the staging of acoustic technologies, through attachments of the law and stagings of the City Beautiful, through detachments of statistics and stagings of the psyche, through the disenchantments of a Second Industrial Revolution and its staged management of time, noise slipped out of the range of the simply sonic and won the privileges of metaphor.

1. Night (h)Owls

Metaphoric noise emerged with one of Nadar's intimates, Charles Baudelaire, under the influence of the German fabulist and composer E. T. A. Hoffmann and of fabulous hashish, whose hallucinations caused "the most inexplicable transpositions of ideas.... In sounds there is color; in color

SOUNDPLATE 5

there is a music." Reviewing the art exhibitions of Paris, with their "turbulent, noisy, riotous bedlams," Baudelaire began to judge painterly color in terms of sound: his friend Eugène Delacroix's palate was "often plaintive"; Horace Vernet's well-received *The Battle of Isley* badly substituted "charivari for color." The pivotal word, the word that swung between the optical and acoustic, was *tone*. According to the German philosopher J. G. von Herder in 1802, painting and music were arts mutually grounded in gradations of tone; his friend Goethe in 1810 suspected that tone would one day be found to have something scientific to do with color as well as sound. Henry Fuseli, the Swiss-English painter remembered for his darkly erotic *Nightmare*, defined tone as a "union of tint and hue spread over the whole" and told a Royal Academy audience that Titian's eye had been "as musical, if I may be allowed the metaphor, as his ear." During a course of lectures in 1821, the royal drawing-master William Craig, a watercolorist, identified tone with luminosity and "discordant" tones with bad art. By 1843 John Ruskin was defending the "dazzling intensity" of the canvases of modern painters by reference to their tones, like those effected by J. M. W. Turner in *Connecticut Steamboat in a Storm*. Tone, explained Ruskin, was "the exact relation of the colors of the shadows to the colors of the lights," which led him to feel the whole of a picture "to be in one climate, under one kind of light, and in one kind of atmosphere." With art-critical precedent, then, and seconded by music theorists who had been homing in upon tonality as the heart of composition, Baudelaire could call the paintings at the 1846 Paris Salon "a tohu-bohu of styles and colors, a cacophony of tones."[3]

Such noise-keen vision was neither unanticipated nor unique, although *Les Fleurs du mal* (1857) became famous for Baudelaire's celebration of "perfumes, colors, and sounds that answer to each other" and infamous for the gleam of his naked darling's "sonorous jewels" whose precious metals dance out a reckless noise as the poet confesses to a passion for all that marries light and sound. He had come upon a similar *son et lumière* in the prose and verse of Edgar Allan Poe, whose work he translated over a period of sixteen years: the "unquiet lights" of chintzy chandeliers in "Philosophy of Furniture"; the relentless beating of a "Tell-Tale Heart"; the murmurs, like the rush of subterranean waters, from the ghastly water-lilies of "Silence." Did Baudelaire know, as Poe would have known, that a "tell-tale" could also be a device to alert engineers when the water in a boiler was too low and the boiler about to explode? Was he not touched to the quick, as an avowedly "irritable" poet, by that passage in "The Fall of the House of Usher" where Poe describes the abstractions painted by Roderick Usher, whose "morbid acuteness of the senses" was so intense that "the odors of all flowers were oppressive; his eyes were tortured by even a faint light; and there were but peculiar sounds, and these from stringed instruments, which did not inspire him with horror"? And could he have failed to recall, during his own tour of the sewers (in usual rose gloves and violet boa?), Poe's account of one of Usher's paintings that revealed "the interior of an immensely long and rectangular vault or tunnel, with long low walls, smooth, white," deep below the earth with no evident outlet or source of light, yet flooded by rays that "bathed the whole in a ghastly and inappropriate splendour"?[4]

I mean by these questions to drag noise-keen vision back to brick-and-stone tunnels that were no less sensational than literary fictions of the underground, and much louder. Back past the iron-and-glass arcades of Milan, London, Warsaw, New York, and Paris—commercial tunnels of reverberant hardwood and marble where Baudelaire strolled for hours, the city outside still buzzing in his ears "louder than an American forest or a beehive." Back past tunnel vision, which was coupled with extreme sensitivity to noise in *migraineurs* like Poe, in syphilitics like Baudelaire, and in those born with Usher Syndrome, a loss of hearing and peripheral vision first documented in 1858 by a German ophthalmologist but named for an Englishman (no, not Roderick) who determined that it was congenital. Judging by letters from the likes of Jane Carlyle and by shelves of popular nostrums, a large though largely unknown number of nineteenth-century adults suffered "sick headaches" and associated phonophobias. Judging by public health data long before penicillin, many among the 10 percent of urban males who contracted syphilis (17 percent of Parisian men, 25 percent of German university students) must have suffered the vertigo and tinnital deafness of its later stages. Arcade passages and tunnel vision, however, afford but the narrowest preview to the new incessancy of noise during the later nineteenth century, an incessancy loudest inside tunnels of stone and brick that had been gouged through mountains, burrowed beneath cities.[5]

Tunnels had roots and routes in paleolithic caves and castle keeps, in the mining of ore and confining of prisoners, but as everyday structures in an industrial West they were products of steam and nitroglycerine: steam engines to pump out water or pump in air, steam-driven boring machines and pneumatic drills (1860s), sticks of dynamite (1866). Steam made longer tunnels possible in the absence of such expendable prison labor as had built the four underground miles of Napoleon's St. Quentin Canal (1810); dynamite quickened the murderous pace. Constructing mineshafts had ever been a loud, perilous occupation; now tunnels of all kinds engaged perhaps a million laborers and killed or maimed as many as a thousand at each of the major projects of the 1800s, while hundreds at a blow were lost to mining disasters, boiler explosions, rail collisions. Pushing through most of the projects were rail companies hungry for connection, although the first long tunnels were canal tunnels and the first rail tunnel was built in 1826 for French miners whose workhorses pulled ore-cars along a track beneath Terrenoire, "Black Earth." By then the mile-long Wapping Tunnel was underway, with rails laid for steam locomotives that hauled raw materials and manufactures between Liverpool docks and Manchester factories. A journalist there heard "the busy hammer, and chisel, and pickaxe, the rumbling of the loaded wagons," then "the frequent blasting of the rock, mingling with the hoarse-sounding voices of the miners," tangs and rumblings repeated at each new incursion into the earth. A passenger of 1836 found the transit through those tunnels "very electrifying," by which he meant very scary: "The deafening peal of thunder, the sudden immersion in gloom, and the dash of reverberated sounds in confined space, combined to produce a momentary shudder, or idea of destruction." That shudder was transposed into a *piano rondo* by a composer who caught the tempi of a locomotive powering through the tunnel into Liverpool. Less musical but more controversial was the tunnel at Box Hill for the Great Western Railway between London and Bristol, due in part to its numbers (2 miles long, 4,000 men and 300 horses at work in 1840 among the swirling dust raised by blasts of gunpowder moving 247,000 cubic yards of spoil, with 100 deaths and hundreds of casualties), in part to passenger fears of a sulfurous darkness, and in part to its acoustics. The noise of two trains passing each other in such a long tunnel, claimed a popular lecturer, would be intolerable; the vibrations from the trains, warned the learned

Rev. William Buckland, Oxford geologist and biblical catastrophist, would bring unlined rock walls tumbling down. Undeterred, rail companies kept tunneling: at Woodhead on the Manchester-Sheffield line (1838–1845, 3 miles long, 157 tons of gunpowder, 32 dead, 700 hurt); at Hoosac in the Berkshires of Massachusetts (1851–1875, 5 miles, first use of dynamite, 195 dead, hundreds hurt); at St. Gotthard in the Alps (1872–1880, 9.5 miles, 310 dead, 877 invalided). In addition to sewage tunnels beneath cities and rail tunnels through mountains, there were drainage tunnels for the scalding waters of deep-shaft mines (the Sutro Tunnel draining Nevada's Comstock Lode, 1869–1877, 4 miles) and shorter, shallower pedestrian tunnels (under the Thames, 1825–1843), then utility tunnels, and tunnels retrieving times past (Layard's Nineveh, 1845–1854; Schliemann's Troy, 1868–1873), tunnels for commuters saving time present (starting with London's Underground, 1863). The total length of man-carved subterranean passages would multiply at least six-fold during the century, twenty-fold if counting man-sized sewer lines, two-hundred-fold if counting each bend in every coal, tin, iron, copper, gold, and diamond mine.[6]

Aside from tunnel vision as a co-symptom of afflictions that entailed tinnitus or phonophobia, and aside from an engineer getting "to learn the present condition of his machine even by the noise it makes as it echoes through the cuts or tunnels," what did tunnels have to do with the constancy of noise? They had to do with the enlivening of darkness, which had been a place and time of quiet. Quiet, mind you, not utter silence: in his camp below the summit of Mont Blanc, moved by the "deep silence" of the night to imagine himself the last soul in a dying universe, Bénédict de Saussure was awakened by the noise of an avalanche; in the depths of the Parisian night, by lantern light, lance brigades of rag-pickers stabbed their hooks into the city's garbage bins for remnants of metal, paper, bone, bread—a hard-scrabbling corps of ten thousand with foraging caps over their ears and donkeys braying behind them. But (cocks, pusses, and asses excepted) darkness had for millennia been relatively quiet: quiet enough for Bostonians to protest the "loud vociferations" of watchmen crying the hours of the night off-key, quiet enough for householders in London to come awake at "a trifling noise" from clumsy burglars, quiet enough for Parisians to make out the groans of bakers at the "nocturnal slavery" of their ovens.[7]

Darkness was being transformed into a louder place and a noisier time; the tunnel was metonymic for what was happening to the dark. Inside mines where grown men had to stoop, eight-year-old boys and teenage girls in harnesses "hurried" tubs of iron along rough floors or pulled carts of coal along screeching rails through ten to fourteen hours of Davy-lamp dusk. When mines were brightened and hours lightened near century's end, miners would use compressed-air drills, more jolting than pick-axes and nervier, higher-pitched than the whomp and whoosh of pumps. And as mines were drilled deeper, miners were constantly "jowling," banging the planks above them with a wooden rod, to assure that the roofs were solid:

> Jowl boys, jowl,
> Jowl boys and listen,
> There's many a marra misten man [many a mate missin'],
> Because he couldn't listen.

Approaching the St. Gotthard rail tunnel (world's longest in 1896), the engineer and fireman donned their gas masks and all of the passengers rolled up their windows, but Oscar Wilde's son Vyvyan found so much smoke sidling in through cracks that he could not see across his own compartment. During the half hour of this ordeal, the heat and dark were so awful and "the noise of puffing and clanking" so loud that children in a nearby compartment panicked, began to scream. In the London Underground of the 1890s, smoke gathered "in great woolly clots" and the noise, said one tabloid, was like "the shrieking of ten thousand demons above the thunder of the wheels." In sum, the experience of tunnels by those indentured to them and by those in transit through them was of darkness *and* noise. As it became harder to avoid tunnels, miners' nightmares of being buried alive would resurface as white-collar panic disorders about being confined in close places ("claustrophobia," 1875–1879) or trapped in crowds with no exit ("agoraphobia," 1871).[8]

Noise inside tunnels arose not only from the grinding of gears, the clash of metal wheels on metal rails or, in coal mines, detonations of fire-damp set off by a "penitent" who crawled on hands and knees with a torch held above his head to rid shafts of build-ups of carbon dioxide. Noise arose also from tunnels of the bronchi, throat, and sinuses as an ominous cough—the deep cough of black lung disease, the barking coughs of the croup, the "harsh ringing cough" of pseudo-croup, the phlegmy cough of broncho-pneumonia, the dry cough of asthma, the wet cough of coalmen's emphysema, the hoarse cough of laryngeal diphtheria, the wracking cough of smokers with throat cancers, the whoop of deadly *pertussis (tussis clangosa)* in children, and everywhere the hollow cough of consumption. In 1898 the general manager of London's Metropolitan Railway cited medical opinions that fumes from the Underground's coal-fired engines were therapeutic for asthma and bronchitis, but within the noisy darkness people were coughing to no good end. Coughing from clouds of rock dust and blasting powder, then dying of black lung at forty-five. Coughing from dust in factories producing lead carbonate for painting the world white, then dying of lead poisoning. Coughing from years of inhaling the sulfur of household coal fires and industrial chimneys, until they had (as every urban correspondent seemed to have) a chronic catarrh, an infection lodged in the sinuses but active over a wider cultural territory: "the whole metropolis," wrote Dickens about London in 1864, "was a heap of vapour charged with muffled sound of wheels, and enfolding a gigantic catarrh." Coughing from dust clinging to the heavy draperies of fashionable parlors, or from the soot of less fashionable rooms filled with tobacco smoke that clogged the nostrils and led to the noisy habit of mouth-breathing, which led in turn, it was thought, to deafness and early death. Coughing from influenza and bronchitis, like most of the Pima children at Tucson's Indian Training School during the cold wet January of 1907. Or from tuberculosis: "We have taken their lands," wrote the school's headmaster, "and we have given whiskey and the White Plague." Coughing then from the tuberculosis bacillus that spread across desert reservations as quickly as through doss houses and sweat shops, into servants' basement quarters that shook with each lorry rumbling past and felt/sounded much like mine shafts, so that "as in that lower world one walked with one's ear nearer the ground," wrote the novelist Henry James, "the deep perpetual groan of London misery seemed to swell and swell and form the whole undertone of life."[9]

Brian Clapp, an environmental historian, estimates that during the 1880s Londoners lost a sixth of their sunlight to the city's "fog," then cites John Evelyn's *Fumifugium* of 1661 to show that smoke and fumes had long been troublesome: "For is there under heaven," asked Evelyn, a Fellow of the Royal Society, gardener, vegetarian, and critic of the burning of sea-coal at the hearth, "such coughing and snuffling to be heard as in the London churches and assemblies of people, where the barking and spitting is uncessant and importunate?" There was: by 1835 Manchester, ringed by foundries, lay under a heaven of "half-daylight." Thus the philosopher-tourist Alexis de Tocqueville, who heard more of the city than he could see: "A thousand noises disturb this dark, damp labyrinth, but they are not at all the ordinary sounds one hears in great cities." Instead, "the crunching wheels of machinery, the shriek of steam from boilers, the regular beat of the looms, the heavy rumble of carts, these are the noises from which you can never escape in the sombre half-light of these streets." Bouts of coughing in Manchester would have been as endemic as in nearby Leeds, where Dr. Charles Thackrah, listening for the insalubrious effects of industrial trades (e.g., the bronchitis of flax workers), had his auscultations upset by workers waiting to be examined, their coughing so strong "as continually to interrupt and confuse the exploration by the stethoscope." In Pittsburgh, it was the fierce coughing of glassmakers, ironworkers, coal miners. Across the industrializing world could be heard at home, at work, and in public spaces the bass cougher who "sounds very much like a young thunder-clap, grumbling at its own noise" or the soprano "sounding like the rolling cry of a locomotive at a public crossing." An invalid at a Denver boarding house, hoping like others to reclaim health by inhaling the bracing air of the West (contaminated though it was by the acridity of cigars and the branging of brass spittoons), found that "everyone coughed—dyspeptics, rheumatics, nervous wrecks, heart patients, kidney patients, ear patients, Keely cure patients—all coughed," and not a one turned her face away, not a one covered his mouth, all afraid to be thought to have the infectious "graveyard" cough of a consumptive. Simulated fits of coughing, behind gloved hands or handkerchiefs, were *de rigueur* in polite society to muffle impolitic yawns, evidence that coughing was more acceptably unavoidable than signs of boredom and more acceptably organic than the sneezing incited by earlier generations' anti-drowse snuff. Doctors diagnosed physical illnesses by the stridor of the cough; close friends and family prognosed consumptives by their coughs. "One felt," wrote the water-cure advocate Mary Gove after visiting Poe's raven-haired wife Virginia, "that she was almost a disrobed spirit, and when she coughed it was made certain that she was rapidly passing away." Psychiatrists certified hysterics by the constancy of their coughs. Freud in 1900 analyzed Dora Bauer's coughing fits as passive-aggressive "symptom choices" for a repressed image of her father being fellated by a woman not his wife. Puccini set the cough of the tubercular Mimi to the score of *La Bohème*. And in Moscow in 1881, at the breathless climax of Sarah Bernhardt's performance as *La Dame aux camélias*—based upon the life of the consumptive courtesan Marie Duplessis—the "Divine Sarah" coughed, and the audience coughed with her, and coughed, and coughed.[10]

Coughing and tunneling did not cease when curtains fell or the sun set. This was painfully clear to consumptives whom drugs had failed and whose nights were racked with that "Noise in my throat." It was clear to Friedrich Engels as he studied the reports of the Children's Employment Commission, which showed that Scottish children, women, and men were forced by mine owners to labor "below ground twenty-four, and even thirty-six hours at a stretch." It was clear to Jules Verne, who toured the "Black Indies" of Scotland's coal region in 1859 and came away with a romantic notion that the depths of mines might someday "provide a refuge for the working classes," the nub of his later utopian thriller about a city laid out fifteen hundred feet below a lakebed, haunted by a vindictive "penitent" at last defeated by happy families of miners working night and day under discs of electric light.[11]

Light of a sudden had a second circularity to it. Although the rhythms of farmers and tradespeople were still geared to sunlight and moonglow, strong were the economic pressures toward a more steadily lit circadian round as factories, offices, shops, taverns, and theaters began to be illuminated by oil lamps with effective glass mantles and metal reflectors, then by gas, electric arcs, paraffin and kerosene incandescence. The move toward light had begun in the 1700s when larger glazed windows and interior mirrors were fitted to middle-class houses. With the "industrialization of light" in the 1800s, not only miners and tunnelers would hack through the darkness. The greater the demand for the night-lighting of commercial thoroughfares, of public buildings and private pavilions, the greater the demand for the ceaseless operation of gas works and oil refineries; the more ceaseless these operations, the more regularly Capital could put Labor to work around the clock; the more ceaseless the labor, the greater the investment in powering up and the greater the disinclination to power down; the greater that disinclination, the greater the push for successive, unending shifts. So industries of light became the epicenter of a new rhythm of sight and sound.[12]

Of sound? Yes. According to George Sala, in *Gaslight and Daylight*, gas itself had a voice, "and I can hear it—a voice beyond the rushing whistle in the pipe and the dull buzzing flare in the burner. It speaks actively, to men and women, of what is, and of what is done and suffered by night and by day." In gaslight, the "shrill whoop" of factory whistles that had signaled the start and stop of work began to mark off one shift from the next. The "supernatural shriek" of train whistles would be heard through winter nights subserving systems of distribution that paid little heed to darkness or seasons. The pounding of paper mills (whose hours in Geneva had been curtailed by a noise ordinance as early as 1539) was now compounded by James Nasmyth's twenty-ton steam hammers, by the roar of blast furnaces, by "chucks a chuckin', drills a drillin'," and all the other sounds that Marietta Holley, the "female Mark Twain," reported in the persona of Samantha Allen after a tour of Machinery Hall at Philadelphia's Centennial Exhibition of 1876: "think of takin' the noise of seventy or eighty thunder-claps, and a span of big earthquakes, and forty or fifty sewin' societies (run by wimmen), and all the threshin' machines you can think of, and fifty or sixty big droves of lions and hyenas a roarin', and the same number of strong, healthy infants, under the influence of colic, and several hundred political meetin's and deestrick schools jest let out and several Niagara Falls; take the noise of all these put together and they don't give you any jest idee of the noise and distraction."[13]

Distraction was no minor affair when growing numbers of men and women had to work through the night and sleep in the day and when cities entertained more activity in the darkness so as to sustain their liveliness in the light. Where before, silence and darkness had been "Solemn sisters! twins / From ancient Night, who nurse the tender thought / To reason," as in the first stanza of the first of nine nights and forty editions of Edward Young's *Night-Thoughts on Life, Death, & Immortality* (1742+), a century later Søren Kierkegaard despaired of finding a tender silence anywhere, anytime, "And man, this clever fellow, seems to have become sleepless in order to invent ever new instruments to create noise, to spread noise and insignificance with the greatest possible haste and on the greatest possible scale." Where before "night birds" (hooligans, harridans) had been creatures akin to screech owls and night walking had been "unfit for honest men, and more suiting to the Thief (the night Whistler) and to Beasts of Prey," now nightworkers obeyed factory whistles that overrode Saint Monday, that post-Sabbath sabbatical of the laboring class. Where before, as A. Roger Ekirch has shown, Europeans of all classes had been accustomed to two periods of sleep with a couple of hours of wakefulness between, hours in which to meditate or make love, card wool or go poaching, nineteenth-century nightwakers were apt to be on the job: rag-pickers, millers, bakers, rubbish men, and prostitutes as before, now also foundrymen, gasworkers, railroadmen, and pressmen putting the news to bed so boys could cry the headlines at five a.m., that "booming sound that fills the city every morning and, to use the words of Mr. Walter Whitman, 'utters its barbaric youp over the house-tops of creation.'"[14]

Not everyone nightwalking was nightworking. With the lapse of curfews, cities had literally loosened up: iron-and-timber gates no longer banged shut to the roll of drums at nine of a summer evening; four hundred sets of chains were no longer dragged across the streets of Nuremberg at nightfall; "Lords of the Night" no longer restricted the burning of lanterns after dark in Venice. With better lighting (and, for some, a diet that included leafy vegetables and fresh fruit, enabling better night-vision) came later curtain times for shows, later closing times for chop houses, dance halls, pharmacies. At eleven p.m. in London in 1859 one could go for punch and hobnob to a *Conversazione*, a "suite thrown open for the reception of a miscellaneous mob of fashionables," what George Sala in his *Twice Round the Clock* called "a vast menagerie" where "the birds, beasts, and fishes crowd and elbow each other, and roar, or yell, or howl, or bark, or low, or grunt, or squeak, or crow, or whistle, or scream, or pipe."[15]

Certainly the night at mid-century seemed newly occupied, and noisier, with multiple voices. "About this time," writes the historian Louis Chevalier, "the night really begins to exist as something particular, no longer as a simple darkening of things that otherwise do not change their nature": too stark a claim. Once Chaos had given birth to darkness—to Erebus, the lower darkness of the dead, and Nyx, an upper darkness for the living—and Nyx had then confected Nemesis, Eris (Strife), and the Fates, night had a field of maneuver all its own, fateful and transformative. It had been the province of sorcerers and wisewomen, of banditti around the bend and physicians at the bedside, of radical pamphleteers and reactionary cabals, of monks and mourners at their vigils, of couriers with secrets and young lovers with assignations, of all who hoped unseen to turn the course of events to their own purposes, and of ghosts, sleepwalkers, and mad people, all those driven by unseen forces. The difference now was not so much that night was keeping new company as that it ran flailing into the day at both ends, like the cock-crowing that upset the Carlyles in Chelsea or the late-night sewing circles of maidens in *Lichtstuben*, "lighted rooms," whose lewd jokes and love-schemes alarmed authorities in rural Germany. The volume of sound occupying Paris in the daylight, suggests historian Simone Delattre, conferred a new value upon nocturnal silence, but when Etienne de Jouy in 1846 described *Paris au clair de la lune*, the mix of sounds he heard by the light of the moon put him in mind of two oxymorons, noisy silence (*un silence bruyant*) and tumultuous slumber (*un repos tumultueux*). From Montmartre, Paul Féval in 1851 heard a profound murmur abroad in Paris, "a noise made up of a thousand noises, a noise without end, inarticulate, indefinable," which, like the breath of a great creature, could be heard all the better at night when the giant was sleeping. An English court in 1860 was hard put to draw a line between day and night; in the absence of curfews and the presence of gaslamps, with factories ever-illumined and noonday skies grim with smog, neither the passage of an obscured moon nor the tumbrels of dustmen gathering horseshit could be inarguable delimiters of night.[16]

If anything distinguished the night, as Joachim Schlör has suggested, it was that life at night seemed more concentrated. "Since the invention of gas light," wrote Robert Springer, co-founder of a German Association for Harmonious Living, "our evening life has experienced an indescribable intensification, our pulse has accelerated, nervous excitation has been heightened." Springer was writing in 1868, by which time the traditional night-watchman had lost command of the darkness, whether by voice, clapper, or tin whistle, and the shift from day to night had become the domain of workers headed toward late shifts at loud machinery. Novelists, urban missionaries, reporters, and reformers also went into the night, investigating a darkness that seemed more knotted than ever with hunger and homelessness, yet fantastic with blast furnaces and sparking locomotives. Once contrasted with day, night now had its own internal contrasts, its own pent and spent energies. Astronomers had long used the night as a filter for stellar clarity; now the night was also a philter, a potion to rouse one from apathy. You could learn something from the night that was unavailable during the day: "if a man never allows the night to whisper something in his ear," wrote a nightgoer in Berlin, "if he never deals with the night side of life, he is in danger of becoming one-sided—the heart remaining dulled to the suffering of the world."[17]

Some of that nocturnal suffering had physiological causes, systemic problems that arise rather from the body's periodicities than from society's heartlessness. Ekirch notes that heart attacks and strokes are much more frequent at night, as are flare-ups of gout, ulcers, heartburn, headaches, and toothache, due to circulatory, digestive, or hormonal rhythms. Pain becomes more acute, so that the night is louder with outbursts of agony and the gasps of the half-asleep, grimacing. Other sounds, more painful to hear than to make, arose particularly at night: the wheezing of asthma, snorts of sleep apnea, whimpers and moans of nightmare. Then as now, gravid women may also have gone into labor more often between midnight and dawn and lost more newborns to darkness than to light; if so, the night would have groaned with mothers at birthing stools and households in the first hours of mourning. Industrial shift-work disrupted the body's clocks, aggravating (or instigating) cardiovascular, neurological, and gastrointestinal disorders, so night would have seemed on all sides to concentrate grief and amplify its noise.[18]

Actually, night concentrated and amplified all sounds, or so it was thought. The astronomer John W. F. Herschel, in an 1830 encyclopedia article on sound, observed that echoes had more repeats at night. W. Mullinger Higgins, a prolific popularizer of science, explained that the clarity of nocturnal sound was due to the homogeneity of the atmosphere, citing the renowned German explorer of South America, Alexander von Humboldt, who in 1803 had found the Orinoco Falls thrice noisier at night than during the day and wondered why. Why the amplification, and why was it so hard "to explain a phenomenon observed near every cascade in Europe, and which, long before our arrival in the village of Atures, had struck the missionary and the Indians"? It was a bit colder at night even in the Venezuelan tropics, and along the Orinoco the breeze picked up as the sun went down, two good reasons why noise should be *less* intense, since scientists had begun to prove that the velocity of sound decreases with lower temperatures and higher winds. Along the Orinoco, Humboldt had felt the calm of Nature at noon, and "Yet, amid this apparent silence, when we lend an attentive ear to the most feeble sounds transmitted by the air, we hear a dull vibration, a continual murmur, a hum of insects, that fill, if we may use the expression, all the lower strata of the air." He listened further: "A confused noise issues from every bush, from the decayed trunks of trees, from the clefts of the rock, and from the ground"—the noise of lizards, worms, burrowing insects. And he had been moved: "Nothing is better fitted to make man feel the extent and power of organic life." Could it be, then, that at night the ear no longer had to contend with the "confused noise" of active life, a sonic background so constant during the day that it went unnoticed but, diminished after dark, no longer masked other sounds? No. Although such logic and listening were impressively subtle, the Venezuelan air abounded at all times with mosquitoes, whose hum seemed much louder by night than by day. The jungle itself came alive after eleven p.m. with "a noise so terrific . . . that it was almost impossible to close our eyes." Tutored by the locals, Humboldt could make out "the little soft cries of the sapajous, the moans of the alouates, the howlings of the tiger, the cougar, or American lion without a mane, the peccari, and the sloth, and the voices of the curassoa, the parraka."[19]

Humboldt pondered. "It was indeed a strange situation, to find no silence in the solitude of woods. In the inns of Spain we dread the sharp sounds of guitars from the next apartment; in those of the Oroonoko, which are an open beach, or the shelter of a solitary tree, we are afraid of being disturbed in our sleep by voices issuing from the forest." The question of the peculiar intensity of noise at night, guitar or cougar, hung in the air. Back in Europe, he ransacked the ancients, the works of Aristoxenes, Seneca, Plutarch, and lit at last upon Aristotle, who had asked in his *Problems*, "Why is sound better heard during the night?" Pseudo-Aristotle had answered, "Because there is more calmness on account of the absence of caloric (*of the hottest*). This absence renders every thing calmer, for the sun is the principle of all movement." Humboldt elaborated: shining on a planet of hills and valleys, the sun heats our atmosphere unevenly, creating currents of air of different density that block or split the paths of "sonorous undulations" and weaken each soundstream. Night's unheated air, less troubled, more uniform, allows each sonorous undulation to proceed at full bore.[20]

Early radio operators, listening through a nearly constant buzz, would be puzzled by nocturnal spikes in the reception of distant stations, which led twentieth-century physicists to plot the atmospherics of the night and half-confirm Humboldt's Greek hypothesis. The physics of sonorous undulations, however, could be no more than half a theory for the felt intensity of noise at night, particularly of city nights, which were becoming so soundful that any lull intrigued. "When I have been out in the night, and return home at one or two in the morning,—the town buried in the deepest repose," wrote a Bostonian in 1832, "I reflect that sixty thousand people are immersed in forgetfulness, and are unconscious of the bright moon and the still, breathless atmosphere." This was a prophetic reflection for John Collins Warren, who in 1846 would be the first surgeon to operate on patients put under general anaesthesia, by which time such night-quiet was being reduced to a brief collective unconsciousness, or what a correspondent to the *London Times* would call a "brilliant flash of silence" amid all the "abominable noises" of the night—those of carpet beaters, hurdy-gurdy players, rowdy cabbies, garrulous codgers set to scraping paving stone. The *Times* editors believed that in Anno Domini 1869 it would take a second Joshua "to bid modern civilization stand still" long enough for a good night's rest. In 1879 the *New York Daily Tribune*, editorializing against noise during a hot August of open windows and the clunk of carts as truck farmers hauled vegetables to market in the moonlight, bent its ear to "a period of the night, when the late ones have gone to bed and the early ones have not yet risen, when we can discover, if we please, how charming is silence." That silence was "positively delicious," but "Alas! The pleasure is brief; for though we may sleep at last it is only for a little while. At dawn, or just before it, the din will begin again." By 1902 the novelist Robert Machray would be moved to exclamation when he found, on *The Night Side of London*, that "From two o'clock to four there is a lull, a quiet, a hush, a vast enfolding, mysterious awe-inspiring silence.... The city sleeps!"[21]

In that "night-hush" (<1849), the wife of a trainman working "night-shifts" (<1839) could not sleep for worry: "Alone, with naught to break the monotony and silence of the night save the low song of flames as each stick sends forth a blaze of brightness, or now and then vibrating on the stillness of an echo of silvery sweet-ness, as the treasured time-piece strikes in measured tones the lonely hours." The hush also discomfited Walt Whitman as he stamped "like a great wild buffalo" through the hospital tents of the Army of the Potomac afternoons and after-dinner for eighteen months, feeding the sick, hearing them out, and reading aloud to them night-long. Whitman's "youp," heard a few pages back, had been rather a *yawp*, not of newsboys but of a spotted hawk at dusk, swooping to accuse the poet of "gab and sloth," to which he responded at the end of his "Song of Myself":

> I too am not a bit tamed...I too am untranslatable,
> I sound my barbaric yawp over the roofs of the world.

That had been in 1855, with the first edition of *Leaves of Grass*. By 1863 he was responding to the "immense roar" of the musketry and cannon of the Battle of Chancellorsville fifty miles from the Capitol, and up close to groans "or other sounds of unendurable suffering." The groans came only "from two or three of the iron cots" and "in the main," wrote Whitman, "there is quiet—almost a painful absence of demonstration," soldierliness suppressing the screams that should have issued from men whose bones had been shattered by cannonballs' "crash or smash" or from amputees, the pain from whose phantom limbs was too powerful for laudanum or injections of morphine. Louisa May Alcott, down from Concord, flung up windows each morning at Georgetown's Union Hotel Hospital to clear the air for the sick and wounded whose fortitude by day "seemed contagious, and scarcely a cry escaped them, though I longed to groan for them, when pride kept their white lips shut, while great drops stood upon their foreheads, and the bed shook with the irrepressible tremor of their tortured bodies." Herself "fond of the night side of nature," she made rounds during "the haunted hours" after the gas lights were turned down; "owling," as she called it, she found that some patients "grew sad and infinitely pathetic, as if the pain borne silently all day, revenged itself by now betraying what the man's pride had concealed so well." Mary Foushée, a nurse at a military hospital in nearby Virginia, remarked a similarly proud silence among Confederate soldiers with lost limbs and awful wounds; "a groan has rarely reached my ears," she wrote, then persuaded herself "that the heroism of our men has developed itself more strongly and beautifully in enduring bodily suffering."[22]

Doctors of divinity and medicine had used a like logic to argue against the use of ether (1846) and chloroform (1847) during labor, on grounds that a woman's birth pangs were foreordained or obstetrically helpful and must be soldiered through, but the night-hush of anaesthesia was widely induced in Civil War surgeries, enabling 30,000 quicker quieter amputations in the North and perhaps as many in the South, on the model of thousands of French casualties who had been treated to "the death of all sensibility" during the carnage in the Crimea. A senior Boston physician, William Ingalls, asked later to recall the era before anaesthesia, remembered when "the surgeon was obliged to operate upon a screaming and writhing being—to feel an almost equal strain upon his coolness, while conscious of the evidences of suffering, the clutched hands, the grinding of the teeth, the sharp catching of the breath, even from those who had the courage to suffer in silence." Like Whitman, Alcott, and Foushée in the wards, Ingalls had appreciated the pain beneath the stoic silence: "One sees the hands tightly closed or strongly grasping a friend's, holding the breath with occasional noisy exhalations; one longs to hear loud outcries, they being the natural expression of pain." The night-hush of the city was becoming uneasily like that, less a respite than an indrawn breath.[23]

Civil War ordinance with its screaming Parrott cannon and whining Whitworth shells, its "Whistling Dick" siege guns and booming 32-pound howitzers, left little in the way of silence that was not unnerving. Like civil wars since the Roman civil war of 49–48 BCE as recounted by Lucan, the sounds of battle were heard with an edgy acuity. As sharp in their war cries as in their shooting, Confederate battalions at Cold Harbor put forth "such a withering fire and yell that the enemy's line broke and fell back immediately," but Gordon Bradwell did not think his Union foes cowards, "for the noise made by our shooting and yelling and the destruction wrought by our 'buck-and-ball' cartridges was enough to frighten any man." Wilbur Fisk, on the Union side at Cold Harbor, heard bullets come at him "like striking a cabbage leaf with a whiplash, others come with a sort of screech, very much as you would get by treading on a cat's tail." John William De Forest described shells soaring overhead as "an amazingly loud, harsh scream" and the musket fire as "a long rattle like that which a boy makes in running with a stick along a picket-fence, only vastly louder; and at the same time the sharp, quick *whit whit* of bullets chippered close to our ears." Men cussed and swore, "the most vehement, terrible swearing I have ever heard," wrote a New Yorker at Antietam; their obscenity was an acoustic mirror for the noise of solid shot "cracking skulls like eggshells," of "bullets hissing, humming and whistling everywhere; cannon roaring; all crash on crash and peal on peal." In the Union camp at Fredericksburg, Louis

Hamilton understood the battle as one noise after another: "At one time we could hear the musketry cackling, like numerous packs of firecrackers set off at once in a barrel—and then—the Artillery would join in like the deep baying of a great hound," then "a great screaming, roaring, round shot like a fiend," and then "*Phiz, Phiz, Phiz*—went the Minié bullets over our heads." Skies were often so dark with smoke and men with mud that even noontime battles took place in a disorienting murk, infantry feeling their way by ear. Thomas Evans of the 12th U.S. infantry could tell "the grating sound a ball makes when it hits a bone instead of the heavy thud when it strikes flesh"; afterwards, "the din of the engagement was still in my ears, and kept up a perpetual buzzing that I could not drive away." Around Atlanta late in the war, Union artillery felt like the beating of sheet iron (familiar, in shorter bursts, from stage thunder and charivari), and a soldier reported that "For two days our hearing was almost gone; it was several days before it was normal again."[24]

Fusillade, cannonade, and the temporary deafnesses of battle led to errors in the relaying of commands and the direction of fire: a Confederate regiment at Chancellorsville, on edge at night after an "infernal and yet sublime combination of sound and flames and smoke, and dreadful yells of rage, of pain, of triumph, or of defiance," replied to intermittent volleys and noises in the darkness and killed one of their own, Stonewall Jackson, who had been out reconnoitering. Anxious were the "death-like" silences before battle, as men crouched for hours straining to hear signal guns, and the night-hushes in forests "full of half-heard whispers—whispers that startle—ghosts of sounds long dead" (Ambrose Bierce, 9th Indiana). Worse were the silences of the aftermaths, for "Out of that silence from the battle's crash and roar rose new sounds more appalling still; rose or fell, you knew not which, or whether from the earth or the air; a strange ventriloquism of which you could not locate the source," wrote Joshua Chamberlain of the 20th Maine Infantry, "a smothered moan that seemed to come from the distances, beyond reach of the natural sense, a wail so far and deep and wide, as if a thousand discords were flowing together into a keynote weird, unearthly, terrible to hear and bear, yet startling with its nearness; the writhing concord broken by cries for help, pierced by shrieks of paroxysm; some begging for a drop of water; some calling on God for pity... some with delirious, dreamy voices murmuring loved names, as if the dearest were bending over them... and underneath, all the time, that deep bass note from closed lips too hopeless or too heroic to articulate their agony." Such was the lull after Fredericksburg; Bierce knew the lull after Chickamauga, the "chill influence of the awful quiet" after the "most infernal noise," a quiet in which a deaf-mute child sees a "clumsy multitude" dragging itself "slowly and painfully along in hideous pantomime—moved forward down the slope like a swarm of great black beetles, with never a sound of going—in silence profound, absolute."[25]

"It was all like a pantomime," recalled a Confederate general, Evander M. Law; "not a sound could be heard, neither the tremendous roar of the musketry, nor even the reports of the artillery." He was referring not to Chickamauga but to Gaines's Mill at the start of the Seven Days' Battles. From 175 miles west farmers could hear the booming of howitzers, and from 110 miles away housewives recoiled at the volleys of fifty thousand rifles, but to the Confederate staff within sight of the mill on 27 June 1862, the fighting seemed to take place in silence. The outcome of that battle was unaffected by the acoustic shadows that haunted the valley of the Chickahominy that afternoon, but at Five Forks, Seven Pines, Chancellorsville, and to some extent at Gettysburg, acoustic shadows were determinative. Where cannon smoke, musket haze, cavalry dust, smoldering brush, exploding shells, muddy bulwarks, and swarms of midges reduced visibility to a few yards even on the sunniest of days, cues were given by bugle boys, drummers, or signal cannon; when certain constellations of humidity, temperature, windspeed, wind shear, ground cover, and topography conspired to produce atmospheric inversions and refractions of sound, commanders and their troops could be unaware that flanking maneuvers had been completed or an attack begun. Regiments could be cut off from each other optically and acoustically, resulting in lost chances, tragic delays: another of the dark, unnerving, sometimes fatal silences of the Civil War.[26]

Nearly four million soldiers, a third of all free American males between the ages of fifteen and fifty-nine, fought in that war. One-quarter were killed or badly wounded or died of yellow fever, typhus, malaria, dysentery (with morbidity and mortality rates among the Union's black regiments nearly double that of white soldiers). Of those who made it back home, a tenth were amputees or otherwise disfigured. Thousands—artillerymen, foot soldiers, gunship "powder monkeys"—had permanent hearing loss from the firing of cannon, the explosion of shells, the crack of Enfield or Springfield rifles at their ears, or typhus. Many more, ostensibly unscathed, could not shake the noise of battle, the shells that had come at them like "fell demons flying through the air." Capt. William J. Seymour of the 6th Louisiana would be haunted by "the unearthly cries of wounded horses" (3,500,000 died during the war, 10,000 at Gettysburg alone), Capt. Robert McAllister of the 11th New Jersey by the screams of whippoorwills trapped in meadows by "the lightning flash and roar of our artillery and musketry." For most, what hung in the air was the noise of dying. "The thunder of battle has ceased," wrote Alva Guest of the 72nd Indiana, "but, oh, a worse, more heart-rending sound breaks upon the night air. The groans from thousands of wounded in our front crying in anguish and pain, some for death to relieve them, others for water. Oh, if I could only drown this terrible sound." (Vultures, frightened away by cannonade, did not circle the fields.) For forty-eight months, women and children in the Middle States had heard the war from their porches as armies passed and re-passed; years later they could still hear the drone of musket fire in the distance, "like heavy wagons being rapidly driven over a stony pike." In the South, women and children could not forget the burnings and bombardments; half a century after, Sara Pryor distinctly recalled the shelling of Petersburg: "To persons unfamiliar with the infernal noise made by the screaming, ricocheting, and bursting of shells, it is impossible to convey an adequate idea of the terror and demoralization which ensued. Some families who could not leave the besieged city dug holes in the ground, five or six feet deep, covered with heavy timber banked over with earth, the entrance facing opposite the batteries from which the shells were fired." Tunnels again: darkness and noise.[27]

Before the war, James R. Randall had composed "My Maryland" in favor of secession. Its last stanza, sung to the tune of "O Tannenbaum," ran:

> I hear the distant thunder hum,
> Maryland!
> The Old Line's bugle, fife and drum,
> Maryland!
> She is not dead, nor deaf, nor dumb;
> Huzza! she spurns the Northern scum!
> She breathes! She burns! She'll come! She'll come!
> Maryland, my Maryland!

But the sounds of the war had hardly been sexy. As they hardly ever were in other wars of the era, from the largest of civil wars, the Taiping Rebellion (>20,000,000 Chinese dead), to one of the smallest, the Federal War (>20,000 Venezuelans dead); from Leopold II's twenty-four years of terror in the "Belgian" Congo (>5,000,000 Congolese) to the Ten Years War (>100,000 Spaniards, >50,000 Cubans); from the War of the Triple Alliance (>120,000 Brazilian, Argentine, and Uruguayan soldiers, >200,000 Paraguayan soldiers, and >500,000 Paraguayan civilians) and the Russo-Turkish War (>120,000 Russians, >120,000 Turks) to the Crimean War (>90,000 French, >100,000? 400,000? Russians, >35,000 Turks, >20,000 British) and Franco-Prussian War (>140,000 French soldiers, >40,000 German soldiers, >100,000 French and German civilians); from frontier wars in Australia and New Zealand (<3000 whites but >80 percent of the aboriginal population) to the United States (<20,000 whites but >60 percent of Native Americans). Not to mention the nineteenth-century wars in Mexico, Nicaragua, Peru, Italy, Austria, Tunisia, Morocco, the Sudan, South Africa, Lebanon, Persia, India, and the revolt of the Communards in 1871 during which, after two months of barricades and bold lyrics, more than 20,000 Parisians were killed on the spot by French government troops, while 40,000 were later hanged, shot, or imprisoned to die of jailhouse fevers.[28]

Tireless, tiresome numbers. Industrial numbers, accentuated near century's end by that most industrial of weapons, the machine gun. Numbers that tailed the survivors of fourscore wars as they returned to find their hometowns industrialized and exploding: Chicago had thirty thousand residents in 1850, a million in 1890; Melbourne, Australia, had one hundred thousand in 1845, half a million by 1900. A planet that had begun the century with twenty-three urban centers over 100,000 ended the century with six times as many, a new metropolis each year. Since the growth of most cities was due rather to in-migration from outlying regions than to urban fertility rates, the ears in cities were often, if not preponderantly, rural ears, and given the high urban mortality rates, at each decade of expansion the cities had substantial populations accustomed from childhood to the acoustics of a village. But villages were becoming towns, towns cities, and cities metropolises, each of which could turn at the drop of a word into a war zone: if not a Communard revolt, then a Manhattan draft riot or gang war, a New Orleans anti-immigrant riot, anarchist and counter-anarchist violence in urban Spain and Belgium, police cavalry crushing Socialists on London's Bloody Sunday, or the Salvation Army riots in sixty English towns (1878–1890), where "Skeleton Armies" of an unrepentant working class hounded General Booth's bands, banging tea kettles and heaving dung at the Christian brass. By 1881, almost three-quarters of England's population had been corralled into cities and four times more people across the globe were urbanites than a century prior. Up against the grid-laying, brick-laying, rail-laying, and tunnel-driving of new towns and ballooning cities, the night-hush could be but a chloroformed moment, a sound shadow obscuring an ever-present noise.[29]

Among the millions of veterans of foreign and domestic wars, a noticeable contingent had palpitations, sweats, deep fatigue, sighing respiration, a "persistent quick-acting heart," and an "aching sensation in the left praecordium." First noted among British troops in the Crimea, the symptoms recurred among Union troops examined by a Philadelphia physician, Jacob Mendez DaCosta, who by 1871 had compiled a file of three hundred cases authorizing a new diagnosis, "irritable heart." Before long, the syndrome was known as "soldier's heart," and recovery from this post-traumatic stress disorder,[30] which had no visible lesions or audible cardiac murmurs, required quiet—the sort of convalescent quiet which a young Florence Nightingale had known at home and upon which she insisted during and after her stint in the Crimea. A member of the English elite (second daughter of reform-minded Unitarians), Nightingale had coughed through a bronchitic adolescence at the family manor in Hampshire. In her twenties she unsettled her mother, her older sister Parthenope, and twenty-seven well-born first cousins by determining that she had a divine call to work with "the sick poor," this at a time when hospital nurses, uneducated, ill-paid, and short-lived, "slept in wooden cages on the landing places outside the doors of the ward, where it was impossible for any woman of character to sleep, where it was impossible for the Night Nurse taking her rest in the day to sleep at all owing to the noise." Deterred by most of her family, and by her own frequent collapses, from answering that call until her thirties, Nightingale briefly attended the Fliedner Deaconess Institute at Kaiserswerth near Düsseldorf in 1850, again in 1851, and twice went to Paris in 1853 to learn the nursing system of the Sisters of Charity. Her French forays were aborted by two attacks of measles and a summons to return to care for a dying grandmother in Derbyshire, where Florence herself had spent the first five years of her life and most summers since. The Nightingale estate at Lea Hurst overlooked the lovely Derwent Valley but lay within a crow's caw of the lead smelter and hat factory that were steady sources of the family's wealth, sources also of the environmental lead and mercury that may have been responsible for Florence's weak ankles as a child, her bronchitis as a teen, and a lifelong sensitivity to sound so pronounced that a former suitor at a soirée chose as his opening line, "The noise of this room is like a cotton mill." Richard Monckton Milnes, M.P., advocate for the rights of married women, had persisted in his attentions for a decade and entered her dreams, but in the end Nightingale rejected him and two other suitors who cottoned to her intelligence, seriousness, tall slim figure, "soft, clear intonation" in five languages, "effortlessly resonant actressy voice," and a "peculiarly beautiful floating walk."[31]

She wanted, as her sister observed, to blow her own trumpet. "Parthe says that I blow a trumpet—that it gives her an indigestion—that is also true. Struggle must make a noise—& every thing that I have to do that concerns my real being must be done with struggle." In August 1853, Nightingale accepted a position in London as Superintendent of the Institution for the Care of Sick Gentlewomen in Distressed Circumstances, a small sanatorium whose capacity she expanded to twenty-seven beds while running up expenses and running out the former matron, surgeon, chaplain, and the "fancy-Patients," or "rather, the Impatients, for I know what it is to nurse sick ladies." It was she who was impatient, for a role of greater significance. When *The Times* and *Daily News* published reports of the conditions at Turkish hospitals, where thousands of soldiers from the Crimean front had been sent to die of cholera, dysentery, typhus, and typhoid fever *before* hostilities had begun, she campaigned to be put at the head of a flotilla of nurses who would sail toward the Black Sea and tend to British troops as the Sisters of Charity were already tending to French troops. Her pen-eloquence, persistence, and patrician connections paid off: in October 1854 she set out for Scutari as commander of thirty-eight sturdy women whose efforts would soon be backed by large public contributions to a Fund in her name. Shortages of all supplies, including ether, plagued her twenty months in the Crimea, so the noise of men in pain should have been grim, but as with Whitman and Alcott, she found that "As I went my night-round among the newly wounded that first night, there was not one murmur, not one groan[;] the strictest discipline, the most absolute silence & quiet prevailed." She liked that silence, and enforced a similar sound-discipline among her female nurses and male orderlies. ("I don't like the word discipline," she wrote later, "because it makes people always think of drill and flogging but, if they would but associate it with the word disciple ----!") Devoting much of her time to administrative issues of sanitation, provisioning, and accountancy, she maneuvered for control of mile-long corridors of enlisted men, morbidly infected or mortally wounded, whom British military surgeons had generally abandoned to rats, lice, unpotable water, and overflowing cisterns of waste. ("The whole reform of nursing," she would write, "both at home and abroad, has consisted in this: to take all power over the nursing out of the hands of the men, and put it into the hands of *one female trained* head and make her responsible for everything.") Omnivorously responsible, she herself worked long hours in her quarters, known as the Tower of Babel for its continuous flux of suppliers, nurses, male orderlies, and soldier-aides speaking dozens of English dialects, good or bad French, Greek, German, Turkish, Arabic. She also introduced the practice of writing letters of consolation to the families of soldiers who had just died. As a result of these letters, of her perseverance in the teeth of military scorn, of her lamp-lit excursions through the wards at night, and of her refusal to leave the Crimea until the last soldier had been sent home, she was praised in poem and song upon her inconspicuous return to Lea Hurst in late 1856:

> So she, our sweet Saint FLORENCE, modest, and still, and calm,
> With no parade of martyr's cross, no pomp of martyr's palm,
> To the place of plague and famine, foulness, and wounds and pain,
> Went out upon her gracious toil, and so returns again.

Her modesty a spiritual postulate, a literary foil, and a political weapon, Nightingale parlayed her fame into a series of hospital, public health, and nursing reforms, but behind the scenes, as an energetic invalid. In the Crimea she had been hit with a two-months' fever, then by sciatica, dysentery, ear ache, and irritability. She came away from the Crimea with something like "soldier's heart" and after a breakdown in 1857 was left in a chronic state of fatigue, rheumatic pains, and hypersensivity, due probably to brucellosis contracted in Scutari from the goat's milk she preferred over the wine taken by other Europeans in Asia Minor. From one of fifteen bedrooms at Lea Hurst and then from her Rheocline Spring Bed in

London, beneath prints of Raphael's Sistine Madonna and Murillo's Virgin, she composed the better half of her life's work of 14,000 letters as well as dozens of books, reports, and plans. Not once during the last fifty-two years of her life did she appear at a public function.[32]

Instead, she invented the public health questionnaire, introduced medical graphing, and published *Notes on Nursing* (1859) and *Notes on Nursing for the Labouring Classes* (1861), with sections on noise that drew as much from her own medical history, her resistance to physical closeness, and her current situation ("entirely a prisoner to my bed") as from her time in the Crimea. "Unnecessary noise, or noise that creates an expectation in the mind, is that which hurts a patient. It is rarely the loudness of the noise, the effect upon the organ of the ear itself, which appears to affect the sick." Those with brain injuries might be bruised by "mere noise," but for every patient "intermittent noise, or sudden and sharp noise... affects far more than continuous noise." Patients were most wounded by sounds that drew attention to themselves without making sense or arriving at closure: whispered conversations, hallway chats just out of range of intelligibility, "the fidget of silk and of crinoline, the rattling of keys, the creaking of stays and of shoes" that signified an ineffectual busyness. Like Bierce's "ghosts of sounds" that triggered a frenzy of firing during the night-hush of battle, such small noises kept the ill from the sleep they needed and intensified their pain, as did the affectation of walking on tip-toe. Her concern for noise was no afterthought; it was anchored in a Christian aesthetic that she had expounded after a visit in 1848 to Santa Maria degli Angeli, a church in Rome built to a design by Michelangelo, whose architecture had impressed her as barren yet strangely harmonious: "That noiseless (if you may use the word) growth of one part out of the other, which reminds one of the growth of the kingdom of heaven from a grain of sand, becoming a great tree, that want of bustle and glaring effect and impudently forcing itself upon one's notice... which is so like the works of God Himself." The spiritual and ethical as well as technical progress of nurses being trained under the auspices of the Nightingale Fund was assessed in reports that had columns for Trustworthiness, Truthfulness, and Quietness, since "A nurse who rustles... is the horror of a patient." Offended by fashions that demanded that stylish women wear crinolines, Nightingale was something of a Babbage when it came to the swish of fabric, but was it peevish to insist that "A good nurse will always make sure that no door or window in her patient's room shall rattle or creak; that no blind or curtain shall, by any change of wind through the open window, be made to flap"? Wasn't it rather good manners and good medicine for nurses to sit and listen without interruption to patients who needed to talk? It could also be therapeutic to read aloud to patients, if done slowly and articulately, without an (urban) hurry and plunge. "These things are not fancy. If we consider that, with sick as with well, every thought decomposes some nervous matter,—that decomposition as well as re-composition of nervous matter is always going on, and more quickly with the sick than with the well," then to make a patient work through interruptions, surprises, and half-silences would hamper mental and physical recomposition. "Irresolution is what all patients most dread," for with it came strain, fuss, noise. "One thing more:—From the flimsy manner in which most modern houses are built, where every step on the stairs, and along the floors, is felt all over the house; the higher the story, the greater the vibration." So never install a patient where she will hear medicine carts overhead, or where he may feel "every step above him to cross his heart." Keep in mind, she wrote, "that every noise a patient cannot *see* partakes of the character of suddenness."[33]

City folk abed at night in flimsy houses might therefore feel themselves under siege, especially those who winced at a footfall: husbands in the grip of migraines, "neurosal" wives, neuralgiac elders, all those so sensitive as to resemble "a finely strung musical instrument." Thus J. Milner Fothergill, M.D., who regarded urban settings as battlegrounds where humanity was at war with itself. In 1887 he lectured the British Association for the Advancement of Science on the fact that town dwellers from embryo are prone to an abnormal growth of the epiblast, which drains the mesoblast and starves the hypoblast, resulting in adulthoods of weak livers, dyspepsia, and an Usheresque sensitivity. An epiblasted woman in bed with a headache finds the tick of a clock intolerable, "and it is scarcely exaggerating to say she can hear the cat walk across the kitchen floor." Or the late-night crank of a barrel organ. Or a milkman's "Koo!" at blink of day. As for the epiblasted man, he "gets more of his neighbors' noise than is good for him": one neighbor rising at dawn to practice a little on the fiddle; another, back from an evening concert, singing snatches of tunes that surge "through his brain in symphonic unrest." Most people adjust, said Fothergill, but the high-strung are sound-tortured, above all at night when clutching for sleep. Around every corner and around the clock, cities took their acoustic toll, for "Man in a natural state lives in the country."[34]

Lutz Koepnick, a media historian, has suggested that rural sounds were so layered as to be intelligible and therefore unproblematic, while the new layerings of industrial urban sounds were unintelligible and overwhelming to town dwellers, who lost their psychic balance amid what could only be heard as noise. This was not quite how Susan Fenimore Cooper heard things in her *Rural Hours*, where folks had to block up their chimneys against the ever cheerful songs of swallows, who would otherwise nest and sing therein from three a.m. to dusk. Rufus Usher, a botanist in pastoral Oxfordshire, observed that all singing birds abide in friendly proximity to humankind, urban or rural; the rest "utter a monotonous cry, or a shriek." But what then of grackles? And what of wars conducted on the gorgeous shores of the Black Sea or in the wooded backcountry of the American South? In a sermon on "Noise and Silence" that he delivered seven times in the years just after the Civil War, the radical Unitarian minister Francis Ellingwood Abbott told congregations in rural New Hampshire and urban industrial Massachusetts that "All barbarous nations and tribes are fond of noise. The greatest of their divinities is the god Racket" and "in their war-councils every campaign is planned under the inspiration of wild whoops and terrific clamor." Yet if "Noise seems to be the universal expression among uncivilized men, for any strong emotion of joy, fear, rage, or grief," he had found that "Even in this highly-enlightened America, we can find no better expression of our public happiness than the snapping of torpedoes and fire-crackers, the clangor of bells, the thunder of cannon, and the still noisier demonstrations of Fourth-of-July orators." What was missing, in town *and* country, was an inner quiet, for "The howlings of the passions make worse confusion in our souls than the shouts of a hundred hack-drivers in a New York railroad station." Unsecured by the "stillness of a soul at peace with itself," we lose our equipoise and are everyone confounded by "The keen chafferings of trade, the loud rivalries of business interests, the din and clack of vociferous tongues, the shrill pipings of penny-whistles blown by boys of all ages from seven to seventy years, the noisy scrambling after a pile of greenbacks," and on and on. "Friends," said Abbott, "this noise of the world would drive us crazy, if we could not flee away at intervals into the absolute solitude of our own souls." Best then "In any season of quiet in our outward surroundings," rare as that might be, or while lying awake at night, to listen very intently for "a mysterious sound which seems to be a mere singing in our ears—a very faint, still, tiny sound probably produced by innumerable small vibrations of the atmosphere confusedly reaching the tympanum of the ear from all directions." What was this "perpetual sound, unnoticed except in hours of still attention"? Abbott could give it "no better name than the *voice of silence*," which "is to our corporeal hearing what the 'still, small voice within' is to our spiritual hearing." It was a silence and stillness neither of idleness nor inaction, "but the stillness of a life that moves without noise or friction to the accomplishment of divine aims"—such as, for Abbott, the abolition of slavery. "Unobtrusive, unostentatious, yet full of power, such a life revolves about the great thought of Duty as the Earth sweeps around the Sun,—with a motion resistless as gravitation, noiseless as a planet in its pathway through the sky."[35]

Wrote a contributor to *The Colored American* in 1837: "The silence of night is the friend of contemplation.... While the multitude sleep, and the sound of their occupations has ceased, and the streets are solitary, and there is no noise but that of the cricket within, or the murmuring of the winds, the pattering of the rain, and the howling of dogs without;—then the soul is self-collected." Just three years later a second contributor demurred that the soul was too distracted in the city, day or night: "The din of the crowded street, the noise and excitement of the public assembly, the bustle and hurry of commerce and amusement, too often, alas! repress that still small voice within." It was "far different in the country," where every scene could be heard to evince God's grace, whether the quiet unfurling of a tender shoot or "the thunder storm, the rain, and the sheeted lightning, and the torrent descending from the mountain's side." Potentially louder than any city, could the countryside have been so idyllic for North America's four million black slaves, what with the pre-dawn clanging of strips of metal to start the day, the crack of whips in the fields under a hot sun, and growling mastiffs in the dark?[36]

Iconically, night had been to day what the country had been to the city: reverie, romance, respite, remedy, recreation.[37] Sonically, the terms were being reversed: night would be to day what the city was to the country, for it was in the industrial city, as in the illuminated night, that sound seemed most concentrated. Visiting the smoky row-houses of Pittsburgh's steelworkers, F. Elizabeth Crowell found that the "Noise pressed in from every quarter—from the roaring mill, from the trolley cars clattering..., from the trains on the through tracks above the topmost row, and from the sidings." The acoustic density of the city was a function of demographics: Manhattan's Tenth Ward in the 1860s had 196,000 people to the square mile and fourteen to a tenement room, the greatest ruckus arising, claimed one reformer, from the digs of organ grinders with their chattering sapajous (Cebus monkeys) and bands of children practicing. Acoustic density was also a function of traffic, of many strangers in almost Brownian motion, as with "the insufferable noise made night and day by twenty thousand private carriages on the streets of Paris" or the "*Chicago unrest*, noise, racket, hurry, bustle, no repose, no particular center, outside people pouring in, the family pouring out," more furious than the "ceaseless tumult" that had affronted Jane Austen's heroine on her return from the calm of Mansfield Park to a thin-walled house in Portsmouth where the "doors were in constant banging, the stairs were never at rest, nothing was done without a clatter, nobody sat still, and nobody could command attention when they spoke." Acoustic density was as well a function of the compression of time in which urban sounds seemed rather to collide than to elide, street cries crushed by "the general roar and rush and rattlety-bang of a chaotic mixture of vehicles, iron-shod horses, and tramping men—a wild orchestral pandemonium, injurious to the delicate organs of hearing, ruinous to the nerves, and subversive to the true ends of civilization," wrote Kate Clark in 1897. Finally, acoustic density was a function of the very air of the nineteenth-century city as a heat sink, its large brick or stone buildings and pavement retaining heat, raising the local temperature and speeding sound along. "Who can bear the thought of those labyrinths of brick," asked Thomas Babington Macaulay in the 1850s, "overhung by clouds of smoke and roaring like Niagara with the wheels of drays, chariots, flies, cabs, and omnibuses?"[38]

At night the air cooled (though not as much as in the country) and city life contracted, but in its contractions the night appeared to compress sound, intensify noise. Compression and clarity, wave and particle: as if the night stripped sound to raw physics. In defiance, poets and musicians constructed nocturnes, dark dreamy evocations of what was, in urban evolutionary terms, vestigial. Gothic novelists had relied upon a framing silence to invest the night with terror; as that frame cracked, Victorian romantics reinvested darkness with recuperative solace, as in the opening lines to one of Kate Clark's sonnets, "An Autumn Night":

> What mellow radiance wraps the slumbering world,
> Soothing its harshness with benignant hand!

At the hands of the Irish pianist John Field, the nocturne also became a full-fledged musical genre in which "an expressive melody in the right hand is accompanied in the left by broken chords." As spun by Chopin, the nocturne took on the qualities of a musical hash(ish), far from Mozart's playful *Eine Kleine Nachtmusik* and headed toward transfiguration in Arnold Schoenberg's *Verklärte Nacht*. Their days and nights broken by discord, urban audiences welcomed the nocturne as an audible anaesthesia, aetheric and etherizing. But the left-handed overtones of the nocturne, like the arching fires of Whistler's "Nocturne—Black and Gold—The Falling Rocket" (1870–1877) or the waves coursing over van Gogh's "Starry Night" (1889), suggested something different, a turbulence almost oceanic. Percy Shelley in 1820 may have been free-associating when he equated the ocean with London,

> that great sea, whose ebb and flow
> At once is deaf and loud, and on the shore
> Vomits its wrecks, and still howls on for more.

By 1857, however, it was common for a poet to use the figure of an ocean for a city, as did Alexander Smith for Glasgow with its 380,000 inhabitants, where

> Instead of shores where ocean beats,
> I hear the ebb and flow of streets.

Engulfed by mud-yellow smoke from a stack rising 450 feet over the St. Rollox chemical works, eddied by the Clyde shipyards that built the majority of the world's iron steamers, Glaswegians found that, day in and night out,

> All raptures of this mortal breath,
> Solemnities of life and death,
> Dwell in thy noise alone.

Walt Whitman, back in New York, rather exulted in "the oceanic amplitude and rush" of cities and their "heavy, low, musical roar, hardly ever intermitted, even at night." Civic boosters, guided by the principle that "The whole business of life is to make a noise in the world," gloried in that oceanic roar, for it was "a beautiful spectacle—the industry of a great city waking up in the morning light" with "smoke puffing afresh from forge and factory; the rattle of wheels here and there breaking the early silence; the strokes of labor commencing from roofs and workshops; the steamers panting at the wharfs," until "one by one, all these energies slip the leash, and one by one these waves of sound swell into the universal roil of activity and toil." Played or painted or spoken amid such swells, "hardly ever intermitted," the nocturne was thin insulation against urban, oceanic noise.[39]

2. Lulls

Other sorts of sanctuary promised other sorts of relief from noise: ruins, the East, opiates, the Rest Cure, cemeteries, the Sabbath. Europeans on their grand tours had associated the picturesque with an optical and acoustic stillness that etched a place as both untouched and foregone. Travelers sought this stillness among classical and medieval ruins, such as the Roman amphitheater in Verona, where "nothing moved, save the weeds and grasses which skirt the walls and tremble with the faintest breeze." But, as the literary scholar James Buzard has shown, such stillness was achieved through

SOUNDPLATE 6

acts of staging that obliterated every evidence of incursion, every steam whistle or stamp of horses impatient at the traces of enormous touring carriages—and it took heroic effort to stage such stillnesses as urban life encroached on ancient spaces. Already by 1817, for twenty-seven-year-old Henry Matthews, these stagings seemed out of reach. Seasick on his way to Malta, the waves "roaring in his ears" and "the noises of Pandemonium all around him," he listened for a picturesque ocean but heard "no 'song of earliest birds;'—no 'warbling woodland;'—no 'whistling ploughboy;'—nothing, in short, to awaken interest or sympathy" from an English squire. Touring Iberia, he seethed at the creaking of ox carts, their noise "as disagreeable as the sharpening of a saw." In Florence he was put out by "the tingling bell of the Mass," in Rome by morning serenades from an "odious squeaking bagpipe," in Naples by "people bawling and roaring at each other in all directions." On the fifth night of Carnival he fled "to the silent solitude of the Coliseum, where you can scarcely believe that you are within five minutes of such a scene of uproar." He would arrive at a more resolute stillness after publishing his *Diary of an Invalid* in 1820 and shipping off as Advocate-Fiscal to Ceylon, where he lived (and died, quite young) in the bungalow of a Colombo suburb. Colombo itself felt like a ruin when visited by Samuel Baker, a big-game hunter depressed by a colonial capital where "The ill-conditioned guns upon the fort looked as though not intended to defend it; the sentinels looked parboiled; the very natives sauntered rather than walked; the very bullocks crawled along in the midday sun, listlessly dragging the native carts." Although the Irish art historian Anna Jameson in her 1826 *Diary of an Ennuyée* was able to stage Venice as a nocturne in which "the silence was broken only by the melancholy cry of the gondoliers, and the dash of the oars; by the low murmur of human voices, by the chime of the vesper bells, born over the water, and the sounds of music raised at intervals along the canals," her diary was a literary fiction, her Venice a set piece. The quest for stillnesses less fictive drew the leisured classes farther and farther afield—to the Sierras, the South Seas, Japan.[40]

Japan especially. Pried open in 1853–1854 at the insistence of U.S. warships led by the coal-powered frigate-steamer *Powhatan* with its giant side-wheel, black smokestack, and sixteen cannon, Japan was staged as a place of Zen silences, the Japanese as monk-adjutants. Tea houses with gravel paths that wound through bamboo gardens were erected at one international fair after another, exaggerating the contemplative side of Japanese life while Western ceramicists, painters, interior decorators, and philosophers idealized the tea ceremony, holding Japan captive to impressions of a stillness no less audio than visual. "They are a quiet people," Mrs. Caroline B. Lawrence assured the New England Women's Club as late as 1907; "one never hears any quarreling or squabbling and the children play quietly." Japanese babies rarely cried, and at sumo matches the spectators maintained a "perfect order and quiet," added Edward S. Morse, a Salem naturalist who had spent five years in Japan collecting shells, ceramics, floor-plans, and ethnographic details. The Japanese colluded in this staging, for they heard most Westerners as bewilderingly noisy. At a dinner for a Japanese diplomatic mission that came to the United States in 1860 aboard the *Powhatan*, the visitors were fêted in San Francisco with toasts that seemed "as noisy as the drinking bouts of workmen in Edo." After dinner, the mayor of San Francisco stood up and, reported Murigaki Norimasa, "banged on a plate. Another man, standing off in the distance, catching the signal, also banged on a plate. When the barbarian music stopped and the place became quiet, the mayor talked about something in a loud voice and at extremely great length. When he finished, everybody stood up and shouted three times in a loud voice, *ippinwā* [hip-hip-hooray] and everybody drained his glass and applauded or else beat on a plate. The barbarian music started again, and the sound of people's voices was such as to make one want to cover his ears." The United States Congress, with its impolitic shouting, sounded to Murigaki like a fish market, and the parlor solos of American women reminded Ono Tomogorō of the "gibberish scream of a bewitched woman." Shibata Takenaka, sent to Europe to study heavy industry, was sure that a visit to Japan would teach Europeans "with their shrieking tongues" to sing like nightingales.[41]

Everything in Japan was not birdsong and hush. Morse himself was annoyed by the clatter of wooden clogs on sidewalks, echoing like "a troop of horses crossing a bridge," and ambivalent about "a plaintive sort of shrill whistle" made in the road by blind men and women advertising their skills at massage. Unambivalent was a Japanese girl who had come from the countryside with her parents to set up shop in Yoshiwara, a district of Edo, only to be stunned by the traffic. "Words cannot describe what it is like after dark, what with the noise of rickshaws thundering by and the glare from their headlights.... Never in all my life can I remember anything like the feelings of the first night I slept here after the move from the profound silence of our house in Hongō." Throughout most of Edo, or Tokyo, the noise was just as bad, for by the 1700s the city had become the most populous on earth, with well over a million people. Bathhouses offered temporary refuge from thundering rickshaws and from the "squeaky shoes" that came into vogue early in the Meiji era, but the adoption of Western sewing machines and growth of heavy industry began new layerings of urban sound louder than rickshaws, more annoying than fancy shoes of "singing leather," more strident than the noisemakers used by plebeian audiences, more constant than the fireworks of Shinto festivals.[42]

From across the Sea of Japan came another, steadier stillness. Ironic though it was that a sedative should provoke gunboat wars to force China's hand in the narcotics trade, Westerners relied upon opium from the East to quiet teething babies, soften the croup, mitigate cramps, settle stomachs and nerves. Unlike chloroform, which would be confined to dental and medical surgeries, opium in all of its forms—pill or powder, alcoholic tincture (laudanum), camphorated tincture (paregoric), alkaloid derivative (morphine) or, later, acetylated alkaloid (heroin, in the 1890s)—was sold over the counter, diffused in patent medicines, and used widely to counteract diarrhea during epidemics of cholera, malaria, and dysentery. Working-class mothers spooned out laudanum cordials of Infant Quietness or Mrs. Winslow's Soothing Syrup (750,000 bottles annually, 1830–1910) to hush young children and suppress their appetites, so that fathers back from night shifts could get some sleep; day laborers swigged balsams of opiated alcohol as short-term stimulants and longer-term deadeners of muscle pain and hunger. The middle classes took it on advice of their doctors to relieve rheumatism, put a quietus to insomnia, blunt migraines. The poet Elizabeth Barrett Browning, suffering from spinal agonies and much else, took it for "life & heart & sleep & calm." England's Prime Minister, William Gladstone, took it in his coffee as a daily supplement to boost his eloquence. Mary Chesnut of South Carolina used it to keep herself on an even keel: at the commotion of "a mad woman raving at being separated from her daughter," Chesnut became so agitated that she "quickly took opium, and *that* I kept up. It enables me to retain every particle of mind or sense or brains I ever have, and so quiets my nerves that I can calmly reason and take rational views of things otherwise maddening." This was in 1861, at the start of the Civil War, during which the Union Army's Medical Purveying Bureau issued ten million opium pills and almost three million ounces of laudanum, paregoric, and morphine. Few confessed to using opium for its psychoactive properties, few became addicted until morphine could be injected by syringe after mid-century, few died of overdoses until heroin was sold in large quantities as a cough suppressant near century's end, but opium was swallowed so steadily by people of every age and condition that in 1868, the year that English law required opium to be labeled a poison, the *Pharmaceutical Journal* noted that "it has been seemingly as reasonable to buy a quarter of an ounce of opium at the village shop as a quarter of an ounce of tea." (A quarter of an ounce of opium was a penny's worth.) Britain that year imported about 300,000 pounds, the United States 135,000 pounds—enough, when cut with inert powders and dissolved in spirits, for seventy single-grain doses for every Briton, a dozen for every American. The opium of the masses *was* opium.[43]

With an ounce of laudanum (its alcohol and single grain of opium), the noise of the world and pangs of the body could be damped; with two or three grains, dulled and distanced; with the six to ten grains of a morphine injection, buried alive. Morphine was the night-hush of the moneyed, a panacea prescribed by 1860 for most chronic diseases, for migraines, and for such "female complaints" as cramps and dysmenorrhea. Louisa May Alcott took it to counteract the effects of the mercurial medicines with which physicians treated the "typhoid pneumonia" she had contracted at the Union hospital, then against insomnia, catarrh, and loss of voice. As with opium, the majority of morphine addicts were better-off women whose spells of anxiety, sleeplessness, and nervous prostration often invoked crushed poppies—or a regimen of silence and immobility that aped the effects of those poppies. This regimen was the Rest Cure, originated by a novelist, poet, and neurologist from Philadelphia's gentry, Silas Weir Mitchell, who had learned the value of enforced bed-rest for malarial Union soldiers under his care. More exactly, the Rest Cure dated from his treatment of Mrs. G of Maine in 1874. Depleted by a series of pregnancies and by postpartum visits to spas, Mrs. G was too tired to sleep, too dizzy to read or write, too weak to climb stairs, and could not think without getting a headache. She could digest food "if she lay down in a noiseless and darkened room" but was too anemic to be left on her own, too jumpy to be tended by family or friends. Her jumpiness was an altitude higher than that of the typical "nervous woman" whom Mitchell saw and whose aural discomforts he described matter-of-factly: "Noises, especially sudden noises, startle her, and the cries and laughter of children have become distresses of which she is ashamed." Mrs. G's jumpiness was akin to that of the other Jumpers of Maine, for whom "any sound, from any source, that comes upon them with sufficient severity and suddenness, for which they are not forewarned and forearmed, may cause them to jump and to cry." The crash of a tree in a forest or the slam of a door set them off "like a piece of machinery," flailing their hands, striking out, tumbling. "In a certain sense," wrote George M. Beard—a New York physician himself prone to claustrophobia, agoraphobia, and a "fear of everything," or so he said—"we are all Jumpers; under sudden excitement, as of a blow, or a violent, unexpected sound, any person, even not very nervous, may jump and cry." The difference was that, like the jumpiest of Mitchell's clientele, the Maine Jumpers were infinitely alarmable. Beard supposed that the (mostly) French Canadian workingmen he had interviewed were half-tranced or epileptic, while Mitchell thought that his more genteel jumpers were wholly exhausted and had lost control of their emotions. In a larger context, the jumpers and Jumpers, all of whom freaked at the screech of a steam whistle, were taking flight from the alarming soundshifts of industrial capitalism and the abrupt timeshifts of a heated market economy with its contrary demands for immediate consumption and brokered futures.[44]

Mitchell's solution was gravitational: continual bed-rest with massage (by a masseuse or by faradic current, to preserve muscle tone), overfeeding in placental darkness (milk and mutton chops, to enrich the blood), and solitude accompanied by "moral medication" (a doctor's daily visits, to reduce tears, twitches, and other emotionality). The treatment went on for eight to twelve weeks until the society matron, housewife, schoolteacher, financier, or railroad agent settled back into herself, caught up with himself. It was a form of ritual *couvade*: a male physician took into confinement a weak, trembling, thin-skinned body, feeding his charge under cover of silence and solicitous darkness until the embryonic figure gained strength and weight, whereupon an adult newborn would be released, stable and resilient, into a world wild with tell-tales—automatic fire alarms, elevator bells, telephone ringers, office buzzers. It was, to the London obstetrician W. S. Playfair, "the greatest practical advance in medicine made in the last twenty-five years." It was, to Peter Dettweiler, director of the Falkenstein Sanatorium in Germany's Taunus Mountains, half the story of healing for those with consumption and lung disease, who required rest, quiet, *and* pure air, lying in silence on lounge chairs on open balconies, reading (as one patient found in the 1880s) Friedrich Nietzsche's *Thus Spake Zarathustra*: "Flee, my friend, into thy solitude! I see thee deafened with the noise of the great men, and stung all over with the stings of the little ones. Admirably do forest and rock know how to be silent with thee.... Where solitude endeth, there beginneth the market-place; and where the market-place beginneth, there beginneth also the noise of the great actors, and the buzzing of the poison-flies." It was, to a Rhode Island painter of trade cards, a tale less of providence than of poison, as Charlotte Anna Perkins Stetson (later, Gilman) found herself in a Poe-like chamber, choking on the milk of mankindness. Before her marriage, Perkins had been running laps and swinging on rings at the Providence Ladies' Sanitary Gymnasium, whose proprietor promoted Exercise Cures, and she had underlined in her diary that "I *hate* to spend an evening in enforced torpor." Yet two years after her wedding in 1884 and a year after the birth of a daughter whose howling gave her crying fits, there she was, committed at twenty-six to Mitchell's therapy of torpor. She had lost her "old force and vigor," was "very sick with nervous prostration, and I think I have some brain disease as well," and "went to bed crying, woke in the night crying, sat on the edge of the bed in the morning and cried." For her, however, as for other women whose politics and callings led them beyond the home, the remedy for such fatigue and depression would have to be aerodynamic, the imagination soaring. Although the Rest Cure afforded women rooms of their own and a period of privacy in which to steel themselves for defiance of old-line patriarchy or new-line capitalism, Mitchell expected the cure to lead his women patients back home. He advised Charlotte to "Live as domestic a life as possible," with only "two hours of intellectual life a day" and no writing or painting. For months she followed his prescriptions for mindless domesticity, and went almost out of her mind. Regrouping, she abandoned the cure and left Providence, her husband, and her household for Pasadena, the arms of women friends, a utopian fiction entitled *Herland*, and a sharp study of political economy.[45]

His own imagination led Mitchell to write novels of the American and French Revolutions, but he kept returning to Civil War battlefields, whose torments had touched him from the moment he began to walk the wards of the wounded who had been carted from Gettysburg to Turner's Lane Hospital in Philadelphia. In his fiction, as in his papers on gunshot wounds, nerve injuries, and the phantom pains of amputated stumps, Mitchell could not shake the sound and feel of that war. In "The Case of George Dedlow" (1866), the hero, like Mitchell a doctor and son of a doctor, suffers the successive amputation of all four of his limbs and the phantom pains that haunt each loss, until he fears that he himself is no more than phantom. In *Roland Blake* (1886), "There was no escaping these persistent echoes from distant battlefields." In *A Psalm of Deaths and Other Poems* (1890) Mitchell still mourned the Union fallen and heard them in "The Waves at Midnight" off the cliffs of Newport,

> Evermore saying——
> Hush—hush—hush——
> They fall, and they die,
> Break, and perish, without reply.

In *Characteristics* (1891), Mitchell's alter-ego Dr. Owen North applauds the heroism of army surgeons during the Civil War, who heard death on every side but kept to their grim task. In 1893, speaking to the graduating class of Jefferson Medicine College, he referred to the students as "young soldiers" and praised "the courage and heroism of [medicine's] myriad dead." In the Civil War scenes of *Westways* (1913), "What most horribly disturbed Penhallow was the hideous screams of the battery horses. . . . he still heard the cries of the animals who so dumb in peace found in torture voices of anguish unheard before—unnatural, strange." On the eve of the First World War, at the age of eighty-five and mostly deaf, Mitchell fell into the delirium of a fatal flu during which he returned one last time to the battlefields of Gettysburg, crying, "That leg must come off—save the leg—lose the life!"[46]

Does it come as any shock, then, that after his first wife and eldest child died during the War, Mitchell had had a nervous breakdown, and in 1872, as author of a tract on the national epidemic of nervous exhaustion (*Wear and Tear*), he had another breakdown, with "grave insomnia" and keen pain in the presence of high notes? This second breakdown was not a case of "soldier's heart" but, as he wrote, neurasthenia, a diagnosis promulgated in 1869 by George M. Beard, who called it "the most frequent, the most important, the most interesting nervous disease of our time, or of any time." Neurasthenia emerged from the maelstrom of "modern civilization, which is distinguished from the ancient by these five characteristics: steam-power, the periodical press, the telegraph, the sciences, and the mental activity of women." From steam-power came the rush of railway travel, the rise of manufactures, and a hundred-fold more business proceeding a hundred times faster. From the periodical press came pages of sensational news that played havoc with the emotions, as well as the incidental "noise of the tearing of newspaper" which, when exhausted from overwork, Beard himself found "unpleasant in the extreme." From (100,000 miles of undersea) telegraph cable and from telegraph lines (running alongside 100,000 miles of railway track) came 33,155,991 messages sent to or across America in 1880, a flurry of tapping and a rain of random intelligence. And from the sciences came much too much to want to know—"The experiments, inventions, and discoveries of Edison alone have made and are now making constant and exhausting draughts on the nervous forces of America and Europe, and have multiplied in very many ways, and made more complex and extensive, the tasks and agonies not only of practical men, but of professors and teachers and students everywhere." As for the mental activity of women, fired up by all of the above, it was unclear whether that proved society's advance or undid its prospects by making women too fragile to cope with the rigors of childbirth and the inevitable ruckus of children. What was clear was that noise characterized modernity: the whistle of steam and the grating of elevated trains, the cries of newsboys, the "nervous noise" of millions of telegrams arriving in soundbites punctuated with breathless STOPs, the stertorous "suspense and distress" of promised discoveries and inventions. "The relation of noise to nervousness and nervous diseases is a subject of not a little interest," wrote Beard; "but one which seems to have been but incidentally studied." Nature's sounds were rhythmical, often pleasing, rarely distressing; "the appliances and accompaniments of civilization, on the other hand, are the causes of noises that are unrhythmical, unmelodious, and therefore annoying, if not injurious; manufactures, locomotion, travel, housekeeping even, are noise-producing factors, and when all these elements are concentred, as in great cities, they maintain through all the waking and some of the sleeping hours, an *unintermittent* vibration in the air that is more or less disagreeable to all, and in the case of an idiosyncrasy or severe illness may be unbearable and harmful." The italics are mine: although Beard asserted that only the tenderest sensibilities were undone by noise, the case he made in his *American Nervousness* (1881) was that the number of such people was mounting in proportion to the degree to which civilization was modernizing, a process that was most ferocious and, thanks to Mr. Edison, most *unintermittent* in the United States.[47]

Which meant that ordinary Americans should take pause. The premonitory signs of neurasthenia were so multifarious (indigestion, tooth decay, hair loss, chronic catarrh, fatigue, noises in the ears, excessive yawning) that few of any class were beyond its reach. Brainworkers, artists, machinists, and mothers of small children might be more susceptible to an earlier onset, but even those with "coarseness of nerve-fibre" (manifest, for example, in "loudness of dress") could contract neurasthenia. The publishers of a new and revised edition of Beard's *Practical Treatise on Nervous Exhaustion* were happy to blurb from a review in the *Alienist and Neurologist* that Beard, more than any other, had "contributed to establish the fact that grave appearance of local disease may exist without, in fact, having a local organic habitation." It was a syndrome that expressed the social disjointedness of a Second Industrial Revolution that was shifting from mere mechanization to precision machining, from steam engines to dynamos, from a translation of forces (linear to rotary) to the transduction of energies (electrical), and from piecework to timework. Americans in particular, observed Beard, were surrounded by the tick and whirr of timepieces (Elgin pocket-watches, Waterbury mantle clocks, Waltham alarm clocks), as well as by anxiously higher-pitched voices and popular music centered on a higher, more fretful key, evidences of a mercantile creed of hurry and worry. Cautioned as they had been against the dangers of sleep deprivation by William A. Hammond, the former Surgeon General, Americans would do well to heed Beard's evangelical call to slow down: "The gospel of work must make way for the gospel of rest." Now that conservation of energy was a law of the thermodynamic universe, it should be obvious that human bodies demanded a similar conservation of personal energy. No need to put corks in both ears and shut yourself up in a dark closet (as did one of Beard's patients whose nerves were "much exalted"), but everyone needed time (out).[48]

Everyone. Before Beard was born, American and European workers had begun to agitate for reduced working hours, marching behind the banners of the Ten-Hour Movement, from which mine children and mill women were among the first to benefit. Between 1865 and 1880, while Beard spent summers relaxing in Europe and American industrial laborers won a ten-hour workday, trade unions put their shoulders behind an Eight-Hour Movement, singing,

> We want to feel the sunshine;
> we want to smell the flowers;
> we're sure that God has willed it.
> And we mean to have eight hours.

After Beard's death from dental malpractice in 1883 and after the international strikes of 1 May 1886—the original May Day—shorter workdays were sporadically negotiated. These successes should have ensured a calmer, healthier society, but that was not how it appeared to liberal critics at the time. Leaving to one side the nutritional deficits that kept infant and child mortality high among laboring families and the industrial accidents and disastrous living conditions that kept their life expectancy short, it was exactly the working-class mode of leisure that gave the genteel world the jitters. Workers seemed hell-bent on spending their "free time" in pursuits noisier and less punctilious than what reformers or pastors had in mind for them. Avid though they might be to improve themselves intellectually and morally, they seemed to end up in saloons, dance halls, brass bands, raucous ethnic festivals, and uproarious union meetings, not in libraries, schools, and churches, or in quiet communion with Nature.[49]

Public parks, inset as "lungs" to give city folk breathing room, became one of the fiercest battlefronts between modes of leisure. Once workdays were shortened and Saturday hours cut in half, wage-earners and artisans had more daylight in which to take deep breaths among these re-visions of royal walks and pleasure gardens. Advocates for city parks, free and open to all, saw them as "public pulmonary organs," then as spas that would draw men and women away from saloons and sordid amusement. Parks would bring parents and children together as families, and families together in one community, the struggling poor, middling, and high society mingling so harmoniously that class antagonisms would wilt and civility return to public life. The wealthy approached city parks as preserves of fresh air and natural beauty, inherently valuable places meant to be as restorative as the well-aired sick rooms of a Rest Cure; accordingly, they did little that might disturb the real estate—a stroll, a smoke, a botanical tour, a carriage ride, a modicum of courting, a stretch of rowing. Their approach agreed with that of America's premier landscape architect, Frederick Law Olmsted, who pictured his Central Park as a soundproofed arcade/arcadia in the middle of Manhattan, where "people may stroll for an hour, seeing, hearing, and feeling nothing of the bustle and jar of the streets, where they shall, in effect, find the city put far away from them." The working class approached parks differently, as neutral spaces open to possibility, free from traffic, from garbage, from industrial cinders, and from railroad tracks that slashed through their neighborhoods from every angle. Clean, open, and safe, parks were expanses where they could flex their muscles and clear their throats, rollick and frolic; they were spaces for sport, for band practice, for flirtations and, in the shadows of arbors or on warm evenings, for sexplay more private than they could expect back home. Bristling at such violations of decorum, designers shunted playing fields and playgrounds to the outskirts of parks to restore the quiet and tranquility that had been upset by "men shouting under the excitement of the skittle ground or the unrestrained merriment of factory girls who used the swings." On aesthetic, health, and acoustic grounds, park commissioners planted signs against rough-housing, required permits for group picnics, fenced in the flower beds, regulated "regulation" playing fields, and posted attendants who, keen as Nightingale matrons, scolded noise-makers. KEEP OFF THE GRASS was no mere assertion of a gardener's prerogative; the signs demarcated a new urban precinct intended to be at once a gallery and convalescent garden.[50]

Beyond the parks, on hillsides worn smooth by hunters and gatherers, and on seashores and lakesides dotted with cottages, lay other sites of convalescence. Those who had the time and money could hold at bay the "New World's Babel, louder with each hour," by indulging in the restful sounds of a mountain farm, "the silver note of the unseen brook, and the clanging of the cow-bells fitfully in the dark, and the deep breathing of the cows." Far beyond the city and its parks, in the North American and Australian West, in the Alps and Hindu Kush, were quieter, less trafficked places yet, and these too could be therapeutic. Deep inside the granite gorges of the Grand Canyon, at the end of two months of exploration and two weeks of rafting the "Great Unknown," John Wesley Powell in August 1869 thought not of geology or geography or his wife back home but of his own return to health. A veteran of the battle of Shiloh, where he had lain in delirium and lost an arm, Powell compared his relief at having survived the pounding white-waters of the Colorado River to his "first hour of convalescent freedom," a freedom as much from the cries of other soldiers as from his own pain: "When he who has been chained by wounds to a hospital cot until his canvas tent seems like a dungeon cell, until the groans of those who lie about tortured with probe and knife are piled up, a weight of horror on his ears that he cannot throw off, cannot forget, and until the stench of festering wounds and anaesthetic drugs has filled the air with its loathsome burden—when he at last goes out into the open field, what a world he sees! How beautiful the sky, how bright the sunshine, what 'floods of delirious music' pour from the throats of birds, how sweet the fragrance of earth and trees and blossoms!" Men of means with "shattered nerves" would more often be prescribed the restorative of wilderness than weeks of Rest Cure. Owen Wister, for example: son of another Philadelphia doctor and an aspiring composer who had left off music to take a clerkship at a bank vault only to succumb to vertigo and fatigue, Wister in 1885 (months after his mother had taken the Rest Cure) was advised by Weir Mitchell to go out to Wyoming, where "Each breath you take tells you no one else has ever used it before you." There, 8,200 feet up at the Point of Rocks, and later at Lower Yellowstone Falls in America's first National Park, he reclaimed his health and purpose in life, which turned out to be chronicling the West. Near century's end, when "tired, nerve-shaken, over-civilized" Americans were touring Yellowstone or the newer National Parks of Sequoia and Yosemite (1890), they expected to snap a Kodak at each primordial vista and, "amid the calmest, stillest scenery" (hoped John Muir), "be brought to a standstill, hushed and awe-stricken," before geysers high as Broadway billboards. They came to the Parks for an assured but contained excitement and may not have gloried, as Muir had, in the "exuberance of light and motion" during a High Sierras storm, where, from his temporary sanctuary in the crown of a Douglas fir, he had heard "The profound bass of the naked branches and boles booming like waterfalls; the quick, tense vibrations of the pine-needles, now rising to a shrill, whistling hiss, now falling to a silky murmur."[51]

Vacations west or anywhere else were not within hearing of the bricklayers, cigar-makers, machinists, or millworkers who by 1900 had won workweeks of 54 hours in Great Britain, 64 hours across Europe, and 56–59 hours in the United States (64–72 hours on farms, more in bakeries) yet worked, so long as they had work, fifty-two weeks a year. Unlike urban elites, they could not elope to lakeside cottages during the summer heat when each household had its windows open and all lives sounded more loudly, nor could they head for the hills at the first sneeze of hay fever, an epidemic (claimed Beard) that had grown together with neurasthenia like ragweed with golden rod. Charitable "Fresh Air" funds did send thousands of mothers and children to farms or seasides for a few days in July or August, when the tenements' poor were most taxed by "the stifling heat, and the noises, and the unpleasant odors, and the lack of privacy"; for the great majority of urban families, however, parks were the only country they knew, the only vacations they took.[52]

There they had space to play, time to think and read and snooze, and the space-time in which to listen—to small talk or love talk, to preachers, poets, or political philosophers. Listening to (and baiting or debating) open-air orators was as much a pastime as attending open-air concerts, football matches, or baseball games. Those who had been to school had themselves been trained in elocution, with Demosthenes as their ancient model, he who had gone to the roaring shore and spoken "his oration there with a loud voice, to accustom himself to overcome the noise of a great assemblage of people." They had learned to declaim from anthologies which, in America, included passages about the vast and vanishing flocks of passenger pigeons with their (*practice those tumbling u's*) "perpetual tumult of crowding and fluttering multitudes, making a noise like thunder." If they made it through grammar school, they would have memorized classical elegies, dramatic monologues, and colloquial verse, like this, about riding on a trolley:

Woman with her baby,
Sitting *vis-a-vis;*
Baby keeps a-squalling,
Woman looks at me;

Asks about the distance;
Says 't is tiresome talking,
Noises of the cars
Are so very shocking!

Schooled or unschooled, people were veteran auditors of hours-long sermons, stump speeches, wedding toasts, union exhortations, lyceum lectures, revival harangues. Connoisseurs of such holdings forth, they appreciated delivery and style and listened for sound as well as soundness of argument. At an inquiry into presidential election fraud in 1876, Col. William Pelton made a bad impression worse when, while confessing to efforts to bribe election officers on behalf of his uncle, Democratic candidate Samuel J. Tilden, he spoke with "a loud, hard, and rather grating voice, and delivered his answers with a quick, jerky, nervous utterance, which often jumbled his words so as to render them partially inaudible." In a world without public address systems, nervous syllables were suspicious, grating and jerky were bad, loudness a necessary evil. Women, who had (been educated into) softer, calmer voices, learned to project their words once they entered the public forum to speak on abolition, temperance, dress reform, or a woman's right to a voice and a vote. Like Amelia Jenks Bloomer, they resorted to passages in song that would carry to the back of a crowd and, like male orators, they chose plain syntax and soundbites that would not drift out of reach. Patrick Conway's band of fifty brass players was praised in Philadelphia for a concert at Willow Grove Park where the band's "most tremendous fortissimo" did not "stun the ear nor distort the composer's meaning," and a critic of band music was willing to concede that "An indiscriminate crackling and tearing of bugles, trombones, and ophicleides, etc., may sound very well in the street." Mark Twain, who struggled to love opera with its "intense but incoherent noise which so reminds me of the time the orphan asylum burned down," confessed to a guilty pleasure in band music. "I suppose it is low grade music—I know it *must* be low grade music—because it so delighted me, it so warmed me, moved me, stirred me, uplifted me, enraptured me, that at times I could have cried and at others split my throat with shouting." Brass bands, brash sports, and brassy orators: not quite what Olmsted's gentry had in mind.[53]

"Standing beneath this serene sky, overlooking these broad fields now reposing from the labors of the waning year, the mighty Alleghenies dimly towering before us, the graves of our brethren beneath our feet," began Edward Everett, former governor, senator, president of Harvard, and at sixty-nine the North's most venerable orator, "it is with hesitation that I raise my poor voice to break the eloquent silence of God and Nature." At his request, the dedication of seventeen acres for a National Cemetery in rural Pennsylvania had been deferred from late October to mid-November so that he could muster his thoughts about the recent battle. (For Bill Bayly, a thirteen-year-old who had watched from town, "no shifting of scenes or movement of actors in the great struggle could be observed. It was simply noise, flash, and roar.") The President was also asked to come, and he did, patiently awaiting his turn through the music of brass bands, a prayer, a hymn, and the two hours of Everett's oration. Lincoln knew what it was to speak for hours, having done so himself on occasion, but this day, when at last he rose to make his "dedicatory remarks," he spoke for three minutes, maybe four. The brevity of his address was fitting for a place that had already been hallowed, as he said, by the blood of fallen soldiers. The tone of his address, delivered in a "sharp, unmusical, treble voice" that carried well into the crowd of fifteen thousand, fit his own mood: his third and favorite son Willie had died of typhoid fever a year ago; his fourth son Tad, back at the White House, was sick. And the text of his address fit the new sense of what a cemetery must be: a "final resting-place" for the dead, a muted soundscape of re-dedication for the living.[54]

Epigraph to an 1863 prospectus for the Cypress Hills in Brooklyn:

> There are the dead respected. The poor hind,
> Unlettered as he is, would scorn to invade
> The silent resting-place of death.

This was a quotation from Henry Kirke White, an English poet who died, as did perhaps half the dead at Gettysburg, before the age of twenty-two. The precocious poet asks in blank verse to be buried in the Wilford churchyard, a quiet place where he may lie at peace in a shaded nook—"a silly thought,"

> as if the body,
> Mouldering beneath the surface of the earth,
> Could taste the sweets of summer scenery,
> And feel the freshness of the balmy breeze!

Yet it's human, isn't it?, to want a "decent residence" for something held so dear,

> And who would lay
> His body in the city burial-place,
> To be thrown up again by some rude sexton,
> And yield its narrow house another tenant,
> Ere the moist flesh had mingled with the dust...?

Too awful to contemplate, a crowded grave in an uproarious city,

> where e'en cemeteries,
> Bestrew'd with all the emblems of mortality,

Are not protected from the drunken insolence
Of wassailers profane, and wanton havoc.

So,

No, I will lay me in the village ground;
There are the dead respected. The poor hind etc.

Composed around 1800, White's poem prefigured the transformation of burial grounds into cultivated gardens and the transformation of graveyard affect from the Gothic or melancholic to the sentimentally contemplative. Where before epitaphs had reminded the living of the wages of sin and the shortness of life, tombstones by 1863 referred to a restfulness that was the ideal state of the virtuous departed and the ideal situation for those who came to lament.

Place for the dead!
Not in the noisy City's crowd and glare,
By heated walls and dusty streets, but where
The balmy breath of the free summer air
Moves murmuring softly o'er the new-made grave.

At best the cemetery had been a wild-flowered country churchyard or a "Quaker Grave-Yard" (Weir Mitchell) in a quiet city square; at worst a boneyard or plague pit to which corpses were carried in carts whose wheels were swathed in bagging "to smother the noise." Soon it would be a gardener's arboretum and a sculptor's gallery, with figurative evocations of the dead, sculpted not because they had been heroes, poets, or duchesses, as at the first of the scenic cemeteries (Père Lachaise in Paris), but simply because they were missed. Or because a full-bodied memorial for the less-than-famous had a chance now of being staged and tended as carefully as a Weir Mitchell neurasthenic or a Nightingale invalid, in a respectful quiet.[55]

Call it a Reform Cemetery, with plots aligned on a grid for ease of access, or a Hygienic Cemetery laid out for proper drainage, or a Garden Cemetery planted for heart's ease, or a Forest Lawn situated on a city's wooded outskirts for a "more natural" bereavement—in each case the new cemeteries were silent cemeteries, inspired as much by desires for quiet communion as by disgust at unkempt burial grounds. *Oration Delivered at the Opening Services of the Rosehill Cemetery Association on Thursday, July 28th, 1859*: "From the noisy town we have come out into the silent woods, from a city of the living to found a city for the dead." The bustle of Chicago behind us, said Reverend Noah Schenck, "today we stand in this secluded grove, awed into silence and solemnity." Rosehill had been designed by William Saunders, who had designed Mount Auburn and the National Cemetery at Gettysburg, and its silence, like that at Gettysburg, was for the living. It makes no difference to the dead where we put them, wrote the poet Annie Parker: "Their rest will be unbroken, though the din of busy life is never hushed around them." We may say we seek a Sabbath stillness on their behalf, but the subdued surroundings in which one can hear "the songs of birds, the spirit voices of the wind, and the rippling of gentle waters" are solely for us: "Away from the noise and bustle of the city, though near enough to be easily reached, the mourner may bring unseen the graceful offering of affection and weep in solitude over the grave of the beloved."[56]

Parker had caught the song of a new profession, funeral directing, with its practiced whisper. She had caught the spirit of a resurrected art of embalming that prided itself in being able to give a corpse "a look of quiet resignation." And she had caught the ripple of a newly hushed ritual that began, for the better-off, with heavy caskets in the stillness of funeral parlors among the rustle of silks, crepes, and tulles, a rustle that continued on to the cemetery. Instead of "half-maniacal" keening, silent tears; instead of hired women wailing, a bugler's "Taps"; instead of torrential weeping, a low slow organ postlude. As funeral directors in their new black suits stifled the sharper outcries of grief by imposing a somber formality and enclosing distraught families in thickly draped carriages, so keepers of the new gated communities of the dead succeeded where urban parks failed: they secured the peace. The rule of quiet implicit in the conduct of funeral directors was made explicit within the walls of the cemeteries to which they brought the coffins. *Regulations Concerning Visitors, Allegheny Cemetery, Pittsburgh, Pennsylvania, 1873*: "Any person disturbing the quiet and good order of the place, by noise or other improper conduct... will be compelled instantly to leave the grounds."[57]

Bells still tolled at the passing of the illustrious, but more briefly and, given the acoustic density of cities, less audibly. A "deep-toned bell" that was "forever tolling its knell" over the archway of a busy cemetery in Whitman's Brooklyn was less a cue to collective grief than a sonic guardian of the hush within. Named in 1838 to invoke "rural quiet, and beauty, and leafiness, and verdure," Greenwood became famous for the visual and aural serenity by which its groves assured a modulated mournfulness at graveside. In the convening of what Rev. Schenck called a "community of sentiment," Greenwood ran a close second to Mount Auburn outside Cambridge, established under the aegis of the Massachusetts Horticultural Society as the first and most elegant of American garden cemeteries. "Why should we," asked the poet and Supreme Court Justice Joseph Story at its dedication in 1831, "measure out a narrow portion of earth for our graveyards in the midst of our cities, and heap the dead upon each other with a cold, calculating parsimony, disturbing their ashes and wounding the sensibilities of the living? Why should we," asked thereafter many a studious young lady elocuting from *Hemans Reader for Female Schools*, "expose our burying-grounds to the broad glare of day, to the unfeeling gaze of the idler, to the noisy press of business, to the discordant shouts of merriment, or to the baleful visitations of the dissolute?" Story himself was interred at Mount Auburn in 1845 under "the shadow of forest-trees" in a family plot among four of his children whose lives had been cut dreadfully short and for the first of whom he had written,

So quiet and so sweet thy death,
It seemed a holy sleep,—
Scarce heard, scarce felt, thy parting breath
Then silence fixed and deep.

His own epitaph read, "He is not here—he hath departed." Edward Everett, also on Mount Auburn's Board of Managers, was laid to rest there on 15 January 1865. Three months later, on the evening of 14 April, Everett's fellow orator at Gettysburg was shot in Ford's Theatre by "the youngest tragedian in the world." A single bullet entered Lincoln's skull behind his left ear, tunneled through his brain, and lodged behind his right eye. The following morning, under the greenish light of a hissing gas jet, Lincoln drew his last breaths, which were heard as "a deep snoring," "a wild gurgling," or the chords of an Aeolian harp plucked by the wind. Over the next fortnight, as the train carrying the body of the President (and the disinterred body of his son Willie) reversed the route by which Lincoln had come from Illinois to the White House, it was met everywhere by lines of mourners. In Philadelphia, New York City, Buffalo, Cleveland, Columbus, and Indianapolis, the lines went on for miles and continued in processions to the beat of muffled drums, the crack of minute guns, and the tolling of firehouse bells. In Chicago, thirty-six young women in white gowns led thousands of dignitaries and ten thousand schoolchildren with black armbands in a four-hour-long procession, at the end of which came the ranks of Coloured Citizens. A local newsman summoned the sounds: "Hushed be the city. Hung be the heavens in black. Let the tumult of traffic cease. Let the streets be still. Let the lake rest. Let the winds be lulled, and the sun be covered up. The bells—toll them. The guns—let their melancholy boom roll out." By 4 May, when the remains of father and son were escorted down a country road to the strains of Handel's "Dead March" and nestled in a hillside tomb among the wooded acres of Springfield, Illinois's new Oak Ridge Cemetery, a tenth of all Americans—one-fifth of the North—had glimpsed Lincoln's casket. More people had gazed upon his face, claimed the Methodist Bishop Matthew Simpson in a final eulogy, "than ever looked upon the face of any other departed man." After the graveside benediction came a few brief thunderclaps. Drops of rain fell in the heat. Lincoln's journey home with Willie had been a last full measure of devotion, and the first national silence.[58]

Quiet held its own in the new cemeteries because they were bastions of the past. Idylls of "immortal" statuary and weeping willows for the remembrance of the dead, they were also refuges from the stir of time. For all of their ordered lanes and reconnoitered spaces, the new cemeteries were rearguard emplacements against the noise of modernity, defending an unruffled permanence that had no public parallel except in museums. And as companies of art lovers made worshipful excursions to gaze upon paintings of the Renaissance that had been varnished to a golden quietness unintended by the Old Masters, so visitors to the new cemeteries honored the recent dead by sharing in a silence that was now their defining quality. Stones were chiseled not with warnings against a mutable world and inconstant flesh but with assurances of the indelible, of names that would not be lost, good deeds never forgotten, smiles only slowly worn to a blank. Through the solidity of their coffins, the standfast of the granite memorials, and the sanitary regimes of groundskeepers, cemeteries willed silence as they willed a resistance to any change other than that of the seasons. If, as Peter Fritzsche has argued, the feeling of being modern was inextricably bound up with a melancholic nostalgia for what had been lost in becoming modern, then the new cemetery was as much a time-out-of-mind as a pastoral preserve. Its stillness was geologic and temporal, its fixity straining, like Lincoln at Gettysburg, toward something more than loss... Under the heat and vibration of seventeen hundred miles of rail traversed at no more than twenty miles per hour and the close breaths of the multitudes who came for a last look at a visage that had been fixed in their minds by the circulation of four score and more of photographic poses, Lincoln's already asymmetrical visage began to droop and discolor. His familiar seriousness, which had been an artefact of having to remain motionless for forty seconds while his face surfaced on a glass negative, was exaggerated by the black bruising around his eyes, left unretouched by order of Secretary of War Stanton as evidence of the assassin's bullet that had shattered the oddly uneven orbital plates of Lincoln's eye sockets. His countenance, which according to one viewer "still preserved the expression it bore in life, though changed in hue, the lips firmly set but half smiling," had to be revived at each stop with brush, chalk and amber—the stillness, like the faint smile, a mortician's fiction. Funeral directors, embalmers, stonecutters, and landscapers of the new cemeteries worked hard to preserve and sanctify an inorganic stillness. A truly modern person would accept the physiology of death, the statistics of public health, the economics of rational land use, and opt instead to be cremated.[59]

Resurrection was the issue: whether Christians had need of bones or other teguments to rise, be recognized and rewarded. Prevailing images of Judgment Day showed gaunt skeletons, sunlit figures, or corpses in winding sheets rising from graves at trumpet's call, their faces no less distinct than those developed by spirit photographers after séances. A similarly apocalyptic physicality attached to the custom of visiting cemeteries on the Sabbath. In Jewish life, *Shabbat* echoed the seventh day of Genesis, the day of the Eternal's rest after the work of Creation and, by divine directive, a day of His people's rest in honor of Creation; by rituals that gathered force after the Roman destruction of the Second Temple in 70 CE, the Sabbath was also a day for communal study and prayer; by rabbinic and later kabbalistic Midrash, which caressed the Sabbath into the figure of a voluptuous bride, it became as well the most blessed time of the week, during which a husband and wife should pleasure each other in bed. In Christian life, the Lord's Day was the First Day of the week, the day on which God had raised His Son from the dead and by which, symbolically, He had shifted the week's holy day from last to first; it was also, prophetically, the day reserved for resurrection at the Second Coming.[60]

None of this ancient history would bear repeating had not certain Anabaptists of the 1530s–1540s determined that there was no scriptural mandate for moving the Sabbath from Saturday to Sunday, and had not millenarian Baptists of the 1830s–1840s determined that restoring a seventh-day Sabbath was critical to salvation. Reading Daniel and Revelation with an eye to the prophetic calendar, the New England Baptist minister William Miller figured that the Second Advent would occur sometime between the vernal equinoxes of March 21, 1843 and March 21, 1844—or, for sure, seven months later, by October 22, 1844. Nothing earth-shattering happened by or between either of the 1844 deadlines, except that Annie Ellsworth in Washington, D.C., did have her father's friend Samuel F. B. Morse use his dot-dash code to wire Baltimore with a quotation from Numbers 23:23, Siyyid Mírzá 'Alí-Muhammad announced himself in Persia as The Báb or "Gate" to a new revelation, Marx met Engels for the first time, the last breeding pair of great auks was killed off the coast of Iceland, and Robert Chambers finished *The Vestiges of the Natural History of Creation*, anticipating Darwin's theories of extinction and evolution. Fifty thousand Adventists, intent on a comprehensive ending and new beginning, expected more to be wrought by God than electrical telegraphy or a survey of fossils; they expected to see "crowns, and kings, and kingdoms tumbling in the dust," expected to hear "carnivorous fowls fly screaming through the air" and "seven thunders utter loud their voices" as the clouds burst asunder. Quakes struck Italy and Indonesia, but these could not shake off the Great Disappointment. Five years later Miller died, unascended but unrepentant of his prophetical arithmetic and, like other Adventists who had been advocates for Temperance and Abolition, still frustrated by the walking-stick pace of the moral transformation of the world. Among those who kept the millennial faith were several who reexamined Miller's numbers and found that something of infinite but invisible significance *had* occurred on October 22, 1844: in keeping with Daniel 8:14 and in anticipation of the Second Coming, Jesus had begun cleansing the Heavenly Sanctuary and its Holy of Holies, which held the tablets of the Law given at Sinai. Before Christ's return and the building of a New Jerusalem, the faithful must show themselves obedient to the Law, above all to the Fourth Commandment, which meant remembering that the Sabbath referred to the fulfillment of the First Creation and must therefore be observed on the seventh day. Catholics and most Protestants except Seventh-Day Baptists had erred in preferring a pagan Sun Day over the Old Testament *Shabbat* (sundown Friday through sundown Saturday), which had grave consequence for redemption. Thus testified Joseph Bates, James White, and Ellen H. White, whose Saturday worship was central to their Seventh-Day Adventist message that the Second Coming was keyed to the keeping of the true Sabbath.[61]

By the time, then, that botanists, Transcendentalists, and young courting couples were strolling through the new silent cemeteries, the American Sabbath was caught up in a scrimmage that was ceremonial and sonological. Ceremonial and sonological? Yes, in the sense of a contest over the performative logic of the Sabbath, whether Seventh Day or First. Should the Sabbath be honored by the sober intonation of God's Word or a joyful noise in celebration of the Lord's world? A private inquest of the soul, or public exultation in Creation? Tiptoed errands of mercy, or clodhopping enjoyment, at God's grace, of a day off? The Catholic Church considered the Lord's Day a time for both collective worship and personal refreshment. What was reasonable by way of refreshment, that was the stickler. Unreasonable uses of the Sabbath had driven Calvinist Reformers to draw up stern rules of observance, especially in England, where precisionist Presbyterians began asserting a much longer and quieter Lord's Day against the games and sports that attracted gentry and commoners of a Sunday afternoon. Charles I, a High Church king married to a French Catholic, scorned Sabbatarianism as "a superstitious seething-over of the hot or whining simplicity of an over-rigid, crabbed, precise, crackbrain'd, Puritanical party," and it was precisely on this issue that Laud, his Archbishop, went after the Reformers. Challenging the historical and theological ties between the Old Testament (Jewish) Sabbath and the New Testament Lord's Day, he intensified the divisions between a crabby Puritan Parliament and a moody King, bullying both toward civil war, which broke out in 1642. And it was precisely the Puritans who would minister to much of colonial New England, where the firing of muskets, beating of drums, blaring of trumpets, blowing of conch shells, or ringing of hand bells (later, churchbells) woke each township to a Sunday Sabbath morning.[62]

Strict observance of the Sabbath dictated a highly aural experience: long prayers "intoned in dragging minor intervals"; the ruling-out or antiphony of long psalms that, during the half hour it took to quaver a response to each of 130 lines, could result in "a horrid Medley of confused and disorderly Voices"; the drone of sixty-page sermons unevenly amplified by wooden sounding boards that rose up behind meetinghouse pulpits. "Deaf Pews," closest to the pulpit, were reserved for those hard of hearing, as was the pulpit staircase itself, whose steps might bristle with elders wielding ear-trumpets, a visible encouragement to silence among the assembled. Assuring an attentive silence was the job of a tithing-man who stalked the room with a hardwood staff for cudgeling rude boys and a dangle of hare's feet for snittling sleepers awake. Back home, the faithful discussed the sermon, entered private meditations into religious diaries, and went off to works of charity or church business. There was no excuse for "*dozing* about with folded arms"; the Sabbath was a day of rest from man's work, not God's.[63]

Whence the reassertion of Sunday laws, which sifted God's work from man's while folding civil authority into religious. The first such edict, issued by Constantine in 321 CE, had closed the courts and forbidden tradesmen from working and merchants from trading on Sunday. In 469, Roman edicts held the Lord's Day aloof from theater, circus games, and judicial investigations, in order "that the shrill voice of the cryer shall cease, that litigants shall have rest from their disputes, and have time for compromise." Quiet and peace, not churchgoing, were uppermost in these and later (medieval) ordinances, whose penalties for Sunday murder, rape, and battery were stiffer than for weekday mayhem, while penalties for missing a mass were slight. Puritans tightened the reins, fining transgressors of the Sabbath (sunset Saturday through sunset Sunday) for hanging laundry, posting notices, singing profane songs, hunting, traveling, or skipping a sermon. These "Blue Laws" never forbade running or kissing one's children (as critics insinuated), but the Sabbath was meant to be spent apart from "distractions and entanglements in the businesses and occasions of life," and it took the ever-earnest Samuel Willard nine months of Sundays to explicate the Fourth Commandment for his congregation. In eighteenth-century New England, Sabbath wardens still watched and listened for illicit traffic and "rude or clamorous behavior, in or within hearing of the assembly."[64]

Gradually, as the Lord's Day was divorced from Saturday evening, and as socializing became the norm for Saturday night and Sunday afternoon, the Sabbath lost its chokehold on noise. The relaxation came as civil authority trumped religious in the making and enforcing of Sunday law and as the appeal of a country ramble ("sermons in stones and running brooks") transcended pulpit hortatory. It came in tandem with the more boisterous worship of Methodists and Baptists, of camp meetings and urban revivals. It came in reaction to Sabbatarian extremism, parodied in 1805 in the figure of a sanctimonious man harrumphing at the violation of the ban on

Sabbath travel: "Even in holy time do impious droves of licentious oxen trample on our laws, and daring swine grunt defiance to the Statutes of the commonwealth." And the impious working class, with greater defiance, availed itself of cheap fares for Sunday excursions to amusement parks, beaches, patches of countryside. For those who labored sun-up to sundown, the "refreshments" of a day off took precedence over pews; for the leisured classes, Sunday rectitude loosened into a genteel recreation that children took to heart. Reprimanded by her father for "playing and making a noise on Sunday p.m.," Lucia Wheaton replied, "with a very serious look, 'Why, we are not playing on Sunday, we're only make believe playing.'"[65]

Her mother in 1851 thought the young Lucia "a natural Sophist," but such sophistry had been learned in Sunday School, which for working-class children was often their sole schooling. Its origin myth had Sunday School being born from noise: in 1781/1782 an English printer, Robert Raikes, had gone to the outskirts of Gloucester to hire a gardener, who was not at home but would return shortly. Waiting, Raikes espied "a group of very noisy boys" in the street. What, he asked the gardener's wife, was "the cause of their being so neglected and depraved"? She: "If he were to visit the place on the Sabbath, he would sympathize with the people of the neighborhood, for the noise and confusion were so great as to deprive them of the quiet enjoyment of that day." Whereupon Raikes, a prison reformer, hired teachers for these little rascals, dragging them to class hobbled with logs of wood so they could not scamper back to Sooty Alley, and caning them when they turned unruly during hours of Bible memorization. Departing soon from modes of instruction that assumed infant depravity, Sunday Schools opted for less terrifying theologies and more Romantic pedagogy as they expanded to handle millions of working-class children who had to learn the rudiments of good behavior: "Make no noise with your tongue, mouth, lips, or breath, in eating or drinking.... Sing not nor hum in your mouth, while you are in company." By the 1850s, principals in Scandinavia, Great Britain, and the United States were leaning toward the views of the *Scotch Sunday-School Teachers' Magazine*, whose editors advocated the singing of secular songs—"good, noisy songs with easy choruses," for "There is no sin in shouting, and cheering, and yelling, and jumping (by permission) in a schoolroom." And if "the children interrupt us with noise, we would stop, and give them leave to create even greater disturbance" until, their animal spirits spent, they are open to "a few words of a serious, religious tendency; and so you will naturally slide into a hymn, and terminate with a prayer."[66]

"Sunday Our National Defence," declared a sub-head in a book of *Manners* prepared by Sarah Josepha Hale. She would know. A mother of five, a novelist, a poet, and editor of *Godey's Lady's Book*, the most widely subscribed magazine in the world, Hale was a force to be reckoned with. Almost single-handedly she had persuaded Lincoln to proclaim a national day of Thanksgiving, first celebrated in the North one week after his address at Gettysburg. Yet even Hale, who believed that full Sabbath observance would go a long way toward securing national civility and personal self-government, protested that "There is something monstrous in... linking Sunday and sadness in the brain of a child." The Sabbath should be for refreshment, not recrimination; practices that burnt a feast of love to the crisp of hellfire could lift up no child's soul. Lincoln himself, who had issued a General Order in November 1862 enjoining "the orderly observance of the Sabbath" by the Union forces, must have known that his soldiers (like Robert E. Lee's, who received a similar order) would be less than reverent in their Sunday demeanor, as he and Lee must also have known that the war would rarely pause for the Lord: troops fought on Sunday at Chickamauga, Chattanooga, Spotsylvania. What did stop on Sunday were the horse-cars of Philadelphia, angering thousands who in 1865 hoped to reach Broad Street and its three-mile-long double lines of mourners waiting for a glance into Lincoln's coffin. Philadelphia's embargo on horse-cars would have pleased the American evangelist Dwight L. Moody, who demanded that public conveyances and popular amusements cease on Sundays, but the Sabbath immobilization caused near-riots in the City of Brotherly Love.[67]

Sabbatarian enterprises everywhere had a way of backfiring. Allied with the American and Foreign Sabbath Unions, Temperance advocates made several pushes for national Sunday closing laws, hoping to dissolve a time-honored contract between local saloons and Sabbath refreshment. Their attempts got tangled up with draconian efforts to ban most secular occupations and public recreations on the holy day, against which came resistance on all sides. Business and government wanted to ensure the daily mail, rail freight, coach and steamer service, and ever-ready telegraph service. Socialites, literary sorts, and middle-class children eagerly awaited the delivery of the Sunday newspaper, with its pictorial supplements and, soon, comic strips. Physicians and health reformers wanted fresh milk available, pharmacies and clinics open, bath houses heated. City fathers needed firehouses manned, wharves watched, parks supervised. Trade unions wanted horse-cars and trains running so that workers could get to marches, lectures, concerts, dances, melodramas, and sporting events, each of which had its own Lord's Day workforce. Altogether, a modern Sabbath appeared to require the labor of actresses, bakers, barbers, carnies, doctors, engineers, firemen, guards, librarians, milkmen, musicians, nurses, policemen, pressmen, stablemen, ticket-takers, trolleymen, undertakers, and, god forbid, saloonkeepers. In 1853 the novelist Harriet Beecher Stowe had an Elderly Gentleman wondering forlornly whether anyone else could remember the "unbroken stillness" of distant Sabbaths, or "recall the sense of religious awe which seemed to brood in the very atmosphere, checking the merry laugh of childhood, and imparting even to the sunshine of heaven, and the unconscious notes of animals, a tone of gravity and repose?" Thomas Lackland, who lived in "quiet rural retirement" and on Sundays heard "no roar of car wheels or shrill whoop of the steam-whistle even in the distance," could have been one of the few in 1867 who experienced unbroken Sabbath stillness—had he not been the figment of a *Boston Post* editor who meant to recapture homespun times "five and twenty years ago." Sunday "rest" now meant any activity but paid work—mending of clothes, shopping, playing kick the (newly proliferant) can. For blue-collar laborers who had muscled through sixty-five hours between Monday dawn and Saturday dusk, and for harried typists, salesgirls, telegraphers, and telephone operators, part of the Lord's Day had also to be spent on laundry and errands. In England, following his fellow working stiffs around on a Sunday, Thomas Wright found wives hard at work preparing Sabbath dinner, husbands rising late to smoke pipes and talk politics in barber shops, single men and women off a-courting, and only the youngest at church, or Sunday School.[68]

"Noisy trade goes on in the quarters where the foreigners live, and the Sabbath is filled with noisy, wanton, and drunken violators." To blame immigrants for the decline of Sunday stillness, as did the Congregational minister Matthew Hale Smith in his tour of New York City, was as a-historical as to press for the return of a Sabbath more circumspect than any Puritan or Seventh-Day Baptist ever managed. If foreigners with their strange inflections and alien modes of emotional expression tended everywhere to be heard as noisy and seen as loud, the imperfectly restful Sabbath acted as an acoustic lens amplifying difference. That many of the "noisy" foreigners were Roman Catholics contributed to Sabbath xenophobia, for Catholics were habituated to the festive Sundays of southern and eastern Europe—despite an encyclical of 1891 "On the Condition of the Working Class," in which Pope Leo XIII stressed that "The rest from labor is not to be understood as mere giving way to idleness; much less must it be an occasion for spending money and for vicious indulgence, as many would have it to be; but it should be rest from labor, hallowed by religion." Jews could not be faulted for resting on a Saturday hallowed by the sacred texts that Christian children still learned by heart, but Jews could still be prosecuted, as were Hindu, Muslim, and Buddhist shopkeepers, for doing business on Sunday. When the strongest of the nineteenth-century American campaigns for a national Sunday-rest law was mounted in 1888, it was defeated by a Congress that had received testimony and a petition of 260,000 signatures from Seventh-Day Adventists who argued that federal recognition of a Sunday Sabbath would make a mockery of freedom of religion and Separation of Church and State.[69]

Local entities across the country did pass Sunday laws, usually in the name of public order or the public weal, like the Colorado town of Argo in 1900, which prohibited "football, baseball and other games and sports assemblages having a tendency to create noises or disorderly conduct on Sunday." A venerable corpus of Anglo-American jurisprudence supported the general intent of the Argo ordinance. After 1885, municipal attorneys were also able to cite, when challenged on the Sabbatarian specificity of ordinances such as Argo's, the opinion of Supreme Court Justice Stephen Field in *Soon Hing* v. *Crowley* that Sunday rest was a worthy cause, since the government had a right "to protect all persons from the physical and moral debasement which comes from uninterrupted labor." Was turning a double play or legging out a triple in the same league as turning chair legs at a lathe? If, as Sabbatarians said, the Lord's Day was "not merely the day of religious duty and rest, but the restoring, the awakening day—the day of recovery and reformation," it could be argued that the exertions of sport were more laborious than reformative. This argument lost its force once health reformers, physiologists, and women's clubs began to prescribe "physical education" as a cure for insomnia, constipation, menstrual problems, neurasthenia, and the many ailments of city life. Sunday recovery, for those debased by hard labor or flustered by a social whirl, would best be effected, they insisted, by exercise in playgrounds and parks, by competition on playing fields, and by hikes in the open air. Let President Benjamin Harrison make federal moneys for Chicago's Columbian Exposition contingent upon its closing on Sundays; people had enough to do elsewise on the "week-end."[70]

"Week-end" was a word new to the 1870s. Granted a half-holiday on Saturday, a practice initiated at George Westinghouse's air brake factory in 1871, Americans came to enjoy what European workers had long claimed for themselves by way of a potable Lord's Day and a recuperative Saint Monday. In England, the Metropolitan Early Closing Association began organizing in the 1840s for shorter workdays and in the 1850s toward a Saturday half-holiday for shop assistants and clerks, in hopes that a less fatigued army of blue- and white-collar workers would sit up in church on Sundays and heed God's word. Paradoxically, the more leisure that some workers gained, the more services other workers had to provide for them, from a two-bit shave to a Sunday spread. For the majority of families who had neither servants nor savings, a Saturday paycheck and half-holiday turned the weekend from stocktaking toward restocking, i.e., shopping. "Goodness knows it is not because their tongues are not loud enough . . . that the chimes in the belfries fail to cause the various churches in the city to be full to overflowing," observed James Greenwood in 1881. "It is exactly at that time when the ringers, fresh to their work, are setting the bells clanging their loudest that the streets below swarm thickest" with shoppers and hawkers at London's Old Clothes Market. So the week-end undermined what was left of Sabbath stillness, even for the staunchly

Methodist son of a Baptist lay preacher, J. C. Penney, whose Golden Rule emporiums in Wyoming in 1902 were kept open on Sundays to do "the good deed" of supplying ranchers, miners, and cowboys who could not make it to store shelves on any other day. By then the half-Saturday was spreading across the business landscape as the Saturday "matinee" spread from vaudeville theaters to cinemas. When the *Virginia Law Register* in 1904 published an article asking, "Should the Sunday laws of our country be changed to meet the demands of our cosmopolitan population?" the question was almost moot.[71]

3. *Bells*

Whatever Sunday-closure customs held tight at banks, business offices, and government bureaus, a thoroughly quiet Sabbath, like a thoroughly hushed night, appeared no longer within the realm of metropolitan possibility and was almost as implausible in the vicinity of mines, mills, and railroad junctions. How explain then the contemporary silencing of churchbells? Of bells that for a millennium had told the time—rising and curfew, or matins/lauds, prime, terce, sext, none, vespers, compline—and tolled the end of one's time, stroking out the number of years the dying had reached. Bells, baptized and blessed, that had rung in each New Year and Lord's Day, feast and fast. Bells that had thundered magisterially against plague, demons, and the havoc of thunderstorms, true to their inscription, *Fulgora,*

SOUNDPLATE 7

Grango, Dissipito Ventos! (Lightning I shatter, great winds scatter!) Bells that had called the populace to fight fires, strap on weapons, mount the parapets, witness beheadings. Bells that could be heard miles off proclaiming treaties, triumphs, and weddings. Bells that had announced queens and bishops in their "progresses" through the countryside, and by whose tones coachmen knew one town from the next, pig drovers one market day from another. Extraordinary bells whose powers of redemption were underwritten by legends of virgins or babes thrown into the molten metal and crying out ever after at each convulsion of the clapper. Bells around which had grown up an architectonics of towers, an archive of poetry—"the poor man's only music" (Coleridge) that "lends the warning voice to fate" (Schiller) as "a heart of iron beating in the ancient tower" (Longfellow). So many bells, said a sixteenth-century Protestant, that if "all the bells in England should be rung together at a certain hour, I think there would be almost no place but some bells might be heard there." Bells bells bells, five thousand in Moscow alone in 1812, each rung fifty times at five a.m., forty times at five-thirty, thirty times at six,

> The swinging and the ringing
> Of the bells, bells, bells,
> Of the bells, bells, bells, bells,
> Bells, bells, bells—

which begin in merriment but end for Poe in alarum, out of tune and shrieking "in a mad expostulation with the deaf and frantic fire," or groaning from the rust in their throats, ghoulish, throbbing. Ancient churchbells, as in *Les Fleurs du mal*, where the poet admits to nostalgia for those bells still vigorous enough to ring through winter mist but is most deeply moved by "*La Cloche fêlée*," a cracked bell that rasps like a wounded soldier struggling for breath at the bottom of a grave-pile.[72]

Over time churchbells could go out of tune, lose their sweetness of tone, crack from age or drunken tinnabulation. In the close company of other belfries, they could clash, cast as they were with different strike notes and partials. They could be badly hung or badly rung, as indicted by a monk in a poem of 1769, when the campanile of his abbey shared the soundscape of Reims with 137 other bells struck so indiscriminately that the poet could neither sleep nor think. In England, upset by all the noise, Laurence Huddlestone of his own accord went from church to church with hammer and chisel to restore country bells to their sonority, and Rev. Hugh Haweis worried that parishioners could be "driven mad with the hoarse gong-like roar of some incurably sick bell." As of 1870, when Haweis was composing *Music and Morals*, he doubted that there remained in Britain a single "musically true chime of bells." Around that time, the McShane Bell Foundry of Baltimore would advertise its five-thousand-pound bells as "a real pleasure to a musical ear when compared with the harsh, unmusical sound of ordinary bells of its size" and guaranteed prospective customers that "If all the churches in the city would gradually adopt *McShane's Bells*, every Sunday morning would greet us with the harmonies of orchestral music, instead of the dissonant noises we have now to endure in some quarters." By 1896, even Boston's Old South Church bell, cast ninety-three years earlier by Paul Revere, was "approaching the end of its harmony" and "its discordant tones grated harshly upon the ear."[73]

Those most distressed by the rasp of old bronze or the chaos of dozens of steeples sonically out of sorts heard churchbells as more than unmusical; they heard them as unmodern. When Quasimodo the deafened hunchback exulted in the bells of Notre Dame, Victor Hugo was sheathing the bourgeois-revolutionary Paris of 1830 with the metal of the medieval, where great cathedrals were worlds unto themselves and great bells with names like Marie and Jacqueline awakened the soul as they shook the rafters, their roar audible for leagues around. When Quasimodo rode Marie, spurring her into a fury of bourdons like "a spirit clinging to a winged monster, a strange centaur, half man, half bell," he was a mythic figure for readers of historical romances; for modernists such as Nadar, the mythic was monstrous and *les cloches* maddeningly bestial: "Every other hour of the day . . . suddenly, brutally, and without any provocation, a storm of noise bursts out in minor and major keys, tolling, chiming, whining, belling [like a deer], groaning, or bellowing. At times, the diabolical bells all seem to agitate each other, one after another and, like madmen, start to howl together in a most frenzied hullaballoo (*tintamarre*)." Or, like grizzled old *clochards* and *bourdons* half-drunk or homeless on Paris streets, the bells brayed defiance at a world passing them by.[74]

Alain Corbin has studied the fall and rise of bells in nineteenth-century rural France, where those churchbells that had not been melted down for coin or cannon during the Napoleonic Wars had to be defended against secular town hall bells and the attacks of republicans who distrusted a Gallican church that was only half-disestablished. Before the Revolution, village churches and abbeys had been almost as likely as cathedrals to muster large numbers of bells; if Rouen and Paris each had hundreds of belltowers, a peasant within four miles of the hamlet of Grandcourt (northeast of Rouen and northwest of Paris) could hear fifty bells sounding from all directions. With the Revolution came the suppression of monasteries and churches and, in short order, a national inventory of churchbells to expedite their removal and re-smelting into copper coinage. Political as was this design to disable the power centers of the *ancien régime*, the silencing of the bells was framed in terms of noise. "There are belltowers that make more noise by themselves than do ten thousand citizens," wrote the novelist, botanist, and revolutionist Bernardin de St. Pierre in 1791;

"and as there are in Paris more than two hundred belltowers, imagine the appalling tumult made by these monuments, above all on holidays. Indeed, it is a monstrous thing... to hear the bellowing of these swollen towers and the barbaric sounds that issue from [what should be] temples of peace, even at night." In 1794, a fortnight after the first national decree abolishing slavery, the Directorate of Trévoux (north of Lyon) declared, "The republican ear must not be struck by the sound of bells, which were invented by priests to stun the people and dispose them to servitude." In 1797, when the Anglo-American radical Thomas Paine was in France and got wind of a proposal to restore the churchbells, he wrote in a white heat: "Religion does not unite itself to show and noise. True religion is without either. Where there is both there is no true religion. A Religion uniting the two [noise and show] at the expense of the poor whose misery it should lessen, is a curious Religion; it is the Religion of kings and priests conspiring against suffering humanity.... As to bells, they are a public nuisance." In 1802 Chateaubriand, the original Romantic, praised the aesthetics of Catholic ceremony and the emotional draw of its bells that evoked "enchanted reveries of religion, family, fatherland," but thirty years later, as an aristocratic liberal, he would claim that he had always thought the blithe ringing of churchbells to be "devastating to the tympanum and wrenching for the soul," neither lifting the spirit nor ravishing the imagination and yielding nought but noise, "an obnoxious brouhaha, or a bitter-sharp and false charivari." Therefore, you bells, hold your tongues! (*Cloches! Taisez-vous donc.*) Not until 1856, a generation after Hugo's creation of the hunchback Quasimodo, did Notre Dame regain its bells and reclaim its "old sonority," but as in Catholic Italy, Belgium, and southern Germany, and in Orthodox Russia, churchbells in France were hinged to political debates that kept refreshing their symbolism, so that their peals, though restrained, were never entirely silenced.[75]

Elsewhere, in Protestant nations and above all in the United States (lacking an established Church), bells whose notes had soared for the Sabbath and knelled for the dead seemed to sour once Lord's Days were overtaken by weekends and churchyards bypassed in favor of garden cemeteries. Even bells in good form appeared miscast for the times. "The sound of the sabbath bell," wrote Thoreau from the bank of a Musketaquid pond, "is as the sound of many catechisms and religious books twanging a canting peal round the world." Nathaniel Hawthorne, who had likened a churchbell to a hanged man and made a horror story out of a bell that tolled for the wedding of an old widower "with the same doleful regularity as when a corpse is on its way to the tombs," described a "Sunday at Home" when "At last, and always with an unexpected sensation, the bell turns in the steeple overhead, and throws out an irregular clangor, jarring the tower to its foundation." In Portland, Maine, beset with peals at seven and eleven a.m., at one p.m., and "toll, toll, toll, from two till six—with the conclusion at nine," the editor of the *Advertiser* could find no excuse for further rounds of tolling at funerals. "The practice is so common, the sound so often repeated, that it has lost all solemnity, all effect upon the healthy and sound, and can only serve to depress and alarm the sick, the nervous, and the over-fearful." Someone signed "X. Y. Z." followed up with a letter asking the town to outlaw funeral bells, which could cast a pall upon the most robust Down-Easter. Imagine "what must be its effect upon the sick and diseased in mind and body—how much pain this sad sound must inflict—how much mischief occasion and for what? To proclaim to the world that Death is, where he has been these six thousand years, in the midst of life."[76]

Longer than that, longer than the six thousand biblical years since Creation, had Death been in the midst of life, or so claimed a new generation of geologists, (paleo)botanists, and Darwinian zoologists who were busy unearthing prehistoric extinctions in the fossil record (and sifting through the mounds of dinosaur bones unearthed by railroad tunnelers). X. Y. Z.'s six thousands were enough, however, to make it clear that funeral bells were atavisms of a sadder age whose human inhabitants had resigned themselves to high mortality and constant mourning. As maternal and infant mortality began to drop in response to obstetrical antisepsis, immunizations, the pasteurization of milk, and the protection of water supplies, and as lifespans began to lengthen with better diets and sanitation, funeral bells began to seem pessimistic—when they could be heard at all, for the noisefulness of large cities was proving too much for church steeples hemmed in by tenements and dwarfed by office buildings. Overmuch or overwhelmed, death knells appeared obnoxious or redundant, as did, to an increasingly vocal choir of critics, churchbells in general.[77]

Competing privately with alarm clocks, cuckoo clocks, wall clocks, and chiming pocket watches, publicly with large clocks mounted on the towers of town halls and train stations, churchbells were losing their monopoly on telling the daily time. A poet at Amherst College in 1852 could still pen a "Song of the Bell" that repeated the old trope of bells tracing the ritual arc of life from birth through marriage to death ("I chime! I chime! / Holy time! . . . I ring! I ring! / Mirth's on the wing! . . . I toll! I toll! / God rest his soul!"), but churchbells were also losing force as makers and markers of sacred time. Bells that had commanded the Lord's Day and commandeered a person's last moments now contended with architectural and legal reconfigurations that diminished their holy *imperium*. Rev. Walter Blunt of Nether Wallop noticed such a reconfiguration in the floor plans of English churches: the belfry, which had traditionally communicated with the nave as a visible, audible element of worship, "thus yielding evidence that the Ringing of the Bells was still esteemed a very sacred thing," by the 1800s was being shut out from the body of the church, or omitted entirely, a single call-bell hung in its stead. Peals of bells, when still in use, were too often noisy amusements "made continually to wake the echoes of the neighbourhood, in no more noble cause" than the competitive sport of bell-ringing, taken up by the most "profane and profligate persons in the parish" and no more sacred in purpose than a steeple-chase. Blunt proposed in 1846 that parishes reinstitute a regime of bells for morning and evening prayers, weddings, Sabbaths, fast days, deaths, and burials, yet he would have the ringing held to a quarter-hour or less—after death "a steady, solemn, though cheerful peal, for five minutes and no more," not the hour-long tolling that had earlier been the custom. Already in parts of colonial New England death tolls had been limited by law to six minutes.[78]

Six minutes, or "five minutes and no more": the brevity and precision beg for comment. The brevity nodded toward a reconfiguration of the auditory experience of sacred time. Where before sacred time had been heard as a time *off and apart*, away from the babble of the mundane, now—in a world of night shifts and unceasing dynamos—it could be heard more simply as *time off*, away from work and worry. Where before it had been heard as a time *out and unhinged*, as with the slow sighs of dream or the glossolalia of trance, in a world of high-speed presses and charging locomotives it could be heard most simply as a *time out*, enjoying the pure freedoms of parks and Nature. Where before it had been heard as a time *in*, for meditation and keeping a spiritual journal, in a world of thin-walled tenements and piercing steam whistles it had to be a time *in-between*, unscheduled moments of peace and quiet that required no further epiphany. And the precision of the six minutes, or the "five minutes and no more," iterated a machine-tooled environment of infinitesimal tolerances, where tenths of seconds were as critical to the achieving of chemical reactions as to the inner workings of Maxim machine guns; where wires were drawn by the mil and the gears of mechanical harvesters had to mesh within one ten-thousandth of an inch; where a twenty-ton steam hammer could crack a nut without bruising the kernel and flour was milled more finely than a hand could judge; where electric alarms rang more instantaneously than eye and ear could tell; where, in brief, the scales of congruence—spatial and temporal—exceeded the capacity of the unaided senses.[79]

Piano tuners could discriminate among the subtlest flats and sharps. Singers in concert choirs matched their voices to pitch pipes that had come into use a century earlier, although the standard pitch had crept upward from an A3 of 406 vibrations/sec in 1713 to a Steinway piano's 457 in 1879, distressing composers, conductors, organists, and orchestral musicians who pressed for a less "squawky" international pitch of A3 = 435. Otologists had begun using finely calibrated tuning forks (invented by a lutenist in 1711, improved by Chladni, perfected by Rudolf Koenig) to quantify degrees of deafness by the audibility of smaller and smaller amplitudes of sinusoidal vibration. In lay circles, the exactness of acoustic tolerances showed up more obliquely in the form of listeners who registered their environments from the soundpoint of the invalid. "For a long time it has been well known to the medical profession," wrote one such listener in 1880, "that in various critical states of the human system absolute silence, or the nearest possible approach to it, is not the least important condition to be secured." Which is why door knockers were muffled in the neighborhood of the deathly ill, and the street muffled with straw. Yet the equally grievous effect of noise on the general public goes unacknowledged, he wrote, and although the French chemist Marcellin Berthelot had not yet proven the internal changes in bodies subjected to storms of sound, noise is indisputably "exhaustive" and distressing: "The strongest man, after days spent amidst noise and clatter, longs for relief, though he may not know from what." People pass their holidays at the seaside, in the country, or at mountain retreats in search, they say, of good air, but it's also an effort to escape noise, "the most unnatural feature of modern life. In our cities and commercial towns the ear is never at rest," what with crowds of vehicles and "the more positively annoying and distracting elements, such as German bands, organ grinders, church-bells, railway-whistles, and the like."[80]

Any seventeenth- or eighteenth-century ear would have been amazed to hear churchbells lumped in with street tubas and hurdy-gurdies as "positively annoying" sounds. The nineteenth-century ear was less in awe of churchbells, so much so that the polymath and ophthalmologist Alfred Smee in 1850 favored a dog of his who "had great objection to the sound of the bell at Lothbury church which tolled at eight o'clock on Sunday morning," the dog howling "in the most hideous manner during the entire time of the unmeaning performance." Whose performance was unmeaning, dog's or church's? "In this case we can but admire the taste of the dog in testifying its disapprobation at the barbarous custom; although the bell-toller did assure me that his bells were charming ringers." In 1882 a *Harper's* travelogue described a Spanish bell-ringer wrangling the ropes of "Twenty bells on swinging beams... One after another their tongues rolled forth a deafening roar, in a systematic disorder of thunderous tones." That same year and closer to home, Jared W. Bell complained of the "distress, agony, and annoyance" caused by peals from the nearby belfry of St. John the Baptist Episcopal Church in Manhattan. "This hideous noise," testified Bell, "is utterly unnecessary to the worship of God and... is simply a relic of the times when there were few if any watches or clocks in the community whereby people could learn the hour of repairing to the sanctuary." Worse, the bell was in disrepair and probably cracked, "of great annoyance to the well, and of serious injury to the sick." Twenty-nine others, including the French Consul, co-signed his affidavit, to which St. John's pastor responded that "he had previously had no idea how objectionable the bell was to the residents of the vicinity," but that on news of any serious illness in the area he had always ordered the ringing stopped, and each Sunday henceforth the church's two-thousand-pound bell would be rung only four times, at intervals of five minutes, to call the faithful to services. The bell, he insisted, had a rich tone, in the key of E, that many loved, yet in the interests of neighborliness he would give instructions "to use the tolling-hammer only, and not to ring the bell at all. The sound in that case is soft and subdued." The technical problem, as the pastor found upon clambering into the belfry, was not a crack; rather, the leather cushioning around the clapper-springs had worn away, so the bell rang more loudly than it did when installed twelve years before. Simple repairs thus might solve the problem, but nothing would satisfy another complainant, Mr. Waters, who stated at a hearing conducted by New York City's Sanitary Commission, "I cannot see the necessity of a bell on any church in this City."[81]

How had bells come to lie within the purview of a Sanitary Commission? According to Charles Chapin, M.D., Superintendent of Health for Providence, Rhode Island, for half a century (1883–1931), "Disturbing noises are certainly a nuisance, if not a direct and appreciable injury to health." He continued, in his influential survey of *Municipal Sanitation in the United States* (1901): "It is claimed that the constant irritation of the nerves produces a very appreciable effect in bringing on that nervous exhaustion, which is the end of many a busy man's career." (In middle age, Chapin's father had had to abandon the practice of medicine due to increasing deafness.) Stories began to circulate of a man here or woman there who, unable to prevent the "ear-splitting noise produced by the bells of a neighboring church . . . sought relief by committing suicide." Eight thousand in Philadelphia were "made ill each year from unnecessary noises," or so it was "authentically stated," wrote Dr. J. M. McWharf of Ottawa, Kansas. Noise, it followed, could be as proper a focus of public health as of police inquiry, whether it proceeded from steam whistles, or (lumped in again) churchbells, as with a Boston ordinance of 1890: "No church bell shall be rung to disturb any sick person, if forbidden by the board of health." Cities had obvious health concerns about animals, who could be noisy when ill, especially dogs whose "perfect bark, ending abruptly and very singularly, in a howl, a fifth, sixth, or eighth higher than at the commencement," was symptomatic of rabies, an epidemic of which broke out in Providence in 1906. But what was it about the "noise" of churchbells *per se* that warranted notice by sanitation officers on alert for the more ominous sound-signatures of whooping cough and tuberculosis, or the bang of water pipes in tenements from which spread typhus, cholera, dysentery?[82]

Was it that churchbells seemed so grievous because superfluous in an era of alarm clocks, their untimely tolls serving "no purpose except to rack people's nerves and harrow up their souls" (*Milwaukee Sentinel*)? Was it that churchbells went on longer and more insistently than other sound-systems, interrupting the quiet that invalids had always needed and the absolute silence demanded now on behalf of those undergoing the Rest Cure? (Each day ten thousand in Milwaukee were sick and "in a condition to be shocked and irritated by the noises," said Milwaukee's Commissioner of Health in 1896. Himself laid up for weeks with a nervous condition, Dr. Walter Kempster had found that "The ringing of bells, blowing of whistles, and rumbling of the street cars passing the door were simply agony to me." In his tremulous state churchbells were the most egregious of tortures, since their ringing had "come to mean nothing.") Was it that modern citizens were unprepared for reverberations off the steep canyons of dense urban avenues where the ringing of bells could be felt as well as heard? ("And these noises," wrote Dr. Robert Hastings of Boston, "are intensified and repeated by the high walls against which they strike and from which they rebound again and again.") Was it that a rising population of infants, elderly, and night-shift laborers had to try to sleep through exactly those hours during which bells were being rung in obeisance to some habit of the Dark Ages? Was it that sleep was no longer typically segmented and routines of bell-ringing that harkened back to medieval liturgical hours no longer made civic, physiological, or sacerdotal sense? Was it that interrupted sleep meant deflected dreams, and that dreaming itself was being reconceived as a physiological defense mechanism? ("One midsummer morning in a Tyrolese mountain resort," Sigmund Freud woke "with the knowledge that I had dreamed: *The Pope is dead.*" What did that mean, considering that Leo XIII, though in his eighties, was yet among the living. Freud's wife asked him, "Did you hear the dreadful tolling of the churchbells this morning?" *Aha.* "I had no idea that I had heard it, but now I understood my dream. It was the reaction of my need for sleep to the noise by which the pious Tyroleans were trying to wake me." In Freud's sleep, and in defense of sleep, his dream had avenged him by killing off the Pope: "*The dream is the guardian of sleep, not its disturbing.*") Or was it that churchbells went so slowly out of tune that they swung for years on the griding edge of dissonance, raising hackles? Had the ringing of these "metal monsters," in other words, become a kind of Death Metal, akin to the "fiendish, hideous clangor" of trolleycars and swept up in the same industrial process by which the "screeching, whistling towboat" was supplanting "the romance of sailing craft"?[83]

Most of these arguments were enlisted by George L. Harrison and eleven other complainants who owned "handsome and expensive residences" within a block or so of St. Mark's Episcopal Church in Philadelphia. Its cornerstone laid in 1848, the church had erected a belltower by 1851 but left it empty until 1876, the year of the Centennial Exhibition, when four bells (of a prospective peal of eight) were at last installed. The bells were rung for as much as half an hour before each weekday service and four times on Sundays (7:00 a.m., 10:30 a.m., 4:00 p.m., 7:30 p.m.), as well as on Saints Days and festivals. Their ringing, claimed Harrison et al. in February 1877 in the Court of Common Pleas, was incessant and "moreover, devoid of every quality which tends to allay the annoyance which overpowering noise produces. It is harsh, loud, high, sharp, clanging, discordant.... It shakes the walls of the houses...disturbs rest and sleep, especially that of children and infants—distracts the mind from any serious employment...lessens or destroys social and domestic intercourse." With the windows open in summer, the noise from the bells was worse, but at 80–94 strokes per minute and 1500–2000 strokes per session, the bells were never less than obnoxiously insistent. Given also that the street was narrow, the elegant houses high, and the bells mounted lower than the roofs of nearby dwellings, the "echoes which vibrate back and forwards between the walls of the church and the walls of the adjacent rooms are most distressing to the senses." St. Mark's had prospered for a quarter-century without bells, and few of the city's other churches any longer had bells, so the four bells of St. Mark's could not be crucial to divine worship; "whatever may have been, in former times and under different circumstances of civilization, the use of bell-ringing as an admonition of the hour of prayer and worship,...the introduction of watches and clocks has entirely superseded such use for all practical purposes." Furthermore, as vouchsafed one MacGregor J. Mitcheson, "The first result of the sudden awakening from sleep by the noise of the bells, was to give my wife and others very violent headaches;—wholly preventing their going to church on Sunday." Nor was the ringing demonstrably "a work of benevolence, of charity, or of education," or, as the church rebutted, "a source of distinct and positive gratification to invalids." On November 3, 1876, a vestryman had received this handwritten note: "Some of my nervous patients are driven wild by the early bells of St. Mark's. If you have ever been in the way of ill sleeping, and have looked to the quiet of Sunday with anticipated comfort, you would understand how grave a matter this may seem to one of these folks. Pray help us to get rid of this annoyance—the early bells. Yours, faithfully, Weir Mitchell." Mitchell's associate, Jacob DaCosta, who lived nearby, also spoke up, citing a recent article in the *Lancet* that "Church-bells, which in the country undoubtedly have a charm, become in the crowded city a positive distress to many sick persons," then detailing the medical impact of "unwelcome noise," which renders "an attack of migraine an unendurable punishment; it aggravates delirium; it may make the difference in the sleeplessness of a fever, between recovery and death." Passing by, Samuel D. Gross of Jefferson Medical College reported that "The church seemed almost to shake with the disgusting sound"; in their frequent repetition, said Gross, churchbells were worse than street organs, worse than the "harsh and discordant clatter" of parrots or the shrieking of newsboys. Mitchell and twelve other physicians signed a letter to the church, protesting "the *early*, the *frequent*, and the *prolonged ringing*" of St. Mark's bells. George Harrison himself deposed that his horror of the bells' vibration, "resembling at all times the effect of a continuous electrical current throughout the system, or the whirr of a circular saw-mill," was exacerbated by anticipation of each next onslaught, whose frequency and regularity made the churchbells "much more objectionable than the occasional and uncertain noises" of, say, fire-engine bells or factory whistles. "Upon the whole," he concluded, "the home-feeling of immunity from intrusion, and the home privilege of exemption from unnecessary outside disturbance, are, in a great measure, lost." His neighbor Mary F. Parsons took the home invasion most personally. Already discommoded by loud conversation on the street, by slammings of doors and rumblings of carriages, she experienced the bells of St. Mark's as a sonic rape: "The force of the metallic masses beaten without any interval of repose almost stuns my senses.... The impression of the metal upon me is brutal force, which I can neither resist nor escape from."[84]

Against such assault, the defense countered with testimonials from bishops of churches whose bells, long in place, aroused no furor. St. Mark's also submitted technical data on the bells, which were "mellow, soft, well-pitched, sweet and harmonious" and could not possibly cause offensive vibrations, since they were stationary chimes struck by small hammers, not swinging bells banged by a clapper. Diminutive in comparison to other beloved bells, those of St. Mark's were neither as disturbing as those of a nearby woollen mill that rang long and loud at five a.m. nor as disruptive as the steam whistles of local refineries. Perhaps Messieurs Harrison et al. had "excessive nervous sensibilities"? Bell-ringing had been "part of the ordinary and usual sounds of city life" for twelve hundred years; proximity to a church belfry in all that time had not caused real estate values to plummet, as the complainants feared. What's more (quoting Rev. Schenck), "a church is not perfect in its equipment without bells." Pocket watches and house clocks had been available for three hundred years, yet during that time belltowers had been erected in many a town whose poor could not afford private timepieces.[85]

On acoustic, psychological, medical, and meteoro-legal grounds, the decision went against St. Mark's. In a residential district, observed the Court, the path of sonic vibrations was unpredictable and could undermine the foundations of a house or of one's health. The repeated ringing of bells, to which people were differently sensitive, could cause "real and substantial" injury to infants, the aged, the sick, and the nervous. As one could not "rightfully infect the atmosphere with noxious smells," so one could not send dangerous vibrations out into the open air. The Court issued an injunction restraining St. Mark's, pending an agreement with the plaintiffs on the ringing of its bells "in a moderate manner."[86]

Just the hint of a gift of bells sent Boston spinning thirty years later. In memory of her late husband, Sarah C. Wheelwright proposed to donate "a peal of heavy tower bells" to King's Chapel, the oldest Unitarian congregation in America. The vestrymen declined the donation, feeling that "the tower is too low and that the [Beacon Street] neighbors would seriously object to the use of the bells." Besides, the Chapel already had a bell, the largest and "sweetest" cast by Paul Revere and still Sunday-serviceable in 1908. Wheelwright then offered the bells to Harvard, which turned a deaf ear, and in 1910 to the city of Boston for the tower of its new Custom House, at which a lawyer protested that "We have long since suppressed the needless whistling of railroad trains within the twenty-mile limit, the German street bands, hawkers' cries, the [tin] can nuisance in the hands of the small boy, etc., etc., and should, indeed, go much further in the avoidance of late night and early morning noises in the street," into which mix he again lumped churchbells. "Long tradition has sanctioned the use of bells to call the faithful to church on Sundays, but many of the leading churches are now without these, as witness St. Paul's, Trinity, etc., and there is certainly no need for a noonday jangle in the busy part of town, where business presses hard and people's minds are strained to the utmost, and as much quiet and concentration is wanted as is consistent with the demands of modern life." Churchbells were anachronisms, case closed: "The idea would be fitting in a monastic age, say, in mediaeval England, but it seems out of place in busy, twentieth-century Boston." The seventy-year-old physician Arthur H. Nichols, an antiquarian of bells and go-between for Wheelwright, was exasperated by the public ridicule of her offer, which when mentioned "at any assemblage produces an effect not unlike the appearance of a fat man at a wake;—everybody laughs." Few seemed to understand that her "chief object was to encourage the art of scientific change ringing," an art which, as practiced by British guildsmen or by students ringing the peal at Groton's chapel northwest of Boston, produced tonal sequences that in full pull (each bell rung by hand by a separate rope according to a rigorous schema) could last hours. Nichols tried to explain: "The experience of two centuries has developed the fact that the sound of bells is most pleasing not when employed in producing hymn tunes, but in the rapid production of musical changes, the effect of which upon the ears is similar to the ever-varying picture of colors presented by the Kaleidoscope." Wasn't a delight in rapid variation the mark of the new cinematic century? Maybe, but change-ringing, more mathematical than melodic, was mocked in much the same way as would be the Cubist paintings shown three years later at New York's Armory. A "Musician" confirmed that Bostonians "seem united in their preference for a melodious chime upon which tunes may be produced." To them, "change"-ringing was meaningless.[87]

No one seemed impressed that the peal would be a full-scale reproduction of the famous bells of St. Mary-le-Bow, whose resonance had inspired John Donne's *Meditation*, "Never send to know for whom the bell tolls; it tolls for thee.... Who bends not his ear to any bell which upon any occasion rings? but who can remove it from that bell which is passing a piece of himself out of this world? No man is an island, entire of itself; every man is a piece of the continent, a part of the main." With the constant "rattle, and roll, and thunder of our modern vehicles" drowning out street criers, it took no tolling of bells to remind a man, or a woman, of the sounding main. "In the town, bells are not always a joy forever," admitted Arthur Hughes, a master ringer in 1904 and soon the master founder at London's Whitechapel Bell Foundry. "Even a well-rung peal on well-tuned bells loses its effect when it has to struggle to rise above the din and jar of street traffic; and bells that are not in tune and are not well rung are often simply a nuisance.... A single noisy, tuneless bell in a crowded town has no charm, and may be a nuisance, just as bells that loudly chime the quarters all night through and murder sleep are a serious affliction." London's *Daily Chronicle* conceded in 1905 that the suburb of Fulham's two-century-old set of ten bells was a "mellow and fine peal," yet "To listen to a really long peal being practised during three or four hours several times a week is positive torture to a musical ear." Nor did it do much good to argue the converse, as did the McShane Foundry: since "the noise of steam and horse cars, manufacturing establishments, carriages and carts rattling over the pavements, etc., is so great, that Bells are not expected to be heard at any considerable distance," it was necessary to install *more* bells, so as "to be heard above the thousand and one noises incident to every large place." Instead, churchbells were being heard as one more of those thousand and one noises. By 1898, the author of an article on Boston bells had no sympathy whatsoever for their historical extent. Finding that colonial churches rang their bells daily at five and eleven a.m., at one and nine p.m. and also for town meetings, funerals, and the opening of markets, he was taken aback: "What a clangor, and for every day of the year, too! The racket must have been intolerable, even worse than that of Independence Day."[88]

Caught between retrospects of ancient bell-racket and the prospect that bells could go unheard in a modern city, churchbells were not so much out of place as out of time. Herbert Spencer, the principal English expositor of Social Darwinism who suffered a nervous collapse in 1853 and thereafter had recourse to a spring-bound pair of ear muffs against a ruffian, possibly regressing world, argued in his *Principles of Ethics* in 1893 that people could worship as they liked, "so long as they do not inflict nuisances on neighboring people, as does the untimely and persistent jangling of bells in some Catholic countries, or as does the uproar of Salvation Army processions in our own." (A decade later Pope Pius X issued a bull that forbade the use of a piano in church, "as also of noisy or frivolous instruments, such as snare drums, bass drums, cymbals, bells and the like," and any semblance of a brass band.) Three years of mockery and disdain of her bell-proposal proved too much for Sarah Wheelwright, who in her own day had sung with the Handel and Haydn Society and helped found the Boston Conservatory of Music. She wrote to Nichols in the fall of 1910, "I never wish to be bothered with the subject again. I am too old to push against the indifference of my fellow creatures as I felt last winter." To be sure, Wheelwright was of an elder generation, old enough to have been called home in her childhood by the dinnertime blowing of a conch shell that could be heard two or three miles away in the quiet Boston of an earlier era. The lesson Sarah said she had learned from the bell fracas was "that it was difficult to do good with one hand without doing harm with the other." The lesson for this Round, as we shall hear time and again, is that each generation inhabits a different acoustic universe, constituted by different musics and memories of sound, by different thicknesses of walls and densities of traffic, by different means of manufacture and broadcast, by different diets and ear-damaging diseases, by different proportions and preponderances of metal rattling in kitchens, clanging on the streets, or ringing in the (differently polluted) air above. Although children, parents, grandparents, and great-grandparents may share the same spaces for much of the same time, their acoustic worlds coincide unpredictably and often oddly. Instructive in this regard was the denouement of Wheelwright's story: at last, in 1912, the Perkins School for the Blind gladly accepted a set of bells from Sarah, who was a granddaughter of that China merchant, Col. Thomas H. Perkins, he who had bequeathed his first mansion to the School when he moved his family to a much grander edifice in Boston. Another granddaughter, Caroline, would never forget that first mansion's massive doorway, hewn in the 1830s from a discarded timber from the hull of the *Constitution*, the thud of whose closing she recalled at the turn of the century as "resounding through the court[yard] in the quiet of the evening." Wheelwright's bells chimed for the Perkins School on the quarter-hour, and still do.[89]

4. Tells

Alexander Graham Bell in 1886 had recommended the Perkins School to the parents of an Alabama "wild child," six years old, deaf and blind. In March 1887 Perkins sent them a teacher, twenty-one, who had just graduated as valedictorian. Annie Sullivan had lost her sight to trachoma, then partially regained it after several surgeries. Helen Keller would never be other than totally deaf and totally blind, but with Sullivan's tutelage and

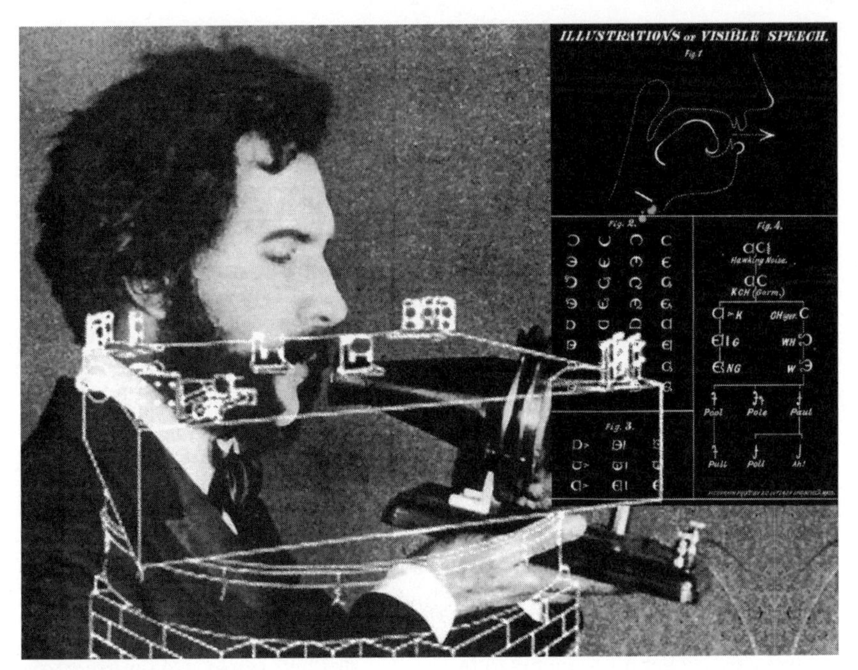

SOUNDPLATE 8

lifelong companionship, she became literate, articulate (nay, eloquent, as befitted a cousin of Edward Everett) and politically active, an outspoken Socialist. What Sullivan heard and saw, Keller felt: from tremors of the earth beneath her feet, clash of knife and fork on china, or water gushing through a pump, she had always been within reach of waves of sound; at the subtle pressure of Sullivan's rapid finger-spelling, she became a person of fluent words who did not hesitate to use them as would a fully hearing person. When thirteen, Keller wrote to Bell's wife Mabel, herself deaf: "We are staying in a lovely place with tall forest-trees all around it; and today the wind is rushing through the trees with a noise something like the noise which you feel when the waves are running high. Then it dies away into a gentle murmur, just as the sea does at low tide." These noises were more than literary figurations; she had been to the Atlantic shore and that summer to Niagara Falls, where she felt the water "rushing and plunging with impetuous fury at my feet." And it was more than figuration when, twenty years later, she put churchbells at the center of a New Year's message. Among the deaf, large bells may be as sensational as strong winds or ocean waves, and whether or not Keller had felt the vibrations of the Wheelwright bells installed at her first alma mater in December 1912, a coincidence of ideology, Swedenborgian theology, and campanology led her to emphasize the ringing of changes in January 1913: "Hear, oh, hear!

The Christmas bells are ringing peal upon peal, chime upon chime! Full and clear they ring, and the air quivers with joy." In the exultant wake of the successful strike of January–March 1912 by the women and children at the Lawrence textile mills, with the support of organizers from the International Workers of the World, Keller had embraced the millennial vision of these Wobblies. "For a Great Change is coming, a wondrous Change, a World-change that shall fulfill all joy in a happy humanity. Ring the Great Change, O Bells! Hear, oh, hear, all people!" An audio-vision: the IWW was known as the "singing union," for wherever Wobblies marched, they sang. Keller, who could not sing, had just finished three summers of voice lessons to strengthen her vocal cords so that she could go out on the lecture circuit with a voice that was warm, personal, and powerful. "The bells and I are strong with a new hope, vibrant with expectancy of this Great Change."[90]

Twenty months later, Europe's remaining churchbells tolled instead a Great War that would recast the scale of noise and the magnitude of silence. In sonic terms the Great War was the Great Amplifier, but, as always, uppercase obscures context. In order to work up to the Great War's amplifications, I must trace the lowercase paths by which audibility and awareness had been reconceived in the decades before 1914. As noise came to feel ubiquitous *and* unceasing, engineers and psychologists inquired into the limits of hearing and the temporality of sound. As deafness came to be understood as a common occupational hazard of noisy industrial work and a probable consequence of noisy urban life, acousticians and physicians worked to quantify hardness-of-hearing while astronomers and psychophysiologists exposed the "personal equation" that made for differences in reaction times and sound sensitivities. Their researches called not for long peals but for ticks, whispers, and brief shocks. As the churchbell, like the ram's horn, became iconic/audionic of millennia past and a future Millennium, the still-small voice of buzzers took up the burden of the present.

Sound, more than it is "here," is just-about "now." Aside from the classical mockeries of Echo, the Renaissance trope of the frozen word, and some Baroque and early Industrial mechanical re-players (barrel organs, automaton flautists, music boxes), sound until the 1870s was of the present, bound by its own *diminuendo.* In Late Romantic language, what the ear receives is the breath of newness: air pressure changing at least fifteen times a second. Our minds, our brains, our "neurons" (so named in 1891) take the shape and amplitude of waves of air pressure and work these into distinctive sounds. As Aden Evens has written, "To hear is to hear difference." Musicians, who have a finer sense of the gradations of time than all but jockeys and magicians, are masters of difference. Their "present" may feel more extensive as they struggle with tempo and timbre to play what is set down for them or to improvise what is in them to improvise, but unlike barrel organs or Edison's phonograph (1877), musicians in company cannot be re-players. Live music, writes Evens, is lived music, sounded differently according to the acoustics of the site, before audiences variously fleshed out on upholstered seats or folding chairs, or passing along the street. Each of us, even she who feels it primarily through her bones, knows the sounding world by its oscillations in the nearly present.[91]

Exactly what the present is, and how to tell it, had been made an issue of more than philosophical moment by the dit-dot of the telegraph and the shriek of the locomotive—or, more exactly, by the sounds of telegraph and rail networks, which fed off each other. With their speed and need for connectedness, trains moved through multifold times as they sped east and west across the meridians, where for each mile of longitude (at 40°–42° N, latitudes of Chicago and New York, Madrid and Barcelona), the solar noon differed by five seconds. A passenger on a train moving at 60 m.p.h. would encounter five-minute discrepancies between solar time and clock time for every hour of travel; stationmasters who were fifteen minutes apart by hurtling locomotive might have station clocks set thirty-one minutes behind or forty-seven minutes ahead. With its electric speed, a telegraph system could relay a time signal well ahead of the sun, but it also ran up against the jangle of 144 official local times and 80-plus railroad times in the United States alone. During the Civil War, officers had their own timepieces geared to noons back home so various that attempts at coordinated attacks could be mortally confused, as happened at Gettysburg to the army of Robert E. Lee, who relied upon his commanders to execute a synchronized offense using pocket watches that ticked with different rhythms. In 1876, waiting at a station in Ireland while on leave with a case of nerves stemming from two rail accidents, the engineer-in-chief for the Canadian Pacific Railway missed his train because an *a.m.* on the schedule had been misprinted as *p.m.*, whence began Sandford Fleming's international campaign for a twenty-four-hour clock and standardized time. Standardized time also made sense to Samuel Langley at the U.S. Naval Observatory, which sold cities, railroads, and clock shops subscriptions to the time signals telegraphed from its own star-transit instrument; how else regulate the unending busyness of modern life that threatened chaos at every new juncture or rail junction? "During the day the rush of passing cars is incessant," Langley wrote in 1878 for *Harper's New Monthly Magazine*, "and when night comes on, the noise of panting engines, which pass the window with a crescent shriek, makes it seem as if our broken sleep had been passed in the vicinity of interminable rushing trains." Five years later, U.S. and Canadian rail managers reorganized regional station times into four continental time zones, which were given global extension and imprimatur in 1884 by the International Meridian Conference in Washington, D.C., after which the industrial world began (with pockets of national resistance) to operate upon twenty-four one-hour zones centered on the 0 Meridian of London's Greenwich Observatory. By the 1890s the shriek of trains racing to stay "on time" to the very second commanded more authority than bells that lackadaisically tolled the quarter-hours from village steeples. America's own commodore of time, the civil engineer and travel-guide publisher William F. Allen, argued that "railroad trains are the great educators and monitors of the people in teaching and maintaining exact time," though standard time could estrange communities that straddled the borders of zones, where defenders of local, familiar, natural time had their strongest support.[92]

Marking the minutes by the whistles of speeding locomotives put time into a zone that was non-local, industrial, and uniform, but acoustically the steam whistles gave a local twist to time and presence. It was not that sound and time were newly twinned; horns, drums, rattles, or conches had sounded ceremonial and calendrical rhythms from "time immemorial." Nor was it that steam whistles made the passage of time feel more dynamic than bells or, say, post horns, which had been blown every quarter mile by national "post-boys." Eulogizing the English Mail-Coach in 1849, Thomas De Quincey deplored "the new system of travelling," where "iron tubes and boilers have disconnected man's heart from the ministers of his locomotion. The trumpet that once announced from afar the laurelled mail, heart-shaking, when heard screaming on the wind, and proclaiming itself through the darkness to every village or solitary house on its route, has now given way forever to the pot-wallopings of the boiler." He who had shared the calm of the Lake Country with Wordsworth and the clamor of London with Carlyle celebrated the "Glory of Motion" at the old Post Office on Lombard Street, where each evening the coaches had been lined up, their mail-bags tied down, their horses readied, and then "What stir!—what sea-like ferment!—what a thundering of wheels!—what a trampling of hoofs!—what a sounding of trumpets!—what farewell cheers!" as they headed out across Great Britain "like fire racing along a train of gunpowder." No, the twist in time produced by steam-driven locomotives came from the relationship of listeners to sounds now crossing regularly in front of them at a velocity that had previously been witnessed only on the rarest of occasions: a person falling from a cliff, screaming; a meteor sizzling across the sky. At these instants, and for hours on battlefronts when rocket squibs screeched overhead, speed and direction had been heard to affect the pitch of the sound. The same held true optically for the color of stars, claimed a Czech mathematician, Christian Doppler. His equation of 1842, inspired by analogy to sounding bodies in rapid motion, predicted that a star should appear bluer (of a shorter wavelength) as it moved toward Earth, redder (of a longer wavelength) as it receded. A Dutch scientist tested Doppler's analogy. In February 1845, C. H. D. Buys Ballot had two musicians calibrate a bugle tone, then asked one to listen beside the track as the other blew past him on a locomotive. Although the test was cut short by a blizzard, its results were intriguing enough to be printed in a magazine for music lovers, who learned that on approach the bugle's tone was heard half a note higher, on departure half a note lower. (That must have been what Langley had in mind when he wrote of a train's "crescent shriek.") Repeating the test in June calm, Buys Ballot took more precautions: he put three teams of three (a coordinator, a musician, and a visual observer) within a yard of the tracks while a fourth team (its musician bugling) sped by on a flatcar pulled at 40-odd m.p.h. by locomotive. Here too there were problems: the engineer uncorked his whistle in the middle of one trial; the background noise of wind and engine overrode the sound of the bugle. Switching to a stronger signal-horn, the musician blew a tone that was heard to climb toward a shorter, shriller wavelength as he neared each stationary team; heading off, the tone dropped down the scale toward a longer wavelength. Conclusion: Doppler was right. Discussion: how we hear even a single steady tone depends on the relative velocities and positions of soundmaker and receiver. Listening for the time as told by the whistle of a locomotive at a country crossing, a farmwoman on her porch in a trestle-back chair could not escape the relativistic twist of the Doppler effect.[93]

Doppler's prediction of optical shifts in the color of stars would be confirmed later in the century by work in spectroscopy. His basic assumptions—that light waves require a universal, transparent medium for their transport and that motion must be measured relative to this "luminiferous aether"—comported with a nineteenth-century physics that clung to the ancient axiom that motion is impossible without a supporting medium. The silence in the vacuum of a bell jar had sustained that axiom: no air = no medium = no motion of particles/waves = no sound. What then enabled light to shine across such putatively empty spaces? A light-ferrying aether (or some plenum no less invisible, intangible, and ubiquitous) seemed necessary to subserve the blithe travels of light waves, whether through the vacuum of a bell jar or from stars through space to Earth. Though the speed of light was great, it had been shown to be finite and subject to the constraints of a material world, such that it lost velocity when passing through bodies of water. Was light also slowed by an aether that (as Thomas Young thought, years earlier) "pervades the substance of all material bodies, with little or no resistance, as freely, perhaps, as the wind passes through a grove of trees"? Would, for example, the speed of light be slightly greater in line with Earth's orbital motion through the aether than when cutting at 90 degrees across that line of motion and the stagnant aether?[94]

Measuring the speed of light had required a gear wheel with 720 precisely machined teeth, a carbon-arc light, two unblemished mirrors, assorted fine lenses, and a tachometer. Measuring the "aether-drag" would require, as Albert Michelson calculated, a device even more sensitive, capable of detecting a change on the order of one in a hundred million yet insusceptible to the smallest extraneous vibration. Having graduated from Annapolis in 1873 as a better musician and mathematician than a midshipman, Michelson was assigned to teach science at the Academy. In the spring of 1880, winning approval from the Navy for postgraduate study in Europe, Michelson sailed for Paris, where he examined the apparatus that had been used by Hippolyte Fizeau and Alfred Cornu to measure the speed of light. At the University of Berlin that fall, he designed an "interferometer" that would similarly split light into two beams on a roundtrip course of identical length; with mirrors astutely aligned, his device would direct the returning beams to the same endpoint, so that he could tell by the interference pattern at their intersection how much the speed of the right-angled beam had been affected by aether-drag and, reciprocally, how fast Earth was moving vis-à-vis the stationary aether. He asked Hermann Helmholtz for advice on experimental design and got a grant to build his interferometer from Alexander Graham Bell, who as a practitioner of a system of "Visible Speech" had a lifelong interest in sound imaging and as an inventor of the telephone had become concerned with issues of interference. Early the next year, his instrument complete, Michelson proceeded with his experiment. It failed. Despite (or because of?) the exquisite sensitivity of the interferometer, which shook in response to "stampings of the pavement about 100 meters from the observatory," it revealed no substantial pattern of interference. Did this mean that the aether was not stagnant but, like a rarefied atmosphere, moved in tandem with Earth (and was thus unserviceable as a still ground against which to calculate the relative speed of Earth in orbit), or that his interferometer was too vulnerable to other sorts of interference? Michelson admitted that it had been "impossible to see the interference fringes except at brief intervals when working in the city, even at two o'clock in the morning," an incidental proof of the enlivening of the urban night but an irksome problem with experimental noise. In 1882, quitting the Navy for a professorship in Cleveland at what would become the Case Institute of Technology, Michelson refined tuning-fork frequencies so that they could be used to calibrate rates of mirror rotation for further trials with light. Three years later he began to collaborate with Edward Morley, a chemist at nearby Western Reserve, who would be the meticulous keyboardist to his high-strung violin. In duet the two amateur musicians redid Fizeau's work on the speed of light in moving water, with a pause to accommodate a nervous collapse by Michelson. In 1887 the two improved upon Michelson's 1881 aether experiment, mounting the interferometer on a stone slab one foot thick and five feet square resting on a wooden ring afloat on a circular trough of mercury, which "massive" slab they set in slow imperturbable revolutions that put a light-beam at right angles to Earth's orbital motion. Again, failure. The two men had engineered an amazing instrument that could not be thrown off-kilter by local vibrations, that was capable of detecting a change of one part in ten billion, and yet it had registered no displacement of light waves by the aether.[95]

Either the aether was so fine a wind-through-a-grove as to effect not the slightest tremor, or its impact was so fine that it would always be hidden beneath margins of experimental error. It took Michelson forty years, and other scientists at least twenty, to adopt a field-oriented, space-time-relativistic physics that gradually dismissed the aether as a suppositious plenum of no consequence to any equation. The aether experiments fit within a cultural history of sound because they drew upon acoustical analogies common to nineteenth-century investigations of light; they fit within a cultural history of noise because Michelson's aether-seeking device, like the psychic antennae of spirit mediums, required thick cushioning against extraneous vibrations. The experiments appear in this chapter because they were paradigmatic of a wider search for "just-noticeable differences" in sensation and perception that defined the limits of discernment and established the constancy of a degree of delay or interruption that became analogical to noise. As much as it was an optical timer, the interferometer was a silent mechanism built to screen for the minutest of vibratory interruptions from an imperceptible substance just-noticeably delaying intangible waves traveling at light-speed. That the experiments turned out to call into question the existence of the aether is less important here than that Michelson and Morley presumed that they could achieve the resolution requisite to catch aether in the act. That physicists of the caliber of Hendrik Lorentz, impressed by the "wonderfully delicate experiment" of 1887, felt compelled to offer relativistic solutions for the startlingly null results is less germane than the fact that the raw data appeared actually to attest to shifts in wave phase and frequency if not in amplitude. "Reducing" their data for publication, Michelson and Morley ignored readings that were (probable) artefacts of observational error (i.e., experimental noise), then converted the results into scalar graphs that damped the remaining oscillations. Debate continues into the twenty-first century as to whether they had good reason to so finesse the infinitesimal; in the 1880s, their finesse was of a piece with the stop-action photography of Eadweard Muybridge in San Francisco and Philadelphia, who caught galloping horses and leaping humans at intervals of 1/5000ths (.0002) of a second, and the physiological sensors of Étienne-Jules Marey in Lyon, who caught "the noises and cracklings" of muscle twitches at intervals of 1/720th (~.0014) of a second, i.e., *before* corresponding flickers arose in the cerebellum. "Precise experiment and exact measurement," announced Marey, who had been inspired by Muybridge to construct a photographic gun that caught successive motions on a single plate from a single point of view, "have begun to appear even in the phenomena of thought."[96]

Between a painful prick and the perception of pain, between spark and idea, there was, as Helmholtz at mid-century had proven, a nervous comma, an intrinsic, ineluctable delay. Using an electrical device that graphed nerve impulses in muscle, Helmholtz in 1849 had discovered *eine messbare Zeit*, a measurable time, between the impinging world and perceptions of it—an interrupt of ~.015 second between contact and cognizance or (in a phrase that first appeared in Paul du Bois-Reymond's French version of Helmholtz's 1850 paper) a *temps perdu*, a "lost time" that would intrigue Marey and, decades later, an asthmatic Parisian novelist who pursued his own *recherche du temps perdu* from a soundproof, cork-lined room. Since "we just cannot perceive any faster than our nervous system works," we are unaware of that delay, but delay there is/was, and in scientific perspective it is/was large, on the order not of a ten-billionth or a hundred-millionth but of a hundredth or—as he would determine in 1867 with regard specifically to human neural pathways—a tenth.[97]

Friedrich Bessel, a professor of astronomy at Königsberg, had already found obstinate full-second discrepancies in individual perceptions of the transit of a star across the hairlines of a telescope. From 1821 to 1833, as Helmholtz knew, Bessel had squinted at hairline grids to map the position of 75,000 celestial objects. Subsequently he spent sixteen months determining the parallax of 61 Cygni so that he could make a good measurement of the distance from Earth of a fixed star: ~66,000,000,000 miles—a number so large it stunned his contemporaries (and it is in fact a bit larger). Since Babel, or the first Egyptian pyramids, priest-astronomers had extrapolated long distances from hair-width differences; modern secular astronomers, understanding that a blink, a wince, or a squint of astigmatism might turn into galactic gaps, tried experimentally and probabilistically to anticipate and compensate for visual errors. Bessel compared his transit data with those of colleagues, then watched them as they watched the stars. Each, he discovered, had what would be called a "personal equation" that set his observations apart from the next; these equations were personally unsteady, varying according to the rhythms of timepieces and (as Babbage had found) to cycles of alertness and fatigue. In tandem with the French astronomer François Arago's insistence in the 1840s on the intransigence of observational errors and Helmholtz's disclosure of a neural *temps perdu*, Bessel's disclosure of systematic discrepancies in fine observation put a permanent pause between the self and the present, a pause both visible and audible.[98]

Audible, because astronomers had been noting the instant of a star-transit by the ticks of pendulum clocks beside them. The swiftness of the correlation between what they saw, a transit, and what they heard, a tick, differed according to individual reaction times, although they trained themselves to estimate, to tenths of a second, the time it took to glance from the eyepiece to the face of a clock. When, from 1846 to 1851, the

United States Coast Survey introduced semi-automated electrical apparatus for the "exchange of star signals," the results still had margins of error no smaller than a tenth of a second, sometimes half a second. Meant to bypass the errors inherent in the old "eye/ear" method, this American Method turned out to be an eye/hand/ear method that was plagued by discontinuities. Unpacking a telescope, a telegraph, a sidereal clock, an astronomical clock, a chronograph, and a slew of batteries, a coast surveyor would press a circuit-breaking key the moment he saw a "clock-star" crossing the hairlines of his eyepiece. Linked to stations at other meridians by telegraph wire, the circuit-breaking key electromechanically recorded the transit time (i.e., the click time) on the smoked paper of a revolving cylinder, eliminating the need to listen for the tick of a pendulum. Yet, to be cėrtain that one had pressed the finger key, one listened for its click. And—because the click was so minimal? or the task so mechanical?—one learned to listen for everything *else* as well, as did Lt. Edward Ord, surveying Los Angeles in 1855. Evening, 14 February: "the plain around the hill...is so quiet, and if it wasn't for the little ground owls—that begin to cu-cu-cooo about in spots—and that coyote laughing such a squeelabelloo—over the next hill—it would seem lonesome—though there's—the big billowing—everlasting—ro-o-a-a-r-r-r-r—onto-the-end-of-the-world of the ocean close by...everything has gone off from the plain—which makes it look so lively by day—the straggling troop of Sand hill cranes that have been gaogle-gobbling and babban-saying to each other near the foot of this gentle eminence have lazily spread their wings—and slowly flapt away—the uneasy & noisy geese have settled in the marshes & ponds—to be quiet (unless the coyotes pitch into them) till morning."[99]

So locales might acquit themselves sonically over and against more generic clicks and taps under a kind of relaxed but attuned listening that would become as ideal for subjects of psychoacoustic experiments as for coastal surveyors. Although the American Method did improve the precision of terrestrial and celestial readings, one of its warmest promoters, Ormsby Mitchel of the Cincinnati Observatory, still thought it worthwhile in 1856 to construct a device that promised to measure an observer's "absolute personal equation," including the "personality of the eye" and the "personality of the ear." Predictably, all measurements using this or other, European apparatuses were of course affected by each operator's personality of eye and ear. Adopting the American Method around 1862, the Swiss astronomer Adolf Hirsch began a decade of studies that showed that determinations of longitude varied according to such personal traits as diurnal changes in attentiveness and the degree to which one actively anticipated a star-transit. Meanwhile, Ernst Mach, the Austrian physicist and psychophysiologist, in an 1865 study, showed that the ear was quickest on the pick-up, the eye slower, the skin slowest. For any sense-sequence (eye → ear, eye → hand → ear), the present tense was, it seemed, both complex and compound. In 1865, too, F. C. Donders and his students in Holland began attaching electrodes to the soles of subjects, then timing how long it took them to indicate by a hand signal whether the left or right foot had been (gently) shocked. The signal-time difference between those who had been told which foot would be shocked and those who had not been told was, so Donders reasoned, the duration required by the brain to render a simple decision (left or right?): one-fifteenth of a second, or one-eighteenth of a second when testing ear → voice. Through such experiments it was becoming indisputable that the human *now* was screened from the world's *now* by two delays, first for the slow race of nerve impulses, next for the mental processing by which these impulses obtain meaning, inspire action. While standard time zones were being etched onto the globe at the behest of railroad and financial consortia, astronomers, and government telegraphers, a new form of intimately local time was coming to the cultural fore and making itself heard.[100]

Heard, literally, because reaction-time experiments, conducted for example in the famous laboratory of Wilhelm Wundt in Leipzig, clocked responses to the bang of a hammer, to bullets dropped on a hard floor, or to steel balls falling onto ebony plates. Wundt had begun his career in 1855 as a physician in charge of the women's ward of a public hospital chaotic with peasants, streetwalkers, and servant girls. He reacted to the noise ("an unsteady, unruly, varying sensation") and the class-awkwardness by turning from medicine to experimental psychology, a field for which, as an assistant to Helmholtz, he wrote up the agenda in 1862 and through which, as director of his own laboratory by 1879, he came to dominate the Western science of perception. His first experiments had contrasted the personal equations of vision, audition, and thought, finding in the latter case that for "each person there must be a certain speed of thinking, which he can never exceed with his given mental constitution." This he proved with a *Gedankenmesser* or "thought meter," a pendulum clock so linked to a bell that he could measure the "shortest time intervening between two successive perceptions" by the ring of a bell. For himself, the just-noticeable interval between sound and sight was an eighth of a second, a little longer than the seventh of a second delay reported by Hirsch for subjects who had to press a telegraph key upon hearing a noise (sound → touch-click). The more complex the stimulus, the more discrimination required and the longer the reaction time, but *ceteris paribus* the ear was faster than the eye.[101]

Did the ear do better with noise or with, say, music? How much of listening to music, whose audiences were at once receptive and discriminating, was bound up with psychophysics, how much with social taste or education? In Bonn, "native town of Beethoven, the mightiest among the heroes of harmony," Helmholtz told a lay audience of 1857 how his researches had been drawn away from the nerves of frogs by "a wonderful and peculiarly interesting mystery, that in the theory of musical sounds, in the physical and technical foundations of music, which above all other arts seems in its action on the mind as the most immaterial, evanescent, and tender creator of incalculable and indescribable states of consciousness, that here in especial the science of purest and strictest thoughts—mathematics—should prove preeminently fertile." Each of his adjectives deserved a treatise, but he rushed on: "Mathematics and music! The most glaring opposites of human thought! And yet connected, mutually sustained! It is as if they would demonstrate the hidden consensus of all the actions of our mind, which in the revelations of genius makes us forefeel unconscious utterances of a mysteriously active intelligence." Verging on god-talk, Helmholtz was marveling rather at the psychophysiology of sound and time. In his own brief time at the podium, he would direct his audience's attention to the most valuable of his discoveries, "the foundation of concord."[102]

Music, as the ancients suspected, was an agitation of the air that recurred "with perfect regularity and in precisely equal times"; irregular agitation generated "only noise." Sound traveled in waves, which most ancients had not known, and the "more uniformly rounded the form of a wave, the softer and milder is the quality of tone"; angular waves gave rise to more piercing tones, or shrieks. Even with perfect regularity and equal times, there were lower and upper limits to the waves a human being could hear: ~16 vibrations, or ~8 cycles per second (the bass rumble of a pipe organ); ~32,000 vibrations, or ~16,000 cycles (the high notes of tuning forks excited by the highest range of a violin). The outer ear funnels these vibrations toward the inner ear and the thousands of nerve-tipped arches of its cochlea, which Helmholtz theorized were aligned "like the keys of a piano," each tuned to resonate at a specific frequency, lowest to highest. Inundated day and (increasingly) night by compound waves from diverse sources, the ear in effect performed mathematical operations (Fourier transforms) upon the mass of sound, analyzing it into individual sinusoïdal wave-shapes, then restoring the whole by continual summation that included the harmonic *Oberpartialtöne* accompanying each fundamental. It was to this world of "upper partials" that Helmholtz staked his explorer's claim. Ever-present and ever affecting timbre and rhythm, upper partials were rarely heard for themselves without "a peculiar act of attention" from trained ears, such as those of musicians or bell-founders. By habit, founders shaped bells so that the stronger, deeper partials were in accord with the bell's fundamental tone; unbeknownst to many, a bell's higher partials were always *out* of harmony, "hence bells are unfitted for artistic music." A congenial flow of upper partials was what created "an artistic bond of union" from measure to measure. The secret of concord lay not with the subtle, nearly-ethereal phenomena of cancellation, augmentation, and interference in the upper regions of every series of sounds. Tones that were just fractionally different from each other, and unstable or fluctuating, produced the wickedest interference, or arrhythmia, as disagreeable to the ear as were unevenly flickering lights to the eye. Dissonance was bad timing; noise was to sound as stuttering was to speech, incoordinate. What a person heard as pleasing was driven by a hidden law of consonance and continuity among the upper partials. "We have now reached the heart of the theory of harmony," Helmholtz concluded. "Harmony and dysharmony are distinguished by the undisturbed current of the tones in the former, which are as flowing as when produced separately, and by the disturbances created in the latter, in which the tones split into separate beats." Dysharmony hurt "the spiritual ear" as well as "the material ear," both of which longed "for the pure efflux of the tones into harmony." Artfully composed, "the stream of sound, in primitive vivacity, bears over into the hearer's soul unimagined moods which the artist has overheard from his own, and finally raises him up to that repose of everlasting beauty, of which God has allowed but few of his elect favourites to be the heralds."

Generous as was this bow to great composers, it solved none of Helmholtz's questions about the physics of sound or the anatomy of listening. In the four editions (1863–1877) of his authoritative *On the Sensations of Tone*, theophany vanished before acoustics and mathematics, although he made it clear from the start that because "Music stands in a much closer connection with pure sensation than any of the other arts," it was the shortest of leaps from physics and physiology to aesthetics and homiletics. Noise too huddled close to pure sensation, Helmholtz knew, whether the "soughing, howling, and whistling of the wind" or "the rolling and rumbling of carriages." The distinction between musical tones and noise consisted in the pace of acoustic variability—"generally, a noise is accompanied by a rapid alternation of different kinds of sensations of sound. Think, for example, of the rattling of a carriage over granite paving stones, the splashing or seething of a waterfall or of the waves of the sea, the rustling of leaves in a wood." Each of these "noises" featured rapid, fitful alternations of diverse sounds, while "a musical tone strikes the ear as a perfectly undisturbed, uniform sound which remains unaltered as long as it exists." It followed that noise was of the instant, while musical tones relied upon elongations of experience to be heard as uniform and undisturbed. Noise could succeed noise, in fifteenths of a second, for hours on end, as in war or caterwauling, but what made noise noisy was its non-periodic soundfulness, each burst registered by (he speculated) the "auditory ciliae in the little bags of otoliths," for if every cochlear nerve fiber resonated to a certain pitch, so squeaks, hisses, and crackles must have their own dedicated co-respondents within the ear. Richard Wagner did cultivate patches of dissonance, but his technique was part of a musical scheme that teased then pleased the ear, which hungers always, said Helmholtz, for harmonic resolution. Crackling in the ear's private reserve, noise resisted resolution; it was no more, and no less, than an intemperate interruptiveness.[103]

Sigmund Exner, an Austrian physiologist who had studied with Helmholtz and coined the phrase "reaction-time experiment" for his own research in the 1870s, found that responses to sparks, clicks, or bells were usually involuntary reflexes. Differences in reaction time were attributable solely to whether a subject had been informed of an impending shock or click. The greater the anticipation of a stimulus, the weaker the focus upon the task of responding and the slower the reaction—as with a young Union soldier awaiting first blood in Stephen Crane's *The Red Badge of Courage* (1895), where cowardice and bravery turn out to be qualities less moral than temporal and the stimuli of war are so overwhelmingly aural ("The noises of the battle were like stones; he believed himself liable to be crushed") that in the crush of noise-time he flees the front in a state of acoustic startle. Or as with astronomers, who sit for hours waiting for a star to cross the spider hairs of an eyepiece, which vigilance, suggested Hirsch, is so taxing that fatigue itself may explain the variability of star data. Wundt kept searching for split-second delays due to acts of mental processing; the international cadre of graduate students who trained in his and in Helmholtz's labs refocused reaction-time studies from the nerve impulses to acts of attention. Noise in this context was peculiar: it drew attention to itself but was easily dismissed as inconsequential or inadvertent. Helmholtz himself had noted that the playing of music was accompanied by "characteristic irregular noises, as the scratching and rubbing of the violin bow, the rush of wind in flutes and organ pipes, the grating of reeds," which irregularities were helpful in picking out an oboe from the sound-mass of a symphony, but "Those who listen to music make themselves deaf to these noises by purposely withdrawing attention from them." At his own laboratory bench, Helmholtz ignored the buzzing of his electrical equipment while in pursuit of the complete *klang* of a musical tone, and he exalted the "peculiar power of mental abstraction or a peculiar mastery over attention" that was vital to hearing the firmament of upper partials in all its glory. Acoustic attentiveness could be so well-directed because the ear was "eminently the organ for small intervals of time," capable of perceiving "132 intermissions in a second" and, thus, the full range of partials.[104]

Quick as the ear was, it could be confounded by a succession of short swift tones like those produced by a siren, whose wail was heard as continuous, or by the syncopations at a ragtime piano. Syncopation, the delay of an expected beat, as in "The Cascades" or "Maple Leaf Rag," produced a psychophysical displacement that could be exploited, at the hands of Scott Joplin, for musical gambol ("It leaps and plunges, laughs and cries, commands and pleads, soothes and inflames, in the most bewildering fashion"). Playing with fractions of time that had become so telling in psychology and physiology, syncopation had as optical counterpart Marey's and Muybridge's stop-motion photographs of creatures running and leaping. The optical counterpart of the siren was the cinema, whose illusion of fluid motion relied upon a succession of after-images lingering long enough to fill in the visual blanks between each of the thousands of photographic stills projected slightly faster than the eye can think. To mistake rapidity for simultaneity was a psychophysical consequence of the *temps perdu* along neural paths; it was also a cultural consequence of life among networks of telegraph lines (on which Helmholtz had modeled the neural system), railways (on which philosophers modeled the conflation of space and time), and newspapers with their Extra! Extra!'s (on which stock traders and insurance agents modeled "news" as overlapping incidents that blur pastpresentfuture, so much so that amateur philologists speculated about an etymological link between "news" and "noise"). Thoreau had anticipated the psychoacoustics of simultaneity in a journal entry of 1838: "How can a man sit down and quietly pare his nails, while the earth goes gyrating ahead amid such a din of sphere music.... And then such a hurly-burly on the surface—wind always blowing—now a zephyr, now a hurricane—tides never idle, ever fluctuating, no rest for Niagara, but perpetual ran-tan on those limestone rocks—and then that summer simmering which our ears are used to—which would otherwise be christened confusion worse confounded, but is now ironically called 'silence audible'—and above all the incessant tinkering named 'hum of industry'—the hurrying to and fro and confused jabbering of men—Can man do less than get up and shake himself?" By 1880, Thoreau's sense of a universe in motion, with all of its concomitant noise, was a commonplace of thermodynamic theory with its concomitant entropy—and a common plaint from such as the Quiet Observer at the *Pittsburgh Dispatch*, whose list of "Petty, yet Painful Annoyances" included cats yowling "just when the world and its cares are fading from your sight" and fat men coughing just "as the prima donna twitters and warbles her sweetest." By century's turn, Thoreau's pell-mell hyphens would be integrated by novelists into stream-of-consciousness narratives, intuited by Henri Bergson into a philosophy of multiple immediacies, and inveighed against as "mere busyness" by the editor of *Harper's Weekly*, who bemoaned "the dissipation" of incessant activity.[105]

Telegraphers and telephone operators at central exchanges in particular had had to deal with the acoustic commotion of busyness—the fracas of keys, garble of crossed signals, curses at bad connections, tap-tapping of typewriters, and clicking of plugs in and out of switchboards from which arose "a dull roar like that of locusts on a sunburnt prairie." Quieter than post-horns or town criers, telegraph messages demanded a heightened alertness to small sounds. "We may hear it breathe, listen to the pulsations of its mechanisms; but we pause in vain to catch any other sound when it addresses its language to the four corners of the earth," wrote an observer in 1854, while telegraphers at rows of receivers had to listen through the cicada-like racket for those systematic pauses that made Morse Code intelligible. Their later counterparts, telephone operators, had to listen through electric growl for consonants that could ground vowels on party lines. Originally a male profession, tele-operation became preponderantly female, in part because women could be paid less, in part because they had, it was said, a naturally "quick" ear, a dexterity at fine motions, endless funds of social patience and, for telephony, a tonal range suited to the capacities of transmitters. Their modulated voices would exercise "a soothing and calming influence upon the masculine mind, subduing irritation and suggesting gentleness of speech and demeanor" during episodes of cross-talk, premature disconnection, or "busy" signals ("simply the reflex of the busy town"). Women also were supposed to have special powers of attentiveness that enabled them to hear beyond the chit-chat and illicit gum-chewing of neighboring operators and to ignore the hum of noisy lines. Finally, they were by company training, if not also by Victorian upbringing, ideally inconspicuous listeners, briefly heard but never seen (bearing up under six-pound harness headsets), and ethically astute, attending more to the sound of a call (its busyness) than eavesdropping on its specifics (its business), so more discreet in their tele-mediation.[106]

Privacy and quiet had been rather assumed than assured in psychological laboratories, with their funnel horns, resonator boxes, tuning forks, harmoniums, organ pipes, sirens and, literally, bells and whistles. For telephony, privacy and quiet were even greater issues, since tinny echoes, crossed lines, and common sense led customers to speak more loudly the longer the distance. While otologists held ticking watches at measured lengths from their subjects' ears to determine the limits of the just-audible, and while psychologists let fall leaden balls onto iron plates to track sensitivities to loudness, ad hoc tests of auditory response would be held every day after 1876, at or near a telephone. The very legend of its invention, as disseminated by newspapers and school texts, was of acoustic oscillation between the loud and the just-noticeable. Experimenting in a "noisy machine-shop" one hot June afternoon when all windows would have been open to Boston's street criers, Alexander Graham Bell heard "an almost inaudible sound—a faint TWANG" from a contraption over which he had been laboring, "a sort of crude harmonica with a clock-spring reed, a magnet, and a wire," of which a working twin sat in the next room, connected by a wire. "Snap that reed again, Watson," cried Bell to his assistant. It was "no more than the gentle TWANG of a clock-spring," capitalized the socialist journalist Herbert Casson, "but it was the first time in the history of the world that a complete sound had been carried along a wire, reproduced perfectly at the other end, and heard by an expert in acoustics." Casson paused for another repeat: "That twang of the clock-spring was the first tiny cry of the newborn telephone, uttered in the clanging din of a machine-shop and happily heard by a man whose ear had been trained to recognize the strange voice of the little newcomer." It had required not only the trained ear of a professor of speech but the sympathies of a suitor of a well-spoken deaf woman (Mabel, later Ma Bell) to make out a tiny newborn cry amid the machinery and street noise of a hot day in 1875. Then, within nine months, as if that first cry had been rather of conception than parturition, Bell (in a different, more private workroom) would shout, "Mr. Watson—come here—I want you!"—a birthpang at once of anguish and command. He had, so Watson said, spilled battery acid on himself.[107]

Mr. Thomas Watson, twenty-two, must have started at the telephonic shout. Watson's ears had always been, as he wrote in his autobiography, very sensitive: unlike other boys, he had grimaced at the firecrackers on Fourths of July and found steam whistles "a horror." Brought into the world behind a livery stable and trained in his teens as a mechanic, he had learned to make and fix call bells, alarms, and hotel "annunciators," sonic devices whose volume he could adjust and control, but his sensitive ears were put most to the test by the séances he had attended from the time he was eight. Listening for just-noticeable raps, knocks, and whispers, he would in his maturity explain these phenomena not as messages from discarnate beings but as manifestations of the power of mediums "to transform some subtle bodily radiation into a mechanical force," just as steam engines transform heat into mechanical motion—or as Bell's telephone used speech to generate a "*sound-shaped electric current*," reconstituted as heard words at the receiving end. For Watson, though, the real fascination of telephones lay in the silence on the line and the mysterious noises that percolated just behind. This was before electric trolleys and grids of electric lights sent "*current noises*" through many a telephone wire, so the noises he heard could not have been mere interference—although it was true that the silence on the line was greatest, and just-noticeable noises most audible, at night, for "the air of the quietest city room during the day is vibrating with a complex of sounds blended into a loud hum to which we are so accustomed that we don't ordinarily notice it," a hum that could overwhelm early, "feeble," telephones. So Watson hung about in the lab well into the night "listening to the many strange noises in the telephone and speculating as to their cause." Was a snap, followed by a grating sound, the aftermath of an explosion on the surface of the sun? Was something like the chirping of a bird a signal from a far planet? What occult forces were just-noticeably at work in the recesses of telephonic sound?[108]

Bell himself had tried out Spiritualism in 1870 when his older brother died, then went on to a more tangible Vocal Physiology, a system of speech training for stammerers, thespians, and the deaf. Vocal Physiology joined his father's system of Visible Speech, his grandfather's methods for removing speech impediments, and his entire family's Scottish sense of propriety with his own researches into the dynamics of mouth, tongue, throat, and lungs. For the grandfather, Alexander Bell (1790–1865), a "Corrector of Defective Utterance," stuttering was basically a matter of "disorderly respiration" that had become more common in society as general levels of elocution and enunciation dropped, leaving the "poor tongue" open

> to the assaults
> Of mimic ignorance and ribaldry—
> The clownish jargons, far and wide diffus'd,—
> The howl and whine of rustic dialect,
> And all the discords of corrupted speech.

In this cacophony, with modern oratory impoverished and modern music no better than "compounded and confounded noise,"

> The general ear, perverted by the drones,
> Seeks nought beyond the modulated drawl
> So often heard in wail of mendicants,
> And crude declaimers at the tavern feasts.

The solution, musically, was "the natural and simple air"; oratorically, schools had to implement a better training of the ear and vocal cords. For the father, Alexander Melville Bell (1819–1905), such training would progress via systematic transcription and recognition—that is, through a visual alphabet of sounds (1867) that could be taught properly and consistently regardless of the timbre or native tongue of any elocutionist. Alexander Graham Bell (1847–1922) learned so well his father's symbols for positions of the lips and tongue and motions of the throat that he could encode "the sound of a cough, or a sneeze, or a click to a horse." He accompanied his father on tours around Great Britain promoting this "Universal Alphabetics," and then, when the family moved to Canada in 1870, on tours around Ontario and New England, in each new town challenging audiences to come up with "the most weird and uncanny noises." Bell would leave the stage while audience members made noises they thought odd or unrecordable; for each, his father penciled a few symbols on a pad for the son to decode and re-sound upon his return, e.g., a loud rasping that Alexander Melville would freeze into Visible Speech and Alexander Graham would convert back into the familiar sound of sawing wood. Such a theatrics of replicated noises made of the father a phonetic stenographer (or "phonographer") and of the son a "graphophone," the term Bell and his cousin Chichester would choose for their version of Edison's phonograph. Having spent his boyhood Saturdays repeating to his near-deaf mother, a devout

Sabbatarian, what he remembered of the morning's sermon, Bell's first hopes for telephony were to optimize Visible Speech through an apparatus "that would enable my deaf pupils to see and recognize the forms of vibration characteristic of the various elements of speech," by which visual cues they would appreciate the distorted vibrations of their own grating speech and learn to correct them. For this reason he was fascinated by "singing" flames that danced before a mirror to the soundwaves of speech or song, then by electromagnetic forces that might carry these waves through wires and cause an eardrum to vibrate sympathetically.[109]

Historians may opt for other points of origin for telephony: Austria, Canada, Germany, Cuba, New York. At the Aural Clinic of Vienna's General Hospital in 1869, a young Boston physician, Clarence J. Blake, did postdoctoral studies with Adam Politzer, an aural surgeon who had recorded the action-curve of tympanic membranes using a phonautograph that rendered sonic vibrations visible. Back at Harvard, in 1874 Blake met Bell, the new Professor of Vocal Physiology and Elocution at Boston University and helped him understand how the ear processes vibrations, after which the two collaborated with Charles Cross of MIT in probing the dynamics of the eardrum, which research persuaded Bell to use a resonating diaphragm not only as a sound receiver but as a transmitter.... Or Nova Scotia, where a younger and tubercular Bell, resting at the family's Canadian home, had trained a Skye terrier to growl continuously while he manipulated its lips, tongue, and muzzle until it could be heard to ask, "How are you, Grandmama?"—prelude to using a continuous current to carry the vibratory shape of words.... Or Frankfurt am Main, where in 1860 Philipp Reis applied the word "telephone" (<1790s megaphone system) to an electromagnetic sound-transmitter and in 1862 exhibited a device with a collodion membrane on one end and a reed on the other that could relay sung notes and vocal inflections but was too congested with rattles, hisses, and buzzes to convey distinct sentences (coughing, voiceless, and unrewarded, Reis died of consumption in 1873).... Or Havana and Staten Island, where an Italian engineer and stage designer, Antonio Meucci, built a "talking telegraph" in the 1850s but lacked the technical advice and social connections to patent and capitalize his invention.[110]

Wherever the point of origin, telephony, like stuttering, was a problem less of sound capture than of release: how to achieve an articulate conveyance of sonorants and a durable, distinct conveyance of obstruents—fricatives, affricates, plosives. Victorian families were privy to a parlor game called "A Secret that Travels," in which a message gets hilariously distorted as it is whispered from person to person. That this game came to be called Telephone tells us more than Bell and Watson might have wished about telephony's origins in noise (and in Bell's mistaking what Helmholtz had done with the production of vowel sounds through tuning forks and resonators). What remains remarkable about the first years of the telephone and its compeer, the phonograph, is how impressed were people with the verisimilitude of the sound despite calls and playbacks that were fraught with hiss and glitch, chirp and scratch, snap and pop. Mark Twain, whose Hartford house had not only an intercom but a "screaming room" where he could curse to his heart's content, mounted his telephone in its own closet, where he suffered enough setbacks to devise a set of checkmarks for commentary on his monthly bill:

✓ Artillery can be heard on the line
✓✓ Thunder can be heard on the line
✓✓✓ Both can be heard on the line
✓✓✓✓ All conditions fail

T. P. Lockwood, an electrician for the American Bell Telephone Company in the early years, categorized the five classes of disturbance on telephone lines by kind of noise: (1) frying, hissing, and bubbling noises; (2) screaming and whistling noises; (3) jerking and rasping noises; (4) dot-dash telegraphic interference; (5) cross-talk. For the latter three he had technical fixes; the first two, prevalent at night due to causes probably atmospheric, were harder to resolve. Then there were the noises that came from the vibration of wires in high winds, which generated interfering currents, and from lightning, and then from the stringing of more overhead wires for Edison's new incandescent lights in homes and offices, so that by 1882 "Circuits that were practically noiseless are now so bad that subscribers cannot receive messages, as the receiver makes audible every variation in the current." The editors of the *American Telephone Journal* scoured scripture for prophetic texts: Psalms 31:2, "I said in my haste, 'I am cut off'"; Job 11:10, "If he cut off, and shut up, . . . who can hinder him?"; Job 13:22, "Call thou and I will answer, or let me speak, and answer thou me"; and Job 15:21, "A dreadful sound is in his ears."[111]

But the laboring and urban middle classes had had fifty years of practice in listening day and night *past* the thunk of three-ton pile drivers sinking piles eighty times a minute for higher and higher office buildings, *past* the blast of furnaces, the grind of freight trains, the careen of traffic, the cries of newsboys, and in private *past* kitchens clanging now with cast-iron pots on cast-iron stoves, the din of mass-produced china and glass that rang at more penetrating pitches than did pewter or earthenware, and, most recently, *past* the immodest grumble of indoor plumbing. If, as Blake acknowledged in an 1878 talk before the British Society of Telegraph Engineers, "In listening at the telephone we are conscious of a considerable loss in the volume of sound," that loss was a minor handicap, for as with the "drummy or boomy" sound of long-distance voices and the narrow band of vocal frequencies tolerated by early systems, callers compensated "automatically" (an 1853 scientific term with regard to physiological reflexes) by speaking more loudly and listening more intently, as if noise were no less to be expected at the phone than in any other conversation across social space. Walter Benjamin, recalling his childhood in Berlin around 1900, wrote that his father, "whose outward manner seems to have been almost always courteous and pliable, possessed perhaps only on the telephone the bearing and decisiveness corresponding to his sometimes great wealth," and only in his "vociferous" altercations with telephone operators did he manifest the forcefulness and energy (and volume) of a more commanding persona. Blake, who had removed a tympanum from a corpse for use in audio experiments, organized Bell's invention: "The mouth-piece of the hand telephone may be compared to the external ear, the metal disk to the drum membrane, the air-chamber to the middle ear cavity, the damping effect of the magnet to the traction of the tensor tympani muscle, and the induced current in the coil to the sentient apparatus." His organ concert turned out to be uncalled-for. Telephones with their line noises were integrated with other naturally noisy ears-full such as the back-chatter of gossip and the patter of sales pitches. Indeed, early commentators were often impressed by the power of the telephone—like the phonograph—to pick up (writes the literary historian Steven Connor) the "individuating tones and accidents of speech," from a brogue to a cough. Bad connections were compatible with otological findings that there were "but few people who have not some defect in one or both ears, whereby their natural range of hearing is to some degree lessened," so telephones could be represented as aids that extended the range, if not the quality, of hearing. Noisiness on the line might even be beneficial, following the logic of the old notion that the hard-of-hearing hear better in noise, as when "riding in a railway carriage, in an omnibus, amidst great noises, or loud, rumbling, roaring sounds." For those with aural catarrh (or sinus infection), claimed one Scottish aurist, an engine's roar could restore tympanic tension to drooping drum-heads and so relieve a temporary deafness. Might not spikes of noise on the line also spike one's hearing?[112]

Dramatically clear were not the phonics but the bell and the *tel-* of Bell's telephone. The intensity of the bell was so confounded with the striking distance (Greek *têle*) from which a call was coming and the importance of its purpose (*télos*) that the ringing of a telephone was granted an imperative urgency that had formerly been reserved to churchbells. Unlike churchbells, however, telephone bells were engineered to have an unpleasant ring; as a company acoustician explained, "A noise that is somewhat unpleasant will undoubtedly attract one's attention more than [a sound] equally intense but harmonious." The dedicated unpleasantness of ring tones was augmented by the dark overtones of being rung up. The telephone, wrote Rudyard Kipling in the rumpled pose of a newspaperman, "becomes a tinkling terror, because it tells you of the sudden deaths of men and women that you knew intimately." That, from Kipling's *The Man Who Would Be King* of 1888. By then, the majority of commoners (who would not have home phones for another sixty years) already feared being called to a public telephone whose ringing so often brought tragic news. Prosperous farmers, society matrons, and city bankers who already did have phones at home found the devices a bit less ominous but more demanding and frustrating: *noisily* demanding, because the unpleasant bell could ring at length if unanswered (in as many as six houses at a time on a party line), and because "howler" circuits had been devised to alert subscribers whose phones had been left off the hook, unintentionally or intentionally; *noisily* frustrating, because phones were designed to make a shrill hum when malfunctioning and because, in working order, they prompted conversations conducted in loud voices. Bell himself refused to have a telephone installed in or near his own study, for he could not abide overhearing half of a SHOUTED conversation. Companies tried to teach their operators and subscribers the correct way to talk on a phone: hold your mouth an inch from the mouthpiece and speak "in a tone slightly above the ordinary conversational pitch, about in the same way as if speaking to a person across the room." Companies also tried to convince users that, just as phone conversations should be treated as face-to-face meetings, so phone receivers should be treated as audio-impressionable devices: "The instrument is not to be compared to a deaf man, but, to the opposite, is exceedingly sensitive." Lockwood, in his *Practical Information for Telephonists*, cautioned that "the upper lip should just touch the mouthpiece, and the tone of voice must be moderate. The resulting sounds will be much clearer in articulation and much louder than if the speaker were to get inside of the mouthpiece (which, speaking hyperbolically, many persons do) and then shout loud enough to awaken his slumbering ancestors."[113]

Still, callers tended to SHOUT, because they knew they were speaking to someone miles away, because they were excited, or because a connection might be broken, without warning. "To talk into a hole in the wall and receive a more or less intelligent answer is enough to fill the soul with awe," said a spokesperson for the National Telephone Association of America in 1904. "But what if the girl [operator] cuts us off in the most interesting part of the conversation?" Being cut off—common in late-nineteenth-century society with its corps of hecklers and henchmen paid to disrupt union meetings, its deafening traffic, and a new curtness aboard jostling trolleys or on the bargain floors of department stores—was most distressing when on the phone, whose noisy line and weak pick-ups required a strict attention to the spoken word that was both aural and postural: "To Listen," read a Canadian notice in 1896: "Place the telephone fairly against the ear, with an upward motion, so that the lower extremity or lobe of the ear is gathered in, into the cavity of the telephone," the lobe thereby "acting as a cushion and at the same time closing out all exterior sounds, thus enabling the voice to be heard with clearness and precision." Given such an intimate acoustics, the shock of being cut off must have been strong and, who knows, may have contributed to Telephone Illness, a syndrome reported by French doctors who found that young anemic women, allured by telephonic vibrations and holding the receiver far too amorously to their ears, became dizzy at the slightest of external noises.[114]

Perhaps the necessity of monaural concentration—of listening, oddly, with the left ear (the left hand holding the earpiece so that the right hand could take notes)—intensified the trauma of being suddenly cut off. And the urgency of each call intensified that trauma. Not long ago we were resigned to waiting a full day for replies from distant offices, wrote a contributor to the *American Telephone Journal*; then, with telegraphy, we expected replies within the hour; now, in the telephonic era of 1894, "response must be instantaneous and uninterrupted, or there is trouble." An insistence on the instantaneous had been with telephony from the start: when conducting the first conversation over outdoor wires in October 1876, Bell had shouted, "[a]Hoy! [a]Hoy! Are you there?" to which, after a brief delay, loyal Watson from across the Charles River had shouted a "Hoy!" back. Bell then responded, "Where have you been all this time?" And while the telephone forced a new breathlessness upon traditional call-and-response, the cost, coupled with anxieties over being cut off, sped up the pace of conversation itself, which before 1900 averaged less than two minutes per call. Whereas large public venues like churches, convention halls, and Victorian theaters were acoustically inhospitable to rapid-fire speech, telephony encouraged it. Demanding prompt response to its ringing and fervid attention to its message, the telephone invoked the timeworn oxymoron of an impatient calm. "Everyone around should be perfectly quiet," directed an Australian electrician in 1878, for "the sound comes in a singularly weird-like manner," and "when a ghostly 'co-o-o-e' is heard, coming from fifty or a hundred miles away, the receiver is almost awe-struck."[115]

Celebrated for its instant embrace of a world of remotely intelligible voices, the new medium was a parlous mix of public and private. Trembling with fears of being summarily cut off and, obversely, of being overheard, users tried to draw curtains of privacy around speech acts that, as telephones became more sensitive and telephone systems more extensive, seemed more potent. Unlike the codes of telegraphy, the language of telephony was open to any eavesdropper; unlike the broadcasting phonograph, telephony was idealized as a closed system which, when perfected, would be immune to party lines and bad weather. The "silence booth," or public telephone cabinet, emerged as another oxymoron within telephony's olio of public and private. There was, wrote one commentator in 1904, "a crying need for a 'private' public pay station" that would "keep out foreign noises and make conversation easy and pleasant. This is particularly true in large cities, where heavily laden trucks and wagons, rattling over the cobblestone streets, the hum of street car motors and the clang of their bells... make the talking in a public place well-nigh impossible." And, he added, since "all, or nearly all telephone messages are of a strictly private nature," silence booths would encourage the use of public telephones by the constitutionally timid as well as by small-town folks who knew how quickly an overheard word could make the local rounds.[116]

"Proofing" a space for sound was a hiss or miss proposition before 1900, whether in psychological laboratories lined with cocoa matting or in public booths barricaded against street noise. People suffocated in telephone booths that were rather airtight than soundproof; when engineers added ventilating fans, these could be so loud that the booths were noisier than music halls. "In these days of the telephone and microphone, sounds are transmitted to enormous distances, or the slightest sound is magnified so that it is clearly audible," wrote James Colling, an English architect, but when it came to the "non-transmission of sound," he could find no reliable guidelines in 1882. Of the 1,364 pages in a major encyclopedia of architecture, three short paragraphs were devoted to the topic, under the rubric of "pugging," or plastering the boards beneath floors and behind walls, an approach that turned sound-deadening air pockets into sound-transmitting solids: "Now this word 'solid' is simply a deception; for it is a fact that the more solid a building is made, the more sonorous it becomes.... A fireproof house, which is built with concrete floors or brick arches, rings like a bell." Colling had a point, above all for North Americans who had been building in a frenzy of cheap wood. After a series of devastating fires in Manhattan, New Orleans, Pittsburgh, St. Louis, Philadelphia, St. John's (Newfoundland), Boston, and Chicago, builders had begun to mortar the spaces between wall joists and fill in the spaces between floors with a layer of lime, clay, and sand, a process called "deafening" because formerly (and mistakenly) used to muffle sound. Encouraged by insurance companies and by new building codes, the filling-in did slow the spread of fires; it also made for less acoustic privacy, more overheard overhead spats, and more irksome bells when telephones rang on and on and on unanswered in neighboring flats.[117]

Outside, on the street, fire departments were putting in place another new communications network that promised speedier, more accurate reaction times. Churchbells, courthouse bells, watchtower bells, and market bells had traditionally alerted communities to fire, but as the high-frequency whistles of factories, trains, steamers, and policemen reached the suburbs and the gongs of streetcars and the grumble of elevated trains resounded in metropolitan canyons, it became harder to hear the bells, let alone identify by tone and number of peals the site of a fire. Firefighting amid smoke and flame had always been a highly aural act, for which the icon was neither a water bucket nor a ladder but a long brass trumpet, emblematic of the tin trumpets through which chiefs shouted commands to their companies. Hearing was so crucial that in 1842 the Union Fire Company No. 2 of St. Louis held a special meeting to consider the new headgear of hose-men, who were sporting caps with protective capes down the back. The capes, which reached around to cover the ears, obstructed the hearing, "and at fires we all know that it is highly necessary that our *ears* as well as our eyes should be open, not only to hear orders that are given by our officers" but to hear the crack of falling rafters and the cries of people hoping to be rescued. When the noise of cities themselves became obstructive, telegraphically linked networks of call-boxes were laid out in major cities—Berlin in 1849, Boston in 1852, Philadelphia, St. Louis, and Baltimore by 1860, elsewhere after the invention of an automatic repeater that enabled an individual signal box to cue a set of alarms in the neighborhood. By 1871 New York City, whose residents had been kept awake by "the most sonorous and ponderous of bells" tolling uptown and down throughout the night and "iterated and reiterated" for minor fires miles away, had installed 919 local signal boxes, operable solely by patrolmen, who had keys. The keys were soon shared with a few "responsible citizens" nearby, and by 1883 some boxes were keyless, awaiting a simple crank, or prank.[118]

False alarms had to be part of the alarm equation, just as false observations were part of the personal equation in astronomy and physiology. Before telegraphic systems were in place, alarms had been raised for trivial sparks, for heat lightning, for the juvenile joy of scampering after an engine company as it jangled out of the station, or as decoys for criminal skullduggery. Electric alarms aggravated the problem: the crank on the call box had to be worked with unpanicked steadiness in order to transmit a decipherable location code to the central office; electrical shorts produced "false rings" as obnoxiously as did boy-gangs looking for excitement on a dull day. In 1882, Police Inspector Byrnes arrested a group of "fire-alarm fiends" and broke up "an organized plan of systematic and malicious annoyance" in New York City; of the 1,195 alarms rung up from London call-boxes in 1887, a third were false alarms, and half of these were reckoned to be malicious. Ellen E. Tyndale thereupon invented an automatic alarm post that would alert the nearby fire station but also "grips the wrist of the alarmist with a steel bracelet, blows a police whistle, and presents the man with a shilling for his trouble."[119]

Electric burglar alarms, coming into their own in the 1870s and 1880s, contributed to the aggravation of false alarms in much the same way as did car alarms of the 1970s and 1980s. They also added to the tone of urgency that was being culturally fixed to the *brrrang* of electric gongs and bells. As improved by Augustus R. Pope in 1853, the burglar alarm dispensed with clockwork and used "burglarious" activity itself—the jimmying of a locked window or door—to complete a circuit that rang a gong or bell in the home or shop, so that the noise of crime would call attention to itself, in counterpoint, for example, to the tactics of thieves involved in the "noisy-dog racket" of stealing brass knockers from fancy doors while an accomplice barked like a dog to mask the noise of the theft, a racket that vanished as electric doorbells replaced heavy knockers. By 1872, through the offices of the Holmes Burglar Alarm Company of Boston and New York, alarm circuits were wired into a central exchange where operators noted the time of each alarm and contacted the police. Electric burglar alarm exchanges (studied by Bell's chief investor, his father-in-law), along with electric fire alarm systems (a major patent for which came out of the machine shop where Bell and Watson had worked), served as the models (and in Boston, the locus) for the first telephone exchanges. All three operated by analogy to the neurological network as understood by Helmholtz and Wundt: an external stimulus—a shock, a break-in, the crank of a call-box—produces a local sensation and noise (a zap, smash, or clack) and not-quite-simultaneously an impulse travels to a control center that takes its own bit of time to ascertain the site and strength of the impulse and direct a meaningful response. "The analogy with the functions of the motor and sensitive nerves of the animal organization is complete," explained a company brochure for the American Fire Alarm and Police Telegraph, proposing a dense urban network that would put a signal box within three hundred yards of any house. "The Central Office is the *brain*, the wires the nervous system."[120]

5. *Spells*

Internal signals—could they also be false? The question was most apt with regard to sound. Tinnitus, or "subjective noises in the head and ears," had been known since antiquity and interpreted as malady or as message. St. Teresa of Avila, experiencing painful noises in the "upper part" of her head while composing her *Interior Castle* (1577), understood the inner rush of hurtling rivers and the whistling of birds as evidence of the Holy Spirit's roundtrips between "the superior part of the soul" and Heaven. Others felt the unrelenting noises as a hellish torment, and after millennia of ointments and leeches applied in or around the ear, an Irish medical professor would admit in 1891 "how unsatisfactory is our special knowledge of the entire subject... and how disappointing are the results of treatment." Henry Macnaughton Jones had himself suffered from a "vascular pulsating tinnitus in my left ear, which accompanied the cardiac systole" and disrupted his sleep. If the pulsation "was exactly like an aortic regurgitant

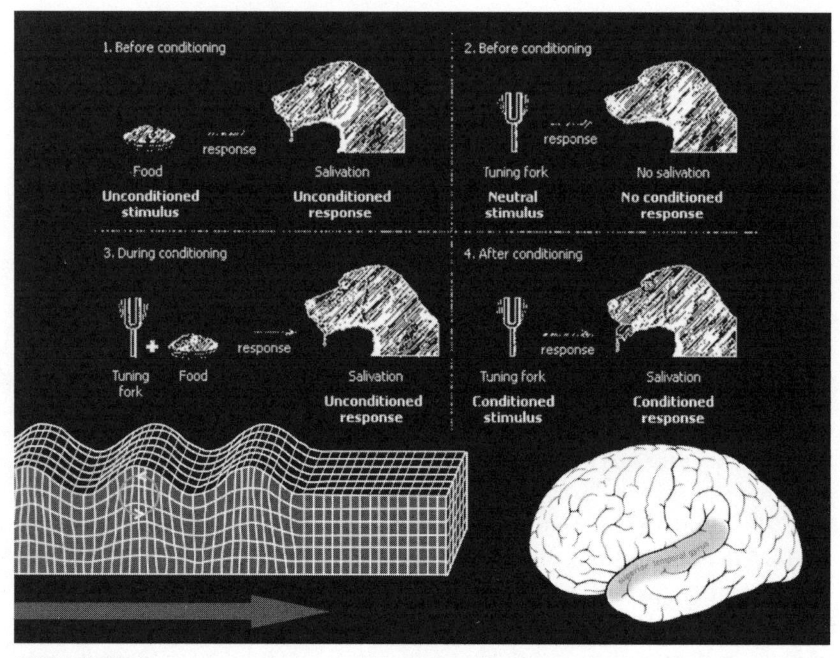

SOUNDPLATE 9

murmur," stethoscopic simile neither explained its origins nor excused him from the noise, which persisted for five years. Some of his patients went almost crazy with the surround-sound of bombs exploding, engines shrieking, hammers hammering, millwheels grinding or, recently, telephones ringing. The correspondences between environmental and head noises were close: "the old lady, fond of her cup of tea, compares the noises to the singing of the boiling kettle, while the servant in the kitchen has a constant ringing of bells in her ears." Correspondences between kinds and causes of noise were speculative: sounds resembling a pump or the hiss of a snake, did they come from impacted ear wax or the chewing of tobacco? In any case, the corresponding prognoses were poor. Macnaughton Jones summed up the perplexity about tinnitus by listing some of its proximate causes: fever, gout, drunkenness, pharyngeal catarrh, cardiac weakness, menopause, mental strain, irregular catamenia, Bright's disease, quinine overdose, toothache, syphilis, smoking, sea bathing, sleeping with the mouth open, a family history of deafness, retroflexion of the uterus.[121]

Experimental neurology, not gynecology, promised to resolve the confusion. After Paul Broca identified the left frontal convolution of the brain as responsible for articulate speech (1862+) and Theodor Meynert tracked the acoustic nerve into the cortex (1867+), Carl Wernicke (1874) and David Ferrier (1876) located functional hearing in the superior temporal gyrus of the temporal lobe, and Meynert's star pupil Sigmund Freud traced the fetal acoustic nerve deep into the medulla oblongata (1885–1886), "noises in the head" might be taken as symptomatic of "disconnection syndromes." Like problems on telephone lines, disconnection syndromes involved problems with neural-signal interference and poor hook-ups, or with lesions in the brain's functional centers and its expanse of neural reticulations. Ferrier, for example, had found that men who suffered injuries to upper parts of the temporal lobe might hear noises so strong as to trigger epileptic fits. Responding to a letter from Macnaughton Jones, Ferrier noted that "tinnitus may occur in both ears as the result of unilateral irritation, the effect of the intimate connection of the auditory nerves and centers." Yet, though investigators reported irregular sensory cortical discharges that might account for tinnitus, and though it was probable, as Macnaughton Jones knew, "that lesions of the adjoining parietal and occipital lobes may encroach upon, invade, or inhibit the hearing centre and cause both deafness and tinnitus," the cause of tinnital *noise* remained elusive. "There is nothing to show that we can strictly apply the hypotheses as to the *probable routes by which mere noise, as distinguished from tones or other regular series of vibrations, are transmitted to the brain*." His italics stressed the "feebleness" of the finest aurists in their hunt for the source of "noise in the ears" and the "exquisite sensitivity of the auditory nerve apparatus," bound up with the brain's intricate convolutions.[122]

"Your trouble is due to a form of dry Catarrh, which has created an inflammation in the auditory canal, and an irritation of the nerves of hearing, as well as a thickening of the membranous lining of the parts," wrote Health Specialist Sproule of Boston to Meriwether Jones of Virginia twenty years later, collapsing a century's worth of guesses into one confident bundle. "In consequence of this, the sensory nerves convey a wrong impression of sound to the brain, the vibrations from the outer soundwaves are not so easily felt, while there is a painful exaggeration of the noise of the blood coursing through the large vessels of the brain." Sproule claimed to have been a surgeon to the British Royal Mail Naval Service, so "coursing" and "vessels" came fluidly to his pen, and there was an ingenious neuro-nautical logic to ascribing tinnitus to the pathologically amplified sound of blood coursing through the brain. "These head noises, described by some as sounds like escaping steam, by others as ringing and buzzing, throbbing or crackling, are, alike to all, the source of keenest suffering" and "considered the bane of the medical practitioner, because so many have tried to cure them and have failed utterly." For $9.00 a month he could "lift this blight" from Jones's life through a combination of "internal toning," nasal inhalations, and a germicide against poisoned mucus. Jones did not jump at the offer (a second letter from Sproule made it $7.20), and I do not know whether the blight was ever lifted. Still, Macnaughton Jones, who would have scorned such a regime, had little better to offer. In Paris, Georges Marage had begun using a patented siren to send low-level sound-waves into the ear to shunt the stirrup one-thousandth of a millimeter back into proper alignment with the hammer and anvil, thus eliminating the hissing or whistling sound caused by its displacement, but this sonic surgery was no sure remedy for those who heard noises like the chirp of cicadas, the crackle of voices on a telephone line, the shuffling of a pack of cards, the ringing of cracked churchbells, or the shouts of an infuriated mob—and to this day much about tinnitus remains puzzling. It will recur, as it must, in the next round; in 1911, awareness of its prevalence and of its poor prognosis confirmed the interminableness of noise from yet another quarter and complemented the study of other "subjective," or incorporeal (eventually, metaphorical) noises.[123]

Convolutions of grey matter could produce auditory illusions apparently unrelated to malfunctions of ear or brain. Theophilus Hyslop, an English physician devoted to the treatment of the insane and the study of *Mental Physiology* (1895), knew Ferrier's work but was more intrigued by sounds that could not be attributed to neuroanatomy. Having a patient who needed injections of morphine every three hours for two weeks, Hyslop's "expectancy and mental preparedness to hear the night-bell, even during sleep, became so powerful" that after the patient recovered, the good doctor kept hearing the bell though an alarm no longer went off every three hours, an "illusion of bells" so vivid that he woke again and again to hie to his patient's bedside. This, said a friend, the London alienist George H. Savage, was a "'lost button' condition of thought" where, for weeks after a waistcoat has lost a button, one goes through the motions of buttoning that button. On second thought, there had to be more to the phantom bells than mere habit, for Hyslop knew of deaf people who had recurrent auditory hallucinations, and how could habit lead one of Savage's own patients to respond to voices that were ringing him up on an inner telephone? Although the "central pathology" eluded science, the hounding sounds probably had something to do with suggestibility and imagination, since auditory hallucinations were most common among impressionable young women.[124]

Vice versa, obnoxiously real sounds might fail to register with people whose ears were fine and whose hormones were less rampant: parents oblivious to their offspring's incessant off-key piano practice, or the bachelor who sleeps through "a rattling thunderstorm" yet "loses all somnolent composure while his neighbor is filling a bath-tub." Such auditory illogics, wrote Dr. George F. Shrady, editor of the *Medical Record*, would be inexplicable until "some enterprising neurologist can locate the noise-centre" in the brain. Meanwhile, cracksmen and rum dubbers (i.e., housebreakers and lockpickers) could escape detection by purposely making a racket so as to calm the suspicions of Manhattan householders resigned to the ruckus of urban nights yet alert to silences broken by the footfall of a "dromedary" (a clumsy burglar) in the quiet of early morning. And how could one be certain that people who slept "soundly" through the thunder of rattling trains or the rattle of thunder had not insensibly reverberated with those noises? Since the 1870s, Rest Cure psychiatrists and respectable physicians, such as Ernest A. Hart, editor of London's *Sanitary Record* and the *British Medical Journal*, had maintained that years spent toiling amid ever-present noise do in time take their toll, if not in nervous collapse then in a loss of mental focus, as with businessmen who come to "feel that the adding up of columns of figures requires more attention than it used to." Or as with factory workers who must shout to be heard over their machines yet swear that they are unaware of the noise: after years of laboring to assert their own rhythms against the heavy blows of steam hammers or drop forges, they suffer extreme fatigue and its mortal consequences.[125]

Marx had construed this fatigue in the context of the political economy of capitalism. Once labor unions succeeded in reducing the hours of factory workers, Capital intensified their labor by enchaining them to faster, higher-powered (and louder, more sonically penetrating) machines in ever-greater arrays. Already in 1844, when Parliament was about to vote on a shorter workday, the masters of textile mills had protested that the "extent of vigilance and attention on the part of the workmen was hardly capable of being increased" and that one more hour's rest would only hurt the national economy, for "to expect in a well-managed factory any important result from increased attention of the workmen was an absurdity." Yet, as Marx pointed out, when workdays were reduced to eleven hours, and in 1847 to ten hours, textile mill output did not decline, for mechanical innovations repeatedly sped up the power looms, whose operations could be attended by fewer if more harried workers. This "systematic heightening of the intensity of labor" heightened not only the strain on the worker but the risk of deadlier accidents. Marx cited a report of 1861 on the excess of mortality from lung disease as a result of that "exhausting state of excitement necessary to enable the workers satisfactorily to mind the machinery, the motion of which has been greatly accelerated within the last few years." In addition to losing limbs and lungs, exhausted laborers became mental cripples, unable to focus on anything and plagued by "false judgments respecting the objective universe," by emotional distortions, romantic illusions, political hallucinations—"all errors, all superstition," wrote Max Nordau, "the consequence of defective attention." From opposite ends of the philosophical and political spectra, Nordau, author of an influential 1892 jeremiad on *Degeneration* and Marx (who died in 1883 of lung and liver problems) worried over displaced and misplaced foci of attention, for the real battle of the industrial world was not for muscle but for minds. At century's turn, a prominent German social critic and political journalist,

Willy Hellpach (with his own grab-bag of nervous problems) lay the blame for this fatigue and resultant "soullessness" on the mechanization and electrification of the workplace, which shattered age-old bonds between workers, their tools, and their products. From that Marxian analysis, Hellpach spun on his heels to blame the severity of industrial fatigue on Marxism itself, for in their campaign to win over the proletariat, Marxists had emasculated labor by splitting its focus: while workers listened with a "fanatical fatalism" for the inevitable collapse of Capitalism, they shared a "more hopeful resignation" in their immediately woeful state, which made them complacent if not also complicit in their downtrodden lives. Their attention divided between unlikely futures and the onerous, ill-paid work at hand, they were deadened to workplaces that pummeled at their nerves and lungs and at their very sense of balance. No wonder they claimed to be deaf to the devastating crunch of machinery.[126]

Lively awareness of sound—whether subjective or objective, oppressed or repressed—could only be part of the story of the hearing of noise. There had to be a more fundamental receptivity, such as that explored by Bernard Pérez, co-founder of the field of child psychology, who noted that infants from three to ten months "are less often alarmed by visual than by auditory impression," so that "fear comes rather by the ears than by the eyes, to the child without experience"—and, wrote William James, to older, less naive folk like himself: "The writer has been interested in noticing in his own person, while lying in bed, and kept awake by the wind outside, how invariably each loud gust of it arrested momentarily his heart." Pérez had concluded in 1881 that "there are hereditary dispositions to fear, which are independent of experience"; James in 1890 did not know exactly what to make of the infant disposition to be more alarmed by sound than by image. Had it to do with shock, with instinct, or with qualities of attention? Since Charles Darwin, who had described attentiveness as central to man's mental evolution, philosophers and scientists had been drawn to the study of attention, dividing it into passive (reflex) and active modes, sensorial and ideational paths, selective, expectant, strained, scattered, or effortful states, each division illustrative of a theory of mind, nerve, and/or motor activity. For John Dewey in 1896, these divisions obscured the congruences by which mindbrainbodymuscle operate. Consider a noise. "If one is reading a book, if one is hunting, if one is watching in a dark place on a lonely night, if one is performing a chemical experiment, in each case, the noise has a very different psychical value; it is a different experience. In any case, what precedes the 'stimulus' is a whole act, a sensori-motor coördination." The organism at each instant is thoroughly but differently prepared to deal with that stimulus. "What is more to the point, the 'stimulus' emerges out of this coördination; it is born from it as its matrix; it represents as it were an escape from it." A noise therefore cannot arise "absolutely ex abrupto from the outside, but is simply a shifting of focus of emphasis, a redistribution of tensions within the former act." To repeat, in case your attention has wandered: "We do not have first a sound and then activity of attention, unless sound is taken as mere nervous shock or physical event, not as conscious value. The conscious sensation of sound depends upon the motor response having already taken place." Noise, in sum, must be a product of the full self in joint session with the world beyond.[127]

Implicit in such reasoning was a deepened understanding of noise as a phenomenal presence rather than an epiphenomenal distraction: noise did not so much erupt or disrupt as return to consciousness. In his *Outlines of Psychology* (1893), Oswald Külpe argued that higher life forms had to be able to narrow their focus, for without such a capacity, "thinking would be made impossible by the noisiness of our surroundings." Conversely, experimenters had also shown that "If a stimulus of minimal intensity (a watch-tick at some distance, flow of sand, tuning-fork tone, liminal smell, &c.)... is steadily attended to, the sensation is found to disappear and reappear at irregular intervals," which made of attention itself a process of discovery, loss, and recovery. In 1904, Knight Dunlap at Berkeley demonstrated that inaudibility itself might be a function neither of acoustics nor of otology but of attentiveness. Placing subjects near a telephone receiver in a "silent room" with padded double doors and soundproofed walls, he sat in a separate room with a "noiseless key" connected to an electric coil that could induce an intermittent or constant "snarl" on the closed telephonic circuit. Shifting from silence to snarl to the pure tones of an electric tuning fork, Dunlap proved "that a minimal sound which has become imperceptible through the so-called fluctuation of attention, may yet be 'heard to stop,'" and that a continuous sound was "inaudible at a much higher intensity than an intermittent sound." Noise, sound, and silence were no longer independently stable entities; they were functions of each other and codependents of the brain. When Dunlap wrote of the "second death of a sound" (lost to lapses of attention), it was not to discourse on deafness but to argue, from a professorial lectern at Johns Hopkins, for some autonomous neuromechanical operation that puts sound through its (s)paces.[128]

John Tyndall, the premier expositor of acoustics to Later Victorians, had confessed in 1867: "How it is that the motion of the nervous matter [in the ear] can thus excite the consciousness of sound is a mystery which we cannot fathom." Since then, neither Helmholtz nor Wundt nor Ferrier had been able to fathom the neuromechanical process by which the voices of those who had experimentally inhaled hydrogen came to sound "unearthly," or the shaking of a toolbox filled with nails, chisels, and files came to sound like, well, noise. Nor even John William Strutt, the Baron Rayleigh, whose two volumes on *The Theory of Sound* (1877) completed the basic mathematics of physical acoustics. Its first chapters composed at the age of thirty as he recovered from rheumatic fever on a houseboat sailing up the Nile, Rayleigh had returned to health and to England to spend another four years on his opus, in whose preface he represented himself as a consolidator. In truth he was a fine "applied mathematician" with a sure experimental hand and a full purse; his theory of scattering, published in 1871, had furnished the first fully correct account of why the sky was (if not in London then at his family's Essex estate or under an Egyptian sun) blue. His interest in acoustics lay with the laws of vibration "to which the sensations of the ear cannot but conform." Beginning with his first paragraph, he removed the theory of sound from the biology of the ear and put it squarely within the field of mechanics; in his ninth paragraph he paused to address the nature of noise before moving on to pages of equations expressing more regular vibrations. Noise was too common and messy a phenomenon to merit more than a few words, for if in extreme cases it was easy to tell noise from music ("everyone recognizes the difference between the note of a pianoforte and the creaking of a shoe"), in most situations it was "not so easy to draw the line of separation," even with ordinary musical instruments, whose tones had unmusical outliers. "But although noises are sometimes not entirely unmusical, and notes are usually not quite free from noise, there is no difficulty in recognizing which of the two is the simpler phenomenon": music. Musical notes had a "certain smoothness and continuity" which, when badly combined and played, could be made to imitate noise, "while no combination of noises could ever blend into a musical note."[129]

That boded ill for sufferers from tinnitus. Although a patient well-versed in calculus might find clues in Rayleigh's analyses of resonance and bells, the physics of vibration had little to say about noises in the head, except to Dr. Marage. If tinnitus eluded the grasp of physicists, physiologists, and psychologists as it eluded the treatments of otologists and psychiatrists, science had had no better success at clarifying the process by which sounds persistent or intermittent or unique became known in the brain's auditory center as music, as noise, or at all. Acts of hearing had something to do with beat and rhythm, with the ear's resonating chambers and membranes, with labyrinthine fluid and cochlear fibers, with the acoustic nerve and the temporal lobe, but neither neurologically nor physiologically was it clear why the readers of Tennyson's "Maud" should agree to hear not a scuffle but "the scream of a maddened beach dragged down by the wave" when "a rough tide rolls in upon a pebbled beach [and] the rounded stones are carried up the slope by the impetus of the water, and when the wave retreats." And why do dogs, who hear what humans cannot, growl at our music? And why was Alfred Smee, who asked this question in 1850 and coined the word *ousaisthenics* for knowledge that arrives by way of the ear, why was Smee so very unmindful of the havoc caused by the bunch of keys he jangled all the time? "Oh, those keys!" wrote his daughter shortly after his death in 1877. "I cannot think of them without a horror. What have my nerves suffered through ye, o keys? If he was thinking, jingle went the keys; if he was writing, again jingle went the keys; whenever an opportunity was afforded him to jingle those precious keys, they were jingled."[130]

Suppose, as the physiologist Ivan Sechenov had supposed in 1863, that all complex behaviors and ideas were "reflexes" of the brain. Suppose that the brain were just a center for organizing or inhibiting reflex action. Suppose that purpose and meaning were not products of an intangible "mentation" but of sensory-motor activity. Suppose that attentiveness were not a quality of "mind" but a quantity of neuromuscular repetition. Rigorously organized, punctual, and focused, Ivan Pavlov was the best man to advance Sechenov's suppositions. As a left-handed youth, he had trained himself to be ambidextrous. As a scientist, he entered his laboratory at nine-thirty a.m. every day but Sunday and left at six p.m. At home, a husband and a father, he ignored the ringing of a telephone between seven and nine a.m. so as to keep (his mind?) clear for the day ahead. In 1904 he won the Nobel Prize for his work on the physiology of digestion, having shown that dogs could be trained to salivate in response to sham feedings—the sight of food or sound of its preparation—exactly as if they were eating, their digestive secretions controlled by nervous reflexes. After 1904 he turned in earnest to the study of "conditional" (conditioned) reflexes, using lights and, famously, bells and buzzers to cue patterns of salivation in response to stimuli that were sensorially unrelated to food. Comparing reflexes to "the driving-belts of machines of human design," he thought he was expanding upon the tradition of Descartes, who had conceived of the nervous reflex. Unlike Descartes, however, Pavlov left no room for that other drive-belt, Spirit, by which to distinguish humans from other animals. Pavlov spoke of a "Reflex of Purpose," an investigatory reflex, even a "freedom reflex," attributing ambition, curiosity, and freedom-loving to higher-order instincts which, like lower-order vomiting or self-preservation, were integral to the organism. So it was that the body, through its network of quick-response systems, could be trained to react to a bell or the tick of a metronome in a standard way and to assign to rings, buzzes, or beats a meaning pre-designated (and alterable) by a purposive investigator.[131]

Attentiveness and alertness were essential for effective conditioning. Dogs drowsy or distracted meant days dog-gone lost. A significant side effect, then, of Pavlov's decades of experiments using metronomes, buzzers, electric bells, whistles, crackling sounds, bubbles, table-scratching, glass-tapping, and hand-clapping, was to underscore the psychophysical power of noise. If sounds of all sorts could be used to establish and elicit a conditioned response, noises could interfere not only with response times but with conditioning itself. Rayleigh had contended that "In the daily use of our ears our object is to disentangle from the whole mass of sound that may reach us, the parts coming from sources which may interest us at the moment"; Pavlov's research depended upon restricting his dog-listeners to what *he* wanted them to hear, and no sound else. Otherwise the association between sound and slobber would be neither manifest nor manipulable. "Footfalls of a passer-by, chance conversation in neighbouring rooms, slamming of a door or vibration from a passing van" would "set up a disturbance in the [dogs'] cerebral hemispheres and vitiate the experiments," as would street cries or the click of a lab worker's shoes. Wherefore a special laboratory, isolated by trenches, was built for Pavlov in Petrograd, each testing room soundproofed and the rest of its environs made as mild as could be, for reflexes, "elemental units in the mechanism of perpetual equilibration," were simply the efforts of an organism to keep itself aligned with its environs. Whether it disturbed a sonic association about to be established or intruded upon silences that were themselves intended as conditioning stimuli, noise was not necessarily unmeaningful sound; it was, in Pavlov's lab, untimely or overbearing sound. Overbearing, when the noise was so loud, as with the "violent ringing of a bell," or so discordant, as with "the blare of a toy trumpet," that panic or fright overrode other visceral responses. In those cases where the ringing of a bell meant to a dog that it would receive food while the buzz of a buzzer meant that it would be electrically shocked, sounding a bell and a buzzer together made for a third sort of noise: contradictory sound. This produced a state of anxiety and, in some dogs, resistance to further conditioning—as if, suggests Peter Pesic, "the scales had fallen from the dogs' eyes" and they heard (bell = food) and (buzz = shock) for what they were, arbitrary constructs of a laboratory world.[132]

Beyond the dog cages of the laboratory, workers in factories and large offices had also been "conditioned" to rise, to work, to rest, to eat, to resume work and to cease working in response to bells, whistles, gongs, and buzzers whose arbitrariness was that of industrial capitalism. Capital, according to Marx, invariably objectified labor and estranged the senses—in much the same manner as Pavlov objectified appetite (= spit) and milled the rich sensations of mouth-feel, taste, smell, vision, and audition down to the bran of reflex behaviors. (The fastidious Pavlov had been unmoved by the efflorescences of Art Deco Paris in 1900; he also cut short his tour of the Jewel of the Adriatic, smelly dirty Venice.) Capitalism put a dastardly spell on the body, such that it had no mind of its own and no idea of what might be shared between bodies for the greater good. It drew attention away from our "inner wealth," making of each of us a lesser being: not a lover of fine music but a consumer of piano rolls; not a freely engaged moral agent but a vehicle for surplus value stockpiled by the bosses. "Every organ of sense is injured in an equal degree" by factory labor, wrote Marx, "by artificial elevation of the temperature, by the dust-laden atmosphere, by the deafening noise," all of which stood in the way of the true work of humankind, which was to realize an equable world and affirm all others in it through the use of educated, reasoning minds and each of our educated senses. Especially the ear. For Marx often elected the love of fine music as the exemplar of sense-development, and in his youthful poetry it had been music, of all the arts, that most expressed desire, longing, and "the inner suffering of the human heart." Like a magical Aeolian or wind harp, he had written,

> My own heart plays the lyre
> That I am myself, that is my pain,
> That echoes out of my soul.[133]

Predictably, what industrial laborers did learn to appreciate with their "crude ears" (Marx) was a kind of noise. The longer they worked alongside machinery, the more expert they became at understanding anomalous sounds: the hoarse gasp of a rusty steam press, the wheeze of a worn belt on a high-speed lathe, the guttural gripe of an industrial stitching machine aching for oil. Alive to such telltales of breakdown and dangerous ejecta, they were, despite their denials, always "in hearing" of the noise of their machines, therefore vulnerable to the mental and physical erosions caused by near-constant loudness and vibration. Where earlier it had been presumed that white collar "brainworkers" were uniquely liable to the fatigue of attentiveness, by the time of Marx's death, physicians and political economists had begun to acknowledge that any prolonged attentiveness could take a wider toll: "the unintermitted action of machinery, though it greatly diminishes physical toil, brings into many forms of labour a vastly increased mental strain through the constant watchfulness and attention it requires." Thus W. E. H. Lecky, the Irish M.P. and historian of rationalism, democracy, and "the natural history of morals." Labor and health advocates discovered signs of mental strain not only in factories but among rail workers, farmboys at mechanical reapers, and sweatshop women at steam presses, a strain manifest in a dramatic upsurge of accidents due to momentary losses of focus during long hours of tensed alertness.[134]

Under such pressure, who would not feel worn out? Who would not need to be "galvanized"? Electrical currents had been applied to the vaporish, arthritic, and impotent a century before neurasthenia was defined, but George Beard, who himself had spells of mental fatigue and a throbbing, buzzing tinnitus ("common with cerebral exhaustion"), also espoused electrotherapy, so the fortunes of ionized medicine were revitalized with the diagnostic fortunes of all forms of prostration. Extending the work of Golding Bird, who in the 1840s had prescribed electroshock to stimulate secretions and galvanic currents to restore the breath of London's asthmatics, Beard and his friend Alphonso Rockwell used electrotherapy not only as an adjuvant to the Rest Cure but as a universal nostrum. Every bodily system had a corresponding current: the "primary" (direct current) for blood and lymph; induced, for muscles; galvanic (electrochemical) for eyes and head; "superinduced" (?) for the nerves. Each of these invisible currents was accompanied by a baseline of electrical noises that patients (like factory workers at their machines) learned to distinguish, for the immediate effects of electrotherapy were often so subtle as to be otherwise unnoticeable. Along with prickles or slight burns, these low hums, buzzes, thrums, and whirrings became persuasions toward a cure, building upon an "electrical sublime" that strung analogies from nerve impulses in the bodily economy to the flow of power in finance, commerce, and industry.[135]

Sonic and haptic, the vogue for electrotherapy between 1870 and 1914 was homeopathic for a Second Industrial Revolution of machine tools with ever-finer tolerances and electrical currents that were not yet, admitted the directors of the Baldwin Locomotive Works, "the silent and reliable servant desired." Let the poison of electricity, which had lit up the workplace, sped up its labors, set up an endless round of shifts, and sent workers into spirals of exhaustion, be transformed into a therapeutic gift that would strengthen frayed nerves, smooth out knots in the digestive tract, and return the entire system to fluency. If a German physician of 1908 knew "very few electricians who don't suffer from overstimulated nerves" and if telegraph and telephone operators during thunderstorms could be knocked out by the shock of excess electricity in the air, then why not reverse the direction of those natural currents and feed them into dyspeptic, fatigued workers? The more often the brain was depicted as a central nerve exchange, the more logical would seem the use of electrical energy to restore alertness. The more frequently reported were the tics, agitations, and mental depletion of railroad switchmen who had to stand for eight to twelve hours a day "at a post that requires unrelieved attention and responsibility," the more logical it would seem that doses of faradic current could re-equilibrate the switchboard that was the mind. And since the demands of attentiveness did their damage by increments, it would seem apt to prescribe weeks' worth of electricity to bring bodies and minds back into focus.[136]

"Focus" belongs to visual frames of reference, which may once have been fitting, when ideas about attention had been rooted in thinking through the eye. By the 1890s, however, and certainly with Pavlov's work after 1904, the bulk of studies of attention involved not thought but muscle, and not seeing but listening, so it would be more fitting to speak of electricity restoring "tone" (*tonus*) to bodily fibers. A word half-muscular and half-musical (and, at the start of this Round, painterly), "tone" was more truly the tonic sought by those who willingly underwent electrotherapy, or any other physical therapy, for fatigue. To have good muscle-tone was to be firm and effectively tensed, one's body at the wait yet never aggrieved at the waiting, neither unsteady nor unnerved. Already I am slipping into emotionally charged and mentalist rhetoric, for wherever one heard tell of muscle-tone at century's turn, the ramifications were positively charged: toned muscles, because they put a body into a posture of easy readiness, resulted in emotional balance, sexual vigor, and mental health where "just-noticeable differences" were at their finest and life could be led with a minimum of self-interference. This at least was the idea, but it was limited by the problematics of attention as posed late in the century. Once visual evidence began to appear (to painters, physicists, physiologists, and philosophers) to be unreliable, and once psychophysiological experiments had determined that other media of experience and memory had their own irreducible deficits, attention itself had to be approached, concludes the art historian Jonathan Crary, as an "essential but fragile imposition of coherence and clarity onto the dispersed contents of consciousness." One could, with novelists, celebrate the flux or, with ragtime composers, put shape and rhythm to the hither-thitherness of perception, but the more that researchers attended to attentiveness, the more diffuse it seemed to be, fogged up by tag-ends of remembrance, by personal equations, by nervous tics, by anxious anticipation, by repetition, echo, exhaustion. Helmholtz himself was responsible for stating one of the paradoxes of attention when he wrote in his *Physiological Optics* that "It is natural for the attention to be distracted from one thing to another," as if the saccades or eye-jumps dictated by our optic nerves in turn dictated that to be truly attentive we had to be distractable, and that the more fully toned were our muscles and minds, the quicker would be our reflexive responses to stimuli coming at us from all directions. Our attendance upon affairs of the intellect, the heart, or the flesh could be no more constant than our pulsating bodies and the vibrating world through which we ∿∿∿∿∿∿∿∿ [137]

Allow me to interrupt. What's been happening is this: on one horizon after another, noise was no longer being heard, defined, or theorized as inadvertent. Noise was now increasingly and *necessarily* everywhen, despite a scattering of soundproofed rooms and silent booths. A phenomenal presence as unavoidable in systems of observation and communication as in band concerts or mechanical equipment, noise had come to seem so inevitable as to be intrinsic to the neurological and psychological engagement of each individual with the larger world, from the first months of life to the last, as hearing hung on to the very end. In the language of the chapel and the laboratory, the ubiquity of noise was the Call, its relentlessness the Response. In the language of the poet and the electrician, the ubiquity of noise was the Gap, its relentlessness the eternal Spark. Consider a device designed by Dr. John H. Girdner of New York, who sang Whitman's "body electric" in a new mode by constructing the body itself as an audible battery. Squinting past those "quacks and charlatans who deceive and rob the public by promises to cure disease with one electric device or another,"

Girdner saw promise in more substantial forms of medical electricity that could be used by surgeons to burn away tumors, cauterize wounds or, most immediately, sound out the viscera. In 1887 he had improved upon a "telephonic probe" that could do what Alexander Graham Bell's induction-coil version had failed to do in the days after the assassination of President Garfield in 1881: locate a (fatal) bullet deep inside a breathing body. For his version, Girdner ran a wire between a receiver and a bolus of aluminum placed in the mouth, then a second wire between the receiver and an aluminum sensor that could be moved along the body; with this set-up he could hear "a perceptible clicking grating noise" each time the sensor came into contact, through the skin, with any (non-aluminum) metallic object buried inside a live patient. The contact closed a circuit that had been generated and constituted by the body itself, so "The current of electricity that operates the telephonic bullet probe is derived entirely from the patient operated upon." By 1903 x-ray machines had assumed the job of locating alien objects and broken bones within the body, but Girdner that year had not yet given up on his probe, proposing that with regular use it could yield equally valuable physiological insights, or *insounds*. "For instance, a man who has fasted for twenty-four hours produces a feeble sound in the receiver; but a few hours after a full meal, the same subject will cause much louder sounds," as would a person with a temperature two or three degrees above normal, none of which activity was detectable by x-rays. Because the whole body sustained the circuit to whose crackling the receiver was privy, his probe might take acoustic stock of the extent and character of electrochemical changes throughout a living body. It would have required years of effort, comparable to the efforts of five generations of stethoscopists, to parse the sound and syntax of each internal electrochemical change—an effort never made. So the noises Girdner heard through his receiver remained noises, clicking grating noises to which people were ordinarily insensible—yet noises all the same, all day long and all night, in every body, in times of vigor, of sleep, and of exhaustion. To be alive—joyfully alive, or deeply troubled, or floating under the spell of a hypnotist—was to be noiseful: another evidence, as personal as it was impartial, of the fall of noise into nature.[138]

6. *Shells*

Back to attentiveness, fatigue, and audition. There is attentiveness, and there is attentiveness. There is an attentiveness that is fully voluntary, as of a violinist auditioning a Guarneri. There is an attentiveness less voluntary, as of a soldier on sentry duty on the edge of a foul swamp. There is an attentiveness less voluntary still, as of a child kept awake by angry voices in the next room. Then there is an attentiveness absent all volition, as that of the ear. You can shut your lidded eyes against sights too gruesome and

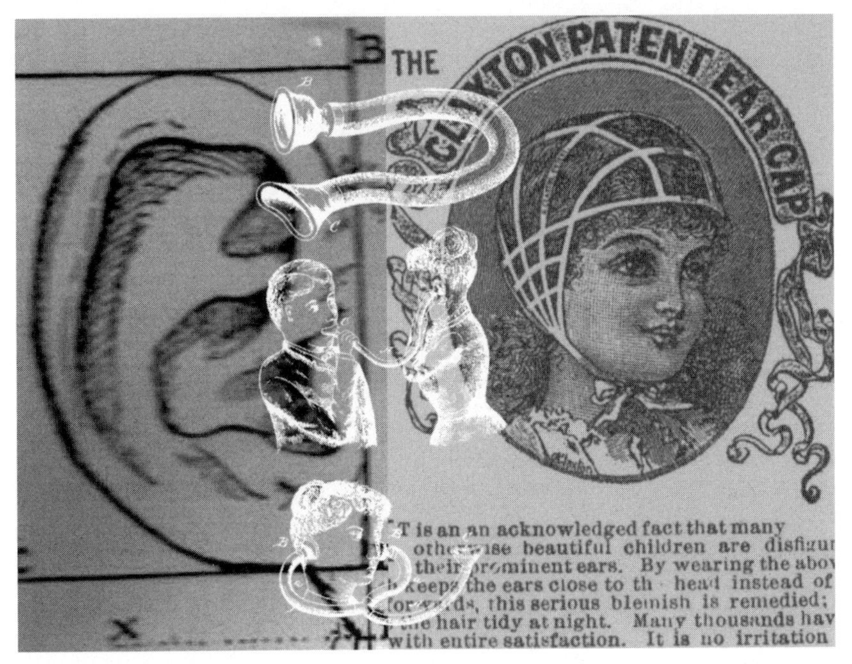

SOUNDPLATE 10

clamp your lips against the unsavory, but, as the sanitary reformer Fothergill announced in 1887, "the ear is ever open." Lack of ear-flaps was no revelation, but in the context of concerns about attention and fatigue, the ears were now exposed as supremely vigilant and vulnerable. Ears had no Sabbath from a universe of vibrations. "There is no absolute repose for the ear," wrote a Philadelphia aurist, Laurence Turnbull; "it is in action from birth till death. If we compare other organs with it, we can see at a glance how liable it is to injurious influences." While the other sense organs relaxed during sleep, the ears had been shown to remain on call night and day, which was how a mother could hear through her dreams an infant about to cry for the breast, or night passengers of a steamboat come awake the moment its paddlewheels shut down. Nor could you shut out "a hoarse or screaming sound, such as that of a siren," by stopping your ears with a finger or a wad of waxed cotton, as city folk knew from experience and any schoolgirl might learn from an Oxford text of simple home experiments in practical acoustics. A sophomore at Brown University in 1860 recorded in his lecture notes on physiology that "In regard to the purpose answered by the semicircular canals, we are entirely ignorant," but his professor did presume that the ear was so tireless a sound-gatherer that it needed a permanent (anti-tinnital) arrangement to prevent it from listening in on its own amplifications, so that the "threads of the cochlea are supposed to be connected with a provision for putting a stop to the vibrations, communicated to the internal ear, and thereby preventing the prolongation of sounds from the mere action of the original."[139]

Otologists would soon prove how crucial were the ears to a personal sense of place and of space: subjects fitted with "sound helmets" or put into "sound cages" demonstrated the ears' skill at locating sound within any of twenty-six sectors above, below, or around the head, so that the ears were at once ever-open and ever-orienting. They kept you balanced, too, for the vestibular system of the inner ear—was it the semicircular canals?—rescued you from vertigo upon lying down, rising up, or making your way across the uneven surface of a revolving planet. Unless attention were prolonged, in which case—wrote one French psychologist—"a constantly increasing cloudiness of the mind, finally a kind of intellectual vacuity, is frequently accompanied by vertigo." Or unless, like Vincent Van Gogh, you suffered from an inner-ear disease that Dr. Prosper Menière in 1861 had linked to vertigo, a disease that reached "fairly epidemic" proportions in New York City in 1872. Open, orienting, and counterpoising, the ears were ever-active, sending impulses to the brain and comparing the phase relations among different sounds (Rayleigh, 1907) to arrive at a sense of where and how you were.[140]

And who. You were, "with a little blunt point, projecting from the inwardly folded margin" of the ear, an evolved primate, sporting the "vestiges of the tips of formerly erect and pointed ears." According to Darwin's revisions of this segment of *The Descent of Man*, "The whole external shell may be considered a rudiment, together with the various folds and prominences (helix and anti-helix, tragus and anti-tragus, etc.) which in the lower animals strengthen and support the ear when erect." Could you also waggle your ears, you were in receipt of the "transmission of an absolutely useless faculty, probably derived from our remote semi-human ancestors." Darwin intended these comments as impartial contributions to a natural history of bodies. Nor was any moral judgment intended (or so they said) by phrenologists who identified the area just above and behind the ears with the faculty of Destructiveness, itself a kind of evolutionary impulse. As the Scottish lawyer George Combe explained in the fourth edition of his *System of Phrenology* (1842), Destructiveness was the natural desire to eliminate life's obstacles while maintaining the fortitude to deal with life's tragedies: "We are surrounded by death in all its forms, and by destruction in its every shape; and Nature, by means of this faculty, steels our minds so far as to fit us for our condition, and to render scenes which our situation constrains us to witness, not insupportable." Given that "the auditory nerve has a more intimate connection with the organs of the moral sentiments, than with those of the intellectual faculties," it was strategic that Destructiveness should be so intimate with the ears, for

"A certain degree of obduracy of feeling, regardlessness of suffering, and indifference to the calamities of our race, is absolutely necessary to render existence tolerable in this world of mingled joy and woe." Yet it was also the case that Destructiveness, strong in carnivores, could make the voice harsh, and that cold-blooded killers might mistake the strength of their own faculty for the harsh voice of Satan whispering in (or from just above) their ears, "kill him, kill him." Destructiveness, after all, had been si(gh)ted by the Franco-German anatomist Franz Joseph Gall only upon scrutinizing the skulls of parricides and homicidal highwaymen, the openings of whose ears rode low on their skulls under a cranial expanse he called "*le penchant au meurtre*." Although Gall would rue having chosen such an inflammatory phrase for such basic needs as self-defense and killing-to-eat, and although Darwin was never as cocky as Social Darwinists with regard to what Herbert Spencer called "the survival of the fittest," by the 1870s it would be almost natural for that new subspecies of *homo sapiens*, the criminologist, to link physical deformities or protuberances to moral deficits or deviances. Among the foremost of those tells, those subtly revealing irregularities, were misshapen ears.[141]

Ears had a thicket of associations in European folklore and systems of face reading. Large ears denoted foolishness, vanity, stupidity, folly, loquacity, and/or long life; miniscule ears, or ears too close to the head, denoted insanity, cunning, deceitfulness, and/or smallness of spirit. Flat scaly ears betrayed a constitutional coarseness, grisly ears a melancholy nature, hairy ears a "good and ready" hearing, gently squared-off ears a greatness and/or purity of soul. Imbeciles, who tugged at their lobes in slack-jawed ignorance, had drooping ears; the feeble-minded had incomplete ears, with narrow openings or missing lobes; the guilty, secretive, and/or ashamed had inflamed ears, and so on. How one thought and/or felt was communicated by and through the muscles of the face until habits of expression (such as envy, which lengthened and narrowed the ears) shaped every inch of the countenance. The physiognomist Joseph Simms took this logic to its extreme in the 1870s with his discovery of such features as Characterioscopicity (the power of discerning character, located at the inner corner of the eye next to the slope of a long nose), Sentinelitiveness (the disposition to watch, with a long nose and full forehead), and Tonireceptionality (the ability to appreciate music, with a rounded ear adapted to catch round and wavy musical tones, as opposed to "an ear lying flat on the side of the head, or angular and pointed in form," which could only receive noises, "square, angular, rough, uneven, or of no describable form at all"). Less extreme but no less confident was Miriam Anne Ellis, who claimed in 1900 that for the typing of personality and identification of persons there could be no surer bet than the ear, since "there is but one feature that cannot be altered at will, and that is the ear. Any attempt at disguising it can be detected and of course it cannot be pulled out of shape and place." Boxers and their sidemen might demur (the spongy or "cauliflower" ears of pugilists had been familiar since Greek times), as might explorers who had encountered tribes who stretched their lobes with heavy stone rings. Ellis's claim for the superior steadiness and remarkability of the ear was distinctively modern, credible only when earless lepers no longer begged on the streets, when slaves—ears notched like cattle—no longer worked the fields, when libellers and thieves no longer had their ears lopped off in public squares, and when wigs no longer hid ears scarred by smallpox or ravaged by syphilis.[142]

Jonathan Swift's 1704 promise of a "general history of ears" had gone unmet due to his bouts of deafness from Eustachian obstruction or, more probably, to the general opacity of wigs, which flapped over the ears of journeymen and masters, shopkeepers and merchants, lawyers and gentry throughout the eighteenth century. Patrician ladies and Jewish wives also wore wigs; most other women and older girls wore lace caps or linen bonnets that covered their natural hair and ears. As wigs vanished from all but the stage, the courtroom, and orthodox Jewish households, and as bonnets disappeared from all but churches and cemeteries, the flowing tresses favored by the Romantics continued to obscure ears male and female alike. By the 1870s, however, women's fashions had swept hair up and back while the industrialization of war and work made shorter hair a saving grace for infantrymen and for factory laborers of both sexes who battled to keep clear of large, rapidly spinning gears and rollers. Earrings returned to the ballroom and salon, leafed in gold, calling attention to the neck and ears through oddly industrial or broadly imperial themes (shovels, hammers, exotic hummingbirds). Ear tubes or trumpets for the hard-of-hearing, which since Swift's time had had to wind their way around and through layers of hair of one sort or another, could now be reduced in size to that of the auricle itself, or hidden in clothing or furniture so that the visible calm of the ear was not compromised. Listening at the telephone, or wreathed by headsets at telephone switchboards, or leaning toward the horn of a phonograph, or waiting upon spirits to speak through dancing trumpets at séances, or sitting in darkened theaters where musicians and magicians demanded one's whole attention, it was best if the ears were unimpeded. Standard eyeglass frames, unlike monocles or lorgnettes, had to be hooked around unencumbered ears, an imperative that took on force in the 1890s once lenses could be ground for degrees of astigmatism as well as far- or near-sightedness. So optometry and otology, fashion and physiology, spectacle and psychology converged on behalf of an open, exposed ear.[143]

Newly exposed, the auricular became more roundly oracular, confirming identity, revealing inner states and pre-destinations. Of the forty-four illustrated heads in an 1832 pocket reference on physiognomy, more than thirty had ears obscured by hair, including the frontispiece of the Swiss physiognomist, Johann Lavater; the three pages devoted to ears had recourse to Aristotle for evidence that thieves, like monkeys, had small ears, and dullards and fops, like asses, large ears. By 1880, the *Annual of Phrenology and Physiognomy* was using the unobscured ear as a centerpoint from which to graph the normative skull, and the Florentine critic Giovanni Morelli was attributing works to Old Masters based on how each had rendered a subject's hands and ears. "For two, yea, three centuries ears were hidden, covered up out of sight by a mass of hair, powdered and otherwise," wrote a self-professed "aurognomist" in *The Strand Magazine* for 1893; now that ears lay open to sight and photographic comparison, it was clear that "Twins, as much alike as the proverbial two peas, invariably differ aurally," leading one to suppose "that there must be some especial sensitiveness in the organism of an ear which is affected by the mind." That same year a resident of Baker Street easily solved the mystery of "The Cardboard Box" after glancing at the freshly severed ears within said box. "As a medical man, you are aware, Watson, that there is no part of the body which varies so much as the human ear. Each ear is as a rule quite distinctive and differs from all other ones." With two bloody, unpaired ears as clues, Sherlock Holmes quickly deduced the backstory, enfolded motives, and helical plot.[144]

Holmes, or Dr. Watson, or their literary executor Arthur Conan Doyle, had been following the publications of the French police clerk Alphonse Bertillon and the Italian criminal psychologist Cesare Lombroso, who identified and classed criminals, imbeciles, and degenerates by their "atavistic stigmata," such as ears too low on the skull, enormous or tiny or shaped like a handle. On Bertillon's photographic grid and Lombroso's moral compass, the ears stood out precisely because they were psychologically inconspicuous. "[P]rincipally on account of its motionlessness, which prevents it from participating in the play of the features, no part of the face attracts less notice," wrote Bertillon; "our eye is as little accustomed to observe it as is our language to describe it." As a result, neither the innocent nor the culpable nor the hysterical thought to disguise the ear, and none—unless beset by scrofula, insect bite, ear ache, or leeches—gave the flesh of the ear a moment's thought. Ears were in this sense more pristine than other facial features, their exposure all the more expositive. Though the lobes be pierced for earrings, ears remained intact—their silhouettes basically unchanged since birth, their unique concha fixed since puberty. On this stasis, and on a mounting anthropometric archive of thousands of faces taken in profile against a standard backdrop and measured in each dimension by the millimeter, the new scientists of the (ab)normal almost went so far as to claim, as had Amédée Joux for his doctorate in medicine, "Show me your ear, I will tell you who are you, where you come from, and where you are going." The ear, declared Bertillon, "owing to the many hollows and ridges which furrow it, is the most important means of identification in the human visage." From the physiognomies of 832 delinquents, 390 criminals, 868 soldiers, and sketches of the heads of "savages" and animals, Lombroso concluded that "28% of criminals have handle-shaped ears standing out from the face as in the chimpanzee," and numerous other culprits had "misshapen, flattened ears, devoid of helix, tragus, and anti-tragus, and with a protuberance on the upper part of the posterior margin (Darwin's tubercle), a relic of the pointed ear characteristic of apes."[145]

Calipers in one hand and a jeweler's loupe in the other, criminologists looked for lobules, pinna, tragus, and external meatus to be, when not leftovers of the past, indicators of a definite present (identity) and predictors of an unmistakable future (character). Because they changed so little while receiving so much, ears were federated with memory and indelibility; because they were at once motionless and labyrinthine, they could be heralds of connivance and conspiracy. The director of an Italian insane asylum, Louis Frigerio, summed up what was known to the charactearological sciences in 1888. Prisons and asylums abounded with inmates whose ears were so large that their heads seemed to have wings. Monkey ears—shaped like handles—prevailed among thieves and ruffians. The auricles of degenerates were situated asymmetrically vis-à-vis the skull. Like the ears of infants, the ears of malefactors and madpeople had incomplete helices, a conical tragus, and small or imperfectly detached lobes. The angle between the auricle and the temporal bone diverged widely from the norm among the crazy and criminal. All told, the structural defects of external ears had functional consequences; they made it hard, Frigerio implied, to attend to the Good Word and the rule of law.[146]

Plug-ugly ears, to be blunt, impressed upon the face such a "disgraceful" countenance (wrote the psychiatrist Bénédict Morel) that they put the moral graces out of reach. They were both source and stigmata of deviance. "In the answers to my Questions issued to medical officers of prisons," reported Havelock Ellis in *The Criminal*, a text that warranted five editions between 1890 and 1914, "I found that the prominent ears of criminals were more generally recognized than any other abnormality." Better known for his sexology, Ellis was one of the few to attempt a causal explanation for the nexus between irregular ears and irregular behavior. Faced with the surprising fact of the "extreme frequency of anomalies of the ear among all classes," Ellis reasoned with Darwinian logic that our pinna, immobile and of negligible acoustic value, were fading away. During this awkward process of "retrogressive dissolution," our outer ears had become "very sensitive to the slightest nervous disturbance," disturbances most common among the wayward and the woeful. So the ear was at once the most impressionable of features and the most outwardly projective.[147]

Centuries of ear-pinching by schoolmasters, ear-twisting by governesses or great aunts, ear-pulling by bullies and big sisters, and ear-boxing by angry parents or foremen had now to be put to an end, for these were no longer deemed so fleeting a violence. Even were that outcropping of flesh, *in toto* or in tubercle, an atavism, it had an intimate bond with character and the character of sound. Thomas Edison attributed his own near-deafness to having been hauled up by the ears into a baggage car of the Grand Trunk Railway, for which he worked as a newsboy and typesetter. Or to having had his ears severely boxed for spilling some chemicals in that car. Either way, by the age of twelve he had been left with just enough hearing to make out the click of Morse Code at a subsequent job as a telegraph operator but with deafness aplenty to shield him from small talk and street noise, so that he had "no doubt that my nerves are stronger and better today than they would have been if I had heard all the foolish conversation and other meaningless sounds that normal people hear." Manhattan was for him "rather a quiet place," his bad ears sheltering his nerves: "Most nerve strain of our modern life, I fancy, comes to us through our ears." Was this why, among his patents for microphones, megaphones, an aerophone to project sound for several miles, and a phonomotor to convert the vibrations of speech into mechanical force and "talk a hole through a board," he never perfected a hearing aid or "aurophone," despite bundles of imploring letters? He himself could hear, he believed, anything worth hearing. Asked to investigate the sources of the noise of elevated trains, he ascertained that the worst offender was the screech of wheels crossing joints in the rails, a sound obvious to him but hidden from other experts who "heard too much general uproar to make it possible for them to make sure of details." Instead of devising a sound-insulated headset for telegraphers, he invented a sounder that responded only to incoming signals, significantly reducing the general uproar in central offices. At phonograph recording sessions, he listened through his teeth or with an ear trumpet for plain melodies and simple harmonies played or sung only with "pure notes, without extraneous sounds and the almost universal tremolo effect" that, like tones too high or loud or complex, made for unwieldy vibrations and, in his head, a general uproar that turned him against the blues, Rachmaninoff, ragtime, and operatic vibrato. He admitted that he had not heard a bird sing since he was twelve, but from his deafness he drew the comfort of a good night's rest, since he could not be aroused by footsteps on creaking stairs or the "demonic laughter of a cat" and so "Slept as sound as a bug in a barrel of morphine."[148]

Facts be known, Edison slept little, taking cat naps on a cot; anyway it was one thing to be the Wizard of Menlo Park, another to tend to a mechanical crusher ten hours a day or to a crèche of young children at home, where good hearing could save the day. Physicians and pedagogues rushed to defend the outer ear because it was the parapet to the most vigilant of the senses. While the interior had to be kept clear of invasive bedbugs, roaches, and earwigs, of children's gardens of cherry pits and ear-shaped lima beans, and of the hairpins/earspoons used by nurses to fish out the bugs and beans, the exterior was being reconceived as the tip of one of the body's tenderest parts. Blushing or "burning ears" were now to be understood as "a common experience among emotional and neurotic persons," and everyone recalled having had their ears set upon in childhood by bullies, teachers, or parents. A box on the ears not only stung and stuck in one's memory; it could cause tinnitus and rupture an eardrum, leading to inflammation and chronic middle-ear suppuration, then to blood poisoning, meningitis or, at least, days of deafness and periods of autophonia. Autophonia was "a condition in which the patient feels as though his cranial cavity contained air instead of brains" and became an "echoing cavern" for a "wild uproar of tuneless noises" from the flow of blood through carotid and temporal arteries, "violent sounds, consisting in part of slamming, cracking, snapping, knocking, clashing, rumbling, humming and hissing noises." Parents, aunts, uncles, teachers and headmistresses were to be cautioned against any affectionate pulling or disciplinary pummeling of so tempting a grab-hold.[149]

Given the exposure of, and heightened concern for, the outward ear, it stood to reason that it would become an object of cultural anxiety. Before photography, wrote Miriam Anne Ellis, "No one had known, apparently, how big even the smallest ear can be. 'What ears!' is the common exclamation on looking over family treasures of photographs of the different members of all ages." Such exclamations were hardly innocent when adjoined to fears of feeblemindedness, or to character assessments based on ear size and shape, or to stereotypes of "jug-eared" drunken Irishmen, droop-eared slothful Blacks (from a continent shaped like an auricle), and "bat-eared," vampirical Jews. The social complications and physiognomic implications of having ears that stood out could prompt a mother of any stripe to send away for "Claxton's Patent Ear-Cap for Remedying Prominent Ears, Preventing Disfigurement in after life," by which was meant, presumably, adulthood. "My baby (a girl) is five months old and her ears sit out so far from her head," wrote Mrs. Robert E. Robinson of Virginia to the U.S. Children's Bureau in 1915. "I am anxious to know some way to train them to lay flat. I tried her cap but she is such a lively child she gets the cap all away." The Bureau advised her to try wide ribbons, a cheap if temporary solution for what an Italian psychiatrist, Enrico Morselli, had identified in 1891 as "dysmorphophobia," the shame of feeling disfigured. This new diagnostic category would justify more permanent intervention once the ears began to reach structural maturity around the age of ten. With antisepsis, anaesthesia, and electric light at their disposal, aural surgeons had become bolder, repairing torn ear drums, performing mastoidectomies to drain middle-ear infections, and contemplating maneuvers to deal with otosclerosis. Once dysmorphophobia reared its ugly head, they were ready with new, less scarifying techniques for corrective or "aesthetic" surgery, taking their first swipes at out-sized ears in 1881 when a young New York physician, Edward T. Ely, excised two square inches of skin from each ear of a twelve-year-old who had been ridiculed "on account of the prominence of his ears." Ear jobs, nose jobs, lip jobs may have appealed most to wealthy minorities who hoped quietly to face down discrimination, but aesthetic surgery attracted as well the settled majorities who found themselves comparing their own features, reflected in household mirrors or off the plate glass windows of department stores, with icons of the handsome, the ugly, the sick, and the criminal. During the last half of the nineteenth century, these icons multiplied with the diffusion of lithographed cartoons, photographic blow-ups, half-tone posters, large advertising sheets, and mass illustrated dailies, then with billboards, magic lantern shows, and early motion pictures, all of which magnified the face and its features.[150]

Crow's feet, tip-tilted noses, inverted lips, more conspicuous now despite the application of a burgeoning variety of facial powders, creams, and glosses, could also be corrected, especially by enterprising cosmetic surgeons like Charles C. Miller of Chicago, who in 1905 was touting the removal of wrinkles, the untipping of noses, and the upturning of downcast lips. Still, much of his practice was taken up by "Operations for Correcting Outstanding Ears"—tucking them in closer to the skull, uncurling the cartilage, shortening the lobe. His earwork was mostly on women, for their hair fashions made them more self-conscious about ear size and shape, and they had frequent need of repairs from the aggravations of earrings. In the flurry of his surgeries, Miller never worried about upsetting the acoustics of the ear, for the cartilage of the auricle, as he understood it, had negligible functional importance, and one could hear "just as well, almost, without it."[151]

Almost. For millennia, people had been cupping hands to auricles to catch at sound, a technique no less effective than metal ear trumpets. To this day, a palm cupped around an ear may be iconic of distance or deafness, a signal for attentiveness or a demand for repetition. As contoured by criminal anthropology, fashion magazines, cosmetic surgery, and the optics of consumerism, the outer ear had become a visual and tactile site—plotted and photographed, pricked and sculpted, bedecked and bejeweled, and, in private, love-bit and nuzzled. But the auricle was also a funnel, and in response to the visual complexities of industrial life and the visual perplexities of the city, the ear—outer and inner—came to be viewed as a catch-all, ever-open, ever-sensitive, ever-solicitous, and increasingly in peril from what city folk were starting to call the Age of Noise. "It goes without saying," wrote Whitelaw Reid in 1878 from behind the thick brick walls of the new home of the *New York Daily Tribune*, "that the ear, in this conflict with modern life, is the most defenseless of all organs.... But all our progress, in this city, has been toward an increase of the general clamor." As editor-in-chief, his own ears were safe so long as he remained in his palatial quarters on the top floor of the tallest office building in the world. Should he descend some two hundred feet, by elevator, to the beer saloon in the basement, his ears would be blasted by "the harsh roar of the streets, the grinding of ponderous wheels over flinty pavements, the loud rattling and creaking and jingling of vehicles of all ages and conditions of decrepitude, the clatter of horses' hooves, the cries of drivers, newsboys" (his own newsboys among them) "...and these noises can never cease to be discordant until Mr. Edison, say, invents some sort of a musical funnel which shall gather up all the sounds of the streets, and, rearranging their clashing waves, shall pour into every home the sparkling airs of the last successful opera." But Edison invented no such funnel, and down at street level the worst of the discord pounded until three a.m. at the ears of his pressmen who worked beside the paper's new, faster, louder, steam-driven rotary presses.[152]

7. Ills

Egyptian workers, wrote Pliny the Elder, had gone deaf after laboring for years beside the cataracts of the Nile, and Egyptian medical texts had anciently associated occupations with their ailments: weavers gasped for breath, dyers went blind. Greek and Roman physicians described the lameness of tailors, the varicose veins of priests, the bad eyes of blacksmiths, the asthma of miners, the stunted lives of metalworkers, and lesser afflictions of fishermen, farmhands, horsemen, scholars, wet nurses, wrestlers. Administering to a continent loud with new construction, Renaissance physicians remarked upon the hernias of carpenters, the congested lungs of plasterers, and the debilities among tin, lead, coal, silver, and iron miners. During the 1690s, Bernardino Ramazzini at the University of Modena offered the first courses on occupational disease and in 1701 wrote the first full-scale treatise on *Diseases of Workers*, yet in four millennia barely a paragraph had been published about the acoustic trauma of workers. To his credit, Ramazzini discussed the hearing of copper-workers who, bent double over their collective hammering, became both hunchbacked and deaf. "To begin with, the ears are injured by that perpetual din, and in fact the whole head, inevitably, so that workers of this class become hard-of-hearing, and if, they grow old in this work, completely deaf." He compared this to the experience of Pliny's Egyptians: "For that incessant noise beating on the eardrum makes it lose its natural tonus; the air within the ear reverberates against its sides, and this weakens and impairs all the apparatus of hearing." Of the deadened ears of artillerymen at their cannon (earlier noted by Ambroise Paré), of quarrymen at their blocks of stone, or

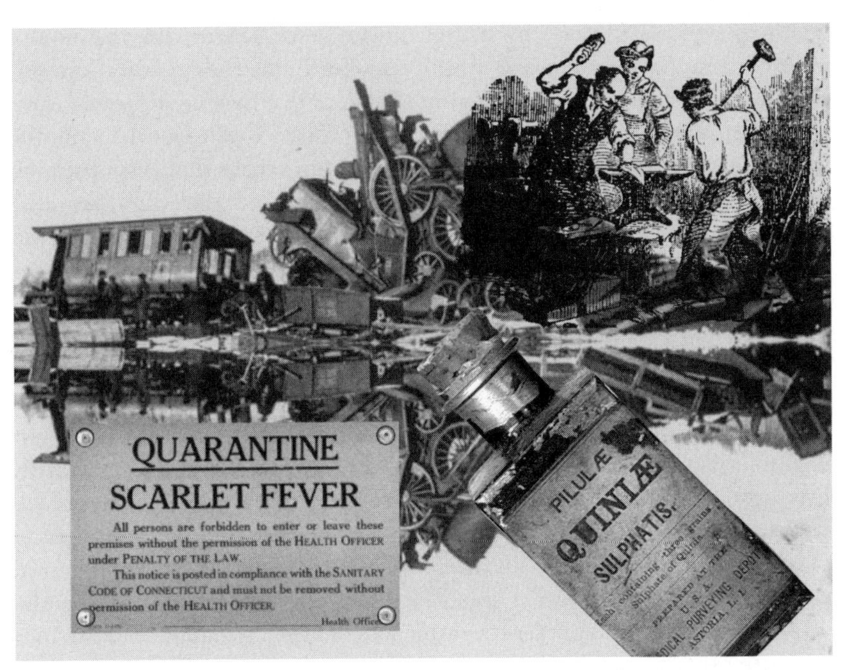

SOUNDPLATE 11

of blacksmiths at their anvils, Ramazzini said nothing. The poorer class of Jews, bent over patchwork in dark rooms, did develop weak eyes and ears, but their deafness came, it seemed, from bad air and bad posture, not from the incessancy of needles. That was all there was to say, even in a second edition, about noise.[153]

Why? Pre-modern communities had not been oblivious to noisy trades, which were exiled to the outskirts of town or subjected to laws restricting the hours during which, for example, grain could be ground, oil pressed, soap boiled, tin and copper hammered, swords forged. And it was not that manmade sounds were innocuous in a world without gunpowder, steam engines, or dynamos: anvils had branged with ear-splitting force; fulling hammers thudded with much the same vehemence when driven by waterwheels; the screak of old windmills could be heard for leagues across Dutch polders. Rather, mortality conspired with otology to hide the numbers of the deafened: workers in noisy trades were usually in dangerous trades and died young, before their loss of hearing became as obvious as among a kettle of coppersmiths. It was also that aging was bound, in emblem and tale, law and medicine, with loss of sight and hearing, so that doctors would not think to listen beyond the typical debilities of age to account for the deafness of an older worker. It was also that standards for the assessment of degrees of hearing loss were lacking, while work-related deafness typically arrives by degrees, over years. It was also that adults who became hard-of-hearing did not make themselves readily known. Half-aware of half-missing sectors of sound, of dullness or slippage rather than stark silence, they compensated by repositionings toward the better ear, by featural and gestural reading, by deft anticipation of responses, by pretences of comprehension. When these tactics failed, they often assumed a protective surliness that masked their sonic isolation. Descending to the bottoms of mine pits during the 1840s, British reformers found miners impatient with questions about their working conditions, an impatience that the paper-and-pencil men attributed to a lack of vocabulary, to underclass gruffness, and to the constant tussle of the tunnels; the recalcitrance was also, I suspect, a problem of hearing.[154]

Many were the causes of deafness, according to the Cheltenham physician John Fosbroke in 1831: cold air, variable climate, nasal polyps, tonsillitis, fever, bladder infections, measles, catarrh. His blunderbuss approach to the ear would not have been memorable had he omitted his discussion of blacksmiths' deafness, a "deafness from extreme noise" that "creeps on them gradually, in general at about forty or fifty years of age." Theirs, he wrote, was a slower, more irretrievable loss of hearing than that of marines who fire off twenty-four-pounders during naval battles and are temporarily stunned by the explosions' echoes among the thick low timbers of the gun decks; less dramatic, the curtailed hearing of blacksmiths had gone medically unremarked except for a Swedish thesis unknown to Fosbroke. Charles Thackrah, who had studied medicine in the company of John Keats and, like the poet, died of pulmonary consumption, was understandably preoccupied with the lungs of the laborers and artisans around Leeds whose work-related ills he studied; even so, in 1832 he added to the ranks of deafened coppersmiths and blacksmiths the grinders who sharpen textile shears to the screech of large revolving stones. Thackrah also proposed, against the grain of industrial capitalism, that the noise of machinery in general could be injurious to the ears of workers.[155]

Yet not for another thirty years—not until the issue of attention had arisen in experimental psychology, and not until life itself seemed to demand unswerving focus—was the noise of the workplace constellated with other threats to workers' health. "In this age of progress and improvement, it is necessary to have our senses perfect, even in the most ordinary business transactions," wrote August P. Lighthill in 1861, and the deaf found themselves at a disadvantage; in his *Popular Treatise on Deafness*, Dr. Lighthill repeated the old prejudice that the blind were "generally cheerful, frank, and sociable," while the hard-of-hearing, constantly at a loss for words and frustrated by fogs of sound, became "suspicious, morose, melancholic, and unsociable," which was no way to keep (a) company. And the number of irritable half-deaf citizens seemed to be growing. Undefended by fat or cell tissue, the ears were exposed to greater vicissitudes of temperature than the rest of the body, so not only concussion but extremes of cold and heat might compromise the hearing of gunners, of workers in machine shops or foundries, and anyone in the vicinity of steam-powered trip hammers. Noise on the job became less of the joking matter it was in 1854 with the Irish novelist Charles Lever's *The Dodd Family Abroad*, where the Belgian industrial town of Liège was depicted as "the noisiest, most dinning, hammering, hissing, clanking, creaking, welding, smelting, and furnace-roaring town in Europe," and where, when the Liègois left off their clanking, they hied to concert halls to listen with dulled acoustic nerves to the newest "din and clangor" from the trumpets, drums, and bassoons of Verdi operas. At century's end, when investigators went out as participant observers into industrial settings, the noise they heard was grim. Walter Wyckoff, a political economist who did stints in stockyards, lumber camps, and machine shops, found the working class speaking with "gruff voices in antiphonal sentences replete with strange oaths." That gruffness and profanity he understood as reactions to the loudness of blue-collar work and the "turmoil of production" in factories, where he heard "the deafening roar of far-reaching machinery, as, with wheels whirling in dizzy motion and the straps humming in their flight, it beat time in deep, low throbs," and buzz saws "rising to a piercing crescendo and then dying away in a sob." Bessie and Marie Van Vorst, two gentlewomen from New York who tried out life north and south as factory girls, heard the "rasping and pounding" of machinery as a "burning din" that workers struggled to outsound, drawing "vocal inspiration from the bedlam of civilization." They themselves were dazed, their "ears strained to bursting with the deafening noise." At a shirt-making plant, "Nothing was as fatiguing as the noise," which stayed with workers long after leaving work. At a southern cotton mill, louder even than a northern shirt factory or a middle-state carpet factory, the Van Vorsts stood stock still: "This is chaos before order was conceived: more weird in that, despite the din and thunder, everything is so orderly, so perfectly carried forth by the machinery": noise as the warp of working life.[156]

Risks of accident, which seemed graver and greater with the arrival of each new piece of "self-acting" machinery, inspired keener concerns for hearing deficits, since mistaken orders or unheard alarms might result in crippling injuries, civil suits, and (first in Germany, then across Europe and North America) demands for worker's compensation. Without good ears, how could engineers hear the crackling noise in the condenser as an engine heated up? Coal miners, the cracking noise at the onset of a "bump" when coal slips suddenly from the walls? Railroad firemen, the hints of a boiler about to explode? Teamsters at train crossings, the hoot of an oncoming express? Most alarming were the accident statistics for railroads, America's largest industry in 1900 and, as tabulated by Mark Aldrich, the largest peacetime cause of violent death. Union records between 1882 and 1912 showed that accidents on the job accounted for at least half of all deaths of blue-collar rail workers. Thousands of passengers and pedestrians were also killed or maimed each year aboard trains, or walking beside them, or crossing in front; in the early 1890s, 600 Chicagoans a year were being killed at grade crossings. Nationally, between 1887 and 1900, more than 86,000 had been killed and 469,000 injured in rail accidents. In 1913 alone, 25,000 would be killed, 300,000 seriously injured, another 2,000,000 hurt: a fatality rate of 263 per million Americans, an injury rate of 1 in 50. (For comparison: in 1909, horse-and-buggy accidents accounted for 3,860 deaths; in 2005, auto fatalities occurred at a rate of 147 per million, injuries at a rate of 1 in 110.) Casualties, to use military language, were much greater than on European tracks, which had crossing guards and gates; across prairie and plateau at half a mile a minute or more, American trains relied rather on bells and whistles to warn off the foot and horse traffic. As large towns sprang up around depots, train whistles began to be heard as nuisances, and trainmen also began to have problems with them, for the whistles, which had no consistency from one line to the next, might be blasted at any time for any number of reasons. Four short blasts on a whistle might mean "recall the flagman" or "note the flag on the engine" or "call for a switchman" or "call for fuel." Since about 1880, German otologists had correlated the years of a railman's service with the severity of his hearing loss, which they attributed to engine noise and "the shrill tone of the steam-whistle." Could older workers hear the danger signals of their own equipment? A 1914 U.S. report claimed that half of railroad engineers had lost some of their hearing to the blast of whistles, "the roar of boiler fires," the squeal and hiss of air brakes, and the "general pounding of the train on the rails." Such hearing loss had a strong bearing on the number and severity of rail accidents. The causal arrows for accidents thus turned in tight circles: the noise of rail (as also mining) operations could cause the very deafness and fatigue that lay behind the high numbers of casualties.[157]

Life insurance companies in 1895 denied coverage to seventy occupations, including miners, railroad men, and steelworkers (in 1907, one quarter of U.S. Steel's full-time employees were injured on the job). These were among the noisiest of occupations, the most deafening of which was boiler-making, where men worked long hours around and within sharply reverberating cylinders of sheet iron as boiler plates were beaten and riveted into shape. The longer he stuck to his profession, the less a boilermaker could hear and the more vehemently he would profess to be able to hear better in noise. A physician of Maine found the staunch profession an earful; after examining forty of this "superior class of intelligent mechanics," Dr. Eugene Holt determined that most could not hear the tick of his watch from the two meters audible to normal ears, and not one had a special talent for hearing a tick, a tuning fork, or a finger-snap in noisy places. Amid the banging, boilermakers had simply grown accustomed to shouting to each other at penetrating pitches that they mistook for ordinary speech. The men knew, to be sure, that their ears were under siege and had tried cotton wool and hard rubber earstoppers to no avail, for these irritated the meatus while scarcely quieting the high-pitched metallic noise around them. Holt concluded that "all persons working at the same occupation in a noise would finally become more or less deaf." Thomas Barr, a Glasgow lecturer on ear diseases who focused upon the "injurious effects of loud sounds upon the hearing," concurred. Of the hundred boilermakers he tested in 1884–1886 at two shipyards, few could hear as well as his control groups of shipyard letter-carriers and of "normals" who had never passed a day in a shipyard. Like riveters and locomotive engineers, the boilermakers were "particularly liable to labyrinthine mischief, from undue irritation of the auditory nerve," which produced permanent hearing loss and terrible tinnitus. "Imagine," wrote a boilermaker of Cincinnati, "a gang of men under a sooty old boiler, often while the furnace is still hot, cutting out or driving rivets overhead, sometimes lying on their backs, or cramped up in a small uncomfortable space, with dust, soot, or red hot scales filling their eyes every now and then, and by way of variation have a brick or tile fall on their heads occasionally." Then imagine an irregular, almost continuous banging of hammerheads on metal and the driving of metal rivets through metal sheets. The men might kid each other about the noise ("DOCTOR: What you need is livening up. You want noise, stir, rattle, bang. Think you can find time to go somewhere and find it? PATIENT: Don't know. I'm a boiler maker and we're pretty busy now"), but they knew that in the long run they would lose much of their hearing. By 1901 their union was pushing for legislation to reduce and regulate workplace noise.[158]

Meanwhile the list of ear-nasty employments grew, some of them adjuncts to the shipbuilding which was driving so much of nineteenth-century industrialization, some of them adjuncts to the capitalism and imperialism that were driving the shipyards. As occupational medicine and public health became fields of their own, journals and international congresses began to fill with reports on the irreversible hearing losses of sailors buffeted by cold winds; of artillerymen sound-blasted by cannon; of divers whose eardrums collapsed under swift changes in air pressure; of telephone operators, their ears shocked by electrical shorts; of shoemakers hammering at their lasts; of cellists rehearsing in front of symphonic brass; of machinists stamping out metal; of coopers, carpenters, and carriage-makers with skreeling band saws, belt sanders, and power planers; of hunters and infantrymen concussed by shotguns and rifles; of linemen at steam-powered printing presses, linewomen at textile or papermaking plants. In many of these employments the damage to hearing was explicable according to a theory invoked by the American aurist D. B. St. John Roosa with regard to telegraphers: "There can be, I think, no hesitancy in believing that the continual recurrence of one kind of sound, having no musical, but, on the contrary, an unpleasant character, must at last cause a congestion of the ultimate nerve fiber in the cochlea. The incessant shock or concussion made upon the drumhead by the blows from dozens or even hundreds of hammers upon vibrating plates, must agitate these fibres in such a manner as to put them out of tune, as certainly as the constant use of the piano will at last loosen its strings." While society was learning to refrain from boxing the outer ear, industry was remorselessly punching out the interior.[159]

Roosa, who had treated seventeen hundred cases of ear disease by 1874, knew that not all damage to hearing was the result of physical blows or electric shock. Icy air and cold water were prime suspects in nineteenth-century etiologies of hearing loss, as were the common cold and the sinus problems of catarrh, whose mucus could clog the Eustachian tube or snake into the middle ear, with lasting harm. Roosa was also alive to the many infections internal to the ear itself, above all *otitis media*, infection of the middle ear, which, if untreated or poorly managed, could leave a patient with permanent hearing loss. There were as well some antagonists whose assaults upon the ear were less immediate and, in Roosa's day, less understood: lead, mercury, solvents. Lead, which at high blood concentrations is now known to attack the cochlear nerve and cause auditory processing deficits, was everywhere: in the water (pipes), the air (white paint), food (yellow coloring, tin cans, glazed pottery, foil wrappers), drink (wine, rum, apple cider, and moonshine aged in lead-hooped casks, milk delivered in cans with soldered seams), cosmetics (eye shadow, astringent lotions, Laird's Bloom of Youth), aphrodisiacs, and a few medicines (nipple creams, suppositories, eye salves, abortifacients). Already high among the general populace, concentrations of lead in the blood would have been highest among lead miners, glassmakers, pottery glaziers, calico printers, pipe fitters, paint mixers, welders, and workers in tin can factories. But the gradual ototoxicity of plumbism was masked by more flagrant symptoms: blue gums, wristdrop, palsy, anemia, confusion, colic. Long before a sufferer complained of hearing loss, she would have been screaming from abdominal pain, so it was rare for an otologist to mention, let alone diagnose, lead poisoning. As health officials and reformers began a century of agitation for laws limiting industrial and domestic exposure to lead, the setting of safe blood levels came under debate; every decade since, research has proven the neurotoxic and ototoxic effects of smaller concentrations.[160]

Mercury was less ototoxic in the form of calomel (mercurous chloride, for constipation) or blue pills (elemental mercury, for syphilis or depression) than in the vapors of methylmercury, which turned Lewis Carroll's Hatter (and many another hatter, gold miner, and dental technician) Mad with tremors, headaches, short-term memory loss, and insomnia. Through mining and industrial runoff, methylmercury flowed into lakes and streams and entered into drinking water and fish at market. Again, as with plumbism, the ear-harm of mercury poisoning was obscured by gaudier, deadlier symptoms, as was the case for another heavy metal, arsenic. "King of Poisons" and weapon of choice against tyrants and tyrannical husbands, arsenic had been used in smaller doses against unwanted pregnancies, to cope with syphilis and, in Fowler's Solution, as an anodyne for skin ulcers, neuralgia, and fatigue. By 1900 physicians were warning of arsenic poisoning from too-regular use of Fowler's, in the wake of earlier alerts about green wallpapers that emitted arsenical vapors. Less evident was exposure through tobacco smoke from leaves sprayed with arsenical insecticides or through the leaching of arsenic from embalmed bodies into cemetery soil and thence into groundwater. Ingestion or inhalation of small but cumulative amounts of arsenic, whether from curtains or wallpaper, from water, wine, or beer (during a British epidemic of 1900), or from cigarettes, cosmetics, concert tickets, or Christmas tree ornaments, could cause hemorrhage, lung cancer, and, we now know, devastation of the auditory nerve.[161]

Subtler still were the ototoxic effects of the toluene, benzene, and carbon disulfide that were being produced in enormous quantities by the chemical industry during the late 1800s. Essential as catalysts and to processes for degreasing, stripping, and printing, such solvents were not thought to harm the ear unless by caustic burn; recent evidence suggests that heavy physical labor increases the tissue uptake of solvents, leading to larger concentrations in the blood that may damage the outer hair cells of the organ of Corti. Over time, exposure to solvent vapors also increases the ear's vulnerability to intense sound and accelerates the loss of higher-frequency hearing. As an expanding chemical industry poured its dregs into pools and rivers whose waters were used downstream for brewing and drinking, the ingestion of such adjuvants of deafness may have been an invisible co-factor in the hearing deficits of urban working-class families.[162]

Wretchedly obvious, by contrast, were the ototoxic effects of many common childhood diseases. Roosa mentioned mumps, measles, meningitis, typhoid fever, and scarlet fever as causes of sensorineural deafness, partial or total. Fever, which was slowly being demoted from syndrome to symptom, had long been known to attack the ears, and it was fever that, in 1848, ranked as the chief cause of deafness among those residents of Hartford's American Asylum for the Deaf and Dumb who had not been born deaf: scarlet fever above all, then "brain fever," "spotted fever," "lung fever," "inflammatory fever," typhus, measles, and mumps (with their high initial fevers). Statistics for all U.S. "deaf-and-dumb" institutions in 1854 showed a like preponderance of scarlet fever, measles, and typhus (confused with typhoid until the 1870s) as causes of non-congenital deafness contracted at an early age: the famous deaf-blind scholar Laura Bridgman lost her hearing to scarlet fever when she was two; Helen Keller caught "brain fever" (meningitis or scarlet fever) at the age of eighteen months. American colonists had regarded scarlet fever as a mild illness; since Independence, observed a Baltimore physician in 1874, it had grown malign and assertive. Five percent of all deaths in England and Wales between 1859 and 1875 were due to scarlet fever, which at that time had no known etiology or remedy. Although Pasteur in 1880 identified the *streptococcus bacillus* by its bent-back bacterial chains and A. F. Eklund in 1882 described the "scarlatina microbe," not for three generations would it be clear that strep throat, tonsillitis, mastoiditis, and scarlet fever were all Group A streptococcal infections and that meningitis may be found among Groups A and B. Nor could Group A or B infections be challenged at their source until the advent of antibiotics in the 1940s; before that, it was almost impossible to mitigate streptococcal infections so as to escape their systemic consequences. When, inexplicably, the virulence and mortality rates for scarlet fever began to diminish in the 1870s, the morbidity rates remained high; epidemics continued into the next century, leaving many more to survive with impaired hearing, even when kept in hospital and overseen for fifty days, as was the rule in Boston. In 1886 one of the founders of the American Pediatric Society, Job Lewis Smith, described a familiar aftermath of scarlet fever as a "thin purulent discharge" that could last months or years, with otitis media as "an obstinate, dangerous, and even fatal sequel." A Cleveland physician, scrutinizing the practices of life insurance companies who took "chronic suppuration of the middle ear" as grounds for rejection, cautioned the medical community against treating the suppurations lightly or not at all: if discharges continued unchecked, they could cut short one's life through cancerous polyps, cerebral abscesses, or meningitis. Those who survived such complications were invariably deafened.[163]

According to a Boston physician reporting on 454 autopsies and 2,232 cases of diphtheria, scarlet fever, and measles, it was the latter that most menaced the ear. When fatal, measles proved to have attacked the ears of all victims, compared to 94 percent for scarlet fever and 82 percent for diphtheria; when patients survived (as they did, increasingly, in 1912), 3 percent of diphtheria cases had residual hearing problems, 11 percent after scarlet fever, and 28 percent after measles (which spread with public education and the large grammar schools that became endemic reservoirs of infection). Such statistics intimated the extensive, enduring impact of measles, scarlet fever, diphtheria, typhoid fever, typhus, and mumps on the otological health of adults, most of whom would have had bouts with several of these illnesses in childhood. The numbers likely underrepresented the percentages of men and women on both sides of the Atlantic who had to cope with hearing losses that stemmed from one of these bouts, since young children with high, contagious fevers were not readily examined for hearing problems, and it took months after a fever had subsided to determine how, or how much, an infant's or toddler's ears had been affected. "If old enough, our patient will tell us that there is pain in the ear and side of the head, that the ear feels full, that the hearing is impaired," and often that there is a tinnital roaring, hissing, or crackling; if the patient is "of tender age," advised a doctor in Detroit, one had to watch for moaning and rolling of the head from side to side, and aggravation at noises "of ordinary volume." Note that last phrase: these diseases could leave a child—or a young man (measles, typhoid fever, and typhus were epidemic among soldiers in the Civil War, the Boer War, and the Great War)—with moderate-to-severe hearing loss, lifelong tinnitus *and* hyperacusis, a shuddering responsiveness to noise and irritability in the presence of high-pitched sounds.[164]

Now consider what physicians had on hand to reduce the fevers and aches of contagious diseases, inveterate catarrh, and other ear-wickéd afflictions.[165] They had quinine, a febrifuge accompanied by noises in the ears that are dose-sensitive. They had morphine, an opiate so efficient at calming the system that it reduces blood flow to the cochlea. They had cocaine (its alkaloid isolated in 1855, available as a powder or spray by 1880), an analgesic and topical anaesthetic which, when taken as a "tonic" by a pregnant mother, makes its way to the fetus and attacks in utero the inner ear and the auditory neurons of the brainstem. They had, by century's end, aspirin, an anti-inflammatory that disrupts the function of a motor protein in the outer hair cells of the cochlea. In short, four of the most widely used drugs between 1860 and 1914 were ototoxins. Had they been taken at sporadic intervals in small doses, their ototoxicity would have been minor and transient; because they were imbedded in patent medicines, cordials, and elixirs used *ad libitum*, and because they were administered—wittingly or unwittingly—to ailing children whose systems more readily absorb and retain the toxins, the amounts consumed from infancy on into adulthood were often massive enough, and frequent enough, to produce more severe, and more permanent, hearing loss.[166]

Quinine's iatrogenic effects were well-known. The tinnitus that almost always and instantly accompanied the use of "Peruvian bark" had been remarked a century before "strong tinctures" were introduced against the more virulent strain of scarlet fever. Once quinine sulfate began to be manufactured in industrial quantities and quinine compounds became cheaper, the aural side-effects of dosing oneself with quinine became more pronounced. If quinine "even in minute doses" starts in on the ear before it gets to reducing fever, physicians also began to report that at continual high dosages quinine may produce vertigo, permanent deafness, and blindness, so prescriptions for repeated doses were sometimes issued with the proviso, "till ears buzz." That ensuing buzz did not deter millions worldwide from taking quinine several times daily for much of their adult lives for rheumatism, sciatica, heart palpitations, "intermittent fever," and headache, or as a preventive against endemic, recurrent malaria. With steady self-medication, quinine could rise to ototoxic levels in the blood, especially among those who, due to quinine's bitter aftertaste or its neurochemical effects on appetite, developed anorexia.[167]

Cocaine, or rather its processed alkaloid, was first applied in eye surgery and dentistry, but its virtues as a topical anaesthetic were overshadowed by its power as an appetite suppressant in dieting powders, as a vasoconstrictor against hay fever and swollen sinuses, as a stimulant in tonics against fatigue, and as a nerve-strengthener for the melancholic. Ironically, it also became a treatment of choice for opiate addiction and for ear ache. Through coca wines, colas, and chewing pastes, cocaine in relatively high doses would have reached a substantial number of pregnant women and the ears of the children to whom they gave birth.[168]

Salicylic acid and soda of salicylate had been prescribed since the 1880s for their analgesic, antipyretic, and vessel-dilating powers. Even then it was known that large doses had much the same side-effects as quinine: a "roaring in the head and ears, a feeling of constriction in the head and deafness, with vertigo and pain in the ear at times" and, at times, hearing loss. Once acetylsalicylic acid, or aspirin, entered the marketplace in 1898 and became available over the counter, sales soared. The result was more regular, concerted self-medication with aspirin for fevers, headaches, muscle ache, and ague. After just five days on what were then considered moderate doses, the average adult would develop narrow-band, high-frequency, bilateral tinnitus. After a few weeks, hearing loss, possibly for life.[169]

To arrive at a reliable estimate of the number of Europeans and North Americans who were walking around at century's turn with an ototoxic buzz in their heads, a roar in their ears, and diminished high-frequency hearing would be no easier than estimating the number who had bad teeth. And this is not to change the subject. We receive sonic vibrations through the bones of our skull as well as through our ears; the teeth and jaw are key acoustic structures. Bad teeth, which can be responsible for indigestion, constipation, and poor nutrition and thus for dosing with ototoxins or for mineral deficiencies that weaken the hearing, can also cause tinnitus and hearing loss. Dental caries, or cavities, were known to doctors to be a prime source of "intense subjective noises" and ear ache: "That carious teeth often give rise to otalgia is so well known," wrote a surgeon at the Manhattan Eye and Ear Hospital in 1898, "that a mere mention of the fact will suffice." Dentists themselves, who were just beginning to acknowledge the importance of microorganisms in creating cavities, had long spoken of "the intimate sympathy existing between the teeth and the ears," such that "nothing was so good for the toothache" in 1820 "as rubbing behind the ear... a mixture composed of equal parts of hartshorn, opium, and sweet oil"; vice versa, they had advertised their services with headlines like "Deafness cured by the extraction of teeth." The surgeon-dentist to the Archbishop of Canterbury had gone so far as to claim that "Cases of true ear-ache are comparatively rare"; violent pain in the ear originated in the wisdom teeth and was displaced upward through the "exposed nerve of the inferior *dens sapientia*." The carotid artery, it seemed, also served as a "direct channel" between aching gums and a sore tympanum, just as the sounds of chewing, or the clicks and crepitus of the jaw, might be amplified into a dreadful grind and headache when the temporo-mandibular joint was out of whack.[170]

Courtesy of otology and bacteriology, dentists could position themselves as guardians of a vulnerable portal. Where before they had been mere mechanics or roving tooth-pullers whose street operations were accompanied by assistants beating drums or blowing trumpets to mask the yowls of clients, they now had fixed locations, framed credentials, and claims to being, as Sarah Nettleton writes, "tooth judges," with a say in daily hygiene. Drawing from an arsenal of laudanum, quinine, and cocaine as well as ether and nitrous oxide to hush the yowling as they tugged, drilled, and filled, dentists acknowledged that theirs were still quite noisy offices. The calm of opiates and local anaesthetics had been offset by the noise of foot-pedaled, and then electrified, drills whose shrillness penetrated the twilight sleep of clients well after "quiet-running" engines were manufactured with parts cushioned to damp the vibration and counteract the pounding of the motor. The higher the speed of the drill, the easier it was on the teeth but the harder on the ears of dentist and client. If modern dentistry was another profession prone to hearing loss after years spent leaning close over patients with a loud vibrating tool in hand, the use of anaesthetics and electric drills directly and indirectly affected the ears of their patients as well. Able to proceed more swiftly and painlessly, dentists could remove teeth more easily. Rather than laboriously filling crumbling molars with tinfoil and lead leaf, mercury amalgam, or silver-tin alloys, they now preferred wholesale extractions—of upper-class teeth blighted by more plentiful sweets from advances in sugar refining; of middle-class teeth blighted by constant physicking with mercurous chloride (calomel) as a purge for constipation; of working-class teeth whose enamel and gums had been eroded by industrial grit. This led to the laying in of more artificial teeth and the replacement of carious ranks of uppers and lowers with full sets of dentures. Given that artificial teeth (no longer of painted wood or ivory but of porcelain or "composites") were poorer at transmitting vibrations, and that people were squeamish about having the natural teeth from a corpse planted in their own mouths, and that dentures were being fixed within a hard-rubber base that further reduced vibration, the complementary acoustics of bone conduction were less available to those already suffering from sensorineural deafness. Given, too, that even the finest dentures were difficult to keep clean and became a locus of oral-aural infections, and that false teeth for the masses could be so poorly fitted that "overclosing of the lower jaw sometimes brought on deafness," dental interventions might themselves injure the ears.[171]

"Set the end of the world on the one hand against a man with acute toothache on the other," hypothesized Marx: physical pain preempted mental anguish.[172] For sub-acute toothache, however, and misaligned jaws with their addled speech, and the industrial grides that set one's teeth on edge, mental anguish prevailed. As it did, too, for intermittent noises in the ears, slow decreases of acuity, and the compensatory "recruitment" that leads a somewhat-deaf person to shudder at higher or intense tones. A distressing insecurity about the stability of one's own voice and the audibility of others, of men shouting incomprehensibly at railroad crossings or toddlers impossibly quiet in the parlor... how many were living like this, with residual hearing loss from childhood infections or ears pinched and pummeled? With tinnitus from doses of quinine or cocaine, from impacted wisdom teeth? With hearing loss after years of pounding at metal, tending power looms, or working most anywhere, ears unprotected, amid the bray of machinery?

8. Scale

More than we can know. Before electric audiometers became available during the 1930s, the calibration of hearing was contingent upon good aural-oral relations: an examiner stood at measured distances behind a patient or applicant and spoke in a low voice or held up a pocket watch, awaiting a response. For aurists like John H. Curtis in 1839, such tests had yielded lucrative proof of how few enjoyed good hearing: "How many, for instance, can hear the insects in the hedges, or a watch tick at the distance of twelve yards? which all ought to be able to do." These were ambitious standards, as profitable as they were problematic. Studying the value of inhalations of sulfuric ether on the ears of a young man with a history of "Nervous Deafness from Infancy, attended with some obstruction of the Eustachian Tube from Mucus," an anatomist at the Philadelphia Hospital recorded on June 9, 1838: "Hearing distance with my watch on right side, 18 inches; ditto, left side, 12 ditto." After five days of ether treatment, hearing distance on the right had increased to 3 feet, on the left to 2 feet 9 inches, and by 26 June the distance on each side was 7 feet, yet despite his vast improvement in hearing, the next day there was "So much noise in street, no test would

SOUNDPLATE 12

answer." In the absence of sound-shielded rooms, tick-and-whisper tests were compromised by echo and interruption; their results were determined as much by the pitch of an examiner's voice, the acoustics of his office, and the peculiarities of his timepiece as by the ears under examination. The tests were also suspect, said the otologist Clarence Blake, on the grounds that examiners had no "simple and universal formula for recording the hearing distance, which may serve to do away with the prevailing uncertainty as to the value of such records dependent upon the imperfect means afforded by a series of tests with watches, not only of different intensity of sound, but of varying tone." Watches ticked more loudly when recently wound, more softly after cleaning and oiling; worse, they pulsed with rhythmic tones that were "easily caught and held when really no longer audible." So wrote B. Alexander Randall, a Philadelphia eye-ear-nose-and-throat specialist who defended whisper-tests well into the 1890s: "Great as are the variations in the loudness, pitch, and distinctiveness of different voices, there is in the whispered speech as much uniformity as in the tick of various watches," always remembering "that the faintest whisper which is at all articulate ought to be heard a half meter in a quiet place, and the full stage whisper at least fifteen meters." For conversation, 15 meters, in a quiet place, seemed about right to Roosa. Catching the drift of two people conversing 40 to 50 feet away in a narrow room with one's back turned was a more practical test, he argued, since people listen for talk rather than tick. As for that tick, hearing a watch at 40 inches, "in a moderately quiet place," was fine for the normal ear. Still, Roosa mourned the fact that "For the ear we have as yet no accurate tests which are adapted for the records of the consulting room," nothing as reliable as the Snellen test for vision adopted in the 1860s.[173]

Impatient with less exacting generations of aurists, Adam Politzer of Vienna constructed "a new acoumeter, giving a definite volume of sound"—a sharp click at the fall of a small hammer released from a short, fixed height—usually audible at 15 meters. Even in its more compact versions, his device required a steady hand and a long room; Politzer's counterparts in Munich and New York found his acoumeter less adaptable to small examining rooms than a whisper and no better than the voice at reproducing tones. Preferable were several complementary tests, conceived in Germany, that exploited the dead-on tones of tuning forks. The first was the Weber test: place a vibrating fork in the middle of a subject's forehead; if the sound is heard as louder on one side than the other, there is some hearing loss. Next, the Rinné test, which contrasted the hearing of airborne sounds through the ear canal with sound heard through the bones of the skull: place a vibrating fork on the mastoid bone and mark the length of time a subject hears the dwindling tone; ditto for that fork vibrating in the air an inch from the ear canal; any difference in duration establishes a person's conductive hearing loss at that frequency. Then the Bing test: strike a tuning fork, place it on the tip of the mastoid bone, then plug up an ear; a person with normal hearing, or with sensorineural hearing loss, will notice a change in intensity; a person with conductive hearing loss will notice no change. Finally, the Schwabach test: if a person stops hearing the tone of a fork before the examiner does, this suggests a sensorineural loss; if the patient hears it longer than the examiner, this suggests... what? a conductive loss? more acute hearing? some obstreperous noise closer to the examiner than to the patient? The forks clearly had their own problems, one of these unavoidably psychophysical: the inconsistency, from one otologist to the next, or from one day to the next, of the "standard blow" that set a fork vibrating for *x* number of seconds. Which technique was best? In 1897 Edgar Holmes of Boston tried them all while testing the hearing of 234 typhoid fever patients; he used his own pocket watch, Politzer's acoumeter, the Rinné tuning fork test, Francis Galton's tone-shifting high-frequency whistle, and a whisper, "but this as well as some other tests were made somewhat difficult as they were conducted in the general wards which were at times somewhat noisy."[174]

Into the noise nonetheless went the inspectors with their watches, whispers, tuning forks, and whistles, for it had become a matter of some urgency to know how well sounds and words were being heard in factories and schools, on the battlefield and in the lives of the life-insured. With the arrival after century's turn of Taylor's scientific management, Ford's assembly lines, and Frank and Lillian Gilbreth's efficiency regimes, workplaces had to meet a new code of aural clarity. Foremen had to be assured that workers could attend step-by-step to the stopwatched sequencing of their jobs and, word-by-word, to a new command structure that had laborers "under load" 42 percent of the workday, "free from load" 58 percent, and "soldiering" (i.e. loafing) 0 percent. At the speedier, though arguably less tiring pace, it was crucial for safety, efficiency, and profitability that there be no aural confusion about points of starting or stopping, wherefore the 23rd General Rule of the Gilbreth Field System:

> Blow *one blast of whistle* at 5 minutes before starting time.
> *Two blasts* at starting time.
> *One blast* at quitting time.
> *Blasts of whistle* to be not over *4 seconds long*.[175]

School-bells were never as loud or confusing as factory or railroad whistles, but classes in urban schools were taught, as the *Brooklyn Daily Eagle* complained in 1901, within range of "The ringing of bells and blowing of whistles with occasional siren variations in all parts of the borough at every hour and half hour of the day and night." As public kindergartens, grammar schools, and high schools multiplied, in tandem with a parallel universe of parochial schools, to accommodate a growing urban (and, in North America, immigrant) population whose children were being encouraged to come to school earlier and stay in school longer, progressive educators insisted on periods of open-air recreation each school day, with more sensory stimulation and better hygienic situations in class—in contrast with the "wasted 'street and alley time' of the masses of city children." Architects were asked to incorporate effective systems of heating, lighting, and ventilation in classrooms that held fifty desks and in auditoriums that also served the adult community. School furniture was redesigned for cleaner lines, sturdier legs, more sanitary surfaces, and anatomically apt curves for children's spines, which led to acoustically livelier desks and chairs. Advised by such as William G. Bruce, editor of *The American School Board Journal*, that "Floors and partitions should be made as sound-proof as possible, so that pupils will not be disturbed by noises in adjoining rooms" or by the scraping of chair legs from rooms above, even the most noise-sensitive of architects were constrained by economies of scale to ink their blueprints with rows of flat-ceilinged rectangular rooms of reverberant wood and plaster (and long slate blackboards), with narrow corridors (and metal lockers) echoing between them. Whatever soundproofing was implemented, architects had no control over what happened once a teacher opened the wonderfully ventilating windows. In Brooklyn, for example, half of the public schools were "surrounded on all street sides by rough, uneven cobblestone paving over which light wagons rattle and heavy trucks rumble throughout the year." In 1902 Brooklyn's Commissioner of Public Works solicited from fifty-three school principals an estimate "of the time lost in classroom work on account of the din outside." The minimum was five minutes an hour, or two hours per week per classroom, and the principals were unanimous that quieter paving, such as asphalt, would make quite a difference, for "a truck load of steel rails passing over a cobblestone street can do more to break up a recitation than anything short of cannonade." It was dandy and progressive to suggest that enthusiastic group recitations would help youngsters memorize their schoolwork, "the impression of so large a body of sound upon the ear" being "very strong." Educators had to distinguish, however, between a reform pedagogy of constructively channeled movement and noise ("To make a young child sit still and keep silence for any great length of time, is next door to murder," wrote the principal of the State Normal School in Trenton. "I verily believe it sometimes is murder") and that predicament in which, "at frequent brief intervals during the school day, neither teachers nor pupils can make themselves heard." Such a din, from brewery wagons no less, put 6,000 students each day out of hearing of the education they ought to receive, costing the city $50,000 in wages (so calculated Commissioner Redfield). The din also thinned the ranks of teachers, who were driven out of Brooklyn or the profession itself by the nervous strain of doing battle with "Truck, Cobblestone, & Co." Add to this din the residual ototoxicity of childhood diseases (and of their palliatives), and the ear aches and tinnitus linked to tooth decay—and what exactly *was* being heard in school?[176]

Or heard on other battlefields, and aboard battleships? Infantry and naval environments had become so loud that Hiram Maxim's silencers for rifles, machine guns, and artillery were indexical of the deafening noise of modern warfare rather than evidence of success at a general deadening of sound. "Noise from the six large guns fired like a salvo," wrote a Berlin military doctor in 1899, "is indescribably tremendous, ear-splitting and unique." Further, the new laminated smokeless powder was thrice as powerful as the old granulated black powder, and because artillerymen used the same weight of the new powder as of the old, detonations could be thrice as loud. Pathological anatomy had at last revealed that gunners suffered lesions in the organ of Corti and degenerative processes in their cochlear nerves. In response perhaps to the pathology reports and certainly to the stringing of front-line telephones that demanded more acute hearing than did in-your-face directives, infantry manuals suggested that soldiers use cotton plugs or other ear protection. For the officers and men of the Royal Navy "exposed to the powerful concussion of the monster cannon of modern introduction," an English physician in 1881 offered to replace filthy, feeble cotton-wool plugs with a "little conical cap of vulcanite," flesh-colored and washable, that would neither tickle nor infect the ear; his prophylactic rubber was widely neglected but well suited to protect the hearing of sailors amid ocean salt and spray. All ranks had to be able to heed commands on the larger steel-sheathed ships whose 10,000-horsepower engines below decks thundered more constantly than the guns above. Throughout these ships ran "voice pipes" that were basically quarter-mile-long ear trumpets made of metal or flexible tubing; officers practiced speaking slowly and distinctly through the pipes so as to make their orders clear and audible, their mouths never touching the pipes for fear of distortion; those on the receiving ends learned to echo each pipe-order back through the hollow tubes, until drowned out by boiler hiss or gun thunder.[177]

War or peace, a warship on a coastline was no silent presence. Captains relied on steam whistles and "screeching, whistling" tugs to make their way through harbors, on foghorns and bell-buoys to navigate coastal shoals, and on whistles and bells in all weathers to keep the crew to its shifts. On the high seas, the men took artillery practice; in port, a warship's guns could be stunningly active with salutes. Touring the Caribbean in 1895 aboard the armored cruiser *U.S.S. New York*, ship's engineer Ernest L. Bennett noted in his log that upon entering St. Thomas, the *New York* fired a salute of twenty-one guns as it came to anchor, then thirteen guns in salute to two French men-of-war also at anchor and thirteen to a German man-of-war, each of which returned the noisy honors. Such friendly fire was multiplied when the *New York* arrived at Kiel in 1895 as envoy to the opening of the Kaiser Wilhelm II Canal between the Baltic and the Black Sea, where more than a "hundred steel floating fortresses" from eight nations paraded up and down, saluting each other and the Kaiser. Their coordinated Salute to Peace entailed the firing of 33 rounds from a total of 2,500 guns, an irony and expense not lost on the world press. Snarled the editor of the *Minneapolis Times*, "The fact that the opening of this magnificent waterway . . . was celebrated by the booming of ordnance from the assembled war fleets of the world, is an indictment of civilization. . . . If such navies are necessary, then liberty is impossible and despotism is a condition necessary for the human race." In 1911 the U.S. Navy had a public festival of its own, with one hundred lead-white warships, including the *New York* (renamed *Saratoga*). "It was when thunder to thunder spoke, as from each deck roared the nineteen guns, which are due for the Secretary of the Navy, that the wonder which this fleet inspires was at its height," declared the *New York Herald*, impressed more by the cumulative roar of the fleet than by its length—six miles of ships steaming down the Hudson. The Secretary, George von Lengerke Meyer, had been consulting experts in scientific management to improve the efficiency of naval gunners so that rounds could be fired not every five minutes, as during the Spanish American War, but every thirty seconds. He hoped also to abate the greater noise amidships by covering the decks with a new sound-absorbent flooring, linoleum, which would never be laid thick enough to keep up with guns that grew louder by the millimeter.[178]

Louder, and deadlier. What were the odds that an infantryman or an ensign would emerge unscathed from battles on land or sea? The odds touched directly upon the assets of life insurance companies, who wanted data not only on mortality and maimings within each occupation but on the hardiness of individual applicants, whether boiler-tenders or stenographers. By the 1860s in England and the 1890s in Germany and North America, companies were requiring initial physical exams and paying a network of local doctors to complete detailed forms that helped determine an applicant's eligibility. In 1911, half of all U.S. and Canadian physicians were on insurance payrolls, noting gait and countenance, collecting urine specimens, and wheedling histories out of applicants. Unlike appointments at compensation offices and military stations where injured workers and reluctant draftees had debilities to prove, a life insurance exam turned vociferous citizens into recalcitrant witnesses, prone to errors of omission or equivocation. Since hearing tests relied upon oral testimony and aural responsiveness, otologists devised ear-deceptions that tricked applicants into revealing that they were less deaf than they claimed (for denial of deferment or compensation) or deafer than they knew (for denial of insurance).[179]

Precision may have been wanting, prejudices rampant, patriotism too much in question, pensions too greatly at stake; even so, "a system of periodic biologic exams" was advocated as a vital part of public and personal health surveillance. Just as "the machinist gives his engine a thorough going-over at regular intervals," wrote ophthalmologist George M. Gould in a widely quoted article of 1900, our bodies deserve a similar, regular going-over. "War has devised a rough and crude system of physical examinations for the would-be soldier; insurance companies have more accurately examined the bodies and life-prospects of their policy-holders to estimate their financial risks; through the Bertillon system, criminology has still more perfectly fixed the anatomic measuring of the bodies of the lawbreaker," and now that dentists were requiring annual check-ups, athletic directors charting the yearly progress of college students, and psychophysicists recording age-related changes in muscular strength and response, the momentum of industry, penology, physiology, and psychology was headed directly toward a "science of man." The power of the sciences lay in their prescience, and medical prescience had to be grounded in the consolidated data of regular exams, for "There are a hundred known intimations and auras of oncoming disease, but there are a thousand undiscovered ones" begging to be found out. Physicians had to be alert for "pre-symptoms," indiscernible except through the study of series. Prevention and "pathogenesis, not therapeutics," were to be medicine's "ultimate study," so it was crucial to track bodily states long before there were complaints, and to plot individual trajectories against a baseline.[180]

Vision and audition, known to change with age, must be checked frequently. Advances in the measurement of, and grinding of lenses for, astigmatism as well as for myopia/presbyopia meant that poor eyesight could be effectively corrected. Their boldfaced handbills notwithstanding, aurists could make no equally effective corrections for permanent hearing loss; ear trumpets and bone-conductors could compensate for some of the lost volume, never for lost tones or the thrum of a crowd. People with auditory deficits were therefore less likely than those with visual deficits to sport the stigmatizing equipment of disability. Without hearing tests, then, the somewhat deaf would be more invisible than the somewhat blind. Given the military and industrial demands of the later nineteenth century, visual and auditory impairments were too disruptive to be left incalculable or undetected.[181]

Testing began in the primary grades, where physicians were on first alert for lice and communicable diseases, then for bad teeth, weak vision, infected ears. The more indigent the district, the greater the number of children who presented with all of the above. Although physical exams took no more than six rough-and-ready minutes each school year, their regularity allowed for the tracking of populations that otherwise might never have seen a physician between infancy and senility. Onsite exams revealed a host of untended, acute cases of scarlet fever, measles, mumps, whooping cough, flu, and diphtheria; annual exams enabled the discovery of more chronic complaints—anemia, bronchitis, asthma, *otitis media*. In Boston in 1899, a seventh of the 17,449 students examined had to be sent home on the spot; in Chicago in 1900, a quarter of 16,805 students. Medical inspectors in New York State in 1908 estimated that a third of schoolchildren were a year or more behind in their studies due to ear, eye, nose, and throat problems. Regular exams helped script a slow-motion datadrama of the alarmingly high percentage of children who were *growing* too slowly or *going* slowly blind or deaf. Going deaf: if 25 percent of Germany's schoolchildren had hearing difficulties, and 40 percent in Scotland, and if only 5 percent of New York City's schoolchildren had unimpaired hearing while 13 percent had pronounced aural defects that made them appear to be dullards, what would happen as they grew older? If three million American children were losing their hearing, what would happen when they all grew up?[182]

One in three would hear poorly, with an "otopathy more or less ignored": that was the rule of thumb in the years before the Great War. Thousands were turned down for military service on the basis of a quick test, a thumb pressed alternately against the tragus of the left ear, then the right ear, as young men strained to hear and repeat words spoken in a "low conversational tone" from at least ten feet away and behind them. Some 16,000 were rejected for poor hearing by German military physicians between 1870 and 1900; in 1897, it was reported that one-quarter of French conscripts had been rejected for hearing problems; 30,000 were similarly rejected by U.S. physicians during the Great War, for "The military man must be able to hear commands, and in time of warfare a keen sense of ear discrimination is useful, since by this means gas shells are differentiated from high explosives while still moving through the air." Draftees who arrived with ear infections were assigned to desk jobs when they were not sent packing—back to the big cities whence they (and their *otitis media*) usually came. Defective hearing, reported the Surgeon General's statisticians, was "of great importance in the draft, constituting 9.9 per 1000 of the defects found or nearly 1% of all"; much larger, and untabulated, were the percentages accepted with hearing poor in one ear but fine in the other or half as good as normal in both ears. Further, while 27,000 were rejected for impossibly bad teeth, another 10,000 with poor teeth were put in uniform, many to endure intermittent ear aches or tinnitus for the duration of the war. Across the Atlantic, the numbers with weak teeth were still more alarming among English musters.[183]

Discolored enamel or a few abscessed teeth were of themselves no cause for rejection but, like shortness or bone-crookedness, they could be signs of another widespread disorder, rickets, that had implications not only for strength and stature but for hearing. In 1901, the Canadian physician William Osler noted that half or more of the children seen in clinics in London and Vienna had signs of rickets, and the deplorable rate of rejection for "poor physique" of working-class youth during the Great War would be attributed by British medical board examiners to cases of undiagnosed rickets. A similarly high incidence of rickets was found in 1921 among urban American children, according to Elmer V. McCollum, the biochemist who had isolated Vitamin A in 1913. A Kansas farmboy who had suffered from scurvy and bad teeth, McCollum was now a professor at Johns Hopkins and, though still tortured by toothache and looking "hollow-eyed, thin, and unwell," was studying the stunted growth, bowed legs, and frail long bones of the children of Italian immigrants living on little food and less daylight in New York's back tenements. Following up on the work of others who had found that liberal doses of sunshine and/or cod liver oil could cure such problems, McCollum determined that rickets was caused by the lack of another vitamin, which he called D. Processed through the skin and the renal tubules, D is vital to bone mineralization in children and adults. It promotes calcium and phosphorus uptake throughout the body, including, we now know, the ears. The otic capsule, densely woven with calcium and phosphorus, is peculiarly sensitive to insufficiencies of D, as are the myelin sheaths protecting the cochlear nerves. During the war, 42,000 American men were rejected for service on the basis of defective bones and joints, 15,000 for curvature of the spine, problems stemming from childhood rickets or the osteomalacia of adult D deficiency, both associated with hearing loss.[184]

Brown or yellowing teeth might signify no more than addiction to cigarettes. Rare and hand-rolled in the 1850s, cigarettes by 1910 were machine-rolled to meet a demand that had risen, in the United States alone, to nine billion a year, and in 1917 to forty billion. Unlike pipe tobacco, chewing tobacco, and cigars, all vying for the throats and lungs of consumers, cigarettes during the Great War were thought so essential to morale that packs were added to front-line rations. It was too early for taut connections to be drawn between nicotine, tar, and emphysema, heart attacks, or lung cancer, and not until the 1990s did data implicate smoking as synergistic to hearing loss from other causes, but the surge in smoking must be considered in any account of the numbers coping with degrees of deafness.[185]

Scrupulousness would also demand a paragraph on diabetes, more often diagnosed around the turn of the century than ever before, due perhaps to dietary changes, to the regular taking of urine samples, to campaigns against adulterated foods, malnutrition, and obesity, or to longer average lifespans (after the first five years) and, in consequence, a larger cohort with adult-onset symptoms. Suspected since the early 1800s of causing hearing loss as well as blurred vision and bouts of fainting, diabetes mellitus has to this day a close if biomedically controversial relationship to sensorineural deafness.[186]

Add it all up—the endemic diseases, epidemics, and childhood "fevers" with their otological after-effects, often permanent; the ototoxic drugs used to treat those afflictions; the boxing of schoolchildren's ears and the familial tugging or cuffing at home; the injury done by industrial noise to the inner and middle ears of working adults, year after year, and more swiftly by the cannonade of battle to the ears of soldiers and sailors; the tinnitus and earache from impacted wisdom teeth, dental decay, and gum disease; the cigar and cigarette smoke, sulfuric ash and coal dust, lead-laced paint and arsenical wallpapers in the most genteel of homes, and the soot and smog outside in the thick city air, responsible at the very least for purulent ear infections and recurrent catarrh—add it all up and the heard world was widely compromised.[187]

9. Spills & Thrills

Widely but differently compromised—according to who you were and how you were faring during a Gay Nineties and Belle Époque of labor lock-outs, transatlantic depression and local recessions, massive migrations and colonial depredation. After decades of exposure to workplace noise,

SOUNDPLATE 13

environmental trauma, and ototoxins, the aged probably heard less well than the young. Ill-fed, sun-starved, crowded into thin-walled rooms close to train tracks and heavy industry, the urban poor probably heard less well than upper and middle-classes anywhere. Making war or steel, hammering at boilers or the chassis of newfangled autos, hunting for sport or for meat, men may have heard less well than women, though women had to feed squalling infants, soothe screaming children, tend to clanging pots and coal scuttles or, at work outside the house, endure the fury of power looms or the squeal of switchboard headsets bristling with interference. So perhaps women simply heard better out of their left ears (shielded from the cries of children held customarily against the left breast while their right hands, trained from infancy to be stronger, took care of a dozen other tasks), men better out of their right ears (pressed to the stock of a rifle and so shielded from the muzzle blast of guns fired customarily with the right hand at the trigger). Whites, having less melanin in their inner ears and more access to higher-powered guns, may have heard less well than African-Americans, who were not so likely to have battlefield experience, to live in large cities, to work in machine shops, or to attend the public schools from which spread ototoxic diseases and (wrote a physician at the Central Dispensary in Washington, D.C. in 1887) "dry catarrh, which is the cause of so much of the incurable deafness in the white race." The blue-eyed, statistically more susceptible to hearing loss, may have heard less well than the dark-eyed, although regression analysis would probably show that the blue-eyed lived longer and had more years to suffer the indignities of noise, or drank more and heard less, as do alcoholics who register sounds more slowly while hearing voices out of reach of teetotalers.[188]

"Do you hear noises of any kind? (Ringing, buzzing, bells, hissing, fire of guns, etc.)," asked the U.S. War Department's medical examiners. "Do you hear voices? Are they real voices? Whose? What time of day do you hear them most? What do they say? Are they pleasant? Unpleasant? Of what do they accuse you? Do you answer these voices? Are any of them friendly to you? Are they real voices or only your own or some one else's thoughts that you hear? Can you hear yourself think?"[189] To the examiners, "hearing yourself think" was not so much a question of noise as of obsession, of a mind too furious, fantastic, untethered or beset; for minds less put-upon, it was still no mean question in 1914. With the cross-talk of telephone party lines, the wash of odd syllables from homemade wireless sets, the scratch and stutter of newly mass-produced gramophone records, the huff and fluster audible through the walls and ceilings of apartment buildings and boarding houses, who would not hear voices hitherto implausible, disembodied?

"Ecological psychoacoustics" is the study of how and what humans hear outside the laboratory, where sounds are almost never pure, isolated, steady tones and come at us instead in storms, with "dynamic changes in frequency, intensity, and spectrum." Out there we hear in soundstreams that we parse by subunits or in parallel. Our "echology" is as temporal as it is spatial, attuned as much to the events that constitute vibrations as to the separation of figure from ground. We have our own inbuilt Doppler effect, overestimating (for safety's sake?) the unevenly increasing intensity of an approaching sound-source, underestimating the decrease on departure; we tend to hear the gradual ramping up of sounds as more dynamic than the decay of the same. Low frequencies must be louder for us to hear them at all, since otherwise we would be disturbed by our own chewing and heartbeats; for like reasons, the stapedius muscle in our middle ear contracts just before and while we speak, reducing the bombard of our own talking. The more time we spend indoors in reverberant rooms, the less adept we become at judging the distance of sound-sources outside; the more uneven the acuity between left ear and right, the less expert we are at judging the origin of a sound. With hearing loss comes a reduced capacity to register shifts in intensity, frequency, and duration and, so, weaker abilities to resolve sounds into streams or events by source or timing: more sounds are unmeaning, more sentences incomplete. Noise then comes at the partially deaf from *out there* and everywhere, but also from *in here* and everywhen. "Can you hear yourself think?" was a telling question for everyone working through an industrializing, ototoxic world *out there* and listening *in here* through a labyrinth of bone, tissue, and filament occluded or inflamed, structurally out of alignment or neurologically out of sync.[190]

Maybe people did not want to think, and noise was one more opiate of the masses. The most egregious of moralists and most empathetic of novelists shared this explanation for the prevalence of noise—of idle wrangling on street corners, of shouting and weeping in theaters hot with melodrama, of brass bands and beer songs in parks and pavilions, of honky-tonk in taverns and ragtime in saloons, of large "torpedoes" or smaller squibs set off in back alleys—among the diversions favored by laborers, assembly-line workers, journeymen, farmboys. Their weekdays dulled by hunger and work, their voices suppressed by dynamos or tractor engines, they would "naturally" prefer, in their free time, the liveliness of a collective noise to the hush of a library or the harmonies of barbershop quartets. Twenty-five hundred were in the stands at Crush Station, between Hillsboro and Waco, to watch a prearranged head-on collision of two old Texas locomotives pulling six freight cars each: "Nearer and faster came the flying trains, the whistles making demoniac music." At impact the boilers of both locomotives exploded, sending shattered bolts, shrapnel, and slivers of oak flying a hundred yards away: a "success," reported the *Boilermaker's Journal* in 1896, equal to that of collisions staged outside Kansas City and Cleveland.[191]

Holidays and weekends, the millions rode the rails of rollercoasters whose sudden turns and steep descents thrilled them with the trappings of disaster. Originally a European recreation of Russian toboggan rides, then a cable railway amusement that by 1846 had a Parisian loop-the-loop, the American rollercoaster was much more intimately bound up with catastrophe, descending as it did from the Lehigh Coal Company's Switch Back Railroad, whose full cars rolled down eighteen miles of track from the adits of Summit Hill on Mt. Jefferson to "shutes" on the Lehigh Canal at Mauch Chunk, seat of Carbon County, Pennsylvania, and home of anthracite coal mining. Mining companies first picked off the outcroppings of coal on the heights of these Pocono Mountains, then drove deep and deeper into the rock, at great and greater risk to miners and their families. Tunnel Nine, which struck the Mammoth Vein on Christmas Day, 1857, after half a mile of drilling, was rich enough to become the longest continually operated coal mine in North America (†1972), with a long history of fatal accidents. As for the Switch Back Railroad, whose steam-driven cable systems pulled empty cars back up Summit Hill, its operators had taken to charging a dollar a head to those eager for a nighttime ascent with a descent reputed to reach speeds over 100 miles per hour. In 1872, after the summit mine closed, the Switch Back Railroad was devoted night and day to tourists, so many of whom converged on Mauch Chunk (35,000 in 1874) that for the next fifty years the area could claim to be America's second-most-popular honeymoon destination, after Niagara Falls.[192]

Coney Island by 1884 had a fifty-foot-high Switch Back Railway of its own, two more in 1885, and then the boom was on, with twelve hundred coasters erected in North America by the start of the Great War. A visit to one of the four hundred new American "amusement parks" would be incomplete without a ride on the (Brooklyn) Flip Flap, (Rockaway) Thriller, or (Coney Island) Rough Rider, where you yelled your lungs out all the way down and half the way back up toward another near-death experience. Edvard Munch's *Skrik* or *The Scream* (at first *Der Schrei der Natur*), painted in Berlin in 1893 and then lithographed for circulation far beyond, had different tonalities ("When the clouds at sunset . . . seem to be a bloody blanket—it is surely not enough to paint ordinary clouds") but the same set of distraught undulations as the steel floor designed by Theophilus Van Kannel for his "Witching Waves" ride of 1908. And the same disorienting waviness as that of rollercoasters: "I felt a huge scream welling up inside me—and I really did hear a huge scream. The colors in nature—broke the lines in nature. The lines and colors quivered with movement," recalled Munch. "These vibrations of light caused not only the oscillation of my eyes. My ears were also affected and began to vibrate. So I actually heard a scream. Then I painted *The Scream*." WILL SHE THROW HER ARMS AROUND YOUR NECK AND YELL? asked the promo for the Cannon Coaster at Coney Island. WELL, I GUESS, YES! In 1910, John A. Miller's safety ratchet, which stopped cars from rolling back while being hauled up the next incline, added an ominous clanking to the sonic suspense of these "scream machines," and in 1912, when Miller's under-friction wheel held the cars more securely to the tracks and his safety bar held riders more securely in their cars, the coasters (the rides, the riders) would shriek with sharper, steeper curves. By then, however, there were other banshee entertainments, namely motordromes, where motorcycles without brakes were ridden at top speed around a banked wooden oval.[193]

Too many motorcyclists died riding the boards for the 'dromes to hold their own against the drone of automobile raceways at Le Mans (1906) and Indianapolis (1909), but "The shrill scream of a woman followed by a chorus of frightened cries and the screech of metal on metal"—the sounds of a fatal accident in September 1911 on Coney Island's Giant Racer—were no dissuasions against climbing into the next higher, steeper rollercoaster. Something about an engagement with risk, and immersion in noise, beckoned. Perhaps it was that, unlike the risk of life-ending accident in coal mines, factories, or trains, the risk and noise of rollercoasters and motordromes were engaged voluntarily and temporarily and were essential to the event: what would a rollercoaster be without moments of mortal uncertainty? what would a motorcycle race or locomotive collision be without the gunning of engines, the crash and blast on impact? There was, too, the appeal of a practically guaranteed dodge from apparent disaster, as with the first cliffhanger serials, *The Perils of Pauline* and *The Hazards of Helen*, that opened in theaters in 1914. Gasp and scream were, alright, a "letting off of steam" for workers who had been bottled up in offices, shops, mills, and factories, but gasp and scream also followed what Bernhard Rieger has called the "uneven contours of risk societies," where ordinary people enjoy witnessing extraordinary destruction (a warehouse in flames, a giant demolition) from oddly unsafe distances, and in hair's-breadth escapes from a falling brick, a rabid dog, a beer barrel slipping off the back of a dray. So Coney Island made a bundle from reenactments of the Johnstown Flood; so the Airdome Theater in Kendalville, Indiana, in August 1911, advertised a "great program of fire pictures" including film of the Triangle Shirtwaist Factory Fire and the "Destruction of Dreamland" at Coney Island. In 1912, when the safest ship afloat collided with an iceberg with "a noise like a cable running out," filmmakers rushed to reproduce, in bathtubs and back lots, the wreck of the *Titanic*, based on thousands of column inches of interviews with the seven hundred survivors.[194]

Strikingly, what survivors remembered was as much aural as visual. While newspaper artists sketched the ship's officers, megaphones in hand, directing passengers into lifeboats, A. H. Barkworth, J.P., was haunted by the *Titanic*'s steam whistles; when the ship's boilers began to explode, "The death cries from the shrill throats of the blatant steam screechers beside the smokestacks so rent the air that conversation among the passengers was possible only when one yelled into the ear of a fellow unfortunate." Edith Haisman, less grammatically stiff, told the press, "It was terrible, lots of shouting and people crying as she went down, people were so upset, never heard anything like it, you could hear the screams of all the people that was on deck, it was really terrible." From one of the twenty lifeboats (too few by half), Joseph Scarrott heard "cries from the poor souls struggling in the water," sounds "in the stillness of the night" that "seemed to go through you like a knife." Laurence Beesley, watching the *Titanic* tilt on one end, recalled the sound of its machinery roaring "down through the vessel with a rattle and groaning that could be heard for miles, the weirdest sound, surely, that could be heard in the middle of the ocean a thousand miles away from land." Mrs. Washington Dodge insisted that "The most terrible part of the experience was that awful crying after the ship went down. We were a mile away, but we heard it—oh, how we heard." It sounded, said John Thayer, "like locusts on a midsummer night, in the woods of Pennsylvania. This terrible continuing cry lasted for twenty or thirty minutes, gradually dying away . . ." Lady Duff-Gordon had "never heard such a continued chorus of utter despair and agony. It was at least an hour before the last shrieks died out. I remember next the very last cry was that of a man who had been calling loudly: 'My God! My God!' He cried monotonously, in a dull, hopeless way. For an entire hour there had been an awful chorus of shrieks gradually dying into a hopeless moan until this last cry that I spoke of. Then all was silent." Eva Hart and her mother spoke the coda: "And finally the ghastly noise of the people thrashing about and screaming and drowning, that finally ceased. I remember saying to my mother, 'How dreadful that noise was' and I'll always remember her reply and she said, 'Yes, but think back about the silence that followed it'... because all of a sudden the ship wasn't there, the lights weren't there, and the cries weren't there."[195]

Who heard any of this when they went to see the silent newsreels and slide shows of the Sinking or, within a month, "Saved from the *Titanic*," a cinedrama starring Dorothy Gibson, a young actress who, in white evening dress, gloves, and black pumps, had been among the seven hundred to survive?[196] And what do I mean by jouncing together rollercoasters 'dromes icebergs locomotives and silent films? I mean to suggest that, given the hearing deficits common to the laboring poor, and to those of every rank who survived ototoxic childhood fevers, and to all adults who used ototoxic remedies, the world of sound would have been heard as similarly jumbled and tinged with a tinnital edge. I mean to suggest that the particular difficulties of the hearing-impaired with high frequencies (shrieks, gasps) underlay some of the jumble of trauma and thrill that greeted disasters impending or replayed. I mean to suggest that the sounding environment was more often than not being heard as out of phase and, at the same time, as piercingly loud, a psychophysical artefact of that hyperacuity which comes as the inner ear tries ever harder to recruit sound from an acoustically and otologically noisy world.

Terror, I am speculating, as it heightens the senses, turns up the gain on the auditory system. Before the sales of cigarette-pack-size battery-operated hearing aids in the late 1940s, few would have warmed to this electromechanical analogy, but the experience of gain would not have been foreign to boys struggling with their crystal sets amid atmospheric squeal or to those elders who, clutching ear trumpets to their auricles, trembled when friends with high voices shouted into the metal horn. Actual or ersatz, terror might turn up the acoustic gain in a way that the hearing-impaired could best accommodate. And at the silent movies, it came through with an amplification at once thrilling and bone-chilling.

Nothing in the cinematic dramatizations of the sinking of the *Titanic* was visually "documentary," as there had been no photos or film of the event; in fact, J.P. Morgan's White Star Line had so few pictures of its ship in dock or at sea (or coming within an arm's length of colliding with the *New York* as it steamed out of Southampton) that images of a sister ship, *Olympic*, were often substituted. But the sounds of silent film in nickelodeons and movie houses were of a different sort, real, anchored, and voluminous. The ruck-and-whicket of film projectors competed with the unwrapping of sandwiches, gulping of beer, slurping of soup. The winding of a phonograph to play a Cameraphone record, the pumping of a pianola, the grinding of an automatic music machine or "photoplayer" (with buttons for "sob music," doorbells, and galloping horses), the playing of a Mighty Wurlitzer pipe organ (with keys for steamboat whistles, fire sirens, quacking ducks, marching bands), or accompaniment by a full-fledged theater orchestra masked the whisk-&-whirr of ventilating fans. The voices of actor-narrators behind the screen or of captions read aloud by friends of the illiterate, of young men whistling with (or at) the (female) piano player and of communal sing-alongs, melded with the calls of chewing gum vendors in the aisles. The sound-effects of a percussionist with "one cymbal struck with a wire (an ignoble sound), one set of bells, hammer for same, one xylophone, resonators for same, a pair of Turkish cymbals, a Chinese crash cymbal, a tom-tom drum, a bird-whistle, a rooster-crow, a dog-bark, a lion-growl, rosin for same, a tambourine, castanets, hand-clappers, a ratchet, a triangle and a beater, a cuckoo, a scrubbing brush for mill-wheel effects," etc., or the blaring of auto horns, striking of anvils, croaking of frogs, and blowing of noses from a Noiseograph machine, seconded the guffaws at comic scenes and screams as danger loomed. So active were "effects men" that a critic complained: "If the picture shows a man in the throes of death on his bed, [the effects man] rises to the occasion by making the bed creak. If by any chance, a horse appears on the screen he feels in duty bound to make us believe that a troop of cavalry is riding at full gallop over a hundred blocks of macadamized pavement." True, there were "scores of pictures that without effects are dull, insipid and meaningless, and with effects, thrill, delight, and please an audience," but sound-effects ought to be "few, simple and well rehearsed." Such sage advice suited deluxe theaters that got films well ahead of time and had staff to preview them, but elsewhere the effects were spontaneous local symphonics of doors slamming, (dogs) barking, horns blowing, and pianists improvising. A theater's score for most silent films was therefore its own; sound-effects, as André Gaudreault has written, situated each showing of a silent film in a distinct cultural geography, with accents added by the audience. More comprehensively aural than would be the Talkies, "silents" were also more democratically oral, for anyone was free to engage with anyone else as images flickered, music played, and effects men struck their gongs, while at the Talkies everyone learned to hush and hang on the next words from loudspeakers synched to the titanic lips on the screen ("The Management of this theatre makes a definite and sincere effort to project Sound Motion Pictures in the most perfect manner possible. We respectfully request Patrons entering and leaving the theatre to avoid loud talking and making any unnecessary noise"). For the hard of hearing, the projection of silent films would have been more sonically alive yet more comfortably audible than Talkies or even vaudeville, since audiovisual cues inside the theater were usefully redundant: captions, captions read aloud, live or/and recorded music, sound-effects, audience gasps or screams or laughter and, in between, running commentaries from those who had seen the film before.[197]

Children were there too, howling, hooting, screaming, singing along with the bouncing ball of a song, throwing peanut shells, booing villains, filling up the slack times on weekday afternoons at "Bijous, Pictoriums, Theatoriums, Electrics, and Dreamlands." In 1914, two-thirds of Cleveland's children went to the movies twice a week, unaccompanied. Did the collective sound-play at the Pictoriums appreciably increase the gain for them too? Even when their acuity had not been affected by scarlet fever or any other of the ototoxic infections of childhood, or by rickets and poor nutrition, or by beatings around the ears, children had years before their hearing matured—low-frequency resolution not before age ten, temporal resolution (the sequencing of multiple sounds) at twelve, full and accurate detection of tones through noise at sixteen.[198]

Logically, one could draw different conclusions from the prevalence of hearing loss among adults and delay or deficits in the maturation of hearing among children. One could argue, phenomenologically, that in the decades before the Great War the world was to many an ear a quieter place, its upper registers in particular damped by partial deafnesses. One could argue, otologically, that the world was a roughly louder place, as inner ears recruited all sounds indiscriminately to compensate for intensities lost, frequencies inaudible. One could argue, neurologically, that the world was a noisier place, filling with exactly those shrill, piercing industrial sounds that pain the half-deaf ear and provoke episodes of tinnitus or feedback. One could argue, ecologically, that the combined forces of steam-driven, internal-combustion, and electrical power expanded the acoustic spectrum—from the rumble of drays to the deeper rumble of subway cars; from the scream of steam whistles to the higher screech and gride of elevated trains; from the sharp blows of hammers at anvils to the sharper, more complex impulses of pneumatic rivet-guns; from the buzz of sawmills to the rasp-and-razz of spark-gap generators—such that daily auditory experience was not so much louder or noisier as more spectrally rich and, for the hard of hearing, more stressful. One could argue, psychologically and kinesiologically, that the sounds that moved people were, in a Progressive era of progressive hearing loss, more thoroughly vibratory: that the longer freight trains and heavier trucks, more strident jackhammers and relentless pile-drivers, new steamrollers and steeper rollercoasters established a more bone-shaking, more deeply *felt* acoustic environment. How to reconcile or negotiate these differences?

10. Squalls

Meteorologically, to begin with. When the Jesuit director of the Geophysical Observatory at St. Louis University stretched nine hundred feet

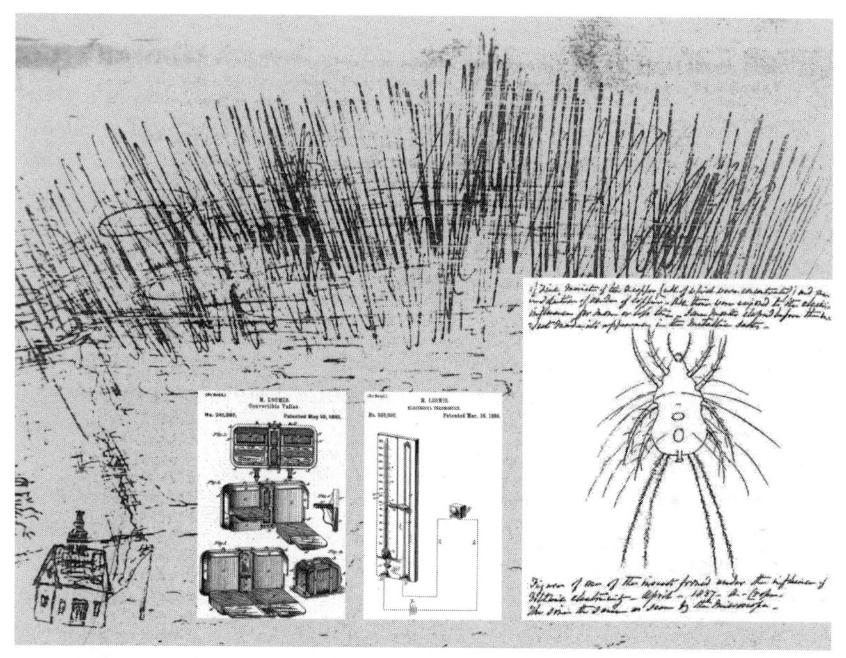

SOUNDPLATE 14

of copper wire in a triangle between the tower of Divinity, a smokestack near Philosophy, and the top of Arts and Sciences, this circuit was as much a signal shift in the cultural sense of sound as a symbolic gesture, in 1911, toward reuniting physics with theology and the humanities. The wire was the aerial for a ceraunograph, or "thunder detector," devised in 1899 by Frederick L. Odenbach, S.J., a Cleveland physicist. Basically a radio receiver that recorded electrical disturbances from distant thunderstorms on a revolving drum, the ceraunograph made the invisible visible and, with its companion ceraunophone, the inaudible audible, by intercepting electromagnetic waves from atmospheric activity hundreds of miles off. Telephone and telegraph operators had heard these "sferics" for decades as awful pops and cracklings on the line; now meteorologists would make something meaningful of that static.[199]

Greek philosophers had been among the first to steal thunder from the gods. Aristophanes had his fun with them in the *Clouds*, where the character of Socrates attributes thunder not to Hesiod's "loud-crashing Zeus" but to the bumping of clouds saturated with water and tossed about by an aerial vortex. Like a man with a full belly who has no end of wamble, waterlogged clouds tumble into each other in the same manner that Strepsiades, after a feast, totters, rumbles, and roars,

> gently at first, *pappax*, *pappax*, crescendo
> Then, *papapappax*, and when I have eased myself *papapapapappax*

and so, asks Socrates,

> Is it not likely, then, when your tiny tummy so loud can roar
> That infinite ether roar louder? Both alike are breaking of wind.

Yet thunder, above all rainless thunder, was ominous enough to send the Roman Senate packing; Emperor Tiberius, tonitrophobic, wore laurel wreaths to ward it off. "If you see thunder and lightning without rain," wrote Rashi, the eminent Franco-Jewish commentator on the Law, "your generation is sinful and God is angry with them." That was about the time of the First Crusade; six centuries later, Puritans and Quakers still equated *The Voice of Thunder* that made the ears ring with *The Sound of a Trumpet Giving a Certain Sound, Saying Arise Ye Dead, and Come to Judgement* (1659). Lest Judgment come aforetimes, seventeenth and eighteenth-century churchbells continued to be rung as counterthrusts to tempest, though statistically neither bellows of thunder nor flashes of lightning were subdued by the bells, and bellpullers had a uniquely high mortality rate during storms. Newspapers and diarists continued to record the sonics of the more dreadful storms, such as that in Bristol (England) on July 15, 1759, whose claps were "the loudest that have been heard in these parts for many years." The loudness was momentous: had not Burke just written, in his essay on the Sublime, that "Excessive loudness alone is sufficient to overpower the soul, to suspend its action, and to fill it with terror"?[200]

Colonial Americans, as Richard Rath has shown, conflated thunder& lightning, experiencing them together as primarily acoustic (if divinely directed) phenomena. At once aural and vibrational, thunder&lightning fell into the same sonic realm as earthquakes, hissing fireballs, and "Moodus noises" in the Connecticut Valley that sounded like "heavy muffled thunder." Europeans, who had their own quakes, meteorites, fog-belchings, and brontidi, also tended to conflate thunder&lightning. Although Jean-Philippe Rameau bound thunder into forty bars of his music-drama, *Hippolyte at Aricie*, beyond the French court thunder&lightning were unrestrained, shattering mirrors, shaking chimneys, startling eels from their lake beds, making lads "run mad," and sending the unlucky to their graves—"the violence of the Thunder and Lightning kill'd four dead immediately" near Colchester in 1708. Such was the power of loudness. The Enlightened tried, as had Socrates, to reason thunder&lightning away from the supernatural, this time by identifying it with something more humanly manipulable: electricity. "In our hands," explained a French physician, "electricity is what Thunder is in the hands of nature; if the effects of electricity are less prodigious than those of Thunder, that is only because art has less to draw upon than nature." Ben Franklin, for his part, would not romanticize Nature; "thunder gusts" were concussions of electrified clouds and, because immune to the seduction of copper rods, more dangerous than lightning.[201]

Further evidences of thunder as electrical did not instantly sunder it from the supernatural or undercut its intimidating voice. In folklore, peals of thunder at a deathbed meant "perdition for the soul"; at Massachusetts law in 1829, a man was found guilty of wanton mischief after discharging his gun in the presence of a woman known to go into convulsions upon hearing a "gun, thunder, or any other sudden noise, or by hearing the words 'gun,' 'ammunition,' etc. mentioned." Worse than the plight of this high-strung Mrs. Gifford was that of blithe little girls who roamed too far from home and were overtaken by storm,

And the thunders crash astounding,
As thou ne'er wert used to hear,
And from cloud to cloud resounding,
Rack thy brain and stun thine ear

until they turned feverish and were suddenly no more.

Thoughtless thus how many wander
Far through Pleasure's luring way,
Till some cloud with flash and thunder,
Crush them in its stormy way.

This was scarcely the reassurance that children needed, in 1841 as now, at the boom and crack of a world overhead that adults themselves did not always take in stride. "We had a tremendous thunderstorm here last night," wrote Charles Titus to his wife in 1843 from Indiana. "I think I never heard such incessant thunder. It was peal upon peal every instant, and that in fearful proximity.... [S]oon I felt an indescribable awe creeping over me, a sense of the majesty and power of God, that I have rarely felt before." Let that majesty be a comfort to you, a mother had told her frightened son in 1830; it stands as surety for the ultimate beneficence of rain, lightning, and thunder. So,

My darling boy, when storms arise,
Thou hast no cause for dread;
For God, who lives above the skies,
Protects thy little head.

As Calvinist atmospherics and Brontëan romanticism gave way to weathers more Unitarian or sweetly transcendental, there was reassurance aplenty for children like young Henry, caught berry picking by a thunderstorm in Lesson LVII of McGuffey's *Fourth Eclectic Reader* of 1866. Crouched in the underbrush, Henry mutters, "How I wish there were no such thing as thunder." His older friend James comforts him: All God's works are blessings, including thunder&lightning, which burn up poisonous fumes in the air. Let us read aloud a poem about thunder as another sign of God's omnipresence, then practice articulating these words: *clang, clang'd; twang, twang'd; cringe, cring'd.* Hold the child in your arms, advised the editor of the *Sunday-School Times* a quarter-century later, while you explain that the clouds are making loud music; soon, the calmed child will be "ready to believe" that thunder is "the very voice of God, which she could listen to with reverent gratitude."[202]

Secular mothers and fathers would define thunder as "a great harmless noise that travels much more slowly than electricity or light" or "the noise which usually follows the flash" from electrified clouds, but nineteenth-century meteorology was too shaky to stand up to that "trembling and dismay when the heavens are gathering blackness!" This was not because theology always trumped physics but because physics was stumped. "We might teach a child who is afraid of thunder and lightning that the lightning flashes are absolutely necessary to purify the air," wrote Miss B. C. M. of Kissimmee, Florida, in a homemakers' symposium on "Calmness During a Thunderstorm" in 1913; her secular, City Beautiful notion of electrified cloud-cleaning was one of many competing hypotheses. Was thunder an echoing concussion among bloated, wind-blown clouds? Was it the "crowding, cramming, squeezing, crushing and rubbing of the thick-packed and close jammed" aerial globules? Sounds of battle between temperamental updrafts and downdrafts? The sonic differential between a positively charged atmosphere and negatively charged earth? An inrush of air to fill the vacuum produced as lightning splits the heavens? A salvo sparked across columns of air as they suddenly expand? Explosions of combustible gases (hydrogen and oxygen) in the lower atmosphere, flung into a chemical tizzy (water vapor "dissociated") by the zigzag of lightning? Bursts of celestial steam along channels superheated by electrical flashes? Or, as a twelve-year-old girl explained to the *Scientific American* in 1908, "a chemical reaction like that which bursts the bell jar, lightning being its liberated energy and thunder its detonation?"... And which of these theories might account for rolling thunder, which for staccato claps, which for successive peals, or, following the sequence laid out by a Forest Expert, the sharp crackle, the heavy rumble, then the earth-shaking explosion of thunderstorm? The possibilities were not so much endless as they were archetypal, embodied in Hesiod's three cyclopean weather-masters, *Brontë* of concussion, *Sterope* of the flash, *Arges* of gleaming whirlwind.[203]

Everywhere that cities were a-building, with their furnaces, steam boilers, fireplaces, and the warm breath of hundreds of thousands within walls of heat-absorbent wood, brick, and stone, heat islands grew up around them. The heat islands of thousands of new cities and a hundred new metropolises became zones of convergence for convective thunderstorms that formed in greater numbers and struck Earth with greater frequency than anyone had known since Babel. Or something like that. Dropping the Victorian bombast or "thunderation," let's just say that the increasingly public and convoluted efforts at a convincing theory of thunder must be attributed in part to its greater urban presence and constancy, eighteen hundred storms (it was said) thundering around the world at any given moment, everywhen.[204]

And everywhere, in attics and basements, on lecture circuits and music hall stages, amateurs of the spark and impresarios of the coil were conducting electrical experiments that showed concussion, flash, and gleam to be hilariously, intriguingly, shockingly—and audibly—inseparable. An example was William Sturgeon's rotating electric sphere, whose whirling sparks were meant to simulate the thunder&lightning of the tropics; the editor of *Annals of Electricity* (1836–1843), Sturgeon had arrived at his calling during "a terrific thunderstorm" in Newfoundland, where he had been serving with the Royal Artillery, whose cannon were no match for a Nor'easter. Later on, with telegraphy and transatlantic cables, telephones and incandescent bulbs, electricity may have seemed less volatile, yet thunder&lightning remained terrific. To Alexander McAdie of the U.S. Signal Service, "the crackling and hissing of sparks, with the attendant phenomena, were intensely exciting" in the summer of 1887 as he stood near the top of the recently completed Washington Monument, the world's tallest manmade structure, measuring the electrical potential of the air during storms of thunder&lightning. Himself more avid than livid, he had to acknowledge, in an essay on "Needless Alarm during Thunder-Storms," the "keen suffering which many undergo just in advance of or during a thunderstorm." The suffering was double: firstly, a sense of impending danger, irrational considering that lightning flashes were "almost always distant" and that "thunder can do no more than the low notes of a church organ"; secondly, "a depression of spirits, which is physical and real, brought about by some as yet unknown relation between the nervous system and conditions of air-pressure, humidity, and purity." To make matters worse, scientists in 1878 had convened an international Lightning Rod Conference because Franklin's copper rods and the "paratonnerres" or lightning arresters on telephone poles no longer seemed to do the trick; the English physicist Oliver Lodge would go so far, said one critic, as to label "every lightning protector with a great notice, 'Beware of this; it bites; it frizzes; it spits, it does all sorts and kinds of dangerous things.'"[205]

"You are startled," wrote a visitor to Fyne Court atop the Quantock Hills, "by the smart crackling sound that attends the passage of the electrical spark; you hear also the rumbling of distant thunder. The rain is already plashing in great drops against the glass, and the sound of the passing sparks continues to startle your ears." It is a dark and stormy day back in 1836 and you are at the manor of Andrew Crosse, who is "in high glee" up in his organ gallery, lit by the gleam of conducting rods and knobs. Known to Somerset folk as "The Thunderer" or "The Electrician," Crosse may not be cackling as sparks leap "with increasing rapidity and noise, rap rap rap—bang, bang, bang," but he does enjoy a good storm. He has composed a poem "To the Winds," for

> Nature ne'er meant this vast creation
> To lie one dull lethargic whole,
> But, mistress of her great vocation,
> Gave to the mass a glowing soul.

Crosse is all about "eternal motion" and electrical fluids that flow through all of life, promising communication "*instantaneously* with the uttermost ends of the earth." He tells you how electricity "broods in the air, rides on the mist, travels with the light, wanders through space, attracts in the aurora, terrifies in the thunderstorm, rules the growth of plants, and shapes all substances, from the fragile crystals of ice to the diamond." Above the Quantock whortleberries, across treetops and hundred-foot-tall masts erected on his estate, sixteen hundred or eighteen hundred or three thousand feet of insulated wire escort aerial electricity to the aerie of his "electric room." There, the "most awful" effects, says Crosse, arise on window-rattling days when the room is shaken by "bursts of thunder without and noise within, every now and then accompanied with a crash of accumulated fluid in the wire, striving to get free between the balls." Those brass balls, one conducting, one receiving, would be cinematically enlarged by set designers for movies of mad scientists, and some now look to Crosse as the model for Dr. Frankenstein. Mary Shelley may have heard him speak in London in 1814 about atmospheric electricity, but his most Frankensteinian experiment occurred years after Shelley made the last revisions to her story. In December 1836, several large mites took shape in, then clambered out of, a sample of iron oxide from Vesuvius that Crosse had saturated with potassium silicate and electrified for twenty-eight days between the poles of a battery. This amazing gestation, in lieu of the crystals he had hoped to grow within the pores of the lava, was replicated five years later by England's other "greatest dealer in thunder and lightning," William Henry Weekes, who had strung a thousand feet of wire over the town of Sandwich and described the "loud cracking reports" from the "torrents of electrical matter" produced during thunderstorms. None beside Crosse and Weekes ever again managed such a spontaneous generation.[206]

Crosse died in 1855, remembered more for the *Acarus horridus* spawned by his "obstetrico-galvanic apparatus" than for his accurate estimates of the amount of electricity in the atmosphere, his progress toward assaying the conductivity of plant soils, or his attempt to make a battery out of roses. He had hoped briefly to pass on his work to Ada, Countess Lovelace, who had studied mathematics with Charles Babbage and recently finished the translation of an Italian report on his analytical engine, to which she added prescient notes and the earliest of computer programs. Now she thought to search out, via the "vital spark" of electricity, a "calculus of the nervous system" that might explain the phenomena of mesmerism. Faraday, who adored Ada, was too ill to take her on, so she turned to her neighbor Crosse, who believed that "Metaphorically speaking, electricity may be called the right hand of the Almighty" but considered spirit-rapping and mesmerism as "absurd and vile, and a rejection of all that is sublime and noble." The two excitable rationalists got along well, but Ada was stirred less by Crosse's unsystematic science and unkempt laboratory ("*Nothing* is ever *ready*. All Chaos & Chance") than by the horserace-betting consortium led by his son John, with whom she had an affair and through whom she twice pawned the Byron family jewels when all the bets went south. Although Crosse's wife Cornelia carried on a few of his experiments after he died and compiled his writings electric and poetic, his scientific heir-apparent would arise on the other side of the Atlantic, in the guise of a dentist.[207]

Fourth of nine children of Nathan Loomis, a civil engineer and a "computor" who helped start the *American Ephemeris and Nautical Almanac*, Mahlon Loomis was five foot eleven—"good proportion and firmly built. Hair eyes and nose have an unconquerable tendency to be red.... Conscience clear as glass, and on the whole," he wrote in 1852, "as happy as any one of my size." He was also ambitious ("I *will* be known to the world.—I *will* do good") and democratic, with a knack for improving dentures so that they could be made available to all: "No more shall the purple clad gently masticate their white bread and venison, with substitutes for natural dentures elaborately wrought from virgin gold, while the hot beans and gruel of his laborer hiss over a plate of base alloy." Distant cousin to Elias Loomis, an astronomer and meteorologist who invented the synoptic weather map and made a fortune in royalties from textbooks on physics and mathematics, Mahlon took an interest in electricity about the same time that his cousin (as a professor at Western Reserve) was trying out telegraphy. At age twenty-one, in 1847, Mahlon had written excitedly to his younger brother Eben about the Drummond or Lime Light, whose white-heat he thought comparable to the power and brilliance of "thunder and lightning," suggesting to him that factories might eventually generate and store the light needed for night shifts, especially when aided by a "perpetual motion galvanic machine." In 1852, while Eben was auditing courses at Harvard and preparing to join their father at the offices of the *Ephemeris*, but also testing his prowess as a medium, Mahlon tried fertilizing a vegetable plot with underground wires and metal plates hooked up to a battery—a form of electrobotany that had also intrigued Crosse. Like Crosse, too, Mahlon had an ear and appetite for thunder. In 1853, he had attended the rehearsal of a German band only to realize that he liked "a different order of Music," to wit: "When thick darkness shuts in upon the earth and nature wears a boding silence as if the elements hold council for demonstrations of their might, and when from the murky distance is heard the muffled coming of the storm battalion—there, *there* are notes of Heaven-made music." The "heaven's cannonading," that's what he loved. He wrote in his diary: "Give me, O Night, those iron tones again! Come in your might, ye winds, playing wild strains of melody, and oh ye lightnings, once more awake the drum-beat of that thunder whose returning echo shall have compassed earth—that I may again hear the execution of that music whose composer is our common God." When he moved to Washington, D.C., in 1856, he was a newlywed whose wife, Achsah, was a professed Christian, "And yet I like this little Dear . . . for *she has a mind*." He settled into a genial deism (unable to believe in "the divinity of Christ, or a Devil, or a Hell") and a congenial dental practice, with shades of Spiritualism and faith in the world electric: "I believe," he wrote in 1858, "as soon as something can be found to bear the same relation to the earth that a wire does, a telegraphic communication can be made without a wire."[208]

Came the solar flares of August 27 and September 1, 1859, and his faith was justified. "One of the most intense coronal mass ejections in history, very near the maximum intensity that the sun is thought to be capable of producing" (say our modern astronomers), the flares produced a "Great Auroral Exhibition" that lasted a full week. Its green and yellow spires, columns of bronze and scarlet, and arches of purple and copper lit up the skies from Havana to Helsingfors (the aurora borealis), Santiago to Sydney (the aurora australis). The lights were so bright at night that a congressman under a dark moon could read the fine print of a bill by the "streams, sometimes of a pure milky whiteness and sometimes of a light crimson," above the Capitol dome in Mahlon's own precinct. From accounts gathered by the editors of the *American Journal of Science and Arts*, which began with a report from Elias Loomis, it was manifest that this was the grandest of auroras in memory, with tremendous deviations in lines of magnetic force as the lights flashed and pulsed, often blood-red, as if cities were burning on the horizon. Telegraphy everywhere was disrupted, sometimes with sparks of fire leaping from the wires; operators scurried to disconnect the batteries. But there was also a report from telegrapher Sarah B. Allen of two hours' contact between Fall River and South Braintree using solely the "celestial batteries" of the aurora to send dots and dashes, and from a telegraphic superintendent in Boston, who found that the aurora induced a current along the line so sustained as to allow for two hours of Morse-Code conversation with Maine. "We are working with the current from the Aurora

Borealis alone," tapped the Boston operator. "How do you receive my writing?" "Very well indeed," replied the operator in Portland; "much better than with the batteries on. There is much less variation in the current, and the magnets work steadier." The conclusion was inescapable, wrote Elias Loomis in 1861 at the end of the third of his four articles in the *American Journal*: auroral light was electric light. Just after the auroral exhibition and before Cousin Elias had ventured his conclusion or written up a treatise *On Electric Currents Circulating near the Earth's Surface, and Their Connection with the Phenomena of the Aurora Polaris* (1862), Mahlon went to the source, examining the results of experiments with kites that had been wired to receive charges from the atmosphere even on fair-weather days. He cogitated over the curious fact that a wired kite overhead could disturb the flow of current to another such kite a distance up and away. "It is the plainest thing in the world," he wrote in his diary on March 8, 1864, "that there is a reciprocating tendency between earth and air, for a constant struggle for static electricity to become dynamic, as proved by lightning, etc., and that it only remains for man to arrange the proper apparatus to conduct it into useful channels." Thereupon he sketched a wireless telegraph to channel "this great electrical ocean, slumbering with giant power."[209]

Others shared his oceanic sense, notably the former captain of the *HMS Beagle*, Robert FitzRoy. A friend of Darwin's until he came to espouse a literalist reading of scripture, FitzRoy would rage at the impiety of any *Origin of Species* other than what he found in Genesis. In 1854 and now an Admiral, FitzRoy had taken the helm of the new Meteorological Office of Great Britain's Board of Trade, devoting himself to what he was the first to call "weather forecasting." In his 1863 *Weather Book*, FitzRoy described electricity as an "all-pervading agency . . . so intimately engaged in every atmospheric change, opposition, movement, or combination, that it should never be left out of mind." Electricity came as close to the divine as a Christian could afford: "Imponderable, intangible, ubiquitous—nay, *materially*, almost omnipotent, this most marvelous of all the elements of our wonderful world, is under control, subordinate to man, and yet as unknown and even as mysterious as 'his glassy essence.'" He did not try to harness this soulful force beyond linking up coastal telegraph stations into a storm-warning network. The network had its highs and its "depressions" (another of his insights), but the prevailing winds were vicious. British attitudes toward weather prediction were so hemmed in by traditions of astrology and farmers' almanacs that FitzRoy's forecasts were ridiculed by the press and his peers. FitzRoy took the raillery as he took much of life in his devout years, with nary a grain of salt, and he obsessed over his fateful role, as captain of Darwin's *Voyages*, in nurturing the demon seed of Evolution. Depressed, run-down, and hard of hearing, he was headed toward collapse and, he feared, total deafness. On doctors' orders in early

1865 he withdrew with his family to a London suburb where he might escape the noise and grief of character assassination, but when he read of the murder of the American president, he became deeply troubled about the state of the world, his health, and his reputation. On 30 April 1865 he slit his throat.[210]

Lincoln's body lay in state in Indianapolis that day, on its way to Springfield, Illinois. For the duration of the Civil War, Loomis had had to defer any trial of his wireless telegraph. At last, in October 1866, he and a passel of friends, including a senator and a congressman, split into two groups and stood on neighboring peaks in the Blue Ridge Mountains, flying kites which each had a square of copper gauze attached to its underside and to six hundred feet of insulated wire. Wire from the first kite ran down through a telegraph set to a galvanometer that would measure any current received; the wire of the second kite, the sender, was grounded and ungrounded (completing/interrupting the presumed circuit) at pre-set intervals to deflect the needle of the galvanometer eighteen miles away "with the same vigor and precision as if it had been attached to an ordinary battery." The needle did deflect, aptly and vigorously, and "although no 'transmitting key' was made use of, nor any 'sounder' to voice the message," wrote Loomis, "yet [the signals] were just as exact and distinct as any that ever travelled over a metallic conductor."[211]

Radio this was not, but it was *wireless*. "A solemn feeling seemed to be impressed upon those who witnessed the little performance," wrote Loomis, "as if some grave mystery hovered around the simple scene." The solemnity was in good measure fatigue: it had taken two days atop Bear's Den and Furnace Mountain to get "the line" to work, and then for less than three hours. Loomis explained that higher regions must be sought, "where disturbing influences cannot invade, where statical energy is stored in a vast, unbroken element, enabling a line to be worked without interruption or possible failure." The kites had not ascended to what he would call the "home of that mystic Needle which points the way to all other lands and space." (In 1861 a Scottish physicist, Balfour Stewart, of Loomis's age and with a like openness to Spiritualism, had inferred from the 1859 "Great Magnetic Disturbance" that the Earth could be viewed "as the iron core of a Ruhmkorff's machine"—an electromagnetic inductor and static generator—separated by a dense insulating lower atmosphere from the conducting medium of the thinner upper atmosphere.) If only he could launch his kites into the air over the Rockies! "There is *no doubt* whatsoever that this plan will work to the utmost nicety," he wrote to his brother Joseph in 1868, and it would be "the greatest thing in the world. I tell you, Joe, it will make the sterile regions vegetate with tropical verdure, and thaw out a 'northwest passage' by God's law . . . it will fertilize the earth and reclaim the heathen. *Telegraph*! Why that is the *least important* result I expect to attain in establishing it." There could be, for example, interplanetary conversation, a new music of the spheres.[212]

Loomis's techniques were rather those of a dentist grappling with plates and amalgams than of a physicist coming to grips with J. Clerk Maxwell's *Dynamical Theory of the Electromagnetic Field* (1864), and his language was more millennial than mathematical whenever he encountered resistance. "The morning comes slowly," he would prophesy, "but the hour is about passed and gone when men may slight with impunity or turn indifferently away from the new born revelations that toiling, pioneering minds garner from rugged fields planted by the living God for our higher harvest and refinement." Such rhetoric attracted New York investors, whom he lost to the financial panic of Black Friday, 1869, and Chicago investors, whom he lost to the Great Fire of 1871. Friends in Congress, impressed by the economics of "one subtle Element, tractable and obedient, inexhaustible and utterly without cost, or tax or tariff," moved to fund his research to the tune of $50,000 but managed only a bill of incorporation for the Loomis Aerial Telegraph Company of the District of Columbia, signed in 1873, one year after Loomis had received a patent on his "Improvement in Telegraphing" that used "natural electricity" to establish "an electrical current or circuit for telegraphic and other purposes without the aid of wires, artificial batteries, or cables to form such electrical circuit, and yet communicate from one continent of the globe to another." He never had the wherewithal to build, as he proposed, a string of towers in the Rockies and Alps, high above the clouds, winds, and other "dissipating influences," where he could "tap the store-house of the mighty thunder and make it whisper glad tidings over the seas."[213]

Three problems, other than money, stood in the way of an Intercontinental, yea Interplanetary Loomis Radio Network: the absence of an appropriate sound transmitter; the absence of a "coherer" or detector; the presence of noise. The first lay within reach, the second was a stretch, the third was an obstacle of the first magnitude. While Loomis in his fifties was still crowing over the wirelessness of his telegraph, Emile Berliner, in his twenties and not long off the *S.S. Harmonia*, was selling the Bell Telephone Company his patent on a carbon microphone. At the same time, David Edward Hughes in his forties had devised a multiple-contact microphone through which "the mere breathing of a fly is heard almost like the trumpeting of an elephant," tested on London streets as part of a fully wireless system with his device at one end and Bell's telephone at the other. (Edison, in his thirties, was preparing to challenge both for priority of invention.) Loomis may have tried out a Berliner microphone, but he failed to imagine a receiver other than a standard telegraphic set; nor did Edouard Branly and Oliver Lodge, who both arrived at the principles for a "coherer" or receiving unit (1890–1893), think to receive other than

—•• •— ••• •••• • ••• •— —• —•• —•• — — — — •••

although engineers' dreams of long-distance telephony were (as we shall hear) not so distant from the practice of wireless telegraphy. The third problem, "disturbing influences," was a failing rather of theory than of imagination or finance, for if Loomis imagined a valise that could be hung on seatbacks and converted into a game-space with fold-down shelves and checkerboard, he was not up to the more abstruse dynamics of electromagnetic fields, and in 1882 would struggle with the technical implications of the telephone. A news article exhuming Loomis in 1902 would claim that "the notion of waves filling all space and yet not lost in the general flux, was too staggering for the imagination of the '60s." This was a misleading excuse, since in Loomis's day air and sound were both understood to travel in waves, with "undulations, or pulsations" that FitzRoy had tracked within an "aerial ocean," and sonic vibrations that were easily discriminable through the atmospheric "flux"—thunder too, although it was mostly flux, a "compound mixture of many discordant sounds." Thus the Yale scientist Benjamin Silliman, Jr., a coeditor of the *American Journal*'s series on the aurora and author of *First Principles of Physics* (1859), who devoted seven pages to detailing the range at which sounds could be heard through specific media: the noise of cannon from 250 miles, by putting an ear to the earth; a powerful human voice in frosty polar air, 1¼ miles; a strong man's voice through a speaking trumpet, 3 miles. By 1873 the notion of ever-vibratory soundwaves had reached the ears of young Florence McLandburgh of Chicago, who at twenty-three published in *Scribner's Monthly* her story of "The Automaton Ear," a science fiction based upon the premise that "sound is never lost. A strain of music or a simple tone will vibrate in the air forever and ever, decreasing according to a fixed ratio" until undetectable by human ears but re-collectable by an ultrasensitive machine that can search for era and select for voice amid the noise, "like the heavy rumble of thunder," of millions of sub-aural sounds lingering for eons in the atmosphere. When the microphone appeared, barely four years later, it was presented as an ultrasensor that made manifest, as the stethoscope had done for the lungs and heart, the perpetual sonicity of the world. "[N]ot only is the air of this room in this marvellous state of motion," William H. Preece told the Society of Telegraph Engineers of which he was vice president, "but every piece of wood, every wall, every picture, everything in this hall at this moment, is almost, I may say, alive, trembling away, moving backwards and forwards, forming what are called sonorous vibrations," and the "great secret of Prof. Hughes's discovery is that sonorous vibrations and electrical waves are to a certain extent synonymous." So the idea that there were atmospheric waves everywhere and everywhen, accessible to any ear that could, with technical help, parse that noise, was not staggering; it was, almost literally, on the tip of the tongue. Loomis was unable to draw an efficient connection between electricity and waves because he knew (and drew) the "great electrical ocean" as one amorphous zone of static "struggling" to become dynamic. His last patent, an electrical thermostat that rang an alarm at a defined temperature, did link electricity to sonic vibrations via a sliding contact: ingenious, simple, logical. In contrast, the wirelessness of aerial telegraphy seemed too grave a mystery and too magical an instance of action at a distance, like the "Cassadaga Propaganda" of magicians who set banjos to playing by themselves inside locked cabinets.[214]

Abandoned by his wife, who thought he had become unwired, Mahlon Loomis passed his last years in the West Virginia countryside, at work on his "Safety Thermometer" and new mineral dentures that would do away with the need for "miserable rubber." In April of 1886, after a stroke that affected his vision, he contemplated cutting a small hole in the top of his head that would improve his sight by admitting more light. He died in October and was buried in a grave near his brother George's home in Terra Alta, "overlooking the haze-crested mountains" whose heights had promised so much. That year, 1886, the oldest son of a lawyer, a physicist in Karlsruhe, Germany, began to show how electromagnetic waves could be produced by human devices, and the oldest daughter of a lawyer, a poet in Amherst, Massachusetts, died of severe primary hypertension, a disease of atrial electrical instabilities. Daughters first, for Emily Dickinson was meta+physically closer to the Loomis saga than was Heinrich Hertz. Her connection reached back to Mahlon's brother Eben, who in 1867 had moved to Washington, D.C. with the offices of the *Ephemeris*. There, despite "the city's ceaseless noise," he found quiet intervals in which to compose poetry, much of it about the sky whose events he tabulated and whose cloud-portents he watched with a sharp eye, having twice been struck by lightning. He did not, however, shy from storms, which after all had led him to his wife:

A low deep groan,
A swelling moan,
The storm's majestic monotone
Breaks on the ear,
And growing clear,
Brings to the heart a chill of fear,
A blinding flash,
A bursting crash...

in the aftermath of which he caught a glimpse of his beloved. And he welcomed, in his *Wayside Sketches*, "the low monotone of the distant thunder" as it shook the air, "as in some dim cathedral the air is shaken by the subbass of its mighty organ." Lecturing on Picturesque Astronomy, he made much of the recent spectroscopic fact that "we are brothers of every star and sun in the universe," a fitting faith for one who had walked the Concord woods with Thoreau. Against the "Christian Juggernaut," he wrote that "The whole universe is but the shadowing forth of the Thought of its mighty Author. So when an artist invents a machine, the (so-called) machine is but the half-abortive attempt to convey to other minds the Idea, which is the Real," which is how he would have appreciated the aerial works of Mahlon. The "sweet, mysterious voices" of the natural world kept him sane, as when, despite the "city's myriad noises" that "weary heart, and soul, and brain,"

> Through the din, and dust, and bustle,
> Comes the forest's whispering rustle

or when, at sunrise, leaving behind

> the city, and the crowd,
> The restless life, the noises loud,

he climbs a hill where "the broad sky bends to speak to me" and

> The cosmic morning's growing blaze
> Rolls outward o'er chaotic maze.[215]

Eben had one child, Mabel, herself a poet and a chronicler of astronomy, and a painter and pianist. She met and married David Peck Todd, an astronomer and erstwhile organist, while he too was working at the *Ephemeris*. Taking a position at Amherst College in 1881, David moved with Mabel ("his perpetual blue sky") down the road from Emily Dickinson and her sister Lavinia at the Homestead and their brother Austin and wife Susan next door. Emily by then was fifty and living in near-total seclusion, but after Mabel concertized at the piano in the Homestead drawing room, Emily would send her a short note or a glass of wine. In 1882, while David was in California observing a rare transit of Venus (when "earth's young sister," rhymed Eben, is "robed in blackness like a nun" against "the splendor of the sun"), Mabel and Austin began a love affair that lasted until Austin died in 1895. Their liaison was abetted by Lavinia, accepted by David and, said Mabel, approved by "the rare, mysterious Emily, who listened silently in the dark" and was "glad that Austin had found some comfort," for "Emily always respected real emotion." (Austin to Mabel, in electric trope, December 1883, days before consummating the affair: "Written words alone are too insufficient now. We have come too near together. We want the true quick speech of the eye—the personal presence which allows the wild life current to throb back and forth from one to the other, with its subtle messages of information and response in no other wise to be imparted or conveyed, as we stand just before the veil of closest intimacy.") Mabel in turn was an advocate for the real emotion and genius in the few poems Emily had shared with her before she died. Afterwards, when Lavinia happened on forty-three "fascicles" into which Emily had hand-stitched fair copies of her poems, it was Mabel's help she sought in publishing the oeuvre.[216]

Here's Emily in May 1852, a young woman, cumulo-nimble, writing to brother Austin, away on business in Boston, about the first thunder shower of the season—"you know just how the birds sing before a thunder storm, a sort of hurried, and agitated song," and "then came the wind and rain, and I hurried around the house to shut all the doors and windows." She had trembled rather with joy than fear: "I wish you had seen it come, so cool and so refreshing—and everything glistening from it as with a golden dew." Next morning she woke to find the day "delightful—you will awake in dust, and amidst the ceaseless din of the untiring city, wouldn't you change your dwelling for my palace in the dew?" And here she is, still in Amherst thirty years later, on thunder&lightning:

> The farthest Thunder that I heard
> Was nearer than the Sky
> And rumbles still, though torrid Noons
> Have lain their missiles by—
> The Lightning that preceded it
> Struck no one but myself—
> But I would not exchange the Bolt Flash
> For all the rest of Life—
> Indebtedness to Oxygen
> The ~~Chemist~~ Happy ~~ransomed~~ ~~A Lover can~~ may repay,
> But not the obligation
> To Electricity—
> It founds the Homes and decks the Days
> And every clamor bright
> Is but the gleam concomitant
> Of that waylaying Light—
> The Thought is quiet as a Flake—
> A Crash without a Sound,
> ~~That~~ How Life's reverberation
> Its Explanation found—

And here's the point. Surrounded by astronomers, herself sky-keen and grounded in a *Compendium of Astronomy* and *Introduction to Astronomy* by Yale's Denison Olmsted (another Loomis cousin, a collaborator with Elias, and theorist of a cosmic origin for auroras), Dickinson nonetheless framed the atmospheric "obligation / To Electricity" in terms of *sound*. There is the "clamor bright" of thunder&lightning, at once flare and brigand, then the soundless crash (~~thunder~~&lightning) whose vibration is more than mere sensation. Her lines—dashed—vibrate when sounded, their heat—and heart—ever in the hearing—

The saddest noise, the sweetest noise,
The maddest noise that grows,—
The birds, they make it in the spring,
At night's delicious close....

It makes us think of what we had,
And what we now deplore.
We almost wish those siren throats
Would go and sing no more.

An ear can break a human heart
As quickly as a spear,
We wish the ear had not a heart
So dangerously near.

And here's the point again, honed for ears so near the heart: the resonances of the world electric in the late 1800s conduced toward a sense of noise that verged on the figurative as it approached the finely audible.[217]

Finely audible: "There is no satisfactory evidence that the aurora ever emits any audible sound," wrote Elias Loomis. "It is nevertheless a common impression, at least in high latitudes, that the aurora sometimes emits... a rustling, hissing, crackling noise." This was an aur(or)al illusion, Loomis argued, a confusion of wind over ice with visions "Of bronze—and Blaze" (Dickinson, on the aurora of 1859). If hiss and rustle were illusions, they were compelling illusions, persistently reported. John Franklin and his English crew, exploring the Polar Sea beneath shimmering auroras in 1819–1822, "imagined, more than once, that we heard a rustling like that of autumnal leaves stirred by the wind; but after two hours of attentive listening we were not entirely convinced of the fact." Decades later, William Ogilvie, a surveyor in the Yukon, took a true auditor out beyond the clatter of a wilderness camp, blindfolded him, and asked to be told whenever the man heard anything from the sky. "At nearly every brilliant rush of the auroral light, he exclaimed: 'Don't you hear it?'" Ogilvie could not. Inuit hunters called bright auroras *Neepeealo!* ("very noisy!"): had they heard or inferred the sound? "The Eskimos say it's noisy when the aurora moves rapidly," reported one ethnographer. "They say that there must be noise up there, because it moves like a cloth in the wind!" Must be noise up there: as if auroras had to be aural, accompanied at least by a sound "like the swishing of a ship," said Capt. H. P. Dawson after at last hearing the lights, months into his stay at the British Polar Station, or "the noise produced by a sharp squall of wind in the upper rigging of a ship." Wrote William F. Butler about upper Canada, "So quickly run the colours along these shafts [of the Northern Lights, resembling harp strings] that the ear listens instinctively for sound in the deep stillness of the frozen solitude, but sound I have never heard." Was it "a soft, slithering whisper"? A sound like "the rustling of silk," or "the crumpling of stiff paper," or flames running through the branches of a pine forest? Marietta Dingle, in Winnipeg at century's end, observed that "When the sound was noticeable the aurora seemed to travel rapidly, in waves of light." Even then, she had to listen hard: "The sound was not at all loud, but was decidedly arresting to music-loving ears, and is difficult to describe as it is so distinctive—neither rustling, nor crackling, nor swishing, but a mingling of all three sounds, faint and distant."[218]

Electrophonic noise, atmospheric scientists call it now. Some had already associated the noise of the aurora with the swish and crackle of electrical activity. Crossing the Canadian Rockies near the Arctic Circle in 1897, Glenn Green heard the Northern Lights make a sound "like one heard in an electric power house." It also resembled the static that radiomen would listen through, and to, as if the atmosphere were not only "the great natural reservoir of sensible Electricity" but of a sometimes thunderous, sometimes hissing, sometimes teasing noise.[219]

11. Electricals

Hooking up Geissler tubes whose gases "perform the maddest antics under the influence of discharges and produce the strangest, most varied, and most colorful phenomena," Heinrich Hertz had brought into his lab, by way of cathode rays, the glow and swish of an aurora, but in 1882 he was in pursuit of an electrodynamic at once more far-reaching and down to earth. He wanted to know, and to show, how electrical conductivity worked. Along the way, Hertz would become an unwitting accomplice in the process by which physics expanded what psychophysics had suggested and industrial capitalism confirmed, that noise *had to be* both ubiquitous and endless.[220]

Throughout its eighteenth-century infancy, electricity *streamed*: it was a fluid. In its early nineteenth-century adolescence, jumping between voltaic piles, or leaping from electrostatic generators, or from lightning bolts into Leyden jars, it *hurtled*: it was a force, acting (like gravity) at a distance across the aether. In its young adulthood, induced to follow a wire, it *flowed*: it was a current, a hybrid of fluid and force. How exactly a current arose was uncertain, in part because time had not been looped into the analysis. Once Faraday proposed in the 1840s that electricity was rather an event than a thing, time became a critical variable and electrical phenomena could be conceived to arise from strains in an ambient "field" traced by lines of force. A charge, as the Cambridge physicist James Clerk Maxwell explained in the 1860s, could then be defined as the expression of a local, transitory change in a field. German scientists, seeking to discriminate

SOUNDPLATE 15

among force, current, and field, split into two camps. The first, guided by Wilhelm Weber and Gustav Fechner, both of whom had also done spadework for psychophysics, centered on the velocity and magnitude of individual charges; the second, guided by Helmholtz, studied engagements of pairs of charged objects. As an assistant to Helmholtz, Hertz had the appointed task of investigating non-conductors, or dielectrics, and in particular how aether as a dielectric mediated electrical interaction. He was therefore alert to Maxwell's field dynamics, which postulated that electric and magnetic forces were identical and should pass through the aether in waves, like light.[221]

Much of this was beyond Mahlon Loomis, Eben Loomis, and Emily Dickinson. No matter: the unsettled weather around theories of thunder&lightning refracted, among the lay public, the turbulent transition of Euro-American science from a natural philosophy of electrical fluids and magnetized objects to a physics of fields and waves. And while electricians like Andrew Crosse enthused about batteries and coils that yielded the tiniest miniatures and tinniest echoes of thunder&lightning, scale was no less a challenge to physicists whose equations appeared to indicate that laboratory apparatus was incapable of producing fields or waves well-enough contained to be measured and tested. Although two leading Maxwellians, G.F. FitzGerald in Dublin and Oliver Lodge in

London, had glimmerings in 1879–1883 of how electromagnetic waves might be generated by an oscillating current, they had no idea as to how such waves might be reliably detected.[222]

Resonance, by analogy to sound, was the answer. Due to his own tone-deafness and a correspondingly "monotonous speaking voice" that rendered him incapable, while with the 1st Railway Guards Regiment, of issuing properly military commands ("I cannot get a sharp, loud sound out of my throat"), Hertz was not a gifted speaker, but in compensation was an intent listener. Adept at carpentry and metalwork, Hertz was as much an aural as a tactile investigator, enjoying the talks and company of the Physical Society of Berlin because, above all, it was "more pleasant to get instruction verbally from persons who are thought of as the best informed in each field instead of forever thumbing through those tiresome books." To be sure, Hertz knew how to hit the books, and he knew Helmholtz's texts inside out, but he was happiest when tweaking wires, amplifying isolated effects, and putting standard devices to new uses, as he would do in his position as physics professor at Karlsruhe Polytechnic. His hands-on habits began to yield the happiest of results in late 1886. Rummaging through the school's cabinets of old apparatuses, Hertz had come upon two flat coils of insulated wire: Riess spirals. Hooking these up to batteries, he noticed that sparks leaping a small gap made in the center of the first coil caused a like pattern of sparks to leap across a gap in the second, nearby coil. He saw, and *heard*, the parallel sparking as a type of resonance akin to the sympathetic vibrations analyzed in *On the Sensation of Tone*; indeed, he termed the interaction between the two circuits "symphonic relations." Hertz then made two leaps of his own: a spark-gap, or open circuit, hitherto regarded as an experimental annoyance, could be used as an "exciter of electrical vibrations" whose short pulses were ideal for laboratory trials; the parallel sparking of his "resonator," to the degree that it replicated those pulses, could be used to detect them. After sixteen months more of systematic finagling, using straight wires with micro-metrically adjustable spark-gaps, he had in hand a potent sender and a sensitive receiver of short (9 meters) electromagnetic waves.[223]

Yes, *waves*. Hertz made the definitive switch from currents in wires to waves in air at Christmastide 1887 while his wife and draftswoman Elisabeth was resting in the afterglow of having borne their first daughter. In the dark and cold of the college's main lecture hall, which he had cleared of the obstruction and hiss of gas chandeliers, Hertz watched and listened for changes in the sparking behavior of his oscillator that he could correlate with the sparking in his resonator. He would not have been able to engage in conversation over the noise of the Ruhmkorff coils he was using to power his apparatus, but in the dark he may have adjusted his receiver at first by cues from the crackle of sparks, then by changes in luminosity and length. Just as light waves, coincident, interfere with each other and produce a grillwork of shine and shadow, so the discharges from Hertz's oscillator, reflecting off (and reset at various distances from) a large zinc sheet, produced in his resonator a pattern of sparking and absence of sparking (or interference nodes) that could be explained only in terms of the properties of waves. Exultant, Hertz wrote to Helmholtz in March 1888, "I believe that *the wave nature of sound* in empty air space cannot be demonstrated so clearly as the wave nature of this electrodynamic propagation," which had carried through "all points of the compass" around the hall. Soon after, Hertz concluded that the waves he had found "do not owe their formation solely to processes at the origin, but arise out of the conditions of the whole surrounding space, which latter, according to our theory, is the true seat of energy." Thus: electromagnetic waves in an energy field.[224]

Hadn't Mahlon Loomis been saying this all along? Not so's one would notice. Oliver Heaviside, on the other hand, had every reason to feel vindicated. For more than a decade, as a telegrapher and tester of undersea cables, then as a defiantly independent though indigent physicist, he had been thinking through the forces of electromagnetic propagation and the practical mathematics of waves and fields. His idiosyncratic, sophisticated analyses of transmission problems and energy flux were time and again maligned and blackballed by the same panjandrum who had made fun of Lodge's critique of lightning rods: William Preece, chief engineer and mathematically primitive high priest of the Post Office, which handled Great Britain's telegraph and telephone service. How perverse, harrumphed Preece, to propose, as did Heaviside (and Lorentz in Denmark), that we should increase (rather than decrease!) the leakage and electrical inertia (or self-induction) of cables in order to enhance transmission. How muddle-headed to pretend to prove, as did Heaviside (with vector algebra, no less!), that signals flowed not as currents within wires but as trains of electromagnetic waves that took telegraph lines merely as guides (as had happened during the 1859 aurora). Then came news of Hertz's oscillator-resonator duet: waves and fields were no longer stranded in the limbo of mathematics. Exultant, Heaviside wrote of electromagnetic waves, "Not so long ago, they were nowhere; now they are everywhere," and he could not stop himself from scribbling a bit of told-you-so doggerel:

> Waves are running to and fro,
> Here they are, there they go.
> Try to stop 'em if you can
> You British Engineering man!

His friend Hertz, to be sure, had not gotten everything spot-on: after his 1886–1888 experiments, Hertz still differentiated wired current from the waves his oscillator propagated all around, since his measurements showed that wired impulses moved less quickly than waves. The difference, as two physicists using the great hall of the Rhône waterworks soon proved, was an artefact of waves bouncing off the sandstone walls and iron pillars of the small lecture hall at Karlsruhe and messing with the current "in" his wires; across a larger space, all electromagnetic energy moved at the same light-speed. That electrical transmissions were not necessarily slowed when proceeding "through" (that is, along) a wire was one more vindication for Heaviside, whose years as a telegraph engineer had been devoted to reducing distortion and whose theoretical work culminated in 1887 with a set of equations for ideally distortionless telegraphic and telephonic transmission. If Hertz was the one who sent waves into the "great electrical ocean" so propagated that they could be manhandled, Heaviside was the one who figured out how those waves could be co-opted to work for long-distance communications through "loaded" (self-induced) circuits, an idea that went begging until recast and patented in 1901 by Michael Pupin at Columbia University, whose ensuing riches should have been, as Heaviside always said, his own.

Hertz, who died of bone disease in 1894, spent his last years reviewing classical mechanics in order to remove the last lint of necessity from an aether that had eluded Michelson and Morley. Heaviside, who lived on until 1925, stuck with electromagnetism; in 1902 he calculated that manhandled waves, or "radio" (so-called by 1903), could make it over the horizon by following the curvature of Earth and bouncing off an ionized region "in the upper air" named, when found in 1924, the Heaviside layer.[225]

Solitary and scathing in his comments on "scienticulists" or poseurs like Preece, Heaviside had a vital interest in communication and distortion. Scarlet fever had taken away his hearing in childhood: "I was very deaf from an early age to manhood, and that has influenced my whole life." As a student from the lower middle class (his father and two uncles were marginally prosperous engravers), Heaviside did better on his own than in noisy urban schoolrooms. As a youth of seventeen, his hearing somehow—and somewhat—recouped, he obtained a job as an operator and then cable tester in Newcastle for the Great Northern Telegraph Company (arranged perhaps by his uncle, the physicist Charles Wheatstone) only to resign after seven years by reason of residual deafness and his first run-in with Preece. Returning to London to live with his parents, moving with them to rooms above his brother Charles's music store in Devonshire, then to a house nearby, and finally to Charles's sister-in-law's large house in Torquay (called "Homefield," but known to Heaviside's friends as "The Inexhaustible Cavity"), he never attended scientific meetings and was as leery of admirers as of strangers. Occasionally he played the ocarina for his nieces and nephews, and early on he collaborated with his older brother Arthur, also a telegraph engineer; otherwise he worked at dense technical articles for *The Electrician* and the *Journal of the Society of Telegraph Engineers*. When a genuine admirer, Silvanus Thompson, Professor of Physics and Principal of Finsbury Technical College (and popular lecturer on all things electrical), invited him over for dinner, the Quaker Thompson made sure to write, with friendly understatement in the first part and friendly persistence in the second, "I know that you live a very quiet life... we live very quietly and my household is a quiet one." Heaviside did not come.[226]

Together, Hertz the tone-deaf and Heaviside the half-deaf made the world a noisier place, literally and figuratively. Hertz put into extended play, one might say, Dickinson's "maddest noise"—what radio hams and advertisers inaptly called "static." The word had referred originally to the spark and bristle of electrostatic charges generated by friction, then to atmospheric static generated by Earth as if it were "one large electrical machine" (wrote the American physician James Knight in 1882), so that "static electricity, from the universality of its action, doubtless plays an all-important part in the sustention of health, the germination of plants, the production of ozone, rain, snow, hail and tempest," and could therefore restore the ill and the crippled. "It shocked my susceptibilities," said T. D. Lockwood, rising to speak during a session of the National Telephone Exchange Association's convention in 1882, to hear a colleague talk about "static current," which to his electrician's ears was an oxymoron; one should talk only of static charge and discharge. Really (as lessons from the New Century Correspondence Schools would confirm), static electricity kept to the surface of things. Unfortunately, Lockwood had made his comments from the floor of the convention hall in an era before microphones, "and as the acoustic properties of the hall were poor," noted the N.T.E.A. reporter, "his remarks were at times indistinguishable." Confusion, nominal and phenomenal, persisted, even when it became clear that with Hertzian waves the source of "static" was dynamic, from the motion of electromagnetic waves all around, all the time. Until 1887 this noise had gone unheard except under liveliest aurora and, perhaps, in thunderstorm, when telephone operators heard sounds "of unknown origin." With radio this noise not only grew louder; it became a constant.[227]

While Edvard Munch and his circle of anarcho-socialist friends in Berlin were speculating in 1893 on how to convert atmospheric electricity into free power for the planet, six-foot-six Nikola Tesla had been standing tall in London and New York theaters, dapper and unruffled amid balls of electric flame on platforms charged with hundreds of thousands of volts, proud that his alternating currents, safe and efficient, were about to light up the White City of Chicago's Columbian Exposition. At the naval school offshore from St. Petersburg, Prof. Alexander Popov was studying electrodynamics and preparing a Branly-Lodge coherer for a lightning rod grounded in such a manner that ships at sea could monitor distant storms.

Adapted to transmitting as well as receiving, Popov's wireless set was a fraternal twin to that being readied by Guglielmo Marconi in Bologna. In 1895, the two men independently used their devices to send and receive Morse Code. Despite the Russian's training in physics, the Irish-Italian's entrepreneurial drive, and the showmanship of the Serbian Tesla—who in 1893 had already displayed his own wireless apparatus, using a Geissler tube as a glowing resonator—the air waves at century's end were rather a ragtime of rustle swish and syncopated crackle than a college glee. Although electricians increased transmitting power so that Marconi on the coast of Newfoundland in December 1901 could claim to have heard a Morse "S" (•••) broadcast from two large antennas on Cornwall's Lizard Peninsula, the signals were fraught with such reptilian noise as to make talk impractical. Radio historians now doubt that the rig at Poldhu was able to send any wave across the Atlantic, and certainly not during the daytime or at the frequencies Marconi reported. Had help come from the Heaviside layer and a flaring sun (the former too thin that afternoon of December 12, the latter unusually quiet), there would still have been no way that Marconi in St. John's—his single-wire antenna trailing from a kite bobbing in gale winds, his coherer so weak as to be untunable—could have been certain of a sound as ambiguous as three clicks (not long buzzes but clicks, since the *used* alternator in Cornwall was restricted to the briefest of spark-gaps). The "•••" heard by Marconi was the "zzz" of lightning over the basin of the Amazon, flashed up the air currents to easternmost Canada, or a static "sss" from a cable station right there on Signal Hill.[228]

Enthusiasts hovering over their receivers a decade before and after the Great War found the "static" itself intriguing, especially at night when the air was most alive with noises similar to those on telephone lines that Watson, alone in Bell's workshop, had imagined as messages from other planets, and to those that radioastronomers would later scan for sign(al)s of life in the universe. Just after the Great War, writing for *Harper's Monthly Magazine*, the American statistician and radio buff Buckner Speed would remain as enthralled as he was frustrated by the interruptions to wireless service—and not just from rude lightning. "Electrical exchanges between clouds and earth and clouds and clouds that do not result in outright flashes will produce scratchy, sawing noises, hisses and thumps and all manner of strange manifestations," many of which had been earlier reported by telegraphers but were more common in radio reception. "Some of the so-called static noises heard in wireless telegraphy may, without too much stretch of the fancy, be fairly called the sound of solar whirlwinds and *maelstroms*. Indeed, we must admit that the sources of some of these mysterious interferences may be in vibrations, explosions, detonations, and reverberations proceeding from the uttermost reaches of space." All of this, "all the curious undertones, overtones and intermittent clatter and whisperings that intrude upon, interrupt, and confuse our systems of electrical communication, constitute what the electricians call 'static,'" which is "the sum-total of the vast complex of vibrations from the myriad initiatives and centers of energy, near and inconceivably far—the mighty murmur, sometimes reaching the magnitude of uproar—of the Voices of the Universe."[229]

Should one wonder, then, why radio pioneers recalled so fondly the crack-allure and "divine interference-whistling" of their sets, though much of this was decidedly manmade? John Ashton, "the original radio ham," sat up nights in his Yonkers attic in 1900 waiting for a buzz or hiss, then frantically adjusted his coherer in hopes of retrieving messages: "All that I was actually hearing was the electric generator supply system in the barn" that Edison had presented to his father in 1896. Edwin H. Armstrong wrote in 1951 to another old radioman: "Go back for a moment to the real world that you and I lived in those days, when there were no amplifiers and the signal strength was dependent solely on the amount of energy that could be picked up from 'the ether.' Recall how we held our breath to hear the faint whispers that passed for signals, closed windows and doors on hot summer nights, used rubber air cushions on the head set and pressed the [ear]phones against our ears, one hand on each, as long as we could bear it, in order to hear the sign-off." While the signals were weak, large wireless sets made so much noise in and of themselves that, on the cruiser *Chicago* in 1905, Lt. Stanford Hooper could hear their operations the length of the ship. A young man in Missouri, fond of "the majestic crash and sizzle of the rotary spark gap," built himself a one-kilowatt transmitter whose "noise was terrific, enough to satisfy the most ardent ham"; its "delightful lightning-like noise" could be heard from a half-mile away, and farther at night. "It is almost impossible to fully appreciate the fascination, the mystery, and the impact," wrote Robert Coe in retrospect, "when a person first heard voices and music in earphones connected to a little box which had no connection with the outside world." In 1900, when Reginald Fessenden heard what was debatably the first voice transmission by wireless, "it was accompanied by an extremely loud and disagreeable noise, due to the irregularity of the spark," and for a generation, radiophonics would be framed by intemperate, nearly incessant ssjssss and ^%#>?+<%!! After microphones were wired directly into the antenna circuit (1906) and transmitters had become sufficiently reliable to make radio talk rewarding, the bulky sets themselves still howled and "sang" of their own accord, with specific spark-pitches, or under attack by atmospherics. In 1913, when the United States Navy put into commission its first high-powered radio station, producing "a high-pitched, almost musical note, easily read through the static" courtesy of a huge rotary spark generator so noisy that operators were forced to stuff cotton in their ears, the Navy's wireless expert believed that "the chief obstacle to successful long distance wireless to-day is the disturbance (static) due to atmospheric electrical discharges," of which no radio system was, or could be, free. Armstrong, who in his youth had listened with pleasurable expectancy to the howl and squeal, would then design a regenerator which, as he explained, used the power of the noise itself (the high-frequency currents of the howling, fed back into the circuit) to strengthen the signal. Whether he was using the power of *X*, symbol for all "extraneous electrical disturbances," or something inherent in the circuitry, it would seem from his very description that noise was no peripheral, passing event but elemental and constant. Whatever else Hertz's waves promised, in popular imagination and professional application they were a latent, potent reservoir of noise.[230]

Heaviside's contribution at this juncture was to insist that no matter how adeptly electromagnetic waves, with their fluxes and curls, were manhandled for communications wired or wireless, there would always and necessarily be residual noise. Given unpredictable intensities of atmospheric or magnetic interference, predictable attenuation with distance, unpredictable disturbances in the receiving booth, and each operator's predictable personal equation, even Heaviside's own equations by which to balance inductance, capacitance, resistance, and leakage could not guarantee an undistorted signal. Telegraphers could compensate for weak or crackling messages by tapping more slowly or putting longer pauses between words. With telephony and radiophony, slow-talking was a wacky, unwelcome option, and further, as Heaviside observed, it was not clear how clear long-distance talk needed to be: "we cannot safely estimate what amount of distortion is permissible in transit along the circuit and how much attenuated and distorted we may allow the vibrations to become before human speech ceases to be recognizable as such, and to be intelligently guessable." We might do best, listening hard for voices through the noises of circuitry, to take our lead from the hard-of-hearing, for "some remarkable examples (of interpreting indistinct indication) may be found amongst partially deaf persons who seem to hear very well even when all they have to go by (which practice makes sufficient) is as like articulate speech as a man's shadow is like the man." It was "surprising what a large amount of distortion is permissible not merely on long lines, but on short ones. It is, indeed, customary, or certainly was on the first introduction of the telephone, and for long after, for people to enlarge upon the wonderful manner in which a receiving telephone exactly reproduces in all details the sounds that are communicated to the transmitter," when many of those exactly reproduced sounds were artefacts of the machinery and in fact deranged the voice. "It would be really wonderful," Heaviside sighed, "if we could get perfect reproduction of speech. The best telephone is bad to the critical ear, if a high standard be selected, and not one based on mere intelligibility." The same could be said for radio talk.[231]

Tesla, highly sensitive to sound and patentee of a process for quieting the flicker and hiss of arc lamps, also bore a share of responsibility for encouraging, among the millions who read of his pronouncements from news headlines or visited his fantastic exhibitions, the sense of an electromagnetic realm at once invisible and brilliant, inaudible and clamorous. In Colorado Springs in 1899 at his new lab, researching lightning as part of a project for the wireless transmission of energy, Tesla orchestrated a pandemonium that could be heard ten miles away: he had a prodigious spark-gap generator whose "loud explosive character" indicated "good vibration"; his high-frequency transformer, its massive primary coil buried underground, produced "*very thick and noisy*" sparks and twelve-foot-long streamers that replicated bolts from above; his secondary circuit, trying to manage 400,000 volts, or a "lightning arc" of 12,000,000 volts, discharged with such a "terrible noise" that his ears buzzed for hours. In the shadow of Pike's Peak—his nearest neighbors being the Colorado School for the Deaf and Blind—Tesla discovered that sound, high notes above all, carried for "astonishing distances" through the pure dry air six thousand feet above sea level, and he could hear thunderclaps five hundred miles off. (Did this mean, in turn, that his giant machines could have been heard as far away as Denver or Albuquerque?) Such acoustic clarity was the result, he suspected, of a "strong electrification of the air" from the many lightning strikes (he counted 10,000–12,000 discharges in two hours on the Fourth of July), which is why he had chosen Colorado Springs, and even so he was none too happy with the intrusions: "A bell will ring in the city several miles away, and it would seem as though the bell would be before the very door of the lab. During certain nights when sleepless I have been astonished to hear the talk of people in the streets . . . not to speak of the grinding of the wheels, the rolling of wagons, the puffing of engines, etc., which are perceptible in such a case, and with painful loudness though coming from distances incredibly great." Still, Tesla would persevere, and it was his confidence in powers that he alone seemed to be able to discipline and transform, to step up and step down—"extreme electromotive forces," inaudibly high frequencies, "high tension" currents—more than anything he overdid or hypersensed, that contributed to an enduring impression of a plenum of sound. The noise and vibration that accompanied him wherever he went, whatever he built, seemed so integral to this plenum that it could easily become a *mysterium tremendum*.[232]

Milling around the *fin de siècle* were others already of a mind to exult in aetheric frequencies, physico-spiritual vibrations, and electro-something waves. Take, for example, the Philadelphian John Keely, erstwhile circus performer and carpenter, whose "hydro pneumatic pulsating vacuo engine" drew on sympathetic vibrations of "triune currents of polar flow of force" and "sound-force, set at chords of the thirds" so as to run in perpetual motion on free energy from "the dominant current (which has never been controlled by man and never can be, any more than the lightning that flashes in the clouds)" bearing a "sympathetically attendant force mysteriously associated, which gives that power of propulsion that induces disturbance of negative equilibrium." What was this triple talk? Rev. Albert Plumb of Roxbury tried to explain: "Neither theological science nor any department of physical science, as it lies in the divine mind, is exactly expressed in any human system"; thus "We can apprehend sometimes what we cannot comprehend. If Keely should die, I fear no one could understand his writings." One must rest content, as yet, with the small confirmatory hisses of "Keal," i.e., Keely vapor, rising from his motor each time he touched a tuning-fork to the generator.[233]

Take as a related example Mme. Helena Blavatsky of the Ukraine, Cairo, New York, Madras, and St. John's Wood circuit, who believed that Keely's motor partook of a universal life-force whose principles were more comprehensively, and surely more comprehensibly, adumbrated by Theosophy. (Keely's own rhetoric had verged on the esoteric: "By what term shall we define that force which, when differentiated, expresses itself on the lower planes of manifestation as charity, self-forgetfulness, compassion, and the tendency of all illuminated ones to association in universal brotherhood?") Drawn toward the universal, Blavatsky ascended to a higher astral plane in 1891, seven years before Keely was crushed by a streetcar. Blavatsky's sandals were filled by Annie Besant and C. W. Leadbeater, whose *Thought-Forms* (1905) reproduced photographs and paintings of individual auras, "'luminous' but invisible vibrations" emanating from the brain. Each clear thought emitted a radiant vibration and a color-form. Clear thoughts harked back to Descartes, the vibrations to Chladni's acoustics ("The body under this [thought] impulse throws off a vibrating portion of itself, shaped by the nature of the vibrations—as figures are made by sand on a disk vibrating to a musical note"), and the color-forms were analogs for states of mind ("murderous rage" revealed itself as an aura of red lightning from a muddy cloud). Conversely, "every sound makes its impression upon astral and mental matter—not only those ordered successions of sound which we call music. Some day, perhaps, the forms built by those other less euphonious sounds may be pictured for us." Leadbeater did just that in *The Hidden Side of Things*, begun in 1908: "The sharp, spiteful yap of the terrier discharges a series of forms not unlike the modern rifle bullet, which pierce the astral body in various directions, and seriously disturb its economy; while the deep bay of a bloodhound throws off beads like ostrich eggs or footballs." As for human beings, "An angry ejaculation throws itself forth like a scarlet spear, and many a woman surrounds herself with an intricate network of hard, brown-grey metallic lines by the stream of silly meaningless chatter which she ceaselessly babbles forth." Worse were the "enormous number of artificial noises (most of them transcendently hideous)" that were "constantly being produced all about us, for our so-called civilisation is surely the noisiest with which earth has ever yet been cursed." The screech of a railway engine launched a projectile more horrid than that of any dog's bark. From the scream of a steam siren arose "a veritable sword, with the added disintegrating power of a serious electrical shock." The perpetual roar of a city corresponded to "the ceaseless beating of disintegrating vibrations" that do permanent spiritual damage, especially to "the plastic astral and mental bodies of children" amid "all this ceaseless, unnecessary noise." From a higher point of view, "all the sounds of nature blend themselves into one mighty tone," the distinctive note that Earth adds to the music of the spheres; on more pedestrian levels it was "abundantly clear that all loud, sharp or sudden sounds should, as far as possible, be avoided by anyone who wishes to keep his astral and mental vehicles in good order."[234]

Vibrations, of course, had been around for millions of years, billions, kalpa upon kalpa, mediating between the physical and spiritual, the material and ethereal, and other cultural binaries still in play during the 1800s. Hegel, lecturing on music, had not been the first philosopher to speak of the "inner vibration of the body through which what comes before us is no longer the peaceful and material shape but the first and more ideal breath of the soul." Nor was Berlioz the first musician to wax philosophical about vibration: "We may *hear*, but we do not *vibrate*. And one *must* vibrate *with* the instruments and voices and *because* of them in order to have genuine musical sensations." Nor was his friend Louis Antoine Jullien, who went on to international celebrity as a conductor of huge orchestras and choirs before dying impoverished in a madhouse, the first to "place his fingers in his ears and listen to the dull roar produced by the blood passing through the carotid arteries, and firmly believe that he was hearing the cosmic A given out by the terrestrial globe in its revolutions through space... the genuine A of the spheres! The vibrations of eternity!" (Hindu music theory, wrote Isaac Rice in *What Is Music* in 1875, had anticipated all of this with its Brahmanic breaths and sixteen thousand musical modes, each "governed by a nymph trying to gain Krishna's love.") Nor was John F. Kitto, stonemason's son, dental apprentice, and Christian missionary, the first fully deaf person to realize the "peculiar susceptibility of the whole frame to tangible percussions" such as "the drawing of furniture... over the floor above or below me, the shutting of doors, and the feet of children at play," vibrations that moved him "to the very bone and marrow" as a creature of *felt sound*. Nor were Georges Gilles de la Tourette and Jean-Marie Charcot the first physicians to use vibrations to heal the trembling and nervous, although Tourette may have been the first to put an electric motor in a metal casque and test out his vibratory helmet on patients whose Parkinsonian palsies has been calmed by rough, noisy rides in train carriages. And was Joseph A. Davis of Niagara Falls the first to claim to cure hopeless cases of deafness by applying a vibrating coiled spring to the ear canal, "the vibrations being generated by contact with a phonograph record" in motion?[235]

Phonographs, however, which for their first forty years were fully mechanical devices that froze and unfroze sound at the point of a vibrating needle, and early telephones, which used a slight current to transduce the vibratory motions of a thin diaphragm across thin wires, effected an historic shift in the way that vibrations were perceived. In 1875, just before the (Edison) phonograph and (Bell) telephone were unveiled, R. S. Wyld, a Fellow of the Royal Society of Edinburgh, reminded the readers of his *Physics and Philosophy of the Senses* that "investigation of the laws of vibrations of air involves the most intricate problems of fluxions [calculus], and that in attempting to conceive their nature, even the illustrious Newton failed." This failure left a residual wondrousness to sound: "As we walk the deserted streets, every step or cough is echoed to our ear: we strike our stick upon the pavement, how sharp and hard and immediate is the reply echoed from the opposite houses!" Thinking sound through, we had to marvel at how "The same soft yielding air carries back alike the prolonged notes of the French horn and the harshest and most discordant notes. It is at first"—and, implied Wyld, always—"a difficult problem to apprehend how so slight an impulse upon so yielding a medium can produce the sharp and far-spread effect." We might expect the wondrousness of sonic vibration to recede before the five-dollar Sears phonograph and five-cent public telephone call, but the amazing amplification of the slightest of vibrations ("so slight an impulse upon so yielding a medium") and the swiftness of registration and reiteration ("how sharp and hard and immediate is the reply!") so intrigued poets, novelists, psychologists, artists, spiritualists, and psychoanalysts that they took up vibration as an analog for sensation, memory, public opinion, visual fields, karmic powers, and phylogenetic trauma. In consequence of phonographs, telephones, and popular exegeses of these new vibrational media, sound-as-vibration began to seem latent in all things corporeal and in much that was incorporeal. Further, because the transmission and registration of sound was never flawless (could never be flawless: Heaviside), noise became resident in sound, imbedded rather as a symbiote than a parasite.[236]

Another vibratory medium, of which two dozen models were available by 1904, fused the sonic and tactile with the purr, scream, and sigh of the orgasmic: the "manipulator" or "percuteur" or, more simply, vibrator. Steam-powered in 1869, it was electrified in the 1880s by Dr. J. Mortimer Granville, author of *Nerve-Vibration and Excitation as Agents in the Treatment of Functional Disorder and Organic Disease*. Marketed ostensibly to soothe male skeletal muscles, vibrators were adopted by psychiatrists to deal with female neurasthenics and hysterics who were known to be both calmed and enlivened by pelvic and vulvular massage that could "restore their bright eyes and pink cheeks," and quickly taken up by women of means ("makes you fairly tingle with the joy of being"). At 15,000–30,000 pulses per minute, the noise of the Swedish Vibrator of 1913 (the entire apparatus no smaller than a breadbox) must have been at once a stimulant, a soundmask, and a simile for orgasm. In 1914, when Freud's acolyte Ernest Jones published his classic study of "The Madonna's Conception Through the Ear," sound was noisily polymorphous and fully perverse. Jones could move with newfound ease, analytic, acoustic, and oscillatory, between the harmonious and the inharmonious—thunder, divine flatus, all-too-human farting, angelic trumpets, stomach grumbles, bullroarers, ritual falsetto, baby's breath, and shit by the shit-load—to expose the root fantasy of a father "incestuously impregnating his daughter by expelling intestinal gas into her lower alimentary orifice." If, in Yeatsian language, love had pitched its tent in the place of excrement—so pitch and tone were bedded with noise.[237]

Waves were also resituated during the late 1800s. Oceanic as always, they had come to be, in addition, unending. Listen to John Tyndall explaining waves of light and heat in an 1865 lecture printed in his *Fragments of Science for Unscientific People* (1871): "Darkness may be defined as aether at rest; light as aether in motion. But in reality the aether is never at rest, for in the absence of light-waves we have heat-waves always spreading through it. In the spaces of the universe both classes of undulations incessantly commingle. Here the waves issuing from uncounted centres cross, coincide, oppose, and pass through each other, without confusion or ultimate extinction." As in the heavens above, so in the Earth below: a new science was detecting subterranean motions—seismic waves—that had previously been insensible, such that the Earth appeared no longer sporadically noisy with quake but in constant turmoil beneath and around discernible fault lines. As seismographs after 1875 became more sensitive and less idiosyncratic, and as observation stations were established worldwide (forty-one across the British Empire by 1907, seventy-one across Japan), not only did the number of reported quakes multiply tremendously (8,331 in Japan alone, 1885–1892) but seismic waves became world travelers: a mild quake in Japan would make itself known in laboratories in Potsdam and the Isle of Wight. Geologists and solid-earth physicists in Japan, Europe, and North America set to work on a theory of longitudinal and transverse wave propagation through Earth's mantle; pragmatists designed scales of intensity, some of which were calibrated in terms of comparable sounds, the loudest first:

1. Trains passing over a bridge or through a tunnel.

2. Thunder.
3. Wind moaning, roaring, howling.
4. The fall of heavy bodies or waves breaking on shore.
5. The boom of distant artillery.
6. An immense covey of partridges taking flight.

There were problems with such scales: different frequencies for each instance, different personal auditory equations, different cultural notions of partridges, the variability of thunder. More crucially, acoustic scales were of no more help than were scales of destructiveness in classifying newly detected "microseisms," those frequent—perhaps constant—shocks inaudible to human ears, unfelt by human bodies, yet recorded by seismographs and due, suggested Otto Klotz in Ottawa, to the relentless beating of ocean waves on distant shores. If Henry Riggs's three-dollar "home-made seismograph" of 1908 would have been oblivious to these microseisms, it did "feel the earth's pulse" at Euphrates College, Harpoot, Turkey, and although it was not in place for the San Francisco Earthquake of 1906, it must have swung violently to the Messina quake of late 1908 during which 120,000 Sicilians died in thirty-five seconds due to the "seismologically bad construction" of stone houses. So wrote one of the world's first professional seismologists, Fusakichi Ōmori, who had been trained in Tokyo by British geologists. While country folk in Pembroke and Leicester were learning to point out the direction of small quakes and describing them implicitly as moving in waves (and their noise swelling to a crescendo then dying away, like the rumble "of six traction-engines" or "thirty or forty wagons travelling at a rapid pace" closer and closer, or the bass vibrations of a great organ heaving and subsiding, or more staccato, like "a heavy body falling several times, say, down a short flight of stairs"), English investigators were loathe to acknowledge that architects on another island nation had a tradition so fitted to a history of seismic activity that one could not trace the epicenter and path of an earthquake by its circles of devastation. "Everywhere the houses are built of wood and generally speaking are so flexible that although at the time of a shock they swing violently from side to side in a manner which would result in utter destruction to a house of brick or stone," wrote the geologist John Milne, teaching at Tokyo's College of Technology, "when the shock is over, by the stiffness of their joints, they return to their original position, and leave no trace which gives us any definitive information about the nature of the movement which has taken place." Instead of applauding Japanese carpenters whose complex joinery left most structures standing after the worst quakes, Europeans would continue to insist upon the fireproof solidity of heavy masonry. They associated bamboo screens, matting, and open floor plans with "flimsiness" and a national timidity that had been ingrained by centuries of temblors. After the Great Nōbi Quake of 1891, when the iron, stone, and brick of Western box-framed buildings turned out to be more frail and deadly than the mast-systems, joinery, and "basket weave" of wooden houses and temples, the Japanese could prove once again the virtues of the pliable. They could also show that slippage at fault lines (geological, cultural) was the source, not the result, of seismic activity. They themselves, however, had no defense against the Great Sanriku Tsunami of 1896 that killed 22,000 on the northeast coast of Honshu.... Waves everywhere, moving between galaxies, occupying the sky, traveling through granite, swelling across and *beneath* the surface of oceans, every so often bursting into the open with a still small, or terrible, noise.[238]

Beneath the skin, too. The heart did not just beat; it made waves. *Electrical* waves. In 1842 undulating electrical currents had been found in association with cardiac contractions, though these currents were too feeble to be heard by the best of stethoscopists. The first recordings of cardiac currents were made by Alexander Muirhead in 1869/1870, and with the aid of a capillary electrometer invented by a French physicist, British physiologists in the 1870s began differentiating the "QRS complex" and "T waves" of these heart currents. In 1886–1887 Augustus Waller in London produced the first true electrocardiograms, through which he graphed "the lines of diffusion of the heart's action current," strangely similar to the lines of diffusion of seismic activity from an epicenter. The EKG also showed that electrical waves cued muscular action and were not merely a result of muscular contraction. By 1902 Willem Einthoven in Leiden had substituted a highly sensitive string galvanometer for the electrometer, improving frequency responses and enabling him to track all five electrical deflections, or P Q R S and T waves (and soon a sixth, U, wave) that, in today's language, "propagate through heart tissue to make the atria (the heart's two receptor chambers) and then the ventricles contract." Occupying two rooms and requiring five operators, Einthoven's machine allowed for breakthroughs in the medical understanding of atrial fibrillation and cardiac arrhythmias, but EKGs would remain research instruments until calibrated and condensed, a process initiated around 1910 by the Cambridge Scientific Instrument Company, co-founded by Charles Darwin's youngest son. By 1914, with two hundred EKG machines in Euro-American examining rooms, leading physicians were all in favor of "arming the heart with a pen" so that it could write itself down or, more precisely, electrically vibrate itself down, revealing not only the (in/audible) rhythms but the (in/visible) shape of the forces driving the chambers of the heart.[239]

Like sphygmographs for inscribing blood pressure, myographs for inscribing muscle movements, and pneumographs or spirographs for inscribing the force and pattern of respiration, the electrocardiograph relied upon the autonomous power of the body-in-motion to draw its own systemic portraits. Like phonautographs that inscribed the shapes of sounds, eidophones that registered the intensities of sounds in multi-petaled Chladni "flowers," and research stethoscopes that relayed chest sounds through a long tube to a sensitive pen on a rotating drum of smoked paper, electrocardiographs translated the (scarcely) audible into the visible by way of the vibratory. All of this new instrumentation appeared to escape the pitfalls of the personal equation. Self-observing and self-recording, the body could surely be truer to itself. With the apparent independence of these recording mechanisms came visual confirmation of the apparent independence of a sonic substructure to the world, analogous to the "record groove of uniform depth on the horizontal disc" of the new gramophone of 1895, where "the groove itself produces the motion of the needle across the disc at the proper rate," as Emile Berliner explained. "Earth is always printing on our intelligence through sounds," proclaimed a layman's letter to the editor in the *New York Herald* in 1901.[240]

Then there was the oscilloscope, through which it seemed that practically all phenomena could be translated into waves—noisily translated, because of the nature of the equipment, or how Nature was equipped. Michael Faraday, and then Julius Plücker in Bonn had first investigated "glow tubes," whose cathode rays were so named in 1876 by Eugen Goldstein in Berlin, who found that these rays could be deflected by an electrostatic field, which discovery had prompted Hertz's work with Geissler glow tubes in 1881–1883. At Hertz's urging, his assistant Philipp von Lenard in 1894 contrived a phosphor screen to register the motion of the rays. Two years later, as news came of Roentgen's x-rays, Lumière's cinematograph, and Marconi's and Popov's wireless telegraphy, Karl Braun in Strasbourg—who would patent his own wireless system and share the 1909 Nobel Prize in physics with Marconi—devised an operational "oscillograph" with a phosphor-coated mica screen. The power and clarity of the first screens (called "oscilloscopes" as of 1906) were swiftly improved so that sinusoidal wave forms could be read from the cycles of alternating current. "It was just what I had long wanted," wrote Braun's assistant: "an instrument with which one could see what was happening in electrical circuits." What was happening was waves, as he might have anticipated from Rudolf Koenig's manometric flame-waves

that responded to changing sounds nearby. Oscilloscopes thus reflected the expected, but allowed for a greater scrutiny of phase-relations. When Lord Kelvin sketched out electrical diagrams that depicted resistance as ʌʌʌʌʌʌʌ, the visuals of sound circuitry were reproducing the ancient imagery of noise as something jagged and prickly, music as rippling and rounded. Graphic artists had meanwhile been shaping their own sonography in order to convey the loud and the furious, as had been done by the German cartoonist Wilhelm Busch, who used the repetition of contours 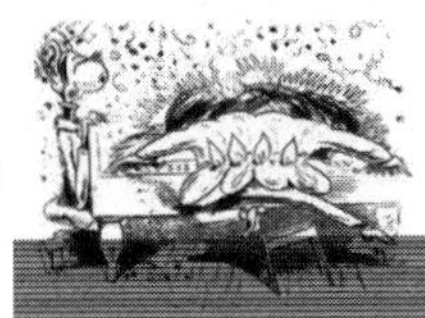 to capture the passionate sonicity of pianistic virtuosity (years before Muybridge and Marey published their photo-sequences of galloping horses or men leaping from frame to frame) and as Busch had also done in 1865 with unevenly radiating interrupted lines for the loud explosion  after Master Lämpel came home to smoke his pipe in peace only to find upon lighting up that Max and Moritz had filled its Meerschaum bowl with gunpowder. The higher the frequency or the greater the interference on the circuit, the more the screens of oscilloscopes would reproduce the ʌʌʌʌʌʌʌ that was iconic of noise. And because of the way the screen was shaped, powered, and illumined (before and after the invention of diode and audion vacuum tubes), it had its own visual noise: ghost images, pattern distortions, stray lines, odd points of entry and exit.... 

And because of the voltages it required, it often hummed or crackled.[241]

12. Visuals

Optical sound has lit up these pages before, in the shape of visibly frozen words, color-projecting musical keyboards, Chladni plates, hashish hallucinations, and impulses toward synaesthesia in critical writing about art and music. What was new—courtesy of physicists, geologists, electrical engineers, and (as Heaviside defined himself) physical mathematicians—was the palette of noises that had emerged in tandem with *and inseparable from* the technological expansion and theoretical deepening of the phenomena of waves and vibrations. These were noises half-visible, half-audible, and increasingly figurative. Theosophists and Impressionists with their spectral

SOUNDPLATE 16

vibrations, Late Romantic poets and early Expressionist painters with their emotive wave-forms, cartoonists with their rays, and art-philosophers like Wassily Kandinsky with his "symphonic" paintings or Marcel Duchamp with his proposals to sculpt a Venus de Milo in sound and "Make a painting of frequency,"[242] all advanced the visuality of sound and the figurativeness of its correlative, noise. Art, poetry, and `pataphysics (that which surpasses metaphysics, a term contrived in 1893 by Alfred Jarry)—were they not the media best equipped to register the upending of the old notion of noise as extrinsic, inessential, and sporadic now that it was proving to be intrinsic, universal, and unending? Where personal "auras" were perceived by illuminati to be shaped and tinted by urban uproar and inner chaos; where atmospheric auroras whispered only to the most expectant or sophisticated of ears; where Helmholtz's "combination tones" were either suspiciously subjective or entirely mathematical;[243] where "static" was as much a statistical as an electromagnetic inevitability; and where subterranean rumbles were almost exclusively seismographic, noise did not have to be audible to be omnipresent. Indeed, once it no longer *had* to be audible, it could be all the more infinite and indiscriminate.

Could, for example, infiltrate the skull of Daniel Paul Schreber, an appeals court judge in Dresden, through rays that only *he* could hear, implanted (as he knew he had been) with filaments (and "female nerves of voluptuousness") that responded like telephone wires. In his *Memoirs of a Neurotic*, published after six years in an asylum (1894–1899) and a return to health, Schreber recalled having been plagued by the "mostly empty babel of ever recurring monotonous phrases" that came, he had believed, from the vibrating nerves of "talking birds." These were transmission lines for the "tone-messages" of souls which, shorn of body, continued to emit their unique sound-signatures, of which he was unique receiver and recorder. The sounds were "projected direct onto my inner nervous system"; were his outer ears to be hermetically sealed, still he would hear them. Breaks from the broadcast came only when he roared back at the cacophonous voices, his own bellowing "drowning with its noise everything the voices speak into my head." He was, in essence, jamming the signals, a phenomena that by 1899 was familiar enough to be practiced by Marconi operators competing with other nascent wireless companies. Schreber also felt that his thoughts were being intercepted, or "tapped," by his psychiatrist Paul Flechsig, a neuroanatomist who framed madness in terms of the electromagnetics of the brain and regarded paranoia as an electromechanical disorder of the brain's sensory and association centers. ("I find," said Marconi in 1907, "that probably the greatest ignorance prevails in regard to what is termed 'tapping' or intercepting wireless messages." True, "No telegraphic system is secret," as Preece had shown; but "It is incorrect to suppose that anyone can at will pick up wireless messages." Promotional guff: Marconi was pressing for laws against wireless eavesdropping. Meanwhile, asylums began to fill with patients who felt that they were wirelessly and unwillingly receiving commands from a world in full broadcast mode, or that their own thoughts, involuntarily broadcast, were being monitored by criminal cabals.) Freud, tapping into Schreber's *Memoirs* in 1911, diagnosed *dementia paranoides*: Schreber had produced a complex delusional structure ("paranoia") to fend off an Oedipal conflict with his father, an eminent orthopedist who had raised his son within an emasculating gymnasium of ingenious apparatuses designed to align the body with the mind and the mind with the middle class. As others in asylums reported receiving radio messages from angry or alien voices through the metal fillings in their teeth or war shrapnel in their bodies, and yet others were assailed by migraines due to cross-talk from a welter of telephonic voices or x-, y-, or z-rays, psychiatry would shift the diagnosis to a new term, "schizophrenia." Paranoid or schizophrenic, the hearing of inaudible voices—earlier understood as mystical or demonic—now had an explicitly electrical grounding and a techno-logic that powered the oxymoron of the inaudibly audible and the metaphor of visual noise. Anywhere that Mary went (as Edison had told his original phonograph), a lamb with oh-so-staticky fleece was sure to go.[244]

Neither schizophrenic nor atypically paranoid, the "scienticulist" Preece also felt hounded by noise. His interest in communications pricked by his "first science lesson" at the age of eight, when his father explained how sound made its way down from blasts at a slate quarry in the Welsh mountains to his ears in Carnavon, Preece had devoted his career to signals and their intelligibility in telegraphy, telephony, and lighthouses. Responsible for instituting systems of electric bells in rail stations and the better sort of hotels, and in his own Gothic Lodge on Wimbledon Common, Sir William was throughout his life acutely sensitive to class and noise, noting in diaries not only his distaste for crowds, Jews, and the "horrid discomfort" of second-class travel but the "doleful sounds" of Belgian rail horns, the "hideous and depressing sounds" of foghorns while he visited with Bell in Boston, and the "horrid howls—more like an elephant's trumpet than anything else" from New Jersey train whistles on his trip to see Edison. In New York City, Preece complained about what we now call visual noise, associated with the expansion of telegraphy and telephony: "Decidedly the most striking feature in New York to my professional eye is the poles that disfigure the streets in every direction. How such an enormity can have been perpetrated is simply incredible. Hideous crooked poles carrying twenty or thirty wires are fixed down the principal streets and sometimes three different lines of poles run down the same street.... [S]ometimes the mass of wires crossing a street form a perfect cloud." This was in 1877, years before telephones or electric utilities had come into their own. In 1885 the chief American journal for the construction trades attacked the "Wire Nuisance." Despite European successes at laying cable underground, American engineers had a horror of the interference that could arise when wires were run together in a cramped conduit, so instead "Every street is crossed, and recrossed, and filled up with these ubiquitous wires, and every severe storm 'floors' them by the hundred." The editors quoted one of Bell's own electricians: not only do poles and wires "disfigure the streets (and just here we may interpolate that all the artistic resources of modern civilization have been utterly unequal to the task of making the telegraph pole a thing of beauty) but seriously interfere with the operations of firemen." Overhead wires on tall poles were also costly to maintain, exposed as they were to the vagaries of weather, road repair, and derangement by the demolition of old warehouses and the construction of tall apartment buildings and taller office towers. Finally, the technical difficulties of underground cables could be easily managed with better wires and insulation.[245]

Fifteen years in the future, (Heaviside's and) Pupin's impedance coils would so strengthen signals as to render negligible the inductive interference from bundles of wires, and the town of Pittsfield, Massachusetts could boast that "Not a wire overhead, except in the outskirts, offends the vision, and not a pole makes an ugly interrogation point on the beautiful avenues." New York City's Mayor Hugh Grant began a program in 1890 to cut down "forests of poles, heavily burdened with scores of cross arms on which depended a cob-web of circuits that, encumbered with the wrecks of primeval kites, literally darkened the sky and actually shut out light from the adjacent buildings." Five mayoralties later, in 1904, the twenty-five arms of the West Street pole, at sixty feet "the greatest ever built," were at last chopped off and hauled away. Such clear-cutting was enabled by the new IRT subway, through whose side-conduits cables could be threaded and wires tucked away; it was supported by a vocal, well-connected City Beautiful movement that had been targeting wires and poles, muddy roads and urban grime, smoke and billboards.[246]

Cities could be beautiful, insisted the women's clubs, civic improvement leagues, and art societies that led the movement, when thoughtfully laid out in "easy undulations," with parks, parkways, and pleasant public buildings. If cities had to grow, let them grow with refinement. Suburban "villa parks" were defended from noise, filth, and congestion; could not the same be done, with good planning and progressive government, for the city itself? Briefs for the City Beautiful, initially aesthetic (a "fairer cityhood"), were framed more effectively in terms of a healthy circulation—of fresh air, pure water, exercised blood, exorcised waste, unimpeded traffic, undeterred commerce. Ease of circulation was quantifiable and inarguably desirable; harder to quantify and more debatable was the reassertion of quiet, a potent but oft unspoken impetus. "The question has doubtless often been asked," noted New England's *Pictorial National Library* as early as 1848 (when Boston, the nation's third largest city, had 130,000 residents), "why Mr. Webster retired thirty miles from the metropolis, to the most quiet of all towns, to find a farm and a country home. It may be replied, that the remarkable quiet of the place was one great recommendation of it, to his own mind." Half a century later, when the population of Boston surpassed half a million, City Beautifiers knew that their cities could never return, as had Webster, to rural placidity, but their acoustic model for the city was nearer that of a village backwater than a hustling metropolis.[247]

Paving materials were tested for their capacity to withstand extremes of temperature and runoff, to reduce mud and dust, and to minimize days spent on repair; they were *promoted* for their cleanliness and qualities of sound. Under the stress of larger stables of iron-shod horses and mules, heavier drays and omnibuses, faster gigs and phaetons with metal rims on their wheels, and miles of train and trolley tracks, road surfaces had been so punished that municipalities in Europe and North America were continually experimenting with new pavements. They tried gravel, cobblestone, granite, macadam, tarred macadam (tarmac), hardwood blocks, vitrified brick, and asphalt (bitumen, sand, and stone) in search of a surface that could take a pounding without cracking, buckling, or overmuch resounding. Geese, ducks, and pigs had been run off the roads, but dogs, donkeys, and horses still took aim at city streets, and under the onslaught of tons of horseshit, dogshit, piss, and spit (all of which did somewhat damp the noise of traffic), pavements had also to resist odor and infectious diseases. By century's end, the durability of pavement was being balanced against anxieties about germs and longer-term desires for quiet boulevards subject to the same "eternal vigilance" as the proper home.[248]

Think of pavement as a rough sound ribbon whose recording and playback mechanisms are shoes, hooves, wheels, canes, pelting rain, hail, and other falling objects. For each groove, rut, or pothole worn into the surface, the pavement returns a vibration peculiar to the shape and depth of the fissure and to the weight, pressure, pace, and contact-contour of each footfall, hoof-beat, or wheel-spin. In this sense, pavement is more sonically dynamic and temporally responsive than any gramophone or (as first conceived by Valdemar Poulsen during the 1890s) magnetic recording wire.[249] In an opposite sense, paving attempts to dismiss the unique sonic signature of a boot-print or cart-wheel in favor of a generic mobility whose sounds result less from the individual traveler than from the local history of travel. In either case, pavement does not loyally reproduce any specific act of trespass; it transduces a history of trespasses (as evinced by ruts or grooves) into a singular vibratory response to the newest trespass. It would be as if, at each playback, the recorded grooves in a gramophone disk returned primarily a distinctive static determined by the number of times the disk had been played and the age-defining dullness of the needle itself—and this analogy is not far off the aural state of early shellacked records after a dozen gramophonic run-throughs.

Granite could last for run-through after run-through, up to fifty years (we're back on the road now), but it was expensive to lay down and slippery when iced over in winter or heat-polished by summer Victorias; it also made a "nerve-rasping noise" when scraped by horseshoes and a "shrill, irregular rattle" under iron-clad wheels. Macadam (granite crushed and compressed to layers of fine gravel) was less costly, less slippery, and less noisy, but when rutted and drenched by rain it became a muddy morass. Further, wrote one keen auditor in 1843, the switch to macadam "had only led to a different *kind* of noise, instead of destroying it altogether; and the perpetual grinding of wheels, sawing their way through the loose stones at the top, or ploughing through the wet foundation, was hardly an improvement on the music arising from the jolts and jerks along [a granite] causeway." So bad was the noise that "Men's minds got confused in the immensity of the uproar and deafness became epidemic." Wooden blocks, like the hexagons of hemlock laid on a base of tarred gravel for Manhattan's Broadway in the 1830s, were quieter still, more easily repaired, less prone than gravel or macadam to a summer "Sahara of dust," and more sanitary than granite sets, whose gaps were quickly caulked with gobs of spit, tobacco, and horseshit. Wood, however, when waterproofed with creosote and rubbed to a gloss by months of heavy traffic, became in wet weather "almost like glass... and accidents are continually occurring, from horses losing their foothold and falling down." Glassy pavement was louder pavement, acoustically alive, while creosote turned wooden paving block into superb kindling for fires.[250]

Road trials continued, and as Dr. Ernest Hart editorialized in 1876 in *The Sanitary Record* with regard to efforts to extend wood and then asphalt pavement across London, "The distraction caused by noise is the reason of the adoption of these costly street coverings. It is not the mere distraction, though, which is so important; it is the effect of continuous loud noise upon the nervous system which renders it so serious a matter." The greater the wear and tear on every pavement, the more far-reaching its physiological effects. "Jars upon the nerves of hearing are communicated to the sensorium like other perturbations of nerve-endings," though a man cannot summon the police at every brutal auditory assault. If a man "were struck upon the cheek, the law would protect him, but if he is subjected to a jar from the noise created by an omnibus passing suddenly from a noiseless pavement on to granite, he must simply grin and bear it." By 1899 the jurors of Manhattan could bear it no longer; street noise had become so distressing that the court had to conduct its business with windows sealed even during heat waves, an intolerable situation that led the Grand Jury to demand that the abutting streets be asphalted. News of this demand reached the ears of a New Yorker who urged the formation of an "An Anti-Noise Society" to muffle not only street vendors and streetcar gongs but the built environment, "the unmuffled and resounding trestles of the elevated railway," the sheet iron of coal chutes, and "the rumble of innumerable vehicles over bad pavement." The city did pass an ordinance requiring teamsters hauling iron beams to wrap them in jute, but its road-quieting efforts, claimed Edward S. Morse (back in Massachusetts after years in Japan and Europe), were inferior to those of Berlin, where the pavements were "so smooth that I heard no noise from the teams [of carriages], but only the sound of horses' feet. In all the large cities in Europe you do not notice the noise of the street." He was exaggerating, though when the English actress Mrs. Patrick Campbell came to perform in New York in 1901–1902, she found the din of that city—its elevators, streetcars, and nasal twang—"demoniacal" and had to quit her first rooms because of the explosive excavation and clank of iron girders being lowered day and night into a new subway tunnel under construction across from her hotel. During her performance of *Beyond Human Power*, a "very quiet tense play in which she stayed in bed from beginning to end," it was said that she was so distracted by the rumble of the three cable-car lines running past the Republic Theater that she got permission from the Department of Street

Cleaning for wagonloads of sound-deadening bark to be spread over the granite pavement outside. In truth, it was the critic George Tyler who had complained to Campbell's manager that he could hear nothing of the play over the clank and jangle of the cable cars, and it was her press agent A. Toxen Worm (no kidding: Conrad Henrik Aage Toxen Worm, a Dane) who played up the tanbark angle for national publicity. The play itself, Norwegian, was among a number of modern dramas favored by "Mrs. Pat," who sought out roles as politically and sexually active New Women.[251]

New Women on bicycles—that's what accelerated the paving of many a road. New Women and pneumatic tires. New Women on "silver steeds" that promised to override "the exasperating noise and confusion of city life." New Women bicycling out to quiet country lanes alongside New Men, their touring clubs loud champions for cleaner, smoother roads. Inflatable rubber tires, explosive in their protests against sharp gravel, uneven brick, and cobblestone. The wire-spoked inner-tubed rubber-tired "safety" bicycles of the 1880s, improved with freewheel gears that allowed for coasting, were so much quieter than other vehicles that ordinances had to be passed requiring bicycle bells to warn unwary pedestrians; since energetic cyclists could reach speeds beyond the power of their brakes, their tendency on crowded streets was toward such constant use of their bells that cities began to rue this new ringing as greatly as that of the new electric carillons in church steeples. High-wheel bicycles had been all the rage in 1867–1869, but by the 1890s the safety bicycle was cheap and reliable enough to be sold by the millions, this when most roads everywhere were still sand or dirt and no more than a third of the streets were paved in the eleven largest cities in the United States. By 1910, in the wake of safety bicycles, streetcars, electric trolleys, automobiles, and the first motorcycles, those eleven cities had paved more than half of their streets, much of these with asphalt because of its reputation for cleanliness and—so insisted Philadelphia's Board of Experts on Street Paving—noiselessness.[252]

Quincy Gillmore of the U.S. Corps of Engineers had put traction well ahead of noiselessness in his ranking of pavements in 1876. Thirty years later, amid what seemed to be thrice the traffic, a civil engineer at the University of Illinois placed noiselessness ahead of healthfulness and traction in a *Treatise on Roads and Pavements*, and found that on "sheet asphalt the only noise is the sharp click of the horses' shoes." Or, as city dwellers found, the noise of jackhammers: "No sooner is a street made smooth and beautiful with this elastic and tenacious surface than the necessity arises for piercing it here and there... until it is seamed and scarred in every direction." In 1902 alone, pavements across New York City were torn up 10,619 times to lay in or repair utility pipes and water mains. Cheaply resurfaceable, blacktop spread across the States and continental Europe, where a greater percentage of the population rode bicycles to and from work. Yet when Robert Saunders left St. Louis in 1911 to ride the "rattlers"—freight trains—as a recruit to the "great fluid army of foot-loose, migratory workers" some called hobos, he saw in five years of tramping just one paved intercity road, California's El Camino Real. Everywhere else it was wagon tracks, packed dirt, mud, and dust.[253]

Dust... philosopher's emblem of the insubstantiality of the material world, preacher's emblem of mortality.... Now it was the street-paver's anathema: sweep it away, oil it down, tar it over. So declared France's Inspector General of Roads and Bridges at a plenary session of the first Congrès International de la Route in 1908. The faster the world seemed to spin, the greater the urge to see dust as flustered dirt, a half-Brownian, half-Darwinian commotion of flakes of skin, specks of seed, infinitesimal flecks of soil. Dust was what aether would be if it had the gumption to make itself known. Deposited on the many new smooth surfaces introduced to the built environment—mass-produced wallpapers, windows, and porcelain; vulcanized rubber, celluloid, Bakelite; cast iron, aluminum, chrome, stainless steel; concrete—and on brighter fabrics dyed with aniline colors, dust became at once more obvious and dubious. In response, domestic economists stripped rooms of sound-deadening but dust-loving draperies, tapestries, and carpets and, hand in hand with municipal sanitarians, turned to new, convincingly loud, devices: street-sweeping machines that sounded like cantankerous tornadoes; huge ceiling fans that shook the dust loose from every corner; corrugated metal cans that clatter-banged when slung into garbage carts at dawn; the notorious "vacuum" cleaner. Electric vacuums were couched very early on within a rhetoric of quiet—the Arco Wand with "no rumbling to annoy home folks or neighbors in apartments"; a "Practically Noiseless" Regina; the Santo so unobtrusive you could use it "in the sick room, in the hospital, the hotel or apartment or anywhere else"; and the Vacuna "Goddess of Leisure" whose "smooth, steady, almost inaudible purr of the turbine... indelibly stamps it as a *mechanical masterpiece*"—but to this day vacuums are louder than they need be, so as to persuade us of their power to devour swarms of the almost-invisible.[254]

Storms of dust, visible or invisible, bore within them particulates of asphalt and cobblestone, motes of glass and arsenic, filings of iron, burnishings of copper. So dust was not only a catalyst of the noises of house- and street-cleaning; like smoke, it was an *agent provacateur* of physiological noise, of wheezing, coughing, snuffling, sneezing. Mixed with smoke, it became a sooty grit that pocked the skin and roughed up medieval stone no less than the white marble and cream-colored brick of the best Victorian enclaves. Smoke on its own was more psychologically disturbing, for it was ostensibly evanescent, yet in its urban-industrial formulations it stuck around, making a nuisance of itself, streaking walls, darkening windows, mucking up clothes. It also mucked up the ears, irritating the canal, impacting the wax. And it aggravated the aural environment beyond, provoking excesses of carpet-beating and, on the gloomier days, spasms of horn-blowing. As it blackened buildings and blighted trees, smoke—or "smog" (1905)—accelerated the absorption of solar radiation and made cities greater heat sinks with more restless residents who fled steaming interiors to cluster on porches or rooftops for relief from blistering summer temperatures and "the smoke and smoulder of London [or Boston, or Warsaw] Brick, hot and choking as one huge Brick-kiln." In consequence, as each new soundcasting device was brought into the home, its volume would have to be turned up during heat spells when windows were wide open. Whence a new round of complaints about pianolas or gramophones (later radios, then stereos) played too loudly.[255]

Upset by palls of smoke since the 1830s, European and American reformers had to learn to shape arguments that would find a hearing among manufacturers unmoved by appeals to pure air or clean linen. If Horatio Allen, visiting from the States in 1828, already found "a canopy of coal smoke" hanging over Liverpool from the chimneys of the Wolverhampton iron district, and if in 1842 the Manchester Association for the Prevention of Smoke was deploring an "eternal smoke-cloud" over the city, wasn't that "the incense of industry," the price society had to pay for a prosperity stoked by coal? If servants and their mistresses fumed over grimy wallpaper and unsightly facades in Pittsburgh as in Sheffield, where "hideous black impenetrable clouds" so blotted out the sunlight that "the very birds cease to sing," didn't artists and photographers cotton to the soft-focus of murky landscapes, wasn't there "a glorious brilliancy of the effects which a soot-laden atmosphere sometimes causes at sundown," and weren't Claude Monet and André Derain drawn to London just to paint its deepest "fogs" (thirty days of fog in 1850, sixty in 1900)? If a union man, arguing in 1896 for a nine-hour day, addressed "Ye iron moulders who spend your lives in the dark steam and gas, and black vapor of the foundry; how will it be to breathe God's pure air for an hour more daily?", how could he dismiss the putatively "antiseptic properties" of carbon that made coalminers immune from tuberculosis and reduced their overall mortality to a rate lower than other industrial workers? And if soot-fall (and acid rain, discovered in the 1850s) forced urban householders to keep their windows closed one day in every seven, wasn't this a more agreeable sabbatical than unemployment? Workers, wrote a Lancashire poet, Mrs. E. J. Bellasis, saw smokeless chimneys as

a signal of despair.
They see hunger, sickness, ruin,

Written in that pure, bright air.[256]

Ornery or smug as they might seem, such apologies for smoke and smog would be reiterated in apologies for noise. Noise too could be defended as the adjuvant of a growing industrial base, a signal of robust economies, backslapping good humor, and crowing vitality. In so far as smoke was the visible residue of commercial, industrial, and urban processes that were ear-sinister when not ear-deafening, noise and smoke were incestuous cousins that reformers hoped to uncouple and deport. Like the anti-noise campaigns that came on their heels, anti-smoke campaigns had to do more than point at the phenomenon with a Ringelmann chart by their side to calibrate the blackness. Activists had to prove, on the one hand, that smoke was hardly as innocuous or fleeting as the wisps rising from cottage chimneys in Romantic landscapes, and on the second hand that whatever "went up in smoke," even as it settled in one's sinuses as a dreadful catarrh, was truly *lost*: tangible evidence of wastefulness, bad engineering, inefficiency.[257]

Commissions had been sent out to investigate pollution from the burning of coal in London as early as 1285, whence centuries of local ordinances against the nuisance of smoke and its undeniable "presumptuousness," a compound of the visual and the auditory. At a fourteenth-century London Assize of Nuisance, Thomas Yonge and his wife complained of both the smoke and the noise of armorers next door whose sledge-hammer blows "shake the stone and earthen party-walls of the house so that they are in danger of collapsing, and disturb the people and their servants day and night." Such nuisance suits were unreliable remedies against traditional practices at the domestic hearth or industries long in place (the armorers protested that the Yonges were newcomers and that they had been there first); verdicts were conditioned by what a person could reasonably expect in the environs of a stable or of a smelter that used eighteen tons of coal and thirteen tons of ore to produce one ton of copper, along with mountains of scoria and emissions of sulfuric acids strong enough to etch a stone wall. Around 1700 shoemakers began putting (noisy) cast-iron pattens on shoes so that the long hems of women's dresses would rise above the muddy soot that covered every major thoroughfare under London's "urban plume." By the mid-eighteenth century, the Welsh were speaking of a "smoke disease" that destroyed the teeth, bones, and hooves of sheep, although workers, sallow and short-lived as they might be, were well-enough paid to descend willingly into "the infernal regions" of smelters. During the Great Copper Trial of 1832, medical men testified on behalf of the healthiness of smelter smoke; in 1895, faced with the linkage of that smoke to deadly levels of arsenic and copper in farm animals and the barren soil of nearby farms, a Welsh jury still failed to assign any liability to the smelting companies in question, although the House of Lords in 1865 had upheld a lower court decision in favor of William Tipping's suit for damages over the ruin of his lands due to "large quantities of noxious gases" from a local smelter.[258]

"Economy Combined with Smoke Prevention" was the slogan of the London Smoke Abatement Exhibition in 1881–1882, visited by 116,000 people now that the "smoke-curse" had turned from a drab grey toward a pasty yellow indifferent to the Smoke Nuisance Abatement Acts of 1853 and 1856; now that galleries had to put their paintings under glass; now that the new Houses of Parliament were being eaten away by atmospheric acids—carbonic, nitric, sulfuric, hydrochloric; and now that statisticians were attributing "excess mortality" from asthma and pneumonia to the "fogs," which "really means fog *plus* smoke." The point of this Exhibition was to convince commoners that "so harmless-looking and quiet a thing could do much mischief," that a smoke-filled atmosphere was unnatural, unhealthful, noisy with coughing and the hawking of phlegm and, as Rollo Russell had written in 1880, an agent of class violence: the unventilated urban poor, unable to escape into fresh country air, would gather in the city's sludge to plot the 4 R's: robbery, rape, riot, revolution. Even modern painters who, like Turner, might once have been inspired by the chiaroscuro of smog, became depressed, said the president of the Royal Academy at a banquet concluding the Exhibition, by "the interminable hours, days, and weeks of enforced idleness spent in the continuous contemplation of the ubiquitous yellow fog." In the years following the Exhibition, the number of full-on "fogs" in London continued to rise, as did national coal production and the feeding of open fires at British hearths; in 1902 every fourth day was smoke-cursed.[259]

Things were little better elsewhere. Although Napoleonic codes had begun the regulation of smoke-belching factories, and Prussia and Saxony by mid-century had legislated against devices "that transmit damaging steam, fumes, smoke or soot to neighboring property," German reformers in 1906 were pressing for a full-on War Against Smoke, *eine Bekämpfung des Rauches*, seeing as how the Prussian bureaucracy had so poorly managed the air quality of a rapidly industrializing nation. Across the ocean, Pittsburgh may have been the first "smoky city" in America, but it was no longer alone in its misery. Any place that had homes, businesses, and factories burning bituminous ("soft") coal with its high sulfur content could end up blanketed with smoke. In Cleveland a woman who lived near the factories of the American Steel and Wire Company could hear at night "the cinders falling like hail on the roof." In Cincinnati, wrote a visitor, "Cleanliness in either person or in dress is almost an impossibility"; from "the smoke of hundreds of factories, locomotives, and steamboats" came a "dismal pall" that stained the linen on clotheslines and streaked each face with soot. In Chicago in 1893, a few miles beyond the white walls of the Columbian Exposition, a visitor found an awful, moral darkness: "Smoke, clouds, dirt, and an extraordinary number of sad and grieved persons," this despite the first thoroughgoing American smoke ordinance and a funded inspection force. The annual deposit of soot in St. Louis at century's end was estimated to be 350 tons per square mile (still less than the 430 in London, 595+ in Pittsburgh). At the New St. Louis Banquet of 1901, a speaker envisioned a city "basking in the sunlight of a smokeless sky" under the leadership of a new mayor who also vowed to rid the city of the offal of boss rule; sixteen years later, smoke still veiled the sun, and John Grundlach of the Chamber of Commerce sent an angry letter to the *St. Louis Globe-Democrat*: "The indifference of the public is astounding. Roofing and paving furnaces emit dense clouds of black smoke in the heart of our shopping district and apparently nobody cares. Street cars pass through the streets trailing a camouflage of dark yellow without objection from anyone in authority." Mrs. Imogen Oakley, chair of the Smoke Nuisance Committee of Philadelphia's Civic Club, had had to admit in 1906 that "Owing to the prevailing discomfort from this nuisance, this Committee will be a center of activity for some time to come." She would also chair the Civic Club's committee on noise.[260]

Unnecessary noise. Committee on *Unnecessary* Noise. *Unnecessary* because the committee had no hope of any out-and-out suppression of noise once it was demonstrably everywhere and everywhen. One could only hope to make a case against that noise for which there could be no earthly excuse. As, for example, screeching streetcar wheels, according to a petition of otolaryngologists urging "the Boards of Health throughout the United States to secure the attention of the Street Car and Street Railway Companies for an early consideration of the sanitary, therapeutical value of noiseless street car wheels," given that "Thousands in our hospitals and homes who are ill and nervous are greatly disturbed by the unnecessary noise of passing cars." This lesson on the rhetoric for the lessening of noise was taught by the anti-smoke campaigners, who had much more going for their cause yet had to retreat from the standard of smokelessness into bivouacs of economy and efficiency. Inspired by Oliver Lodge's speculations of 1907 on atomic energy, Annie Besant and C. W. Leadbeater could envision an Earth with "no grime, no smoke, and hardly any dust," while a noiseless planet was, in science fiction as in show business, moribund. If nuclear power was barely a glimmer in the eyes of physicists in 1913, it was possible that year for a well-heeled theosophist to adopt electric lights in lieu of oil lamps, electric irons in lieu of coal-heated behemoths, clean-burning gas ranges in lieu of wood or coal stoves, asphalt pavements in lieu of muddy gravel, and a White Steamer in lieu of backfiring gas-powered automobiles—a White Steamer, noiseless, free from vibration, "odorless and smokeless and, therefore, . . . the only car which can be driven in Central Park without danger of the driver being arrested because his car is emitting malodorous vapors."

Granted such prospects for smoke-free kitchens and avenues, granted the obnoxiousness of smoke, and granted that physicians bore loud witness to the deadly effects of coal dust, soot, and smog as well as the dangers of "vitiated" air in houses shut up against a dirty sky, anti-smoke campaigners should have made more headway than they did. But they had to contend with the bluster that heating, manufacturing, and shipping were *necessarily* grimy affairs, and they had to prevail against the more treacherous, ostensibly altruistic argument that working-class families would suffer if economies declined or prices rose due to the substitution of more costly (cleaner) fuels, such as hard coal, gas, or electricity, for cheap, plentiful (and highly sulfuric) soft coal. When, after a full year of discussion, the Mothers' Club of Cambridge, Massachusetts, presented a "Remonstrance Against the Smoke Nuisance" to Boston's city fathers in 1909, they found that "The corporations petitioning for *more* time for pouring forth smoke were present and all agreed that nothing could be done to diminish the smoke without greatly increasing the cost of their products."[261]

Declaring, as did Grundlach in St. Louis, that "There is as much sense in smoke production as there would be in open sewers," was a dramatic rebuttal: who would not concede the sensory and sanitary benefits of closed conduits for effluents? Even more persuasive to businessmen were rebuttals in terms of dollars and cents. "Smoke is absolute proof of imperfect combustion," declared the Kewanee Boiler Company in its ads for Smokeless Firebox Boilers. "It is a sure sign that much of the heat in the coal you are using is being wasted—for smoke is nothing but good fuel which has been only partially burned." A cartoon in the *Chicago Record-Herald* postulated that the savings in laundry bills from no longer having to wash besooted shirt collars every night would pay for building the Panama Canal. Frederick Law Olmsted featured this cartoon in his 1908 pamphlet on *The Smoke Nuisance*, in which he declared that "smoke is economic waste in itself," a waste compounded by the expenses of extra laundering, washing of windows, and cleaning of streets, not to mention the cost of treatments for asthma and emphysema (or the noise from new trolley lines taking workers away from the smoke of the center city out toward clearer, cleaner suburbs). Smoke did not merely symbolize waste—it was waste, and it generated waste.[262]

Literal and figurative were all mucked up here—the excretory and excessive, shit and dissipation. As Adam Rome has shown, the word "pollution," which had long denoted moral and sexual defilement, began to be inserted into nineteenth-century debates over sludge-slowed rivers and ink-thick creeks, while "air pollution" did not come into common parlance until 1907 (and "noise pollution" not until the 1960s). Instead, air could be "corrupted," "contaminated," or "fouled" by "noxious vapors," "exhalations," "the smoke evil." Three points had to be made, in any case: as with waste, dirty air could mount up; as with embedded grime, it could not be wished away like a passing cloud; as with noxious weeds, it was by nature invasive. After a few days of smog the unnecessary became annoying, after a few weeks unhealthy, after a few months unrighteous: "Chicago's black pall of smoke, which obscures the sun and makes the city dark and cheerless, is responsible for most of the low, sordid murders and other crimes within its limits," claimed the president of the association of women's clubs in Chicago. "A dirty city is an immoral city, because dirt breeds immorality. Smoke and soot are therefore immoral." And insidious: as a male physician wrote, "Women living in sunless, gloomy houses and attired in somber clothes [to hide the soot] also prove to be irritable, to scold and whip their children and to greet their husbands with caustic speech." It was plain to a smoke inspector in Sheffield that although it was a "*silent* nuisance," smoke had to be called out, for "A proportion of the physical degeneracy of the human race, of which so much has been said during the last few years, is directly traceable to the constant inspiration of impure air." As for the more spiritual side of inspiration, Mrs. S. S. Merrill of Milwaukee feared that smoke could "unchristianize the nation." Campaigns against smoke were perforce smear campaigns through which smoke was indicted for laying waste to land, bodies, buildings, and souls while hiding behind a facade of progress. Like the most blatant of billboards, smoke was at once brazen and opaque.[263]

Where and whenever the air was clear enough to see twenty feet in front of you, what you were likely to see in any town center and extending out along rail lines into ever-noisier meadows and farmland were flyers, placards, posters, painted signs, and billboards. With the advent of web-fed presses, of endless rolls of manufactured paper automatically cut and trimmed, and of photomechanical color lithography, not only the pages of magazines but the sides of construction fences ("hoardings"), factories, commercial buildings, and barns could be cheaply plastered with loud bright advertisements for bread and circuses, pills and ointments, soap and cough syrups and, across America, Uneeda biscuits with the very first "million-dollar" advertising campaign, involving posters nationwide on subways, trolleys, and trains as well as huge boards. Standardized in the States at 8′8″ × 19′6″ on twenty-four contiguous sheets, billboards were often double- or triple-decked, and the bill-posting industry exploited opportunities larger and longer still, painting/pasting monumental advertisements on the sides of grain elevators, along breakwaters, and on the ten thousand square feet of the exterior brick of a Chicago warehouse. An enterprising salesman "coming suddenly upon that almost majestic advertisement" might be overcome by "a feeling almost akin to awe," and at street level there might be some charm to watching "the flaming poster just being pasted by the energetic if not acrobatic bill-sticker," but City Beautifiers,

Municipal Housekeepers, and Urban Planners would have none of it, especially as downtown signs began to flash with blinking incandescent bulbs.[264]

London's *New Review* in 1893 called upon five eminent Later Victorians to inveigh against "The Advertisement Nuisance": the M.P. and historian William Lecky; the novelist and philanthropist Walter Besant (brother-in-law to theosophist Annie); the elite portraitist William Blake Richmond; the comic novelist Julian Sturgis, heir to a Boston mercantile fortune; and the noblest of the quintet, Mary Jeune (Lady St. Helier), reformer and memoirist. Each pursued a characteristic tack. Lecky, who distrusted popular culture and rootless populations, saw billboards as yet another vehicle in the vulgarizing of daily life, which had been so conned by speed and greed that "the green fields along all the chief lines of railway for more than a hundred miles from London have been disfigured by hideous placards." Besant, who had built a People's Palace in London and written on *All Sorts and Conditions of Men*, took the American "Abuse of Public Advertising" as a warning to England, which if not vigilant might find its most idyllic settings over-towered by "huge advertising boards in all colours" promoting Catchall Kodak or Electric Instantaneous Apple Blossom Hair Starter. Hope there was, wrote Besant, for if "the Millions have yet to learn how to unite" against ugliness, Canadians and Americans had recently stripped away decades of "vile advertisements" from the banks of Niagara Falls. Richmond, for whom Charles Darwin and Otto von Bismarck had sat, and who preferred the glossy surfaces of Renaissance masters to the nervy brushstrokes of Impressionists, described the £4,000,000 spent annually on public advertising as a tyranny of vision. What advertisers sought was not admiration for their artistry but attention to their product: "Should sober colour and a rational scale not succeed in claiming the attention of the passer-by, brutal colour and Brobdingnagian proportions must be resorted to." Sturgis, the gentleman wit with quarters in London and a manor in the country, addressed hoardings in global agricultural and local satirical terms. As English wheat and cattle were displaced by imports from New Zealand and Canada, the farmers of Albion would seek out new and lucrative crops: stalks of billboards, stacks of placards, rows of signs. "There is absolutely no limit, as it seems to me, to the spread of this triumphant crop," in ever greater profusion and variety. Jeune, who worked on behalf of the children of sunless slums, acknowledged that publicity was vital to modern enterprise, but when she saw Cheapside placards on stilts and barges lurching into the countryside, she blanched at the desecration, for "Even the poorest person, the most habitual liver in towns, the veriest cockney is warmed into expressions of joy and appreciation at beautiful scenery."[265]

Beautiful World, the journal of Great Britain's Society for the Checking of Abuses of Public Advertising (SCAPA), argued from 1893 through 1909 that its cause was "not a question of sentiment but one of utility." If authorities spent thousands on gardens, museums, and grand public edifices to "train the eye" of the citizenry, why in heaven would they tolerate a "chaos of catchpenny inscriptions," flashing lights, and "sky-signs" atop tall buildings? The Society's secretary, Richardson Evans, drew an explicitly acoustic analogy: "If, in a crowded room, every one talks gently to his neighbour, there is a chance that all the groups will be able to converse; but if the speakers go on competitively raising their pitch, no one will be able to hear in the babel of shouts. Is it necessary to point the meaning of the parable?" It was not. In 1894 a contributor to *Beautiful World* had proposed that "the doctrine that applies bye-laws in defence of the ear, should hold good of bye-laws in the defence of sight." The next year, after publishing another contributor's diatribe on "The Grievance of Unnecessary Noise," the editors confessed that they had "often been urged to make relief to the public ear as much a part of our aim as consideration for the public sight." SCAPA dared not blur its focus, but the editors agreed that the advertising nuisance was no worse than the noise nuisance, which "penetrates into our houses, our studios, our sick-rooms, and our hospitals." Besides, there was no hope that the smog whose acids eventually erode garish colors and giant letters would soften the raucousness of knife-grinders or the howls of newsboys ("worthy of an eighteenth-century madhouse") who also upset that "quiet enjoyment of the world" which the Society meant to defend. So arose an alliance among campaigners against excessive noise and gigantesque advertisement and a merging of rhetorics that expanded the field of visual noise.[266]

A certain Mr. Crackanthorpe suggested that SCAPA use alliteration to drill children in slogans such as "Reckitts ruins rural retreats and renders road and rail repulsive." Clever yet mild, as befitted a complaint about soap powder billboards. The convergence of aural and visual ugliness was more sharply dramatized by Louisa Starr Canziani, an artist mad enough to run the horrors of street music up against those of the "*Eye Grinder*" whose bad commercial art was no better than the mincemeat notes ground out at the crank of a hurdy-gurdy. "A glaring billboard," ruled a judge in California, "is just as offensive to the immediate residents" as a stone-breaking machine "or the chime of hoarse bells." And like the noisiness of a world sped up by electricity and powerful motors with ever-more piercing horns, sirens, and whistles, signage had become more extensive and intrusive in order to keep up with people moving at an ever-faster clip. Commuters hurtling down the road at fifteen, twenty, or even forty miles per hour would never notice a handbill; at twenty-two feet per second (= 15 m.p.h.), they might attend for at least a few seconds to a thousand bulbs flashing down a city block or three hundred connecting colorful sheets trumpeting Barnum's circus. Largeness and brilliance, like noise and other forms of the obvious, were as much a function of speed as of commerce.[267]

Were commerce itself jeopardized, as was the case in the Rhineland, where hundreds of thousands came "every summer from far and near to seek on the sunny heights, wooded hills, and shady valleys rest and delight," there was no stopping a law for "preventing the disfigurement of places remarkable for their natural beauty." The Diets of Prussia, Hessen, and Saxony could not "overlook the pecuniary loss that will accrue if the progress of disfigurement diminishes the profits from the tourist traffic," but such rationales for legislation, like the French taxing *les affiches* by the linear meter, appealed to Mammon to save his own skin and ignored the aesthetical-ethical-medical principles upon which anti-billboardists built their case for loveliness (and visual quietness) everywhere—inside subways, along Asylum Hill in Hartford, on the approach to the Academy of Our Lady of Mercy in Pittsburgh, or on the main streets of Magdeburg. "I don't believe it is possible for a person for his own interest to make a great clamor in the city at any time of the day," testified the muralist and art professor Frederick Dielman in his capacity as president of New York's Federation of Fine Arts. At the hearings of the Mayor's Billboard Advertising Commission in 1913, which had to review the city's 385,955 square feet of boards, Dielman noted that "the German brass band has disappeared from the streets of New York. I believe there is a regulation in the way of the organ grinder, and I think that the offense directed against the ear has been regulated." Should not offenses against the eye be regulated on the same moral—and medical—grounds, given that "We know perfectly well that delicate people are intensely susceptible to the color of their surroundings," especially primary colors? Discussing "Hospitals and Esthetics" in 1915, Grosvenor Atterbury, architect of the recently opened Henry Phipps Psychiatric Clinic at Johns Hopkins Hospital, would explicitly invoke "visual noise"—in quotes—as an idea crucial to his claim of "an analogy between discordant esthetic conditions, in color, line and form, and the corresponding reactions of the senses of smell and hearing when irritated by what we call smells and noises" in hospital wards, which should be to all the senses "a tonic," beautifully therapeutic. The lawyer Albert S. Bard, founder of the Citizen's Union and president of New York's Municipal Art Society, argued at nearly the same time in nearly the same language that the "hot and restless signs" of outrageous commerce should be as unwelcome in the poorer parts of town as would be an iridescent board forty feet high framed in gold at the entrance to Central Park, for "You shriek at us so loudly and thrust forward your twiddling lights and symbols so without proportion or restraint or beauty that we shed you as we shed the street signs and sounds in sheer self-preservation." Outdoor advertisers made a fatally neoplastic noise; they "scream and shout and clamor at the public, visually speaking, each shrieking that his own goods are the best, and the whole effect is one of violent contradiction," not to mention the erection of hideous structures "which deform our cities and scar our open country."[268]

Behind the "painted monstrosity" of each billboard lay rubbish. The bracings of billboards caught, hid, and bred piles of refuse, and "Therein this evil-smelling and epidemic-breeding debris rots beneath a summer sun," becoming (said doctors) a source of typhus and typhoid and "a veritable Frankenstein of death." Hygiene aside, "rubbish" was also vulgar, and in being vulgar, bulked up the notion of visual noise. Lurid billboards for laxatives, burlesque shows, and chewing tobacco were seen-and-heard to "flaunt their loud color, their ugly vulgarity in the face of every passer-by." And what could one say about the slogans painted "with grotesque coloration" on the sides of outhouses, barns, and covered bridges, or the use of stone cliffs to proclaim a new preparation for hemorrhoids? It was as if, wrote Richard Gilder, poet and editor of *Century Magazine*, "at a symphony concert vendors of soap should be allowed to go up and down the aisles and bawl their wares." Speaking in 1908 to the Sphinx Club of Manhattan, whose membership consisted of (quietly enigmatic?) advertising men, J. Horace McFarland of the American Civic Association predicted that outraged public opinion would no longer tolerate the sun setting behind Crystal Domino sugar, rising to Harvard beer, and crossing east to west along roads and rails "begirt with whiskey, phonographs, fly screens, corsets, tobacco, beer and razors, often to the second and third tiers of shouting signs." If the merchants of Berlin, Paris, Vienna, and Buenos Aires got along fine without billboards, there was hope of ending the blight in America.[269]

Ah, but the 8,500,000 linear feet of billboards and $100,000,000 invested in outdoor advertising in the United States in 1908 had an advocate no less optimistic than McFarland: E. F. Trefz of the Gunning (sign) Company and System. Speaking before the Convention of Associated Advertising Clubs of America in Kansas City, Trefz dated his profession to prehistoric times when hunters left "advertisements" of their presence upon the trees and rocks; in fact, the out-of-doors was the "only natural" advertising medium, for Man was still "an outdoor creature, trained to outdoor observation, susceptible to outdoor influence," and in need of outdoor activity. Most men were of the order of athletes, not aesthetes; to prove this Trefz took up the topic of music: "Few enjoy Wagner, some pretend to, and the rest prevaricate, but the most of us can hum 'So long, Mary.'" How do folks everywhere come to be comfortable with a George M. Cohan Broadway tune, he asked? The laws of psychology with regard to effective advertising were fivefold: visibility (or "scale"), stability (of message), and variability (of design) prompt people to take notice; suggestion and reiteration fix the message in the subconscious, from which arises "the resolve to buy." For the masses that every advertiser hoped to attract, it was thus a matter of calling-to-attention, then re-minding. What "supersensitives" considered a "brutal" attack on the eyesight, Trefz knew as relief from monotony; audiovisual flatness was the bane of advertisers indoors and out. More than a credo of advertising, wasn't it an axiom of human biology that "A gray world, without variation, would be deadly in its monotony"? Loud colorful advertisement was not merely a way of life for advance men, politicians, newsboys, and music salesmen who cranked up victrolas playing ragtime on the sidewalks outside their stores. It was the way of all who were truly alive.[270]

Animation was a spectacularly primitive definition of life and liveliness. "We are at present," wrote Sylvester Baxter for Gilder's *Century Magazine* in 1907, "in the childish period of advertising development—the era of Chinese gong and fire-cracker methods." (Modernizers in China, meanwhile, were cracking down on street peddlers, rickshaw runners, and all whose drums, cries, calls, and carts got in the way of commercial traffic or official business.) Baxter's ideal was quiet, orderly Germany, where he had studied in the 1870s; his style was Bostonian, as secretary to Greater Boston's Metropolitan Park Commission and Metropolitan Improvements Committee; his language was optico-acoustic. Confronting the "very babel of discordant appeal" that forced upon the public eye "in unpleasant ways and at inopportune moments, attention to things which may or may not be wanted," he hoped that advertising could evolve from hallooing and huckstering to the equivalent of a concert of classical music, especially with regard to the interiors of railroad stations, "frantic with advertising of the most shrieking character." Like billboards that shouted across verdant valleys, "Just now the chief idea in this sort of advertising is vociferosity. The making of a huge din, however, would render the perpetrators liable to prosecution for being a public nuisance. At present they disregard the fact that the eye is as sensitive to discords as is the ear." Despite the uneven application of nuisance law in cases of annoying sounds, anti-noise suits and legislation had at least the patina of antiquity (how the metaphors mix!) and could be pursued on grounds other than aesthetic. "Had not the common law taken the auditory sense under protection, we might possibly now have the equivalents of myriads of boiler-shops, calling the attention of passers along the highways to the merits of all sorts of wares, with megaphonic phonographs, steam-calliopes, and other appalling acoustic devices. But that is precisely the sort of thing that thousands of advertisers all over the country are doing to the eye."[271]

"Visual noise" was not yet a phrase in circulation.[272] I have begun using the phrase nonetheless, for the concept, if not the tag, had gained a wide hearing by 1900. In the 1770s, the German mathematician, electrician, and philosopher J. G. Sulzer had argued that "the sense of hearing is by far the most effective path to the emotions," and that musical dissonance was "far more easily perceived than is a clash of colors." For example, "Who would claim that clashing or unharmonious colors have ever caused pain?"

The ear, by contrast, could be "so adversely affected by dissonances as almost to bring the listener to despair." But from Baudelaire's criticism of "cacophonous" painting through Baxter's repudiation of "shrieking" signage, the concept of visual noise had grown in reflex to the perceived ubiquity and constancy of noise, its electrical transfiguration, and its inaudible refigurations. The campaign against egregious advertising was vital in this last regard: not only did campaigners conflate uglinesses optical and acoustic; they also depicted the effects of the visually ugly in terms of its painful intrusiveness, a quality that had previously been reserved for odor, smoke, thievery, and noise. "Loud" clothing, acrid smells, London fogs, and oompah bands could be *ob*trusive, offensively obvious in public. When, however, one began to speak of the *in*trusiveness of visual noise, one was entering a new domain: privacy.[273]

13. Personals

"Comeliness" is a neutral state, "neither of charm nor of repulsiveness." Like true gentility, wrote Richardson Evans, it invites no attention. In this new "Age of Labelling," unhappily, the aesthetics of the unobtrusive were being painted over by a splurge of self-promotion: "I am: therefore I announce the fact in large capital letters." If such boldfaced marketing and mass merchandising were "a trespass on the Public," the dare and blare were most awful when penetrating the most intimate of precincts: the

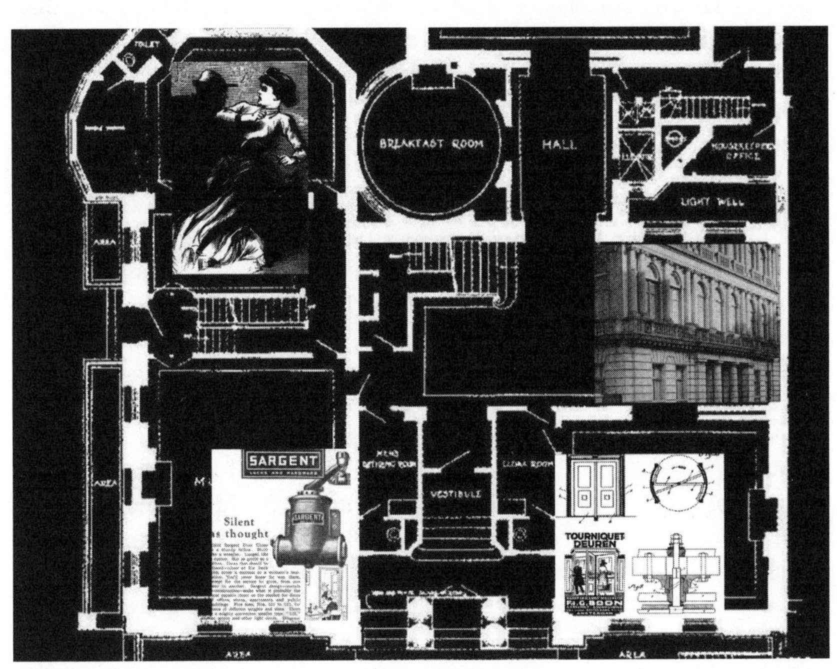

SOUNDPLATE 17

home, the heart, the mind. Evans's circle found such advertising offensive "because it is (through the eye) a vulgar intrusion on their intellectual calm. To them the perpetual iteration of one name or phrase in glaring colours, jars on the nerves as much as the persistent solicitations of the same begging impostor at the same street corner." Nag nag nag nag nag.[274]

Billboard ordinances had been voided in one court after another when towns tried to legislate visual propriety or defend a standard of comeliness that was, wrote Justice Wilkin in *City of Chicago* v. *Gunning System* (1905), "mainly sentimental, [intended] to prevent sights which may be offensive to the aesthetic sensibilities of certain individuals." Outdoor advertisers like Trefz argued that their posters and billboards were "the poor man's gallery of art"—attractive, professionally designed images erected in grim industrial sectors and along the dullest of country lanes through miles of cotton. Justice Barber, in *Commonwealth of Massachusetts* v. *Boston Advertising Company* (1905), agreed that the State could bar advertisements "of indecent or immoral tendencies, or dangerous to the physical safety of the public" but had no right to remove a forty-foot-wide billboard anchored in private land, whose three-foot-high letters stood out, black on orange, at the edge of a Boston parkway, for "the well-being of the ordinary person who uses a public park or parkway never can be so far affected by the visibility of signs, posters, or advertisements placed on the ground as to injure his health." When the Missouri Supreme Court did at last find cause to rule against Gunning (1911), it was on grounds of physical safety rather than visual assault: a person could be imperiled by the fire hazard of trash accumulating at the foot of billboards too close to public highways, by thieves or loose women lurking behind billboards too large to see around, and by slats sent flying from boards too rickety to withstand high winds.[275]

Would that there were another legal approach, one that made invasiveness itself the issue. What if, along with the common law's acceptance of a right to the "quiet enjoyment of one's property," there were "a right to be let alone"? That was the phrase used by two young lawyers, Samuel Warren and Louis Brandeis, in a seminal article of December 1890—a phrase that would advance with Brandeis up the judicial ladder to the Supreme Court, where he sat from 1916 to 1939. The right to be let alone, they argued, proceeded implicitly from a library of other established rights: the right to an unbesmirched name (laws against libel and slander), the right to an uninterrupted spiritual life (laws guarding freedom of worship), the right to the expression of one's thoughts and sentiments (intellectual property law), and the everyday right to be rid of noise, odor, dust, smoke, and vibration produced in one's vicinity (nuisance law). All told, these rights presumed "in reality not the principle of private property, but that of inviolate personality."[276]

Proper Bostonians both, sons of wealthy merchants (grain, paper mills), first and second in the same class at Harvard Law School, housemates and law partners, members of the Dedham Polo Club and of a charmingly anachronistic Society for Apprehending Horse Thieves, Brandeis and Warren had personality enough to become alarmed by reporters and detectives armed with small cameras snapping "surreptitious" and "instantaneous photographs" that "invaded the sacred precinct of private and domestic life." Sapped by the intensity of the daily round in an urban society with a diverse, burgeoning population, citizens had need of new assurances of restorative solitude and secure retreat. Nine months before their brief for privacy appeared in the *Harvard Law Review* co-founded by Brandeis, he had rushed back to his family home in Louisville to attend the funeral of his older sister Fanny, who had committed suicide at the age of thirty-nine. In October, he wrote from the quiet of his Boston lodgings to Alice Goldmark, a second cousin in New York and now his fiancée: "Blessed quiet, blissful absence of people. Oh, Alice, if you were but here, how beautiful it would be. Dear, I cannot tell you how much I long to be with you, to be with you thus alone, far, far away from the bustling, babbling crowd." Some of this was romance, some the pulse of desire. "This week, it seemed as if I must have you." But there was more: like other reformers who led lives proper, public, and thoroughly engaged, he felt himself to be under severe tension from the urban and political whirl: "The noise, the crowd, were almost intolerable. I felt as if I should flee from this wilderness of men." The next days he went canoeing on the Charles. "I thought, Alice, how you will love the canoe, and how well it expresses you: the silent dignity, strong but tender, sensitive to the slightest touch, responsive to every word, listening with bended head to each whisper of nature, with a heart for all human emotions and a soul to grasp the divine." Not everyone had a canoe or the hours to paddle out alone on a river at midday; it was crucial, ergo, that the courts recognize a right to privacy, and that the public lay claim to those rights. "Our hope," he wrote Alice, "is to make people see that invasions of privacy are not necessarily borne—and then make them ashamed of the pleasure they take in subjecting themselves to such invasions." He and Alice were married in 1891 by his brother-in-law Felix Adler, founder of the Society for Ethical Culture. Adler would later address his Society on "The Moral Value of Silence" in which he declared, "Slowly, gradually, with much difficulty, the right to curtain off our inner world has been won in civilized communities," but in 1898 this was a right confirmed rather by the architects of seaside manors and city palaces than by the courts. While anti-billboardists hoped to clear avenues of eyesores, Brandeis and Warren meant to erect a judicial screen against eyesorrows (whence "eyesores"), eyestrain (whence, said his doctor, Brandeis's fatigue), and concomitant I-sorrows. For Warren these sorrows would come to a woeful pass after a decade of bitter and increasingly public disputes with his younger brother

Ned over art, aesthetics, and unfair disbursals from a family trust drawn up by Louis twenty years earlier and administered by Samuel ever since. In December 1909 Ned filed a lawsuit against his brother *in re* the trust; on February 18, 1910, at his house in Dedham, Samuel took up an axe and split some logs, then took up a shotgun and turned it on himself. His heirs, and the newspapers, protective of his reputation, reported that he died of apoplexy.[277]

Shock! lay at the heart of the Brandeis-Warren brief for a right to be let alone. The photographic shock of flashpowder lighting up the most sensitive corners of life. The newsstand shock of sixty-point banners "screaming" from the papers of Hearst, Harmsworth, and Pulitzer, whose new journalism was known for monstrous illustrations, brash typography, and loud headlines (as on October 28, 1890 in the *World*, where just below "PROGRESS OF BRYAN WAS A WHIRLWIND OF FIRE. Hurrahing Democratic Mass at the Meeting Places, While the Central Rallying Point Blazes with Bombs and Resounds with Shouts," came "A SHRIEK, STREAK, YELP AND A BOB-TAILED DOG," an account of the "lightning-express" progress of William K. Vanderbilt, Jr., hurtling through a town in his mud-covered car, as well as color comics, faked interviews, and sensational features: "DOES MODERN PHOTOGRAPHY Incite WOMEN to BRUTALITY?" The billboard shock of three-foot-tall letters "shouting out" their wares in otherwise serene surroundings. The tabloid shock of the *National Police Gazette*, with its ads for Organs Enlarged!, its headlines (on December 6, 1890, when the Brandeis-Warren brief appeared) about "WICKED WOMEN! And Very Wicked Men, Just as Hard," and half-page illustrations of a woman in negligee fleeing a man with a cleaver and of "Mrs. Harvey Sharp, a Camden, N.J. Lady, Attacked by a Burly Colored Desperado in Mount Ephraim," who "ESCAPED BY SCREAMING." Shock, of course, was not new to century's end; it had however gathered technocultural force as an individual sensation and an index of modernity. The shock of combat (< *choquer*, "to clash," the medieval French referent) was no longer that of knights charging on armored steeds; it was the pounding of artillery. The shock of swift motion, another old referent, was not just the rush of cavalry but of racing locomotives pulling trainloads of passengers subject to constant vibration and frequent collision, from which arose "railway spine" and thousands of claims of spinal concussion. The shock of unpleasant surprise (1691) had been intensified by the suddenness of steam whistles shrieking and telephones ringing at midnight. The shock of horror (1704) had been intensified by trick photography, by industrial mutilations, and by glossy garish half-tone prints—enabled ironically by a Warren mills process for coating paper with pigment to be used in illustrated periodicals. The shock of being scandalized (1840) was magnified by *fin de siècle* eroticism, by "shocker" fiction (1907), and by the "brutal" dissonances of the music of Charles Ives (*Central Park in the Dark*, 1906),

Arnold Schoenberg (*Five Pieces for Orchestra*, 1909), and Igor Stravinsky (*The Rite of Spring*, 1913). Mildly intriguing electrical shocks (1746) had been horribilified by the first explicit use of shock for torture (of cats and dogs, in Thomas Edison's crusade against Tesla's alternating current) and for killing (the electric chair, 1890). Medical shocks—nervous shock or "traumatic neurasthenia," shock from the loss of blood during surgery or childbirth—these had multiplied as surgeons armed with anaesthetics and carbolic acid performed more invasive procedures, as medical men muscled their way into obstetrics with a proclivity for Caesarean sections, and as psychiatrists puzzled over hysterical paralyses, testing or treating their patients with electric shock. Eight months after Warren's suicide, during an Ether Day Address at Massachusetts General Hospital, George Crile first enunciated his "Kinetic Theory of Shock" that linked the brain, adrenal glands, and liver to the pathology of shock, surgical and emotional, such that shock had to be understood as the totality of an organism's response to trauma. By now, shock had such momentum that it could be seen-and-heard to arrive in "shock waves," particularly when propagated during explosions, as analyzed by the French physicists Jouguet and Crussard and described in English with language borrowed from geologists who for fifty years had been documenting the "shocks" of earth-waves.[278]

Shellshock lay just over the horizon, in the trenches of the Great War. We are not there yet. Instead we are in the company of "Mr. Never-Close-the-Door" and "Mr. Always-Slam-the-Door" at a more pedestrian crossroads of privacy and shock. Slam-banged and "jarring the nerves," or bolted to bar the nosy, the door itself was under siege. Like interior walls, domestic doors had become thinner as hardwoods and heavy metal fittings became more expensive; doors were now customarily strip-paneled and fitted with flimsier locks. In their flimsiness, they would blow open, wheeze through warped slats, and bang half-shut—or they would flap, like the new wire-mesh doors invented by Hannah Harger in 1887 to let in the breeze but keep out the flies. Whereupon the firm of Yale & Towne began producing pneumatic door checks that "close doors quickly and gently, firmly and silently," like strong grandfathers or (vanishing) butlers, so as to "Protect your health from drafts, your nerves from odors and noises." Goodbye to Mr. Always-Slam-the-Door; hello to "the quiet positive closing" of front doors, lavatory doors, and screen doors. By 1911 the Caldwell Company of Rochester, New York claimed to have sold two million "Dime" Screen Door Checks whose rubber bumper held "banging and jarring" to a minimum.[279]

Pneumatic door checks made a "slight hissing sound" that could be mistaken for a *sotto voce* wolf-whistle. Another kind of door was making its entrance with a whoosh. In 1888 Theophilus Van Kannel, inventor of carnival rides, patented a revolving door, its three radiating wings so hinged and enclosed that they could never be slammed shut by the wind or blown open by interior pressures, as would happen to ordinary doors in the lobbies of buildings with banks of elevators. The revolving door was also "perfectly noiseless," with "no screaking springs to close it, and no whistling checks to counteract the springs." It afforded simultaneous egress and entrance while excluding not only "wind, snow, rain or dust" but the "noises of the street" and was therefore a good choice, insisted a 1904 circular, for courtrooms, churches, theaters, and libraries. You might wish to install it solely on account of its soundproofing qualities, for example in factories where "a room containing some nerve-racking noise from machinery or work adjoins a room where quiet is most essential." Entrance and barrier—"Always Open Always Closed"—the revolving door became an atrractively modern synecdoche (notes the historian James Buzard) for the ambivalences of a threshold, at once inviting and dangerous, transitional and secure. Each traveler passes within a self-contained if transparent cabinet, a turnstile reconstrued for privacy. The revolving door envelops and expels, protects and intimidates, its motion conditional on the number, hurry, awkwardness, and bulk of the people pushing at its vanes. Embraced momentarily by a room of one's own, a person may find herself caught up in a (social) whirl, a three's-a-crowd reel exploited by silent film comedians and personal injury lawyers.[280]

After the Great War, another illustrious brief for privacy would be filed, this on behalf of a writer's need for a more permanent room of her own, with a lock on the door and £500 a year. It would be a room, wrote Virginia Woolf in 1928, dedicated neither to Rest Cures nor changes of costume but to the presence of a reality more vivid than that of "the common sitting-room." It would be the kind of room that Montaigne had proposed, and which three centuries of well-off men had enjoyed: "We should set aside a room, just for ourselves, at the back of the shop, keeping it entirely free and establishing there our true liberty, our principal solitude and asylum. Within it our normal conversation should be of ourselves with ourselves, so privy that no commerce or communication with the outside world should find a place there." Given that in English country houses the size of private apartments had shrunk during the 1800s while demands for sociality had expanded; given that walls had been thinned to the point that no longer was it "almost possible to fire a cannon in one's drawing-room without attracting undo attention from one's next-door neighbor"; given that the sound-deadening draperies and rugs of the Victorians were being stripped away; and given that most anywhere "Dogs will bark; people will interrupt," women needed a place of No Trespass for reading, musing, and self-confiding: "five hundred a year stands for the power to contemplate,... a lock on the door means the power to think for oneself." Back in, on, or about December 1910, when (as Woolf would write, in jest as to the exact month—and why not in

May, with the hullabaloo of Halley's Comet?) "human character changed," she had an entire floor of a Bloomsbury house to herself and at least £500 a year, some of this inherited from her Aunt Caroline Emelia Stephen.[281]

Adeline Virginia Stephen's youth had been neither private nor quiet nor emotionally secure. She grew up in a London family where "silence was a breach of convention," in a house of seven siblings and seven maids a-cleaning, of bedrooms always shared and years of sexual abuse from two half-brothers. Just outside bleated the traffic of Hyde Park Gate, where, as she reported in detail, accidents were frequent: a pedestrian squashed by an omnibus, a hansom cab overturned, a woman cyclist knocked flat. In 1895, her mother Julia, who had tended to a dying uncle, dying sister, dying mother, and dying father, and who had published *Notes from the Sick Room* in which quiet was paramount ("no noise is more exasperating than the scraping up of cinders... the room should be so gently hushed that the patient should feel able to drop off to sleep at any moment"), died quite suddenly; Virginia, thirteen, had her first nervous breakdown. In 1897 appendicitis took her half-sister Stella, and in 1904 her father Leslie succumbed to cancer; Virginia fell into a suicidal depression. Sent to a Friend's (and friend's) to recover, she went on to The Porch, Aunt Caroline's place in Cambridge. An influential Quaker theologian, Caroline Stephen also heard voices, inner voices to which she listened in silence, with a discernment less psychiatric than spiritual. ("A kind of modern prophetess," Virginia called her, and altogether "a remarkable woman," though prone to a "woolly" benignity.) In "ideal retreat" at The Porch, where "I can sit by an open window for hours if I like, and hear only birds' songs, and the rustle of leaves," Virginia found her own voice; with Caroline's encouragement, she began to write and publish essays. She was hesitant to return to London and a crowded house, for she longed, as she confided to a friend, "for a large room to myself, with books and nothing else, where I can shut myself up, and see no one, and read myself into peace." In 1906 her brother Thoby, two years older, died of typhoid fever (because misdiagnosed by the family physician, George Savage, editor of the *Journal of Mental Science*). When she moved the next year to Fitzroy Square in Bloomsbury with her brother Adrian, she was trying at twenty-five to shake off four deaths. The noise of the Square itself came near to driving her mad: she had to install double windows to mute the sharps of carts and the flats of vans. But the Bloomsbury Circle was gathering and she was happily among them, writing, talking art and sex, teaching night classes for workingwomen, attending concerts of new music. The bequest from her aunt further "unveiled the sky to me." Working on a novel, *Melymbrosia* (eventually *The Voyage Out*, where a woman enters the turbulent world and a man hopes to write "a novel about Silence"), in March 1910 she had another collapse, returning that autumn from a Rest Cure—one of many prescribed by Doc Savage—to help her sister Vanessa crusade for Women's Suffrage and her friend Roger Fry curate the first English exhibition of the paintings of Van Gogh, Matisse, and Picasso, a show received in London "with shrieks of misunderstanding and interested horror." In those busy days, luncheons "were accompanied by a sort of humming noise, not articulate, but musical, exciting, which changed the value of the words themselves." Why, she asked in *A Room of One's Own*, "have we stopped humming under our breath at luncheon parties?" Was it the change in human character in late 1910? The noise of traffic? The art? The war? "Certainly it was a shock (to women in particular with their illusions about education, and so on) to see the faces of our rulers in the light of the shellfire. So ugly they looked—German, English, French—so stupid."[282]

Alfred Harmsworth knew the war was coming, and used his papers, the *Daily Mail*, the tabloid *Daily Mirror*, and the more stolid *London Times*, to raise the alarum. His public brashness, however, was counterweighted by a personal secretiveness of manner, quietness of voice, distaste for foul language, and a dislike of noise he had manifested since childhood, when it had been ascertained that his were unusually large "ear orifices." Lifelong, he fled to silent sanctuaries after periods of self-promotion or publicity stunts. By the time he was made Lord Northcliffe in 1905, he owned Elmwood, a manor girded by a nine-acre garden and high walls, beyond which lay Broadstairs, a seaside village still (as Dickens found while working there on *Nicolas Nickleby*) "a little fishing place, intensely quiet." Behind elms and firs, in large quarters of his own, Northcliffe required his servants to wear soft-soled shoes so as not to squeak on their rounds, and although he had a telephone in each room and a pair of live alligators in the greenhouse, Elmwood was sanctuary enough for him to contain each of his hypochondriac fears—of going blind or getting fat, of heart attack, pancreatitis, ptomaine poisoning, or pneumonia (for which he kept a Throat Diary of coughing fits). The private silence, as Brandeis, Warren, and Woolf might have said, sustained and refreshed him, he who in public thrilled to the sound of auto races, airplanes, and anything that accelerated, including his presses, loud and powerful.[283]

Lord Northcliffe's American counterpart, Joseph Pulitzer, was equally schizosonic. In the arena of metropolitan journalism, where dozens of newspapers were hawked on noisy streets and read on noisy trams, cartoonish sensationalism was crucial, or so he said, while in private he insisted upon an impossible quiet, all the more so as eyestrain worsened to near-blindness. He had a "nerve condition," wrote an editor of the *New York World* who knew his boss well, that made him "so extremely sensitive to noise that the clink of a spoon, the sipping of soup, the closing of a door, the cracking of almonds by a guest at dinner, the rattle of a paper—anything in the nature of sniffling, or scratching, or scraping, or any kind of noise whatsoever except such as he could expect and identify and enjoy, such as the rumble of surf, or the wind, or the galloping of horses would react upon the Subject like a cannon shot, precipitating a paroxysm of pain and anger, and a certain degree (we must assume) of fright." Or of shock, for then came "violent cursing, shouting, shaking of the fist, stamping of feet," culminating in fits of sobbing and exhaustion that the ubiquitous Weir Mitchell could not cure, in part because Pulitzer held his insomnia dear while defending his media empire from midnight coups. From 1900 to 1903 he threw his architects into fits of madness, if not sobbing, as they struggled to build him a noiseless palazzo just off Fifth Avenue. McKim Mead and White, who preferred to design grand residences with lively music rooms (and who ignored his command to scratch a music room from their plans), divided each of Pulitzer's four stories of limestone and granite into three acoustic zones separated by thick masonry, with sound-deadened ceilings and floors between each story. They tested the zones by harkening from room to room as assistants pounded on walls, jumped on landings, hallooed through windows. Pulitzer had to agree that even the sirens of fire engines did not reach his private wing, padded as it was from floor to ceiling and separated by a bridge from the rest of the house, but the triple lot at 7 East 73rd Street came with a small natural spring, and he swore he could hear the pumping of the electric sump pump that kept his Venetian palace from floating away. "We have been building a house for Joseph Pulitzer, who is a nervous wreck and most susceptible to noises," William Mead wrote in exasperation to an acoustics expert at Harvard, "and he has discovered many real and imaginary noises in his house. Some of them are real and can be obviated, and," Mead pleaded, "we have great confidence that you can discover the cause and a remedy for them." One solution was to redo the plumbing and pumping so that no pipe communicated directly with the living. A second was to lay cork floors. Inadequate! At night Pulitzer complained of the creaking of the woodwork; he told his cousin, the great Viennese otologist Adam Politzer, that the house was a failure. A third solution was to hire a new firm to design a detached bedroom in the back garden, with triple casements, double walls, double doors, a floor set on ball bearings, and a ventilating fan ineffably muffled. Alas, even "The Vault" was imperfectly silent (what did Pulitzer hear? was it the sound of silk threads "absorbing" stray vibrations in the chimney shaft? a trolley conductor's bell on the edge of Central Park? the throbbing of his own pulse?). He abandoned East 73rd, spending his last years in thickly draped suites on ocean liners or aboard his own yacht, where bells were never rung, engines were always muffled, and the only sound was the lapping of waves at the bow. It was on his yacht that Pulitzer died in 1911. He spoke his last words in his native German: "*Leise, ganz leise, ganz leise*"—"Softly, so softly, ever so softly."[284]

Ganz leise or *lentement, assez lentement*: Proust's motif as well. In the opening paragraphs of the overture to *Swann's Way*, begun in 1910, his alter ego Marcel remembers as a child waking in the early dark to the thought that it's time to go to sleep, as the sound of train whistles intimates at a traveler passing swiftly, expectantly toward home through a deserted countryside. He falls asleep again and wakes, like Pulitzer, to the creak of wood paneling, then sleeps and wakes to more adult, more sexual thoughts, sleeps and wakes again to meditate upon insomnia and its disorientations of light and time, thence to reflections on the motion of thought itself and the presumed immobility of things—of furniture, of chairs and tables. Like that iron table beneath the chestnut tree in Combray, decades ago, when Swann, a neighbor, would make his entrance while the family hears, from "the far end of the garden, not the copious high-pitched bell that drenched, that deafened in passing with its ferruginous, icy, inexhaustible noise any person in the household who set it off by coming in 'without ringing,' but the shy, oval, golden double tinkling of the little visitors' bell." Gathered around the table and speaking in the whispers of a quiet summer's evening in the dark of the garden, the family could tell Swann only by his voice, a voice that sets the story in motion. Another familiar but oddly transformed voice had put Proust on the path of *À la recherche du temps perdu*, a voice heard over the telephone eight years before. His mother had rung him up to tell him of her parents' death and "suddenly her poor shattered voice came to me through the telephone stricken for all time, a voice quite other than the one I had always known, all cracked and broken; and in the wounded, bleeding fragments that came to me through the receiver I had for the first time the dreadful sensation of all things inside her that were forever shattered." Had he mistaken a bad connection for emotional upheaval, static for mortality? After her death in 1905 he became more reclusive. Asthmatic, insomniac, and acutely sound-sensitive, Proust so loved operas, plays, and passionate concerts but so hated the undertow of "stupid" audiences and the clamor of avenues that he subscribed in 1911 to Théâtrephone, a service by which he could listen to live performances transmitted from microphones on stage through telephone lines to a receiver in his apartment, which by then he had lined with sound-deadening cork. The cork darkened under the burning of medicinal powders to fumigate his lungs, powders whose smoke and ash filled in the indentations of the cork so that the walls ironically became smoother and more acoustically alive with every wheezing year; during the War he could hear the buzz of French airplanes overhead and air raid sirens "like the heart-rending scream of a Valkyrie." *Swann's Way*, the only volume of his opus to appear before the War, ends in the Bois de Boulogne, where instead of pairs of horses pulling plush Victorias Marcel finds "nothing now but motor-cars, driven each by a moustached mechanic" and instead of elegant society in feathered bonnets the shrill cries of large birds. Like father like son: in national and international forums, Adrien Proust, M.D., had urged the creation of a *cordon sanitaire* around Europe to block the spread of cholera from the East; his son created a private *cordon sanitaire* of cork and curtains around his rooms in Paris and by the time he died in 1922 had built up seven volumes of ruminations, very public yet personal, around a faded *belle époque*. Imagination, wrote Proust, is like a barrel organ (*un orgue de Barbarie*) which has gone off its pins and plays a tune other than the one you choose.[285]

Sometime back, we heard *un temps perdu* gain its first currency with regard to intrinsic delays in the transmission of nerve impulses from skin to brain; now it resonated with issues of history and subjectivity. Just as Proust was outlining his *recherches*, through which all of his senses, sharp as they were guarded, would contribute to a sinuous reconstruction of a passing era, Georg Simmel in Berlin was presenting his own researches into the sociology of the senses, taking the ear to be the primary agent of the self, its secrets, and its *temps perdu*. Heir to a fortune made in music publishing, so able to thrive as a dynamic lecturer with no vested university appointment, Simmel was a peripatetic philosopher weaving his way through the *agora*, trying to make sense of money and fashions, of the flux of people and power. As the city flung strangers into each other's personal spaces and intensified each encounter with the "unexpectedness of onrushing impressions," so its neural shocks exacted a toll: a half-hearted life. Stressed by the derangements of capitalism, one made one's way in the world less by feeling than by intellect, which promised the safety of pattern and the security of indifference. How else "preserve subjective life against the overwhelming power of metropolitan life" than by keeping one's quickest emotions and deepest feelings to oneself, relying instead upon quadrilles of punctuality, etiquette, and mechanical civility? Accordingly, the city was "the genuine locale of the blasé" and the ostensible, with everything private (everything constituting the uniqueness of an individual) held in reserve as a kind of municipal bond. In this schema, the eyes are media of re~cognition, finding in faces and scurrying figures the sum of what citizens are at their least relaxed. Like daguerreotype cameras, eyes light upon what endures; when eyes meet, people romanticize that sudden reciprocal connection into love at first sight or a glance into the soul. Ears, however, are the true instruments of interior life, for they register successions of vibrations already struck and vanishing, like the flow of our natures. Further, the ear itself is "the egoistic organ pure and simple, which only takes, but does not give," yet paradoxically it commands no private property, for sound is *broad*cast and, in musical concert, can create "an incomparably closer unity and commonality of mood than is the case among the visitors to a museum." A "passive appendage" receiving the liveliest of sense-impressions, the ear is the preeminent organ of mood, of subjectivity. As a receiver of stolen (overheard) goods or keeper of whispered confessions, it is a tender of secrets, a steadier archive than the eye. We remember what has been said more easily than what has been seen, claimed Simmel, because we see generically but hear idiosyncratically, so that in sensorial terms the ear is the marker of personal time and the maker of personal memory. "It pays for this egoism"—returning to an old trope—"in that it cannot turn away or close itself, like the eye; rather, since it only takes, it is condemned to take everything that comes into its vicinity."[286]

Questionable as was Simmel's theory once gramophone records and player pianos had entered millions of "living rooms," and questionable as was the theory of using Maxim Silencers in every corner of every room (proposed by *Puck* in 1909) to insulate a flat against street singers, barking mutts, and braying victrolas . . . even so, sound on all sides was being taken more personally, and noise as a greater personal affront.[287]

14. Walls

"Space, light, air, privacy are natural rights," declared a report of the Social Service Commission of the International Committee of the YMCA at the congress of the Men and Religion Forward Movement in 1912. Not simply a legal right, privacy was evidently invested in the individual as a moral agent. Wherever "the spirit of the home is to dwell, it must have light and air, room and privacy, a chance for quiet and a place for play." Space, light, and air were properties of a commons to which all originally had access;

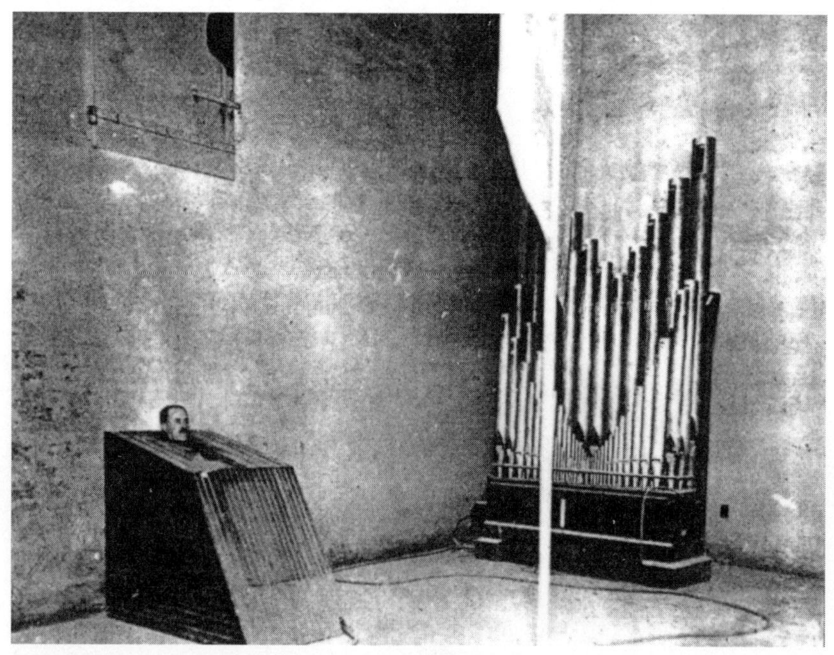

SOUNDPLATE 18

privacy was rather an assertion of the intact self, its senses, and its notion of time and company. Any invasion of privacy was therefore a shock to four systems: the body/mind; the soul/spirit; the commonweal; the moral economy of the world. "With a shock we realize that what is justly called progress has been marked by relentless encroachment upon the home." And modernity, wrote the novelist Theodore Dreiser in 1911, was all about shock: "We live in an age in which the impact of materialised forces is well-nigh irresistible: the spiritual nature is overwhelmed by the shock." Railways, telephones, telegraphs, newspapers, "and, in short, the whole machinery of social intercourse," were too much for our brain-pans. (Dreiser would know: one brother an electrician, another a vaudevillian and publisher of popular music, three sisters in fixes, himself an urban gadabout with an ear for the vernacular.) Under such pressures people in cities or on the open road had need of "shock absorbers," a new term working itself out from the undercarriages of autos and airplanes, and a "sturdy, shock-proof exterior" like that of Lester Kane, the thoroughly modern cad in Dreiser's second novel, *Jennie Gerhardt*, about an innocent small town girl to whom "the jingling sounds and the changing odors which the city [of Cleveland] thrust upon her were confusing and almost benumbing her senses," making her an easy mark for Kane. The problem with a broad claim to natural rights in space, light, air, and privacy was—in acoustic, architectural, and metropolitan terms—that a private quiet seemed technically incompatible with open horizons, good ventilation, and easy circulation, whether for Pulitzer a few steps from Fifth Avenue, Proust on Boulevard Haussmann, Woolf on Fitzroy Square, Simmel on the busy corner of Leipzigstrasse and Friedrichstrasse, or a "silent man" named Sabine in the subbasement of Harvard's Jefferson Physical Laboratory, 17 Oxford Street, in Cambridge, Massachusetts.[288]

Nominated the class prophet of Ohio State University in 1886, and speaking (softly, as always) on "The Power of Public Opinion," Wallace Clement Sabine knew whereof he spoke: his father Hylas was a state senator, editor/publisher of the *Union* [County] *Press*, and, during the years that "Tinto" was in college, Ohio's Commissioner of Railroads and Telegraphs. When Wallace went on to do graduate work in physics at Harvard, where his father had spent two years on a law degree, he came well-prepared, having studied Chladni figures and Lissajous curves with the physicist T. C. Mendenhall, who would later assume the presidency of Worcester Polytechnic and, in 1893, of the American Association for the Advancement of Science. In Cambridge, Wallace left behind his family nickname but not his father's interest in communications, which he pursued in summer jobs with Bell Telephone and in a paper co-authored with his thesis professor, physicist John O. Trowbridge, on "Electrical Oscillations in Air" and electromagnetic tuning for wireless telegraphy. From Hertz and electromagnetism he switched over to Helmholtz and acoustics, from which he gained a reputation as "a wizard on Sound" and a position at Harvard, though he never earned a doctorate. While his brother-in-law Wilbur Siebert, a professor at Ohio State University, began a lifetime of research on the American antislavery movement and Underground Railroad, Sabine moved underground with a purpose almost as liberating for such a benchworker: freedom from errors in measurement caused by an all-too-noisy world outside and above. In the student manual he had created for Physics C, "a laboratory course in physical measurements," Sabine had accented "all the precautions which should be taken in the accurate and proper performance of the experiments," the need for many repetitions, and the importance of precision in statements of results, which should be expressed mathematically in thousandths, not 20 but 20.000. "The student should not ask himself whether his results are correct and in agreement with known values, but whether his method is correct and his observations well taken."[289]

Underground, Sabine meant to determine the principles by which to improve the unspeakable acoustics of the new Fogg Art Museum Lecture Room, a domed half-shell half of whose audience could hear nothing, the other half only noise. Designed by Richard Morris Hunt as one of his last commissions and built at the same time as his crowning achievement, The Breakers (the palatial Vanderbilt villa in Newport), the Lecture Room was open for classes in 1895 only to be abandoned weeks later by a disgruntled faculty. Asked years earlier how much he knew about acoustics, Hunt had assured Rev. Henry Ward Beecher, most illustrious of Northern pulpit orators, "As much as anyone." "And how much is that?" persisted Beecher, whose low, well-modulated, "far-reaching voice" had to contend with the volumes of Brooklyn's Pilgrim Church, which a music critic had described as having "no pretence of architecture save that of the convenient and humble barn" where a thousand and more congregants strove to make out his sermons. Hence Beecher's persistence. Hunt had replied, "Not a damned thing."[290]

The damnable unintelligibility of speech in Hunt's hall most upset Harvard's first professor of art history and a formidable man of letters, Charles Eliot Norton, an advocate of the City Beautiful but not of Hunt's Beaux Arts styling. The Museum and its attached Lecture Room echoed with the problems of democracy as Norton laid them out in private letters and public lectures: in a young country lacking a "tradition of intellectual life" and the "power of sustained thought," the multitudes were drawn to the loud and superficial, easy marks for advertisers, demagogues, and jingoistic politicians. "The rise of democracy to power in America and in Europe," he wrote in 1896 to his friend Leslie Stephen, Virginia's father, "is not, as has been hoped, to be a safeguard of peace and civilization. It is the rise of the uncivilized, whom no school education can suffice to provide with intelligence and reason." Despite such head-shaking and "conscious that I am listened to with a certain sense of kindly, patronizing sympathy," Norton lectured, hectored, and lettered-to-the-editor, for he exemplified the Brandeis-Warren ideal of a cultured individual with uncompromised personal space but an engaged civic spirit. In this he was akin to the hermetic but voluble Thomas Carlyle, from whose friendship and from the editing of whose letters Norton would have known how contests between private and public, quiet and noise, play out. "Future generations will look back upon us with pity for the inconvenience and noisiness of our present life," Norton had told the undergraduates taking Fine Arts IV in 1890. And he had pronounced, in "sonorous tones" that echoed Carlyle: "This is a noisy age." So when he came to address the eight hundred ears under Hunt's new dome only to find each of his shapely phrases tumbling into a gallimaufry of syllables for some, dead silence for others, the barbarous acoustics would have been as symbolic to Norton as was the form of a rotunda, "indicative of lack of culture and good taste." The cacophony bespoke as well the predicament of reason and beauty in a materialistic nation with a good heart but ever on the verge of savagery. "We are corrupted by ugliness," wrote this thin, pale, stooping, balding aesthete, and if Nature itself were imperfect, the fine arts could be instructive where not redemptive. He appeared to one Harvard student in 1898 as "A man born blind, [preaching] the religion of pictures, a deaf man worshipping music, a man devoid of sensuous experience erecting altars to the Aesthetic." None of this could have have been said of him during his early years as a political reformer, devout Unitarian, and passionate husband, but his vision had turned dark, his beliefs agnostic, and his passions reclusive upon the death of his young wife in 1872. Urging a "noble discontent" with all that was ignoble, as the historian Linda Dowling tells us, he was an embittered republican, alarmed at the bellicose patriotism he heard rising around him during the 1890s. So the entire museum, barbarous in its parts, ignoble as a whole, demonstrated to Norton a "lack of civilization": "Had it been intended as an example of what such a building should not be, it could hardly be better fitted for the purpose." He made his grievances known to his cousin, the chemist Charles W. Eliot, president of Harvard, who forthwith asked Sabine to look into correcting the acoustics of the Lecture Room, "which had been found impracticable and abandoned as unusable."[291]

Forty years younger than Norton, Sabine would not have been unsympathetic to elements of the older man's politics—his vocal opposition, for example, to the imperialism of the Spanish-American War, or his concern for intelligent, intelligible public speech. The Midwesterner's inclinations may well have been more strongly populist and progressive than those of the Boston Brahmin, who at century's end was alternately pessimistic and patiently philosophical about the promise of democracy, which, "ideally, means universal public spirit" but in practice yielded to "general selfishness and private spirit." Sabine's maternal grandfather, J. R. Ware, had been a Quaker abolitionist. His mother, Anna Ware, had been educated by Horace Mann at Antioch and in 1863–1864 was one of the very few members of the "opposite sex" admitted to the Harvard lecture halls of Asa Gray (anatomy) and Louis Agassiz (zoogeography); a crusader for Temperance and Women's Suffrage, she lived to celebrate the end of the Great War and passage of the Eighteenth and Nineteenth Amendments. Jane Downes Kelly, whom Wallace would marry in 1900, graduated from the Woman's Medical College of Chicago (now Northwestern University) and won an appointment in Boston as one of the first female orthopedic surgeons; associated for decades with the summer session of Harvard's Department of Physical Education and with the New England Hospital for Women and Children, she directed her energies to the care of crippled children and the health of young women. So, when consulting in England in 1914, it would be Wallace Sabine who proposed that the House of Commons remove the brass grill that had long kept honorable members from being distracted by women visitors in the gallery above and which, in turn, had kept these women from clearly following the chamber's debates.[292]

Legal scholars have combed the lives of Brandeis and Warren to establish the motives behind their unprecedented brief for a right to privacy. Literary scholars have combed the lives of Woolf and Proust, seeking the intimate sources of each passage in their celebrated novels. Historians of science have made little of Sabine's biographical facts, for he was never one to intrude the personal upon the professional, especially given his sense of himself as an experimental physicist in a small, amorphous world (well-drawn by Emily Thompson) where acousticians had trained rather as engineers or contractors. He had also to deal with a physics community convinced that the problems of acoustics had been solved by Rayleigh and Helmholtz, so that further research was merely an application of the masters' work, of interest only to the trades. Nonetheless, his approach to acoustics had correlates with the fine columns of numbers perused by his father's father, John Sabine, auditor for Union County; with the manual dexterity and physical endurance of his uncle, Andrew Sabine, a Union surgeon throughout the Civil War; with the legal training and technical flair of his father Hylas; and with the inventiveness of his father-in-law, Francis V. Kelly, at the firm of Browne & Sharpe, a leading manufacturer of precision-gear cutters and calipers. He was influenced equally by the educated rigor, intensity, perseverance, and politics of his mother and his wife, and considered his discoveries in acoustics a gift to the commonweal, charging hesitantly for most of his consultations. His chief assistant was John Connors, a Harvard janitor; the mathematics he used were within reach of anyone with a jot of algebra. Sabine's professional name would be made after years of finicking measurement and a plotting physics, when he arrived at the principles for what was, at heart, a functional democracy of sound: accessibility, equipotency, flexibility, transparency. His tests, equations, graphs, curves, and materials were meant to help architects design public venues where listeners in the front rows were not privileged with greater clarity of sound; where any singer or speaker, woman or man, could be heard without distortion or loss of volume; where acoustic space honored the diversity of listeners and positions; and where good remedies for acoustic miscues could be adopted at low cost, without mystification.[293]

Centuries of masons, architects, engineers, and more recently, physicists had claimed to be able to (re)design a good room for music or speech and had typically failed. What did Sabine do differently? He took audience into *bodily* account. He thought about rooms as filled not simply with sound but with sound absorbers, not simply with orators but auditors. When asked how to deal with disturbances among students in a lecture hall, the consummate if soft-spoken lecturer had said, "It is perfectly easy; all you have to do is survey the audience and look every man in the eye." Similarly, he attended to each material, each seat, and the clothing and bulk of each woman and man in a prospective audience as of potential significance. One might presume that the science of lingering sounds would require a democracy of measurement, but others before him, with liens on science or a love of harmonies, had looked to geometers' lines of refraction and reflection, then to classical theories of proportion to model the architectonics of sound. They had also, it is true, done some listening: they had clapped or whistled in various halls and timed the seconds it took for the sound to become inaudible, but almost always in emptied spaces, themselves as audiences of one or two. An exception was Théodore Lachèz, an architect who studied reverberation and sonic interference with the physicist Félix Savart in Paris during the 1830s and understood enough to write in 1848 that "absolute vibratory calm scarcely exists," so that architects must be prepared for rooms full of "a thousand small ancillary noises" arising from the audience, whose numbers affect the qualities of circulating sound. Yet despite concerns for volumes too great and walls too smooth; despite his scorn for the "Acoustique fantaisiste" that perpetuated unutterably bad acoustics in the refashioned halls of the National Assembly; despite his own tests with sound-curtains and sounding boards, Lachèz exalted shape over substance, espousing a polygon akin to a wide megaphone. He also doubted that any formula could help determine the design of rooms for good listening, even the isacoustic curve that he claimed to have discovered in 1835 but is usually attributed to a Scottish engineer, John Scott Russell. Studying wave mechanics, Russell in 1838 had come upon an equation by which the seating in any hall could be arranged along the arc of a rising curve from stage to back wall so that all members of the audience had good sight-lines and "equal hearing." Sabine was dismissive of this curve, for it ignored (like other ideal geometries that presumed hard lines and impervious solidity) the absorptive individual. Sabine's democratic vista was oriented toward assemblies of pliable, pored, rustling people and heterogeneities of density and surface, all of which required empirical research to meet his three conditions for good acoustics: sufficient loudness; proper relative intensities; clarity, or "freedom from extraneous noise." And the non-introspectiveness of Progressive thought allowed Sabine to sidestep the quagmires of musical taste to arrive at "perfectly determinate" calculations of loudness and distortion once "the reflecting and absorbing power of the audience and of the various wall-surfaces are known."[294]

Closeted more hermetically than Proust or Pulitzer, Sabine began his acoustic work in something like Hell, which is what MIT students at the Rogers Laboratory of Physics called their own constant-temperature room and which, like Sabine's room at Harvard, had been devoted to experiments with boiling points, steam pressures, and fuel efficiencies. At the base of a central tower resting on its own foundation and separated by a gap from the remainder of the Jefferson Physical Laboratory so as to be free of vibrations from classrooms, boilers, machine shops, and the nearest road ninety meters distant, lay a sparsely ventilated room with one door that let out through two (gapped) eighteen-inch-thick brick walls into an anteroom. The room also had a small, re-sealable opening in the arched masonry ceiling from which a "tall, slender, and rather silent" researcher like Sabine could be lowered to the concrete floor. In that inner sanctum, from half past midnight to five a.m., during which hours a diligent young man not yet married would be bothered neither by society nor the rumble of carts near Harvard Yard (nor by the throbbing of his appendicitis, almost fatal), Sabine graphed the ebbing of the audibility of organ pipe tones as they were absorbed by selected materials. When measuring the absorption coefficients of walls or sections of tile, he would be further enclosed from shoulders to toes in a plain wooden box resembling a steam-bath cabinet, from within which he could silently manipulate an electrical organ console and the key of a chronograph.[295]

Anchorite he was not; Sabine was listening *out*, to a world in need of amendment. After examining the sonic responsiveness of a manometric flame and of photo-manometry, he chose to exclude all sound-sensing mechanisms but human ears—his own, his assistants', and those of volunteer audiences. As anyone who has taken an audiometric test may agree, it's a strange, strained practice to listen for a single sound to vanish, stranger still when listening for the decay of isolated pure tones never heard in daily life. To be precise, as Sabine hoped always to be, he was listening for what was inversely proportional to decay, namely "the duration of audibility of the residual sound," which hypersensitives like Pulitzer and Proust were most adept at registering. The point, however, was to listen at the level of an average concertgoer or undergraduate or, when your hearing was as acute as was Sabine's, to arrive at a standard of calibration indebted to no single pair of ears and to no unique room in which "the buzzing of a stray mosquito . . . produced a noise like a dynamo." So, upon determining that the decay of an organ pipe tone an octave above middle C in the empty Lecture Hall lasted an interminable 5.62 seconds, during which "even a very deliberate speaker would have uttered the twelve or fifteen succeeding syllables," Sabine began filling the hall with rows of cushions from nearby Sanders Theatre so as to assess the effect of their elastic-felt presence on the loud uneven swirl of sound in the aptly-named Fogg. The Harvard cushions stood in for the absorptive masses of clothed human beings, one at each seat; with each additional row filled, the decay was shortened, loudness softened, sound disambiguated.[296]

He could have stopped when each of the 436 seats had its own cushion and the duration of tones was down to a respectable 2.03 seconds. He knew then that enough cushions or equivalently sound-absorbent fabric would restore intelligibility to Fogg-talk, though he had to do some further experimenting in order to figure out the best placements. He had already found, as to reverberation, that an object's absorptive power is independent of location, since sound disperses very quickly throughout a room; he would soon find that the intensity of a sound decays logarithmically, so loudness could be plotted on a steeply sloping downward curve, and that the relative placement of absorptive materials vis-à-vis a sound source affects both the initial distribution and intensity of sound in a room, as well as the probability of echo. But he did not stop with these 436 seats. He went on to stack all fifteen hundred of the cushions in precarious piles rising up into the dome and reducing duration to a deadbeat 1.14 seconds, for he was pursuing, really, a solid, tensile geometry of sound. He wanted to comprehend sound as the active occupier of a space, not just the received quality of a place. His quest would lead him to a triumphant equation defining the acoustics of a room neither by its shape nor its proportion but chiefly by its interior volume (and the mean free path of all sound reflections under a dome, such as Hunt's, by the cube root of its volume). In the meantime—interrupting nightly tests in the Fogg whenever a trolley passed within two blocks or a train within a mile (shades of Michelson and Morley!)—he had to construct a scale for the comparative absorptive power of other seats, fabrics, floor coverings, curtains, and wall materials by substituting these for so many square meters of cushions, then to come up with a universal module for absorption, something other than a Harvard cushion.[297]

Herez rugs, chenille curtains, canvas, plaster, brick, hardwoods, oil paintings, hair felt, carpeting, linoleum, cork, and ornamental indoor plants all had to be calibrated, as well as "the absorbing power of a man and of a woman," for the audience was, he insisted, "a factor of every problem, and in a majority of cases it is one of the most important factors." Victorian astronomers and psychologists had dealt with the problem of the "unavoidable presence of the observer" by computing personal equations for each woman at a telescope or each assistant at a lab bench; the case of architectural acoustics was much the obverse: in lecture rooms, music halls, and theaters, a company of individuals had always to be presumed to be present and meaningful. A company of *clothed* individuals: Sabine had to discard two months' work and three thousand data points "because of failure to record the kind of clothing worn," then went to the trouble of having three separate observers make six determinations of the "total absorbing power" of three different (middle-class, ordinarily dressed) audiences, and of a fourth audience in the same room without chairs or settees, and of the chairs and settees themselves, plain ash or upholstered, cushioned or uncushioned, until he could show that the absorbing power of an "isolated woman" was (due to additional layers of clothing) .54— or .06 more than that of an "isolated man"—and that the absorbing power of an audience per square meter was .96, twice that of a single man.[298]

.96 of what? Of air. The module Sabine chose as the standard unit 1.00 for sound absorption was a square meter of open window. A square meter, in effect, of air. This makes surprising sense if you are listening for decay: a closed window, like any hard smooth uninterrupted surface, is a near-perfect reflector of sound, and with the installation of domestic glazed windows and permanent plate-glass display windows, cities had become that much noisier; in the open air, sounds find no surface off which to rebound, so (excepting strong wind, thick fog, and temperature gradients) they dissipate swiftly and evenly, by an inverse square law, without echo. Let us start then out of doors, wrote Sabine in his article on acoustics for *Sturgis' Illustrated Dictionary of Architecture and Building*, and consider every enclosed room a limiting case of a flat open field with a single listener. Outside or in, the hearing of sound is conditioned by the size and density of the audience, but indoors the total volumes of air and the sizes of audiences are finite, and walls, floors, and ceilings to some degree inelastic. Outside, there can be much more "extraneous" noise, which is why Sabine during quiet summer nights (what with the racket of crickets and birds) preferred to use cushions for his module of measurement rather than an open window; otherwise the natural world deadens sound, unless a grove of hardwoods—or a granite cliff—should happen to reflect the music of a band back to rows of listeners embanked *en plein air*. Inside, there is likely to be more interference, more dead spots; otherwise the built environment promises loudness and liveliness. The trick is to temper that loudness and liveliness in the interests of an audience whose bodies are never equidistant from the source of the music or the talk.[299]

Sabine had begun with surfaces and ended up with volumes; had compared the absorptive powers of dozens of solid materials and ended up with a unit of air; had surveyed all elements animal, vegetable, or mineral that might constitute the sound-life of an interior, from mortar, wood, and leather to men and women, and ended up with an almost generic, usually invisible gas; had begun, in other words, with a sense of the built world as deviously dense and ended up with an ecology of vents. This was of a piece with the contemporary Fresh Air movement and with Sabine's own decision in 1899 to take up rowing: "The long sweep of the oars, the rhythmic clank of the oarlocks on the out-rigging, the swish of the sliding seat," he wrote his parents, "is music." The difficulties long posed by architectural acoustics had, it turned out, been due to thinking of sound as trapped within walls, rather than being welcomed or spurned in every direction by materials more or less elastic, more or less porous. Sound, it seemed, wanted little more than to be taken in, absorbed. Not a man to use such loose metaphors, Sabine followed Rayleigh and Helmholtz in conceiving of sound primarily as vibratory energy, which "will continue until either it is transmitted through the boundary walls, or is transformed into energy of another type, for example, heat." For his purposes it made no difference which; the crux was that sound had rather to be accommodated than captured (another metaphor he never used). For a century, musicians had referred figuratively to the "volume" of an instrument or a voice, by which they meant its compass and power, and in popular usage "volume" had come to mean sheer loudness. Neither gramophones nor telephones in 1898 had controls for loudness, but with his measurements and equations, Sabine was learning how changes in the cubic area of a hall, and the absorbency of its surfaces, could better facilitate the sound of a voice or instrument and affect its tone and loudness. Literally, figuratively, he was learning how to turn the volume up and down.[300]

During the years that Sabine was laboring to clarify the Fogg, Major Henry Lee Higginson, Boston's leading financial and musical broker, was in the final stages of planning a new home for the Boston Symphony he had founded. Its old venue, the Boston Music Hall, was a drafty firetrap at the center of town; each time an audience arrived to fill its 2,600 seats, the hurly-burly of the traffic jam outside played alongside the performance within. The new hall, as the Major proposed in 1892 to Charles McKim, "should be on the street level I think and have perhaps two galleries of small dimensions. I think that it should be lighted from the top only, i.e., from windows in the top or on the highest part of the side walls. I say all this on account of quiet—to keep out sounds from the world." Within, the hall should foster the "full and yet delicate effects of sound," as did Leipzig's admirable Neues Gewandhaus, opened in 1884. How Martin Gropius and

Heino Schmieden had succeeded with the acoustics of the Gewandhaus was unclear, and architects had had many a bad sound day when modeling new halls after favored old ones. So, although McKim studied the shoebox shape of the Gewandhaus, his taste for grandeur (as in the Boston Public Library) and Higginson's insistence on more than the measly 1,560 seats in Leipzig, led him to design an enclosed Greek theater, with "wedge-form stage walls and the succession of rising concentric arches immediately behind and above the orchestra" that would function, he predicted, as a series of sounding boards. But how could he, or the Major, or the Symphony, be sure? Asked in October 1898 to review McKim's plans by President Eliot, the Major's chum, Sabine hesitated, for he had been "floundering in a confusion of observations and results" from thousands of measurements taken in the Fogg, in the Jefferson sub-basement, in lecture halls elsewhere at Harvard and at the Boston Public Library. He was also probably a bit blue, for in September he had at last refitted the Fogg to make it "entirely serviceable," but after three years of research to be sure of the strategic placement of sound-absorbing cloth and sound-reflective screens, the Lecture Hall was hardly an exemplar of good sound. On October 29, however, he embraced the new project, for he had just realized that "the curve, in which the residual sound is plotted against the absorbing material, is a rectangular hyperbola with displaced origin; that the displacing of the origin is the absorbing power of the walls of the room; and that the parameter of the hyperbola is very nearly a linear function of the volume of the room." The implications were aesthetic and prophetic: the neater the curves, the more reproducible would be the acoustics.[301]

Reproducibility was key, for until 1898 the ground rules of interior acoustics, such as they were, had proved more useful (when useful at all) in *ad hoc* corrections than in prior calculation or prediction. Other architects and engineers—Charles Garnier with his Grand Opéra in Paris (1861–1874, when completed the world's largest theatrical edifice, though seating just 2,156), Gabriel Davioud and J. D. Bourdais with their do(o)med Palais de Trocadéro and its Salle des Fêtes (1878, seating 6,000)—had claimed to be "consistently scientific" yet their buildings had fallen so short of their sound ambitions that the Manhattan music critic Henry Krehbiel doubted that "the element of mere chance" could be eliminated from the acoustics of a new hall. Garnier, for example, had spent years reading books on theaters ancient and modern, had visited dozens of concert halls, and had found—instead of a set of acoustic laws with regard to shape, proportions, ventilation, ceiling height, and seating arrangements—a science in its infancy at best, and at worst "a tohu-bohu," a chaos of maxims and practices that worked to his advantage, since he did not have to learn any higher mathematics. "The most successful architects will tell you," so the successful British architect T. Roger Smith told his students at University College,

London, "that their judgment of the probable acoustic effect of this or that arrangement is as much, or more, the result of a sort of instinct—a kind of internal perception of what will do, or what will not do." In 1898, the mechanical engineer Eugene Kelly of Buffalo published what promised to be a rational, mathematical account of *Architectural Acoustics*. As did Sabine, he found that stringing thin wires across ceilings to absorb wayward vibrations was useless, that vast domes (like that of the acoustically hideous Royal Albert Hall, 1870, seating upwards of 7,000) should be avoided, and that walls of brick or stone should not be finished with slick plaster ("How startlingly distinct is the sound of the joiner's hammer in a newly plastered house!") He also explained that "Echoes, crashes, jars, etc. are caused by the infinite flexibility of the air, which is practically incomprehensible to the ordinary mind.... A volume of air the size of a City Hall can be easily set in motion by the chirp of a cricket." Yet he pursued no covering equations or systematic onsite studies of the causes of sound Dead, Harsh, Cold, Rich, Warm, Pure, or Live; instead he drew inferences from his Appendices of Ideal Auditoriums and Existing Failures and succumbed to an idealized geometry, espousing perfect ovals and perfect cubes as perfect solutions to "crash" ("a blind echo and a primary wave in collision without an intervening silence") and "flutter" ("a primary sound wave in contact with a circular wall or ceiling"). And no one seemed to be paying attention to him, anyway. When the Pan American Exposition opened in Buffalo in 1901, August Esenwein and James A. Johnson had given its reception hall, the Temple of Music, a large octagonal dome.[302]

Boston Symphony Hall was a gift to a new century, which began historically with the death of Queen Victoria in January 1901, announced on both sides of the Atlantic by tolling churchbells, police whistles, and the shrill cry of newsboys. Then, in September, under the Exposition's octagonal dome, came the clear echoes of an assassin's two pistol shots, killing President William McKinley. At its premiere in October 1900, the Symphony Hall's acoustics were put to the test most appropriately with Beethoven's *Missa Solemnis*; by October 1901, when the anarchist assassin "Fred Nieman" (as Leon Czolgosz liked to sign himself) was electrocuted with a-c jolts of 1700–1800 volts, all the technical reviews were in: the hall was fine, the sound "astonishingly clear and smooth" if (according to critics and musicians who had grown used to fierce reverberation) lacking "body" and "fulness." Even so, the building contractor, Oliver Norcross, doubted that credit for the clarity and fluidity of sound should go to Sabine's interventions (or the seconding opinions of Charles Cross, a physicist at MIT) that persuaded McKim to return to a rectangular shape, to keep the ceiling low, to shorten the auditorium to avoid tunnel effects, to rake the seats as steeply as possible, to hollow out the space under the stage, and to reposition some of the air vents. The other contemporary masterbuilder of Boston theaters, Clarence Blackall, was also skeptical of the value of Sabine's contributions, ascribing the hall's success to its perfect imitation of the Gewandhaus and declaring, as late as 1908,

> Acoustics is the one baffling problem which has so far set at naught scientific research. There has been some most excellent work done by investigators such as Professor Sabine... but when it comes to determining in advance what the acoustic properties of a given hall shall be, the only guide is experience.... If there is today any workable theory for determining the acoustic properties of a hall of audience, it has yet to be successfully applied in practice, and the most we can do is draw a few lessons from observed facts, and even these must be applied with fear and trembling.

Despite such professional deafness, it was Sabine to whom Mead would appeal, in fear and trembling of Pulitzer, when at wit's end regarding the New York villa, where, as Sabine found, the rapping of his knuckles on a wall of the servants' dining room in the basement could, via an invisible scaffolding of iron angle-bars, be transmitted "telephone-like" up two stories to Pulitzer's bedroom with "astonishing loudness and clearness," the sounds making an end-run around a layered floor of carpet, wood, reinforced concrete, three inches of sand, and three inches of lignolith block. And it was Sabine to whom Blackall himself would appeal for theater projects after 1910.[303]

Year after year, Sabine's list of correspondents and consultancies grew as it came to seem feasible not only to insulate urban interiors from external uproar but to limit—in advance—the transmission of disturbing sounds from within. This was a newly urgent concern as flush toilets were installed in middle-class houses; as well-to-do homes were outfitted with electric alarms, buzzers, vacuum cleaners, and kitchen appliances; and as the managers of theaters and concert halls had to cope with audiences accustomed to the volume and immediacy of vaudeville sketches, band music, vocals, and classical symphonic excerpts projected by the flaring horns of gramophones or the sounding boards of player pianos. In 1909 a young Chicago architect, Alfred S. Alschuler, wrote Sabine for advice on plans for the Sinai Temple, the city's premiere Reform Jewish congregation. Reform synagogues at this time were merging the traditional center platform, or *bimah*, with a platform at the western wall that held the ark and Torah scrolls, so Alschuler was concerned about acoustics as well as sight lines in the new arrangement. A graduate of the Armour Institute, Alschuler had trained for a year in the offices of Dankmar Adler, a structural engineer respected for his genius at acoustics; in partnership with Louis Sullivan, Adler had built the sonically excellent Chicago Auditorium (1889, seating 4,200), a far better hall than New York's Metropolitan Opera House of 1883, with more democratic sight lines and acoustics, as befitted the vision, weeks after Chicago's Haymarket riot, of "a large public auditorium where conventions of all kinds, political and otherwise, mass-meetings, reunions of army organizations, and, of course, great musical occasions in the nature of festivals, operatic and otherwise, as well as other large gatherings, could be held." But Adler had died in 1900, so Alschuler turned to Sabine. He wanted to know how to construct the floor of the *bimah*; whether a certain height would affect the projection of readers' voices; whether wood paneling at the back could serve as a sounding board; how to craft the pulpit and position the choir (other Reform innovations); how best to slope the risers; the acoustic effect of covering a cement floor with strips of carpet; whether to put air registers under the seats and how to configure ventilation; whether it would be acceptable to build a curved ceiling over the balcony, and so on. "It is rare," responded Sabine, "that the inquiries in regard to acoustical questions are so definitely or so well put," but it was his own consultancies and publications that had alerted architects to the value of such detail, because he could prove that some details mattered more than others, and that internal volume (of air) was critical. Sabine could also assume, from the volume of inquiries, that acoustics mattered to people, either for the clear transmission of sound in theaters, churches, synagogues, city halls, and legislatures, or often for better soundproofing, as in libraries, banks, courthouses, business offices (and typewriters), phonograph sales booths.[304]

Strangely enough, the more clearly he identified the sources of reverberation or showed how the sounds of indoor plumbing could be "telephoned" through walls and floors by unremarked metallic fittings, the more he seemed to be a one-man band noisifying the built world. He became alarmed at the tone of correspondents who seemed to be holding him responsible for their sonic predicaments, as if his exhaustive review of variables had made architectural acoustics forbiddingly complex. "In order not to feel myself an outcast in society," he wrote in 1912 to Bertram Goodhue of the firm of Cram, Goodhue, and Ferguson (architects of New York's Cathedral of St. John the Divine, where Sabine found sounds loitering for as long as ten seconds while echoes shuffled from chancel and apse), "I have to stop every now and then and remind myself that whatever I may have done in architectural acoustics I did not invent the problem. Please exonerate me from being the cause of your problem." What was going on was, in retrospect—or, to coin a useful word, "retrocoustically"—understandable. As Sabine at Harvard and Charles Ladd Norton at MIT began to apply general equations to the problems of interior acoustics and to devise sound-absorbent (and fire-retardant) panels for walls and partitions, it began to seem reasonable to expect architects to be able to design, in advance, acoustically acceptable spaces, so the profession was under increasing pressure to conform its aesthetics to principles of acoustics as expounded by engineers and physicists rather than fine artists. The pressure, which could be traced back to the thoroughness of Sabine's and Norton's methods as well as to the extensiveness of their measurements, was threefold: to attend to each structural and ornamental element as if it were a prospective source of acoustic chaos or intolerable silence; to attend to the volume of the whole in terms rather of dynamic sound volume and decay than of the solid, permanent geometry of harmonic proportions; and to regard walls and floors rather as acoustic membranes with distinct qualities of reflection and transmission than as the defining perimeters of differently functional spaces. There was also increasing pressure from clients themselves, as traffic noise and neighborhood gramophones threatened from the outside and desires for full ventilation and freedom of movement arose from within, to consider exteriors rather as absorptive materials than as hard facades. No wonder that McKim, working with Sabine on Symphony Hall, had feared that "the building will look, when it is finished, more like a deaf, dumb and blind institution than a Music Hall." And no wonder that architects in general might hold Sabine responsible for the bewildering number of acoustic variables that had now to be taken into account ahead of time—as if noise were hiding, almost literally, in every nook and cranny of a blueprint, ready at any moment to cry havoc.[305]

15. Phenomenologicals

"Always already extant and on hand," so the philosopher Martin Heidegger would say of our intrinsic entanglement with the world, where what we humans construct unavoidably reveals what has been already if preliminarily present. Noise too now seemed "always already extant and on hand," and if noisiness was not yet another way of talking about our intrinsic entanglement with the world, it was n/ear to it. Cognoscenti may suspect from this language and the slash in "n/ear" that we are about to ascend the slippery oeuvre of the philosopher Jacques Derrida; instead, I am headed back to Heidegger's mentor (and Derrida's *bête blanche*), Edmund Husserl, who had lit upon the "always already" decades before, between the time that Sabine was mapping out the 3,485 golden pipes of the organ in Boston's Symphony Hall and rapping his knuckles on Pulitzer's eighteen-inch-thick walls.[306]

Around century's turn, Husserl had put in the footings for "phenomenology," which was as much an attitude toward philosophy as it was a philosophical system. Having first excavated the underpinnings of mathematics, Husserl then developed the thesis that it did no good—in fact, it did ethical harm—to dismiss as unreliable the sensible world and our sensations of it. Rather than disembarrass philosophy of the objects that impinge daily upon us, our fleeting grasp of them, and impressions of fellow beings in similar flux, philosophy must make the most of our constellation of

SOUNDPLATE 19

immediacies; only through the sweep of moment-to-moment consciousness, carried forward by the intuition and reason of a transcendent ego, do humans achieve whatever solidity becomes us. Music, as Husserl observed in his lectures on flux, consciousness, and internal time in 1905 (just as five startling papers on Brownian motion, time, and relativity were appearing from the hand of an amateur violinist who often thought in music and saw his life in musical forms), was proof of the sweep of the self: "That the consciousness of a tonal process, of a melody I am now hearing, exhibits a succession is something for which I have an evidence that renders meaningless every doubt and denial." With each succeeding phrase, Husserl drew closer in principle to the Gestalt psychology of one of his teachers, Carl Stumpf, and in practice to the physics of Sabine, whose work would soon attract notice in Germany. "When a melody sounds, for example, the individual tone does not utterly disappear with the cessation of the stimulus or of the neural movement it excites. When the new tone is sounding, the preceding tone has not disappeared without leaving a trace." If prior tones appeared and vanished separately, "we would be quite incapable of noticing the relations among the successive tones" and could never hear a melody. Nor could we know melody if the tones were not modified in their moment-to-moment constellation by our presence, our faculties for hearing, our powers of memory, our intentions of coherence.

Otherwise we would hear "a disharmonious tangle of sound, as if we had struck simultaneously all the notes that had previously sounded." Noise, one could say, was the state of the world without us, always at the ready.[307]

Thus the phenomenologist, like the acoustic physicist, listens for—and *so establishes*—clarity and community of sound, reflecting upon the immanent meanings, or reverberations, of each event and act. To Husserl's lectures at Göttingen in 1906–1907 came thirty-four-year-old Theodor Lessing, D.Phil., a poet, dramatist, essayist, critic, activist, aspiring academic, and recent author of *Schopenhauer, Wagner, Nietzsche*, based on lectures for workers he had given at the train station in Dresden. Lessing agreed with Husserl's prior critique of the contradiction between Immanuel Kant's epistemological relativism (what we know is conditioned by who we are) and his moral absolutism (what we do can be judged *a priori* as good or bad). Lessing was now taken with Husserl's proposal as to how a phenomenological stance, which accepts vicissitude and receives the transient, can uphold enduring values. Drawing from the work of an Austrian philosopher, Franz Brentano, Husserl argued that consciousness was braced by intentionality, a forceful reference to the factual, which in turn braces all things-in-themselves, a circular relation through which we not only take the measure of our world but take heart. At least Lessing took heart, for a doctrine of imbedded intentionality, of phenomena as inseparable from purposive act, at once grounded and liberated his own socialism and feminism: a committed philosopher should be ready to protest on behalf of the rights of women, of workers, of Jews whose ambit of action had been so limited by bourgeois-industrial tyrannies as to demean the intentionality of their lives.[308]

Husserl and his wife, born to a prosperous Jewish community in Moravia, had both converted to Lutheranism before marrying; Husserl's career thereafter as a university professor in Germany had proceeded with Christian grace. Born a decade later to highly assimilated Jewish parents in Hannover, Lessing did not convert, despite an escalating number of encounters with antisemitic prejudices and educational policies. In 1906, due in part to his public resistance to such policies, he had neither a secure teaching position nor a secure income. He saw his life as a chain of personal wounds and collective injustices, some of these because he had remained a Jew (a secular Jew, who would later and famously write about the self-hatred of secular Jews), some because he had attacked the indignities of industrial capitalism, some because he had made common cause with workers, some because of the coldness of his father (a physician to the elite) and mother (the daughter of a banker). As the historian Lawrence Baron has shown, Lessing was drawn to phenomenology because it could frame a philosophy of social action that he had been building upon the twinned pillars of suffering and empathy. Empathy for those who suffer could become a kind of suffering in itself, and one's own suffering could make one empathetic with others likewise wounded, so these pillars were twinned to the circumstantial world. And in fact, pain and suffering could be no less than the origins of consciousness, shared pain and suffering no less than the origins of politics. Whereas Husserl was seeking to shape, in a typically German hybrid of naturalism and idealism, a science of philosophy (a rigorous method for thinking about thought), Lessing was seeking an intellectually and emotionally convincing path from personal distress to political energy.[309]

Already he had issued a call to arms against noise. Noise made an intentional life more difficult. No minor blip in the moment-to-moment sweep of our lives, noise was now a persistent disharmony, a counter-presence that violated the first rule of reason (as Schopenhauer had written, using English for effect): "never interrupt." In essays of 1901 and 1902, and in a newsletter, *Der Antirüpel* (Anti-Rowdy, 1909–1911), Lessing called out German society—not just selfish organ grinders but fife and drum corps, gramophone owners, piano-pounders; not just teamsters carting scrap iron but bureaucrats who kept paving streets with uneven cobbles; not just squawking-parrot fanciers and screeching-alley-cat feeders but churchmen ringing bells dawn to dusk; not just loud drunks in the night but shouting children at play in the streets at midday because they had nowhere else to go. He quoted Carlyle at length. He cited the medical literature on neurasthenics and other sensitive souls in order to make his case for a Right to Quiet, *Recht auf Stille*—one subtitle of his newsletter. It had to be a right, not just a class privilege: surrounded by constant uninvited noise, people essentially lose their personhoods, their ability to hear their own thoughts and reflect upon the world. This was an environmentalist argument with a phenomenological spine. As noisemaking can be a means for suppressing awareness or drowning out frustration, so, wrote Lessing, the human response to terrible noise is either psychic withdrawal, madness, suicide, or indiscriminate retaliation against the offending *Zischen, Stossen, Stampfen, Kreischen, Pfeifen und Schreien*.[310]

Should the stepped-up noisiness of life at the end and turn of the century be understood as an auditory hallucination or a modern abomination? as evidence of physiological weakness, psychological frailty, or social discombobulation? To the many readers of the 1892 treatise on *Degeneration* by the physician and novelist Max Nordau, the problem was not noise *per se* but *fin de siècle* waves of hysteria, emotionalism, and "egomania" that resulted in two disturbing if contrary acoustic tendencies: the first, a morbid sensitivity to sounds; the second, addiction to aural excitement, such that, in concert music, "A dissonant interval must appear where a consonant was expected," and a *Katzenmusik* polyphony must confuse the ear, like "the jangle of a dozen voices." To Nordau's ear, noise was all of the above—hallucination, abomination, discombobulation—but it was epiphenomenal, an alarm bell rather than the cause of alarm. Born Simcha Südfeld, son of a Hungarian rabbi, Nordau himself was a healthy cosmopolite in Paris for whom noise warranted nothing like the attention that ought to be directed to body-building, character-building, the building of good houses for workers and (as he reentered the Jewish fold in reaction to the Dreyfus trial and pogroms in Russia) the building of a Zionist movement for the creation of a haven from antisemitism. What then led Lessing—also a Zionist, also an advocate for workers, also a cosmopolite—to concentrate on noise and propose a "coalition of intellectuals" to combat this evil?[311]

Answer: bacteriology. Now that major diseases could be explained in terms of communicable infections, and infections could be explained in terms of invisible but detectable cells, it had to be clear that each success at speeding up travel and transfer was also an opportunity for microbes to make ever-quicker, wider rounds. The same "Malthusian" tornado of trolley-riding, train-traveling, bicycle-racing, and auto-touring that raised dust and spread germs also raised a ruckus and spread noise—another invisible and invasive presence. The same sanitational cyclone that meant to thwart microscopic villains through storms of carpet-beating, vacuuming, toilet-flushing, pot-washing, dustbin-emptying, and (mechanical) street-cleaning produced a sensational ruckus. Like bacteria, the world's newest bogeys, infiltrating everywhere and everyone, noise had the power to sabotage the sturdiest of bodies, derange the best of minds. Nordau would say that noise had such power only to the extent that a population had become so weakened it could not resist the world's customary intrusions; Lessing would say that noise had such power because, in its current amplified state, it wore down all defenses of the blood and nerves. When the Austrian dramaturge Alfred Freiherr von Berger went into his garden in 1907 to take a sounding of the city around him, he explained that he had done his sampling "just as sometimes the drinking water, which one consumes daily, is tested for chemicals or bacteria." An early advocate of a right to quiet, he heard from within his sedate district of Vienna the horns of three street bands and the whistles of three factories, the barking of two dogs, pounding of two pianos, buzzing and hooting of two automobiles, as well as the cry of a peacock, a phonograph playing symphonies, the roar of zoo animals, the screech of train and trolley brakes, the bang of railroad bumpers, the blare of trumpets from a barracks, the squawk of a parrot, and the high pitch of "a scythe being sharpened." Such a mob of noises could only be damped, and effectively damned, by a national defense league.[312]

Second answer: thermodynamics. In the terms that Lessing had used to analyze the thought of Schopenhauer, Wagner, and Nietzsche, noise was the audicon of entropy. Where living beings struggled to maintain their integrity and self-organization against the literal dissoluteness of the physical universe, noise—that "super-abundant display of vitality, which takes the form of knocking, hammering, and tumbling things about," grumbled Schopenhauer—was also what eroded psychic energy, wasted intellectual force, used up emotional reserves, scraped the nerves, and sapped the very effort at a tuned consciousness. Socially it was "the most impertinent of all forms of interruption" (Schopenhauer), physically the most fundamental rudeness. Lessing established the headquarters for his defense league in Munich in 1908, a city of "sensual sloppiness," or so he wrote, that had tripled in population since 1870. Skirted by an expanding industrial sector and rows of barracks for laborers, the center of Munich was overrun by beggars and whores, and its workdays were often interrupted by labor protests and strikes. Lenin had been in Munich since 1900, pacing in his room and whispering aloud each sentence of *What Is To Be Done?* before composing his detailed plan "in which everyone would find a place for himself, become a cog in the revolutionary machine," while his wife Nadezhda Krupskaya washed the dishes as quietly as possible, learned "secrecy technique," and helped publish *Iskra* (The Spark). The Lenins returned to Russia in late 1905 just as Arnold Sommerfeld was arriving at the University of Munich to set up the first institute for theoretical quantum physics. If only he could, wrote Einstein in 1908, he would take the train up from Bern to hear Sommerfeld's lectures on the mathematics of electrons—and perhaps a first glimmer of his revision of Ludwig Boltzmann's 1898 statistical definition of entropy, which, Sommerfeld would show, must be understood not as disorder but as a lack of restraint on the ranges of variables required to describe the thermodynamic state of any system. Lessing had read Boltzmann, was unfamiliar with Sommerfeld, but already understood noise as a lack of restraint.[313]

Der Überbrüller (The Out-Hollerer), one of the names proposed for the house organ of Lessing's league, was too much like the new unrestrained college yells that a certain Professor Schnittfeidel had detailed for the *New York Sun* and that Lessing reprinted with a sneer: Auburn's vaguely Indian "Ki-yi-ya! Ki-yi-yi! Hoop-la-hi!"; Kentucky Central's "Razzle, dazzle! Sis-boom! Ah! Central University, Rah, Rah, Rah!"; Hiram College's magical-classical "Brekekex! Koax! Koax! Brekekex! Koax! Koax! Alala! Alala! Siss-s! Boom! Hiram!" Other proposed titles were no less noisy: *Cri-Cri*, *Der Schreihals* (The Screaming Child), *Das Nebelhorn* (The Foghorn). But *Der Antirüpel* it would be, best understood as *The Counter-Lout*, for Lessing was most concerned with effecting a change in a society whose individual relations with the sounding universe had gone sour: a phenomenological transformation, not a shouting match. Sure, he supported legislation against country animals in cities and heavy industry in residential centers, against steam whistles anywhere, against pianos and gramophones playing in the quiet hours after lunch, against churchbells too frequent and sirens (which were not signals but *Unfugen*, nonsensical nuisances) whenever.

He applauded the substitution of traffic lights for police whistles and of asphalt for granite paving blocks, and lauded Baedeker's introduction of R ratings for hotels that were *ruhig*, restful. But most telling was his high dudgeon in response to a correspondent from Dresden who claimed that the millennial quest for a Music of the Spheres had been trumped by modern appetites for a whole host of sounds, terrestrial or celestial; instead of going on the warpath against noise, musicians were now learning to listen for new sounds, such as the backfire of automobiles, as sources of inspiration. Lessing was outraged: "Is there a music of noise? I could just as well ask, is there a logic to nonsense? Music is music, and not noise. And noise is noise, and in no way music." Otherwise one might end up with a symphony of backfiring autos, rumbling freight trucks, stamping packhorses, and blundering beer wagons—a music of wasted if furious energies. (This exchange occurred in 1910. Had Lessing heard of Filippo Marinetti and the Italian Futurists, whose first manifesto was published the year before? He was up to date on Paderewski's *tonitruone*, an instrument used in concert for imitating thunder, and predecessor to Futurist orchestrionics.) "I fail to understand," thundered Lessing, "how the loudness of tone production or the complexity of a held tone can be confused with noise. Only a musical ignoramus would do this." Or someone who had surrendered his personal boundaries, ceded her ear-hold to the entropic enemy in return for an aesthetic that was an apologia for rampant industrialization, urbanization, and its resultant pandemonium. Another of Lessing's correspondents, Dr. Pudor of Leipzig, upset about the noises audible in a pedestrian subway, did not mistake them for music even though he reported them poetically as "Brrr-bumbum—trab trab-trab—tuuh—s sz s sz s sz—kling-klingkling—tjätjä—janemizeseruzalltakinoschwanz." The goal of Lessing's *Deutscher Lärmschutz-verband* was neither to eliminate noise by redefining all sound as latent music nor to escape noise by constructing sleepatoria at the bottom of the sea (however attractive this latter vision may have been to Lessing, whose first essays against noise had been cued by the student ruckus around his rooms in Munich—or was it because in 1901 his first child had just been born and he could not bear the noise that kept infant Judith from her sleep, and so him from his?). The anti-noise movement must engender in each person a heightened awareness of the deadliness of noise that led beyond whining to municipal action, to more acoustically sane urban planning, and to such innovations as automatic door-closers, quiet typewriters, and sound-deadening walls. *Non clamor sed amor*, read the motto of Lessing's league: "make love not noise."[314]

Egoism was the problem. Private property and egoism. Private property, capitalism, and egoism. Private property, and the tension that builds where private property is felt to be threatened by a multitude of other property owners or by gleams in the eyes of the dispossessed; mass-merchandising capitalism, and the oxymoronic impulse toward having what everyone else has in order to distinguish oneself; egoism, and the competitive ideology of individualism, in which men claim not to care what others do so long as that does not curtail their own freedom of maneuver, which in practice prompts each to maintain a more impenetrable buffer against the other and to assert a wider margin of maneuver in the face of apparent encroachment by muscle or bad music. "Men," because as a socialist and feminist, Lessing was pursuing here a political critique that was leftist, gendered, complexly philosophical, and (said his critics, say his biographers) therefore doomed to a small following, despite a boast in 1910 that his league had a million sympathizers in Germany alone and his insistence that noise was everyone's concern, including American multimillionaires in their mansions in Newport, whose summer idyll was daily broken by gawking tourists in beeping, backfiring cars.[315]

Lessing saw and heard that automobiles, trucks, motorbuses, elevated trains, and flying machines were the vehicles of the future; he was not reactionary and not (as was in vogue) neo-Gothic. If he hoped to enlist the millions in a "War against Noise, Vulgarity, and Incivility in German Domestic, Trade, and Commercial Life" (the longer subtitle of *Der Antirüpel*), he was not retrocoustic. Like Karl Lamprecht, the eminent German medievalist and comparativist who supported the anti-noise campaign at the same time that he pressed for a controversial new historical methodology that encompassed cultural, environmental, and psychological factors, Lessing was considering the cultural, environmental, and psychological aspects of sonicity as the key to a global evaluation of societies that were urbanizing and industrializing faster than they could humanize or empathize. Lamprecht was seeking out the "social-psychic factors" in political change and social unrest or "disharmony"; he had been stirred by the "new historical atmosphere" of modern Germany, where the "increase of city activity, with its nervous haste and anxiety, its unscrupulous abuse of individual energy, progress of the technical arts, and the extraordinary multiplication of the means of communication... produced a great number of new stimuli, which neither the individual nor the community could escape." At the same time that people began to expect constant excitement, they became irritable, with a special irritability, or *Reizsamkeit*, in regard to those urban dissonances he too found insufferable. (Did Lamprecht, who had collaborated with Wilhelm Wundt in the 1880s, follow the work of Wundt's assistant at that time, Oswald Külpe, who claimed in 1893 that people could discriminate 150 shades of color, 694 levels of brightness, but 11,063 audible tones?) The refined and the artistic sought out "indefinite titillation by minimal dissonances of the tones, the gentle rustling audible in the building and the interior rooms, suggestive silence in conversation"; the rest went in search of "deafening music, a life in the streets filled with the greatest discords of movement, of visual phenomena, and of tones." The job of the historian was to identify the "*dominant*" chord (Lamprecht's italics for "the very kernel of personality") of every epoch. Lessing was listening for the binaural implications of lives terrorized not so much by change as by the noise of change.[316]

Groups in Austria, the Netherlands, France, and the United Kingdom had also begun protesting the noise, and sometimes the historical Gestalt (urbanization + centralization + mechanization + electrification) from which the noise arose. Those with a sense of context appreciated the irony that the most distressing noises came from newly powerful or newly extensive steam whistles, sirens, gongs, horns, bells, and buzzers mounted wherever soft human bodies had to be warned of hard, swift machines approaching at inhuman speeds, pounding with inhuman force, or acting with inhuman regularity. The automobile became the exemplar of this irony. In concert with asphalt pavement, motorcars with rubber tires had been heralded as a solution not only to the mess of roads mucked up with horseshit but to the grinding of the "ball-bearingless axles" of wooden drays and the sharp blows of horseshoes striking cobblestone. "When the motor comes in, with its rubber tires and improved bearings, and the sound of iron hoofs becomes infrequent in our streets," predicted the editors of *The Horseless Age* in 1896, "then we shall realize that one of the blessing of the motor vehicle, and one, perhaps, we had not fully anticipated, is that it is a conqueror of noise." So quiet were the first years of humming electrics, chuffing steamers, and putt-putting diesels that ordinances were swiftly passed requiring motorists to squeeze their horns (or "honk," a new usage in 1895) often and loudly, to alert pedestrians and equestrians unaccustomed to the hush of horseless vehicles. With such laws in place, however, it was as hard to restrain automobilists from intemperate honking as it had been to restrain cyclists from bell-ringing, and much harder to stop them from honking deliriously at two a.m., or as mating calls, or in sarcastic salute—or at every crossing, so as to avoid being found liable for accidents because (as in case after court case) they simply had not honked. So the motorist, enamored of ever-more powerful cars, "got a horn which could be heard a mile in order to frighten the man ten feet in front of him." The prefecture of police in Paris had to annotate its honking laws to remind drivers that it was a serious error to suppose that once they had used their horns, they could no longer be held responsible for ensuing accidents. By 1908, as toy manufacturers introduced miniature (pedal) automobiles complete with small squeeze-bulb horns, the larger versions were coming under suspicion as criminal accessories to collisions whose keynote was rather that of metal through glass than wood against flesh: "On crowded thoroughfares the clanging of bells on trolley cars, fire apparatus, police and hospital ambulances and the honks of motor vehicles are often so confusing and clamorous as to bewilder pedestrians who, in trying to avoid one danger, run into another."[317]

Spot on. "There was a time within the memory of men still living," wrote the editor of *Harper's Monthly* in a prose that accelerated from irony toward inevitability, "when those who had been tormented by the tinkling of the horse-car bells promised themselves relief in the coming of the cable-cars, the electric cars. But the cable groaned and growled incessantly when it replaced the horse, and when the trolley-wire, overhead, or underground, replaced the cable, though at first it propelled a car which emulated the noiselessness of the mystical current it bore, soon began to drive through the streets a vehicle with flattened wheels which battered the senses with the effect of innumerable hammers."[318] Thus William Dean Howells, who at sixty-nine had spent his latter years seeking out ever-quieter quarters in New York City, in Bermuda, in Maine, and most ironically in "too quiet" Portsmouth, New Hampshire, where engineers widening the harbor in 1905 created the largest manmade explosion that had ever been heard, blasting a quarter-million tons of rock into the air. Howells continued: "Some forecast the horselessness beginning to characterize our civilization as relief from the clatter of iron hoofs on the stone pave, as a silence which no other sound should break; for who could have imagined the percussion, the whiz, the whir, the honk of the automobile?" In 1911 the *Saturday Review* and *London Times* ran features, editorials, and letters to the editor on "The Hooting Nuisance," which was crueler than the noise of muffler cut-outs, tire blow-outs, or occasional crack-ups. "Instead of the crepitation of thousands of tapping hoofs on the pavement we have the definite mechanical buzz of the motor for ground tone," above which rang the shifting of gears and above that the honking of cars, each of which was now like a busker racing around the city making "loud grunts or sudden hoots, yells, squeaks, other sounds that one can only imagine to be like the death-rattle of a mastodon." Would that these too were on their way to extinction, what with sixty-seven blasts per minute between 7:50 and 7:55 p.m. on an otherwise quiet London street in Mayfair! How dare the Westminster City Council prosecute one of the last of the gentle muffin men for ringing his time-honored bell in the street when there was no decent regulation of these new horns, audible a mile off, sounding a "perfect medley of grunts, whoops, shrieks, squeals, snorts, snarls, and bellows to which no human ear can ever hope to grow accustomed"? While the thud and hurtle of horseless carriages was running off the remaining peddlers' carts and organ grinders, this did not give audimobilists the right to fill in the void with their own shouting and honking. If the Putney and Bayswater buses had been deprived of their coach horns years ago and ships' pilots on the Thames had been barred at night from tugging the cords of their steam whistles, what held authorities back from forbidding motorists from "making noises which have been compared to a dying horse, a tuberculous cow, a famished seal"? The Locomotives on Highways Act of 1896 did require "light locomotives" to give "audible and sufficient warning of their approach," but nowhere was "audible" sufficiently defined, so motorists in their roadster-locomotives erred ever on the side of the raucous. In the United States, where automobile registrations sextupled in the years between 1907 and 1913, with a concomitant rise in the numbers injured and killed in traffic accidents, Chicago's ordinance of July 17, 1911, requiring that every motor vehicle have "a suitable bell, horn, or other signal device," had to be replaced five months later by a more specific ordinance in the double negative that motorists must not use "any device which will not produce an abrupt sound sufficiently loud to serve as an adequate warning of danger." Both ordinances had a proviso that such devices not be used "except to warn," but abrupt and loud were more attended to, on the road as in the codes, than the etiquette, or ethics, of restraint.[319]

Might not the horns emit *un ton grave*, a deep note, as directed by the 1909 International Convention on Motor-Cars, or, as Brown Brothers of London promised for their horn, "Melody *NOT* Noise"? Not if what was required was a sound abrupt, loud, and *alarming*. Besides, motorists like Lady Agnes Geraldine Grove, for whom there was "something in a vibrating, pulsating machine under one's own immediate control that sends the blood coursing lawlessly through the veins," wanted a sound equally antinomian and egotistical. Whence in 1908 the famous "ah-ooh-gah!" of the electrical Klaxon (< New Jersey Greek, "to shriek") that began its triumph over other horns as it triumphed "above the downtown din!" The Lovell-McConnell Company of Newark, which had bought the patent from an Edison associate and fastened upon the name Klaxon, claimed that "accident insurance adjusters the country over ... trust no other signal." The Klaxon "*always* warns, in time, *before* you are confused." It was pricey, a double sawbuck and up, because the company pledged Permanent Satisfaction; by the start of the Great War more than 400,000 had been mounted on cars across the planet. Other horns more biblical or classical tried to compete: the Jericho-o-o-o-o-o ("A Pleasing Tone / An Insistent Warning"); the Gabriel ("the *only* truly musical horn ... signals the approach of a *gentleman's* car"); the Sparton (whose Ya-Hoo-Ta "mingles with the crowing of chickens, and sounds high above the roar of the city traffic"). According to an Austrian official, the car horn attacked "everywhere in the city, at all times and from all sides." Whether at the wheel of a "silent running" Aerocar Touring Runabout or a Northern Limousine with its "wonderful Silent Northern engine," the dashing motorist had little compunction about noising through every intersection. For one German, the medley of horns was "Sharp, cutting, shrill like a whistle, deep like a trombone or a foghorn, shaking us like an explosive charge."[320]

Vienna, like other expanding urban centers, was audience to increased traffic noise from automobiles and their horns, locomotives and their whistles, motorized streetcars and their gongs, unmuffled motorcycles and traffic bullhorns, all amplified by urban canyons lined with taller office buildings and larger apartment complexes, such that the city in 1911 seemed to be a "stone vessel...from which noise cannot escape." The Austrian Health Care Society had by then redirected its efforts from cleaning up grotty sewers to reducing "that other uncleanliness, that loud odor which is called noise." Thus Austria's premier ethnographer, Michael Haberlandt. Sad was it that the ears, most forbearing of sense-organs, loyal wardens, should be undefended in these civilized times against indescribable overtones from the thunder of cracked churchbells and all the new steam-powered and electrical tumult of mindless noise-seeking and barbaric noise-making. Cities were being acoustically polluted (*verpesten*), wrote Haberlandt, not by modern industry alone but by the sonic restlessness of an underclass of wild men and the exuberance of the machine-infatuated. Shame on rotten-good-for-nothing technology ("*Pfui über die faule Technik*") for not laying down the law by laying down noiseless roadways.[321]

Dutch authorities, threading their way through more bicycles and canals than cars, debated a different problem of rotational noise, that of the cheap gramophone, which in the hands of working people might wheeze for hours over scratchy disks of questionable if sometimes orchestral music. This sonic repeatability—was it at odds with a mutually respectful urban life or a pedagogical path toward respectability and better musical taste? The class issues were clear here too, if not as stark as they were across the North Sea, where British activists had been instructed by James Sully in an 1878 essay on "Civilisation and Noise" that would be as widely quoted in English as Schopenhauer was in German. The psychologist Sully had an ear to the imperial powers of noise, which held children and savages in thrall, but was also capable of the "violent capture of the attention" of urbane adults through the trance rhythms of, say, "a chimney-cowl revolving with the wind" or through that wretched state of suspense, say, "in which the irregular barks of certain lawless dogs are apt to plunge a man with a sensitive ear." Since the more civilized the person, the more sensitive the ear, it followed that noise was the more injurious the more it overtook the lives of the social and intellectual elite, for example in capital cities where such persons cluster. The year before, Sully had published a study of *Pessimism*; the question at hand was whether civilized people surrounded by increasingly noiseful environs, their ears "exposed to a whole *Iliad* of afflictions," had reason for optimism due to a strength of character or discipline of mind that could rise above the turmoil of modernity, or to a hardening of the tympanic membrane after weeks of exposure, or to a process of natural selection by which those most put out by noise gradually leave the field to advanced souls who are more aurally resolute. No, he decided: no good reason for optimism in the capacity of even a superior intellect to resist storms of noise, for individual resistance, if not futile, was stressful, fatiguing, and ultimately debilitating. The answer had to be collective resistance that would arise as society became more civilized and a "general growth of sensibility" led to an equally general protest against all those shriekings and crashings that the current populace blithely ignored or, as with oompah bands, seemed to welcome. Such progress, however, could be generations in coming, as sensibility had to be refined by good taste, which was always so far ahead as to bring up the rear. Meanwhile, activists should refrain from arguing that those who suffer from noise "consist of some of the most valuable members of society"; instead, they should appeal to common sympathies for the weak and defenseless, i.e., the ever-open, tremulous ear, and to a not-uncommon empathy for those in pain, since "The pains inflicted through the ear are deep and pervading, analogous to bodily hurts." Moral suasion, medical evidence, legal injunction, and new taxes on dog-owners must do the work that will eventually be done as communities of sentiment act against outrages upon all of the senses.[322]

Scoffers seemed to agree that (with the exception of revolving doors, rubber-soled shoes or "sneaks," asphalt pavements, and electric cars) the modern world was noisier. The crux of the issue, especially in the New World, was how much of that noise was *virtuous* noise, and whether warriors against noise—like London's new Society for the Prevention of Street Noises—were making war on modernity itself. Whoever objected to the purr of a sewing machine, hum of a telephone wire, clack of a typewriter, zoop of a pneumatic dispatch tube, or the gurgle of a flush toilet—undeniable benefits of contemporary life—had better retreat to villages untouched by the modern arsenal of sanitary, laborsaving, and message-speeding devices. According to an anonymous philosopher of noise in 1893, "Within civilization itself, indifference to noise is one of the distinctions of a system rudely healthful, both in body and mind." Anyone frazzled by the minor anxieties of noise was lacking as much in willpower as in tympanic resilience and was, in a word, becoming outmoded. Hollis Godfrey, who as a scientist and high school instructor in Boston found that urban noise had increased "to a point beyond all reasonable tolerance," still conceded that "A really silent city is impossible," and began his chapter on "The City's Noise" in *The Health of the City* (1910) not with a denunciation but a Whitmanesque dithyramb: "Now loud, now low, now sounding in musical humming rhythm, now clanging in sharp staccato or rising in plangent, shrieking chords, the song of the city comes to the listening ear." An editorial in *Scribner's Magazine* in 1909 went so far as to theorize that noise could be a protective presence. Take for example the ease of reading inside a railroad car while surrounded by "the measured rattle and bumping of the car wheels, the hoarse whistle from the engine, the groaning and gasping of starting, the rasping shriek of the rails and the brakes at stopping, the insistent inquiries of the porter, the inarticulate shout of the breakman, the impertinent chant of the train boy." Wasn't there some consolation in realizing that in these days "of [an] alleged progressive demoralization of the nerves," noise itself could form "what may be conceived of as a sort of enveloping enclosure of sound"?[323]

Knowing commentators have accounted as meager the successes of early anti-noise movements, with their memberships minuscule and middle class, and have explained their "failure" as the turning of a deaf ear on all that was modern. But Charles Babbage had been as modern as they come, a man literally at the cutting edge of a Second Industrial Revolution, of social thought and economic analysis. Lessing, although deeply saddened by the effects of industrialization and urban compression, had no nostalgia for rural society or the putative benevolence of a landed (Junker) elite. He despaired of capitalism and its ratcheting up of speed and power on behalf of unrestrainedly consuming and suggestible egos, but he was listening toward further stages of transformation, toward a socialist world and a moral philosophy that would sustain concerted action on behalf of women and workers. He used media old and new to make his points, and he used them well. Like Babbage, he should rather stand accused of miscalculating the prospects for his own era than of repudiating a new century, or of foisting middle-class sensibilities upon a modern sensorium; like Lamprecht, whose interdisciplinary methods would later be taken up by the *Annales* school of historians in France, he was "ahead of his time" in believing that history was rooted in myth, not stone, and that the writing of history was always already a reworking of the past in hopes of editing a half-written future. He believed (too) much in change, and he hoped that the noise of the modern would not be mistaken for its promise.[324]

What made him think that a war against noise could be won? Precisely those sociocultural changes and technical advances that were responsible for heightened levels of noise were those, he thought, that could be turned against noise. The same modern engineering that produced gas-powered automobiles could quiet them with mufflers and lubricants. The same modern efficiencies of manufacture and distribution that put pianos and gramophones in parlors could put hydraulic stoppers on doors, rubber bumpers on garbage cans, double panes on windows, mutes on telephone bells. The same modern rationality that planned new cities and carved up the countryside with roads could situate schools and hospitals in public gardens and pave the streets with asphalt. The same modern medical establishment that administered a serum against ototoxic diphtheria and inspected children's ears so that adults heard more, or better, could demonstrate to skeptics the somatic effects of noise and prescribe earplugs or quiet. The same modern powers that prepared so effectively for war had also introduced social insurance and compensation for workmen's injuries, so administrators should be able to come up with effective means for restricting the hours of churchbells, licensing pets and street musicians, regulating the routes that trucks take through town. The same modern architectural momentum that drove buildings skyward with safety elevators and huge fans could be directed toward reducing the noises that banged through apartment walls or blasted from factories. Modern skills at exact measurement and detection could be used for identifying, measuring, and reducing levels of urban-industrial sound. Legal philosophies that granted personhood to corporations could be redeployed to establish full rights to quiet and privacy in a West with smaller houses, fewer servants, and thinner-walled bedrooms. And the same modern media that were loud with advertisements for circuses, alarm clocks, and Independence Day firecrackers could be used for propaganda favoring the quiet man, the quiet woman, the quiet house, the quiet street, the quiet city. If noise was an inevitable offshoot of industrialization and urbanization (however detested by such avatars of the modern as Pulitzer, Northcliffe, and Proust), communal acts of silencing were reasonable requests to make of the truly modern. If the nineteenth century had been born with the steam engine and the piercing cry of steam whistle, if it had fostered the siren, the dynamo, and the telephone bell, then the twentieth should be ethically prepared and sociotechnically equipped to restore acoustic respectfulness to each lifeworld.

Besides, Lessing was much encouraged by the success of the Society for the Suppression of Unnecessary Noise in New York City.

16. Julia, a Quadrille (in Five Figures)

Quadrille? The following sections play out as a complexly partnered dance, with jigs and pigeon-wings at the edges of the square, and inevitable missteps among those not up to the balletic *jetés* and *chassés* that had been *de rigueur* early in the nineteenth century. But I have also in mind the quadrille as it evolved into a more popular, raucous, flinging affair, a form rather of exercise than of footwork. And also the quadrille as an instrumental genre caricatured as a battle among loud and louder musicians. And the loudest quadrille of the century, Louis Jullien's "Fireman's Quadrille," attempted later by other ambitious conductors who could never equal the 1854 premiere in Manhattan before an audience of forty-two thousand at the Crystal Palace, where Jullien's enormous orchestra was assisted by two military bands, a battalion of sound-effects devices and, with the connivance of P. T. Barnum, three fire companies who rushed into the theater at the climax, ringing brass bells, shouting through megaphones, and hauling lengths of hose up through the galleries to douse "something

SOUNDPLATE 20

like a blaze" in the cupola.[325] And, of course, the most memorable of quadrilles, from *Alice's Adventures in Wonderland* (1865), Lewis Carroll's "Lobster-Quadrille," with its refrain, "Will you, won't you, will you, won't you, will you join the dance?" For my quadrille is at once a dance-card of turn-of-the-century invitations to join an international movement against noise, a set of biographies of those calling the steps or orchestrating the music for this movement, and a family romance that would have acoustic, social, legal, and technological repercussions well beyond the turn of the twentieth century. In order to appreciate the agile footwork and partnerings of Julia Barnett Rice at the head of her Society for the Suppression of Unnecessary Noise, it is certainly important to understand her own medical, musical, social, and familial life, but it is also vital to understand the engagements, interests, aggravations, and international pursuits of her husband and intellectual partner Isaac Leopold Rice. So, in the manner of the liveliest of quadrilles, Isaac's adventures will weave through the patterns established by other callers of the dance and complement, often with grace if not always with economy, Julia's lifework.

1: Morse the Malacologist

"Obviously," declared an editorial on City Noise in 1902, "the first object is to create a more healthy state of public opinion. For this purpose citizens who love quiet might form a society, with a small annual subscription; this being the orthodox British method of starting a reform." From his offices in London and in Auld Reekie (Edinburgh in its heyday of smoke and stink), and as nephew of that Scot who had published *Vestiges of the Natural History of Creation* before founding *Chambers's Journal*, Charles Edward Stuart Chambers allowed that some urban ruckus was unavoidable, such as "the rumble of the loaded wagon, the clang of the tramway-bell, the snort or the shrieking whistle of the locomotive," but would not credit "hardened city men" who claimed to be so evolved as to prefer "the tumult of the town" to the quietness of the countryside. Their minds and nerves must suffer like all city folk from the wear and tear of the rattle of horse-cabs and "thud of shunted trucks." If some noises were unavoidable byproducts of industry and commerce, most were, or were becoming, avoidable, such as the "raucous shout of the coalman and the jangle of the milkman's bell" or church carillons ringing through the night. Since "to live constantly in an environment of noise is as inimical to perfect health as to eat unwholesome food or to breathe in a vitiated atmosphere," Chambers proposed that concerned citizens form a "Society for the Suppression of Unnecessary Noise."[326]

Ungainly and almost apologetic, the title did not immediately take—the SSUN did not rise. It had been forty years since Florence Nightingale had condemned "unnecessary noise" in hospital wards, forty years during which ear activists in Milwaukee and Manchester, London and Manhattan, Paris and Munich could warm to the concessions of that phrase. In the United States, however, the needful was least bound up with the traditional. Where an aggressive entrepreneurial economy, greased by international combines and goaded by national advertising, was hell-bent on making invention the mother of necessity, the very notion of "needfulness" was on the table. Needfulness was put into question not only by Thorstein Veblen's 1899 attack on "conspicuous consumption" but by a sisterhood of household economists and a fellowship of efficiency experts. While Chambers in Great Britain counted among necessary noises "the whooping of sirens on board the steamers," across the Atlantic such stalwarts as Edward S. Morse, Edward A. Abbott, and Julia B. Rice would take issue even with those, and Dr. John H. Girdner was unwilling to concede much of anything by way of noise to modernity or the metropolis.

Morse we have met before, in Japan and back home in Massachusetts. Asked to establish a zoology department for Tokyo's new Imperial University, he had returned to Salem in 1879 at the age of forty-one a world authority on marine biology, Japanese ceramics, and the ethnography of archery. The next year, as director of the Peabody Academy of Science, he began to expand its natural science collections as well as its Asian holdings, already diverse with gifts from whaling captains. The Peabody was no backwater; it welcomed 36,056 visitors in 1883, some to gape at odd snails and naked skeletons, some to stare down a series of life-size figures from China, Japan, and India, some to tremble before an intricate wood carving of "Heaven and The Day of Judgment." But what drew Morse out of his study into the public fray was not the bustle of these museum crowds but the yowl of steam whistles. New Englanders used these whistles with abandon, typifying a nervous monomania too prominent in their character even for a Down-Easter who had collected his first wages as draftsman for a company that built steam engines and locomotives.[327]

Hmmm. Here was a man who had been expelled for his boisterousness from a series of grammar schools and in middle age preferred military music to Mozart; who dreamed of fire alarms as a youth and in adulthood would chase after fires at the clang of a fire-bell. So what had gotten into him? The spirits, perhaps. Upon the death of his older brother from typhoid, Morse in his teens had repudiated the Calvinism and revivalism of his father, turning instead to the *Spiritual Telegraph* and public séances in Portland's Mechanics Hall, where he was moved by the shade of Edgar Allan Poe speaking beautifully through a medium. If the Other Side did not stick with him as a lifelong presence (he would call mediums "arrant humbugs" in 1898), the Spiritualist conviction that God had "no room for Hell or Devil" certainly did; when Morse later encountered Charles Booth and his Salvation Army soldiers with their bass drums, bugles, and hellfire language, he was—despite a passion for marching bands—appalled at "the most marked exhibition of emotional insanity I ever witnessed." So there was his spiritualist predilection for a revelatory quiet and his rationalist predilection for meditative examination, and then there were his bad teeth and his tonsils, chronically infected from 1858 to 1862: cause enough for him to be underweight and rejected for Maine's Twenty-First Regiment during the Civil War, cause enough also for a chronic sensitivity to high notes. Hyperacusis might explain why, despite an accurate ear for tones and a capacious auditory memory that enabled him to chant long passages from Noh drama, he never got the hang of Japanese music or of Japanese singers who made "the most curious squeaks and grunts" accompanied by plucked strings whose overtones ran toward the highest registers. His biographer, Dorothy Wayman, suggests that Morse returned from Japan with an appreciation for aesthetic simplicity and mental tranquility; although he had fumed at the Japanese "slamming of storm blinds as they are pushed in place, one after another, at night," he envied their mental and spiritual calm, and "the fact that they are never seen to drum with their fingers, nor do they whistle or rattle anything in their hands or indulge in manifestations of nervousness as we do." As for Morse's own nervousness, there was the stress of a chronic insolvency: due to his insatiable collecting of mollusks and ceramics, he was rarely out of debt until the age of fifty-four, when Boston's Museum of Fine Arts (at the behest of a good friend on the

Board of Directors, Samuel Warren) purchased his 105 barrels of Japanese pottery and hired him as keeper of the collection. Debt made him irascible, writes Wayman, and Japan made him nostalgic for the olden days, "when rikishas spun noiselessly over dirt roads."[328]

Debt and petulance, aesthetics and simplicity, tonsils and overtones, spirits and silence—these help account for Morse's animus toward steam whistles, which he could hear from freight yards near his house on Linden Street and which he would frame as archetypal of the "savagery of modern machinery," like the siren, which had "no justification whatsoever." Not fully at home with his times, Morse never used a typewriter and never, to his death in 1925, drove a car. He did come to own a phonograph and play classical recordings, but his preference, even as a naturalist, was for an oddly unnatural hush. After his eighty-fifth birthday, hard of hearing and pained by rheumatism and bronchitis, he went off to recuperate in the White Mountains of New Hampshire, where he was happy to hear "Absolutely not a sound. No whistles or puffing of engines. No crowing roosters or cackling hens. No barking of dogs or squalling of cats. No mooing of cows or bellowing of bulls. No shouting of children or crying of babies. Not even a songbird or cawing of crows. Doubtless the insects chirp but I do not hear their high vibrations, so it is simply heavenly." All the more reason why, in handwritten letters, in speeches, and in pamphlets, Morse had spent years trying to persuade boards of health, college students, and selectmen of the medical, mental, moral, cultural, and political urgency of silencing, at the very least, steam whistles. Addressing the graduating class of Vassar College in 1895, Morse contrasted the rudeness of Americans with the politeness of the Japanese, who had learned how to live harmoniously in rows of tenements whose rooms were walled off only by bamboo curtains. Educated women had a special obligation to teach the American masses "the efficiency of good manners in the battle of life." Efficiency, manners, battle: having invented a superb filing system for his collections of brachiopods and stoneware; having mastered the etiquette of daily life in Japan; having split with his mentor, Louis Agassiz at Harvard, who was of a generation and a geological stratum to reject Darwinian theory; and having in 1886 presided over the American Association for the Advancement of Science, Morse could integrate the rhetorics of zoology, anthropology, evolution, and scientific Whiggism which together constituted the intellectual underpinning of his thirty-years-war on noise.[329]

Addressing the graduating class of Worcester Polytechnic in 1900, Morse was steamed at the misplaced efforts by municipal authorities to silence "the hurdy-gurdy, which may be heard for a block or two, and which gives infinite delight to the children, and yet permit a pandemonium of factory whistles in the morning" and railway whistles all day long—whistles that could be heard miles off. Polytechnicians with hopes for livable cities should demand "the suppression of every unnecessary noise," just as hotels, which had once sent a man through the halls to wake patrons for breakfast with "a Chinese instrument of torture—the gong," now rang the rooms by telephone or sent bellboys to tap at doors. Could not a city, like a good hotel, be shaped into an "orderly and temperate community"? Not when people were jolted by steam whistles, a menace to sleep and sanity, he admonished the Massachusetts Association of Boards of Health in 1905. If the morning crying of Sunday papers could be suppressed, as it had been in Boston, why not also the clanging of churchbells, "which is supposed to call to prayer but which often leads to the opposite," and why not the steam whistle, whose shriek had already been forbidden in Berlin and Atlanta? "Just in proportion as a family or community become civilized, just in proportion are unnecessary noises suppressed," for the sake of "invalids, of the sick and dying, of the nervous and overworked," and for everyone else. Morse quoted the *Philadelphia Medical Journal*, whose editors had called American cities "pandemoniac with avoidable noises" and who had asked, "How soon shall we learn that one has no more right to throw noises than they have to throw stones into a house?" According to an issue of *The Philadelphia Medical News*, the deluge of urban noise was tantamount to a slow collective suicide.[330]

Trivial this was not. When Salem's Board of Health passed a rule in 1898 against spitting on the sidewalk, the rule had been ridiculed but was soon admired. Was it Morse who persuaded the Board's medical director to argue in his Annual Report the next year that anti-noise regulations might also be ridiculed but were equally vital, and that "Rigid methods should be taken looking to the suppression of all unnecessary noise"? As Dr. Walter B. Platt of the University of Maryland had written (and Morse had read): "If there is any one thing proving the unutterable thoughtlessness of man, the stupidity of municipal control, the general idiocy of health boards, it is the shameful permitting of factories blowing piercing blasts in the morning for absolutely no purpose." In his files of clippings, Morse accumulated heartening evidence that a crusade against noise of all kinds would have support far beyond the medical community. Since the first century CE, when a Roman candidate ran on a plank to reduce the ruckus in the roads, people had responded warmly to appeals to protect their ears—or so Morse interpreted the discovery at Pompeii of a poster for a political party of Long Sleepers and Deep Drinkers. In seventeenth-century Germany, city fathers had been praised for exiling goldsmiths and "other noisy workmen" to the outskirts of town, and by the 1880s, Berliners could be fined for whistling on the stoop of an apartment house or beating carpets over a front balcony. In Munich, noted Dr. Platt, who had been writing since the 1880s against the injuriousness of noise (which, when continuous, "acts as certainly upon the nervous system as water falling upon a harder or softer stone"), the streetcars had been relieved of their bells, and across Europe milk carts no longer rattled. Even in that "Nation of Noises," China, where "The Celestial's Nerves Are Proof Against Discordant Sounds," creaking wheelbarrow axles had been lubricated by recent order of the police, the coolie-porter's singsong silenced, and factory whistles restrained. Closer to home, the Corporation Counsel of Chicago in 1900 had crafted an overarching ordinance proscribing "the playing of musical instruments or the operation of other sound-producing devices" (i.e., phonographs) in public places before seven a.m. or after nine p.m., the driving of "unnecessarily noisy vehicles," the keeping of clamorous animals, the unloading of coal too early in the morning, and so forth, in accordance, said Counsel, with "a feeling throughout the country generally that, in our cities especially, there is a great deal of noise of which there is no real need." Chicago Aldermen's lists of "useless noises" included yelling peddlers, cackling hens, "rah rah boys," elevated trains, ice wagons, cut-out mufflers, and, mysteriously, a "summer garden." (Was that a beer garden?) Closer still to Salem, the *Boston Herald* had editorialized in 1896 against "The Crime of Noise" on city streets, where "steam whistles shriek everywhere, and then the whizz and thunder of the electric cars begin." Modernity had its benefits, no doubt, but "Electricity is not a silent slave; it is a violent master, and even its commonest forms rend the nerves of all who deem it a power of good. All ordinary sounds of city life have accentuated under the stress of this electrical age." It was high time for readers collectively to defend themselves against the "bustle, turmoil, shriek and tussle" of an increasingly electric day and electrified night. In 1893 the editor of the *Boston Courier* had taken an excursion to Salem and found a once-dignified town "made hideous" by train engines "shrieking and tooting by day and night with a shrillness and persistency that are maddening," and by factory whistles, in their discord, worse still. Village Improvement Societies had sprung up across the country in the 1880s; now was the moment to form City Improvement Societies to restore urban air, urban water, and an urban calm. After his own labors in Salem, workmen tipped their hats to Morse: "Hardly a day passes that a horny-handed son of toil doesn't thank me in the street for my efforts in the direction of stopping the Steam Whistle," he wrote to his best friend, the banker John Gould, in 1904. "The smaller noises affecting the immediate neighborhood; the hand organ I would not suppress, as the children enjoy it and it don't begin at midnight. But the whistles," these were a separate matter, "waking up the invalids in the next county," so he would not relent in his attack on "the hoodlum whistling" of locomotives, particularly in freight switching yards that shuddered with as many as eighty-three whistle blasts in half an hour.[331]

Eighty-three? In the interests of science and the commonweal, Morse had counted the blasts audible from his house each day from September 13 to October 4, 1900—and the eighty-three were exceptional. On a more ordinary day, September 27, he had counted 119 between 7:15 a.m. and 10:00 p.m., an average of one blast every seven minutes. Such exactness was open to the same ridicule as Babbage's punctilious charting of street musicians; also vulnerable was the manic synonymy of an anonymous letter that a friend suspected Morse of sending to the directors of railroad companies: "Gentlemen: Why is it that your switch engine has to ding and dong and fizz and spit and bang and hiss and pant and grate and grind and puff and bump and chug and hoot and toot and whistle and wheeze and jar and jerk and howl and snarl and puff and growl and thump and boom and clash and jolt and screech and snort and snarl and slam and throb and roar and rattle and yell and smoke and smell and shriek like hell all night long?" Morse received, in return, and typewritten for anonymity, "A Plaint & A Warning":

Re-MORSE-less fate
Would relegate
My warning note
To place remote,
Where summer heat
Cannot be beat
And winter chills
Are turned to grills.
If Styx I cross
You bet, Oh Morse,
I'll give a yell
To wake all H---
And by that shriek
Will, so to speak,
Consign to Mars
Off side the stars,
To soundless sound
And peace profound,
All Anti-Noisites.

No closing couplet here; just the void of space for Morse and other old fogies who would stall commerce, paralyze industry, and deny humanity its hijinks. When the "millionaire burg" of Brookline outlawed all noises after three a.m. and Chief Corey in 1904 made silence the watchword of his suburban police force, the *Boston Post* took the Selectmen to task. Shall no milkman drive too loudly over the pavement, slam the back gate, rattle his bottles? Shall no motorman clang his gong in the dark? Shall no dog bark? When Anti-Noisites moved to subdue the mill whistles in Haverhill, "a city of factories," the *Salem Evening News*, no less, demurred: "It is not to be expected that the quietude of a village shall be brought to pass in an industrial center like the 'Mistress of the Merrimac,'" which produced a twentieth of America's boots and shoes. And when the Van Winkles of Mount Vernon, New York passed an ordinance in 1905 that "No dogs shall bark, roosters crow, cats meow, parrots talk, nor shall horns be blown or whistles sounded after 10 o'clock at night," the constabulary was in big trouble with satirists, who drew up chicken muzzles, parrot padlocks, and a list of Sleeping Duties:

Hark! Hark!
The dogs don't bark.
Dead quiet fills the town.
Gum shoes. No booze.
No cackles, cooes,
Meows, yelps, or yowls—
The burg's a-snooze.

All that was left was to build padded chambers around the town's most irritating snorers. Then none a wink of sleep would lose.[332]

Bad rhymes and bitter satire could prove that Anti-Noisites were attracting attention and making headway, but anti-noising was never a snap. In 1900 the *Boston Transcript* commiserated: "Just fancy yourself an anti-noise crusader for a time, if you think the mission of such a body is to be easily won." A cornet player may admit that he'd survive without his midnight practice, but people will ever excuse their own horn-blowing under cover of undeniably more dreadful noises in which they have no part, such as the backing-and-forthing-and-screeching of locomotives in freight yards. Meanwhile the patent office is flooded with schematics for ever-louder alarms. Organ grinders and German oompah bands may be vanishing from the streets, but trolley cars skee-reech around the corner and peanut vendors start up their own shrill katootles. Soap-box orators haranguing on behalf of the urban poor thoughtlessly open for business smack dab in front of Boston's Relief Hospital, deaf to the needs for peace and quiet of the suffering poor inside, and must be asked to move.[333]

"Extremely harrowing," all this, a newspaper critic seemed to agree, in his review of *Troublesome Noises and How to Stop Them*, a pamphlet by "an indignant Scotchman" in New York. Noise, wrote one Malcolm MacGregor in 1903, was like a sneak-thief; it robbed the oblivious of their health. Every noise emits "endless vibratory waves" that constantly bombard us, whether or not we profess to hear them; millions of these atmospheric concussions, roiling the air of cities, can reduce residents to a state of collapse. "A panicky person . . . would probably be seized with a wild desire to flee for life to the depths of the backwoods," sneered the reviewer: the situation was hardly so intolerable. "Indeed, a certain amount of noise has, doubtless, an invigorating and bracing effect on a healthy nervous system," and if noise was one of the "perennial drawbacks of urban life," it helped "keep the population of the big cities within supportable limits."[334]

Up against such a metropolitan Malthusianism, what could one do? One could, as did Henry Pynchon Robinson for *The Chautauquan* in 1894, so wildly praise noise as to explode the logic of the pro-noisites. For those who knew him, a classicist with a Yale education who would be celebrated as a poet of elegies for the "gentle, quiet, true and peaceful," each line of his essay "On the Gentility of Noise" must have been a source of hilarity. Those who knew him not should have been alerted by the outrageousness of his noisifying Darwinism: "The early germ or keynote of our homelier noises is plainly paleozoic... man must be classed as eminently the noise-producing animal.... From all we know of the processes of savage and barbaric peoples, noise is a reserved force, a final and last resort in the ceremonies of war and worship." Evolution somehow dictated that "the leviathan roar of a great city"—"the snap of whips, the bang of doors, the clatter of teams, the clangor of bells, the rattle and rumpus and dash and clash of a busy populace, everywhere colliding and at war"—must be "typical of high development." And just as it "would not need a professed logician to show that the final mission of glass and crockery in the world is to make a crash," so door mats, carpets, cushions, soft-shoed library chairs, and ear-cotton were "all feeble evasions of the unwritten law of an expansive civilization to be noisy." Noise was verily an "elixir," its "broad range of racketing vibrations... sanitary and inspiring." Must not the most sober acknowledge that the stillness of staunchly quiet houses was productive of nothing so much as melancholy? The true models of modernity were Democratic conventions, "well-known to give an almost ideal representation of tumultuous uproar," and stock exchanges, whose "panic-din" was essential "to loosen the mind and produce the abandon of cool daring that is the basis of reckless speculation" upon which the world today was necessarily built. How else was one "to make a noise in world"?[335]

All before 6:30 in the morning: "5.45 baker's cart with aggressive bells, calculated to wake everybody within their sound; 6. Church bells ringing a lively peal as if desiring to make nervous sinners swear—mingled with loud puffing of locomotive; 6.10, more church bells and train starts up again; 6.15, somebody pounding with a hammer, an awful noise, each blow like a bomb. A passer-by whistles 'Hiawatha'; 6.18 pounding still going on and added to it come shrieks of a locomotive; 6.20, another baker's cart and passing at the same time a very rattling cart. The train is puffing with great noise, also; 6.22, locomotive whistles, three piercing shrieks. Another whistle. Another rattling cart," and so on through the sifting of coal and the testing of a fire alarm. That was Morse's response: a journal of diurnal noises. A liberal New York columnist argued instead for "The Commercial Value of Quiet," referencing the high prices willingly paid to recuperate in quiet at health resorts. Dr. August P. Clarke in 1895 asked the members of the American Medical Association to recognize "The Importance of the State Government Control of Artificial Agencies That May Be Productive of Noises." Dr. William C. Krauss, concerned for the public health in 1897, put forward a taxonomy of noises natural and artificial, and among the artificial, noises necessary, unnecessary, or traditional ("a task not easily accomplished, and I trust you will bear with me in my rather hazardous undertaking"). Edward F. Baker in 1902 hazarded a suit for damages to the value of his property caused by the noises of the Boston Elevated Railway Company, and he won. One of Morse's friends contemplated murder, ticked off by reports of a new children's toy for the Fourth of July, a "non-hurtful cannon" that produced a loud explosion with little heat and no burst, invented by a safety-minded physicist at Lehigh University. "My first impulse is to kill Professor William Suddards Franklin and plead 'justifiable homicide,'" wrote C.H. Ames in 1905. "Possibly, however, we ought to accept 'din without danger' as one step forward in the evolution of the race." In lieu of such Social Darwinist stoicism, one could help draft bills, as did Morse in 1906–1907 for the Massachusetts legislature, "An Act to Prohibit the Unnecessary Blowing of Steam Whistles" by locomotives, tugboats, factories, and fire companies. Of Senator Jenks, who fought for years in the Rhode Island legislature for the silencing of factory whistles, the *Providence Journal* in 1907 wrote, "If Dante in his 'Inferno' neglected to put at least one circle of hell under the influence of a howling, shrieking pandemonium of whistles, he let pass a great opportunity."[336]

2: Girdner the Populist

Patience, persistence, and courtesy, said Edward A. Abbott, underlay his own successes in combating "the infernal barbarism of factory whistles." Raised in Indiana, he had come to Chattanooga in 1887, where he went into the hardware business with his brother-in-law, William P. Silva. Silva built an art studio above their store and became a respected impressionist favoring serene gardens, marshes, magnolias, and pines at dawn and dusk. Abbott himself planted a superb flower garden, sowed the seeds for a City Beautiful club, and plotted for public parks. Around 1903–1904, as Silva was deciding to make a life of his oils and leave Tennessee for Paris, Abbott began to obsess over noise, "stimulated thereto by a mill whistle almost over my bedroom window." His house was also close enough to a rail line for steam whistles to disturb his sleep, and instead of sheep he counted at least forty-one "shrieks and yells" of locomotives during a single night. For the next decade he would vigorously protest the "fool uproar" of his no-longer-small town, which over forty years had grown forty-fold into a metropolis of 100,000 that would welcome the new century with an electric trolley grid, a new rail junction, and an airmeet, though troubled by older violences of public duels and lynchings. In 1908–1909 Chattanooga got new rail lines and a new terminal, steeling Abbott's determination to rid the city of the particular "evil of steam whistles." In order to discover "the why and what for" of noise, and "to find a cure for it," he had mailed out hundreds of letters across the nation to "doctors, lawyers, teachers, laborers, city officials, legislators, preachers, manufacturers, police chiefs, and everybody likely to have expert information about noise"; his respondents agreed that avoidable noises ("steam whistles, barking dogs, crowing cocks, church bells, and the like") menaced "health, business, happiness, morality, manners, religion, progress and prosperity." Dr. William J. Morton of New York, the first to use dental x-rays (and son of the claimant to the first use of ether in surgery), assured Abbott in 1905 that his concerns with "the jar, the bustle, the clamor and the confusion of sound which we call noise" were scientifically grounded, given that waves of noise, when projected into the brain through the auditory nerve, advance a line of thought "which leads to weariness and anxiety and resentfulness." Abbott himself had been unwearying in his anti-noise efforts. As a result, the nearby mill had not whistled over him since 1908, and like "A saw mill up in Maine, a cotton mill in Georgia, an arms factory in Connecticut, and an oyster cannery on the gulf coast," all of whom sent words of encouragement, the Chattanooga mill had not gone bankrupt once it left off shrieking. Locomotives, however, were still "yelling" through town in 1911, so Abbott began exposing what he called the "yelling peril."[337]

"Yelling peril" must have been an unsettling pun to a Japanophile like Morse, though he had few kind words for the Chinese, whom he found noisy and filthy. It must have also been upsetting to an anti-imperialist like John Girdner, who admired the Japanese for their meatless diet and opposed American military action in the Philippines. But the phrase was brazen enough to fly in the face of Girdner's bullheaded neighbors, New Yorkers who celebrated noise,

> For I am of the city born and of the city bred,
> And I want the noise around me, underfoot and overhead;
> I'll get enough when I am dead of quiet country joys,
> But while I live I want the town, its life, its rush—its noise!

That rush was symptomatic of what Dr. Girdner termed "Newyorkitis," a kind of Knickerbocker neurasthenia that he described, with tongue in medical chic, in 1901. Five years before, he had published a widely read essay on "the plague of city noises" in which he proposed a "Society for the Prevention of Noise," to be modeled not after New York's notorious Society for the Suppression of Vice but the more modestly effective Society for the Prevention of Cruelty to Animals. This was apt, since city noises assaulted, first, our animal nature—"physical man might properly be defined as five senses mounted on stilts"—then the nerves and mind.

Everyone knew that boilermakers suffered from the noise of their occupation; in the long run, all classes of people in the urban cauldron would suffer from the "continual rattle, roar, and screams which assault their ear drums at nearly all hours of the day and night." That some city dwellers grow accustomed to the noise, and claim to be fond of it, was no proof that it was any less destructive. Was alcohol any less destructive to habitual drunkards who consume cases of liquor without keeling over? Enormous mental energy was required of New Yorkers struggling to listen through a constant Babel. "Observe the confusing and almost stupefying effect on the inhabitant of the way-back rural district when he visits the city for the first time.... His trouble is, that he has no skill in selecting and discarding among the million sights and sounds that rush in upon his consciousness." (Reading this, Morse would have recalled a letter back home on his first trip to the city as a country boy from Maine: "The first thing that would strike you about Boston, yes, strike an awful blow to your thinking apparatus, is the horrid din and incessant rattlety bang of heavy wagons, omnibuses and carts over the paved streets. Every court, lane, and alley is paved... cobble-stones mostly used and causing the most infernal racket.") The racket drove men and women of all ranks toward nervous prostration. "How often," wrote Girdner, "the physician in his daily rounds finds it necessary to prescribe 'perfect quiet' in order that the flickering spark of life remaining in the patient may be brought back to a healthy flame. Yet in nine cases out of ten, that perfect quiet he deems so important cannot be had, owing to the noises from the street." And, added Dr. E. Fletcher Ingals of Chicago, physicians themselves were subject to the same nervous prostration; a local doctor with a large practice that kept him up every day until midnight had recently died of insomnia when each night's sleep was interrupted as early as four a.m. by the noises of the street.[338]

Remember Girdner? We have met him before, inventor of a telephonic bullet probe that he thought might have saved President Garfield from the septicemia of an assassin's bullet in 1881, and which later Girdner hoped to use as a full-body stethoscopic scan. Morse knew Girdner's work, and may have corresponded with him as he had with Abbott, but Girdner thought of himself as a player for higher stakes than steam whistles. Son of a Union Army surgeon and a schoolteacher; A.B. from Tusculum College, Tennessee, 1876; M.D. from the University Medical College, New York University, 1879; lecturer on surgery at the New York Post-Graduate Medical School, and the first to graft skin successfully from a corpse onto a living body, Girdner had been physician to President Grover Cleveland and to the celebrated actor Edwin Booth, a man after his own quiet heart, known for the refinement of a thespian style grounded in the "thoughtful introspective habit of a stately mind, abstracted from passion and suffused with mournful dreaminess of temperament." Girdner was also, by 1900, in regular contact with his good friend William Jennings Bryan, whose New York State supporters needed a platform that Girdner was willing to write and whose second presidential campaign needed a running mate, a position for which friends wished to put forward his name. "I have to listen to, and prescribe for, both the physical and political troubles of a larger number of visitors every day," he told Bryan; as a busy and discreet man, he had thus far refused his consent to be formally nominated for vice president, lest this "prove at all embarrassing to you," as "you know I am not looking for glory nor office." Of course, if Bryan's running mate were to hail from New York, the state was sure to go Bryan's way, and no other New York prospect had been as steadfastly opposed as was Girdner to sending 126,000 troops to put down the "Philippine Insurrection." So why didn't Colonel Bryan, who had been under Girdner's orders to rest his silver tongue, join him on Block Island, "where you can get perfect rest and quiet before the campaign opens"? It would do him a world of good. "You have obeyed my orders *not to talk* so well of late that I trust you will obey this last order," he told Bryan, who did visit and did regain his mellifluent voice but took Adlai E. Stevenson as his running mate. They lost the election to a Republican ticket of McKinley, who stuck to his front porch in Ohio, and the vociferous Theodore Roosevelt, who rough-rode across the country behind posters claiming that "The American Flag Has Not Been Planted in Foreign Soil To Acquire More Territory But For Humanity's Sake."[339]

Newyorkitis was composed as reports were filtering back from Manila of commands to massacre Filipino civilians, of American soldiers torturing guerillas using an ironic "Water Cure" or writing home of their doubts about this "quagmire" (Mark Twain's word: "why, we have got into a mess, a quagmire from which each fresh step renders the difficulty of extrication immensely greater").[340] Girdner's small treatise was meant "as a plea for a wider thought horizon, a more genuine and comprehensive brotherly charity, less materialism, and more cultivation and development of those qualities which distinguish man from the lower animals." He was building upon his articles in the *North American Review* on city noise (1896–1897) and an essay on "theology and insanity" (1899) where he had shown that the delusions of the mad run in parallel with the anxieties of the era, so that "The insane are not now tormented by the devil and his imps, but telephones and phonographs are continually ringing in their ears. Others suppose they have steam engines in their heads, and many imagine they are persecuted by men of large fortunes and of great political power." The last was a projection of Girdner's own anxieties—in part because he had taken on powerful States' Rights advocates and state medical societies in a proposal for a federal Department of Health with a cabinet officer to coordinate health campaigns and medical inspections (since "disease germs have no respect for State lines"). And in part because he was not only a popularizer, writing about germs, nutrition, and noise for *Munsey's Magazine* and *The Junior Munsey*, but an urban populist who, after Bryan's defeat, became physician to, friend of, and briefly business partner with Tom Watson, the Georgia lawyer who headed the Populist Party ticket in 1904.[341]

Akin to Asiatic cholera, Newyorkitis was a communicable disease, wrote Girdner. His inflamed language was in keeping with that of Joseph Howard, one of the first syndicated newspaper columnists, who after sixty-five years in Manhattan had found that "There is no getting away from the ceaseless rumble of elevated cars, surface cars, heavy coaches, burdened wagons.... There is no relief from hand organs, none from street peddlers' cries, none from the shrill whistle of the postman and none from the shameless shoutings of newsboys.... It's death to nerves," so much so that "the general get up and get in hurrah fashion of everybody and everything, conspires in a harsh discordancy to pull down, tear up and drag out by the roots the most callous nerve that ever nestled in bone or muscle." A reviewer for New York's Reform Club believed that other cities were no different and awaited exposés of Bostonitis, Chicagoitis, Minneapolitis—and in fact, Girdner by 1906 had a steamer trunk full of clippings and letters from correspondents around the planet who found their own parts of the world intolerably noisy. Would that Populism were equally communicable. In his private correspondence and published articles it was clear that the inverse notion of communicability for the good of the people was never far from Girdner's pen. The *élan*, rapidity and technological wizardry of Manhattanites would be best applied elsewhere than in spicing up gossip, slurping oyster cocktails, dodging cable cars, and whistling for Mrs. Leslie Carter in her scandalous role as "Zaza," the music hall singer and siren. Meanwhile, as soundwaves reverberated between the solid walls of their tall buildings, New Yorkers were going almost as deaf as boilermakers. To resist this deafness, of a piece with their flat-footedness and near-sightedness, New Yorkers had need of a group of Anti-Noisites who would promote new asphalt pavements and rubber tires, push for the enforcement of ordinances against teamsters who neglect to wrap loads of iron in old carpet, and persevere against "the useless postman's whistle,... the yelling 'rags and bottles' man, the horn-blowing scissors-grinder and four-in-hand driver." Girdner concluded with a short ode to New York by the anti-war poet Ernest Crosby, publisher of the *Labor Prophet* and president of the Anti-Imperialist League of New York, who lived up the Hudson on a country estate in much the same manner of his friendly correspondent, Tolstoy. "O sprawling, jagged, formless city!" Crosby began; if only your "seething energy" could be set on conceiving "something nobler than a full stomach" or a swollen bank account: "Become now at last conscious of the germ of soul that is in you, and stake your overweening energy on that!"[342]

Girdner did stake some of his own overweening energy on the

germination of an anti-noise society, of which for a spell he was president and of whose fruits little record remains. While Morse's anti-noise program was essentially a cross-cultural critique of American urbanism and industrialism from the perspective of a small-town Down-Easter and hyper-acute naturalist, Girdner's anti-noise program was a political critique of American violence. While Morse concentrated on the inanimate, on imprints and residues, geologically slow accretions and patiently fired clay, in contrast with the awful impetuosity of steam whistles and the whine of street cries, Girdner focused on an animated sociology of working bodies threatened by war, germs, and the selfish aggressiveness of capitalism. His attachment to Populism was so strong that it was seen by the *New York Sun* as having stigmatized his efforts at noise suppression: "As long as the Anti-Noise Society was alone in the fight for quiet and its president insisted on inviting the Hon. William Jennings Bryan to town at frequent intervals, the campaign gave little promise of success." In neither Lessing's nor Morse's nor Girdner's case was Anti-Noisite activism a bourgeois defense of privacy or property against the encroachments of the urban or transient poor, and if Morse, Abbott, and perhaps MacGregor bespoke elements of a nostalgic anti-modernism (as in Morse's obsession with archery), these were balanced by an embrace of the logic and rhetoric of modern science, ethnography, archaeology, social statistics, and a philosophical concern with equity.[343]

3: Rice the Pianist

We come then, and therefore, to Isaac Leopold Rice and Julia Hyneman Barnett Rice. Together and independently they were modern and media-savvy. Isaac we have met before, adumbrating on Brahmanic breaths and modes. He had been born in Rhenish Bavaria just as the Bavarian Parliament was debating a bill to give Jews full legal and economic equality. When the bill was pulled down to defeat by a strong undertow of antisemitism in the wake of the suppressed revolutions of 1848–1849, Mayer and Fanny Rice emigrated to Pennsylvania, as did twenty thousand other Jews from southwestern Germany.[344] The Rices arrived in Philadelphia in 1856, when Mayer was forty-eight, Fanny thirty-two, and their son Isaac six, by which age he may already have been tickling the ivories, for at thirteen he made his debut as a pianist and at fifteen received a warm national notice in *Dwight's Journal of Music*, his performance so "creditable" that his friends and the reviewer alike "confidently anticipate for him no mean position in the ranks of our native virtuosos." After high school he went to the Conservatoire National in Paris, financing his studies as a foreign correspondent for the *Philadelphia Evening Bulletin*. Finishing up at the Conservatoire, he gave concerts briefly in Germany but lacked the funds to return home until he taught school for a while in England, where he also did well in chess tournaments. During the 1870s he set himself up as a music teacher in New York City, supporting his aging parents while publishing piano exercises, commercial pieces ("Wild-Flowers: Five Waltzes"), and a treatise, *What Is Music?* Music was "a great part of the cosmos, and not a human contrivance." This was something of a Brahmanic proposition, and Rice would be known thereafter in musicological and theosophical circles as an expositor of "Hindu" tonalities, which to Western ears bore little kinship to music. For a cosmopolite and virtuoso who had concluded that the "difference between light and sound is not in kind, but in degree" of vibration, and who had found to his delight in Benjamin Silliman's *Principles of Physics* that "The aggregate sound of Nature, as heard in the roar of a distant city, or the waving foliage of a large forest" could be resolved to a single tone, a piano's middle F ("the key-note of Nature"), it was no grand feat to hear the music in vibrations from the East. For a chess aficionado, it was equally a matter of course to explain *How the Geometrical Lines Have Their Counterparts in Music.*[345]

Non-Euclidean in the geometry of his own life, his careers ran in parallel lines that every so often converged, much like his famous Rice Gambit, which relied upon the centrality and sacrifice of a knight, the only chess piece that flies across space and turns corners. In addition to his music teaching, composing, performing, and chess-playing, he enrolled at Columbia Law School in 1878 and received his degree in 1880 with prizes in constitutional law, international law, and the history of diplomacy, and then studied under the German-trained legal historian John W. Burgess, who at that time had in his classes "three very extraordinary men, all of whom have exerted much influence in shaping my own life's course": the brilliant Nicholas Murray Butler, later president of Columbia; Theodore Roosevelt, fresh out of Harvard, always in the front row with "a dozen prepared and sharpened pencils bristling out of his upper vest pocket"; and Rice himself, "one of the most ingenious minds that ever came under my tutelage." Hired as a lecturer in political science and as librarian for new acquisitions in economic, social, and historical analysis, Rice introduced subject cataloguing before Melvil Dewey (of the Decimal System) was asked over from Amherst to direct a larger, revived, Columbia library.[346]

Roosevelt had come for the law, stayed for political theory, then leapt at a chance to enter the New York State Assembly. By 1884, Rice was teaching at the Law School and publishing monographs on constitutional, interstate, and international law that were, wrote Burgess in retrospect, "among the soundest and best expressed to be found in any literature." Rice's arguments anticipated Roosevelt's programs for conservation and trust-busting. "We have seen," wrote Rice, "vast combinations springing up everywhere and in all branches, seizing for themselves that which ought to benefit and belongs of right to the whole community, crushing out all competition with relentless cruelty." Rice attacked discriminatory freight rates that favored monopolies and defended federal action to break up such combines when they reached over state lines. "The proper function of a government is to protect the liberty of the people against aggression from within as well as without." During an era when the business corporation was gaining its legal identity and claiming the liberties of personhood, Rice defined civil liberty as "the result of the restraint of the legally acknowledged and vested private rights of the more powerful individuals and classes of the community," provided "that this restraint be exercised by the sovereign people." Unrestricted rights were "nothing less than instruments of oppression in the hands of the more powerful individuals and classes." He upbraided Herbert Spencer for exalting private interests above public needs, since private interests were usually those of the rich, while the public needs that were neglected were usually those of the impoverished. The State did not exist, as Spencer had maintained, solely to enforce private contracts and mete out police justice (against obstinate churchbells, for example, and Salvation Army bands, two of Spencer's peeves). The State, wrote Rice, had more positive contributions to make, all of whose engagements and expenses Spencer had decried: public education; public libraries; public hospitals; regulations protecting public health and the health of workers in factories and mines. According to Spencer's Social Darwinism, the State had "no right to prevent the unlimited exploitation of the weaker by the stronger, and it must allow the former to be crushed and destroyed in the struggle for existence, so that the latter may have full opportunity to survive." How sadly mistaken, observed Rice. Mr. Spencer had misread his Plato as badly as he had bungled his history of music by which he hoped to show that progress is always from the homogeneous toward the heterogeneous, from the choral mass toward the sovereign aria. In musico-political fact, Rice argued, any Government that is not a tyranny must be an agent of the choral masses, and its aim must be to carry out the will of the people, not the interests of bedizened soloists. Given that each citizen has co-ownership in the State, it behooved the State to have citizens who are healthy, educated, and secure enough to make their preferences known; in turn, citizens must realize the public spiritedness required of them to assure an effective Government, one that overcame class hatreds and corruption.[347]

Side by side at Columbia with other highly assimilated Jewish lecturers, Rice would never seem as radical as the Marxist-Socialist Daniel DeLeon or as influential as the progressive-tax economist Edwin R. A. Seligman. Nor would he become as distant from the Jewish community as "Ed," youngest son of Joseph Seligman (another of the 50,000–60,000 Bavarian Jews in America by 1857), who established one of New York's foremost investment banking firms, held together Northern financing during the Civil War, founded the Hebrew Orphan Asylum, and gravitated toward Felix Adler's Society for Ethical Culture, later led by Ed. Nor would Isaac become as uneasy with his origins as Daniel, who upon entering the political arena—as editor of *The People* and Socialist Labor candidate for governor of New York in 1891, 1902, and 1904—repudiated his descent from a venerable Jewish family in Curaçao and invented out of whole cloth the alb and tunic of a Catholic heritage. In 1883–1884, however, when Rice, DeLeon, and Seligman were all at Columbia, they met regularly at the house of Emma Lazarus and at Rice's house nearby to consider the plight of the Jews. Specifically, they mulled over (to no resolution) the creation of a society that would help Russian Jews fleeing the pogroms stirred up by the assassination of Czar Alexander II to emigrate rather to agricultural colonies in Palestine than to the twenty thousand slum tenements on the Lower East Side.[348]

Cynics might suggest, and Marxist critics have insisted, that such concern was proof not of loving-kindness toward fellow religionists but of anxious defensiveness on the part of prosperous, cosmopolitan, nonobservant Jews who felt themselves threatened by millions of poor, insular, orthodox immigrants. If Emma Lazarus penned the poem that would be inscribed on the pedestal of the Statue of Liberty in 1886 to welcome "the huddled masses yearning to breathe free," she was also a wealthy urbanite who preferred Friday evening salons to the closeness of synagogue air, who corresponded on transcendental topics with Ralph Waldo Emerson, and who in 1882 had begun a friendship with the artistocratic Henry James. Wasn't she, and weren't Rice, DeLeon, and Seligman, fearful that the reputation of the upper strata of the Jewish community, and in consequence their own ability to negotiate ethnic, social, and cultural borders, would suffer if too many Jews of a lesser ilk took up cramped quarters in their city? (Already in 1877, in a public scandal, Joseph Seligman & Family had been denied rooms, as Jews, at their customary retreat, Saratoga's Grand Union Hotel.) Weren't the plans of Lazarus et al. an effort at rerouting the Exodus and detouring a presence that could undermine their own security and mobility as tasseled, bearded men pushed wooden carts through streets, called for rags and scrap, and peddled old clothes at the tops of rasping, nasal voices? How much different, at heart, were the meetings at East 57th Street from other uptown German-Jewish salons in which the typical *-ky/-ki* endings of Slavic names were, it is said, mocked and slurred into "kike"? And weren't Lazarus et al. as aggrieved by the loudness, the dialectal harshness, the essential *noisiness* of Old World Jews as that masterful modernist, Richard Wagner, was offended by the barbarous, noisily Oriental modes of "Jewish music"?[349]

Ambivalence there surely was, and class suspicions on all sides, but Isaac Rice left no evidence of antagonism toward Easterners whether of Benares or Lvov. It would be dangerously clever to turn the anti-noise movement upon itself in the same manner that anti-Semites have blamed Jews for antisemitism. Historians of sound have examined conflicts over the ringing of churchbells that set rationalists against the religious, the lax against the pious, and Protestant against Catholic, but they have avoided, even with Lessing (the first to describe "Self-Hating Jews"), the Jewish Question. Was sensitivity to noise, and a readiness to protest sounds too sharp or repetitious, characteristic of a Jewish bourgeoisie alert to phenomena that might, through some ethno-religious resonance, jeopardize a carefully earned status in Christian society and a studiously fixed, if unhealthily neutered, quietness? For Isaac, the personal nexus of noise-suppression, ancestral Jewishness, and strict positionality was drawn tight in just these years, prompted much less by steerage-loads of Russians than by one red-bearded, sharp-tongued, arthritic Jew from the Prague ghetto (via Vienna and London) who toured Philadelphia and New Orleans in 1882 then returned to settle in New York in 1883 and become, within three years, the first (self-proclaimed) "world champion" of chess. From Wolf/Wilhelm/William Steinitz, Rice learned a style and philosophy of play that was computational enough to be called, as his mentor insisted, "scientific." In lieu of romantic French forays, Hibernian raids, or headlong Crimean assaults, Steinitz moved his pieces as would a corporate lawyer, calculating the value and equivalences of each piece, looking to gain small but measurable, permanent advantages. A Steinitz match was rarely exciting to watch. It had none of the boldness of the barbarian marauder or the rapacious capitalist; it was a game of defense, of fine formulae and middle management, as deliberate as the Bessemer process and just as steely. If he took risks, these came in eccentric openings by which he challenged himself, like Houdini, to escape from ostensibly fatal situations; a contemporary called Steinitz "the Champion Wriggler of the Known Universe." The rest of the game was about consolidating positions, exploiting margins of profit. His matches were so slow (eight hours, or ten, or twelve, in an era when timers were introduced to tournament play) as to be called "eternal." It was exactly the game for "even-tempered" Isaac Rice, "quiet, kind and lovable, but most persistent and thorough." And in the tense silence between moves (when Steinitz might hum a few bars from *Tannhäuser* or, nearer checkmate, a funeral dirge) you were never alone with your thoughts; you were in mute conversation with another who might admire your slyness and forethought but had always your demise in mind. Analogies to the status of an intellectually gifted, politically articulate, nonobservant Jew playing for high stakes in a Gentile world would not have escaped Rice.[350]

4: Julia Barnett and Isaac the Venture Capitalist

When exactly during Isaac's jockeying for position in the realms of music, academia, law, chess, and international Jewish affairs he met Julia Hyneman Barnett, "a handsome, refined, and cultured woman of the Castilian type," I do not know. Julia's father, Nathaniel Barnett, belonged to a family that had been in the New Orleans wholesale and auction business for seventy years. How Nathaniel met and, in 1859, happened to marry Annie Hyneman of Philadelphia, I do not know. I do know that when the Civil War broke out, Annie fled North with one-year-old Julia to stay with her parents Sarah and Leon in relative security. Returning to Louisiana at war's end, Julia studied to be a concert pianist; on the eve of her debut she found that her hands had cramped from over-practice, and she never did perform in public. By 1878, exploiting connections through the family firm of Katz & Barnett, Nathaniel had moved to Manhattan with his wife and (now) seven children, in pursuit perhaps of a more congenial religio-political climate than that taking shape in postbellum New Orleans.[351] It could have been as early as 1881 that Isaac was smitten with Julia, if it was she who inspired his music for an atypical song, "The Spring Gave Me A Friend." In any case, Isaac did have Julia as a music student around that time, did ask for her hand in marriage, was rebuffed by her parents (what were his prospects as a piano teacher?), and the two drifted apart until they met again at a summer hotel "quite by accident," just when Isaac had a law degree under his belt and a position at Columbia. Julia's parents now consented to the marriage. Isaac, thirty-five, and Julia, twenty-five, wed in December 1885, a few months after she had earned her M.D. from the Women's Medical College of the New York Infirmary.[352]

Established by the Blackwell sisters, the college was run by Dr. Mary Putnam, a resourceful tactician—with her husband, the German-Jewish physician (and 1848 revolutionary) Abraham Jacobi—for the new field of pediatrics and a new generation of suffragists. Julia's mother herself was a public advocate of women's rights. So was Grandfather Leon, a Grand Mason of the Shekinah Lodge, publisher of the *World's Masonic Register*, editor of *The Masonic Mirror*, and a philosopher whose *Fundamental Principles of Science* (1876) spelled out the equality of women in every sphere. A force in Julia's life through to her majority, he spent his last decade in the Barnett household on East 60th Street, dying in 1880. During those years Julia's Aunt Leona was a Broadway actress with a penchant for commanding roles—Lady Macbeth, or Pauline Deschappelles, a merchant's daughter scheming for a title in *The Lady of Lyons*. Julia's great-aunt Rebecca was a poet known also for a novella, *Women's Strength*. Her Aunt Alice was an essayist on women's organizations and "Woman in Industry"; she would embrace the politics and in 1893 the life of Charles Sotheran, a Yorkshire man who had come to her father with Masonic credentials and gone on to make a name for himself as an antiquarian bookseller, bibliographer, literary historian, and co-founder of the Social Democratic Workingman's Party of North America. This became the Socialist Labor Party, from which he was expelled in 1894 through the maneuvers of another Knight of Labor, Daniel DeLeon, whom Sotheran had scorned as a mere lawyer and who in turn dismissed Sotheran as a "250-pound perambulating scrap book and historic junk shop." Sotheran then took up writing, criticism, and muckraking for the *New York World*, dying in 1902.[353]

Both sides of Julia's family, the Barnetts and the Hynemans, were of Dutch Jewish extraction, secularly educated, internationalist, nonobservant but not disaffected. The names of her aunts and uncles hewed to the Sephardic line of Moses, Rachel, and Samuel; among her brothers and sisters a Bertram and a Maurice held hands with an Isaac, a Sarah, a Miriam. Grandfather Leon would have nothing to do with "a personal God"; her uncle Sotheran, a Rosicrucian, introduced Madame Blavatsky to the term "theosophy"; Julia's brother Leon Elias late in life consorted with Rudolf Steiner's Anthroposophy; her children raced around after everything under the sun except the sacred; and her own leanings were toward a diffuse humanism, but neither Julia nor Isaac held themselves aloof from the Jewish community, however easily they moved between worlds and cultures. Isaac's younger brother Joseph, a physician, came to the wedding from his offices at the Montefiore Home for Chronic Invalids, a recent addition to the Mount Sinai Hospital complex; when Isaac left Columbia the next year for private practice at law, he partnered with Nathan Bijur, who had sat in on the Lazarus discussions of Jewish emigration and in 1911 became a founder of the American Jewish Committee. Isaac contributed to Jewish charities and helped subsidize the Jewish Publication Society's translation of the Tanakh, the Hebrew Bible; Julia would give $1,000,000 for the building of a convalescent home allied with Beth Israel Hospital.[354]

Julia herself never formally practiced medicine. Her career after marriage—in addition to raising six rambunctious children born almost every other year from 1888 to 1902 and nicknamed Dolly, Polly, Tommy, Molly, Lolly, and Babe—was "helping other people. Anyone who had troubles came to her," wrote Polly (i.e., Dorothy); "just to be with her seemed to straighten out the world." She was also occupied for much of her life in publishing *The Forum: A Magazine of Politics, Finance, Drama, and Literature*, which Isaac had begun as his wedding gift to her and which evolved into a leading outlet for Progressive thought. Isaac had the wherewithal to start up *The Forum* because his lawyering was so profitable: he innovated instruments and gambits in corporate finance to assist bondholders of rail companies struggling to hold off Gilded Age raiders and plunderers. He could sustain the magazine, which was rarely in the black, because his lawyering led him from the restructuring of debt to the raising of millions from Anglo-Jewish and European financiers for syndicates hoping to salvage the smaller independent railroads from corporate mergers, thence to the funding and running of companies on the technological forefront. "I am very much interested in new inventions," he would tell a congressional committee in 1908. "That is what might be called my 'fad.' If a man has a new enterprise and cannot find capital, sometimes he comes to me and I raise the capital for him, and in this way I have been the founder of new industries in this country."[355]

Boxes of "pickled energy" were the first to intrigue him. While Sabine at the Columbian Exposition had been impressed by its architectonics but preoccupied with courting the M.D. who would become his wife, Rice, already coupled to an M.D., came away from Chicago in 1893 with another gleam in his eye. Much of the glow of the White City relied on storage batteries that could be reliably charged with the lower-cost energy now available through increasingly powerful dynamos. The most reliable storage battery was the Chloride Accumulator, tested in France by Clément Payen before he emigrated to America, where he expanded his invention into several series of U.S. patents (1889–1890), then for a goodly sum assigned them to a tailor-made subsidiary of a Philadelphia gas company. That subsidiary, the Electric Storage Battery Company (ESBC), was making scant progress in its marketing to electric streetcar and lighting companies, telephone exchanges, and electric fan and sewing machine manufacturers due to a slough of infringement suits, most of them groundless, a few swampy, all mired in legal costs. Edison's own storage batteries a miserable failure, the Wizard had publicly discouraged further investment in these utopian devices that, like the back half of a perpetual motion machine, seemed to solve the problem of sequestering and mustering energy at will—if only their acids did not eat through metal casings and rubber insulation. When Isaac Rice arrived on the scene in 1894, buying up shares in ESBC and assuming the role of corporate counsel, the best recourse was to buy out some five hundred competing patents, regardless of technical merit, so as to clear daylight for Payen's battery and related improvements made overseas by engineers less cowed by the Wizard. This Rice proceeded to do with four millions in capital from German investors and promises of technical aid from a triumvirate of British, French, and German electrical firms. As major stockholder, then as president, he put ESBC back on its feet, or rather on its wheels, for in addition to supplying energy for lathes, drill presses, automatic fire alarms, automatic pianos, and automatic phonographs, he heavily promoted (900-pound) batteries for electric streetcar systems and the electric automobile.[356]

Likely he had seen a young man driving a nine-seater electric carriage at upwards of seven miles per hour through the White City, a running advertisement for a storage battery sold by the American Battery Company of Chicago. (The driver was eighteen-year-old Edgar Rice Burroughs, son of the company's vice-president and an adventurous "flop" until he conceived of Tarzan; the "wild and terrible cry" of Tarzan's ape-call, which rose above the "lonesome noises" of the jungle in 1912 and was first heard on celluloid in 1932, was no far cry from the ah-ooga! horns on early automobiles.) Rice himself was among the first civilized men to drive an electric hansom into the parks of Manhattan. Awash with sailor-suited children under the genial sway of Julia and the pelican eyes of her mother Annie, Rice's car was a running ad for the hundred Electrobat taxis of his Electric Vehicle Company, bought in 1897 and linked with his Consolidated Rubber Tire, Lindstrom Brake, and Quaker City Chemical Companies: he had learned from railroads the virtues of vertical integration and always had a stake in "cognate industries." When well-heeled competitors bought out both ESBC and the Electric Vehicle Company in late 1899, Rice had millions to invest on another technological frontier: the submarine.[357]

Editors at London's *Electrical Review* had considered the principal defect of the storage battery to be "its modesty." Unlike gasoline and steam engines, the electric engine did not "spark, creak, groan, nor slow down under overload.... It stays where it is put, and will silently work up to the point of destruction without making any audible or visible signs of distress." It had also the qualities desirable for powering a submarine, so long as the engine did not up and quit. Rice knew the history of submarine design from an 1887 article published in *The Forum* by Lt. Edmund L. Zalinsky, who made it clear from the first paragraphs that a practical system of energy storage was critical to the success of any manned vessel meant to move underwater. An inventor and manufacturer of pneumatic guns, Zalinsky would befriend John P. Holland and invest in his Holland Torpedo Boat Company, whose submarines used a large storage battery to drive an electric motor for safe quiet progress underwater. Electric motors eliminated the danger of gasoline fumes and the racket of cylinder noise pounding at the ears of seamen in a confined, metallic space. After years of odd competitions, safety tests, and stalled decisions, the U.S. Navy in 1895 at last commissioned Holland to build a submarine to specifications so inflexible and infeasible that the *Plunger* was sunk long before it was finished. Out of funds by 1897 but not yet out of breath, Holland at fifty-five had greater hopes for *The Holland VI*, which would have an internal combustion engine for surface maneuvers, an electric motor for power below, a fixed center of gravity for swift calm descent, and a propeller aft of the rudder for stability. What he needed was a venture capitalist. After a visit on the Fourth of July, 1898 and a trial descent in September, that's what Rice offered: capital, and connections.[358]

Capital, connections, and a clear conscience. The British Admiralty was dubious of such an unethically silent weapon, "underhand, unfair, and damned un-English"; Clara Barton, who accepted an invitation for a ride in 1899, deplored the implications of such silent warfare. Holland, however, and Rice (an admirer of Tolstoy and supporter of the Peace Society of New York) appeared to be operating under the same rationale as the industrial chemist Alfred Nobel, whose first Peace Prize was shared in 1901 by a society for international arbitration and the founders of the International

Red Cross: "My dynamite will sooner lead to peace than a thousand world conventions. As soon as men will find that in one instant, whole armies can be utterly destroyed, they surely will abide by golden peace." So while Rice's Electric Boat Company (now cantilevered by his Electro-Dynamic, Electric Launch, and Industrial Oxygen Companies) might have a patented corner on the best design and technology, he counted on selling submarines to the United States, where Teddy Roosevelt in 1898 was Assistant Secretary of the Navy, *and* then to other nations, for a balance of power and profit that would make Isaac one of the world's major arms dealers. By 1900, after *The Holland VI* outdid itself on trials along the Potomac, he had in hand U.S. Navy contracts for six more subs. Over the next decade, as his brother-in-law Maurice (Secretary and Treasurer of the company) oversaw U.S. contracts for another twelve, Isaac made overtures to the navies of Japan, Russia, Turkey, Portugal, England, France, Germany, Austro-Hungary, Sweden, Norway, Denmark, Mexico, Peru, and Venezuela. John Holland thought he was jumping the gun, as there was much yet to be done to engineer a good submarine—better ventilation for longer periods underwater, safety servo-mechanisms, and quieter engines that did not sound to the crew like rock crushers. Underpaid, underappreciated, and undermined, Holland left Rice's employ in 1904 under decree not to divulge trade secrets to prospective customers or competitors, although that hero of children's fiction, Tom Swift, seemed to have had no difficulty in 1910 constructing his own submarine with an electrical propulsion system and gasoline motor. Holland died in August 1914, on the verge of the Great War, by which time the Electric Boat Company was on the verge of bankruptcy.[359]

True, Rice had managed to sell three hundred submarines, many contract-built by English shipyards, but sales for years were sluggish and the company had not issued regular dividends since 1911. Worse, U-Boat 9, adapted from Holland's designs as patented in Germany and infringed by the Krupps (against whom Electric Boat had won a judgment), sank three British cruisers in one hour on 22 September 1914. Sick and weary after a long sales trip to Russia and troubled by the sinkings, Isaac saw his fortunes further decline as shares of Electric Boat stock fell to a low of $8. His stock began to rise again in the spring of 1915 when it became clear from news on all fronts that the war would not be ending anytime soon; orders flooded in not only for subs but for Rice's subsidiary 380-horsepower Elco launches, whose shallow draft was ideal for eluding submarines and whose speed helped chase them down. Now Isaac would recoup his fortunes, selling 28,000 shares of Electric Boat at prices rising from $60 to $200; never did he recover his health or regain the good conscience that decades of arms peddling had made the world safer. In the summer of 1915, after U-Boat 20 sank the magnificently fast *Lusitania*, on which he, Julia, and the children had sailed more than once, Isaac relinquished the presidency of the company. By then he had realized $2,000,000 or more from the sale of stock, some of which he had a mind to reinvest that August in a new gas-powered refrigerator to be marketed by his Car Lighting & Power Company. He died in November at the age of sixty-five, leaving an estate valued at $40,000,000. Had he been able to hold off on the sale of his Electric Boat shares for three more months, to October, he could have left his family six times as much.[360]

Speaking of hypotheticals, had Rice in 1899 held off for nine months from his investment in Holland's submarines, he might have had the funds to make good on yet another venture: the Marconi Wireless Telegraph Company of America, for which articles of incorporation were filed in New Jersey in late 1899 with August Belmont as treasurer, Rice as president, and "authorized capital" of $10,000,000. One of the profit sectors would be ship-to-shore communications: "We expect to equip the signal and life-saving stations along the coast with the wireless system, that they may warn approaching vessels in time of fog or storm," Rice told the press, adding:."The uses to which the wireless system may be put are almost unlimited." Sadly, his resources were most limited; deeply invested in the Electric Boat Company and digging still deeper to buy the Casein Company of America (milk derivatives for paint, coatings, and adhesives), he had to leave the wireless enterprise to others (and to his son Isaac L. Rice, Jr., who "has mechanical talent and has experimented considerably in wireless telegraphy," reported one paper). In 1907, though, Rice assured the press that the United States (i.e., the Electric Boat Company) led the world in the design of submarines *and* of a "new signal apparatus attached to these vessels" that could warn of the approach of enemy ships. He was intimating at wireless, but the apparatus was a different sort of device, with a longer history.[361]

Since the 1850s, as tonnage and traffic kept pace with the size and speed of steam-driven ships, and as hurricane activity, storm severity, and sound amplitudes of bad weather peaked in the North Atlantic due to the Multidecadal Oscillation of 1870–1899, nautical engineers and coastal authorities had been struggling to devise signals more distinct than bells, whistles, horns, rockets, or cannon. Between 1900 and 1913, while Rice was promoting the safety and silence of submarines, at least one in eight British commercial vessels sank or foundered and the peril of a sailor in the best-equipped fleet in the world was twice as great as that of a coal miner. Meteorologists establishing America's first hurricane warning service after the tragedy at Galveston in 1900, and physicists tracking wind currents and air viscosity discovered what sailors already knew, that neither foghorns nor Joseph Henry's "very large steam siren" nor eight huge megaphones, seventeen feet long with mouths seven feet wide, broadcasting the shrill of a steam whistle to all points of the compass (as tried on Falkner's Island off the coast of Connecticut in 1904) could be effectively heard under foul conditions. "The mariner has long since learned to be exceedingly cautious about depending upon aerial sound signals, even when near," stated the U.S. Navy's Hydrographic Office. "Experience has taught him that he should not assume that he is out of hearing distance of the position of the signal station because he fails to hear the sound; that he should not assume that because he hears a fog signal faintly he is at a great distance from it, nor that he is near because he hears the sound plainly." Years before Marconi and other, better radio men overcame the static and crossed signals of shore-to-ship wireless, it had occurred to more than one inventor, including Alexander Graham Bell and his nemesis, Elisha Gray, that an underwater signal might solve the problem, given the common notion that within the "deep repose" of the ocean "storms, no matter how terrible, have no power." In 1901 the Submarine Signal Company was organized in Boston to market Arthur J. Mundy's version of a subsurface microphone, which Gray had dubbed a "hydrophone," in tandem with an electromagnetic subsurface bell whose tones, issuing from within a water-filled box attached to the hull of a ship or from a tripod anchored to the floor of the continental shelf, could be heard ten miles distant, even during an ocean snowstorm. Each of its strokes controlled from the shore or from the wheelhouse through electric cables, the bell could be made to send distinct signals or Morse Code through the water at four times the speed of sound in air, furnishing information about latitude, longitude, and looming dangers. A hydrophone could pick up these bell signals forty-five minutes before a deckhand would hear a fog whistle, and "The smash of paddlewheels, the deep, slow pound of the heavy merchant ships or battleships, the clack and whir of the higher speed machinery on destroyers or torpedo boats" would become identifiable to experienced operators listening in. By 1907 submarine bells on tripods were set along stretches of North American, French, and German coastlines, and bell+hydrophone systems were being installed on ocean liners, dreadnoughts and, perhaps, submarines.[362]

Isaac Rice had no share in the Boston company. Given his penchant for vertical integration, he had clearly missed a beat with his Electric Boat Company by not also being a coxswain in the wireless world, so he may have put his submarines on the leading edge of electrical signaling, which, said the *New York Sun* in 1906, made the sea "safer now than the streets of Manhattan." To which streets we must now return, keeping in mind the foghorns and boat whistles that would have been audible from the southeast corner of West 89th and Riverside Drive, overlooking the Hudson. There Isaac and Julia had built a modest mansion, finished in the fall of 1902, months after the dedication of the Soldiers' and Sailors' Monument directly across in Riverside Park. Like the Monument, the frontage of Villa Julia had flights of marble steps leading up to a marble arch over formal doors, all of which, in style and proportion, made the Villa's western approach seem of a piece with a Monument honoring the Union dead. The mansion itself was Renaissance and Georgian, with a dark red brick facade lined in white and crowned by the tiles of a sloping attic that hid from view a gymnasium and servants' quarters on the fourth story. The formal entrance was always locked; everyone came around instead, as today, under the arch of the *porte cochère* on 89th Street, distinguished by a stone relief, still *in situ*, depicting the six Rice children at play. The children, of course, had rooms of their own. So did Isaac, a very special room, rock-solid and soundproof, built into the foundation, with a door as thick as that on a banker's vault, quieter than Pulitzer's "Vault" on the East Side. This was Isaac's chess room, where his patient tactics and marble chessboard would be undisturbed by the clamber of an indoor elevator, the hum of an electric refrigerator (another Rice enterprise), or the whoosh of indoor plumbing... undisturbed by the rumpus of "Six Children Who Never Hear Don't" (a pedagogy that yielded, wrote daughter Dorothy, children "not really bad" but "what is called 'uncouth'")... undisturbed by the dozen tutors who daily mounted the mahogany staircase to the third-floor warren of children's bedrooms... undisturbed by the menagerie in the colonnaded south garden, a haven for lost or convalescing animals... undisturbed, finally, by the whistles of tugboats on the river.[363]

Elevated trains, screeching overhead on rails put in place by 1880, had driven the Rices from West 73rd to an apartment on 79th and West End. From there they moved one block west to a "suburban villa" on 88th and Riverside Drive, originally built for Egbert Viele, a civil engineer, Civil War general, and urban planner who had sketched the first plans for Central Park and directed the Commission that worked on Riverside Park. His villa had been constructed when Riverside Drive was an escape (like Prospect Heights Park in Brooklyn, according to Viele's prospectus of 1859) from "the glare, and glitter, and turmoil of the city," and though the Drive remained remote from most traffic, the villa proved too small for a household always expansive. Deciding to build their own, Isaac and Julia in 1899 invested $225,000 in a lot one block north, next to that held but never developed by the Jewish department store magnate Benjamin Altman. The Rice's first architect conceived a baroque edifice with an observatory ninety-six feet high, an indoor pool, and an infirmary for the children, as if his clients were of the same mold as Charles Schwab, who in 1901 began erecting a seventy-five-room chateau a mile down the Drive, the largest private dwelling ever seen in Manhattan. But the Rices had no wish to match pinnacles with the President of U.S. Steel or to outglow the new red-brick mansion on the opposite corner of 89th, called Bishop Potter's Residence once the widow Clark married Henry Codman Potter, Episcopal Bishop of New York and founder of the Cathedral of St. John the

Divine. So the Rices turned to Henry Beaumont Herts, who, like Isaac, was a member of the Harmonie Club, New York's elite German-Jewish philanthropic, dining, and singing society.[364]

Something of a prodigy, Herts as a Columbia undergraduate had won an open competition to design the Columbus Arch in Central Park. With Hugh Tallant, whom he met while studying in Paris, Herts in his twenties set up a firm that would become famous for its designs of vaudeville theaters and a rebuilt Polo Grounds, validating the enjoyments of common folk. Herts and Tallant also became known for their skill at acoustics, most applauded in their Brooklyn Academy of Music of 1908. "The effect is so startling," gasped a reviewer for the *American Architect and Building News*, and "the ability to hear well in any part of the different auditoriums and halls so remarkable, that we were led to call on Herts and Tallant to ascertain if this result was deliberately achieved or as is so often the case, due to fortunate combination of circumstances." With nary a nod to Sabine, the two assured the reviewer that they had studied the matter for twenty years and "were able and willing to guarantee the perfect acoustic qualities of any building they undertook." Villa Julia, commissioned when Herts was thirty and a new father, was too acoustically lively; reverberations from hardwoods, marble, and smooth plaster perpetuated such havoc in the upper stories that Dolly/Muriel, the poet of the family, would hide out in the elevator and stop it between floors to achieve the solitude she needed to write her verses, as in "The [Soldiers' and Sailors'] Monument at Night":

> Thy very beauty sends unto the soul
> A heavenly silence, sweeter than the pause
> Of passing sound, is to the wearied ear.[365]

Commanding what *The Real Estate Record and Guide* in 1903 called "the finest view of the Hudson that is obtainable on Manhattan Island," Villa Julia was punished for the splendor of its vista. Sun or fog, morning and night, workday and holiday, the whistling of tugs rose to the villa's bedroom windows. If the noise and congestion of street traffic in New York was already notorious in the 1870s, the noise of river traffic was also rising to a crescendo with "the shrieks of passing steamers, the discordant notes of harbor craft, the puffing and wheezing of tugs, the din of escaping steam, clanging bells, howling men." So wrote the art historian and nature-writer John C. Van Dyke in 1909 about the "new New York"—where those on the banks of the Hudson lay at the mercy of compressed-air horns, sirens powered by 108,500 footpounds of steam energy, and O'Brien Electric Boat Whistles with "NO WEAK BLASTS, but a perfect signal at all times." O'Brien whistles would blow "whether engine is running or not" because, ironically, they ran on (Isaac's) electric storage batteries. And blew with no less fervor on weekends and holidays, when tugs took landlubbers on harbor cruises or shuttled them to Coney Island. Captains used their megaphones, bells, airhorns, steam whistles, electric whistles, and high-pitched "peanut whistles" as communication fore to aft, as signals to the ships they were maneuvering, as warnings or greetings to nearby craft, as halloos to crewmen ashore and, on excursions, for the sheer hallelujah of it.[366]

"Horrid beyond endurance" the *Milwaukee Sentinel* had called the outrage of a steamer that had upended the whole city one night in August 1861 "with a series of the most deafening, defiant, discordant and diabolical noises that ever split the tympanum of humanity or disembowelled the peaceful air." Trying to get a drowsy bridge-tender to raise the Spring Street Bridge, the captain had used every one of his many steam whistles, so that it seemed "as though all the furies dammed were let loose upon the night. Everybody woke up with a shiver—cold perspiration stood out upon several thousand foreheads. Everybody immediately thought of the day of judgment." In a sense it was, for the Civil War was upon them. Any protest by an Anti-Noisite would be drowned out by the ensuing apocalypse, and it would take another Milwaukee generation before Health Commissioner Kempster in 1896 had the support to mount a campaign against "the superfluous noises of the city," under which rubric fell the blowing of tug and steamboat whistles, whose "outrageous noise" could be heard for miles. Hearing of Kempster's campaign a thousand miles off, editors at the *Boston Herald* and *New York Evening Post* praised his plans, as did Morse up in Salem, who quoted Kempster in his own tirade, *The Steam Whistle A Menace to Public Health* (1905). Neither Boston nor Milwaukee, to be sure, had the tireless chug and whoop of New York harbor, through which, in the single year 1896, steamed or sailed 4,460 vessels from overseas and 10,229 from coastal ports. On the Hudson at all times were also hundreds of barges laden with stone, sand, coal, garbage, excavation debris and, daily, 400 tons of horse manure and 200 dead horses. All of these barges and many of the seagoing ships had need of the six hundred tugs operating in the harbor, an "Irish Navy" of primarily Irish crews.[367]

Never one to seethe in silence, Julia Rice would muster her resources against the indiscriminate whistling of tugs on the Hudson, "horrid beyond endurance." It is tempting to explain what followed as part of a larger tug-of-war between the city's Jews and Irish Catholics. That tug-of-war had been politicized since the election of 1886, when Eastern European Jewish voters first jeopardized Tammany's grip. The tug-of-war had stiffened in 1897, when the German-Jewish elite became publicly engaged in electioneering, throwing their support behind the Fusion candidate Seth Low. A wealthy Progressive who had been mayor of Brooklyn and was President of Columbia University, Low would finally if briefly succeed in 1901 in wresting the mayoralty of the city away from the Irish machine, which had disregarded the condition of roads, schools, and hospitals—all of personal concern to Isaac with his electric taxis, Isaac's brother Joseph with his research into education (of which more anon), and Julia with her medical degree. The tug-of-war had become most tense in July 1902 at the funeral for Jacob Joseph, Chief Rabbi of the city's Orthodox Jewry, as fifty thousand mourners wound through the Lower East Side past the printing press factory of R. H. Hoe and Company, whose workers hurled wood and metal scrap down on the procession. Mourners, outraged, returned fire. After volleys of stones and bricks and tussles on the street, the "riot" ended with the bludgeoning of scores of mourners by a squad of (chiefly Irish) policemen under the baton of an Inspector far-too-aptly named Adam A. Cross. Jewish leaders across the religious spectrum united this once to demand an inquiry, and Mayor Low appointed a committee of reformers, among them Isaac's friend Nathan Bijur, to look into the mélée. It turned out to have been spurred more by the antipathies of German Lutherans (employed at Hoe but being displaced from lodgings on the Lower East Side by Eastern European immigrants) than by rivalries between the city's 700,000 Jews and 1,100,000 Irish Catholics, but the fact that nightstick brutality was common to the quelling of "disturbances" of all kinds, whether labor meetings, suffragist marches, or gang wars, did little to excuse New York's finest.[368]

Low had promised to root out corruption in the police force—and (after one of his campaign speeches was interrupted by a street organ, by the honking of an automobile, the rankling of a truck carting iron rails, the gong of a streetcar, and the jankling of a milkman's cart) to suppress unnecessary noise. Unfortunately, his first police commissioner was timid and conservative even in response to the riot; his second commissioner was so active that he alienated tenured Irish interests and then the German Protestant population, 726,000 strong and irate when its beer gardens were shut down on Sundays by honest (and Catholic, and zealous) cops ordered to enforce the State's Sabbath closure laws. Low accordingly lost his bid for re-election and Tammany returned to power late in 1903 just as Villa Julia was finished, so Julia Rice would have to face down not just the sonic traditions of an "Irish navy" but the vindictiveness of an Irish City Hall that associated the Rices and *The Forum* (for which Low had written) with the stench of reform. What's more, it would have been public knowledge by 1905 that Isaac Rice was pursuing in court the discontented John P. Holland, a native of County Clare, a teacher with the Christian Brothers in Limerick, an instructor at St. John's Catholic School in Paterson, New Jersey when he came over to America, and until the 1890s a Fenian sympathizer whose first submarines were built with funds from Irish nationalists. All this may explain why, once retaken by Tammany, the city suddenly retracted the Rices' permit to construct a seven-foot-high terrace wall (for security? privacy? greater quiet?) along the corner of 89th. Pained relations between the Irish and the Jews might also account for the recalcitrance of municipal authorities to respond to Julia's complaints (her "foolery") against the tootling of the tugs of Shamrock Towing.[369]

Disenchanted Irish maids or chauffeurs from the Rice household would have to be subpoenaed to deepen this indictment of Anti-Noisism as a soundscreen for haughtily Liberal, anti-Catholic prejudices. But none would be forthcoming. On the contrary, Julia had a talent for enlisting in her ranks such stellar Irishmen as the Tammany bigwig Charles F. Murphy and his brother, John J., President of the New York Contracting and Trucking Company. Further, the City Improvement Society had reported in 1899 that the large majority of complaints it fielded from all sides about life in New York City had to be filed under the category of "Noise," so neither the Progressive Dr. Julia Rice nor the populist Dr. John Girdner had a peculiarly political sensitivity to sound. And should a researcher someday locate that 1899 Noise file, she would find, as elsewhere, that those accused of noisemaking were most often the underclass, Catholic (Irish, Italian) *and* Jewish (Russian, Romanian, Lithuanian, Polish), whose days had to be spent in thin-walled tenements or on the congested streets hustling their wares—in a metropolitan area that had grown by more than a million in ten years, with a doubling of passengers on local buses and trolleys every decade since 1870. If it was not their street cries, street music, street games, or rent strikes that raised the hackles of neighbors, then it was the c/row of their dogs and roosters and the pitch of their public entertainments.[370]

Xenophobia? Was it xenophobia, then, and class fears, that were the real issues behind Anti-Noisism? Julia Rice was four generations into an American identity; Isaac, a first-generation immigrant from Central Europe, had already begun to worry about a city and nation inundated by Jews from Eastern Europe. Both Isaac and Julia were easy enough with their wealth from the moment of their marriage to consider themselves thoughtfully rich, neither greedy for the patinas of old gentility nor consumers of the opulently second-rate but publishers of the best in literature and ideas and lifelong advocates of artistry and intellect. They hired as tutors for their children some of the most promising young musicians and chess-players (Emanuel Lasker, for instance, who in 1894 at twenty-five defeated Steinitz for the world title in chess). They were present at the creation of Barnard College for Women (Julia), sponsored collegiate chess tournaments (Isaac), and helped start the Poetry Society of America. They had learned the postures of attentive silence at the concert hall and chess table, and although they enjoyed the rambunctiousness of their children, the family entire knew when to hush, how to listen for nuance. Isaac's command of music, law, languages, and finance, Julia's command of music and medicine, and their exalted position on Riverside Drive: did these inflate

Julia's overblown indignation at the workaday whistles of those socially, and intellectually, and literally, beneath them?[371]

5: Julia and Everyone Else

Self-righteous or self-confident as she may have been, Julia was well aware that a campaign for quiet issuing from a woman in a mansion on Riverside Drive would be open to all such innuendos—from the Left, of classism, ethnic prejudice, bourgeois maternalism; from the Right, of hysteria, busybodiness, economic naïveté; from both sides, of motives darkly Jewish. With other Progressive women who opted for Reform over Revolution, she also held to an ideal of social harmony that would be classless and raceless though brokered by an educated white elite. Like her counterpart in Pennsylvania, Imogen Oakley, she therefore deflected the debate from her person and position toward those less blessed: it was no wealthy matron at her high window who had to suffer the indignation of tugs noisily plying the waters around Manhattan, but the indigent in hospitals along the Hudson, and above all the hundreds of sick poor children sent to contagious-disease hospitals with whooping cough, scarlet fever, typhoid fever or diphtheria who could neither sleep nor be sung to sleep while tugs blatted to crew members drunk in saloons at two a.m., their horns, sirens, and whistles criminally indifferent. Noise everywhere and everywhen, vexing everyone: this was more than an historian's chronology; it was what the Anti-Noisites knew they had to prove, across classes and generations. Oakley—a veteran of the anti-smoke wars in Pittsburgh, daughter of a prominent surgeon and professor of medicine, sister-in-law to a chemist who came to manage a steel company, second cousin to the nation's best optical instrument maker and director of the Carnegie Institute of Technology, wife of a successful stock broker, mother of a leading illustrator and art educator, and in her own right a national advocate for women's clubs as agents of municipal housekeeping (in Japan as well as America)—Oakley positioned herself not as the D.A.R. she was but as the voice of the people. She told a reporter that "in her desire to find out how the city could best serve its working class," she had gone into the slums and asked women what they deemed "the greatest evil in their crowded tenement life." That was noise: "It never stops. It is killing us.... You can get away from the noise during the summer; but we can not. We are right here in the middle of it all our lives." Was it she who was behind a petition in 1907 to the Philadelphia Civic Club from "a large number of working women in the Eleventh Ward, asking the assistance of the Club in their endeavor to restrict the shouting and yelling of the street venders and push-cart men"? They needed help in abating "the useless noises of the streets" which began at four a.m. and continued through midnight, making sleep impossible for workers returning from night shifts, infants at their naps, and mothers around the clock, "by reason of the din kept up by street vendors, peddlers, newsboys, and organ grinders," as well as the cluck and clatter of nearby poultry yards.[372]

Old nuisances, a new polity—a constituency not just of the thin-skinned leisured classes with so little to do that they could afford to be annoyed by the sounds of an industrious world, but hard-working city women and those "horny-handed sons of toil" who had come up to Morse on Salem sidewalks to thank him for reducing the steam whistling in freight yards. Oakley also learned from Morse to keep a daily log of "useless noises coming to her ears in her own house on a 'quiet' residential street." She bracketed the quiet because her elegant home was itself "bounded on the front by trolley cars and on the rear by a densely populated backstreet," and from her parlor she could remark "a different, unnecessary, and preventable noise on an average of every five minutes between 4 a.m. and midnight." And what of women far more badly situated than a broker's wife? Should they not also rail and fume at continual disruptions to their families' peace and health? "The people of the slums do not like noise, as popularly supposed; they hate it." Responding to the petition with a series of noise abatement initiatives, Oakley got "a shower of grateful letters—all from hard-working men and women in the tenement districts, districts whose inhabitants, if we are to believe the noise apologists, would actually suffer if deprived of the familiar and stimulating noises of the street." The letters confirmed that the "toiling poor" were neither strengthened by noise nor immune to it; if they had borne up for years under terrible racket and industrial "bluster and brag," that was heroism in the face of poor housing, bad urban planning, worse politics. More than society matrons or suburban housewives, they knew how hard it was to maintain their humanity and preserve their respectability. (Did Oakley have glimmerings that union membership across the nation had quadrupled between 1899 and 1904?) In the ideal city, she wrote, "All unnecessary noise will be silenced, and much that is now deemed unavoidable will be found to be quite unnecessary." Newsboys would be off learning a trade. In lieu of barrel organs on street corners there would be pianos in public schools. Trolleys would be shunted underground, and no citizen would tolerate for a minute "the sputterings, snortings, and shriekings of the present-day car. The engineers of the ideal city will have evolved a noiseless motor."[373]

She had been in touch with Julia early on, but evidently had not gone for a spin in one of Isaac's silent electrics. Or had Julia first been in touch with her, as she was with Morse, with Abbott, with the Drs. Girdner, Shrady, Weir Mitchell, and Kempster? The eloquent Anti-Noisite (and anti-Tammany man) E.L. Godkin, editor of *The Nation*, had died in 1902, but she could still enlist William Dean Howells, who in his seniority as editor of *Harper's* was finding "that years which dull the senses to so many things do not bring that thickness of hearing which would save the sage from the sharpest tortures of modern sound. In fact, it seems to render him more alive to them." Would that there were "sound-brakes," he wrote in 1906, akin to windbreaks on the prairie. Through Howells she would recruit Samuel L. Clemens, now a New Yorker domiciled "in one of those dignified old mansions of lower Fifth Avenue that have set their kindly, patrician old faces sternly against the marauding march of skyscrapers and business loft and hotel." Through Isaac she knew President Butler of Columbia. During morning constitutionals she could stop to chat with her neighbor, Bishop Potter, who had been outspoken against police corruption "in the name of these little ones, these weak and defenseless ones, Christian and Hebrew alike," children of all races and religions. However, it was Morse to whom Julia sent packets of clippings and copies of her far-flung correspondence; it was Morse's anti-noise career that she, like Oakley, took to heart; and it was Morse's advice she most sought.... Or was it simply his benisons? Like Mrs. Imogen Brashear Oakley, Mrs. Isaac Leopold Rice had at hand models of successful activism advanced by women of a certain social register and persistence: the rise of the City Beautiful movement and revitalization of Temperance and Suffrage campaigns; the passage of laws against child labor and "white slavery"; the spread of kindergartens, settlement houses, playground associations, and pure milk depots; the promotion of anti-smoke ordinances and urban planning; the underwriting of working-class health surveys; the emergence of a Women's Trade Union League; the shaping of new concepts of municipal economy and public health.[374]

Unlike Imogen, Julia had close at hand four additional models:

- *The Brooklyn Daily Eagle*. Under the critical eye—and ear—of St. Clair McKelway, grandson, son, and son-in-law of physicians, the newspaper since the 1880s had had a penchant for detailed articles and blunt editorials about the noisiness of what had become the fourth largest industrial city in the United States just before its consolidation with Greater New York in 1898. In 1901, McKelway's staff studied Brooklyn's "roaring, sizzling, humming, gurgling, banging, clattering, clanging conglomeration of sound"; they determined which noises ought to be tolerated as necessary, and which must go, at least from the sonic prospects of prosperous businessmen and upstanding home owners in houses of resonant terra cotta and yellow brick. Necessary were the noises incidental to the operation of elevated trains (which in the 1880s had their way smoothed by Isaac Rice, counsel to the Kings County Elevated Company); unnecessary were newsboys shouting "news of impossible events" at intolerable hours.[375]

- Dr. Sara Josephine Baker. An 1898 graduate of Julia's alma mater, Baker three years later passed the city's civil service exam for medical inspectors of schoolchildren. Such positions had been a Tammany racket, but when Mayor Low appointed a more professional Board of Health, Baker began serious work in Hell's Kitchen to reduce high infant mortality rates due to adulterated milk, poor maternal nutrition, and a host of infectious diseases. It fell upon her to inspect for unsanitary conditions and also to field countless complaints about noise within and around tenements. Herself alert to every sort of sound, from the "tick-tack" of pebbles dangling on pins rapping peskily on window glass to the fare-collection bells on trolleys, Baker was acutely aware of the crying of babies, which she understood rather in a sociological context than as an "exercise" (said some male pediatricians) necessary "to expand the lungs" of newborns. When she founded the Little Mothers' League to train older girls to care for younger siblings, she promoted it with playlets, one favorite plot of which involved "a crying baby, screaming its head off, and a harassed mother who has no notion whatsoever of what to do about it. Enter several neighbors, some sympathetically distressed about the mother's troubles, some furious because the baby is keeping them awake, all making a terrific hubbub with unintelligent suggestions" until a Little Mother arrives, sizes up the situation, removes a prickly safety pin, and all returns to quiet: infant hygiene as the riddance of unnecessary noises.[376]

- The West Side "anti-noise contingent." Organized by Girdner, whose own brownstone was on West 71st, this contingent had been active since 1900 in crafting ordinances against street noise and the screech of those Els whose tracks had no sound-deadening devices set between the rails. Loosely organized, the group drew inspiration from *Newyorkitis* and the writings of such as Philip Hubert, Jr., a New York music critic, journalist, and dramatist who had proposed in the *North American Review* that there be formed "a society in New York and every large American city for the suppression of unnecessary noise." Dins of Iniquity, cities had associations to suppress vice, poverty, and cruelty to animals; why not a SSUN? Hubert's essay appeared in 1894, weeks before he withdrew to a small farm on Long Island where he and his family lived the life of Thoreau, raising chickens and bees: "The real secret of the greater endurance and longer life of the man who lives in the country is that he is not subjected to the ceaseless noises of the city that keep the nerves constantly alert," though surely, like Thoreau, he heard loggers with buzz saws clearing the forests "with an agony of sound." Twelve years later, a little worse for wear, Hubert was lured back to avenues clucking with car horns and abuzz with suicidally repetitious street organs playing "Bill Bailey" or "Wait till the Sun Shines, Nellie." He returned to his desk at the *Herald* just as Julia launched her SSUN.[377]

- Her brother-in-law's successful campaign for educational reform. By 1888, Joseph Mayer Rice's interests had shifted from childhood diseases to the hygiene of school buildings, thence to the general issue of public education, so he left off his medical practice and went to study psychology and pedagogics at the universities of Jena and Leipzig. On his return in 1890 he investigated teaching practices in classrooms across America, collecting data from more than one hundred thousand elementary school students in thirty-six cities, from which came twelve influential reports first published in *The Forum*, then as *The Public-School System of the United States* and *Scientific Management in Education*. More devoutly than America's foremost child psychologist, G. Stanley Hall, who had also studied at Leipzig and written for *The Forum*, Joseph took a clinical, statistical approach to comparing the effectiveness of modes of teaching. He factored in (and out) the amount of time spent teaching a subject, the style of teaching, the corresponding achievements of students (on his own standardized tests), their ethnicity, and their family's economic status. Widely reported by the press, his conclusions put "the new education" on legislative agendas: "mechanical" teaching did not work; verbatim recitation from schoolbooks was a waste of time; poverty of pedagogical invention among teachers was more determinative of weak student performance than poverty at home. "Mechanical" or "unscientific" teaching meant all education "conducted on an antiquated notion that the function of the school consists primarily, if not entirely, in crowding into the memory of the child a certain number of cut-and-dried facts." By "the new education" he meant Progressive education as demonstrated by teachers who established a bond of sympathy with each child as "a frail and tender, loving and lovable human being" and who used their knowledge of the laws of mental development to help children learn to observe, to reason, and to make moral choices.[378]

Joseph saved his harshest language for the teaching of reading when it "consisted of nothing but empty words, silly sentences, and baby-trash," and for the "spelling grind" at all times, whether too long (fifteen minutes a day was plenty) or arcane ("exogen," "coniferal"). Improvement in spelling, he could show, was a function neither of drill time nor of a student's native tongue but of maturity. And given the absurd orthography of English, many of the ways that students misspelt "physician" were not unreasonable: sounded out, what was wrong with fasition, fesition, fisition, fisision, fiseshon, or fazishon? The phonic, sonic elements in his critique drew less comment from the Committee of Fifteen appointed by the National Education Association to respond to Rice than did his emphasis on independent thinking and individual growth. As pianists and humanists, Isaac and Julia would have taken Joseph's implicit point that bad teaching was basically unmusical: it ignored laws of development, was mindlessly repetitive, arrived at coherence only through loudness (group or sing-song recitation), was stressful for children, distressing for auditors. It was, in short, noise.[379]

Each subject, wrote Joseph, had its saturation point each day, after which it too became (he did not use this phrase) mental noise, and it was a waste of effort to push past that point. "Scientific management" à la Taylor and the Gilbreths would study the patterns of physical fatigue in order to set more productive work regimes for laborers at each factory or brickyard task; Joseph used studies of mental fatigue to prescribe better classroom conditions for learning. His own Society of Educational Research lobbied for the standardization of educational procedure and reduction of class sizes in the interest not only of efficiency but of political reform, sustained by a citizenry that would demand of its government, as of its schools, an ethical, "rational leadership."[380]

"Macerior begs the aedile to prevent the people from making a noise, disturbing the good folks who are asleep." Thus began Julia's first anti-noise essay, borrowing from Morse the Pompeian graffito that foretold her own reforms, which would steer from the unnecessary toward the unjust. She would be more effective than Morse, Girdner, Abbott, or Babbage because she had more tools at her disposal: a musician's (and a Southerner's?) sensitivity to tone; a physician's confident use of medical literature and logic; her own and Joseph's close relations with medical men and educators; her own and Isaac's social, industrial, financial, academic, and Jewish contacts; *The Forum*, its ties with Progressives and writers; Isaac's legal perspicuity, technical expertise, and tactical patience; her own political talent, deftness at controlling the direction of interviews, and skill at articulating an anti-noise program from the perspective of children, the working poor, and the "poor, lone woman" (with which she first identified herself: a "lone woman, with no one to help me," she told the *New York Times*).[381]

Returning in August 1905 from one of many tours of Europe, where Isaac had been scouting contracts for submarines, Julia heard tugboat captains tugging on their ever-more powerful whistles more often, more stridently, and at greater length—as long as 38 seconds a toot. Her children would crawl out of bed at one in the morning and creep down to the art studio to sculpt or draw, "simply because they could not sleep." She too was sleep-deprived; the electric fan she kept near her bed for its continuous buzzing could not mask the piercing sounds. Learning that some captains were using their whistles to serenade servant girls on Riverside Drive and sure that other tootling was equally unnecessary when it consisted of "30 blasts followed by a long ear-splitting shriek," Julia called upon the Board of Health, the Police Department, the Warden of the Port, the Dock Commissioner, and the Port Collector's Office for respite. No one seemed to know who had purview over the boats once they left the dock. Julia had stationed men along the shore to count the frequency of the whistles; they found that there was "scarcely a minute in the night" when whistles did not sound, some loud enough to be heard four miles away. She had also hired Columbia law school students to count the blasts audible from the front and rear of Villa Julia: 1,116 whistle-blasts on a brilliantly clear Sunday night when hardly a toot was called for, triple that on cloudier, foggier evenings, up to six blasts a minute. Now she went out in person to the docks to talk with tug captains and toured the Gouverneur and Bellevue Hospitals on the East River, the Children's Hospital on Randall's Island, and the Metropolitan Hospital on Blackwell's Island, all situated near barge and tugboat moorings. She called again upon each city bureau, armed this time with three thousand signatures petitioning for the suppression of unnecessary whistling, testimonials from hospital superintendents "representing the sufferings of 13,000 poor people," and letters from neurologists about the harm done by the shriek of whistles to the nervous, depleted, contagious, and chronically ill. ("The bacilli which haunt the air are not worse enemies than the sound waves that quiver in the air and beat in upon our brains," editorialized the *Tribune*, which counted among its readers many college graduates and Reformers. "There are benign bacilli, but there are no benign noises.") At Bellevue, the staff could not sleep for the "incessant whistling of passing steamers," and apprehensive patients, if not whipped into a frenzy, mistook the whistling for their death-knells. "Heretofore we have been compelled to stand this disturbance, sometimes without a murmur, as we would thunder from the clouds or a pestilential visitation of Providence," wrote the resident alienist; could anything truly be done? New York's Chief Medical Officer, the pathologist Hermann M. Biggs, who had expanded the city's program of school medical inspections to test for diminished hearing, was "thoroughly in sympathy" with Mrs. Rice's efforts and told her that the Health Department had fielded similar complaints from the city's own Scarlet Fever Hospital at the foot of East 16th Street and the Riverside Hospital on North Brother Island, but all he could do was submit to Corporation Counsel the question of his jurisdiction, if any, over river traffic. Julia hied to the Counsel, who dithered. On the strength of letters from such as Capt. George A. White of the Hudson River Day Line, who thought that tugboats sporting whistles powered by 150–200 pounds of steam pressure were engaged in nothing less than "pure rowdyism," she got the National Association of Masters, Mates, and Pilots to pass a resolution "urging all its members to abstain from unnecessary use of the whistle, 'especially the so-called siren whistles.'" She went to Washington, D.C., to attend an annual meeting of the Board of Supervising Inspectors of Steam Vessels, her briefcase bristling with statistics and bulging with affidavits from tug and scow captains aggrieved at the confusion of whistles so irresponsibly tooted that the "amount of whistling often makes the one signaled-to angry, and he pays no attention, and so the whistling continues." The Board, created to deal with unsafe boilers, deplored the situation but threw up their hands; they had no power to silence such "hoodlum whistling."[382]

An impasse, was this, or a signal moment? It is common these days for "noise" to be characterized as liminal, as sounds that teeter on the edge of meaning, fitful of place and fretful of time, all the more unsettling for eluding definition or resolution. Liminality was not what Julia Rice had in mind when she campaigned against noise, nor was it what decades of Manhattanites had had in mind when protesting against fire-bells and bellowing firemen, churchbells and peddlers, organ grinders and newsboys, "ruffians with throats of brass" and nighthawk cabs, construction workers and the grind of rails. They knew what the noises meant; they knew well where/who/what the noises came from; they knew very well how much earlier than dawn or later than dusk the noises arose. Knowing the time, place, and meaning of the noise was crucial, for only then could they argue, as they usually did, that the sounds came at the wrong time, in the wrong place, or in excess, or as atavisms no longer to be abided. Not liminal: rather, obtrusive and inexcusable. Once noise was admittedly everywhere and everywhen and jangling everyone—Oakley in her well-appointed home and the poor in their tenements, Rice and her children in Villa Julia and the indigent in riverside hospitals—a campaign against frustratingly untouchable whistles or sirens had to be an assertion that no humanly engineered sound should be beyond human command. If tug whistles resembled "the wails of lost souls in torment, or children in terror, or wounded horses, or distant Scottish bagpipes, or the Banshee," one ought not grant them a legal limbo. True, there were as yet no standard units for the loudness of a sound, and although Julia was briefly in touch with Sabine, the intensity of aural sensations had no generally acceptable metric. Wrote a young polymath, Norbert Wiener, in a short meditation "On the Measurement of Sensory Qualities" just before the Great War: as it would be "impossible to say that the sound of a foghorn is, say, two thirds as loud as that of a boiler factory," so "we do not measure the loudness of a watch-tick in thousandths of a boiler factory." Still, attention had to be paid and arguments made if banshees were not to raise the hackles of the thousands each night who needed to regroup after the stampede of the urban day, or haunt the sleep of the thousands in hospital beds by the shore who were trying to recover their strength.[383]

"Cell exhaustion, fatigue cannot be avoided, but recuperation is a right of nature, the wanton interference with which constitutes an infringement upon the personal right of the individual, and which should be protected by laws." A right. Stirred by psycho-legal claims as definitive as this (from George W. Jacoby, a leading neurologist), and with a Brandeisian sense of the lines to be redrawn between public and private life, Julia approached her congressman, William Stiles Bennet, a lawyer, jurist, and public defender of immigrant "South European and Hebrew peoples." Noisemaking was apparently a federal issue when on interstate or coastal waters, and if she could get Bennet to ask the Secretary of the Treasury to send a revenue cutter to report offenders or, better, to introduce a bill giving local officials the authority to police the acoustic waters, and if Congress would pass it and President Roosevelt sign it, the problem of this defiantly "free soundway" might be settled over Tammany's muddled head.[384]

Oakley never operated at this level, though in her work on national civil service reform she came close. Nor did she have the ear of a President who had published five articles in *The Forum*, who knew Isaac Rice as a political scientist, a Columbia professor, a corporate lawyer, a sponsor of international cable chess matches, a Navy contractor, and an escort for his daughter Alice, who had been taken for a courtesy descent in one of Rice's submarines in 1903, and who in 1905 had just wed a congressman from Ohio and was best of friends with Mignon Critten of Staten Island, who was betrothed to an eloquent congressman from Kentucky, J. Swagar Sherley, who happened to sit on the House Committee on Merchant Marine and Fisheries and who would speak up (after hearing tugs skirling near a Critten family charity, the S. R. Smith Infirmary?) in favor of HR 17624, through which the Secretary of Commerce and Labor would give the Supervising Inspector-General of Steam Vessels, and local representatives, including revenue cutters, the authority to order tugs to cease and desist their unnecessary whistling, since "Much confusion has arisen in some of the large harbors of the country by the use of sirens and other loud whistles for purposes other than signaling, making it difficult to hear the regular signals of other vessels." That's a long, involved sentence for the long, involved politicking that must have lain behind Julia Rice's triumph during the 59th Congress, when Section 4405 of the U.S. Statutes, with regard to whistles on steam vessels, was revised, passed, and in February 1907 signed into law.[385]

Months before, Julia had been contemplating the next steps. "Encouraged by the success with which my little crusade against tug whistles was rewarded," she wrote Morse in October 1906, "I now want to devote myself to the task of forming a Society for the Suppression of Unnecessary City Noises." The tug whistles were 5 percent of what they had been before, she noted, and although they would become more directional as furious captains would "shiver the evening air whenever they pass her home," she had what appeared to be an open invitation to steer her course toward broader lanes. Dr. Thomas Darlington, the city's popular Commissioner of Health, had that month addressed New York's Society of Medical Jurisprudence on "the desirability of suppressing a large proportion of the noises which now combine to make the city a pandemonium." The *Tribune* welcomed his concern but was disappointed in his "hopeless tone," as if none of the existing laws against noise could be enforced. "The only hope," wrote the Republican editors, would lie "in a popular movement for reform and the arousing of public sentiment on the subject." A real estate man, Granville Nicholson, responded to the editorial with a letter claiming that he was, "as far as he knows, the only person who ever had anybody arrested here for making an unnecessary noise," viz., a teamster who had not properly secured the iron girders he was transporting. Proud to have accosted the perpetrator of the clanking, even if newspapers at the time ridiculed him as a "crank," Nicholson suggested that "what is wanted in this town more than anything else is some person of wealth, leisure and ability (like the late Henry Bergh, who founded the Society for the Prevention of Cruelty to Animals) who, oblivious to ridicule, will, with unflagging zeal and everlasting persistence, take hold of and pursue the matter until the infernal noises which we now endure are suppressed."[386]

Cue Julia; SSUN rises. On 3 December, as Roosevelt was delivering a State of the Union address in which he urged the passage of a law that would give his trust-busters sharper legal teeth, since "the real efficiency of the law often depends not upon the passage of acts as to which there is great public excitement, but upon the passage of acts of this nature as to which there is not much public excitement, because there is little public understanding of their importance," Julia Rice laid out her plans for the SSUN to a *Tribune* reporter: "I have long felt that there has been an urgent need for such a society, but as nobody else appeared willing to take the initiative in the matter I have finally had to do so myself." She went on: "The noise question grows more important every year in this city, and it has become more aggravated of late than some people imagine." Hers would not be "an anti-noise society, but only against unnecessary noises," such as the noise made by boys scraping sticks along metal railings a pebble's throw from hospital wards, or milk cans improperly secured, rattling through the streets on wooden carts well before dawn. In preparation, she would be gathering letters of support, selecting a board of directors, and drawing up a constitution and bylaws, with the help of Francis Hamilton, Solicitor for the Collector of the Port. She would also be drafting an appeal to the public to relieve "the intense suffering of our sick poor from the noise-evil" and to remove "one of the greatest banes of city life" in general, viz., "unnecessary noise, which first, wrecks health and then is chief torment of illness." Like Oakley, she solicited support from all quarters, believing that "those whose sympathies and efforts are devoted to ameliorating the condition of our congested tenement-house districts will willingly aid a work which will strive to render conditions there more endurable. To the sensitive, noise, even amidst spacious surroundings, is disturbing; in confined quarters it is torture." The SSUN's success, she wrote in the *New York Times Sunday Magazine* later that month, "depends on the molding of public opinion and united general work."[387]

Darlington himself was "very glad" to be on Julia's Advisory Board, as were Felix Adler, William Dean Howells, Samuel Clemens, Nicholas Murray Butler, Richard Watson Gilder (Progressive editor of *Century Magazine*), Luther B. Dow (General Manager of the Association of Masters, Mates, and Pilots), John M. Farley (Catholic Archbishop of New York), and the director of the Hanover National Bank, the dean of Columbia Law School, the president of the Academy of Medicine, the president and director of the Metropolitan Museum of Art, three congressmen, six physicians, and six professors. Among the latter was John Bassett Moore, Acting Secretary of State in 1898, drafter of the treaty ending the Spanish-American War, advisor on Latin America to Roosevelt, Professor of International Law at the Naval War College and Columbia, advocate of arbitration to advance corporate interests and "displace war between nations," contributor to *The Forum*, and a founder of the Hague Tribunal. Each of these men was nicely positioned to help promote the SSUN, supported by a Board of Directors composed of Julia and Isaac, John Girdner, T. C. Martin (editor of *The Electrical World*), Morton Arendt (an electrical engineer at Columbia), James T. Shotwell (historian, diplomat, and peace activist), John Jerome Rooney (lawyer, patriot-poet, jurist, and Irish Catholic politician, his wife the head of the Civic Education League), and the superintendents of sixteen hospitals.[388]

Name-dropping and networking were invaluable to a Society that meant to go after unnecessary noise across the urban grid: the horns of impetuous or obstinate automobilists, the gongs of insensitive streetcar conductors, and all uproar around hospitals. Julia promised that the SSUN would also address the "whirring of motors, the grinding of brakes and the pounding of flattened wheels" of the electric trolley system, run (coincidentally?) by the syndicate that had bought out Isaac's taxi company. That whirr, grind, and pound, and the "whistles of the starter, the shouts and jokes of the idle conductors and motormen, the altercations with drunken passengers" had driven insane not one but two women in the Sloane Maternity Hospital. Then there were the milk vendors, who should have rubber tires on their wagons, rubber shoes on their nags, and stuffing around their cans; she herself, she vowed, would write "10,000 letters to people and advise them to deal with only those who do business in a quiet way. I shall request them always to do this in the name of the poor who are trying to get their much needed rest."[389]

Writing on behalf of the exhausted poor desperate for a night's sleep, the sick poor desperate for a fortnight's peace, and the immigrant poor desirous of an audible education in a new language, Julia was also working implicitly on behalf of metropolitan harmony. The traffic at city centers was more nerve-wracking than before or since: horse-drawn carriages with jangling harnesses competed for rights-of-way against streetcars with their brass bells, electric trolleys with their gongs, pushcarts with their peanut whistles, ambulances and fire engines with their sirens, teamsters hauling loose steel beams, demolition teams with skreeling jackhammers tearing up cobblestone to make way for quieter asphalt, locomotives pulling out of depots with their whistles wailing, and electric hansoms, Stanley Steamers, or gas-driven Curved Dash Oldsmobiles with their double-toned horns, hisses, or chugs—all as yet untempered by standard traffic signals or practical speed limits. (Overhead, New Yorkers also heard the screech of Els, and from street gratings the rumble of subway trains below.) It was as well an era of labor strife: twice as many strikes nationally between 1900 and 1905 as during any previous five years; in New York City, strikes by housepainters, streetcar conductors, and four days of rioting in 1907 when striking longshoremen fought off scabs. A time, too, of epidemic and class-borne fears of infection: after a bad season of typhoid fever, "Typhoid Mary" Mallon, an Irish cook, was dragged from a Park Avenue home in March 1907 and thrust into isolation in Riverside Hospital; during the summer came an outbreak of polio that struck children in tenements and brownstones alike, killing twenty-five hundred, paralyzing thousands more, and leaving many survivors severely deaf. There could be nothing innocent or apolitical about a SSUN that had to negotiate between the demands of industry, commerce, and public health while protecting the poor, the sick, the newborn, and the Newyorkitic from the clamor of boats, trains, trolleys, churchbells, and factories. Across such a wide soundstage, Julia had need of both Tammany and Reformers, Irish Catholics and German Lutherans, Episcopalians and Ethical Culturists, behind-the-scene bankers and public health commissioners, diplomats and neurologists, physicians and philosophers. And engineers.[390]

Victor Hugo Emerson in particular. Director of the record department of the Columbia Phonograph Company, Emerson had been experimenting since mid-November, free of charge and at Julia's behest, "to discover whether or not records can be made of these distressing sounds around hospitals for the humane purpose of trying to help the sick poor." On the cutting edge of sound recording technology, Emerson had filed fourteen U.S. patents in the field since 1893; his newest patent would be granted on 18 December for a "Sound-Record" whose blend of shellac and celluloid was hard enough to obviate the need for a change of stylus after each session and whose smoother surface allowed for less hiss at playback. Sound disks and cylinders had already been adapted to advertising and to language instruction. With the addition of electric motors and a reliable storage battery, gramophones were also being used as a commercial broadcasting medium, set playing on the sidewalk to attract passersby to music shops and amusement parlors—a loud form of publicity which Julia meant to outlaw. Meanwhile, she and Emerson were the first, aside perhaps from apostles of Temperance, to cut records for the purpose of engaging an urban reform, and the first intentionally to record sounds that were to be foregrounded, and detested, as noise. That Julia should have been in touch with Emerson months before the first meeting of the SSUN, and that she set the stage for his recordings in a set of interviews published weeks before the meeting, was equally extraordinary, as if she had studied how best "to make a noise," in the parlance of the advertising world, and how to use new media to her advantage. On January 14, 1907 the first meeting of the SSUN was called to order in the library of Villa Julia, where everyone listened "with pained faces" to graphophone recordings of "the whirr and bang of flat-wheeled surface cars, the shriek and rumble of the elevated expresses, the honk of the automobile, and the various discordant street cries, tooting of whistles, yells of street vendors, &c." Those were examples of unnecessary noises, said Julia. "They are an abomination, and we shall do away with them if it takes us all Summer."[391]

Elected president "in spite of my piteous appeals," Julia wrote Morse, "I wish that you lived here, to head our Society." Morse kept his distance. (Julia to Morse: "You owe me at least a million letters.") Morse would never bask in the glow of such vice-presidents as Julia had rounded up: Howells, the international man of letters; Moore, the international arbitrator; the Very Rev. George M. Searle, Superior General of the Paulists, mathematician, astronomer, and Smithsonian scientist. Notwithstanding her prestigious board, it would take more than a summer for the SSUN to abolish, abate, or otherwise dull the exiguous noises of New York. Nineteen years later, in 1926, Julia would be at it yet, protesting that "There is still much to be done.... There is no reason why New York should be a Babel." In 1927 came the first blockbuster Talkie, *The Jazz-Singer*, more noise from New York; in November 1928, had Julia bought a ticket to the 79th Street Theatre, she would have been less than amused by Walt Disney's *Steamboat Willie* with its scratchily animated, sonified scenes of a demonic mouse swinging cats by their tails for an experimental *Katzenmusik* and tugging on steam whistles in manic defiance of HR 17624. Julia died the next year, at the age of seventy, two weeks after Black Tuesday but far from the defenestrations of Wall Street, at her daughter Dorothy's country estate in Deal, New Jersey. Honored in obituaries as "A Crusader for Quiet" from days of yore when "the demand for a layer of asphalt over the surface of the Belgian blocks became a political issue of citywide importance," she died just as another crusade against noise was charging up and a black van packed with engineers was monitoring the streets of New York with a new device that measured sound levels by small, precise, logarithmic increments.[392]

Hers were not the decades of the Talkie, the audiometer, or the decibel, nor, for most of her life, radio broadcasts. What she had at her command were calling cards and petitions, letters and cables, posters and flyers, medical statistics and diagnoses, essays and editorials in the national press, newspaper interviews and news-clip services, local ordinances and federal law, legal briefs and testimony,[393] surveys of hospital staffs, teachers, and students, and meticulous enumerations of whistle blasts, horn honks, gong bongs. And, most dramatically, Emerson's open-air recordings of Manhattan streets, which sounded to one reporter like this:

> Errrr-rags! Grrrrr umpfh, errr-rags! Rags! Strawrbries! Casholclo's! CHEAP casholclo's! Eeeee-yip! Freshlivecrabs, all alive! Errr-rags! Gooowarrtumpfh! Errrr-rags! Ta-ta-ta-----ta-raaaah! Ding dong! Scissorsgroun'-----lineup overthere? Ding-dong-ding! H-A-double R-I-G-A-N spells Harrig-----A robry in the park; seated alone in the Y.M.C.A.-----errrr-rags! Casholclo's-----Cud-dle up a little closer, Baby mine! Cud-----errr-rags! Want a line up? By Killarney's lakes and dells; by Killarney's-----Sullivan! Sullivan! Big Tim; y'all know of him;John L, y'know of him as well. Oh, there never was a man named Sull-eye-van who wasn't a damned fine Irish-----errrr-rags! Strawbries!

His transcription missed the more mechanical of Emerson's "canned noises of the New York brand": trolleys scraping around bends in the track, ragtime songs from Tin Pan Alley leaping out from double-horned phonographs wheeled onto sidewalks in front of penny arcades. And who was that Sullivan!? That Big Tim y'all knew? It was Timothy D., who with his cousin Timothy P., the Little Feller, ran the Tammany machine in the Bowery and Lower East Side. Big Tim had begun in saloons, moved on to tenements and rings of thieves and prostitutes, served in the State Assembly and U.S. Congress, and prided himself on having friends in every crook and nanny of the island. The two Timothies had more clout than the boxer John L. and a well-publicized concern for the welfare of working women, whose hours they fought to reduce and whose children they fitted each spring with new shoes. As aldermen, they would be the ones to put forward and assure the passage into law of Julia's most distinctive, and lasting, accomplishment: quiet zones.[394]

"Zones" had been on the lips of astronomers for a century, botanists and geologists for eighty years, military strategists and coal miners for fifty, neurologists and psychologists for thirty, rail executives and diplomats for twenty, timekeepers, trolley conductors, and Isaac's taxi company for ten or less. While towns since Jericho had segregated noisome occupations in precincts just within or beyond their walls, and while Frankfurt and Berlin in 1891–1892 had designated types, heights, and functions of buildings in each ring of development around the city center, urban planners in America in 1907 had still to battle retailers and manufacturers over the legality of residential, commercial, and industrial zoning. Los Angeles led the way, as California courts upheld its restriction of slaughterhouses, laundries, and saloons to specific districts, and Chicago was close behind, but the federal bench, which had approved the assertion of eminent domain for decades of rail lines that cut audibly and visually through the centers of one town after another, was reluctant to constrain real estate developers, shopkeepers, or factory owners. However, once cities began improving their streets, they were *ipso facto* zoned, as much by environing sounds as by echelons of class, according to which parts of the city were paved and which had asphalt, brick, wood blocks, or cobblestone. And once cities began to foster protected pastoral spaces, a pedestrian would observe a tacit zoning for "calm and restfulness" in the vicinity of garden cemeteries and central parks, as well as explicit signage and policing for quiet and orderliness within. Then there was the example of the Four Mile Law: in 1877 Tennessee had forbidden liquor sales within four miles of any chartered school in unincorporated areas, a statute quickly adopted by incorporated communities; this made each schoolhouse the nucleus of "an alcohol-free zone eight miles in diameter"—a zone also sonically free of the hilarity and brawling common to areas with saloons and liquor stores. In the same mode of moral surveillance, New York City had passed a law in 1893 that forbade saloons within two hundred feet of churches and public schools; reformers hoped to extend this law to "the dirty little candy stores and cigar shops that are found adjoining the girls' and boys' entrances of the public schools." By 1905, the city had in place, among a number of other unenforced noise regulations, a Section 530 that forbade peddlers from bawling out their wares within 250 feet of a church during services, a courthouse in session, a school during the schoolday, and a hospital at all times.[395]

Two years later, in June 1907 with the passage of the aldermen's ordinance, everything would be different, just as it was that summer at Coney Island, where barkers for one day were denied their ten-block-loud megaphones. Now official signs would be posted around the city; now alert patrolmen would insist that horses be walked, newsboys hushed, and street vendors rerouted around hospitals; now the Sullivans would be looking on, and Mark Twain would prowl for quiet in his white flannel suit. Right. The signs were small ("NOTICE—HOSPITAL STREET"), the police force too small to keep vigils in Quiet Zones, and Twain, when at home in his dark, cathedral-vaulted premises on Fifth Avenue near Washington Square, was dictating his autobiography, caring for sick daughters, and listening to synthesized music piped in over his telephone line, "as loud as if an orchestra were on the spot," from Thaddeus Cahill's Telharmonium, a two-thousand-switch electric keyboard occupying the entire floor of a building a mile and a half uptown.[396]

For Isaac Rice, whose reputation as a lawyer and venture capitalist had become entwined with the intricacies of rail, auto, and banking trusts, 1907 *was* different: Wall Street had been growing bearish as high-interest short-term railroad obligations became due then overdue and deposits in trust companies contracted. By autumn, after the collapse of a scheme to corner the copper market and the failure of the American Ice Company owned by C. F. Morse (who had used stock in one New York bank as collateral to become a controlling shareholder in another bank, and another), a panic had set in that so compromised Isaac's investments that he and Julia accepted an on-the-spot offer for Villa Julia, which they sold for $600,000 to Solomon Schinasi, a Jewish-Turkish cigarette mogul. The family took a suite at the elegant St. Regis; Julia thought it was "wicked" to spend a hundred dollars a day on a hotel, but Isaac had to keep up appearances during the panic, lest the noises of rumor, innuendo, and economic insecurity jeopardize his financial standing. Within a year, the panic was over and his finances were steadied, but he had to withstand a month of highly publicized congressional hearings on charges of influence peddling and fraud with regard to Navy contracts. His name finally and entirely cleared, Isaac moved the family to eight rooms in the Ansonia, a Beaux Arts utopia on West 73rd and Broadway designed by one of the architects responsible for the Soldiers' and Sailors' Monument. The idea and funding for the Ansonia, a seventeen-story colossus of hotel suites and rental apartments laid out, it would seem, with artists and musicians in mind, came from W. E. D. Stokes, heir to the fortunes of Ansonia Brass and Copper, a company that provided the copper wire for America's telephones. Stokes had been lobbying for years for quiet asphalt paving on Broadway and bridle paths in Riverside Park, so he and Julia would have known (of) each other. As the Rices, their servants, and the several immigrant families they supported spread out into a dozen and eventually twenty-two rooms near the top of the Ansonia, they could rely on its three-foot-thick walls for soundproofing against sopranos down the hall and car horns sixteen floors below. In the Rice foyer sat a gift to Julia from a European admirer, a statuette of Silence, "the forefinger of the right hand placed gracefully on the closed lips."[397]

Across town, the 1907 panic stirred the Fifth Avenue Association to resist the march of skyscrapers that blocked out the sun, congested traffic, and threatened to further depress real estate. In Lower Manhattan, Florence Kelley of the National Consumers' League worked with the Fabian socialist Benjamin Marsh to form a Committee on Congestion of Population; they mounted an exhibit in early 1908 that dramatized the viciousness of crowding and the virtues of planning commissions that controlled for density and land use. Over the next eight years the Fifth Avenue Association, the Committee on Congestion, and the City Club of New York (which had been active against dirty streets and noisy pushcarts) pressed for local zoning laws that would not be vetoed by the Governor or struck down by the courts; this they would achieve in 1916. Zoning, then, had a momentum separate from Julia's Society, and an appeal not exclusively acoustic. It was a strategy, like Roosevelt's new National Parks, at once territorial and temporal, a means for certifying the value of a place and securing it from encroachments industrial and demographic, mass-mediated and political. It was also, in literature, art, and commerce, a means for making the most of terrains that spilled over traditional borders or experiences that outsped them.[398]

Nabbed for motorcycling through Manhattan at high speeds (aboard a somewhat "noiseless brand of machines"), teens Polly and Tommy were lectured in the fall of '07 by a judge who advised them to abandon their fast life and embrace the efforts of such a one as Mrs. Rice, who was doing so much to stop unnecessary noises. Their mother was *that* famous? The Rice kids did not cease their racing and were understandably less impressed than were Health Commissioner Darlington and Police Commissioner Bingham, who at Julia's request had joined forces to silence "honking autos, tooting boats, and metal-laden trucks," and the barking mutts of the Bide-A-Wee Home for Friendless Animals. Within months the commissioners concluded that a more enforceable set of regulations was needed before most noises could be muted. A new anti-noise ordinance was passed in the fall of '08 under the aegis of Little Tim and two other aldermen, Samuel Marx and Max Levine, after hearings during which "the Downtrod and Uppressed" had their say about bowling alleys, blacksmiths, tomcats, and streetcars. At the hearings Julia sat "all in black net, a black hat and occupying [Alderman] Reggie Doull's own mahogany desk," at the head of a company of thirty other powerful women from the Equal Suffrage League, the Legislative League, and the Mt. Vernon Women's Republican Club, to assure that the noise nuisance was taken seriously and to direct all unraptured ears to another playing of Emerson's records of the streets of New York.[399]

Adults to one side, the SSUN was smiling on an associated project to which Twain had lent his name. Twenty thousand children in public schools and the Free Synagogue were now engaged in composing pledges of respectful quiet around hospitals—no pounding of fences with baseball bats, no roller-skating, no kicking of cans while hanging around waiting for the excitement of an ambulance, its sirens at full roar. "I *promise*," wrote a child eager for a membership card in The Children's Hospital Branch of the SSUN and a big button promoting HUMANITY in blue on white, "*not* to play near or around any hospital. When I do pass I will *keep my mouth shut tight*, because there are many invalids there. Nor will I make myself a *perfect* nuisance." The children may have been "annoide," but her brother-in-law Joseph would not have minded the misspellings. It was the purity of the sentiment that counted, and for Julia, who was "eager to leave nothing untried to help the sick," there was the additional reward of "teaching the young the beauty of thoughtfulness for others."[400]

Young and old everywhere: Pittsburgh, Philadelphia, Boston, Baltimore, London, Paris, Berlin, Munich, Rome. Each harbored a faithful remnant of earlier Anti-Noisites; each had a nucleus of prominent citizens and physicians with whom Julia had been corresponding and from whom would arise new anti-noise cells, secondary SSUNs. Pittsburgh and Philadelphia, of course, had Imogen Oakley, whose concern with noise would not let up even at her death in 1933; a posthumous collection of her free verse featured "Midnight in the City," whose hush was broken by pounding trolleys, by giant trucks "leaving in their wake / A piteous wreckage / Of wakened children and wan-eyed sufferers for sleep," by motorcycles "Cracking, snapping, exploding, on their nerve-destroying way." Boston had Morse, Charles Eliot Norton, and a cadre of other noise-naysayers: Edward Everett Hale, chaplain to Congress; William Everett, former congressman; the MIT professors W. T. Sedgwick, biology and public health, Charles Ladd Norton and Charles Cross, physics; the Harvard professor James J. Putnam, neurology and psychiatry; Dr. Agnes Vietor, public health; Dr. Clarence J. Blake, otology. Among the most vehement was Hollis Godfrey, a graduate of Harvard and MIT, science director at the all-girls' High School of Practical Arts in Dorchester, and all-around popularizer of science, who described metropolitan noise as beating "on the wearied ear with whips of strident steel"; he made the muting of noise a priority for *The Health of the City* (1910). In 1908, at a downtown hotel, the leaders of Boston's women's clubs had listened in consternation to samplings from Julia's trunk full of recorded New York noises, but by 1911 the wealthy women of the Back Bay had formed a Women's Municipal League that hoped from the start "to induce women servants, poor wives and mothers of the tenement house districts, and women who work in stores, shops and factories" to join with them "on equal standing" in efforts to improve the milk supply and to eliminate such unnecessary noises as empty cans rattling in milk carts trundled over cobblestone, revolving chimney pots "squeaking incessantly for want of a little oil," and the "weird cries of cats." Marian Peabody, chairwoman of the League's Committee on the Abatement of Noise, would admit in 1914 that "an impression seemed to start almost simultaneously with the starting of our Committee that we did not and could not accomplish anything," but they had been in touch with Julia Rice, with Imogen Oakley, with Anti-Noisites in London and Berlin, and people all over Boston knew already "where to come for advice and assistance about noise." Meanwhile, the Committee's thirty members were "patiently and tactfully" laying the groundwork for an ordinance erecting quiet zones around hospitals. If his Honor the Mayor had spent several weeks in a hospital "and liked the noise, and [said] that he thinks as people have always got well in hospitals in spite of the noise they will continue to do so," Elizabeth Lowell Putnam and others of the Boston aristocracy were certain that "the great majority of sick people are irritated and their recovery delayed, perhaps even prevented by the needless clatter of an empty tipcart trotting homeward over cobblestones—the shrieking of a car wheel on an unoiled curve—the wild bellow of some motor siren—the shouts of children fresh from school right under a window where life or death perhaps depend on sleep or waking." Alleycats likely would not cooperate with the Committee's efforts, but the Lowells, Putnams, Peabodys, Minots, Bazeleys, Porters, and Crowninshields were all on board.[401]

Baltimore had Dr. William T. Watson, chair of the Anti-Noise Committee of the City Medical Society, which, energized by the SSUN, began to inventory the city's noises and anti-noise ordinances. "Many old-fashioned citizens," commented the *New York Times* in 1912, had joined the campaign; so had the American Federation of Labor and the Jewish Educational Alliance, whose new-fashioned members rallied against the loudness of factory machinery and the frequency of factory whistles. Inventories were dodgier here, for while Julia in New York had marshaled her allies and evidence so as to defuse suspicions of ethnic bias, there was more than a touch of anti-Catholicism in the Baltimore "crusade" against church and convent bells, more than a touch of racism in the rank(l)ing of grievances about noise. Among the patients at the Union Protestant Infirmary, for example, first came crowing roosters and cackling hens, then barking dogs and yowling cats (owned or fed, it was assumed, by blacks), then hucksters (black and white), then "Noisy schoolchildren, black and white," "Negroes quarreling in the alley," the "Noisy play of Negro children," and "Negroes singing until after midnight." Although Watson tidied up the list for a talk on noise to the Medical and Chirurgical Faculty of Maryland, there remained the problem of "Noisy Negroes in alleys" at night and "Colored street evangelists" on Sabbath afternoons. Subject to the Jim Crowing of Mayor James H. Preston, who fought as hard to segregate black and white block by block as he did to improve the streets, air, water, and parks, Baltimore in 1912 had half a million people now, a sixth of them African-American, a sixth foreign-born, and it was more difficult than ever to keep them from crying their "Wa-a-termillon" or blowing fish-horns to announce the catch of the day. Thanks to Mrs. Rice, the port officers of Baltimore, which had thirty-four steamship lines and the largest Atlantic fleet outside New York, could call upon waterway supervisors to enforce the whistle-blowing provisions of the Pilot Laws. Thanks to Morse, Rice, and Abbott, residents who lived near Union Station with its four hundred trains a day or within wailing distance of freight yards knew that something could be done about "the bells, the blowing off of steam minutes at a time, the poof, poof, slow and measured, then faster and faster; then the unearthly shrieks of the whistle, which awakening one from sleep, give the impression," wrote Mrs. Clarissima Mabbett, "that the engine has determined to come into the room and take possession." Thanks again to Julia, Baltimore as a dense metropolis of one hundred thousand dwellings—the center rebuilt and paved after the Great Fire of 1904—had by 1911 established quiet zones around its hospitals, sanatoria, and asylums. And thanks to Dr. Watson and fellow physicians, Baltimore in 1914 put Maurice Pease in uniform as the first noise cop in America.[402]

Off now to England on the White Star Line's flagship, *Adriatic*, accompanied by mocking sirens and whistles from tugs, ferries, and launches. "Somehow the skippers had learned that Mrs. Isaac L. Rice of the Anti-Noise Society and her four daughters were on board," although the *New York Times* in this summer of 1907 had to admit that "most of the tugboat men like Mrs. Rice and have very generally fallen in with her ideas about the advantages of quiet navigation." The mocking chorus, and the route out of the harbor, gave Julia the idea of writing to John D. Rockefeller to see what could be done about the whistling of Standard Oil's many tugs that crossed between New Jersey and New York. While her children went down to the engine room to stand "half-hypnotized by the huge silvery throw of the cranks sweeping magnificently round and round in broad flashes of light" (wrote daughter Marion, who in 1913 would be the first woman at MIT to graduate in chemical engineering), Julia could discuss anti-noise tactics with fellow passengers Charles McKim (architect), Herbert Nathan Straus (son of Isidor of Macy's and nephew of the Secretary of Commerce), and the New York financier Harry B. Hollins (friend and frequent host of His Royal Highness the Prince of Wales).[403]

Prince George of the House of Windsor was not at the docks to greet Julia or Harry Hollins, but she could count on the noble remnants of a London Association for the Suppression of Street Noises that had resumed Babbage's prosecution of strolling players. During the 1890s, the numbers and noisiness of street musicians seemed to have grown despite the 1864 Bill for the Suppression of Street Music and a series of subsequent bylaws. The apparatus of justice could not keep up, it appeared, with technical advances in the fatal power of street organs, whose range differed as much from those of fifty years ago "as the range of a Martini-Henry rifle differs from that of the old Brown Bess." The Association had also gone after junk men who, defying the law, roared for "rags and bottles" at all hours. "Distracted" of Blackheath meanwhile complained to *The Times* of the continued "blowing of steam whistles for calling workmen together," specifically prohibited by Act of Parliament. American Anti-Noisites had a habit of citing England for legislative precedent, but Anti-Noisites in Britain found their own laws toothless, esteeming instead the bite of decrees Russian and German, Austrian and Australian, even Italian. When Morton Arendt, Honorary Treasurer of the SSUN, arrived in London that July, he was stunned by the latitude granted milkmen and peddlers, telling reporters, "If we had a street in New York as noisy as the Strand, we should soon take steps to effect an alteration." Despairing of the noise, England's finest writers were reputed to be deserting the metropolis; lesser authors, engaged with "penny dreadfuls," became leading members of the Society for the Suppression of Street Noises: Charles Fox, editor of *The Boys' Standard*, and Max Pemberton, editor of *Chums* and *Cassell's*. Also involved in anti-noise activity had been Laura Ormiston Chant, a purity crusader and suffragist whose attack on music halls and the Empire Theatre's promenade of prostitutes was the fore-end of a campaign against street pianists and organ grinders. Then had come a medical conference on the harmfulness of noise, headlined by the widely quoted Theophilus Hyslop, M.D., and sponsored by the Street Noise Abatement Committee of London, aroused by the visit of Mrs. Rice to act upon protests from such as the journalist and playwright George R. Sims, author of *How the Poor Live*, who had called London's "snorting, rattling, rumbling, horn-blowing motor-traffic" a "Gargantuan-Gilbertian Grotesquerie." Julia herself liked quoting George Bernard Shaw on the worst of noises being those that are "State-aided," such as churchbells and military bands: another reason to lobby for national anti-noise laws. When she returned to England in 1909, she arrived with the bearing of one who knew her center of political gravity: in deference not to Mrs. Cornelia Vanderbilt but to his esteemed passenger Mrs. Rice, the captain of the *Mauretania* foreswore the traditional steam whistles on leaving harbor. Julia arrived as the official American delegate to a conference of representatives of the anti-noise organizations of England, Holland, Denmark, Austria, and Germany; really, she was the presiding muse. Receiving Julia on behalf of the Betterment of London Society were Sir Theodore Martin, the nonagenarian lawyer, poet, and counselor to the late Queen, arm-in-arm with Thomas Bowden Green, an engineer whose earlier cause had been the National Thrift Society. Martin and Green had been scouting out duchesses, dukes, and lesser blood to endorse a petition requesting of Prime Minister Asquith "early legislation in reference to unnecessary and objectionable street noises." Noise, however, had less priority for the Liberal (soon Liberal/Labour) government than did the creation of a system of old age pensions and National Insurance. Even a deputation of London bobbies demanding shorter hours on the beat so as to better recuperate from the noise of traffic did not result in broader anti-noise laws, this despite a surprising call for A Society for the Suppression of Senseless Sounds from Britain's finest avant-garde journal, *The New Age*, upset by the "No-man's land" of "foulness and discord" that lay on the other side of the double windows of the most comfortable, and artistically adventurous, of London abodes.[404]

Paris already stood in the avant garde of Anti-Noisism, or so wrote those who would be ill-prepared for another, contrarian avant garde, Futurism (coming soon!). Years before Julia was on the international scene, the city had thought fit to reduce the *charivari* of urban life by muffling street criers during the day and coachmen at night, licensing the corps of organ grinders down to an armful of accordionists, and recasting the steam whistle so that it "begins with a soft musical note and swells slowly to one sharp, clear intonation that would not awaken a baby slumbering directly under the whistle." In 1905, the symbolist poet and social reformer Adolphe Retté, amid his conversion to Catholic royalism, dreamt of a curious silence on Paris streets and of an invitation to "the funeral of Mr. Noise, who died last night. Killed by contemporaneous excesses, he was held in horror by the Eternal Himself." Julia herself, in her many prior visits to Paris, had heard just one steamboat whistle along the banks of the Seine, and a few months before she returned to Paris with the SSUN also risen, the "ingenious, sometimes seductive" novelist Marcel Prévost anticipated and applauded Mrs. Rice as a "Queen of Silence" who had allied the "fine élite of the scientific world, the political world, and the fashionable world." Prévost had the highest respect for her, he told the readers of *Le Figaro*, but "I have asked myself, long before she did, why the noisy folk, who are so few, victimize with such ease the lovers of silence, who are legion." He knew why beaters of carpets, layers of paving-stones, and drivers of piles chose to make their noise at daybreak: "by outraging the morning stillness they think they usurp for the moment a prerogative of royalty." This made sense to such a social climber as himself, but what school of architects had agreed to fashion apartment buildings so that an upstairs neighbor undressing at a late hour seems "to hurl upon the floor a pair of paving-stones in guise of boots; then a pair of breeches made of zinc, a cast-iron waistcoat, and a leaden jacket"? Prévost was asking in the guise of the engineer he had once been, but also as a self-professed annihilist, that is, as one who believed that "Noise must be killed, annihilated." A man ever dubious of the role of the New Woman as half-virgin, half-dynamo, he looked forward to meeting Mrs. Rice, who, so far as he could tell, had the qualities necessary to conducting "a warfare sublime." As "*la chère prophétesse du silence,*" Julia may have been able to prove to the editors of the *Journal des débats* that noise kills more people than typhus, cholera, and measles combined, but in Paris in 1907 the matter of churchbells had yet to be settled. Prévost himself had complained for years about the "booming" of discordant churchbells that sounded the quarter hours long into the night and woke one at dawn's earliest light. And if Gustave Charpentier, a composer whose friends and feelings ran rather with bohemian anarchism than bourgeois annihilism, flung open his Montmartre windows to the cries of carrot sellers and chair menders that would figure as a collective voice and political noise in his popular opera *Louise* (1900), by 1912 booksellers along the Seine were organizing a chapter of the *Amis de Silence* to protest traffic noise so horrid "that one can no longer saunter with any pleasure...along the quays where the book-booths used to invite to blissful browsing."[405]

Little time for browsing, anyhow, although on their many transits through Paris, Isaac and Julia would acquire a large collection of French first editions. From 1907 to 1914, Julia had as many appointments to keep on behalf of her SSUN as did Isaac on behalf of his subs; both were selling silence (streetside or subaqueous) and security from attack (noise or dreadnoughts). In both cases the sales went more smoothly in Europe than America, for the kinaesthetics were more confined, the sense of motion through space more constrained by centuries of borders and walled cities, so that Europeans were readier to consider a communal regulation of acoustic waves and the construction of navies beneath the waves. In fact, more European than American laboratories had been working to improve submarine design, so Europeans could better appreciate Electric Boat's (i.e., John P. Holland's) advances, and more generations of Europeans had been working to suppress noise, with legal and cultural precedent, so they could better appreciate the successes of the SSUN. Most appreciative was Theodor Lessing, who graciously translated into German the poetry of Julia's daughter Muriel and who, inspired by Julia's relentless letter-writing (three to four hours a day), hoped to maintain a wide network of corresponding Anti-Noisites. Progress was also being made in compiling "blue lists" of quiet hotels; paid-up members of Lessing's Anti-Noise League had access as well to urban retreats, "houses of silence" devoid of pianos and parrots.[406]

Upon each return to the States, Julia had yet another noise-suppression tack in mind. As so often happened when Americans went to Europe to study, negotiate, tour, or confabulate, they came back across the Atlantic recharged, ready to take on a New World that appeared at once more deranged and more daring than the Old. The SSUN in consequence would beat down more fiercely on churchbells and on that "concentrated extract of noise" known as the klaxon. "Honkhonking," which drove Anatole France out of Paris, would lead Julia to count honks by the half-hour (656) in Piccadilly Circus in London and persuade the Association of Licensed Automobile Manufacturers to help her get a handle on the bleating in New York, where the number of registered motor vehicles was multiplying exponentially, from 500 at century's turn to 62,660 in 1910. Out of the shadow of the SSUN would also come Quiet Zones posted around grammar schools to "Save the Children By Killing Noise," an offensive affirmed by S. Josephine Baker, now Director of the Bureau of Child Hygiene, by William H. Maxwell, New York's Superintendent of Schools, by nine thousand teachers and principals, and by the student body of a Brooklyn school surrounded by trolleys on one side, wheelwrights on a second, a scrap iron junkyard on the third, and on the fourth a stoneyard.[407] Most famously, she

SOUNDPLATE 21

put the SSUN at the high noon of a national rescue operation to save little rascals and fearful mothers from the mortifications of the Fourth of July.

17. All in All

All these pages on Julia Hyneman Barnett Rice stand as necessary prologue to a full sounding of her efforts to damp the noise of Independence Day. However great a lost cause, her campaign was neither huff nor hubris. It was a considered response to a widening consensus, in the years before the Great War, that noise—variously defined—was everywhere, everywhen, and adversely affecting everyone. That consensus had arisen from within turn-of-the-century physics and philosophy, science and engineering, literature and law, architecture and medicine. It owed its momentum to notions of and motions toward efficiency and sufficiency, liveliness and civility, education and independence. It had to do with expectations of illness, obligations of audience, and declarations of a right to privacy. More than an earful of perceptions or wavelengths of sound, the vibrations that were noise could be felt literally and figuratively throughout bodies social, cultural, economic, and politic.

Precisely because noise had so many extra-acoustic elements, skeptics could claim that noise grievances were a question of nerves. If Mrs. Rice had found that "all the cities of Europe manage to make more din in one day than New York does in two," then she should learn her lesson from "the placidity of Continental town folk in the midst of their thunderous, screeching, booming, whizzing, rattling, grating, shrieking environment that noise is more a product of the nervous system than of air vibrations," and reorganize the SSUN into a SAWN, a Society for the Abolition of Weak Nerves. Or a question of politics in disguise: the *Detroit Free Press* thought it important to state that the "representative taxpayers" who were eager to suppress streetcar noise around the city were "not radicals nor anti-corporationists but of substantial business standing, fully capable of giving even a street railway company a square deal." Or a question of perception: if three voices are shouting, then "the noise becomes noticeably stronger when a fourth is added, but if thirty are heard, one more, or even five more would not be heard; ten more would have to join to make a perceptible difference." And so on, explained Hugo Münsterberg, professor of experimental psychology at Harvard. If three hundred noisemakers produce a noise, one hundred more would be required to make a noticeable difference, which datum reflects a universal law of accommodation: humans become inured to any sensational extreme encountered on a daily basis. It is a modern superstition, wrote Münsterberg, to blame the complexity of civilization for epidemics of nervous strain, since we all "become insensitive by adaptation to our tumultuous surroundings. When we return from the mountain woods we hear the roaring of the city for a day or two, and then it sinks below our consciousness and no longer harms well-adapted nerves." Indeed, "nervousness itself is an illusion," as is that figment called the "subconscious." Yet when it came to the long-term effects of noise, Münsterberg's own research proved that accommodation was never benign: "The laborer, for instance, usually believes that a noise to which he has become accustomed does not disturb him in his work, while experimental results point strongly to the contrary." That is to say, the noise of machines, "which in many factories makes it impossible to communicate except by shouting, must be classed among the real psychological interferences in spite of the fact that the laborers themselves usually feel convinced that they no longer notice it at all." It followed that those who could not stand the noise around them would do well, for the sake of productivity if not mental health, to press for the removal of offending sources. As did the hero in D. W. Griffith's 1909 silent film, "Schneider's Anti-Noise Crusade," who late one night, struggling to finish a speech for the morrow, catches burglars in the act but lets them proceed once he finds that they are stealing exactly those household devices whose noises had kept him from concentrating on his task.[408]

Still, a patriot might protest that quieting whistles, horns, or even churchbells was one thing, the Fourth of July entirely another. There might be a vogue now for "silence" cloths to put under dinner plates on your sideboard;

for hunting shoes "built on scientific principles, to secure noiselessness;" for Corkolin flooring, warm and "noiseless"; or for the Siwelclo Noiseless Siphon Jet toilet with a flush you won't hear "outside of its immediate environment"...but a silent firecracker? There might be a push for soundproof rooms beyond the homes of neurasthetes and millionaires, as in the labs of psychologists studying perceptible differences and ornithologists studying birdsong, or in hospitals and asylums, or in music stores where patrons could listen to piano rolls before buying them...but what would the Fourth be without hard drumming, shrill piping, cowbells, and festive explosion?[409]

Good question. From the start, there had been two schools of thought about Independence Day. The first reveled in bands and parades, roasting and toasting, and assorted artillery common to earlier civic festivals; the second, in the minority, held out for a more pensive event akin to a Puritan Fast Day, during which to ponder the meaning of freedom. George Washington, who inaugurated the celebrations at his Brunswick, New Jersey, encampment in 1778, had gone for a bit of both: church sermons; artillery. For Manhattanites the two schools harked back to an era before the island was English, to a proclamation read out by Dutch Calvinists at New Amsterdam City Hall on December 31, 1655: "Whereas experience has manifested and shewn us, that on New Year's and May days much Drunkenness and other irregularities are committed besides other sorrowful accidents such as woundings frequently arising therefrom," henceforward there shall be on those days no "Firing of Guns, or any Planting of May Poles, or any beating of Drums, or any treating with Brandy, wine or Beer." Not that the small fines of such ordinances had much effect, but the lines were drawn, and would stay drawn, less by class or party than by creed—Calvinists, Quakers, Unitarians, Universalists, Spiritualists, and some Episcopalians on the quieter side of the Fourth, about which, for obvious reasons, comparatively little is known. Contrast this to the experience of ten-year-old Noah Blake, a Connecticut farmboy who walked into town in 1805 for a reading of the Declaration of Independence and recorded in his diary that he had "Never heard so many bells and cannon shots"; the horse his father bought that day just had to be named Bang. Of an older generation, Leverett Saltonstall in Salem shuddered at such noisy days: "I could never see the necessity of disturbing all sober people to celebrate the independence of the Nation." A lawyer, flautist, Federalist, and soon a Unitarian, Saltonstall found it "horrible at the first dawning of the day, when you are enjoying a morning slumber to be disturbed by that confused sound—and to have your sleep afterwards murdered." At the nation's fiftieth anniversary, in 1826, the Virginia poet Hiram Haines, waiting with everyone else in Petersburg's Court House Square, was stunned at ten in the morning "when off went in the most sudden and unexpected manner the little Iron Cannon—'Bomb! Bomb! Boo'—and such a rattling and clashing of tavern bells as was never before heard on this momentous occasion." Henceforward, the Fourth would be louder longer, a noise of thirty-some hours, with squibs and musketry from sunset on the Third throughout the night and day and evening of the Fourth on into the morning of the Fifth. "This is a most odious day here," wrote the New England novelist and devout Unitarian Catharine Maria Sedgwick in a letter of July 4, 1836. "All our liberty seems made into gunpowder, and noise to be the only expression of rational beings." In New York City, where the "terrible noise and racket" of the Fourth included "Innumerable discharges of fire arms, millions of rockets, and tumultuous enough effervescence of juvenile patriotism," *The National Era* in 1847 was impressed rather by the absence of rioting and "no infringement in the slightest degree of public decorum." Noise without brawl or bloodshed was the liveliest possible evidence of the virtues of democracy, though Maria Dyer Davies in Macon, Mississippi, among her Methodist family of small planters, thought "It would be behind the age [in 1854] to devote a day—all of it—to shooting guns—making speeches—eating barbecues—&c—for the love of those old men who are dead and buried long ago." In Illinois that year, Daniel Brush, the founder of Carbondale, had his own stock of "skyrockets, Roman candles, torpedoes, firecrackers, magic wheels, wriggling serpents, etc.," for the joy that he, like other boyish men, took in such noise as was transcribed by a young correspondent to *Robert Merry's Museum*: "Cr-r-r-a-a-ck!—fizz!—pop!—bang!—'The day we celebrate'—bang!—'our nation's birthday'—fiz-z-z-z!—whis-s-s-h!—'Bunker Hill, and spirit of '76'—bang! . . . 'now and forever, one and inseparable'—siz-z-z-z!—boom!—'who fought, bled, and died in freedom's cause'—E Pluribus Unum forever!—bang!—who-o-o-s-s-s-sh!—pop!—pop!—pop!—crack!—who-o-s-s-s-sh!—fiz-z-z-z-z-z!—s-s-s-spuk!—whis-s-s-sh!—bang!—crack!—whiz-z-z-z-z!—boom!—who-o-s-s-s-sh!—siz-z-z-z! fiz-z-z!—s-s-spuk!—who-o-o-o-s-s-s-sh!—BANG!!!!!!!"[410]

Accidents would happen, as the Council of the New Netherlands knew they would. One of Brush's rockets fizzed into an open box of fireworks. Serpents "began to hiss and up the street they started squirming and jumping . . . The magic wheels rolled and tumbled, the Roman candles shot forth the best they could, the crackers all popped at once, and the torpedoes with loud reports exploded." The inadvertent pyrotechnics came off as a climax to the rites of freedom, but within a decade, after the shellings and maimings of the Civil War, fireworks would seem more ominous, and the quiet school received a closer hearing. In Thomaston, Maine, Hannah Hicks awoke on the Fourth of 1864 to battery guns "popping away right merrily"; long distressed by "the wicked Southern rebellion which is more than I can submit to calmly," she did not feel "one bit jubilated today but very sober," so went on no picnic. Instead she visited the graveyard and Union dead. "Fireworks of *Death*," a Union soldier camped near Petersburg called the army show that Fourth, which otherwise "passed off, quietly, with us here. The men are tired, and are resting. We had no stupid speeches about the Spread Eagle, and the old [illeg.], the nigger, and Independence, etc., etc. Everybody was *sober*, here; no noise, no hurra's." In 1865, less than three months after Lincoln's assassination and guns firing funeral salutes each minute for hours at a time, Benjamin Brown French in Washington had little heart for a rowdy Fourth. He who had been chief marshal at Lincoln's first inaugural heard "constant uproar of guns, pistols, crackers, etc.," and "commenced the day by discharging out of the chamber window the five loads which I put into my revolver on the day that the President died. After that I *was silent*."[411]

Damping down the noise of Independence Day could be a matter of gravitas: if quietness was the *New York Tribune*'s sine qua non for any holiday, quiet was all the more requisite for the "national sound" of the Fourth; as early as 1842 its Unitarian-Universalist publisher, Horace Greeley, had editorialized against the "spirit of Pandemonium" that seized the nation each July and stunned the ears of the crowds, for "When a Society cannot celebrate its greatest day in a more worthy manner, it is a proof that it is destitute of all sentiment of art, unity, and concert of action." Or damping down could be a matter of redefining independence as did one newspaper editor who, "On the morning of the recent Fifth of July,—while the incessant crackling of crackers was fracturing the sunny air on every side," was watching and listening to a new sewing machine whose "flying lever and whirling wheel delivered a music not certainly so inspiring, or so brilliant, as that of the instruments off in the square, but whose cheery strain kept caroling on" as a song for the "Independence-Day for Women," freed as they now were from the drudgery of mending. Damping could be a matter of the diffractions of class: the Fourth was the only summer holiday for most workers. Unlike the privileged already at ease in their summer houses, those getting off work on the evening of the Third tended to whoop it up with drums and fish-horns and the pranks of youth—running up the street randomly ringing doorbells and tossing squibs. Or so ran the stereotype in cartoon and joke, such that the owner of a Massachusetts textile factory in 1870 thought fit to boast about the decorum of *his* workers during the Fourth, "when everything was quiet; when there was a smile on the face of every one and no one was in liquor." Yea verily, damping could be a matter of Temperance: without demon liquor, people would be more circumspect or less fumbling with their fuses, and fire companies need not race, bells clanging, from one blaze to the next. Damping could be a matter of piety: during the 1858 wave of revivals, Methodist moanings trumped all fireworks; Stephen Albro, editor of the Spiritualist *Age of Progress*, would "not mingle with those who rejoice outwardly, by burning powder, marching with banners, and swallowing alcoholic enthusiasm, in honor of National Liberty, whilst their souls are bound with chains the most galling and tyrannous." Damping could be a matter of culture: as "lowbrow" entertainments diverged from "highbrow" after the Civil War, the highbrow hoped to substitute exhibitions for explosions, edifying lectures for flummox, and professional illuminations for the frowsy rockets of amateur squizzileers. Damping could also be a matter of age and personality: "Among people of refined taste and sensitive nerves, 'Going out of town to avoid the Fourth' is a phrase so common in my time that it ceases to awaken attention, and is taken as a matter of course," noted Julia Ward Howe, she of the "Battle Hymn of the Republic," an abolitionist, suffragist, and Unitarian, writing in 1893 in *The Forum*.[412]

Mrs. Peterkin, skittish and susceptible to sick headaches, pyrophobic and of a mind to imagine the worst, would have liked to go out of town. In lieu of fleeing on the Centennial Fourth, she had made her boys promise to usher in the day with only and exactly five minutes of hornblowing but, as usually happened with the Peterkin clan, things bum-tumbled toward comicatastrophe. In the din of bells and cannon at dawn, noon, and dusk and lusty imitations of donkeys braying, pinwheels would ejaculate prematurely, a twenty-six-hour fulminating paste have its day in the sun, and faces get blackened by gunpowder. Great-niece of the patriot Nathan Hale, niece of the orator Edward Everett of Gettysburg and numerous Fourths, sister of the Anti-Noisite Unitarian chaplain Edward Everett Hale, Lucretia Peabody Hale had it in her bones to play off the quiet school against the loud. When she thrust nervous Mrs. Peterkin ("We are all blown up, as I feared we should be") into the hoopla of the Fourth, Hale knew that the many readers of *The Peterkin Papers* (1880) would appreciate both the ebullience and anxiety of a day of spark, flame, whoop, and bang that could turn so quickly from hilarity to tragedy. "I wish people would show their patriotism in a different way," wrote one woman after the hullabaloo of the Centennial Fourth, but then spent much of her letter carefully capturing the rhythms of the noise: "firing of cannon, ringing of bells, &c., which one could stand but the lesser sounds by the boys—bunches of fire-crackers—all going at once—pop-pop-pop—right under our very windows—torpedoes—tin-horns and everything under the Heavens that could by any possible chance be brought to produce a sound, was called into requisition, and this confusion, din, and hurrah was kept up *all night*." Torpedoes, Bombshells, Grasshoppers, Geysers, Serpents, Sky Rockets, Dragon's Nests, Weeping Willows, and Devils Among the Tailors ("a rich stream of brilliant fire, then a fountain of reporting electric stars, ending with a burst of serpents thrown to fifty to sixty feet high and exploding there with loud reports") were readily available from Masten and Wells of Boston; Calliope Rockets (which "fill the air with loud, screeching whistles, that can be heard a great distance") and Screamers (which "ascend with a loud screaming noise, and when they reach the highest altitude explode with a heavy report") could be had from Consolidated Fireworks of New York. Hundreds of such firms relied for their annual profits on the thirty-odd hours of Independence Day, the eight or so hours of New Year's Eve, three-hour political rallies, and a not-so-silent Southern Christmas morning, but the Fourth was always most worrisome for those apprehensive, like Mrs. Peterkin, of cheap explosives causing the most sorrowful of "woundings."[413]

Brooklyn had an ordinance against Independence Day ordnance by 1877, which "Anti-Bumble" indicted as an assault by Dutchmen and immigrants upon John Adams and Samuel Adams, the Founding Fathers who had encouraged Young America "to make all the noise they choose on the Fourth of July"; anyhow, wrote Anti-Bumble, "This stuff about 'carnage, fire, &c.,' is all rot." But when a like law was enforced across New York City in 1878, the *Tribune* was delighted to report that while the mercury had risen into the nineties, no alarming fires had been sparked, and "Rarely—perhaps never—has the 4th of July been checked off the calendar with so little noise and so short a list of accidents." In the reckoning of the cost of the Fourth in terms of property, limb, and life, Chicagoans took the lead, due perhaps to the lingering resonance of the noise and devastation of the Great Fire of 1871:

> And loudly the fire-bells were clanging, and ringing their funeral notes;
> And loudly wild accents of terror came pealing from thousands of throats...;
> But louder, yet louder, the crashing of roofs and of walls as they fell;
> And louder, yet louder, the roaring that told of the coming of hell.
> The Fire-king threw back his black mantle from off his great blood-dappled
> breast,
> And sneered in the face of Chicago, the queen of the North and the West.

Chicago sneered back and rebuilt in brick, stone, and steel, with rails splayed out in every direction but the lake. The sounds of the city grew harder and more piercing as Chicago became huskier and more inflammable. So in 1884 the City Council—which had already forbidden train whistles except when "absolutely necessary," held boat captains to "three sharp, short sounds of the whistle to be given in succession, as quiet as possible, and not to be prolonged," and banned the sale of small packets of firecrackers—passed an ordinance that classified noises as "unavoidable," "tolerable," and "intolerable," then regulated the tolerable and assigned fines for the intolerable. The Mayor vetoed the law. S. J. Jones, a prominent otologist who had moved seven times in two years to escape Chicago's street noise (and who, claimed one wag, had at last made his house "quiet as a graveyard" by filling in all cracks with strips of rubber), encountered similar opposition from another mayor (with a "preference for street noises, as testimonials to the urban character of Chicago") when he began a citywide anti-noise crusade in 1900. No mayor, however, could veto the publication of statistics on the deadliness of the Fourth due to wayward sparks, wryneck fuses, mistimed explosions, concussions, and stray bullets: 3 dead, 1,074 injured, and $149,000 in fire damage across the nation in 1899; 14 dead and 201 injured the next year in Chicago alone. For 1903, the city's totals had risen with the noise level—thousands of caps had been placed on trolley tracks to explode "like the rattle of musketry when the wheels passed over them"; afterwards there were 19 dead and 314 injured, mostly children. Dismayed, the Chicago Amusement Association, which had supplied the *Chicago Tribune* with some of its data, decided to try a startling experiment in 1904: the Association would give unlimited free firecrackers to every child so long as they lit their matches in public parks while firemen, physicians, and policemen stood by. The result as of the sixth of July: 1 dead, 101 injured, and $80,000 less in property damage than the year before. As the *Tribune* cautioned, however, lockjaw from cuts caused by bad blanks and faulty crackers would slowly reap its own gruesome harvest, doubling fatalities nationwide, where the totals stood provisionally at 52 killed and 3,049 injured.[414]

These were moving numbers, on four fronts: they grew from day to day after the Fourth; they grew nationally from year to year; they were larger each year than the casualties after the major battles of the Revolutionary War (as Julia noted, using statistics gathered since 1903 by the American Medical Association, also in Chicago); they were, for the most part, children. As Viviana Zelizer has shown, American valuations of childhood had been shifting toward a sentimental sense of the child as an open, suggestible mind and a tender, vulnerable body rather than a willful spirit in need of discipline and an economic asset in need of a job. Only one-sixth of children between ten and fifteen were working stiffs in 1900; a like percentage from the ages of five to ten were probably earning their keep on the streets, in shops, in mills and mines, on farms, or helping with piecework in tenement bedrooms. The rest, the majority, were no longer so "useful," although better-off girls had "baby" dolls, miniature tea and kitchen sets, and scale-model sewing machines on which to practice homemaking skills while better-off boys had small tools, steam engines, and complex mechanical toys advertised as child "quieters." (These to compensate for the more plentiful variety of noise-making playthings now on the market: air guns, cap pistols, miniature pianos, chiming Rough Riders and, for the fortunate few, electric trains and crying dolls that hushed when rocked not from end to end, "as a ship would toss in an angry sea," but from side to side.) Children became more "precious" as the average number of successful births per mother declined by two (from 5.42 in 1851 to 3.56 in 1900—Julia belonging in this respect to the older cohort). Working-class and middle-class parents were now obliged to attend to their children rather as challenges in the management of time, germs, and sound than as contributors to household income. Where before proper childrearing had entailed instruction in the sacred, now, as Felix Adler declared in his capacity as first chairman of the National Child Labor Committee in 1904, "childhood shall be sacred." The shift was neither clean nor sudden, and more urban than rural. Even so, and as one consequence, farmers and wage-earners began to take out life insurance on their children, to provide for decent, formal burials in a still-precarious world.[415]

Reformers who had supported child labor laws, school medical inspections, and prosecutions for cruelty to children were in the meantime scrutinizing another serious threat to children: "accidental" death. Accidental deaths had usually been agricultural, industrial, and adult; by 1910, as public health programs, cleaner water, better sewerage, and widespread vaccinations took effect, accidents had become the leading cause of death for children. Mortality between the ages of five and fourteen was more likely to come from being run over by trains, trolleys, and cars or blown up by misfiring "toy" cannon than from infectious diseases; and if not death, then crippling amputations. These statistics could be tied to the Fourth, as in a picture postcard of a boy, a girl, and a dog standing on the edge of a circle of exploding firecrackers: "Fotograph your boy before the 4th of July. You may not get a chance after." When, therefore, the art teacher Jeanie Lee Southwick addressed Chicago's 20th-Century Club in 1904 on the topic of "A Rational Celebration of the Fourth of July," her call for a safe, sane holiday was part of an Aristotelian, proto-modernist action against accidentals, i.e., all that does not inhere in the constitutive essence of a thing, being, or event. In the case of business and industry, accidentals were both sign and cause of dysfunction and inefficiency; in the case of architecture, costliness and obstacle; in the case of household economics, clutter and decay; in the case of tugboat whistles, confusion and interruption; in the case of Independence Day, disfigurement and death. It might be of the essence of a child or a Chinaman (wrote Philip G. Hubert in 1894), to love noise; as a mature and prosperous nation, Americans should have outgrown the savage notion of South Sea Islanders that "the louder the tick the better the clock." If Americans continued to import shiploads of cheap Chinese fireworks to celebrate the Fourth, they were dooming themselves to a barbarous vision of freedom. "We must begin everywhere, as regards the Fourth," agreed Charles Eliot Norton in 1904, "to insist that our town authorities should suppress the use of cannon crackers and other similar explosives, of toy pistols and other instruments of like danger, and the ringing of bells, and other noise in the streets before sunrise. If we secure this," he told Edward Morse, "we can then go on to provide the means for a rational and civilized celebration of the day."[416]

Decades before Julia Rice began to advocate "a joyful but sane" Fourth in place of the "old, selfish, strident, senseless method of celebration," the political, medical, social, and sacerdotal rationales for a quieter Independence Day had been laid out. Decades before the Russell Sage Foundation published her pamphlet, *Our Barbarous Fourth* (1908), others had been protesting its terrors and its attack upon the ears. Years before she published her manifesto *For a Safe and Sane Fourth* (1910), other Anti-Noisites had spoken out against the insanity. In 1903, the town fathers of Springfield, Massachusetts, outlawed personal fireworks on the Fourth; in their place arose battalions of marching boys and communal pyrotechnics. By 1906, the SSUN barely arisen, women's clubs were moving to substitute pageantry for pandemonium and the Playground Association of America was planning to launch a countrywide campaign for "A Safe and Sane July Fourth." Soon the Sage Foundation would foot the bill for the production of "a stirring drama of love and danger" for the movie houses, with a young widow as heroine and fireworks dealers as the villains, their hearts black as gunpowder. Why then did Julia feel compelled to take up a cause so well and widely underway without her?[417]

Had she a need, as the "apostle of silence," to speak for the voiceless, for the "tens of thousands of hospital patients" who grimaced at the approach of the Fourth, like so many children in magazine illustrations with hands clapped to their ears, and the millions of anxious, terrified mothers at home? "Yes," declared New York's Mayor Gaynor in 1913, "it makes everyone rejoice that the crippling, blinding, and killing of children and grown people by explosives on Independence Day is a thing of the past."—Did she feel obliged to defend her axiom that "Deep feeling is mute" and that "deep joyfulness and gratitude cannot vent themselves in wild ebullitions of boisterousness whose only feature, whose only aim, is din, din, din"? No doubt, as confirmed by the report of the Sane Fourth Association of Chicago on its Historical Pageant and Army Tournament of July 4, 1910: "the arena lights went out ten minutes after the performance began, and for fifty-seven minutes 100,000 men, women and children sat or stood quietly and without complaint, waiting," rewarded with flyovers by army pilots, rocket and balloon launchings, and troops scaling walls, building pontoon bridges, performing musical calisthenics.—Was it that she sought out a national reputation as "The Woman Who Reformed the Fourth" in order to prove that no noise, once shown to be unnecessary, was unassailable? Perhaps, for as President Taft said on July 4, 1911: "When we think of the way the Fourth of July has been observed in the past, and of the terrible consequences, it ought to make us blush that we have not taken means to stop it."—Was it that the success of reformed Fourths in Manhattan, Cleveland, Chicago, and cities farther west would allow her to draw upon the resources statistical and rhetorical of mayors, health commissioners, and governors across the country in order to advance new anti-noise initiatives? Quite likely, seeing as how, according to the *Boston Herald*, "About five years ago the noise was strained out of the Fourth of July with great success.... Anti-noise societies are now being formed all over the country, and if they succeed in their crusade we may have to pay 50 cents a day to see the home team play ball, instead of standing outside the fence and keeping the run of the game by the rise and fall of the madhouse chorus inside."[418]

Well, alright. But there was more to Julia's engagement with the Fourth than complementarity with the SSUN, or the mention of her name and cause in 12,000 lines of press coverage in the first six weeks after she announced her campaign. ("Mr. Jacob Riis," she wrote to Robert Underwood Johnson, now editing *The Century*, Riis the muckraking photographer whom she had met in Germany, Riis of *How the Other Half Lives*, Riis mayhap a distant relative of Rice, Riis "thought that perhaps you might be interested in hearing about the large number [of clippings] which have been sent me.") There was more even than the satisfaction of a warm letter from Woodrow Wilson, President of Princeton University: "I wish to express my entire sympathy with the interesting work you are undertaking for the suppression of unnecessary noise and the sane reform of our present way of celebrating the Fourth of July."[419] The Fourth presented Julia with a unique opportunity to make a point about noise that she could never otherwise have made so forcefully. Noise hurt. Noise had physical aftereffects. Noise, through its skyrocket surrogates, left a bodily trace, a wound. Noise, sometimes, killed.

Physicians could, and did, refer to cases of insanity, suicide, and heart attack vaguely attributable to noise; they could not point to autopsied trajectories of noise coursing through the body like bullets. "The Coroner's jury is yet to be found," said the *Chicago Tribune*, "which shall render a verdict of 'death from the city noise.'" Physicists could display tracings of uneven soundwaves produced, for example, by Dayton C. Miller's phonodeik of 1908, and Emerson the phonographist could play recordings of street noise; neither could show that visually irregular or auditorily excruciating sound did lasting bodily harm. Bad vibrations might affect one's mood, one's social and sexual relations, and the likelihood of contacting the Other Side through "the voice of the silence"; if, as with the shock of the 1906 San Francisco earthquake ("the air still quivering with the echoes of that stupendous noise"), bad vibrations could devastate buildings and china, no one had evidence of strictly acoustic destructiveness beyond the dogs, "couriers of the cataclysm," moaning and whining, jaws dripping, heading out of the hills.[420] Loud sudden close sounds could, everybody knew, cause temporary deafness and ringing in the ears, but permanent deafness could not be irrefutably, anatomically linked to blown-out eardrums or other otological damage caused putatively by noise. Julia the physician would have known all of this, would have appreciated the implications.

Misfiring blanks, bursting torpedoes, cans of crackers splitting open with flame and fury, serpents hissing at the wrong time in the wrong place—all of the accidents of the Fourth were the ringers that Julia was looking for: things essentially noisy that had, in countable and mounting instances, corporeal consequences; things sought out primarily for their noise that were demonstrably unnecessary and physically dangerous; things intended to be loud and transient that became, too often to be ignored, indelible; things that were the very embodiment of noise and which, flying off in all directions, could blind an eye, lodge in the brain, take the life of a child. A nasty business this, angering even the Fireworks King himself, Mr. H. J. Pain, whose Pain Manufacturing Company excelled in the spectacular reenactment of battles: the celebration of the Fourth, he knew for a professional fact, could be no less splendid for being less flammable and foolhardy. It was sheer hoodlumism to set off giant firecrackers fourteen inches long or "shooting canes" stuffed with potassium chlorate, sulfur, and powdered glass, all of whose noise might register itself, in a flash, on bodies. Not solely by shock ("Crackers' Noise Kills Baby; St. Louis Boys Scare It to Death with Giant Explosives"). By manslaughter: the arc and figure of noise.[421]

Time finally to consider—Julia was quoting the *Minneapolis Journal*—"how our annual worship of the God of Noise is to be abolished. This blatant and death-dealing Divinity long ago usurped the shrine occupied by Patriotism." Time, said Twain, to do away with "the bedlam frenzies of the Fourth of July," which Americans, said Howells, celebrated "like a nation of lawless boys." No clamor overtook the Swiss during their Independence Day: "As night descends, the bells on all the churches are set to pealing in a sublime concert of gratitude, rising with penetrating poetry through the serenity and softness of a summer night." From the City of Brotherly Love Imogen Oakley wrote, "Any celebration that carries with it a train of ambulances, doctors, and nurses is neither safe, nor sane, nor civilized." Thirty hours of BANG!!!!!!! was not even traditional: during the Early Republic—Julia had done her research—people listened to patriotic poetry and readings of the Declaration of Independence, rang a few bells, enjoyed a thirteen-gun salute, watched small rockets go up, and went as families to the theater. And had anyone bothered to ask modern children what they wanted on the Fourth? Baseball, ice cream, trees to climb, pie-eating contests—and the rich serving dinner to the poor, suggested one of the thousands of girls and boys Julia interviewed in 1912. So "Let us, on this day, forget the noise of battle." The alternative—noise shrieking its way like shrapnel through the flesh and bones of the young—was just too horrible.[422]

& now to war.

ROUND THREE

Everyhow

On hearing what had not been heard,
could not be heard, should not be heard.
Calibrating and recalibrating noise. Toward what end?

Rounding the Great Barrier Reef in 1606 and threading his way through "an archipelago of islands without number" off the coast of Papua New Guinea, Luis Baéz de Torres found no gold or spices to redeem the long voyage from Peru, and no sign of *Terra Australis*, the fabled southern continent, so had his crew snatch up a score of Islanders, "black people, very corpulent and naked... that with them we might be able to give a better account to Your Majesty," Philip III. In fact, Torres sailed within view of Australia's northernmost peninsula but mistook its low hills for more of the archipelago. The Spanish admiralty kept his report a trade secret, hoping it held clues to the whereabouts of a new continent as rich as the Americas. Months earlier, Willem Janszoon, sailing from the opposite direction, had sketched the first European map of any segment of the Australian coast, then skirted the western edge of the three hundred sandbars and cays of what one day would be called the Torres Strait. Supplies running low and nine of his crew killed by Islanders, he turned back. Other Dutch ships from the East Indies came after, measuring distances between sea and shore by cannon shot, between bodies white and black by musket. In 1644 Abel Tasman claimed western Australia as "New Holland" but stopped short of the Strait with its treacherous shallows. Not until 1770, after a British fleet took Manila from Spain and its hydrographer fell upon archival copies of charts drawn by Torres, did Europeans return to the Strait. The English came to take possession of those parts of Australia still deemed *terra nullius*, "no-one's land" (like that which Captain Cook named New South Wales), then to offload, over the next eighty years, 100,000 convicts (thieves, forgers, fences, political agitators) sentenced to transportation. William Bligh, who had sailed under Cook and seen him killed in Hawaii, knew the waters best: set adrift by the mutineers on the *Bounty* in 1789, he and eighteen crewmen had "by God's grace" negotiated the Strait in a small launch. In 1792 he came back through as commander of the *Providence,* using his cannon as much for their noise as for the damage that

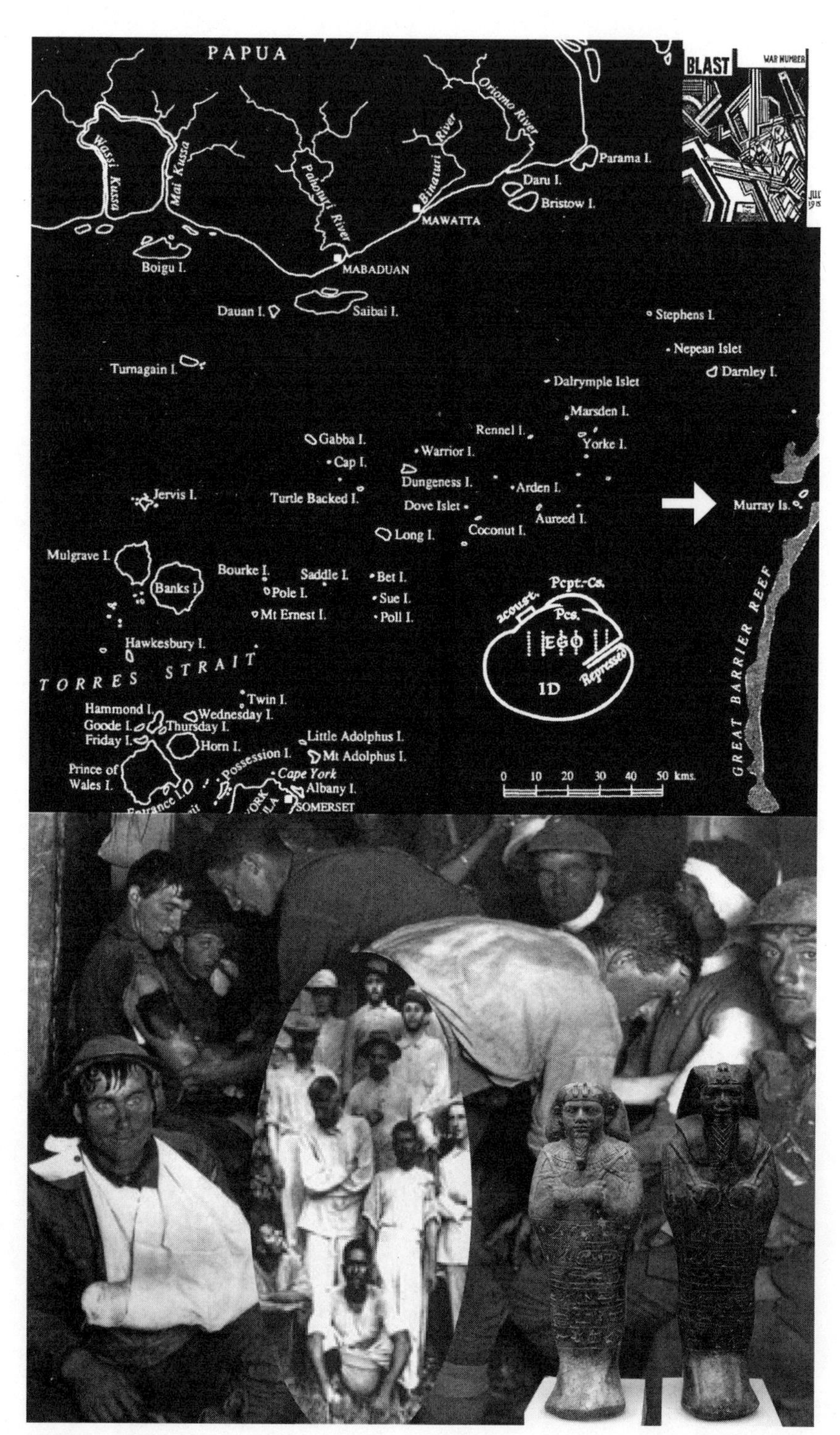

SOUNDPLATE 22

four-pound shot might do to outrigger canoes half the length of his sloop. The Islanders were defending fragile pools of fresh water, Bligh a cargo of breadfruit seedlings from Tahiti destined for Jamaica, to feed African slaves working plantations of sugar cane.[1]

Whoever these whitefellas were, ancestral ghosts or shipwrecked magicians through whose skulls, smoked and cured, the future might be divined, they were loud and careless of the ear. In the Australian languages of the Cape York peninsula and the Western and Central Islands of the Strait, and in the Papuan language of the Meriam of the Eastern Islands, thunder had given rise to humanity and the ear was the seat of intelligence. If the ear was *milka* (as it was near Cape Bedford), then to lack an ear was to be stupid, *milka-mul*; if to be unable to hear was to forget, then to break open the ear, *milka-bandandaya*, was to reconceive one's life. So the Islanders listened to the exhalations of the dugongs, conversed with the Eclectus parrot, and sang what they did not say, with glissando and trill. They danced to lizard-skin drums or *warups*, hollowed-out bamboo *thrums*, and tree-nut rattles. They swung bullroarers to catch the voices of hero-gods. They shouted at corpses to let the dead know how they felt, and the spirits of the dead whistled back, in code.[2]

I am paraphrasing a monograph of 1903, *Superstition, Magic, and Medicine*, by Walter Roth, Northern Protector of Aboriginals. Created in 1897, the Protectorate over the next decades would forcibly remove thousands of "half-breeds" from their Aboriginal mothers to mission "reformatories." Roth himself was partial to the Islanders, whom he considered superior to Queensland Aboriginals and whom he tried to protect from the predations of Asian and English freebooters. By 1897, most of the four thousand Islanders had taken up Christianity as the better part of valor and pidgin as a needful tongue through which to negotiate with Europeans; on these grounds, Roth accorded the Islanders, especially the Meriam with their stately tortoise shell masks and orderly gardens, more respect than his charges on the mainland. It would be decades before Islanders relinquished their totems and stone shrines, their loops of skulls and mortuary cannibalism, but Roth had hopes for them so long as they found other amusement than singing "absolutely filthy" songs.[3]

Easiest with demands to lay aside raids in favor of trade and skulls in favor of mother-of-pearl were the selfsame Meriam, whose 1.5 square miles of Mer, unlike most islands in the Strait, had a fertile soil that inclined them toward the calm of subsistence farming and reef fishing. Named Murray Island by the captain of the *Pandora* (dispatched to capture the *Bounty* mutineers, shipwrecked in 1791 while retracing Bligh's route), Mer lay near the Strait's eastern entrance, convenient to the exchange of trochus shells and bananas for iron blades and fish hooks. After "The Coming of the Light" with the missionary ship *Surprise* in 1871, Mer harbored a Papuan Institute where worship and woodshop throve for a dozen years and a written language was generated to record the stories of Eastern Islanders: *ma kari nasoli*, "you hear me," declarative. The Meriam for their part learned to hide from the eyes of a Samoan convert, "teacher Josiah," agent of the London Missionary Society, who had children and adults flogged with stingray tails for violating the Sabbath or wearing too few clothes: *ma kari nasor*, "you listen to me," imperative.[4]

Though their numbers were declining from the eight hundred they had been at mid-century, the Meriam did not forget who they were, descendants of Malo, the shapeshifter who had come from Papua and whose cult the men secured by the banging of giant clam shells, the shriek of bullroarers, and the loud slapping or sucking sounds of octopus Malo. Nor did the Meriam forget how long they had tended to Mer—seventeen generations and more—with the laws of Malo in mind: "Malo's hands are not grasping, he holds them back. He walks on tiptoe, silent, careful." Nor did they stop lengthening the ears of their children with weights and piercing the margins at puberty with holes for pendants of shell and wood. But they could not avoid the syphilis and measles brought by the outlanders, who by 1880 had a hundred boats operating in the Strait and three depots on Mer; nor could they ignore the abductions of their women and conscriptions of labor at gunpoint. Such violence had been accelerating since the 1850s as beds of pearl shells and sea slugs (aphrodisiac *bêche-de-mer*) attracted boatloads of divers from Japan, Polynesia, and the Philippines; as a gold rush in Queensland drew thousands of Europeans, Chinese, and Malays through the Strait; as Anglican and Lutheran missionaries and their South Sea protegés put local pagan shrines to the torch.[5]

Witness to the epidemics and violence was an English zoologist, Alfred Cort Haddon, who had spent a year in the Strait in 1888 exploring its marine biology—much of that time on Mer as a guest of the missionaries Hunt and Savage. Haddon returned in 1898 on a different and urgent mission, convinced that comprehensive ethnographies must be undertaken before the peoples of the Strait were destroyed by the West in its drive for mother-of-pearl buttons, pearl-handled pistols, sex, and power. His father, grandfather, and uncles had directed a firm that published Baptist tracts before branching out to the brokerage of African produce, his mother had composed *Stories of the Apostles* before branching out to nature writing in *Woodside, Look, Listen and Learn* (1887), but Haddon had quickly shed the piety of his youth to become an agnostic concerned rather with ethics than evangelism. After his first visit to the Strait he was already upset about the decline of "right living" everywhere, in the British Isles as on the islets between Australia and New Guinea. "They have taken away our traditions," he said, in sympathy with "lowly people" East and West, "they have taken away our folk-lore, they have taken away our gods, and I know not where they have laid them." It was Haddon who went in search of what was being taken away, Haddon who transposed "fieldwork" from the natural sciences into anthropology, and Haddon who brought into the field, to Mer, four medical men whose voices would soon be heard grounding British anthropology, founding the discipline of social psychology, and sounding out the terms of shellshock.[6]

1. Ethnosonographies

W. H. R. Rivers, M.D. and Cambridge professor of experimental psychology, measured the eyesight of the Meriam; then, initiating the genealogical method of anthropology, he traced their lines of kinship while his friend Haddon measured their bodies, learned their customs, collected their art. William McDougall, M.D., pricked and pinched them to assess their susceptibility to pain; then, on the trail of social psychology, he went off to observe the head-hunting tribes of Borneo. Charles G. Seligman, M.D., took stock of their general health, their drugs, their categories of illness, and their techniques of healing, inventing medical anthropology. Charles S. Myers, M.D., measured their senses of smell and taste, their reflexes, and their hearing; then, as a violinist drawn toward ethnomusicology, he analyzed their dance and song. Sidney H. Ray, a self-trained philologist, parsed their language. Anthony Wilkin, a young Egyptologist, sketched, filmed, and photographed their dwellings, their tools, the layouts of their gardens, and their rites. At forty-three, Haddon was the oldest. The mean age of the group was thirty—the average life expectancy of the Meriam. All but Wilkin (†1901 in Egypt) would make it through the Great War.[7]

That Rivers had specialized in the physiology and psychology of the senses and that four of the seven—McDougall, Seligman, Myers, and Wilkin—had studied with him at Cambridge, this was apt to a central enterprise of the Expedition: testing the "fairly unanimous" ascription to "savage and semi-civilized races" of "a higher degree of acuteness of sense than is found among Europeans." Like his companions, Myers found no evidence that the Meriam could register phenomena that "escape the most acute European," although "the constant rustle of palm-leaves and the beating of the surf" made tests of hearing difficult. And if by "acute Europeans" were meant himself, Rivers, and Seligman, by whose acuity he calibrated the distance at which an adult should be able to hear the ticking of a clock or stopwatch, well, on Mer the three of them suffered from "below par" health—and dulled hearing, said Rivers, and likely some tinnitus from dosing with quinine to ward off malaria. In addition, the Meriam were such strangers to ticking that it took them a while to localize the sound, let alone to indicate the distance at which the ticking was lost in the rustle of coconut palms. Further, Meriam men had had their ear drums repeatedly ruptured during years of diving for shells, deeper and deeper, to satisfy foreign demands for mother-of-pearl; the repeated tympanic trauma had made many of them hard of hearing or deepened a deafness caused by epidemics of measles brought by those foreigners. Finally, when testing the ability of the Meriam to discern differences in tones, Myers used tuning forks whose scale was familiar to children taught Western hymns in the mission school but not to older folk, who sang their traditional songs in what Myers heard as wavering, indistinct pitches with "scant feeling for tonality." The Merriam could tell a puff of wind from the highest pitch of a Galton whistle (no mean feat), but their hearing as adults seemed collectively poorer than that of Seligman, Rivers, or Myers's control group, a passel of Aberdeen Scots.[8]

In sum—and this was the thrust of all seven folio volumes of the Expedition's reports—how well the Meriam saw, heard, or felt was a function of what they needed to see, hear, or touch in their own world. How they led their lives and mourned their dead was a cogent response to what the sea and the Strait demanded of them and what they as heirs of Malo-Bomai asked of a land less and less insular. (In 1898 Haddon found a new road circling Mer, built by Queensland prison labor.) That Haddon hurtled through a lifelong stammer; that Rivers was the son of a country-curate-hyphen-speech-therapist who conducted sessions with Lewis-Carroll-hyphen-Charles-Dodgson and six of Dodgson's stuttering sisters and was himself a stutterer yet a university lecturer; that Seligman suffered from a weak heart that preoccupied him lifelong with issues of fatigue but did not hold him back from trips to Ceylon and the Sudan—these biographisms informed their reports no less than that Seligman and Myers were Jews critical of racial stereotyping. Like Rivers, McDougall, and Haddon, Myers and Seligman shared a wariness of predestinarian thinking, biological, social, theological, or political. The expedition had "been planned as something far bigger than a rescue party to a decaying culture," wrote Myers; the Meriam were primitives *and* coevals. Their history, environment, and resilient networks of kin had always to be taken into account. For Myers (whose mercantile father was one of nine children and whose musical mother one of fifteen), the senses of the Meriam were neither inferior nor superior to those of Britons; they were differently oriented, differentially favored. Neither surrogates for intellect nor the cause of its suspension, their senses were honed by minds that grasped the ecology of their milieu. "Take the savage into the streets of a busy city and see what a number of sights and sounds he will neglect because of their meaninglessness to him"; on reef-fringed islands, it was another matter, for "the savage will take an interest in, and make an inference from a sound, which may be meaningless to, and perhaps neglected by the European ear." Primitives did have a higher threshold of pain, reported Seligman, friend and then relation to Myers (by Charles's marriage in 1904 to Edith Babette Seligman), and their threshold between dailiness and dream was more "pervious." Yet, as Seligman would conclude, "the fundamental lesson taught by the Torres Straits expedition was that institutions and the ideas underlying them could only be studied adequately"—listen to the weave of coordinating conjunctions—"when the interrelationship of the individuals concerned had been learnt *and* classified, when the part each played in regard to his comrades on any particular occasion had been studied, *and* the relation of ceremonies *and* customs to each other had been understood, *as well as* the purpose *and* relative importance of each in the social fabric."[9]

So too with Jews as a people, or perhaps a race. If there were identifiably genetic components to Jewishness, these were conditioned by degrees of education and complicated by annals of intermarriage reaching back to dark-eyed Hittites and blue-eyed Philistines. These annals were pursued by Redcliffe Nathan Salaman, a Jewish physician who regretted not having joined his friends Myers and Seligman on Mer. Salaman did obtain from them strands of Papuan hair to enrich his comparative data, and his ties to the two became even closer when Seligman in 1905 married his sister Brenda, one of fifteen Salaman siblings. Thereafter, and at length, Salaman discussed with his brother-in-law the photo albums of Jewish faces he spent years compiling. As there was something steadfast to being Meriam (whose features were described as handsomely "Semitic"), there was something steadfast to being a Jew which Salaman found in physiognomy and Myers in a love of the arts, adeptness at science, devotion to children. Neither man believed that Jewishness, physical or cultural, predisposed to virtue, to liberalism, or to socialism, and neither at first worried that a genetic, anthropometric, or aesthetic affirmation of a Jewish "race" would inflame an antisemitism already obvious to Zionists like Theodor Herzl, who had witnessed its virulence in such centers of civility as Vienna, Berlin, and Paris. After the Great War, however, as Seligman, Salaman, and Myers endorsed programs to improve the health of working-class Jews and campaigned to convince Parliament that tighter restrictions on Jewish immigration would disadvantage the nation, they met with an antisemitism strong and persistent enough to require, by the end of the next war, a collaborative study by psychologists and sociologists. Chairing that study and presenting its findings would be Myers's final public achievement.[10]

Haddon had come to Mer in 1898 already critical of racial prejudices, Social Darwinist presumptions, and British imperial practices. His sympathies lay with a regenerative socialism that could restore workers to their right livelihoods and honor a people like the Meriam, whose wants were few and easily supplied and who refused to "work for the white man more than they care to." If the Meriam made mummies of their dead (and not long ago had eaten the hands and feet), their ceremonies for the departed were more moving than the "ghostly mockeries of the funeral plumes of the professional upholstery [of modern undertakers] which have only lately been abolished in England." Unique among contemporary ethnographers, Haddon credited his informants for their help and called them "old friends," to whom he returned time and again with a conviction, shared by Rivers, that colonial outposts staffed by a corps of empathetic anthropologists could protect "primitives" from a marauding West. Rivers came to socialism more gradually, inspired by the social democracy he saw on Mer and the communalism he saw across Melanesia but also by a notion of what it would take for moral sensibilities to regroup in a world when ethics were foundering on the shoals of an abrasive capitalism. Two weeks before he died, in 1922, Rivers defended socialism from critics who sneered that the best the masses could manage was mobocracy. What of Melanesia, he asked, where "all the members of a social group of considerable size are allowed the use of the produce of the cultivated land of the group regardless whether they have taken any part in the cultivation or have contributed in any other manner to produce the commodities"? That was extreme, but anthropologists had come across other peoples working peacefully together, and wasn't the animus against worker solidarity a distinctively European hysteria?[11]

When the Expeditionary Force left the Torres Strait after seven months of intensive inquiry and extensive collection, the Meriam became restive. The Force, wrote a colonial administrator in 1899, had "set them thinking." They began to make noisily explicit those rights (a Western term) traditionally implicit. Meriam resilience had been strengthened by the Force's requests for the remodeling of masks that had been destroyed by or hidden from missionaries, reenactments of forsaken or secret rites, the digging up of lost or buried cult objects, the resumption of abandoned or deferred ceremonial dances. Haddon had fixed upon Mer not simply because its rich soil allowed for permanent villages whose inhabitants would be available to undergo testing but because of the "attachment of the Meriam to the past." He and Rivers relied upon that attachment to gather up genealogies and myths. The Force entire encouraged the Meriam to recover much that was on the verge of vanishing and so, by one of the ironies of a cultural conservatism promoted by moderns, reclaim a politically cogent present. Known as an independent sort, the Meriam from then on were at the forefront of protests when Islander integrity was threatened: they walked off Queensland company boats when they found they could not use them as they wished (1913); killed the Protector of the Islands during a dispute over compulsory medical inspections (1935); led a maritime strike in 1936; joined an Army strike of 1944; and convened Islander meetings in 1988 to discuss secession from Australia. In 1992, Australia's High Court decided for Edward Koiko Mabo, a Meriam litigant, in his land tenure suit against the government of Queensland. Mabo and his lawyers had contended that the Meriam had no concept of vacant land and that each of their eight clans had well-marked garden plots whose boundaries had been settled through inheritances reaching back centuries. Ergo, the principle of no-one's land by which the British had taken possession of Mer (and the rest of English Australia) was inapplicable. The island was not a *terra nullius*; it belonged to the Meriam, and they unendingly to it. The proof lay in 3,489 pages of testimony by living Meriam; in the exercise book given by Sidney Ray to Pasi of Mer, who in 1898 had written down "fifty-nine pages on Meriam myths, custom and classificatory systems"; and in the *Reports of the Cambridge Anthropological Expedition to the Torres Straits*, over which Mabo had pored while a gardener at James Cook University. *Terra nullius*, ruled the High Court, would remain determinative with regard to political sovereignty, but it had no merit with regard to the gardens of Mer. No-one's land was a fiction of the brigand.[12]

2. *Sonotherapies*

Or the fiction of a war to end all war, where no-man's land was another, louder killing ground, a *terra nullius* where self-possession was as much at risk as were regimental banners slipping in the mud. Where high explosives buried soldiers alive next to bodies skewered with shrapnel. Where shards of skull glowed at night under Very flares. Where Blaise Cendrars saw his sergeant "blasted in mid-air by an invisible ghoul in a yellow cloud, and his blood-stained trousers fall to the ground *empty* while the frightful scream of pain emitted by the murdered man rang out louder than the explosion of the shell itself." Where—but that's more than enough just now about the Great War, which began in August 1914 and was sure to end before snow fell that autumn.[13]

Shellshock was not new to the Great War. It became of moment because the war did go on. Because, as it went on, it became entrenched, men on both sides crouching under longer, fiercer bombardments from larger, louder shells. Because, as the trenches lengthened, the war drew down the numbers of the draftworthy and sapped the morale of infantry, so that medical questions about how to treat tens of thousands of exhausted, deafened, or nightmare-riven soldiers became military questions of how best to patch them up and return them to an interminable Front.

Already during the Boer War, which went on not four months, as expected, but four years (1899–1902), a civil surgeon attending English soldiers had reported sixty cases of "general nervous shock, immediate and remote, after gunshot and shell injuries." Similar to "soldier's heart" during the U.S. Civil War and to the "railway spine" of passengers in the aftermath of collisions, "general nervous shock" was a disorder of time and sound, not a lesion of the brain, heart, or spinal cord. As such it shared their problematics, "the absence of evidence of outward and direct physical injury ... the obscurity and insidious character of the early symptoms ... and the very uncertain nature of the ultimate issues of the case." Many of Dr. Finucane's cases arose from the battles of Colenso and Spion Kop, British disasters both, and instances of a new kind of warfare disconcerting in its arrhythmias and invisibilities. "The battles in this campaign," wrote one British officer, "do not consist of a few hours' fighting, then a grand charge, resulting in the rout of the enemy, when men can see the effect of their work. No; this is very different. Think of it, a two-mile march under the fire of an invisible foe, then perhaps eight or ten hours' crouching behind any available cover—an ant hill or a scrubby bush—when the slightest movement on a man's part at once enables the hidden enemy to put him out of action." A battle with ghosts. At Colenso, Capt. Walter Congreve knew only "little tufts of dust, all over the ground, a whistling noise, 'phut,' where [bullets] hit, and an increasing rattle of musketry somewhere in front." Smokeless Mauser rifles pinned down the ill-prepared British, who suffered 143 dead and 1,002 wounded while the Afrikaaners remained hidden, distant, reloading their high-velocity rifles with light, small bullets that, whistling into the air at 2,400 m.p.h., could pass through one body and enter another. Among the wounded was a British corporal, a bullet lodged in his left ear. After successful surgery to remove the bullet, he had some residual deafness and tinnitus, both understandable, but as his ear healed he continued pale, shaky, slow, and headachy, losing weight and memory, "easily upset and put out," shivery and sleepless, becoming the wraith that was his experience of war. A private at Colenso, knocked senseless by a shell exploding a few yards away, was briefly paralyzed on his right side; his paralysis recurred upon removal to hospital in England, along with "noises in the head, dimness of sight, and some deafness," his reflexes "exaggerated with tremors of the hands and tongue" as if he were condemned to reenact the explosion. At Spion Kop, where the British lost 1,500 to the Boers' 350, a private was shot through the ear and blacked out. Making a good recovery from the physical wound, he was beset by "periods of stupidness and unconsciousness," headaches, and vertigo, signs of a "nervous breakdown" that he himself attributed to "shock and panic at Spion Kop, even before receiving his wound, as the result of heavy shell-fire and rifle-fire from the Boers, who were invisible and well under cover whilst the English troops were freely exposed."[14]

One might attribute such shock and panic to poor physical health, an issue during recruitment when a third of the candidates were rejected on medical grounds, and a greater issue after the war, when a commission looking into British military preparedness learned that many regiments had been "physically very inferior and liable to disease—boys and weeds." Or was the shock attribuable to the dismayingly skillful use by the Boers of dynamite, long-range Krupp cannon, Mauser rifles, and machine guns on swivel-mounts, which together made for a sonically terrifying war? "You can't imagine anything so demoralizing," wrote Lt. Frank Isherwood from Colenso. "The awful sinking at one's stomach pit when you hear the nasty buzz, followed by a plomp and burst. Then there are nasty pieces of shell which wander about and are quite capable of taking one's head off." This was not the glory foreheard on a widely-cranked gramophone recording of 1899, *The Departure of the Troopship*: "crowds at the quayside, bands playing the troops up the gang-plank, bugles sounding 'All Ashore,' farewell cries of 'Don't forget to write,' troops singing 'Home, Sweet Home,' which gradually receded in the distance, and the far-away mournful hoot of the steamer whistle." Once the men arrived at the Cape, wrote another lieutenant, "There is so little to describe. The infantry soldier sees nothing except the men on either side of him and the enemy in front. He hears the crackle of the enemy's fire somewhere—he does not know where—and he hears the whit! whit! of the bullets, and every now and then he knows vaguely someone near him is hit—he feels the smell of the powder (cordite) and the hot oily smell of the rifle. He fires at the range given, and at the given direction, and every now and then he hears 'Advance!' And he gets up and goes on and wonders why he is not hit as he stands up. That is all.... No glitter—no excitement—no nothing. Just bullets and dirt." Under the pounding of Boer mortars that could loft ninety-four-pound shells from miles away, Sol T. Plaatje, a court interpreter and one of seven thousand Africans besieged along with British forces in the town of Mafeking, recorded only "Shelling—shelling—shelling—the monotony of the day." The Boer siege and bombardments, said Plaatje, "worried the souls out of us."[15]

"Worrying the soul out": what arrhythmias and dissonances did. Wrote one officer, about the shells, "Though at first they seemed innocuous, the noise they made, first the boom of the cannon shot, then the shriek of the shell two or three seconds after, and finally the terrific—or so it seemed to me—explosion... at last began to tell on my nerves." But this was still a war in which more soldiers died of disease than battle, and in which 70,000–100,000 Boers and black Africans (military and civilian, adults and children) died from starvation or illness in concentration camps. Of the 340,000 British regulars and irregulars who had sailed for the Cape, 72,000 were invalided back to England. The noise at home of a war at last won, the "Pax Pandemonica" of patriotic howls and "blatant senseless music-hall songs accompanied by horrible noise-producing instruments" during celebrations of the relief of Mafeking and capitulation of the Boers, would be antistrophed by Rudyard Kipling's "Lesson" of a war badly begun and cruelly ended. Having burned out thousands of Afrikaaner farms and slaughtered thousands of sheep and cattle, the victors left the veldt overcast with vultures, barbed by 3,700 miles of wire, and bolted into imperial place by 8,000 blockhouses of concrete and metal, so it was hard to accept what the poet Swinburne called a "peace with pride in righteous work well done." It was rather "The Last of the Gentlemen's Wars" and the end, wrote Kipling, of "all our most holy illusions." Some 30,000 British soldiers were hospitalized in South Africa with "rheumatism" or "disordered action of the heart."[16]

Major William Joseph Myers of the King's Royal Rifle Corps, educated at Eton and Sandhurst, veteran of the Zulu Wars and postings to Cairo, died in October 1899 at the Battle of Farquhar's Farm, drilled through the head by a Boer sniper. Charles Myers would learn of the death of his uncle shortly after his return from the Torres Strait and Borneo to assume a position at St. Bartholomew's Hospital in London. Six months later, his own health at dangerously low ebb, it was to his uncle's beloved Egypt that Charles repaired. There he passed a year of recuperation puzzling out hieroglyphics in the company of those "excavators of antiquities" who had helped the Major amass thirteen hundred pieces of ancient Egyptian art of "exquisite workman-ship, beautiful color, and, often, historical significance." While in Egypt, Charles also finished his medical thesis on *myasthenia gravis*, a "premature tiring of voluntary movements" often conflated with hysteria or neurasthenia, since autopsies of those few fatal (and working-class) cases showed no neuromuscular pathology. His interest in the topic stemmed from his own chronic fatigue, and his prescription for treatment was exactly what he prescribed for himself: *complete rest*, "confinement to bed in a large airy room," warmth, quiet, and calm.[17]

He was not one to lie abed. In addition to a trip by camelback to the Great Oasis, Myers took it upon himself to review *Naturalism and Agnosticism*, a series of lectures delivered in Aberdeen (1896–1898) by the occupant of a new Chair of Mental Philosophy and Logic at Cambridge, James Ward, a theologian manqué who had gone to Germany to study the physiology of nerves and returned to teach, first, psychology, then moral science. Ward had leanings toward neither classical logic nor that mathematical logic taking form at Cambridge under the chalk of Alfred North Whitehead and Bertrand Russell; from his lectures it was also clear that he had lost interest in the mind as a bundle of brain cells. He had become an Idealist, and a theist to boot, marshalling evidence for a "hypermechanical directive agency" that would explain why life had not exhausted itself ages ago in the entropy dictated by the Second Law of Thermodynamics. Neither physics nor biology nor Darwinian natural selection could account for, or supply, "rational insight" and "spiritual light." At the center of life must be "God only, and no mechanism." Such a claim, not entirely anathema to Myers the Maimonidean philosopher-physician, was rebarbative to Myers the experimental psychologist and physical anthropologist, who would spend months in Egypt and the Sudan measuring the heads of soldiers from ear to ear—the auriculo-frontal radius, the auriculo-occipital—to determine whether the Lower Nile had been settled, as some claimed, by two distinct races, one white, one black. Against Ward, Myers maintained that "neither the idealism of the teleologists, nor the naturalism of the mechanists is one whit the more real, the more adequate, or the more true than the other." What *was* real was "the unity of states of consciousness which we have called mind or experience" according to how we foreground the subject (the thinker and perceiver) or object (phenomena). The solution to the "world-riddle" was not "'God or Mechanism' but God and Mechanism" and "one undivided continuum of states of consciousness," just as he could show that it was neither a black race nor a white race that had settled Egypt but a single people whose physical traits ran along a continuum of variability that yielded more "negroid" features the further south one went toward the dryer, warmer Sudan.[18]

Myers's last installment of his "Contributions to Egyptian Anthropology" was printed in 1908, when he was back at Cambridge finishing his seminal *Text-Book on Experimental Psychology*. In the fall of 1908 another Jewish physician, moving along an uncannily similar path, was also in England. He who had turned professionally from experimental psychology/neurology to psychiatry, and personally from the Greco/Roman to the Egyptian, was about to broach the anthropological. Having finished his sessions with the Rat Man, Sigmund Freud spent the summer with his family in the Bavarian Alps, reading, writing, and gathering wild strawberries. On September 1, he set off alone for London to pass a week engrossed no longer in a case of obsessional neurosis but in the Egyptian cases at the British Museum. Ernest Jones, his first "Celtic" disciple, who had been impressed as much by Freud's skill at listening as by his theories of infant sexuality and who had attended the International Psychoanalytic Congress in Salzburg that April to hear the Master speak for five hours about the Rat Man, furnished Freud with a bibliography of British anthropology, but he was abroad that fall, so there was little in town to distract Freud from his antiquities. Not the opening of J. M. Barrie's *What Every Woman Knows* ("that when she trusts to her instincts and shuns reason and logic she is apt to be right"). Not the Nineteenth Eucharistic Congress, where six cardinals, fourteen archbishops, seventy bishops, and fifteen thousand lay Catholics celebrated the perpetual miracle—and psychodynamic enigma—of transubstantiation. Not even the "Vision of Salome" at the Palace Theatre, danced in the (ostensibly) nude (from Greco/Assyrian models) by Maud Allan—as teasingly erotic as *Gradiva*, that skirt-lifting woman of Roman bas-relief who had been reanimated in a 1903 novel that in turn had inspired a paper by Freud on delusion and dream.[19]

Since 1900 and the publication of his "Egyptian" book, *The Interpretation of Dreams* (dreams, he said, resembled hieroglyphs), Freud had been drawn toward the Old Kingdom of the hermaphroditic trinity Osiris/

Horus/Isis and, one way or another, toward the land of his fathers: "strange secret yearnings... perhaps from my ancestral heritage—for the East and Mediterranean." Released from an obsession with Rome and Greece and now "intoxicated with things Egyptian," he had become a collector of small Egyptian sculptures, stone reliefs, and funerary figurines, *shabti*, interred with mummies to do the hard labor of the deceased in the afterworld—just as he labored to relieve his patients of the oppressive weight of their past (never promising "that a freedom *from* history will lead to a freedom *in* history": Richard H. Armstrong). Unearthed in heaps by the same "excavators of antiquities" who stripped from the outer wraps of mummies their tempera portraits like the one owned by Freud that looked to be a self-portrait, dozens of *shabti* stood by him as silent auditors on the shelves of his consulting room, on his dining table, and in his luggage anywhere he went. Analogies between archaeology and psychoanalysis were clear and present to him: stripping away layers of memory; digging up occult images; piecing together glyphs and shards to recreate a dark past. He would hang an engraving of Abu Simbel over the analytic couch, invoke Egypt in his analysis of Leonardo da Vinci (1910) and return to Egypt near the end of his life in *Moses and Monotheism* (1937–1939), but his concerns with origins and mourning, fathers and sons, memory and race were leading him now to Myers's other outpost, Australia. There, the mental life of tribes "described by anthropologists as the most backward and miserable of savages" offered "a well-preserved picture of an early stage of our own development" and insight into archaic complexes that gripped modernkind. For this he would dip into papers by Haddon and Rivers; in turn, Rivers, Seligman, Myers, and McDougall would rely on his writings when called upon to treat the psychic wounds of the Great War.[20]

Am I free-associating, as Freud encouraged the Rat Man to do? Am I running us through a maze, like the famous laboratory rats at Clark University in Worcester, Massachusetts, where Freud, Jung, and Sandor Ferenczi lectured on psychoanalysis in 1909 at the invitation of G. Stanley Hall, doyen of American psychologists, who had recently advised a colleague to review the literature on shock?[21] What I am tumbling toward is this: Rooted in the fieldwork of the Cambridge Seven and embedded in the couchwork of Freud was a notion of originary violence whose noise was its most insistent repercussion and active residue. Sonically and psychically, noise was the keenest reminder of fraught beginnings, whether the shriek of a bullroarer or a catch in the throat. An analyst might suggest that the Cambridge Seven mated psychology with anthropology so as to cope with the disturbingly attractive forms of ritual cannibalism they met up with on Mer, and that the Viennese Master performed the shotgun wedding in *Totem and Taboo* (1912–1913) so as to cope with disturbances of totemic memory and violations of professional taboo that faced him each day in his study, where his fathers and his future met at the large ears of the *shabti* Senna, deputy of the deceased.[22] Or that the Seven, aggrieved by the violence of a paternalist colonialism whose effects followed them through densest jungle, and the One, projecting a series of violences physical or fantastic wherever children took on their parents, needed a theory of trauma and transformation in order to make something of the world other than impending collective catastrophe.

While the field anthopologists sought arrows of cultural diffusion and artefacts of cultural conflict among the twists of myth and turns of kinship, the armchair anthropologist sought out the logic by which intrapsychic conflicts are converted into physical disorders, not merely as hysterical symptoms but as constituents of a basic negotiation between mind, body, and family—generational, prehistoric, probably racial, possibly pan-human. Listening as intently as did the Meriam to the sea, the wind, and the voices of ancestors, and as did the Cambridge Seven to drums and tales, Freud was most keen on the noise of talk. If he claimed to be able to make little of the tonal relations in Western music, he made much of the sounds of speech, for his early research was in aphasia and his own memory was "phonographic": what he heard once he could repeat almost verbatim. As phonography took in every yawn and snuffle, so Freud attended to the inadvertent and excrescent, sensing in them things as yet untold that were sharply telling. A preternaturally fluent speaker, he was keen on Anna O's odd coughs and mumbles. Keen on Dora's nervous cough, metallic hacking, and asthmatic breathing. Keen on Frau Emmy von N's "spastic interruptions amounting to a stammer" and her peculiar "clacking." Keen on the racingphrases, and hesitaaaations of the Rat Man. Throughout the "talking cure," tics and tantrums of the tongue told Freud more about what had been buried alive or cannibalized in the psyche than any screen memory or dream displacement, for only the noise of that trauma had the power to pierce the layers of fustian and denial.[23]

Soon after Freud left England, the *Illustrated London News* featured "A New and Terrible Engine of War": shrapnel grenades that could be fired from military rifles. Grenades dated back to the Renaissance but they had burst open with poison gas during the Russo-Japanese War (1904–1905), during which Japan tried out the newest armaments and Russian doctors had to cope with a new group of soldiers whose psychoses were stamped "with a depressive character such as is not noticed in times of peace," patients lying motionless for weeks, staring at the wall, fearful, with "great disturbances of sensation." (Their faces must have resembled that of Oskar Kokoschka's *Self-Portrait as a Warrior* of 1909, a gaunt, harrowed bust in clay and tempera with hollows of deathly orange around the eyes.) Launched like small bombs, the new grenades exploded in fragments of metal that tore at the flesh where they did not kill. They were the physical expression of that nightmarish fragmentariness and noise which Freud had always to puzzle through in dreams, fantasies, phobias; they were also, like analytic *matériel*, intensely personal ("anti-personnel") weapons that could have generic points of origin.[24]

By 1909, fragments and noise were also being integrated into the cultural sound system. From Italy came F. T. Marinetti's first Futurist Manifesto: "We intend to exalt aggressive action, a feverish insomnia, the racer's stride, the mortal leap, the punch and the slap." From Munich came Wassily Kandinsky's "Yellow Sound," a music drama (unperformed until 1982) driven by the psychomotors of synaesthesia, with "intensely yellow giants" and "vague red" flying creatures crossing the stage to the shrieking of tenors. And from Vienna came this letter in the hand of Arnold Schoenberg, writing to the concert pianist and music theorist Ferruccio Busoni:

> I strive for complete liberation from all forms
> from all symbols
> of cohesion and
> of logic.

As it was impossible to have a single sensation at a time ("One has *thousands* simultaneously"), so musical expression had to be wildly concatenated, "And this variegation, this multifariousness, this *illogicality* which our senses demonstrate, this illogicality presented by their interactions, set forth by some mounting rush of blood, by some reaction of the senses of the nerves, this I should like to have in my music." Have it he would, in a 1909 opera for soprano and orchestra created in collaboration with a young lyricist, Marie Pappenheim. A dermatologist and poet, Dr. Pappenheim was a cousin to Bertha Pappenheim—Freud's Anna O. Writing in a white heat, she drew her plot and portrait from the case histories of Anna O and Dora, devising lyrics as broken and elliptical as the language of the women at their most tormented. The atonalities through which Schoenberg strung the lyrics seemed so illogical, so noisy, to prospective producers that *Erwartung* (Expectation) had to wait in the wings until the noise of the Great War had subsided.[25]

Fragments and noise were also critical to the tensions between Jung and Freud in early 1909. Son of a poor Calvinist pastor, Carl Jung had married into a wealthy Swiss family. Adept enough as a psychiatrist to have worked out the protocols of word association, Jung was the ideal vehicle, thought Freud, to lift the psychoanalytic movement out of its Central European ghetto and deliver it to a wider world as a new science that could not dismissed as a "Jewish national affair." Jung had inclinations toward another sort of mediumship, impelled by his mother's ghost-watching and a doctoral thesis on a cousin's paranormal powers. On 25 March 1909 he showed Freud his own command of "catalytic exteriorization phenomena," that is, the power to summon poltergeists or the telekinetic rattling of objects. Freud heard "loud reports" from his bookcases and inexplicable noises from his cabinets, which seemed to confirm Jung's mediumistic dexterity. The experiment, he wrote Jung three weeks later, "made a powerful impression upon me," so he had listened further into the matter, "and here are the results. In my front room there are continual creaking noises, from where the two heavy Egyptian steles rest on the oak boards of the bookcase, so that's obvious. In the second room, where we heard the crash, such noises are very rare. At first I was inclined to ascribe some meaning to it if the noises we heard so frequently when you were here were never heard again after your departure. But since then it has happened over and over again, yet never in connection with my thoughts and never when I was considering you or your special problem." The noises were not at Jung's beck and call. "My credulity, or at least my readiness to believe, vanished along with the spell of your personal presence; once again, for various inner reasons, it seems to me wholly implausible that anything of the sort should occur. The furniture stands before me spiritless and dead, like nature silent and godless before the poet after the passing of the gods of Greece." How he was trying to ennoble the loss of magic! How he was struggling to honor his heir ap/parent, whose mana was as close to magical as any Freud had known, while disabusing him of a free-floating mysticism: "I therefore don once more my horn-rimmed paternal spectacles and warn my dear son to keep a cool head." Sometimes noises were just noises, fragments of sound from a creaky universe, though it may be that "Every noise indistinctly perceived gives rise to corresponding dream-representations." As for unprovoked rattlings of shelves of *shabti*, this was no more proof of a spirit-world than was that number mysticism by which gamblers and primitives bank upon a responsive universe; to speculate otherwise would signal a return to animism, or to Australia.[26]

Jung and Freud blanched at an invitation to speak to assembled psychiatrists at the Australasian Medical Congress in 1911. Though fascinated with Aboriginals as avenues of access to the archaic or archetypal, neither went Down Under, in part because the Australians, oblivious to the impending rupture, expected a joint effort. The invitation would seem otherwise to have been well-timed: Jung in 1911 was amassing world symbols for his *Psychology of the Unconscious* and fixing upon a Central Australian fertility ceremony to prove the universal sexual symbolism of fire; Freud was absorbed in anthropologies of Aboriginals. Yet to both, Australia was less a continent in the Southern Hemisphere than a "phantom of psychoanalysis" in the left hemisphere. For Freud, who only once set foot outside Europe (to speak at Clark University and tour Manhattan, finding America "a big mistake"), Australia was a template of how some thoughts come to be unthinkable, some places unapproachable.[27]

Neurotics and primitives, wrote Freud on the first page of *Totem and Taboo*, have a similar mental life. Not only do neurotics regress to the animistic; their psychic processes return them to the primal scenes: noisy copulation, witnessed at an uncomprehending age; patricide; cannibalism. Freud appreciated that "even the most primitive and conservative races are in some sense *ancient* races and have a long past history," which meant that, like psychiatry, ethnology could never hope for unobscured glimpses of primordial ideas or anxieties. He was nonetheless as sure of his acoustic methods for reconstituting the prehistoric roots of ambivalence and the emotional logic of "hordes" as of his methods for retrieving the root conflicts of neurotics, who "may be said to have inherited an archaic constitution as an atavistic vestige." Although his evolutionary schema for civilization (animism → religion → science) may have been stuck in Late Victorian positivism, what he was saying in 1911 was quite alarming: Europeans at opera or Aboriginals in the outback must struggle every day to keep their psychic balance in the startlingly long wake of a singular trauma—the murder and ritual devouring of a father by his sons. All laws, deferrals of desire, rituals of compression and confession proceed from the inherited memory of that event and its suite of societal and psychological reorganizations. All ghosts and irksome spirits arise from the shared guilt in that cannibalistic killing.[28]

Footnote: "If a spirit is scared away by making a noise and shouting, the action is one purely of sorcery; if compulsion is applied to it by getting hold of its name, magic has been used against it." I mean to make more of this single footnote than it can reasonably bear, since I consider it, in analytic terms, overdetermined, and in archaeo-acoustic terms, undersounded. The footnote occurs in Freud's chapter on "Animism, Magic, and the Omnipotence of Thoughts"—that last a phrase contributed by the Rat Man. Freud considers animism the first "truly complete explanation of the nature of the universe," attractive enough to linger on in idioms and superstitions. Complementary to animism was a "body of instructions upon how to obtain mastery over men, beasts and things—or rather over their spirits." In an earlier footnote, Freud had quoted James Frazer (he of *The Golden Bough*) on rituals for warding off the ghosts of victims pursuing their murderers: "These ceremonies consist of beating on shields, shouting and screaming, making noises with musical instruments, etc." But Freud does not ask why noise would ever have been thought effective against such ghosts as, for example, a murdered patriarch, and in his later note he considers naming to be an advance over noisemaking. What was going on at the bottom of these pages?[29]

Five things. The five violences of echo: muffling, intensifying, truncating, distending, delaying. Psychoanalytically, noise re-minds of an act too terrible to be voiced other than in imitatively loud or choked, delayed fragments, invoking the originary violence. Noise distends by distortion and by introjection, the latter that habit of the neurotic taking serial fragments of the sounded world as onomatopoetic warnings, as did five-year-old Little Hans, Freud's constipated and horse-obsessed case of 1909, when associating the noise of a fallen bus-horse that slipped in the street and "gave me a fright *because it made a row with its feet*" with his own angry stamping before running to the toilet and dropping a "*lumf*" or turd. Truncating noise fends off vengeful spirits by obliterating the final syllables of accusations, aborting events, denying the consequences of desire. Noise repels by intensifying, in the same manner as does obscenity, which his friend Ferenczi was exploring while Freud was laying out *Totem and Taboo*: "dirty" words bare all, making shamefully public what should be private, audible what should never be heard: "the hearing of an obscene word in the treatment often produces in the patient the same agitation that on some earlier occasion had been produced by accidentally overhearing a conversation between the parents, in which some coarse (usually sexual) expression had slipped out." And noise can suppress, by overlay or dissonance, as used by schizophrenics in defense against intruding voices; Freud, who had finished his analysis of Schreber's *Memoirs of a Neurotic* just before turning full attention to Aboriginals, himself muffled the political and social-critical passages of Schreber's delirium, reading exclusively for the sexual, Oedipal, and mythological.[30]

Sorcery, then, was wildly reflective and insufficiently reflexive, colluding too closely with the unconscious. Like the "Antithetical Sense of Primal Words" described by the philologist Karl Abel, the noise of sorcery expressed "contraries by identical means of representation," echoing the primal violence. The double words that astonished Abel and intrigued Freud (e.g., *ken* for strong/weak) came from "primitive Egyptian," a language whose roots had also been collided into compounds "in which two syllables of contrary meaning are united into a whole" (e.g., "outside-inside") and, incredibly, words that reversed their meaning when their sounds were reversed (as "ruckus" reverses to "succour"). Freud's point, developed in yet another footnote, was that such sound reversals bore an intimate relation to dreams and the root ambivalences enacted by taboo and embodied in conscience. The sonic was so much in play that in 1912, completing *Totem and Taboo*, Freud found that "the totem-ambivalence question suddenly fitted, snapped shut with an audible 'click.'"[31]

Magic, in contrast, was farther from raw event, and quieter. The magus did not inflect an originary violence; he knew the secret names of things, and by naming them controlled them, his lexicon a new instrumentality for deturning guilt. If magic relied on a delusional omnipotence of thought and an animism of relations, at least it asked for insight and self-restraint; the next step was religion, which consolidated both and codified Necessity in the form of an inescapable mortality and omnipresent divinity. Last came science, which made of life an ethical and intellectual discipline neither daunted by nor disdainful of death. Harshest of antidotes to narcissism, death was no resolution, philosophical or psychological, to the hard memory of a primal coup and the trauma of cannibalistic murder, whether among a horde of brothers or the citizens of nation-states. While the analyst bravely exposes the incestuous fantasies of early childhood, the analytic process is meant to affirm life and reconcile instinctual desire with the economies of a societal self, not to prepare men for a Great War or a good death. When Freud published "Thoughts for the Times on War and Death" in early 1915, his son Martin was already on the Eastern Front manning artillery in Galicia, the Rat Man had been taken prisoner by the Russians, and all Europe was regressing to sorcery and noise. "Not only is it more sanguinary and more destructive than any war of other days, because of the enormously increased perfection of weapons of attack and defence; but it is at least as cruel, as embittered, as implacable as any that has preceded it. It sets at naught all those restrictions known as International Law, which in peace-time the states had bound themselves to observe; it ignores the prerogatives of the wounded and the medical service, the distinction between civil and military sections of the population, the claims of private property. It tramples in blind fury on all that comes in its way as though there were to be no future and no goodwill among men after it has passed." Freud wrote of "disillusionment"; what he heard, and felt, was shock.[32]

Having dealt with the most extreme of regressions and self-destructive of impulses, he was yet taken aback by the vehemence of a war that "has brought to light an almost incredible phenomenon: the civilized nations know and understand one another so little that one can turn against the other with hate and loathing." So much for science. Or religion. Or magic. It was back to primal ambivalences and a primitive silence: "The state exacts the utmost degree of obedience and sacrifice from its citizens, but at the same time it treats them like children by maintaining an excess of secrecy." The war confirmed for Freud the thorniness of the "ineradicable traces" of the killing "of the primal father by the company of his sons" and the impossibility of weeding out all "evil tendencies," akin to the impossibility of that perfect ear-cleaning about which Freud's disciple Karl Abraham had written in 1913, concerning the erotic fantasies and frustrations of children who stick fingers in their ears so often and pleasurably that they develop eczema or worse. Was it any consolation that "our fellow-citizens have not sunk so low as we feared, because they had never risen so high as we believed"? How attain a "higher plane of morality" when "the primitive mind is, in the fullest meaning of the word, imperishable"?[33]

Just one word: plasticity. Psychoanalytic experience shows "that the shrewdest people will all of a sudden behave like imbeciles as soon as the needful insight is confronted by an emotional resistance, but will completely regain their wonted acuity once that resistance has been overcome. The logical infatuations into which this war has deluded our fellow-citizens, many of them the best of their kind, are therefore a secondary phenomenon, a consequence of emotional excitement, and are destined, we may hope, to disappear with it." We may hope. And shell shock, that too he thought would disappear at war's end.[34]

Charles Myers advanced the phrase "shell shock" in 1915 while Freud was composing his "Thoughts." Myers soon regretted the term, which implied that there was a physical etiology to the constellation of symptoms that Finucane had called "general nervous shock." During his time with the Royal Army Medical Corps in France, it had become clear to Myers that in cases of disorientation and muteness where there were no discernible head wounds, the psychological impact of shelling was paramount and the most effective treatment was "obtaining persuasively the recall of repressed memories." Regardless, the phrase stuck. Co-founder and editor of the *British Journal of Psychology*, Myers's choice of words carried weight, and his four articles on the topic (1915–1919) in Britain's leading medical journal, *The Lancet*, each had "shell shock" in the title.[35] As a shorthand for the experience of modern warfare *and* as a diagnosis of trauma, "shell shock" made direct sense to the ranks at the Front and their working-class families at home, for whom the phrase was at first explanatory, then accusatory—what fools were sending their boys into no-man's land? Wrote a physician who worked for three war-years with a field ambulance, "Young soldiers prepare to become a case of shell-shock almost before the first shell drops near them. The very fact of the noise of the explosion causes a certain amount of emotional upset and that is sufficient to send them over the borderline.... To the soldier's mind it was as much an entity as scarlet fever, with the further addition that, being incurable, shell-shock was more to be dreaded."[36]

Stuttered into shell-shock and collapsed into shellshock, "shell shock" echoed the impersonality of the war, where enemy faces were illegible through the mud, smoke, and phosgene mist. With unheard-of rates and ranges of fire and with bombs dropping from Fokkers, Sopwiths, and Spads, the war played out as the obverse of a silent film: all sound and mostly darkness, with silence at the intertitles and sudden fiery close-ups as if a reel had sprawled off its sprockets and started to burn in the projection booth. Tanks, the newest of cavalry, "a strange regiment of monsters," seemed to Mary Borden, a poet and nurse near the Front, to be "Dragging themselves forward by their ears." (The men within the dark tanks drove half-blind; under barrage, the solid steel shook "like a wobbling jelly" in "an inferno of slammed doors.") From their beds in triage tents the wounded woke at each shellburst, eyes wide, ears erect. If, said a French guard, they fell back to dreaming, theirs was a sleep buffeted by implacable noise, *un bruit implacable*. Day and night what came through the dirt and smoke most distinctly was the explosion of grenades, mines, and mortar shells, the concussion of heavy artillery, the cackle of machine guns, the huff and swoosh of flamethrowers, the roar of biplanes on strafing runs, all reverberating within newly steeled helmets. In a segment on "Steel Helmets for All" in his memoir, *Undertones of War*, Edmund Blunden found "their ugly useful discomfort supplanting our old friendly soft caps" and understood that "The dethronement of the soft cap clearly symbolized the change that was coming over the war, the induration from a personal crusade into a vast machine of violence."[37]

At the Battle of Ypres in 1915, the first 420 mm shell heard by an infantryman "was such a sound as one might hear if hanging by the arms from a high railway trestle while an express train passed at full speed overhead, its emergency brakes set and all wheels locked." At the third Battle of Ypres in 1917, the Allies mounted a preparatory barrage of 4,500,000 shells, 4.5 tons of shells per linear yard of Front. The sound of a bombardment, said an English captain, "was as if on some overhead platform ten thousand carters were tipping loads of pointed steel bricks that burst in the air or on the ground, all with a fiendish, devastating, ear-splitting roar." When a shell burst a hundred feet from Amos Wilder at Salonika, "My soul scattered into as many atoms as the shell," but during a later barrage aggravated by the roaring engines of a convoy of trucks, the worst of sounds was that of a small yellow dog barking near his blanket. "It's the impudence, the Personality in sound," he wrote, "that makes it rasping." If for a minute the air stopped shuddering with shells, then men heard the screaming of wounded horses, which was (wrote Erich Remark, hit by shrapnel on the German side) "the moaning of the world... the martyred creation, wild with anguish, filled with terror, and groaning." Eight million military horses died during the Great War. "The screaming of the beasts becomes louder... ghostly, invisible, it is everywhere, between heaven and earth, it rolls on immeasurably."[38]

Soldiers on leave, far less articulate, gave up trying to explain what it was like at the Front. The Royal Welch Fusilier Robert Graves, wounded at the Battle of the Somme, understood why: "you can't communicate noise, noise never stopped for one moment—ever," and for years after his own screams would wake him from unsound sleep. On the German side of the Somme, "The earth rocked and the sky boiled like a gigantic cauldron" and you could not tell one shell from the next, even though you listened "with that peculiar intentness that concentrates all thought and sensation in the ear," for their "long-drawn howls," wrote the infantry officer Ernst Jünger. "There was nothing but one terrific tornado of noise," with barrages so intense that "Head and ears ached violently, and we could only make ourselves understood by shouting a word at a time. The power of logical thought and the force of gravity seemed alike to be suspended.... A N.C.O. of No. 3 platoon went mad." And if the noise could be told, its toll would be too great. Wrote the poet Wilfred Owen while recuperating from shattered nerves at Craiglockhart War Hospital, where Rivers was on staff:

> If you could hear, at every jolt, the blood
> Come gargling from the froth-corrupted lungs,
> Obscene as cancer, bitter as the cud
> Of vile, incurable sores on innocent tongues——

you would not think it so sweet to die for your country. If you could *hear*: the leitmotif of "war-shaken men." It did not have to be the sound of shells spiralling overhead or of chlorine gas drowning a man in his own saliva; it could be troops slogging through the rain "with a great noise like a scattered herd, dominated by the sharper noise of waterbottles and mess tins jangling together, shouts, orders from somewhere vague, muttered oaths...." Or plodding in the heat, "our lungs sounding harsh like hoarse whispers, dull sounds beneath the metallic ring of jolting bayonets bumping against the steel of rifles," when "the sharp little sound of jingling mess-tin chains rings in the ears gratingly, like everlasting cicadas." Or a "rain of maggots, which all through the night, above our heads, had made a noise like rustling silk as they gnawed their way through some dead man's guts."[39]

Best not to hear. After "months of blood and mud, hunger and cold," wrote Guy Thorne, author of such prewar "shocker" novels as *The Soul-Stealer*, it was "noise which thrust itself between the skull and brain and beat out thought." The noise of the Great War was too great not to be visceral. At a field hospital, the Harvard physiologist Walter Cannon was prepared to manage the rapid pulse, pallor, and cold sweat of wound shock and surgical shock; what he came up against was shellshock, whose similar symptoms and "fixed fears"—which he mulled over with a fellow physician, William McDougall—proceeded from "a lowered threshold between nerve cells." Thorne was right: noise did thrust into the body, did beat out thought, ganglion to ganglion. Cannon himself felt the "real *thumping*" of bombardments that "thud the air and shake the earth"; along the Front the "noise was so continuous that it soon sank into the back of consciousness, as if without significance." That "as if" was misleading: men in trenches listened intently; distinguishing between sounds could mean their lives. "We listen for an eternity to the iron sledgehammers beating on our trench. Percussion and time fuze, 105s, 150s, 210s—all the calibres," wrote a French soldier. "Amid this tempest of ruin we instantly recognize the shell that is coming to bury us. As soon as we pick out its dismal howl we look at each other in agony." And sounds in the air were but half the battle. There were also echoes in the bones: a "continuous hum and rumble and thumping—sounds conveyed through the earth, which the air did not carry and revealing the hideous business going on uninterruptedly. The noise of the big guns," Cannon went on, with such deliberate understatement that he must have been registering the irony of his surname, "gets on your nerves at times, and you wish they would cease—quite apart from any thought of the waste of human labor and of human lives. They become personally *annoying*." The author of *Bodily Changes in Pain, Hunger, Fear, and Rage* (1915), Cannon appreciated better than most physicians the somatics of war. He had special sympathy for machine gunners, whose concrete emplacements were "perfect resonance chambers," so that with every cartridge-belt expended they shook from metatarsal to cranium. Listening across from his medical barracks to the trenches, "I knew in my bones that I should have been mortally and sickeningly frightened if I had had to 'stick it,' as the English say, in that dreadful vortex."[40]

Vortices of sound; of noise. Lying on his stomach, shells exploding fifteen meters behind and beyond his narrow trench at Verdun (1,200 artillery pieces firing 2,500,000 shells during the longest field battle in history, with 337,000 German casualties and 377,000 French), Paul Pireaud scribbled a note to his wife: "I am writing you in this wretched position surrounded by a din the likes of which I have never heard before." After a mortar shell burst near Private G. L., burying him in mud, his "Head became very giddy, with singing noises; he was like a drunken man; had staggering gait," a vortex unto himself. An American corporal would not climb out of the hole made by a shell that had "broken just by him; he luckily had not been hit but his nerves had been torn to pieces from the shock"; though a friend was able to calm him, he remained "delirious and as helpless as a baby." In the shell holes, trenches, and ten-foot-deep sloughs of mud of no-man's land, everything imploded. The physical force of constant shellfire, as the historian Peter Leese explains, was so fierce that after three hours men became nauseated by the noise, lost all sense of time and sky. Huddled under a table at the Front, Ford Hermann Hueffer heard "a shelling just overhead—apparently thousands of shells bursting for miles around and overhead. I was convinced that it was all up with XIX Div. because the Huns had got hold of a new and absolutely devilish shell or gun. It was of course thunder. It completely extinguished the sounds of the heavy artillery, and even the how[itzer] about 50 yds away was inaudible during the actual peals and sounded like *stage thunder* in the intervals. Of course we were in the very vortex of the storm, the lightning being followed by thunder before one cd. count two—but there we were right among the guns too." Author of a novel, *The Good Soldier* (1915) and a writer for Britain's War Propaganda Bureau before enlisting, Hueffer in 1916 took notes on the sounds of war: how in wooded country heavy artillery makes the most noise, on dry down it "hits you and shakes you," and how in hot dry weather it gave him a throbbing headache—"over the brows and across the skull, inside, like a migraine." Three months later he was "blown up by a 4.2 [mm shell] and shaken into a nervous breakdown, which has made me unbearable to myself and all my kind."[41]

Ford Madox Ford was the name he would take after he had come back to himself. Hueffer or Ford, his was already a name to reckon with, as founder of the *English Review* in 1908 and the first to publish D. H. Lawrence and Wyndham Lewis. He had chosen his words carefully when writing from the Front, to Joseph Conrad, of being "in the very vortex" of a storm. Swept up by the passions of Ezra Pound, poets and artists of the English avant garde had been drawn to vortices optical, kinetic, and acoustic: ✱"All experience rushes into this vortex. All the energised past, all the past that is living and worthy of living. All MOMENTUM, which is the past bearing upon us, RACE, RACE-MEMORY, instinct charging the PLACID, NOW ENERGISED FUTURE."✱ In the spring before hostilities began, they fell in with "Vorticist" as the title for their movement, which drew its physics from Helmholtz's "Vortex Motion in Ideal Fluids" (1858), its metaphysics from *The Unseen Universe or, Physical Speculations on a Future State* (1875) by Peter G. Tait and Balfour Stewart, its kinetics from a nervous ambivalence toward Italian Futurism ✱"We do not want to change the appearance of the world, because we are not Naturalists, Impressionists or Futurists (the latest form of Impressionism)"✱, and its psychic drive from Freud ✱"WE NEED THE UNCONSCIOUSNESS OF HUMANITY —their stupidity, animalism and dreams.... WE ONLY WANT THE WORLD TO LIVE, and to feel its crude energy flowing through us"✱. The first issue of their new magazine, *Blast*, edited by Lewis in June 1914, featured black-and-white vortices of "Radiation" and "Plan of War" and a set of manifestoes, a play by Lewis, a story by Rebecca West—who began a novel about shellshock in 1916—and a short story by Hueffer that began, "Permanence? Stability! I can't believe it's gone. I can't believe that that long tranquil life, which was just stepping a minuet, vanished in four crushing days." The second and final issue, the "war number" of July 1915, had a vortex of artillery on the cover and a letter from the trenches by the sculptor Gaudier-Brzeska: ✱"The BURSTING SHELLS, the volleys, wire entanglements, projectors, motors, the chaos of battle DO NOT ALTER IN THE LEAST the outlines of the hill we are besieging. A company of partridges scuttle along before our very trench."✱ At the still center of the vortex lay "a sort of death and silence in the middle of life," as Lewis would write, that allowed an artist in the maelstrom to grasp the unities of the world. The letter from Gaudier-Brzeska was published posthumously; he had been killed in June. That summer, Lewis was on his way to Flanders as a bombardier and then artillery officer, crediting his sturdiness under barrage to encounters with F. T. Marinetti. Performing a *Bombardamento* from a volume of his onomatopoetry re-chording the siege of Adrianopolis that he had rushed to witness in 1912 during the First Balkan War, the father of Futurism had with capacious lungs so outsounded a gigantic drum at the back of a hall that his *ZANG-TUMB-TUUUMB* surpassed, wrote Lewis, "all the 'heavies' hammering together, right back to the horizon." There had also been the Grand Futurist Concert of Noises at a London music hall in June 1914, when Marinetti and his crew cranked the handles on twenty-three cubes, or "noise-organs," out of whose metal funnels had come what Ezra Pound called "a mimetic representation of dead cats in a fog-horn," along with Two Noise Spirals and "A Meeting of Motor-Cars and Aeroplanes": *un succès fou* that to Lewis's ears quite outdid the noise on the Front. Across the Channel, Claude Debussy wrote that "at a time when only Fate can turn the page, Music must bide her time and take stock of herself before breaking the dreadful silence which will remain after the last shell has been fired."[42]

Out from the wings now must come the Master of the Vortex, René Descartes, three hundred years late and just in time. His physics, which proposed tight vortices of colliding particles as the basis for global stability and local change, had been essentially an acoustics, first of music and then of war, insofar as the first principle of music-*cum*-physics and war-*cum*-politics was the striking of solid bodies in a world of tremulous air and impassioned speech. Although his *Compendium Musicae* of 1618 was derivative of earlier Italian theory, it was his first book, at twenty-two, and set the tone for later, more original efforts to account geometrically, and mechanically, for sensations. The *Compendium* appeared at the start of the Thirty Years' War, for which Descartes enlisted in the army of Maurice of Orange, looking for action and finding siege. That year he also apprenticed himself intellectually to Isaac Beekman, a Dutch physicist whose intrigue with the mechanics of the almost-invisible had led him to explain musical consonance and dissonance by means not of Pythagorean ratios but of the vibration of soundwaves, coincident or embattled. For Descartes, who had scant training in music and could not tell apart an octave from a fifth, music was rather a mathematical than an auditory puzzle, but as he worked toward a full-bodied philosophy of nature, sound became an issue of mind, matter, and method.[43]

His method, his famous method, was given mythopoetic articulation on November 10–11, 1619, Martinmas Eve—a time of leases coming due, vocations chosen (he got his law degree on November 10, 1616), bonfires lit for carnival, bellringing "as loud as possible," and lewd rendezvous. Descartes by then had withdrawn from the war to his own winter quarters in a quiet part of Ulm. That night, inspired by his isolation, migraines, melancholia, the stuffiness of a heated chamber, the anguish of a break with Beekman, or exhaustion in pursuit of Truth, he dreamt three dreams so vivid he wrote them down and instantly glossed them as climactic. Freud, asked in 1929 to interpret the dreams, considered them a short step from waking thoughts and would not expound upon them; others since have been less reticent, taking them orgasmically or as images of parricide. I of course take their *sounds* for all their worth, and not without reason: the dreams gave Descartes a vorticist drubbing. In the first he was tossed and turned by a whirlwind and spun away from Law and the Church; in the second he heard a sharp, piercing noise like a clap of thunder during a storm and woke in a fright to find sparks of fire in his eyes. These dreams, he wrote, summed up his past: his Jesuit schooling, his legal studies, the war. The third, confirming the philosophical method he had lit upon that very dusk, forecast his future. Centered on a dictionary of the sciences, at first complete and then in-, and on an anthology of poems, at first familiar and then un-, the dream was a mazy flipping of pages and copperplate figures among Enthusiasm, Wisdom, Imagination, and Philosophy, all of which came into focus on a line from the Latin poet Ausonius: "*Quod vitae sectabor iter?* What path in life shall I pursue?" [The poet continues, "The courts are full of uproar; the home is vexed with cares..."] René's answer: the life of the mind. Of *his* mind. Of clear and distinct thoughts in *his mind*. The dreams had to have come from on high, for through them were answered most distinctly his unspoken prayers and queries concerning intuitions of the intellect. Whether a "reaction-formation to epistemological insecurity," oneiric anxiety over a long "separation from the organic female universe" after his mother died when he was an infant, or the celebration of his discovery of a "Wonderful Science," his dreams later resounded in the trenches of the Great War, where the experience of shellshock was vortical, its dilemma Cartesian.[44]

What is it to hear in a dream a clap of thunder so loud it awakes you? To be spun about by a thrumming whirlwind, half metaphorical and half kinaesthetic—the result, speculated Descartes, of sleeping on his left side, in which position heartbeats are heard more strongly? Questions about the relations of mind, matter, and memory that had been sonically and somatically posed in the classical figure of Echo and the Christian trope of Pentecost were most poignant in the trenches under barrage, where "One's whole body seemed be in a mad macabre dance," wrote a Canadian at Vimy Ridge. "I felt that if I lifted a finger I should touch a solid ceiling of sound, it now had the attribute of solidity." By the ninth day of bombardment at Verdun, the men of a crack French unit were weeping; not that they had been hit or hurt, but the sound had gotten to them, a vortex of concussed air that shook them to the depths, as if caught up in a "loudening tornado." In 1910 Kaiser Wilhelm II had cautioned cadets at a German naval academy, "The next war and the next battle at sea will demand of you healthy nerves," but by 1915 his army was stricken by a "war neurosis crisis." German commanders had to send in a corps of psychologists as a second line of defense, using "surprise attacks" of high-voltage electricity, hypnotic commands, loud noises to drive soldiers out of their hysterical deafness, or a metal ball pressed against the larynx to induce a fear of being smothered and a scream that unblocked the speech of the hysterically mute. (Another tactic: place the shellshocked in beds where they can hear the howling of the insanely violent; when they recoil from the noise and recover their senses, put them to work assembling alarm clocks.) The Cartesian divide between body and mind, never exact, was no easier to maintain along 15,000 miles of trench than was the divide between savage and civilized, and neither Myers (in his review of Ward) nor McDougall (in his 1911 *Body and Mind: A History and a Defense of Animism*) was inclined to defend such divides. There had to be crossovers, with which Descartes himself had struggled when trying to account for the phantom pains of soldiers with amputated limbs. So: was shellshock a phantom pain of the already phantasmic Front, of "the ghostly wail of a shell's screw-cap as it whizzed through the air after an explosion, the plaintive moaning of the men, the buzzing of great swarms of flies disturbed by the bombardment, the high-pitched screaming of rats, and, sublime absurdity, the singing of birds"?[45]

Sandor Ferenczi, Freud's good friend, had been in charge of the ward for nervous diseases in an Austro-Hungarian hospital for only two months when he gave a talk on "Two Types of War Neuroses" in 1916. His first visit to the ward had been bewildering, and deafening. "You see here about fifty patients, who almost all give the impression of being seriously ill, if not of being crippled. Many are incapable of moving about; for most of them the attempt to move causes such violent tremors of knees and feet that my voice cannot be heard above the noise of their shoes upon the floor." From the early work of Breuer and Freud he knew of conversion hysterias through which affective experiences are fixed in some part of a (paralyzed) body. What amazed him was the aptness and immediacy of this conversion—how each localized tremor or hesitation of gait encysted a specific trauma, a single powerful sound or concussion frozen into a resonant, residual tremble. He found also at the root of a common hyperaesthesia (sensorial hypersensitivity) an anxiety neurosis that made it impossible for the patients to act, to move forward without fear and trembling. "The encounter with an overwhelming force, the blast of air from the shell, that hurled them to the ground as of no account, may well have shaken their *self-love* to its foundations," and so they had regressed to unsteady toddlers or fetal securities, all eyes and ears. Most shaken were those with supersensitive hearing, who could not abide even the relative calm of the hospital day room yet could never be at rest even in perfect silence, as one explained, "because I always have to listen so carefully to hear whether there really is not some sound to be heard."[46]

Testified the British military historian John Fortescue: "Of course in old days, noise played a less important part than at present, and men could *see* their enemies, which undoubtedly reduced the strain." What a young medical officer saw during his first days on the Front was "The earth sprayed up in splashes of flame and smoke and soil. What looked like a man with no head—and probably was—leapt into the air" while "voluminous whorls of smoke unwound themselves in evil patterns." But what lay entirely beyond his experience, as he wrote, was the noise: "I do not know how the Infantry stand its continuity... and still stay sane." Was shellshock a psychic failure to assimilate the sonic outrage of battles where "It seemed as though the air were full of vast and agonized passion, bursting now with groans and sighs, now into shrill screaming and pitiful whimperings, shuddering beneath terrible blows, torn by unearthly whips, vibrating with the solemn pulse of enormous wings... a condition of the atmosphere, not the creation of man"? Was it then the "uncanny effect of shells" on the mind, or "normal gun-deafness"? Why make more of the "voice of the guns" than did one soldier-contributor to *Cornhill Magazine* in 1916, who kept his poise by cataloguing gun voices as if no more than bird-calls—the 18-pounder's "ear-splitting crack"; the German 77 mm like the "banging together of two planks in a courtyard where there is some echo from the walls"; the "intermittent, hollow, rushing sound" of a heavy howitzer, "with an ever-deepening note which dies away"; the "whirr" of shrapnel splinters and "miaow" of flying bullets; "the loud 'clump clump clump' of the bursting 'Archies'" fighting off aircraft?[47]

Certainly most men had not been prepared for the noise. Veterans may have had some sense of what they were getting into... or not. William H. Hamilton, a sergeant in the Tenth Cavalry during the Spanish-American War and a leader in the war relief effort in 1916, told the Urban Club of Brooklyn, "Being a military man myself, I threw out my chest when offered the chance to go abroad, believing that it was a glorious and fine thing to be a soldier. I changed my mind after observing what the soldiers have to endure in the trenches and on the field of battle. The roar of the heavy firing machinery is enough to drive one insane." Unless an artilleryman, or a civilian quarryman, locomotive engineer, or metalworker with long exposure to penetrating sounds, an army regular and a new recruit would have been equally unsettled by the constancy and density of the noise of the Great War. Not mortality itself but the noise of dying. Almost every young man had seen death in the faces of brothers, sisters, parents brought down by epidemic, food poisoning, or traffic fatality; factory laborers and trainmen had borne up the mangled bodies of co-workers; city-slickers knew well the smell of dead horses (10,000 died on Chicago's streets alone in 1906). Few soldiers, however, were prepared for the sound-masses of the Great War—where, for example, it took 500 rounds from a 75 mm cannon to blast a 5-by-10 meter gap in the barbed wire of no-man's land. During basic training, men tasted mud; in practice trenches they felt the constriction of space and loss of perspective. At the Front they learned to recognize the odor of mustard gas, the odd loudness of gas-spewing machines, the distinct "plop" of gas shells, and the coughing fits of the gassed. But neither drill nor manual, and neither Germans nor Allies, prepared men for the "rolling deafening noise," such "an infernal noise [that if] you want to make yourself understood by your neighbor," wrote a German deserter, "you have to shout at him so that it hurts your throat." In keeping with the continuous-aim firing of the ever-more thunderous guns of dreadnoughts, navies had devised systems of silent commands and their gunners had adopted "ear defenders" more readily than army artillerymen; infantry, crouched under barrage and caked with blood, blanched at the glycerine-saturated cotton-wool or rubber "antiphones" that promised rather ear infection than protection. William Yorke Stevenson, an ambulance driver, knew enough to stuff cotton in his ears and keep his mouth open to prevent his ear drums from rupturing during an incessant *tir rapide* from a hundred field guns laying down a barrage, but corrosive mustard gas, "with its affinity for fats and oils," ate into vaselined cotton and bore through rubber plugs. Another driver found men in hospital who had "lain for hours in shallow holes with shells bursting every moment within inches of them and inflicting sudden and awful death among their comrades; their ears have been deafened with noise which in itself would produce prostration." On their cots they were "babbling in delirium, some shouting and cursing as they fight their way out of the ether dream in which they are reenacting the horror of the trenches." After weeks in hospital, "They will become emaciated and fretful, calling out querulously, cringing at a touch."[48]

Yes, they were reenacting more than the noise, cringing at more than the memory of shrapnel. They had slept atop rotting corpses; they had dug in as if gravediggers at flood tide. At Verdun, Fernand Léger found "almost mummified heads of men emerging from the mud," several with "fingers in their mouths, fingers bitten off with the teeth" to stifle the pain. "To die from a bullet seems to be nothing. The parts of our being remain intact," wrote Paul Dubrulle, a priest and a private in the trenches. "But to be broken into pieces, torn apart, reduced to pulp, there you see an apprehension that the flesh cannot withstand. That is the worst of what suffering is about during a bombardment." Given that the Great War was the first in which the majority of casualties came from bombardment, and given that two-thirds of wounded British and German soldiers returned to fight at the Front, and that of the 2,800,000 French soldiers seriously wounded, a third had been wounded *twice*, the apprehension was real, the reenactment precise. What Myers called shellshock the French called *commotion*—a vector of the co-motion of pounding waves of sound and of bodies flying apart. Call it shellshock, war strain, traumatic neurosis, hysteria, *Granaterschütterung*, or *commotion*, it was a condition structured by vortices of noise and fragments whose impact lasted well beyond the armistice, though that noise has been neglected by physicians convinced that it has no physiological anchorage and must be a ghost of things more substantial. Not counting civilian deaths or the mortality from the flu pandemic in 1918–1919, 10,000,000 men died during the war; twice as many carried war wounds into the 1920s—amputations, twisted limbs, a hissing in the ears, or the utter muteness of a tongue paralyzed and vocal cords shattered. Thousands of soldiers with what the French called "broken faces" and the English called the "thousand-yard stare" were pensioned off on psychiatric diagnoses, thousands more got nothing but grief, and all bore their *commotion* into the 1930s: 500,000 men or more, on all sides, may have lived for decades with symptoms—restless wandering, nightmares, stuttering, and overreaction to sudden noise. "For verily," wrote a Canadian veteran in 1930, "I (along with all my mature contemporaries) did live one life prior to 1914, and am now living another"—noise the sharpest of dividing lines, noise "that for volume, extent, persistency and general raucousness exceeds anything ever yet heard by man," noise so blaring that it made impossible that prewar life with its "languorous quiet wherein one could find pleasure in one's self, in one's own thoughts." Wracked by the blast of a doorman's whistle ("over the top," again?), shuddering at backfires (grenades?), the shellshocked were ever in retreat, like the stammering War-Crossed veteran Septimus Smith in Virginia Woolf's *Mrs. Dalloway*, who prior to his suicide hears his dead friend Evans singing and is "always stopping in the middle; wanting to add something; hearing something new; listening with his hand up." Was it the music of a barrel-organ that he heard or, as Evadne Price would write, "screams of men growing louder and louder, maddened men, louder, louder, louder, shrieking down the song"? Price's novel, *Not So Quiet*, was drawn from the journal of Winifred Young, an ambulance driver on a Western Front still audible in 1930. "All the ideals and beliefs you ever had have crashed about your gun-deafened ears," and you could "no longer believe in anything but bloody war and wounds and foul smells and smutty stories and smoke and bombs and lice and filth and noise, noise, noise..."[49]

Throughout the debate over whether shellshock was neurological or neurotic, somatic or psychosomatic, there was little disputing the medical fact that the ear itself *could* be functionally damaged by concussion. "It is very difficult or impossible to draw a line between cases of 'noise' deafness and those of 'shell' or 'explosion' deafness," stated two English physicians: the swiftly changing air-pressure of a blast was exactly what created its sharpness and loudness. On other fronts, war was no gentler to the ear. In the trenches, eczema was rampant; a man scratching at itchy ears with dirty fingernails or sticking a grimy fingertip into an inflamed canal could give himself *otitis media*, an excruciating condition that might leave a man years later with tinnitus or hearing loss. Since neither the French Adrian steel helmet, the British Brodie, nor the German Stahlhelm fully covered the ears, metal splinters or chips of stone and bone thrown up by an explosion could lacerate the meatus, enter the mastoid process. Medics applying ear-drops with syringes often pushed into the recesses of the ear canal the urine-soaked sand and bloody wax that had accumulated in the meatus during hours of huddling in the trenches, bare ears pressed to earthworks. Concussions from shells bursting or detonations of nearby land mines could rupture ear drums and crush the outer hair cells of the organ of Corti, as could "salvos of low-bursting shrapnel, which cracked off in our faces, with a cloud of black smoke, at about twenty feet from the ground," so that ambulance drivers and infantry temporarily lost their hearing during fierce barrages. "Imagine a trench, five foot, being blown flat as a billiard table, and 'Crump, Crump, Crump!' the whole time," wrote a Canadian officer at Ypres in 1915: "The air was so thick with high explosive smoke you could hardly breathe, and the noise!" With proper diet and cleansing, blown-out ear drums would heal after a month of buzzing; on the Front, healing was chancier. Special otological centers had to be opened to deal with thousands of shattered eardrums and infected ears. "If you were now to bring from France all the men with chronic middle-ear suppuration," said an English doctor in 1916, "you would have battalions of them all over the country."[50]

Back home, psychics were finding that, like tympanic membranes on the Front, there was a hole in the partition between the world of the loud and lively and the world of the silent and departed, a hole big enough for Raymond Lodge to be heard by Moonstone, a spirit medium listening in with his mother Mary and his father, Sir Oliver, the physicist and radio theorist. Raymond had entered the war in 1915 and written from Flanders in April that "We are all right here except for the shells. When I arrived I found everyone suffering from nerves and unwilling to talk about shells at all." He was eager to try out the ear-defenders his father had sent, vowing to test them as soon as he had a chance to fire off the deafening brass pistol of a Very flare. On June 22, 1915 he wrote, "Well! What a long war, isn't it?" Four days later he was off to machine-gun school. In July he was dodging German "sausages," meter-long aerial torpedoes. In September he was killed, joining "hundreds of his friends" on the Other Side. There, as Moonstone informed the Lodges, missing limbs were restored to the spirit-bodies of the young men, who were speaking out in strong spirit-voices against selfishness and materialism.[51]

On This Side, the war brought home not only soldiers so disfigured that they wore tin masks over blasted faces but the material sounds of something like battle. Something was wrong if Harry Crosby, ambulance driver and proper Bostonian, could write of German artillery sounding like "trolley cars tearing out of the subway by Arlington Street." There should be no easy analogy between sounds at home and on the battlefield, yet Dr. T. Ritchie Rodger of Edinburgh was moved by the war to review Euro-American and Japanese research on "noise-deafness" among railway linemen, boilermakers, and factory workers, whose ears when autopsied showed signs of trauma. He concluded that all men in any jeopardy of hearing loss, whether shipyard riveters or shipboard gunners, should be using earplugs. In Prague it was a lawyer for the Workmen's Accident Insurance Institute and manager of an asbestos factory, Franz Kafka by name, a devotee of Ohropax cotton-wool earplugs and an Anti-Noisite of Proustian proportions, who drew out the parallels between shellshock and industrial stress in a successful petition to establish a hospital for neurological and psychiatric (war/work) disorders. In Boston in 1916 it was Marian Lawrence Peabody, chair of the Noise Abatement Committee for the Municipal League, who made the nexus explicit. "In these days," she said, "the harm done by noise is being brought home to us in new and terrible ways. In Europe hundreds of young men are nervous wrecks and scores more have lost their minds owing to the horrible noise of modern war." Then she brought the noise closer to home: "though we are filled with horror by the tales which come from the war zone, how do we know that the girls, all day behind the newsstands in the subway, submitting patiently to the ear-splitting shrieks of the cars as they follow each other grinding around the curves, are not breaking under the constant strain? I do not think the shells in battle can make a worse sound, nor one more shrill and nerve wracking." Quiet zones, as instituted around Boston's hospitals, were a start; the Municipal League must press for quieter streets. "Is not a procession of our city carts, each shaking and rattling in every part as they trot along, quite equal in noise to so many artillery wagons?"[52]

You would "feel it in your ears more than hear it," wrote Lt. Oliver Lyttelton of the Second Grenadiers about the incessant bombardment. This was comparable to other forms of "traumatism by machine"—by vibration—now being reported, and to the vasospasms of "dead hand/white fingers" found among Italian miners at pneumatic drills as well as gunners at machine guns. "Mechanical Death," as ambulance driver John Hargrave titled it, "whistled and screamed and crashed" over Gallipoli. The noise of Mechanical Death literally got under the skin: there was "psychogalvanic" evidence of physiological changes upon hearing the word "bombs," changes that could lead (via surges of adrenalin charted by Cannon) to muscular fatigue and mental collapse. An English friend of the American efficiency expert Frank Gilbreth broached the war and the noise, the vibration and the fatigue in terms of waste: "Nature knows no waste," declared the mechanical engineer Henry J. Spooner, who would be quoted for decades by Anti-Noisites. "Mankind are the only wasters, and our greatest wastes are those of time, opportunity, health, and life." An early motorist with an ear "delicately attuned to the wide range of detonations," then a lecturer on power plant design, he had become an authority on industrial fatigue once the "urgent call for shells and still more shells" led to workers doing overtime seven days a week. The war could be a "wake-up call" to an "era of retrenchment, frugality, economy and thrift" if workplaces were quieted and simplified, for "Nothing is more fatiguing to some workers, particularly if they be highly strung, than a noisy, unrestful, fussy atmosphere. There is far too much avoidable noise in many industrial works, and when it is realised that the human ear can be affected by vibrations of the air ranging from 16 to 40,000 in the second, it can be well understood how injurious noises may be to the nervous system."[53]

Dan McKenzie, a Scottish laryngologist at the Central London Throat and Ear Hospital and editor of the *Journal of Laryngology, Rhinology, and Otology* throughout the war years, endorsed Rodger's plea for earplugs and other efforts to protect the ears of workers and soldiers. However, as a clinician who at forty-six had reached the acme of his profession, he had more in mind than defensive measures. In 1916

> while Julia Rice in New York was coping with the death of Isaac and the fractured pelvis of her daughter Dorothy, who with her co-pilot Larry Sperry (he of autopilot fame) had spiraled down into the waters of(f) Babylon (Long Island);
>
> & while she was funding a hospital for convalescents "needing quiet and rest" and pledging $500 a month "for as long as the war lasted" toward the purchase of anaesthetics for wounded soldiers on the Western Front (where Dorothy's estranged husband, Waldo Peirce, was an ambulance driver and a painter of battle scenes);
>
> & while the Navy was about to charter a barge owned by the Electric Boat Company as a floating sub base, to be christened the *USS Isaac L. Rice,*

McKenzie in London put the anti-noise movement on a war, and metrical, footing. Known for his powerful voice and spontaneous poetry, he exercised his "remarkable powers of declamatory utterance" in *City of Din: A Tirade against Noise*. His final "Litany of Din" was entirely of the home front, of

> Buzzers busy; factory bulls—
> The screeching of mishandled tools—
> Squealing brakes and clanking rails,—
> Strauss's music (shrieks and wails)—

but his tirade seems to have been cued by a letter from a relative on the battlefront: "The noise, especially during the two hours before the attack, was appalling. It was unceasing, heartrending, brain-rending. It was noise gone mad, out of all bounds, uncontrolled." McKenzie hoped that the War Office would find ways to reduce the artillery noise for "our own men" while blasting the Germans to kingdom come; for his part, he would "assault" another "outwork of evil," viz., the "shocking amount of superfluous sound in our modern cities—useless, dangerous, gratuitous noise." Noise by definition was "anarchical" and "painful" sound. Its painfulness, as Henry Head the neurologist could show and Spooner cite, was *not* all in the mind; the substantial pain of noise could prompt suicide, even murder. Civilization might be impracticable without a modicum of noise; still, civilized life need not be raucous. How ironic it was that Suffragists struggling to further civilize society had to practice their oratory in London by having one among them speak "while the rest of the class imitates the noise of traffic." If Dante never imagined noise as torture, the Florentine had not heard the "unearthly screeches, squeaks, and groans" from the hellish underbelly of a motorcar or a driver pounding a Gabriel horn "startling enough, in all conscience, to awaken the dead." Total silence, on the other hand, felt deathly, as one physician declared after a stretch inside Prof. Hendrik Zwaardemaker's soundproof chamber in Utrecht, so it was not silence that McKenzie sought, but soundness: significance, safety, stability. Citizens suffered less from loudness than from acoustic overlays akin to tinnitus, "as incessant, as persistent, as distracting as that symptom can be at its worst." The needless ringing of churchbells, needless music in restaurants (protested the conductor Thomas Beecham), needless shrieking of train whistles amounted to a "long bombardment by noise" of "unprotected nerve-endings in the ears" and the "auditory nerve-centres in the brain."[54]

Not that McKenzie could find evidence of "damage to the ear itself from the noises of everyday life." At autopsy, the ears of metalworkers, riflemen, and beetlers (finishers of cotton cloth) might reveal lesions due to the sonic intensity of their occupations, but not the ears of an ordinary bloke subjected to

> The fog-bound syren's echoing wail—
> The rigging-shriek of ship in gale—
> The weary bang of flapping shutter—
> The gibberish that madmen mutter...

Among the first to assess the clinical value of a labyrinthine test for diagnosing noise-deafness through evidences of vestibular instability, McKenzie was also among the most vociferous in demanding scientifically controlled studies of the long-term bodily effects of noise, so he always had less robust otological data than he wanted in order to pursue, as he did throughout the 1920s, a public health crusade against noise. Consider, he wrote, the noise of a tram surfaceman cutting through steel rails with a cold chisel; consider therefore wearing earplugs. Or consider (he cited Spooner) extending the efficiencies of quiet operation from the factory to the street. Was it not a reasonable hypothesis to assume that if constant vibration caused cracks of fatigue in the hardest of metals, constant noise could cause mental fatigue in the sturdiest of men? But should you trust what a surfaceman said, or a shellshocked corporal? The labyrinthine nystagmus test, as McKenzie had found, failed to distinguish hysterical deafness from that caused by ear disease or physical damage, and during the war there was growing skepticism about assumptions that shellshock had its otological counterpart in a "labyrinth storm" in which physical imbalance due to distress in the vortex of the ear complemented psychical imbalance in the vortex of war. Specialists argued that even with injury to the labyrinth, total deafness was rarely the sequel unless catalyzed by hysteria; presentations of total deafness at the Front far exceeded the number of cases with damage to the bones, membranes, or nerves of the ear.[55]

Deafness, to be sure, was easier to feign than fever or a broken arm. Total deafness, whether attributed to concussion from exploding shells, dum-dum bullets, or infection, would therefore always be suspect as malingering. So would deafness less than total; hearing tests at induction were perfunctory and audiometry in its infancy, so physicians had no useful baselines on individual soldiers.[56] With its multitude of symptoms (deafness, muteness, stuttering, trembling, nightmare, insomnia, anorexia, loss of affect, hyper-reactivity to sound and touch), shellshock could point as well to cowardice and deceit as to brain injury or neurological trauma. Alarmed at the large numbers of men withdrawn from action during the first two years of the war, commanders on both sides pushed for a medical consensus that the phenomenon was thinly psychological—and not physiological, neurological, or psychiatric. That is, they pushed to divorce the shell from the shock and shock from the brain. Most of the shellshocked had not been hit by a shell, and "the wounded are practically immune from shell shock," claimed a London physician, "presumably because a wound neutralizes the action of the psychological causes of shell shock." Disoriented by the death and dismemberments around them, recruits might well be hellshocked, might need a quick course of mental hygiene, a dose of moral stamina or bucking up, and sessions of electroshock to return them to their patriotic senses, but they did not need months of therapeutic confessions behind the lines or back home, did not need anyone burrowing into their dreams or ferreting out sexual angst. The noise of war, which the shellshocked both resumed and repudiated through episodes of deafness, trembling, or shrieking nightmare, was only noise, and a man could become as acclimated to cannonade on the Front as to the screech and rumble of trolleys and lorries on city streets. Concussions of air from bursting shells did not collapse the labyrinths of the ears or do lethal harm to nerves. A ruptured eardrum healed quickly; a man could live with tinnitus easily enough for the rest of his life.[57]

Questions about the causes and treatment of shellshock were questions about the deepseatedness of the experience and memory of noise. If noise were presumed to be insubstantial, even when constant or *fortissimo* it could not register in the bones or nerves—could not do the harm that shellshock embodied. As a commotion of air, the noise of war was as problematic to medical science as were exclamations (bloody hell!) to the science of linguistics. In the lectures of Ferdinand de Saussure, whose prewar *Cours de linguistique générale* was published posthumously in 1916, exclamations were unaffiliated phonemes, momentary but not momentous, like the sudden obscenities of *ticqueurs* or the foul-mouthings of sailors. I do not force the analogy: under bombardment for the first time, Wyndham Lewis found the "air full of violent sound," and with "the tom-toming of interminable artillery, for miles around, going on in the darkness, it was as if someone had exclaimed in your ear."[58]

Lewis, as we have heard, took this with aplomb, as did Ludwig Wittgenstein, who had spent months helping Myers refine devices for acoustics experiments at Cambridge and by August 1914 was in Austrian uniform, soldiering through the war in artillery workshops and a howitzer regiment until taken prisoner by the Italians. Later he would declare that a person could legitimately ask, on hearing a loud noise, "Was *that* thunder, or gunfire?" but it would be nonsense to ask, "Was that a noise?" To a veteran of the Great War, to a man who spent a week in hospital suffering from "nervous shock" after a workshop explosion, noise was psychically and semantically self-referential, though what one person sensed as degrees of loudness might be sensed by another as different qualities of sound. (For example, whistling: to many, it was a rudeness, loud or louder; to the "acutely sensitive ear" of Wittgenstein, a virtuoso warbler capable of Schubert, whistling was a melodic delight.) Throughout the war, Wittgenstein kept up his early philosophical investigations, drawing out the cat's cradles of semantics and hoping to rise above the trench mentality of insular logics, advising himself, "Don't get involved in partial problems, but always take flight to where there is a free view over the whole *single* great problem, even if this view is still not a clear one." That was a pilot's attitude, an aviator's altitude, up where the war had its poster boys and last romance, though airmen too, aloft over battlefields blurred by smoke, had to listen closely to their engines (were they misfiring?) and learn to identify the yaw of their plane "by the 'singing' of the wind through the wires of the machine." Down below, less philosophical soldiers knew within hours on the Front that the Great War *was* noise, that the noise was dangerous, and if noise of itself was not fatal then it was advance notice, and emblem, of mortality.[59]

And what of the noises made by the shellshocked themselves—the weeping, the strangled voices, the grinding of teeth? Descartes had asked, How can I know that someone who looks to be in pain is not feigning the signs or nerve-tricked by a phantom limb? How can I know what another is feeling? Wittgenstein, who would spend twenty years back at Cambridge (1929–1951) tackling the philosophical problem of "other minds," began a lecture on "private experience" by insisting that "The experience of fright appears (when we philosophize) to be an amorphous experience behind the experience of starting"; how a philosopher got from the physical starting, or startling, to the fright, was a dilemma of commitment, of committing oneself to speaking that perpetual connection. Freud, for whom deceptions (displacement, screen memories) were analytically revealing, resolved the dilemma by treating chronic pain as always both psychic and physical, a resolution of personal moment to him from the onset of his throat and jaw cancer in 1923 to his death, twenty cigars a day and thirty-three operations later, in September 1939. Chronic pain, however stifled or stiff-armed, and chronic fears of painful situations, however cathected, advert to a "return of the repressed" so strongly inevitable that intentionality and pretense could be ruled out. Noises of distress, hyperreactivity to noises reminding of distress, or preemptive deafness, were expressions of an ego mortally afraid of being sucked into the vortex of an "internal enemy," a homicidal, cannibalistic, and obscene libido which had as its war-echo the violences of flesh-rending, shrieking barrages. War neuroses, wrote Freud, arise primarily among conscript armies whose citizen-soldiers, unused to battle, shy from the risks and cruelties required of them. "The conflict is between the soldier's old peaceful ego and his new warlike one, and it becomes acute as soon as the peace-ego realizes what danger it runs of losing its life owing to the rashness of its newly formed, parasitic double." This was the double or evil twin hunted through myth and literature in 1914 by Freud's colleague, Otto Rank, then tracked in 1919 as "the uncanny" by Freud. Shellshock was in effect a fault-line of the uncanny: the civilized man blanches at a savage-within-and-shadow-beside whose tremors he cannot shake off, not with a flailing of arms, not with hypnotic suggestion, and surely not with faradic currents used by German, Italian, Russian, French, and English army doctors to make a cure behind the lines seem more frightful than life in the trenches. Restored to the Front with his parasitic double still quaking beside him, a citizen-soldier would be haunted again by mental conflicts ("opposition to the command to kill other people, rebellion against the ruthless suppression of his own personality by his superiors") that were encapsulated in sound-instants (the shrill whistle blown as a signal to go "over the top," the sirens alerting to gas attacks, the fatal gurgling of soldiers caught without gas masks, the panicked breathing of men no longer feeling like themselves inside their masks, the bursting of shells on all sides) and would again be displaced as paralysis, deafness, daytime babbling, nocturnal screaming. For the psychoanalyst, the problem of "other minds" became a three-body problem: how a third, the therapist—half-archaeologist, half-anthropologist, and all ears—could listen between the modern neurotic self and its primitive doppelgänger.[60]

Id, ego, and super-ego, that intrapsychic three-body problem, had not yet been constellated. In 1923, when Freud diagrammed the structure of the mind for *The Ego and the Id*, EGO asserted its reality principle from a dotted grid within a bulging ovoid of the pleasure-principled ID while a narrow canal, open to the outer world and bordered by the Repressed, angled up from the right like the entrance-trench to a pill box. ("A monstrous Easter-egg," wrote Lewis, "its shell of four foot thick concrete sunk in the earth. Its domed top protruded slightly above the level of the parapet, covered by a thick coat of caked soil," with narrow slits for machine-gun barrels.) Conscious and preconscious "perceptions" had their spots above and inside the ovoid, but super-ego was absent. Atop to the left was a small rectangle representing the auditory lobe, which Freud called the *Hörkappe*, a "hearing cap" the ego "might be said to wear... awry." That *Hörkappe* vanished from a nearly identical diagram drawn by Freud in 1933; in its place presided the super-ego. This substitution, noticed by the analyst Otto Isakower, confirmed the supremely auditory role of the super-ego, which was in a sense the terminal point of that eighth cranial nerve that had been traced by Freud the young neurophysiologist. Transmitting the impulses for audition, postural stability, and orientation in space, the "vestibulocochlear" eighth nerve was functionally homologous with the super-ego as a commanding inner voice, an omnicoustic listener-in (no sound too fragmentary to be ignored), and a strict positioner in ethical space. Freud himself, in *The Ego and the Id*, was at pains to identify all verbal residues in the memory as auditory, echoes of words heard long before one could read or write, and he implied that these remnants, half mnemic and half noise, shaped all subsequent talking-out of early trauma. So the "talking cure," whether for infantile Oedipal conflicts or war neuroses, had a rationale beyond that of sympathy, empathy, or confessional catharsis; it was a scientific way to make noises make sense, and the capitol of Freud's "empire of the ear."[61]

Before the war, inheriting some money on the death of his father, Charles Myers had subsidized a new laboratory in experimental psychology at Cambridge, his own "empire of the ear," where he had the benefit of a soundproof room like that at Utrecht. Using instruments perfected by Wittgenstein, he studied "the influence of timbre and loudness on auditory localization" and "individual differences in the attitudes of listeners to musical sounds," but early in 1915, at the age of forty-two, he placed his folders on the psychology of music in storage and proceeded under commission to Boulogne "to supervise the treatment of functional nervous and mental disorders occurring in the British Expeditionary Force." He had campaigned hard for this appointment, though he had "virtually no experience with psychotic disorders," and he did well. By 1917 a crew of psychotherapists was working under him at Abbeville, though field-generals remained dubious of the diagnosis of "shell shock" that he had been among the first to invoke, and more than dubious of the need (as Myers insisted even after he withdrew the term) to obtain "the recall of repressed memories" to calm the inner turmoil of thousands of mute, deaf, shaking, or paralyzed soldiers. Tired of fighting with military muckety-mucks who regarded such symptoms as a crafty sort of malingering (or "sinistrosis"), and unwilling to treat the shellshocked (as did his medical superior) with a short rest and shots of rye, he requested a transfer back to England, where he spent the remainder of the war inspecting shellshock hospitals and trying out therapies, some psychoanalytic, at Maghull north of Liverpool, where his friends Seligman and Rivers had been sussing out the Freudian approach to war neuroses. All along, Myers was prepared to listen to the mnemes and noises of men of all ranks: prepared by his training in performing and listening to music, by his experimental psychology of audition, by his fieldwork in ethnomusicology and medical anthropology, by his Gestalt reckoning of human experience, and by leanings toward a Liberal-Socialist program for helping factory workers and immigrants. If dizzied, dissociative soldiers did not have cerebral hemmorhages, and if the "high-frequency vibrations caused by an exploding shell" did not produce "an invisibly fine 'molecular' commotion in the brain which, in turn, produced dissociation," it was nonetheless crucial to talk out their "long-continued fear," horror, anxiety, fatigue, insomnia, and constipation, so that they eventually regained a mastery of their emotions and their lives. Those with the worst prognoses were young infantrymen who had been buried by a blast or pitched into the air by a bursting shell, but Myers had hopes that these too could be "reintegrated," a double-edged word in wartime: restoring the whole of a personality to a healthful state (no migraines, deafness, nightmares, or spasms); restoring the soldier to his unit, at the always pathological and nearly deafening Front.[62]

There was still more to do for wartime listening. Myers was asked by the Admiralty's Anti-Submarine Department to devise tests that would select for men best suited to hydrophone work. A prelude to sonar, the hydrophonic detection of submarines demanded a listening below the surface of the sea as sustained and acute as psychoanalytic listening for noises of the un/sub/conscious—or stethoscopic listening for the sounds of the heart and lungs. In fact, "Listeners" looked much like long-range stethoscopists, whether in coastal quarters or shipboard. Each operator wore binaural headsets whose rubber tubes reached to microphonic drifters offshore or over the hull to a microphonic "underwater ear" on a ⊥-shaped bar that could be rotated in all directions. Waves breaking on the bow, as well as engine and propeller noises from surface ships, including one's own, had to be filtered out, mechanically or mentally, which required "great concentration for effective continuous watch," or else convoys would have to stop all engines while operators listened for the enemy below. And because the best defense against submarines was the depth charge, operators were at risk of ear damage from loud sudden explosions magnified by their headsets. Hydrophones were problematically passive, waiting upon the underwater approach of sound-signatures whose direction and distinctiveness in stormy waters would not become clear until too late. More active and effective would be equipment that emitted sounds whose underwater echoes could be identified more certainly and at greater distances. Building on its expertise with underwater bells, the Submarine Signal Company in Massachusetts improved upon echo-ranging equipment that initially had a limit of two hundred yards, less than the range of an average torpedo. Company engineers constructed "rats," gray cylindrical devices with wire tails, placed at the vertices of a triangular rig by which underwater sound-signatures could be detected fifty miles away, but for reasons of political economy, the Company failed to integrate into the rigs their Fessenden oscillator, which produced "a continuous series of undamped waves" and would have made sounding more reliable, reception less susceptible to extraneous noises. Since hydrophone signal amplifiers were themselves loud and staticky, shipboard operators had to be isolated in sound-cushioned cabinets and left to the empire of their own ears. By 1917 listening had become doubly demanding as the Germans made their U-boats quieter and equipped them with exterior carbon mikes to listen in on the Listeners, though then a submarine had to shut down its own engines, which otherwise resounded through the hull.[63]

Landlubbers during the war had their own extraordinary listening to do, aside from identifying the sounds of shells. Everywhere on the Front were listening posts, blasted trees (or painted props) into whose bombed-out trunks were sent men trailing telephone wire, to report on movements in no-man's land. Among these brave amateurs was Wyndham Lewis at Passchendaele, "with the German trenches a few yards ahead, and beneath me," and Ludwig Wittgenstein, who asked to be assigned to a listening post on the Russian Front—"Now, during the day, everything is quiet, but in the night it must be *frightful*. Will I endure it?? Tonight will show. God be with me!!" God was, but most forward Listeners had short lives; anything upright between the trenches attracted snipers. The professionals, called "sound rangers," stood well behind the trenches, helping artillerymen locate enemy cannon by contrasting the flash from muzzles with the "infernal loud crack" arriving so many tenths of a second later, then calculating soundpaths back to their sources. With listening cones that grew larger each year and enormous horns that magnified binaural phase differences, rangers could pinpoint enemy batteries in two minutes. The English also had "listening wells" to pick up the acoustic fronts of warplanes through resonating discs whose trembling could be transmitted microphonically or stethoscopically to Listeners in seven-foot-deep shafts. Army engineers attached microphones to galvanometers that could distinguish the first shockwave of an artillery discharge—a low-frequency gun-wave—from the louder, higher-frequency shell-wave that had obscured the first wave and (because shells traveled faster than the speed of sound) upset calculations of distance to enemy positions. Unlike the human sound rangers, galvanometers were insensitive to grinding clutches, the braying of army mules, the cursing of soldiers. Insects were a more formidable problem: attracted to the heat of the wires, they crept along the lines and nibbled at the insulation, so men in headsets had always to be on the listen—and the look-out, now that they had pictures in their heads of the sound of smaller guns as sharp kicks in lines drawn by the stylus of a galvanometer, larger guns as sloping humps.[64]

To *picture* sound had been a millennial project: geometries of the music of the spheres; illustrated legends of frozen syllables; frescoes of the revilers of Christ and mourners of the Crucifixion; arches incised with the gaping mouth of hell; screaming grotesques guarding cathedral rooftops; palace ceilings sculpted with trumpeting angels or brushed with celestial choirs; woodblocks and etchings of street peddlers; oil paintings of battle scenes; portraits of the well-off at their keyboards;[65] fireworks photoengraved as explosions of asterisks on Christmas cards, New Year's prints, Independence Day programs. Only in the half-century prior to the Great War, though, had the graphics of ₦0↨§Σ gone beyond pen-and-ink riots, pained or shouting faces, a chaos of forms. With the advent of the phonautograph, seismograph, electrocardiograph, and cathode ray tube, the visual oddness of noise as a regular irregularity had made its way in stuttery lines toward the scientific eye; with the advance of theosophical, dramaturgical, and aesthetic theories of vibration, visual noise approached the public eye. Painters just before and during the Great War took up the cross-modalities of the vibratory by figuring chromatic algorithms for "visual music" as a spiritual regime (Kandinsky), building up a "synchromatist" oeuvre (Stanton Macdonald-Wright, Morgan Russell), or entitling their works fugues, sonatas, symphonies, "Sentimental Music" (Arthur Dove), "Blue and Green Music" (Georgia O'Keefe), "Instrument for New Music" (Paul Klee) or, back to basics, "Painterly-Musical Construction" (Mikhail Matiushin). Joseph Stella set up his studio above a frenetic intersection on the Lower East Side and thrilled to the *Battle of Lights, Coney Island* (1913), which critics compared to the explosion of a munitions factory. Teaching Italian at a Baptist Seminary in Brooklyn, Stella crossed and recrossed the Brooklyn

Bridge and stood among its girders time and again to be "shaken by the underground tumult of the trains in perpetual motion, like the blood in the arteries—at times, ringing as alarm in a tempest, the shrill sulphurous voice of the trolley wires—now and then strange moanings of appeals from tugboats, guessed more than seen, through the infernal recesses below." Then he painted the bridge as a vortex of brilliant steel c(h)ords.[66]

Painting sound was the art of the future, wrote a critic for *The Nation* in 1912, and that future was strident, embracing ragtime or worse. At pianos around the San Francisco Bay in 1913–1915, Henry Cowell pounded out tone clusters with forearms and elbows, scoring *Adventures in Harmony* for "arm chords" and finding musical effects for the chaos of *Creation Dawn*, for the vortex of *The Cauldron* (a whirlpool off the Carmel coast), and for the choreic muscle spasms that put him on crutches in *Mad Dance*. In New York in 1915, Aleksandr Scriabin's *Prometheus: The Poem of Fire*, scored for searing music and leaping color, was performed with the aid of a ten-foot-high Color Organ built by A. Wallace Rimington, an Englishman who, had he survived the war, might have founded an institute of color music. Scriabin himself, who had been interviewed by Charles Myers concerning his sound-color synaesthesias, died in 1915 of color changes: pimple → pustule → carbuncle → furuncle → fatal infection. His "dense harmonies, swirling chromaticisms, drum crashes" had often been greeted by the whistled hissing common to premieres of new concert music after century's turn. A Russian critic called Scriabin's Second Symphony the Second Cacophony: "One dissonance after another piles up without a single thought behind any of it." Unfair, if not untrue: Scriabin in retrospect thought his Second banal. The pianist Anatol Drozdov had been closer to the mark when he heard "the sobbing and shouting of social struggle," for Scriabin's assertion of simultaneous melodies, confounding coloration, and ancient scales was as much a political as a (syn)aesthetic statement, more romantic than radical but ever in honor of the plebeian strife of daily life, and defiantly cacophonous. Erik Satie scored *Parade* (1917) with steam whistles, typewriters, and pistol shots, but meant originally to foreground "a barrage of vocal sound" based upon one soldier's vivid narrative of time at the Front; that soldier, Jean Cocteau, at last agreed with Sergei Diaghilev and Satie that mechanical outrages would be more evocative of the noise of war than the most "pyrotechnic" of voices. Other composers, reacting to the *fortissimo* traffic outside symphony halls and anticipating the whistling, seat-rattling, and catcalling of outraged audiences within, countered with battalions of musicians and singers—one hundred strong for Richard Strauss's *Elektra*, one thousand for Mahler's Eighth, twelve separate orchestras with choruses, and a brass band on a barge, as imagined by Charles Ives for his *Universe Symphony*. Invalided out of the French Army and newly established in Manhattan, Edgar Varèse let the "parabolas and hyperbolas" of fire-engine sirens surge through his

Amériques (1918–1921) and haunt his *Hyperprism* (1922–1923) as elements of "organized sound masses"—masses set in visual motion by Marcel Duchamp, also in New York, drafting his plans for the rods, gears, coffee grinders, cowbells, and cogwheels of the *Large Glass* (1915–1923), a metal-framed double window at the instant before ignition✺explosion. *Large Glass* was a transparent working out of his instructions for *With Hidden Noise* (1916): "Make a readymade with a box containing something unrecognizable by its sound and solder the box."[67]

Premature burial? Vampire waking in a coffin? Futurist noise-organ? Thrum through a hydrophonist's headset in a soundproofed cabinet? Or "A Case of Paranoia Running Counter to Psycho-Analytic Theory of the Disease"? Published in 1915, this was one of the rare cases in which Freud "altered the *milieu*... in order to preserve the incognito of those concerned," a strangely apt closeting to a case that had begun as a readymade referral. A lawyer had been contacted by a thirty-year-old woman, single and "singularly attractive," seeking protection from a co-worker with whom she had, after much wheedling, consented to an affair. When at last she went up to his bachelor rooms and they had begun to make out, "she was suddenly frightened by a noise, a kind of knock or tick" from the direction of a writing-desk near some heavy curtains—maybe, said her frustrated lover, from a clock on the desk. But the mood was broken, their caresses cut short, the undressing curtailed. As she left the apartment and (not nude) descended the staircase, she passed two whispering men, one of whom was carrying something under wraps that "looked rather like a box." This, she decided, was a camera that had been hidden behind the curtains to take compromising pictures of her; the tick had been its click. She had received no threats or blackmail notes; on the contrary, the man made further protestations of love. Yet she was sure that he was conspiring to ruin her. The problem for Freud was not in ascertaining the paranoia but in explaining it, for psychoanalytic theory associated paranoia with fears of homosexual bondage, not heterosexual play. As he had to, Freud probed for that primal fear, which he found in the gap between the first and second tellings of her story: the woman had in fact been to the man's room twice and in-between had noticed a co-worker whispering (an account of their tryst?) to the office manager, an affectionate matron who resembled her mother, with whom our subject was living, an only child, father deceased. Aha. "Love for the mother"—really "the infantile relation to the original mother-imago"—"becomes the protagonist of all those tendencies which, acting as her 'conscience,' would arrest the girl's first step along the new road to normal sexual satisfaction." But wait. What was that unrecognizable sound that interrupted the afternoon tryst? Was it the noise from the soldered box, the camera? No, it was the noise from that other soldered box, the mind. It was the noise of the primal fantasy of eavesdropping on parents in the primal act, and (as Otto Rank had suggested to Freud) "it reproduces either the sounds which betray parental intercourse or those by which the listening child fears to betray itself." And it was both, since the woman through a partial regression had smuggled herself into the place of her compromised mother, then heard herself in the act. "I do not believe that the clock ever ticked or that any noise was to be heard at all. The woman's situation justified a sensation of throbbing in the clitoris." It was the noise from "an isolated clitoris-contraction" during lovemaking in which there had been no genital contact but assuredly arousal. The clit tick.[68]

Onomatopoetics never had it so good. For Freudians and for Futurists, sound was exploratory, often revelatory—and always echoic. The voice made for more than lexicality, the world's noise for more than nonsense, with shock as the pivot from a neurological to a psychological understanding of mental life. "Our increasing love towards matter, the will to penetrate it, and to know its vibrations, and the physical sympathy that binds us to engines, incite us to use onomatopoeia," said Marinetti in 1914. Since Futurists were short-form adventurers in love with speed and blur, their literary arena was mostly poetic, their poetry onomatopoetic, and their onomatopoeia mostly noise, whether the *dum-dum-dum-dum* of "the rotative noise of the African sun and the orange color of the sky," the *pic-pac-pum* of fusillade, the *stridionla stridionla stridionlaire* of "the clashing of great swords and the furious commotion of waves," or the "abstract onomatopoeia" of *ran ran ran*, "the noise and unconscious expression of the most complex and mysterious movements of our sensibility."[69]

Consider Futurist writing and theatrics, then, as a disturbing double to the Freudian, equally alert to the automatisms of the spontaneous and the mechanisms of the improvised, attending to the inner coherence of the cacophonous and the precision of the inarticulate voice. These contradictions had to be the very stuff of the modern, where the visible was incredible or discredited. Where the tangible had become impalpably atomic or temptingly indeterminate. Where perspective, as in Guillaume Apollinaire's "Zone," was now split between the pedestrian and the pilot, and from the clouds above, in his graphic poem *"Il pleut,"* came a pelting rain of women's voices and the neigh of "an entire universe of cities of the ear."[70] Only a universe of ears was up to the task of simultaneity that modernity demanded, and the sum of the simultaneously audible was noise. Noise expressed at once the hoarse power and gross inefficiency of machines.[71] Noise exposed the noble outrage of which the human spirit was capable and the torture to which a human body was susceptible. Although Futurism burst onto the scene before the Great War, it was from the start the *volte*-face of shellshock, lurching to trip the wires of Dada and Surrealism.

Proclaimed Marinetti in his first Futurist manifesto of February 1909: "We intend to sing the love of danger, the habit of energy and fearlessness.

We will glorify war, the world's only hygiene—militarism, patriotism, the destructive gesture of freedom-bringers, beautiful ideas worth dying for, and scorn for woman." The next year, Umberto Boccioni, Carlo Carrà, Luigi Russolo, Giacomo Balla, and Gino Severini called upon the Young Artists of Italy to "express our whirling life of steel, of pride, of fever and of speed." Swept up into the whirlwind was Balilla Pratella, a composer whose folk-opera had caught Marinetti's ear. He asked Pratella to write a "Manifesto of Futurist Musicians," which Marinetti instantly rewrote and published in six thousand copies. Pratella-Marinetti decried "that absurd swindle that is called *well-made music*" and summoned young musicians to the standard of the Futurists, who had already sponsored their first "*pandemonio sonore*," a concert of noises in Milan. This was not what music critics meant when they took up "Futurist" as a label for symphonic music that was as difficult to listen through (Scriabin, Schoenberg, Stravinsky) as to play. No, the Italians wanted to shred all musical traditions in favor of "amoral liberty, of action, conscience and imagination," so that everyday noises, like the "crashing down of metal shop blinds," might take center stage. In 1913, Pratella's *Inno alla Vita, Musica futurista per orchestra*, a work of semitones and polyrhythms, premiered to a negligible buzz among the audience but a second performance seemed to "drive the public insane." Fruits and vegetables were thrown, and a few blows, and slices of chestnut cake: a wild success. In the hall was the painter Russolo, son of a church organist. Listening (to the music, or the audience?), "a new art came into my mind which only you can create," he told Pratella, "the Art of Noises, the logical consequence of your marvellous innovations." Or so he wrote. As Douglas Kahn has shown, Russolo had conceived *L'Arte dei rumori* some months before and as a more revvvolutionary music than Pratella's, a music of "the most dissonant, strange and harsh sounds" in sync with a world of collaborating machines. Composers, said Russolo, must henceforward unstop the emotional, acoustic, and military power of the infinite variety of "*noise-sound*" in order to shake the nerves of audiences with a "COMBINATION OF THE NOISES OF TRAMS, BACKFIRING MOTORS, CARRIAGES AND BAWLING CROWDS." Pratella, though he spoke of catching the "musical soul" of factories and battleships, had no heart for mechanical combos, so Futurist music became linked more publicly with Russolo's new *intonarumori*, large Helmholtzian resonators fashioned in crazy parallel to

percussion	brass	woodwinds	strings
"Rumbles"	"Whistles"	"Whispers"	"Screeches"
Rombi	*Fischi*	*Bisbigli*	*Stridori*
Tuoni	*Sibili*	*Mormorii*	*Scricchiolii*
Scoppii	*Sbuffi*	*Borbottii*	*Fruscii*

in cahoots with *ostinato* pounding of metal and terracotta and assorted grides and stridulations, all meant to shock concert halls, "these hospitals for anaemic sounds," back to life. It was time to exult in the manly meatus and the hardy tympanum, and "It's no good objecting that noises are exclusively loud and disagreeable to the ear." Given that "Every manifestation of our life is accompanied by noise," noise was not merely intemperate sound; it was a prime expression of the "fantastic juxtapositions" by which life was enlivened. ("Little girls walk by with their tiny noiselings," wrote Vladimir Mayakovsky.) Let us enjoy, therefore, "the eddying of water, air and gas in metal pipes, the grumbling of noises that breathe and pulse with indisputable animality, the palpitation of valves, the coming and going of pistons." The sounds of the natural world and the old musics of the concert hall were too predictable, too easy to ignore. "Noise, however, reaching us in a confused and irregular way from the irregular confusion of our life, never entirely reveals itself to us, and keeps innumerable surprises in reserve." Noise was essentially inexhaustible, so Futurists could use *Ululatori* (howlers), *Rombatori* (roarers), *Sibilatori* (hissers), *Crepitatori* (cracklers), *Gorgogliatori* (gurglers), and *Ronzatori* (buzzers) to create new musics. As for painting, away with the subdued, all things muted or mitered. Embrace instead "Reds, rrrrreds, the rrrrrreddest rrrrrrreds that shouuuuuuut" and "Greens, that can never be greener, greeeeeeeeeeeens, that screeeeeeam." The art of noise was an art of motion, and whatever was in motion was in disequilibrium, as in Carrà's *Jolts of a Taxi-Cab* or the fragmented grillwork greens of Balla's *Speeding Automobile* or the slashing yellows of Boccioni's *The Noise of the Street Penetrates the House*. Painting, announced Enrico Prampolini, should be "chromophony," a klaxonry of sound and color that displaces the atmosphere like the soundwaves of a car horn breaking up "into a myriad chromatic scales."[72]

Prampolini had been reading up on vision, and color, and Aboriginals. He may well have read Rivers and Myers on the perception of color among the Meriam and others, for he knew that "Animals, like primitive peoples—including some still living in Australia—only distinguish dark from light colors, and have arrived, with some exceptions, at a third level of perception" by which they somehow can see through opaque bodies. Presbyopic Europeans compensate by perceiving "the whole chromatic gamut of bright objects." Someday people should be able to take in all atmospheric vibrations, including the polyphony of colors of different intensity, which Prampolini exalted as a "violent irradiation of evolving vibrations." So, "Destroy, destroy, in order to rebuild consciousness and opinion, culture and the genesis of art." This was the violence of a movement whose politics would arc over from anarcho-socialism to fascism; it was, wrote one critic, "an acromegaly of noise."[73]

Among the few Futurist women, speed and violence were secondary to the reintegration of the senses through the sounded body—a body dancing to musical colors or making theatrical love and redirecting the Futurist strut toward an attack on rigidities Catholic and patriarchal. Marietta Angelini, who as Marinetti's maid had glided across Persian rugs three-deep in his palazzo in Milan (built with the millions his father made in Alexandria), broke gender decorum and violated class deference with free-word collages and polemics about a Wireless Love Room and sex with her employer, "the Red Man." Rougena Zatkova sculpted *Sensibilita, Rumori e Forze Ritmiche della Machina Pianta-Palafitte* (too literally, "Sensibility, Noises, and Rhythmic Powers of the Pilework Machine") as if woodstripping could free the soul. Valentine de Saint-Point, who would move to Cairo and convert to Islam in 1924, performed veiled dances with geometric movements calligraphically swollen with the energy of the erotic. Where Marinetti projected a set of gestures in which voice and body merged in liberating processes of metallification, liquefaction, vegetalization, petrification, or electrification, Saint-Point gestured to a Manifesto of Lust. "Lust," she wrote in 1913, "when viewed without moral preconceptions and as an essential part of life's dynamism, is . . . a powerful source of energy . . . the expression of a being projected beyond itself . . . the sensory and sensual synthesis that leads to the greatest liberation of spirit . . . the panic shudder of a particle of the earth."[74]

Panic shudder: aren't we back to shellshock, and to the prewar shock felt by Marinetti after he drove his car into a ditch in 1908? Racing "through streets as rough and deep as the beds of torrents," he'd been yammering with two pals about shaking the gates of life and testing their hinges and had braked sharply to avoid oncoming cyclists shaking their fists at him. His car in its inertial fury had swerved wheels-up into the roadside muck and pinned him briefly beneath—a shock that made him "sing the love of danger" and "the beauty of speed" in his first manifesto, where he went on repeatedly about not repeating the past and thrilling in a future that was "violently upsetting." Turn and turn about, one literary historian has suggested that the corpus of Futurism as defined by Marinetti exactly embodied traumatic neurosis as defined by Freud. Rethinking the traumas of war in *Beyond the Pleasure Principle* (*Jenseits des Lustprincips*, 1920), Freud characterized traumatic neurosis as "a type of repetition compulsion that indexes a failure of symbolization, an incapacity to render experience into the symbolic language of dreams or mentation. By this logic," argues Lawrence Rainey, "perhaps it is only inevitable that the traumatic automobile accident leads Futurism to declare a historical end to symbolism." In lieu of dream or metaphor the Futurists exulted in onomatopoeia, "the primal language of shock," says Rainey, who proceeds to invoke the vortices of Descartes: "It epitomizes and restages the principle of matter itself, which is now envisaged as animating a universe of indifferent objects in perpetual collison, a continuous concussion that lends 'natural' sanction to the mechanical jolts and vibrations of modern technology, even as it simultaneously recapitulates and replicates them." With onomatopoeia let loose, Futurists no longer need syntax or "wires" between words. In "Contraband of War," Marinetti explained that "the strident onomatopoeia *ssiiiiiii* renders the whistle of a tug on the Mose, and is followed by the veiled onomatopoeia *ffiiiiii fiiiiiii* echoing from the other shore. The two onomatopoeias have saved my having to describe the width of the river." But the *ssiiiiiii* simply reenacts the scene, the echoing *ffiiiiii fiiiiiii* allows no time for reflection, and it is through such intense immediate circularity that the traumatized are left beside themselves in a world of ceaseless shocks, bereft of metaphors, the self as scattered as shrapnel and (argued Freud, and Myers, and Seligman, and McDougall, and Rivers) in deepest need of recognizance.[75]

Wrote Rivers early in 1918, "It is natural to thrust aside painful memories just as it is natural to avoid dangerous or horrible scenes in actuality, and this natural tendency to banish the distressing or the horrible is especially pronounced in those whose powers of resistance have been lowered by the long-continued strains of trench-life, the shock of shell-explosion, or other catastrophe of war." But what is banished, as Freud insisted, never vanishes of itself; the repressed always returns, readymade. For a soldier who has been buried alive in the mud and rock thrown up by an explosion or who has stumbled on a companion "blown into pieces, with head and limbs lying separated from the trunk," the noises must come bursting out. Unless listened to and through, to full catharsis, the noises that invaded a man's dreams, woke him from fitful sleep, and threw him off kilter would in time isolate, emaciate, paralyze. Dreams were the key to knowing what had been repressed, but the secret conflicts they bespoke were not sexual or Oedipal. They were, as Rivers insisted, existential: how to survive in the face of horrible brutality, how not to have to face that brutality again, and how to face oneself in the long aftermath. These conflicts were easily heard in the fragmentary, blatant dreams of the uneducated, those shellshocked enlisted men he had been treating at Maghull Hospital. More skill was required to take stock of the complex dreams of officers at Craiglockhart in Edinburgh, where shellshocked university men enjoyed university comforts by day but "by night each man was back in his doomed sector of a horror-stricken Front Line, where the panic and stampede of some ghastly experience was reenacted among the livid faces of the dead." There, in "Dottyville," Wilfred Owen at twenty-five composed his "Anthem for Doomed Youth"—

> What passing bells for these who die as cattle?
> Only the monstrous anger of the guns.
> Only the stuttering rifle's rapid rattle—

and Siegfried Sassoon took notes for his Sherston trilogy, in which Rivers assumed a heroic role, "always communicating his integrity of mind; never revealing that he was weary as he must often have been after long days of exceptionally tiring work on those war neuroses which demanded such an exercise of sympathy and detachment combined." Rivers had been most struck by Freud's "theory of forgetting" and by Freud's knack at eliciting precisely those memories that were most painful and most revealing. He was much less concerned than Freud about intrapsychic mechanisms, much more concerned, as an anthropologist of kinship must be, with the talking *out*. If a dream is "a solution or attempted solution of a conflict which finds expression in ways characteristic of different levels of early experience," then an anthropologist familiar with the stages of social and mental life in primitive cultures could draw upon his years of fieldwork among cannibals to elicit underlying fears, puzzle out oneiric shapes, trace the genealogies of shock, and recast them in such a way that they could be confronted and their terrors dissipated. Still, as Rivers was quick to admit, the most empathetic of anthropologist-physicians—one who knew the frustrations of the stuttering of the shellshocked, who himself was susceptible to breakdown—would be unable to help a soldier aching to forget an experience "so utterly horrible or disgusting, so wholly free from any redeeming feature which can be used as a means of readjusting the attention, that it is difficult or impossible to find an aspect which will make its contemplation endurable." Such as? Do you really want to know?[76]

Many army physicians—German, English, Italian,[77] and Russian—did not want to know. They were chary of coddling men too cowardly to cope. They were wary of endorsing insomnia and tremor as excuses for the surrender of a deadlier purpose. For some, the repudiation of shellshock was a matter of national honor; for some, ancestry and class; for some, manpower; for many, manliness. None of this would have come as a surprise to a veteran Anti-Noisite. Those who doubted the power of sheer noise to do enduring violence to a body on the Front used the same language and logic as those who doubted the malignity of noise back home. Anyone protesting the hullabaloo of industrial civilization did not appreciate modernity; had neither the moral stamina nor the strong nerves of an evolved citizen; was taking refuge in a hypervigilance that no class could any longer afford to indulge and honest working folk rarely entertained. Men overreacting to shellbursts were like the women's club committees petitioning aldermen to outlaw squibs and peanut vendors. Not only were such whimperers foolishly denying where they *were* (at war or in the city); male or female, they were using the most feminine of wiles to ingratiate themselves with politicians and newspapers while surrendering to the most primitive, illogical of fears. Shellshock, said some hard-line physicians, was male hysteria; worn-out blood lines, said others; congenital predisposition to insanity, said others; "sheer funk," said the neurologist Henry Head. It was not the war that was to blame but inexpert screening for preexisting neuroses, whose wartime deficits could in any case be remedied by hypnosis, intimidation, isolation, discipline, a shot of whiskey, or electroshock.[78]

Was it as simple as that? As simple as reeducating a man "to life as it is and as others are living it" (thus Donald Laird, a young professor of psychology in Wyoming)? "One had little time to think out the many theoretical problems," wrote William McDougall while treating the shellshocked. His experiences with "a most strange, wonderful, and pitiful collection of nervously disordered soldiers" lifted his esteem for psychiatry and confirmed the value of the field of social psychology whose outlines he had drawn in 1908. It was his war service, when "for the first time in my life.... I was giving my whole time and energy to work that was indisputably worth while," that impelled him to finish *The Group Mind* (1920), in which he, like Freud, grappled with issues of mob panic, national character, and "collective consciousness" that both had scouted out before the war. But Freudians clung to "esoteric dogmas" and private communions, while McDougall held to the exoteric, an individual's relations with society and the sway of the multitudes. Politically his sympathies lay rather with "individualism and internationalism" than with Nationalism or Socialism, though insofar as his psychological postulates justified any philosophical ambitions, these tended toward a "synthesis of the principles of individualism and communism, of aristocracy and democracy, of self-realisation and service to the community." His was scarcely the behaviorism of James B. Watson, man of mice and mazes, nor was it the Pavlovian trapshoot of stimulus-and-response; it was an anthropological and psychological theory of the commons, of degrees of "similarity of mental constitution, of interest and sentiment." The implications for treating shellshock were threefold: to restore a man to himself (individualism), to his brigade (communal responsibility), and to a *world* war (internationalism). The implications for the noise of war and the noises of war neurosis were also threefold: as consciousness was never exclusively personal, neither were the nightmarish noises that kept a shellshocked man from sleep; noise was an index to, and a pathology of, the social; healing a man of the noise of war did not mean silencing the battles within but hearing them out and restoring a viable life in the company of others.[79]

"Personally I find that gun-firing stimulates me like the drum-beating of savages," wrote Haddon from the French Front in 1917, "and so far I have not felt the slightest fear." He had made his way through a blizzard to reach Christ's Hut, a dug-out captured from the Germans, and after a spot of tea had given "a lecture on New Guinea Cannibals to about 150 men. It went off very well." A man of sixty-two and still in his prime, he had spent the first years of the war editing further volumes of the *Torres Straits Reports*, revisiting Australia and New Guinea, and devising cat's cradles for engagement with primitives whose languages he never learned. Now he would enjoy his three months in "Cambridge House" five hundred yards from the trenches, acquainting the rank-and-file with anthropologies of lives little more savage and far more comfortable than those on the Front, despite the "hellish orchestra of war... with its discords of booms, bangs, crumps, crashes, rat-tat-tats, whizzes, screechings and whines." He found less stress at Cambridge House than back in Cambridge, where civilian anxiety was rampant.[80]

Home, reworking the *Reports* on the verandah of his neatly-gardened house, Haddon himself had no such anxieties, no compunction about resuming professorial duties. Myers, on the other hand, swung away from the academy toward industrial psychology and the study of industrial fatigue, fields informed by his engagement with shellshock. Postwar factories might be said to have suffered collateral damage from the shellbursts of the war, given the number of workers who returned from the Front with hearing loss, tinnitus, and sensitivity to sharp or loud noises. Myers's student T. H. Pear, an early initiate into Freudian mysteries, had collaborated at Maghull with an anatomist to write in 1917 about *Shell Shock and Its Lessons*, one of which was that the anxieties of most of the shellshocked had existed before the war and would not disappear with peace. Exploding shells and poison gas aggravated these anxieties, but nervous breakdowns stemmed from emotional exhaustion, since "before this epoch of trench warfare very few people have been called upon to suppress fear continually for a very long time." In effect, a soldier in a trench fought always on two fronts: "The noise of the bursting shells, the premonitory sounds of approaching missiles during exciting periods of waiting, and the sight of those injured in his vicinity whom he cannot help, all assail him, while at the same time he may be fighting desperately with himself." Once the war was over and a man returned to the din and demands of factory life, he fought again on two fronts, straining against the soundpower of machinery. Fatigue and unrest, as Myers explained in 1921, might be ascribed to a short-circuiting of "lower" nervous processes by auricular as well as muscular strain; the consequent irritability and restlessness of workers might rise in pitch from gripe to strike, so factory owners would be wise to attend as much to soundproofing as to scientific management. In the 1920s, Myers created an Institute of Industrial Psychology as a culmination of his work in anthropology, experimental psychology, and psychiatry; it was also a response to the societal problem of shellshock. Since his time in the Torres Strait, he had been operating implicitly on the conviction that "The true nature of things is to be found in process—happenings, events, or acts in a space-time continuum." Noise, memorable and evidential, occupied the space-time continuum as fully as any other event, and could be understood as the epitome of process. So it was not surprising that Myers's most devout disciple, Frederic C. Bartlett, should move from studies of *Psychology and Primitive Culture* (1923) and *Psychology and the Soldier* (1927, addressing shellshock as conversion hysteria) to a still-influential inquiry into *Remembering* and a brief account of *Psychological Experiments on the Effects of Noise* (1932), then to a full-scale review of *The Problem of Noise* (1934).[81]

"We come from silence and we return to silence, but in between it is all noise." So George Slythe Street in a "Digression on Sound" for *The New Age* in 1918. His digression was not yet a philosophy of noise, though it was of a piece with the other carpings of a fifty-year-old British essayist and novelist who deplored modern music and/as the uncouthness of coughing in public. There was more of a philosophy of noise in the *Bruitism* of Richard Huelsenbeck, Hugo Ball, Tristan Tzara, and the Dadaism that had arisen in Zurich and Berlin in 1916. Dadaists borrowed *Bruitism* from Marinetti but understood noise more existentially, as "life itself." Bruitism, said Huelsenbeck (a medical student before the war, physician and jazz drummer after), "compels men to make ultimate decisions." Dada was nothing if not political, though resolutely self-contradictory. "Anyone who lives without rules, who doesn't love museums except for their floors, is DADA," wrote a wealthy forty-year-old from the archetypal urb of noise and simultaneity, New York City. "A true work of Dada," he continued, "shouldn't live more than six hours"—though this particular New Yorker would make his name as a collector of modern art. "I, Walter Conrad Arenberg, American poet, declare that I am against Dada, seeing no way but that to be up dada, up dada, up dada, up dada," a Manhattan Upanishadada of the primitivism that Huelsenbeck admired in ragtime and imitated in his "Negro poems," all ending "with the refrain 'Umba, umba,' which I roared and spouted over and over again into the audience" at the Cabaret Voltaire in Zurich.[82]

This was war poetry by those opposed to the war and hoping to rescue from its vortex whatever was primal but not shameful. For Futurists, shells and shocks were nerve-food to brace society for an apocalypse of the old and apotheosis of the modern—"We arrive hearing a shout we throw ourselves to the ground," wrote Boccioni in a war diary months before he was killed in action: "shrapnel explodes twenty steps away and I shout: At Last!" Dadaists wanted their polyglot rockets of words and caravans of typographic marks—Ball's "hollaka holla/anlogo bung/blago bung/bosso fataka/# # # #"—to prepare the ears for primal truths. "Noises (an *rrrrr* drawn out for minutes, or crashes, or sirens, etc.) are superior to the human voice in energy," wrote Ball, who had been rejected for German military service, gone to Belgium to see the war for himself, been horrified ("It is the total mass of machinery and the devil himself that has broken loose now"), joined anti-war protests in Berlin, then decamped for

Switzerland. "The noises represent the background—the inarticulate, the disastrous, the decisive," he wrote, and are "a direct objectivization of a dark vital force," dark and necessarily shocking.[83]

According to Julia Rice's friend, the neurologist George F. Shrady, there were two types of shock, "shock proper or torpid shock" and "excitement or erethistic shock." Torpid shock had to do with neurasthenic depletion and loss of mobility or expressivity, erethistic shock with excitability and hypersensitivity. But a pair of Cleveland physicians had argued just before the outbreak of war that "the essential pathology of shock is identical, whatever its cause. That is, when the kinetic system is driven at an overwhelming rate of speed—as by severe injury, by emotional excitation, by perforation of the intestines, by the sudden onset of an infectious disease, by an overdose of strychnine, by a marathon race, by foreign proteins, by anaphylaxis—the result of these overwhelming activations of the kinetic system is a condition which is identical, however it may be clinically designated, whether surgical or traumatic shock, toxic shock, anaphylactic shock, drug shock, exhaustion etc." Both kinds of shock (here I follow Tim Armstrong, historian of modernism) had to do with compressions of time. The *Bruitism* of the Futurists compressed the future into the present, so that noise was fully "resident" (quoth Douglas Kahn, historian of the arts of sound). Dadaists used *Bruitism* to compress an archetypal past into the present, so that noise was, rather, reminiscent. It was no coincidence that the shapes sculpted by Hans/Jean Arp would recur in the dreams of Jung's analysands, or that Huelsenbeck, after drumming up Dada, established himself in New York as a Jungian analyst. If we accept Armstrong's distinction between "notions of trauma grounded in the catastrophic wounding of the body or the psyche" and "those shock-effects which exist within an economic conception of the body, which simply represent the flow and processing of its interchange with its environment," then the Futurists were catastrophists of noise, the Dadaists its economists.[84]

3. Silences

Under the economy of the wooden horse ("Dada," in French), silence had a more pervasive role. As did silence in the *pittura metafisica* of Giorgio Morandi, who had suffered a breakdown after enlisting in the Italian army in 1915, then withdrawn to his native Bologna to paint still-lifes of glass bottles, coffee pots, and clay vases, objects whose serenity, dusted by "remembered" light, sat in *contrapposto* to the war beyond. As did silence in the architectural planes and solids of Giorgio de Chirico's *Le Muse inquietante* (1916), inspired by his wartime hospital assignment in Ferrara, "city of silence." As surreal wooden figures and lipless dressmakers' dummies, his Disquieting Muses were cut off from conversation yet appeared to be asking, as did the English poet Marian Allen, "And what is war?"

STATIC!
something ought to be done about this
March 13, 1928.
J. A. GLASSMAN
LOUD SPEAKER
Filed Feb. 23, 1927
FILL IT UP
STATIC
STOP LOOK LISTEN
916,885.
Patented Mar. 30, 1909.
[silence 2'00]
SILENT
as the Sphinx...when your transmitter's off
LEAPS INTO ACTION WHEN IT'S ON!
"Drop that Cough"

A growing stillness, many empty places,
A haunted look that comes in women's eyes...
A sudden gust of wind, a clanging door,
And then a lasting silence—that is war.

When, finally, the killing came to an end and armistice was declared, silence invited philosophy; philosophy, silence.[85]

Kafka, in his "Silence of the Sirens" (1917), had Ulysses eluding them at first by orneriness and Ohropax, bound to the mast by chains, his ears plugged. (Homer had allowed Odysseus/Ulysses the heroism of listening to the Sirens, whence the chains; during the Great War, heroic listening was no longer credible, whence the Ohropax.) Then Kafka bethought himself of the new powers of sirens: warning of poison gas attacks along the Front, of air raids in the streets of London or Paris. It occurred to him that a Siren/siren—*die Sirene/die Verführerin*—might be at her/its most irresistible *when silent*. Perhaps the siren songs were memory screens, and all the Sirens needed to seduce a man from rough seas or deafening barrage was the promise of stillness, alluring to sailors who had next to brave the strait of Scylla and Charybdis. This silence of the Sirens, was it perhaps known to Ulysses, who chose not to shame them with his knowledge, so made a pretense of resisting their... musics? As does Leopold Bloom in the Sirens section of *Ulysses* (1918–1920), where James Joyce takes his hear-all hear-o through light opera and music hall songs and a whistled rendition of *Tutto è Sciolto* (All Is Lost Now) from *La Sonnambula* ("A low incipient note sweet banshee murmured all. A thrush. A throstle.... Blackbird I heard in the hawthorn valley. Taking my motives he twined and turned them. All most too new call is lost in all. Echo") to a sketch of shellshock ("Yet more Bloom stretched his string. Cruel it seems. Let people get fond of each other: lure them. Then tear asunder. Death. Explos. Knock on the head. Outtohelloutofthat") and a theory of acoustics as Musemathematics all the while realizing that Blazes Boylan right then is having his way with his beloved Molly ("Her wavyavyeavyheavyeavyevyevy hair un comb: 'd") as the talk in the hotel bar teems with offcolor otology, "Sure, you'd burst the tympanum of her ear, man, Mr Dedalus said through smoke aroma, with an organ like yours." A man and a woman hold a shell to their ears: "The sea they think they hear. Singing. A roar." Shell shocks? "The blood is it. Souse in the ear sometimes. Well, it's a sea. Corpuscle islands." An auricular Brownian motion that keeps the interior monologue going, a kind of towncrier or tinnitus.[86]

"Don't talk.... Spies are listening," cautioned a poster created for the U.S. Army Intelligence Office and featuring the head of the Kaiser on the body of a large spider: a silent presence luring careless words to its web.[87] The war had not been, could not be, all noise. Hydrophonists and other

Listeners demanded silence. The general manager of the Los Angeles Submarine Boat Company, John Milton Cage, who had improved on the surveillance hydrophone and patented a diesel system for underwater propulsion, lost a Navy contract to build his "Submarine of California" because of too-visible exhaust bubbles and too much noise from the diesel engine for a vessel meant to lurk in the silent deeps.[88] "Zones of silence," which divided the inner sound-area of a barrage or explosion from an outer sound-area extending as far as two hundred miles, were more evident than had been the acoustic shadows of the American Civil War, due to the greater noise of high explosives and the intensity of twentieth-century cannon, which made for more dramatic contrast. Attempting to chart and predict these zones of silence, which floated in odd rhomboids behind the Front, meteorologists struggled to line up the variables: wind vectors, temperature changes at times of day and different altitudes, the gaseous composition of lower and upper atmospheres, topography, frequencies of explosion. Since zones of silence were part of the same acoustic phenomenon that *increased* audibility on the rim of the outer sound-area far from the centerpoint of explosion, civilian ears well away from active battlefields would every so often experience the barrages of the Front as surprisingly strong and close, as if a weather front of frozen sounds had unfrozen of an instant and stormed down upon them out of nowhere.[89] So silence was as elemental to the war as was noise, and in 1915, when Francesco Cangiullo presented his terse Futurist drama, "Detonation," silence shared the lead with a bullet:

> Night. Cold. A deserted street. A reverberation. One minute of silence.
> A pistol shot.
> One minute of silence.
> Curtain.[90]

Le cafard, French soldiers called that time of silent waiting in a trench between assaults or bombardments when a curtain of darkness might suddenly descend upon you, "you don't know why, and you start considering all the reasons for feeling depressed." *Le cafard* appeared most often at dusk, when "You feel cut off from the rest of the world," and in those rare intervals of quiet when a sergeant could dig the mail from his pouch, calling out names and listening into the air for the dead to come forward and claim a letter. English soldiers heard in the silences of the Front the improbabilities of natural beauty, which opened old wounds. "You will have a terrific tearing and roaring of artillery and shot in the dead of night," wrote J. C. Faraday to the *London Times* in 1917; "then there will be a temporary cessation of the duel, with great quietness, when lo! And behold and hear! Hearken to his song! Out come the nightingales, right about the guns." Or in a momentary hush of

> the shriek of hurtling shells: and hark!
> Somewhere within that bit of deep blue sky,
> Grand in his loneliness, his ecstasy,
> His lyric wild and free, carols a lark.

Soldiers along the Western Front had also sung, together, during the silence of the Christmas/New Year's truce of 1914, initiated not by the generals but by infantry on both sides of no-man's land sharing photos or smokes, trading harmonica solos and renditions of "Silent Night/*Stille Nacht*." A British soldier remarked then on "the strange unreality of the silence of the night, of the silence of the moon in the sky"; a German noted the "extraordinary hush, which seems quite uncanny." This sort of silence, like all things *unheimlich*, had an oddly independent presence, indebted to no frame of sound "but perceived and pondered on its own." The silence could even be a sounded silence, itself almost unbearable, as at Gallipoli in 1915, at the "most weird concert" that Horace Bruckshaw ever attended. With lightning in the sky and artillery "roaring a solo with a chorus of rifle fire, stray bullets even reaching the spot where we were," his fellows were singing "comic, secular and sacred songs. The limit, however, was reached when Gilbert Wilson, a chum of mind, and who is a professional, sang *Will o' th' Wisp*. He sang it splendidly but the effect was almost unearthly." Too much was passing too quickly, out of range of words.[91]

Almost unbelievable, then, would be the rumors of an armistice that reached a rehabilitation ward near Le Havre in 1918. "November 11th dawns—the skies are dull and grey.... A heavy air of depression pervades the place. The boys lie on their beds saying little, and that little in muted voices. The nurses move silently about." No joking, no bantering, noted an aide, Lena Hitchcock. "Suddenly striking the air the bells ring out, the whistles shriek. We pause—our very breath suspended. Someone makes a queer, throaty sound. Miss Cameron's voice pierces the stillness. 'Well, they've signed.'" A Red Cross worker enters the ward waving flags and shouting *Fini la guerre!* "She goes from one to another with her hateful clack and din. Miss Cameron and I both try to stop her, but it's impossible." The men grimace. One, in great pain, "fears the war will have to be done all over again."[92]

Some *will* do it all over again: the shellshocked. Some, the culture-shocked, will feel that the war in its sonic sense has never ended: could anything be done, a Harvard alumnus will ask in 1925, "to heal the blows of sound"?—blows coming blast and furious from a juggernaut of steam-shovels and cement-mixers commanded by the shrill voice of "Hula-Hula-Ragtime-Beelzebub, alias Jazz, the central figure in the religion of noise." Sober people (like Earley Wilcox, Harvard Class of 1895) demand relief "from the shell-shock of unnecessary noises"—and income tax. To Henry

Hazlitt, apologist for the Free Market, those noises will seem "entrenched" in the modern economy, although "The fact of shell-shock in war should be rather good evidence of what can happen when noise is bad enough; it seems at least a fair inference that the constant rumbles, clangs, honks, shouts, screeches, chitters and roars of the city must finally consign the more sensitive to various mild forms of shellshock." Deafened or dead-fingered, industrial workers will take the wartime referent to heart: "I used to work a pneumatic drill but I cannot any longer; the constancy of the action is so much like a machine-gun." A university psychologist, researching the afterwaves of shellshock in 1926, will find that miners were least able to carry on as usual after the war; they were far too susceptible to "redintegration," a neurotic reaction "wherein part of an original situation calls up the whole original response: a sudden noise, for instance, will recall the entire neurotic display originally manifested by the patient in response to the overwhelming stimuli of the front-line situation of which the noise was an element." Our brains, a French doctor will write, absorb hundreds of "shocks" that are folded into the layers of our consciousness—to rebound (a French reformer will claim) when cued by noise. A psychoanalyst at City College will reveal that New Yorkers suffer from "Peace Shell-Shock," having caught the malady "largely from riveters." In engineering terms, "Peace Shell-Shock" is as inescapable as the piston knock that arose during the Great War with the advent of light, high-speed motors for tanks and aircraft. Such engines made it "very difficult to obtain silent running, for at the end of the compression-stroke the piston is thrust violently from one side of the cylinder to the other." Will anyone draw the analogy tighter? Lead tetraethyl might calm the high-speed engines that had accelerated across the Front; what will cure the ptupomania, "or mad craving for noise," of the Roaring Twenties? "The word 'civilization,' is said to be derived from the word 'quies,' meaning rest or quiet," will write an eye-ear-noise-&-throat man of Washington, D.C., having taken leave of his dictionaries in his haste to attack the mania; "the idea being that through the freedom from the necessity of labor one secures leisure, rest, and quiet; but the unexpected has happened, and there has come instead unrest, disturbance and noise." And he who will take up where Julia Rice leaves off, Shirley Wynne, M.D., Commissioner of Health for New York, will plead for more "civilization" to cope with the din of the city, which is doing to citizens "the same thing on a smaller scale" that shellshock had done to soldiers of the late War. Neurosis, the Noise Commissioner of London will explain, "may be attributed to noises just as in the case of shellshock."[93]

War's end, then, would be best commemorated neither with fireworks nor with the hoopla of a Red Cross greenhorn but with silence.

Two minutes of silence. National silence.

For the first time it was possible for a willed silence to be thoroughly national, even international. Although churchbells had been rung across the realm at the deaths of queens, kings, presidents, or heroes, and although common folk had gathered in shuffling quiet as cortèges moped down the road or a train in black crepe steamed slowly down the tracks, the collective silences of the past had never been synchronous across townships, provinces, or nations. Nor had they been fully silent. At the funeral procession for Umberto I, assassinated by the anarchist Gaetano Bresci in 1900, "the crowd in the streets was extraordinarily quiet" amid "the tramp, tramp of the soldiers' feet" and "the muffled drums of the dead march" when of a sudden "the silence—which till then really had been remarkable—was broken by a sound like the buzzing of thousands of insects." Asked an American visitor to Rome, "Who can these be?" "The lawyers are coming," said her friend Patsy, and come they did, chatting injudiciously above the clank of their ceremonial gold chains. Then someone knocked over a chair and the crowd panicked, shouting "Anarchists!" "A bomb!" The next year a similarly nervous hush enveloped the funeral of President McKinley, whose assassin had been inspired by Bresci. Thirteen years later there was an unseemly hurry to the passage of the cortège of Archduke Franz Ferdinand, heir to the Austro-Hungarian throne, and his wife Sophie, assassinated in June 1914 by a Serb nationalist who thereby set off the guns of August for what would be called by 1918 the "First World War" (not in expectation of a Second but in contrast to all wars previous). So, once the Versailles Treaty was signed, it would be doubly important for memorial silences to be synchronous across borders, signifying both the global community of fallen heroes and an end to nervousness about sudden deaths. And now that the world shared a grid of time zones, with master clocks electrically synchronizing local minute hands, the Allies could effectively coordinate a ritual silence commemorating the end of the war. One hundred twenty seconds of silence for a war during which one-fifth of Europe's population had been killed, wounded, or driven from their homes by the longest, heaviest barrages in history. One second of silence for every quarter-million soldiers killed or wounded by a war in which casualty rates for the major European (and Australian) armies ranged from 44 to 75 percent, and in which young men from every populated continent had fought and died or lost their legs or arms or minds. A war some of whose survivors woke years after in the middle of the night to the sound of shellbursts.[94]

Each year from 1919 until 1939, at 11:00 a.m. on 11/11, trolleys came to a halt, factory workers put down their hammers, schoolchildren shut their books, and the victors of the Great War stopped to reenact the muteness of shellshock. The muteness was in line with a mode of mourning modeled by Later Victorians of the upper class and funeral directors for the middling sort: quiet and composed, with caskets that could be opened "by springs, without any noise." It was a muteness implored of families by soldiers on the Other Side, who directed mediums to "tell the mothers and fathers and sisters and wives to stop crying. No man can stand the sight of tears, the sound of sobs. They feel it much worse here." It was a muteness that had been customary Down Under but was more far-reaching, for Australians and New Zealanders had now to keep company with a "phantom army"—one in five ANZAC volunteers died in the war, and 25,000 went missing; half of all Australian families were bereaved. In 1921, mediums made 4,332 attempts to speak with the dead or missing on behalf of these families, but otherwise the bereaved kept their silence, "For they who mourn sincerely / Mourn silently and low." Muteness, too, seemed to be all that was possible at the end of a war of blast and vanishing:

> When it is over, the noise and the clamour,
> Then come the creeping years, heavy with silence—
> Peace cannot break it, and we who must suffer
> Must, and in silence.[95]

Armistice Day 1918 had itself been greeted with war-whoops and artillery: newsboys crying "Extra!" with sirens wailing, rockets stuttering into the air. There was some looting of the shops of pacifists, some last-minute Kaiser-baiting and newly anxious Red-baiting, and everywhere scenes of bedlam ("Wild Crowds Howl Kaiser's Dirge in Loop/Whole City Goes Mad," read the banner of the *Chicago Tribune*). In London the Fabian reformer Beatrice Webb heard "a pandemonium of noise and revelry," but the day was cold, dark, and wet, a day—as Virginia Woolf found—of loutish, staggering revelry, rather dull-eyed than joyful. Soldiers on all fronts still huddled in their trenches, already grousing about the pace of demobilization.[96]

Terms of peace, less than clement, would not be agreed upon for seven long months, during which time thousands of soldiers died of war wounds and thousands more (like Wittgenstein in Italy) remained as prisoners of war, hostage to the politics of the *post hoc*. After the signing of the Versailles Treaty in June 1919, the South African politician, adventurer, and children's book author Percy FitzPatrick proposed to King George V that Armistice Day of 1919 be dramatized with an "Empire Silence." FitzPatrick's son Nugent, of the South African Heavy Artillery, had been killed in 1917 in France, and FitzPatrick would plant a grove of trees in Nugent's memory; the idea for a public silence arose from one he had witnessed at the reading aloud of South Africa's first war casualty lists in local churches, and/or from the daily two-minute noontime silence promulgated in 1916 by the mayor of Cape Town, inspired perhaps by a businessman who urged the city to set aside a minute of thanksgiving for those who had survived and a minute of grieving for the fallen. And/or from the *London Evening News*, which in May of 1919 published a letter by E. G. Honey, a journalist from Australia, where communal silences were customary as memorials for those killed in mine disasters. Residing in London, Honey now proposed a five-minute Commonwealth silence, "Five little minutes only," for "Communion with the Glorious Dead who won us peace, and from the communion new strength, hope and faith in the morrow." Yet another precedent was more definitely ascribable, and impossibly poignant: on January 8, 1919, at 2:15 p.m., Americans had observed a minute of silence in respect for the passing of Theodore Roosevelt, who in 1918 had been painfully deafened and blinded by a throat infection that spread to his middle ears ("bilateral acute otitis media"), thence to the mastoid process and left eye, and who, in the throes of this agony, learned that his youngest son had been killed in aerial combat over France. The American, Australian, and South African precedents convinced the War Cabinet to press the King to declare a "Great Silence" for the anniversary of the Armistice, "which stayed the worldwide carnage of the four preceding years and marked the victory of Right and Freedom." Upon a trial among the Grenadiers at Buckingham Palace, Honey and FitzPatrick both in attendance, five minutes proved to be insufferable.[97]

Believing that "my people in every part of the Empire fervently wish to perpetuate the memory of the Great Deliverance, and of those who laid down their lives to achieve it," George V then proposed "a brief space of two minutes" on 11/11 during which "all work, all sound, and all locomotion should cease, so that, in perfect silence, the thoughts of everyone may be concentrated on reverent remembrance of the Glorious Dead." Just before eleven a.m., citizens were alerted to the oncoming silence by loud guns, rockets, churchbells, fire alarms, and a bugler on the balcony of Selfridge's Department Store, "a tremendous burst of synchronized noise across the country" and an acoustic irony that few remarked. "Nobody can imagine what a silent London—still for two minutes—is like," wrote the editors of the *Daily Express*. "No traffic, no business, no talking—nothing for a brief but sacred space of time." In Manchester, "The tram cars glided into stillness, motors ceased to cough and fume, and stopped dead, and the mighty-limbed dray horses hunched back upon their loads and stopped also, seeming to do it of their own volition," according to the *Guardian*. "The hush deepened. It had spread over the whole city and become so pronounced as to impress one with a sense of audibility. It was a silence which was almost a pain." Over the years, as the Great Silence was stamped into the calendars of the Allies, that pain would subside for all but the shellshocked, who might go "all of a tremble" at the prelude of cannonry, then at the silence itself, which was too awful a reminder of having been concussed into deafness or buried alive. By 1921 the bodies of Unknown

Soldiers (English, French, Canadian, American, Australian), representing the three millions who had gone missing in the mud of no-man's land or been so obliterated as to have lost all (but national) identity, had been formally interred, and the ceremonies of 11/11 converged upon their tombs.[98]

Here and there, and increasingly during the Great Depression, the silence would be interrupted by down-and-out veterans shouting "Anyone want a medal?" or Communists singing anthems of the working class. To these, the Great Silence was less a reverent nod toward a lost generation than a cover-up of the brutalities of capitalism, a cloaking as transparent as the "golden silences" explored by Ferenczi in a foray of 1916 on analysand silence as a form of anal retentiveness and resistance to therapeutic transference. After all, Hiram Percy Maxim's famous silencers were only cover-ups for the speedier murder enabled by his father Hiram Stevens Maxim's lightweight machine guns that could "mow down an entire army" at six hundred rounds per minute. A suspiciously Freudian cover-up: the father's weapon, totem of modernity at its deadliest, symbolically magnified by his own gun-deafness and then his death in mid-war; the son's taboo, his hatred of noise, and his silencer.[99]

Silencers had caught the public's ear well before Hiram P. came on the scene. Clockmakers since the 1830s had taken to putting a "silent" switch on clocks with chimes, for those (invalids, mothers with sleeping babies) who found the chimes upsetting. In 1861 a London shop began selling a "silent practice drum" patented by John Azémar; *Punch*, influenced by its noise-hating cartoonist John Leech, hoped that Azémar would turn his hand to a silent piano, silent barrel organ, silent brass band. Nothing doing there, but twenty years later Grant Allen, a popularizer of science, would pick up the gauntlet with *A Tale of Love and Dynamite*. Sydney Chevenix, a chemist, has been experimenting with mut(at)ing nitroglycerine and an associated "self-compensating double wave-action, whereby the sound undulations, in their mutual interferences, spontaneously deadened one another so as to be absolutely inaudible." The prospect of a "perfectly noiseless detonating compound" excites him: "Just picture to yourself the use of such an explosive in war, for example! You're fighting a lot of uncivilized enemies, and you send out your sharpshooters under cover somewhere, and they pick off the enemy, one by one, noiselessly, silently, unseen, unsuspected; and the unsophisticated savages don't even know they're being shot, or where the firing comes from, but merely find their men dropping down all around like magic by the dozen..." His assistant, a Polish nihilist, prefers to contemplate its use in "practical politics." Say "You're fighting a lot of enemies of the human kind—emperors and bureacrats and such-like vermin—and you stick a little bit of the new explosive under the chief criminal's bed, and it goes off pop in the middle of the night, noiselessly, silently, unheard, unnoticed.... That would indeed be developing the resources of civilization." Love, wouldn't you know, gets in the way of such-like civilizing, and the love-object herself, Maimie, becomes the first to make practical use of the "anacoustic explosive." She takes up Sydney's experimental noiseless pistol and shoots him in the back, silently.[100]

First on the military-industrial agenda, however, were silencers for rifles. The noisier a soldier's rifle, the greater the risk not only of ear damage but of precise retaliation. (The report of a Springfield rifle could be heard three miles away, and the United States was paying out large pensions to veterans who as artillerymen and gunners had lost their hearing on the job.) Technically, as explained in patent applications, the trick was to regulate the volume of gases escaping the barrel of a gun by means of "a simple, light, yet durable device adapted for attachment to the muzzles of rifles" that will "materially modify or entirely prevent the noise made when a weapon is discharged... without impairing the power or accuracy of a weapon." This was done most profitably by the "silent firearm" patented on February 25, 1908 by Hiram P. Maxim, whose stated purpose was diplomatically divided between the military (tactical concealment) and humanitarian (kindness and cochlear security). His silencer would be useful on the domestic scene, he said, for putting down cats and dogs in animal shelters or cattle in slaughterhouses, where the noise from guns "maddens" the poor beasts. "I do not know how silent the New York S.P.C.A. wants the gun to be used by it to be," he told the papers, "but I gave them enough silence to satisfy them." In a demonstration on February 8, 1908 in the library of a New York law office before sportsmen, scientists, and reporters, he proved that with a section of his patented pipe screwed onto the end of a rifle barrel, the report would be "less than would come from the whizzing snap of a toy air pistol." By then other silencers, or mufflers, were already reducing the noise of automobile exhaust pipes by creating air traps to deal with back-pressures, but Hiram P. always said that he had happened on the principle of his silencer while stepping out of a bathtub and watching the water swirl around the tub down the drain. The secret to quieting a gun must therefore be to inscribe a series of spirals on the inside of the barrel so that the heated gases, like bathwater, would be slowed on their exit and no longer burst from the barrel in a fury. So construed, his silencer reversed the visual acoustics of the Edisonian phonograph cylinder and turned phonography inside out, using spiral grooves not for fusing but defusing tracks of sound.[101]

Hudson Maxim, his uncle, had invented Maximite, a smokeless high-power explosive; used in long-range rifles, it kept soldiers out of the sights of their enemies. Hiram S., his father, had made his fortune by inventing the spring-loaded mousetrap and then by harnessing the energy of gun recoil to a reloading mechanism that made firing nearly automatic. Whence the "machine" or Maxim gun, which distanced ranks of soldiers from hand-to-hand combat and led to the creation of a deadly *terra nullius* between battle lines. Son and nephew of engineers, Hiram P. as a child had been taken with the surge and sophistication of locomotives but distressed by their noise, which "hurt my ears and frightened me." Welcomed onto an engine at the age of seven, he had been "completely overcome by the hissing noise" and appalled by the engineer's voice, "the harshest and most rasping I had ever heard in all my life." He guessed "it had to be this way in order to penetrate the awful noise," but he would not let loud enough alone, whether as superintendent of the American Projectile Company or as an engineer in 1903 for what had been Isaac Rice's Electric Vehicle Company, whose quiet cars he preferred to those gas-powered behemoths for which he too designed mufflers. He would work until he achieved the kind of combustive silence that bore within it the contradictions embodied by shellshock. "I am fully conscious of the awful possibilities of this gun," he confessed, "and my conscience is not at all easy on the subject. The rustling of the leaves in the trees or the noise of passing wagons would completely drown the sound of the [silenced] gun. You would not even know what killed the victim until an examination was made."[102]

"Maxim Robs War of its Thunder," read the headlines in 1909. With silenced smokeless rifles, all else an army needed to advance inaudibly and invisibly across battlefields would be a "chameleon uniform," the kind that would be worn during the Great War once quartermasters abandoned the gaudy for the drab and a cadre of Vorticists and Cubists began to apply "camouflage." If Maxim's invention were embraced by the generals, the silence of war would be "incalculably more terrifying than the roar of artillery and the vicious crackling of small arms fire," for "In the battle, soundless except for the bullet whine and the shell scream, death will seem more sudden, inexplicable, horrible, and panicky than ever." Under such ghastly quiet, predicted the *Minneapolis Bellman*, war would cease, having become "too dreadful for human nerves," especially if, as Maxim thought possible, his silencer were fitted to cannon, and there could be "No noise, no fuss, no muzzle blast." Ambrose Bierce, whose *Devil's Dictionary* defined noise as "The chief product and authenticating sign of civilization," heard silencers otherwise, as morticians burying civilization: terrorism would become unstoppable, as "Anarchy, discarding her noisy and imperfectly effective methods, gladly embraced the new and safe one." The *Cincinnati Commercial Tribune* reacted to this "New Battle Terror" as had Clara Barton to the terrible silence of Isaac Rice's submarine: "To those who desire world peace and who have been through the trial of blood and battle, the new noiseless rifle has a distinct meaning."[103]

Anarchists nor armies, as it turned out, had much use for silencers; both relied on noise as a weapon that could be more concussively sudden than any quiet, hence more startling. And what sort of urchin would seek out a silencer for target practice within the city, as suggested by 1911 Maxim ads: "Many short opportunities to shoot which he has not had because of the noise"? Newspapermen, playwrights, and film directors, extrapolating from rifles to revolvers, could think of a different clientele: criminals. No matter, as Maxim repeatedly explained, that any revolver however silenced would still make a noise "like the cracking of a whip" due to the speed of its bullets and the loose joint between cylinder and barrel. "With the noiseless pistols in the hands of the stickup men," decided a columnist for the *New York Star*, "the man making his way home after dark would have just about as much chance as a pink ostrich plume in a blast furnace." The seminal literary effort was Bayard Veiller's Broadway play, *Within the Law*, produced in 1912, where Inspector Burke catches up with Joe Garson, who shot Eddie Griggs "Because he was a skunk and a stool pigeon.... I croaked him just as he was goin' to call the bulls with a police whistle. I used a gun with smokeless powder and a Maxim Silencer, so it didn't make any noise. Say, I'll bet it's the first time a guy was ever croaked with one of them things—ain't it?" Like Joe the forger, Bayard the writer was surely proud of the neat twist of croaking a squealer with a silenced gat. Ever since, the device has taken on a sinister cast in film and fiction, but when news first spread of Maxim's patent, commentators were ready with more benign applications: a cryless baby, a barkless pooch, a wordless shrew, a frustrated firecracker. A cartoonist for the *New York Evening World* posed "A Modern Paradox" in two frames: above, a soldier asleep beneath a Noiseless and Smokeless Gun, thinking "War is a Dream"; below, an urbanite in bed, hands over his ears, thinking "Peace Is Hell!" as an alarm clock goes off, a telephone rings, tugboats toot, trolleys clang, and roadsters honk up into his window.[104]

Himself "appalled" by the military implications, Hiram P. & Co. stopped making silencers for firearms in 1926 and focused on peacetime noises, which "injure health, increase insanity, decrease production, reduce the value of real estate, and cost humanity incalculable sums of money annually." That year Sir William Joynson-Hicks of the British Home Office heard the noise of motorcycles as an unacceptable echo of war, and in response to a manufacturer who suggested that it was hard to tell apart "an adequately and an inadequately silenced motor cycle," dictated this reply: "Sir William Joynson-Hicks cannot understand how anybody can feel that there is a difficulty in deciding between a machine which is decently silenced and one which goes past a row of houses in the middle of the night like a Gatling gun." Aside from mufflers, the "noiseless" typewriter and file drawers sliding on roller-bearings had been signal successes; eliminating other sonic sources of aggravation and fatigue in the workplace would be a good thing—so good, said an American civil engineer, that "it makes better citizens." Toward that end, and others more plainly commercial, the

Maxim Silencer Company convened a champagne caucus of engineers and scientists on New Year's Day 1928 to discuss the ways and means to make the engines of civilization "absolutely silent." But Maxim, an advocate for amateur radio, was no eunuched Silentiary. A 1925 catalog for *Maxim Industrial Silencers* referred not to silencing the world but to methods of "noise extraction" that Hiram P. hoped to apply to pneumatic drills, air compressors, aircraft engines, and offices. Diesel exhausts at power plants were the worst offenders, their three-foot-wide pipes loud as artillery and audible fourteen miles away. With apt technology, these too could be softened, as could other products now promoted in the pages of *Good Housekeeping*: Hoffman radiator valves that "prevent hissing and leaking"; Studebaker sedans, whose bodies would not "rattle or squeak after thousands of miles of hard usage"; Westinghouse fans, famed for their "Beauty Power Silence." In the meantime there should be "hospitals of silence where persons injured by noise or needing rest from noise, may have quiet," and invalids at home should be protected from the rattling of dishes. While the Hush-a-Phone company was selling "voice silencers" (small parabolic sound reflectors that allowed switchboard operators to whisper and yet be heard), Maxim was working to make industrial gigantism more acoustically tolerable.[105]

"Great flakes of snow were falling as relentlessly and silently as the bullets from a Maxim soundless gun," wrote Wex Jones in his *Stories of the North, South, East and West* of 1908. Among the first instances of a silencer translated into simile, other instances swiftly followed, as in the sports section of the *New York Evening Journal*, where a columnist listed the many "Silencers" before Maxim's: Jack Johnson in the boxing ring, nipple bottles for crying babies, cops bashing demonstrators, and Death, "The Great Silencer." There it was again, that ancient pairing of death and silence which the noisy had always used in their defence (noise = life; silence = death), but more of silence was in the air than Death, and for more than two minutes. As George Steiner has argued, the development of complex mathematics since the seventeenth century and of symbolic logic since the nineteenth had distanced the sciences and social sciences from workaday language; the more abstruse their symbols and operations, the more silent they would appear to the uninitiated, assuming a transcendent power to access the Truth without words, without a sound. And as Harold B. Segel has argued, even the arts of the West had begun to turn away from spoken words toward mime, expressive dance, and shape-poems, while the Belgian dramatist Maurice Maeterlinck "touted the superiority of silence over speech in the drama as the gateway to cosmic mystery." On the postwar stage, a "school of silence" tutored by the French playwright Jean-Jacques Bernard relied on gesture and long pauses. Theater, wrote Bernard, should devote itself to the inexpressible, unconfessed, and unconscious, as in his *Martine* (1922), whose rural heroine is "mute and defenceless, caught in a net of words and actions for which she is not responsible, which she cannot name or understand." The play comes to "a silent conclusion in which nothing is heard but the ticking of the farmhouse clock asserting unbearably the implacable routine of daily life." Inadequate to the task of declaring meaning amid the relentlessness of the quotidian, the furor of the industrial, or the barrages of war, language must cede to silence the powers of refuge and remediation.[106]

"Anything a man knows, anything he has not merely heard rumbling and roaring, can be said in three words," claimed Ferdinand Kürnberger. It was this nineteen-word apothegm from a "disillusioned revolutionary" that Wittgenstein took as the motto for his *Tractatus Logico-Philosophicus*, finished in 1918, published in English in 1922, and ending in a nine-word caution: *Wovon man nicht sprechen kann, darüber muss man schweigen*—"What one cannot articulate, one must silently pass over." But holding one's tongue when one *cannot* speak may be as diagnostic of trauma as of intellectual restraint. Working through whatever traumas led him to his own postwar silence, Wittgenstein would return to words and philosophy a decade later, never again to give up on them, though indifferent to seeing them in print. Some of his contemporaries would more happily engage in that contradictory practice (or "apophasis") of publishing extensively about things so profound or ephemeral that they can only be expressed indirectly or through negation.[107]

During the war, the four hundred apophatic pages of *The Empire of Silence* had already made their appearance. The empire's annalist, Charles Courtenay, was an Anglican cleric with an ecumenical sense of quiet, his volume a chrestomathy of quotations ranging from the Neoplatonist Porphyry ("the worst man, even though silent, offers praise to God through his silence") to the Friend Caroline Fox ("How I like things done quietly, and without fuss. It is the fuss and bustle principle, which must proclaim itself until it is hoarse, that wars against truth"). Silence fell into categories: compulsory, contemplative, belligerent, restorative, punitive, reverent, pedagogical, proverbial. Each silence had its own timbre, mellow or gruff, and the silences of "the world's most stupendous war" were most grating—torpedoes fired from unheard submarines, bombs falling soundlessly through air, poison gas floating over no-man's land. So one did not always welcome silence, nor could one always choose one's silences, literal or metaphorical. "*In stormy times* we must be silent. There are electric atmospheres which a single word will precipitate into a storm," as in the Alps, where climbers must keep "the strictest silence, lest the vibration of their voices should precipitate the threatening avalanche."[108]

Such a silence was an imposed silence, originating in noise or the fear of noise. Courtenay understood the societal implications: "A silence which is noisily created is in constant need of being re-created, since it is made from without." That was how the Armistice silence felt to a charwoman telling off a general in *Two Minutes' Silence*, an Australian anti-war play of 1930:

> The General: . . . I insist that you pay a proper observance to the two minutes' silence, which I have rigidly enforced in my household since the first Anniversary of the signing.
>
> Mrs. Trott: I ain't likely to forget it. And I don't need no silences to tell me nothin'. There is plenty of silence for me when I go 'ome too tired to sleep, and them's the sort of 'orrible silences yer can't stop when yer want to.

Silences scarcely less war-torn and noise-born were meanwhile being embedded in modern philosophy. Twenty-five times between 1914 and 1918, Edmund Husserl at the University of Freiburg had fled with other townspeople into private cellars and public shelters to escape the bombs dropped by British planes in a series of air raids more frequent over Freiburg than any other German city; between raids he would have heard the barrages from a Front just thirty miles off, so loud they could drown out a peal of cathedral bells. Having lost a son and peace of mind during the war ("Can we accept this horrible 'concrete' world as the real world?" he asked a friend in 1917), Husserl began to insist on a self-silence deep enough to hear oneself speak, a quietness of voice through which a person could come to a transcendental experience of the presence of others and an ethical regard for their voices. Husserl's assistant in philosophy, Martin Heidegger, who had sat through the war behind a desk in the Office of Postal Censorship, took silence a step further: the language of being was essentially "the ringing of stillness" through whose "soundless gathering call" the essentiality of others' lives could be shared. Only through an interior silence could language assume its finest functions, attention and conscience. "In talking with one another," wrote Heidegger in 1927, "the person who is silent can 'let something be understood,' that is, he can develop an understanding more authentically than the person who never runs out of words." Not that taciturnity was the optimum. "Authentic silence is possible only in genuine discourse; in order to be silent, *Da-sein* [the self-interrogating being] must have something to say, that is, must be in command of an authentic disclosedness of itself." One's capacities for "hearkening" to others were often impeded by deference to intrusive technologies and by the noise of a man-handled world that interrupts itself with the imperiousness of buzzers or telephones or the bells of Big Ben tolling through the Two-Minute Silence.[109]

Recondite stuff, these ontologies of silence. For the masses there was a snow-white cream to deal with "Something men never speak about that every Girl should know," and Mum is the word! For the middle classes and aspirants thereto, there was guidance in poise and focus, which required an inner quiet. "A great nature," wrote Susanna Cocroft in one of her many pamphlets on physical culture, "is conscious of a sacred stillness, a Divine undertone, too deep for sound"; from the stillness comes serenity; from serenity, surety; from surety, poise; from poise, the power to sway others, for "Thoughts held in the silence continuously radiate through the voice." Theron Q. Dumont, author of *Personal Magnetism* and *The Power of Concentration*, advised his many readers to get in touch with "that deep silence of the subconscious mind, where deep thoughts, and the silent forces of high potency are evolved." Call it "Cosmic Consciousness" or a "wireless station, the message recorded coming from the universal mind," this silent source could help focus thoughts and overcome the flinch of muscles at the slam of a door. And wasn't it a sense of the command of silence that had attracted American voters to "Quiet but Convincing 'Cal' Coolidge," elected vice president in 1920? A Vermonter who perpetuated disconcerting pauses, he claimed that "I've usually been able to make enough noise to get what I want," which in 1919 had been to suppress the Boston Police Strike, but the Kansas journalist William Allen White, probing the mind of a man who became president in 1923 upon the sudden death of Harding, knew he had to "begin with Silence—find its root cause in Coolidge, and its deep need in leadership." If Coolidge's New England type was "taciturn, monosyllabic, crabbed, dry, weaned on a clothespin," Cautious Cal was also of Indian blood and heir to trappers who knew when to make their move. According to his autobiography, he owed his love of brevity and clarity to having grown up in Plymouth Notch, which "was all close to nature and in accordance with the ways of nature." Some said that he spoke so little because his voice had the quack of a duck. Others said that Cal had merely a Northeasterner's twang, girded by such judiciousness that he rarely "stopped thinking long enough to speak" (and when he did speak, he averaged eighteen words per sentence, in contrast to Teddy Roosevelt's forty-one, Woodrow Wilson's thirty-two). Mutt and Jeff, turning on the radio in 1924 to find out what the dopes in Washington were saying about a tax reduction plan, heard instead five minutes of "oppressive silence" and figured at last that "President Coolidge must be talking." Cartoonist Bud Fisher entitled this strip "An Oration by the President Consists of About One Word or Maybe Less," although as the stalwart of a Republican Main Street, Coolidge defended the noise of business: "the rattle of the reaper, the buzz of the saw, the clang of the anvil, the roar of the traffic are all part of a mighty symphony, not only of material but of spiritual progress."[110]

Then there was Meher Baba. If Coolidge was "a spiritual hermit, greedy of his silence," Meher Baba's silence was greedy of attention. Merwan Sheheriarji Irani was born in 1894 in Poona, near Bombay, grandson of the keeper of a Tower of Silence where Zoroastrians (Parsis) lay the corpses of their dead to be picked clean by vultures. Merwan's father had wandered Persia and India for ten years as a dervish, then settled down to marry, run a teashop, study Persian, Arabic, Gujarati, and Marathi, and write hymns for the Parsi community. His mother, also multifluent, encouraged Merwan in his collegiate studies of Urdu, Hindustani, Persian, Gujarati, Marathi, and English. A lover of literature, drama, and music, Merwan appeared to be headed toward a career in the arts, composing poetry and fiction and managing a theater company. When the company folded, though, he returned to help at the teashop and, soon, to take up the spiritual path his father had left off. In 1913, cycling along the Poona Road, he had met an old Afghan holy woman, Hazrat Babajan, who kissed him on the forehead as if a disciple. So he became, and in January 1914 she predicted that he would "shake the world to a great upheaval," whereupon he went into a nine-month trance or "spiritual headache." He revived by November 1914 as India readied 800,000 men for the Great War in a show of loyalty that nationalists hoped would persuade the British toward independence. Himself, he spent the war seeking out other spiritual masters (Sai Baba bopped him on the head to calm him), managing the tea (and toddy) shop, and in spare hours sitting beneath a Tower of Silence, "banging his forehead against the stones to relieve the spiritual agony." Blending paternal Zoroastrianism with the Catholicism of his early schooling, the Sufism of Babajan, and the Hinduism of Sai Baba, by 1922 he was a Sadguru, a Perfect Master, with a lingering habit of knocking his head against floor boards. Known now as Meher Baba, Father of Compassion, he built an ashram and attracted forty-five disciples who traveled with him throughout India and Persia, then built a small model town, Meherabad, at what had been a military camp.[111]

Early in July 1925 this lover of poetry, jokes, and radio commenced a year's silence, partly to mourn the imminent death of Babajan and "partly because of the disturbances, wars, and disasters which were coming to the world and to India in particular." In July 1926 he did not resume his songs or talks, disappointing female disciples for whom "His voice was the *sound* of Love." ("Now what is voice?" Baba had asked. "Really speaking, voice is everywhere in the universe.... When a sound is let out of the mouth, it is lost in the universal Voice.") Baba had not yet left off his silence when he embarked for England in 1931 aboard the same ship as Mohandas Gandhi, released from prison after leading the Salt March and headed to a London conference on Indian independence. Gandhi and Baba conversed, Baba using an alphabet board to whose letters he pointed in rapid succession, Gandhi urging him, "Baba, it is now time for you to speak and to let the world hear." Baba had once identified himself as the incarnation of a Hindu freedom fighter who had battled the Mughal Empire before it was supplanted by the British Raj, but now he remained silent. In his silence, however, he was rarely as ascetic as Gandhi: he laughed long and loud, hummed and thrummed, snapped and clapped; he listened to blissful music and enjoyed the vibrations in each of seven higher planes; he guffawed over Laurel and Hardy, enthused over jazz. In London he attended a Promenade Concert, visited the cenotaph of the Unknown Warrior, then sailed for Istanbul, Genoa, finally New York—where he was met at the docks by Malcolm and Jean Schloss, owners of a metaphysical bookshop. Baba took big chunks out of the Big Apple then denounced American egotism. He stood outside Sing Sing to reveal that "In this prison there is a man who is my agent; he does good work for me; I shall free him when I speak." He hobnobbed with Leo Tolstoy's daughter Nadine; in his capacity as The Awakener he told her, "[The mind] is like a wound-up alarm clock: it will ring at the appointed time." He visited Boston, where Thomas Watson, Bell's first lieutenant of telephony and a veteran seeker of more spiritual media, wept in anticipation of their meeting. The most celebrated of his admirers in America was Princess Matchabelli (née Norina Gilli, stage-name Maria Carmi), whose three thousand performances as the Madonna in Max Reinhardt's three-hour silent drama, "The Miracle" (1911 London premiere, 1924 on Broadway) had led her into the arms of Russian nobility and the marketing of a line of perfumes. Through Norina, Baba hooked up with Hollywood, where in the spring of 1932 he toured the movie studios and had a bothersome tête-à-tête with Tallulah Bankhead, who asked for a love potion. Disappointed that Greta Garbo, "most spiritual of stars," and most notoriously silent, was too ill to come to dinner, he watched her celluloid double the next day in "Grand Hotel," playing an aging, suicidal ballerina who tells her lover, most memorably, "But I want to be alone." The movie ends on a downbeat more Eastern than Western: "The Grand Hotel. Always the same. People come. People go.... Nothing ever happens."[112]

Baba himself was preparing for a stupendous event: he would break his silence on July 13 in the Hollywood Bowl, with Mary Pickford at his side and an international radio audience hanging on his every word. "It has been asked why I have remained silent for seven years, communicating only by means of an alphabet board, and why I intend to break my silence shortly," read a message from Baba. Answer: "Avatars usually observe a period of silence lasting for several years, breaking it to speak only when they wish to manifest the Truth to the entire universe. So, when I speak, I shall manifest the Divine Will, and world-wide transformation of consciousness will take place." And where else but Hollywood, which after thirty years of silent films had broken through to song and speech, just in time to dispel a global Depression?

> Only Hollywood will be able to delineate
> Baba's life story to the world,
> which will listen to his Song's resonance because of it.

> His Song, though echoless, merges within itself
> all the echoes of the world!
> And in him are all the images of the world!

First, though, he had something urgent to do in China.[113]

Off he went and away he stayed, with Mary Pickford left to explain to the tens of thousands of expectators, and to dozens of women wearing *haute couture* "God-realization dresses," why Baba had scurried back to India after his dash to China instead of appearing as promised at the Bowl. That week, as the California Bonus Army lay siege to Sacramento and a national Bonus Army of unrewarded veterans was encamped in Washington, D.C., Pope Pius XI had announced, "The Time is ripe for intervention by God." Premier Herriot of France, after sixty hours of negotiation among European leaders to resolve the issue of reparations lingering from the Great War, announced that the Lausanne conference had achieved "Peace on Earth, good will to man!" Another prophet, Nikola Tesla, coming out of seclusion to meet the press for his annual (seventy-sixth) birthday interview, hinted at two new inventions, the first of which would be "like the 100,000 trumpets of the Apocalypse," the other "like the shout with which Joseph's [*sic*] army brought down the walls of Jericho." Had Baba's silent thunder been stolen by popes, premiers, and master electricians? In any case it was two years before he saw fit to return to California, by which time his Lovers had managed their cognitive dissonance well enough to arrange for the Awakener to star in an autobiographical feature film (never finished).[114]

Henceforth, at each crisis on the path toward a second world war, Baba's disciples at once built up and fended off millennial anticipations of Silence broken. On news of the signing of the Munich Pact in September 1938, Baba indicated that war was inevitable and "When war is in progress, then I will speak.... Spiritually, both war and peace are nothing; but externally war is the most dreadful thing, and unless it were absolutely necessary for the spiritual upheaval, I should never allow the war to be." For the nonce he would help the *masts* of India, "God-intoxicated souls" so overcome by spiritual energies that they appeared to have gone crazy; if the world was also going crazy, well, he had that under control. After Belgium was overrun by German tanks, Baba secluded himself for ten days in order to "decide definitely the length of time this war shall last, and exactly when it will end." If the war continued, he would speak in August 1941, or when "Peace is just on the point of appearing," or after the war had played its part in helping humanity "to inherit the coming era of truth and love, of peace and universal brotherhood, of spiritual understanding and unbounded creativity." In 1947 Jean Adriel (Schloss) published a biography of Meher Baba in which she noted that "This long silence—now in its twenty-second year—has been an enigma to many people and his continual postponement of breaking it has proved a stumbling block to many others."

Aware that skeptics accused Baba of using silence as a publicity stunt, she had asked him to explain. He told her "that from the outset, he knew, of course, precisely how long he would remain silent, and he knew therefore that it would run into many years," but like a wise father he had been developing in his devotees a "capacity for patient waiting" and deepening their longings for the infinite. On New Year's Day 1950 and again in 1954, he was understood to have announced in oddly Christian language that the time was "very near for the breaking of my silence and then, within a short period, all will happen—my humiliation, my glorification, my manifestation, and the dropping of my body." Obstinately corporeal in 1960, he assured the faithful that "I shall not lay down My body until I have given the WORD to the world.... As I am the PIVOT of the Universe, the full pressure of the universal upheaval will bear on Me, and correspondingly My suffering will be so infinitely overwhelming that the WORD will escape from out of the Silence." Lovers collected his sayings on silence: "Silence is the only language of the realized"; "There are some who live in a perpetual hullaballoo, in a tornado of noise.... These will not proceed on the Godward road"; "Loud noise is sacrilege on the sky." Bearded, smiling, he materialized on posters as the "Don't Worry Be Happy" avatar of the Sixties and The Who.[115]

Had he told his Lovers "that his Silence would continue for the rest of their lives—that his Advent, in fact, was a mission of Silence—none would have been able to support the loss," wrote one Lover, Francis Brabazon. "So he egged them on with announcements and promises that he was about to break his Silence; and further-more that when he did speak all people would bear witness to that speaking and his inner Circle would become God-realized—for his speaking would not be mere speech but the Divine Creative Word." That was one explanation: pastoral psychology. Brabazon offered another, more millenarian: the full fruiting of Meher Baba's Brotherhood would be a New Humanity "in which we will not need to converse with God by signs and symbols so that the agents of oppression shall not overhear us; we will talk to the Beloved in our own tongue." Silence could be sacred camouflage. On New Year's Day 1949, Baba asked his Lovers, who had already been observing a Day of Silence on the anniversary of the start of his own (July 10), to keep mum for a full month in the run-up to his Speaking-At-Last: "The year 1949 marks an artificial end to an artificial beginning and the real beginning to the real end." Such a chiliasm of silences almost-broken, suffused with communal suspensions of sound and breath and an apocalyptic sense of the cosmic power of sound unleashed, was grounded not only in the Great Breath of the Vedas, Parsi Towers of Silence, Sufi trance, and shellshock. The silence resonated now from the sands of the Jornada del Muerto, where on July 16, 1945 a huddle of officers and scientists in a desert bunker heard the "deep, growling roar" of the first nuclear fireball. "It worked," someone said quietly. That was J. Robert Oppenheimer, and Sloka 32 of Chapter 11 of the Hindu classic, the *Bhagavad Gita*, came to mind—as well it could to a physicist who had studied Sanskrit, translated the *Gita* while at Berkeley in 1933, and kept the text within arm's reach thereafter:

Now I am become Death
the destroyer of worlds.

Later he would add the implicit prelude, from Sloka 30:

If the radiance of a thousand suns
were to burst into the sky
that would be like
the splendor of the Mighty One,

who was Vishnu, a.k.a. Time, destroyer of worlds. Others who knew the *Gita*, hearing Oppenheimer recall what at first had been a silent thought but became the audible, seminal quotation of the Atomic Age, finished Sloka 32 for him:

View in me
The active slayer of these men;
For though you fail and flee,
These captains of the hostile hosts
Shall die, shall cease to be.

Thus the silences of Hiroshima and Nagasaki. It would be overbearing to fix Meher Baba at the center of this circle of silence had his Lovers not sought it out: "Baba wants to use a simile about the atom bomb. Just as an atom bomb, which in itself is so small, when exploded causes tremendous havoc, so, when he breaks his silence, the universal spiritual upheaval that will take place will be something that no one can describe." What else was left in a world of mushroom clouds but silence as both doomsday weapon and doomsday itself, like the wind-driven newspapers whipping through the silent city streets of the final frames of the post-apocalyptic film *On the Beach* (1959)? Withheld sound was no less than the last word in silencers.[116]

4. Loudspeakers

"Sometimes Baba sat with us on the lawn and gave us thirty minutes 'in silence,'" wrote one of his American enamorata. "This we loved." It was silence that intensified his presence, like the silence to be broken in one of Samuel Beckett's first efforts, "Assumption" (1929), which began "He could have shouted and could not" and ended with a voice apocalyptically released, shaking "the very house with its prolonged, triumphant vehemence." It was an expressive silence like that in the finest silent films and

Hemingway fictions. Like that in the oils of Edward Hopper, from "House by the Railroad" (1925), a quiet we know will soon be broken, to "Nighthawks" (1942), a silence almost unbreakable. Or like that in the prose of James Agee and the black-and-white photographs of Walker Evans in *Let Us Now Praise Famous Men* (1941), the silence of endurance, and in Evans's hundreds of photographs of empty parked cars. Silence, writes the literary historian J. A. Ward, offered "a standpoint of stability and even of sublimity" during the 1920s and 1930s "from which to observe the distressing but never finally threatening chaos of noise that, to Agee, characterized the age in which he lived."[117]

Epitome of that noise was the loudspeaker, an old idea electrified. During the seventeenth century the Jesuit Athanasius Kircher had conceived of speaking trumpets so large they resembled towers of Babel turned upside-down, which in effect they were meant to be, broadcasting intelligible sound. The French novelist Villiers de l'Isle Adam, in a speculative satire of 1874, extrapolated from the acoustic premises of ancient Greek amphitheaters and the modern acoustic promise of gramophone horns to introduce a new invention, a soundsystem that could be built into any auditorium so as to outclap outweep outshout outlaugh and outstomp the most outrageous of claques, ensuring that any performance could be received with a fulsome Bravo! or *Bra-wow-wow* at the press of a button. If a claque paid to whoop or huzzah on command was "a machine made of humanity," and if every theatrical glory owed its success to its claque, "that is to say, its *shadow*, its element of fraud, technique, and nothingness," i.e., in-house Publicity, then why not a Glory Machine that could more reliably and automatically reproduce "the Screams of Terrified Women, the Stifled Sobs, the Genuine Infectious Tears, the Sudden Chuckles" or "the Howls, Choking Fits, Encores, Curtain Calls, Silent Tears, Threats, Curtain Calls with Howls as Well" of a well-lubricated claque? Why not install, behind the mouth of every ornamental cupid or caryatid, small electrical bellows working the speakers of gramophones so as to bring to perfection the *Wow-wows* and "all the noises of the Public," with tear gas and laughing gas concealed in tubes, and seats fitted with heels of hard rubber studded with nails? Such a Machine would be "so powerful that if necessary it could *literally* bring the house down." Glory need never again be left to chance, for the very auditorium, from balconies to backdrop, would be in cahoots with the ambitions of performers, composers, and playwrights, sweeping along its cargo of auditors in one loudspoken excitement, Babel turned inside-out.[118]

Glory was not too grand a word for anticipations of what loudspeakers might effect, considering how "amplifying telephones" were put to use in Madison Square Garden on 11/11/1921 as the Tomb of the Unknown Soldier was being dedicated at Arlington Cemetery. All the way from Virginia, the voices of President Harding and assorted dignitaries "rang out in loud and distinct tones to a crowd of 30,000 guests of the American Legion," and "Except for a slight blurring of deep sounds and the occasional re-echoing of the rolled 'r,' the voices were reproduced with sharpness and clearness," reported the *New York Times*. The sound was "flung out from an amplifier about the size of an ordinary desk concealed behind flags on the speaker's platform" and reached all corners of the auditorium. In the open air of Madison Square, where thousands more had gathered, and in San Francisco, "powerful electric megaphones also trumpeted the vibrant tones of President Harding's address." The commanding presence of the loudspeaker emerged earlier than projected by H. G. Wells in 1899 in *When the Sleeper Wakes*, where Londoners of a distant technotopian future live in spartan, sound-deadened flats while, outside, "General Intelligence Machines" blare the news to thirty-three millions. During the 1920s and 1930s, public address systems were built into convention halls and portable speakers designed for rallies out of doors. By 1932, Western Electric sound systems had been installed in 14,405 Anglo-American moviehouses showing mostly Talkies, and soon the Nazis, who had used mobile vans outfitted with loudspeakers for their successful electioneering in 1932, had hundreds of free-standing "speaker-pillars" blaring across Germany. Everywhere, radios with improved amplifiers filled homes and offices with more continual music and more tuneable voices singing or speaking through more sensitive microphones.[119]

"Nearly three years ago a shy and reticent young man walked into the office of Powel Crosley Jr. with an idea for a radio loud speaker under his arm." C. A. Peterson had conceived of "a marvelous actuating mechanism so designed as to vibrate freely without choking regardless of the heavy electrical impulses applied to it," and in a short time he had "revolutionized the loud speaker field" with a radio speaker that offered "amazing tone," "surprising volume," and "startling fidelity of reproduction." So crowed the Crosley Radio Corporation of Cincinnati in 1927, proud of its Musicone radio that "delivers pure, true tones, without distortion regardless of how suddenly the crashes of orchestras or high shrill notes come through it." Problems of range, power, responsiveness, fidelity, and noise had plagued electro-mechanical speakers from their origins in the 1870s, whether in telephones, phonographs, or radios. Early moving-coil transducers could not handle all audible frequencies, and their "magneto-electric" components had their own loud, defiant hum; the greater the volume, the louder the hum and more distorted the sound. Oliver Lodge in London had fiddled with the coil in 1898, and in Napa, California, Edwin S. Pridham and Peter L. Jensen had so improved upon it that they were happy to call their 1911 device a Magnavox, which GreatVoice ("a voice not of this world... a supernatural colossus") would be wielded in 1919 by an ailing Woodrow Wilson hoping to persuade an audience of fifty thousand in San Diego of the virtues of a League of Nations. Still, in 1921 Norman Ricker found the "quality of the voice and of music from the present loud speaker with horn attached" far more "unnatural" than supernatural, and "very disturbing." A young physicist who had studied acoustics and sound-ranging at Rice University (in whose dormitories he endured the "loud hellraising racket" of fellow students), Ricker had a notion of eliminating that horn altogether, since the quality of sound through a headset attached to a radio or phonograph was always superior to the sound in the open air massacred by the typical loudspeaker, whose horn suppressed low frequencies and distorted the high. Instead of a large spiralling horn hearkening back to biblical rams' horns, why not use a small frame fitted with a conical diaphragm like that patented in 1901 by an Englishman, John Stroh, so as to amplify sounds with less distortion, as done by the diaphragms in stethoscopes? Or Ricker's own paper-cone diaphragm speaker, developed in 1921 during a summer job at Western Electric? Or two or more such paper cones, configured to respond to low frequencies (woofers), the standard mid-range, and high frequencies (tweeters)? And why not couple the moving-coil to Lee De Forest's audion, or vacuum tube, patented in 1907 but just now revealing its full potential for receiving radio signals and amplifying sound? (The three-electrode vacuum tube had "an electric valve so highly sensitive that by its means a very feeble electric current is enabled to control a very much louder one," explained *St. Nicholas Magazine* to its wide audience of children in 1922. With reliable battery power and vacuum tubes, an engineer could "boost" a sound to make it a million times stronger, a billion, ten trillion.) Success! "Gone was the feeling that the orchestra was playing through a muffling blanket," testified Ricker. "*The orchestra was in the laboratory!*" and it was playing "Nothing Could Be Finer Than to Be in Carolina in the Morning."[120]

"Through the ether there will come to you sometime a Voice," read a poetic advertisement by the Jewett Radio and Phonograph Company of Detroit. "A Voice of Friendship—A Voice of Romance. Be ready with a Jewett Superspeaker to catch every revealing inflection of living, breathing, human Personality!" More than the renewed production of phonograph records, which were limited to 3–4½ minutes per side at 78 r.p.m. and seemed to attract scratches from the get-go, it was the burble of activity on the postwar airwaves that led to urgent demands for better speakers, since voices and music were now being broadcast for hours on end from stations powerful enough to maintain news and melody for long patches between static. The Master Radio Corporation emphasized that its 1922 Clarophone was no loudspeaker but a *clear*speaker, enabling a roomful of people to hear every word or note, unlike radios where multiple auditors had to listen in on linked headsets (sitting "silent and lifeless" side by side, wrote a German cultural critic, Siegfried Kracauer, "as if their souls were wandering about far away"). Radio receivers themselves were still dubious: "No rasping, buzzing, breaking or any noises save those caused by your receiving set." By 1925, radio audiences had "developed quite a critical ear," for which the Sonora Phonograph Company was prepared: "It took phonographs years of experiment to progress from shrill tin horns to the perfect mellow tones of today. Now the Sonora Radio Speaker with a single step brings to radio this same refinement of tone." With the amazing Dictogrand speaker one could "hear 'the feeling' in a mother's croon." Few of these marvels incorporated the best design—a coil-driven, baffled diaphragm—as adumbrated in a 1925 paper by a pair of General Electric engineers, and as of 1928, according to a young physics professor in Iowa, there was yet "no complete theory that can be applied to any actual loud speaker on the market." But everyone shared the same ambitions: distortionless amplification of voices and music for larger audiences increasingly remote. "Radio fans," announced one manufacturer in 1923 with unjustifiable confidence,

> If you still
> think Loud Speakers
> are hoarse and coarse
> you haven't heard the
> loud, clear
>
> Herald.[121]

Properly installed, a loudspoken voice could be heard a mile away. Or more: the Christmas carols broadcast for ten hours by loudspeakers mounted outside the eighty-first floor of the Empire State Building in 1930 could be heard across nine square miles of the city. In 1931, New Jersey folks five miles off could hear the world's largest loudspeaker boom a recording of Big Ben from the roof of Camden City Hall; the mayor presumed that in emergencies he too would be heard to the limits of his jurisdiction. These paled in comparison to another "world's largest" speaker, shown at the National Radio Exhibition in London in 1929 and capable, it was said, of carrying a voice thirty miles in the open air. Although public address systems had (still-annoying) problems with transients, tinniness, and echoes, by the 1930s loudspeakers were touted as capable of commanding audiences of half a million, far larger than the forty thousand who had listened "intelligently" to the music from Belgian belfries during the Great Carillon Competition of 1910. Claims to such vast audiences were evidence of a cultural investment in the amplification of sound; put in the context of increasingly regular radio broadcasts, booming radio sales, and fleets of loudspeakered "sound wagons" soliciting votes or faith, the large numbers were credible tall stories. "The Spellbinding Kilowatt," headlined the *Scientific American*; "Microphones, Amplifiers and Loud-Speakers.... Enable the Human Voice to Reach Vast Multitudes." Prewar orators had strained their lungs to be heard by a couple thousand; these days a person—even a meek female—could speak in conversational tones and be heard, indoors or out, by tens of thousands. Were you a devotee of *Popular Mechanics*, you could build a six-foot loudspeaker from plans published in 1927, and "You will be amazed at the volume and tone that will gush forth."[122]

Amazed, or abashed. If the choirs of loudspeakers now arching over political rostrums (and "looking much like the trumpets which are supposed to be part of the heavenly angels' equipment") so extended the audibility of candidates that delegates at last put "a premium on what is said rather than on how loud it is said," they were heard elsewhere rather as too loud than as well-spoken. By 1929, Cecilia Winkler of New York City could think of naught but draconian remedies to "the radio problem through closed windows." Either pulverize the inescapably loud loudspeakers of all home radios or let civilization go deaf and crazy. "The sound of the loud-speakers carries through window panes, floors, ceilings, walls," she wrote angrily to the *New York Times*. "Astonishment has been expressed at the need for more and better psychiatric institutions. I am convinced that radio loud-speakers have been one of the causes." Justice Cardozo, presiding over a noise suit againt the Paramount Mansion Company, was taken with the range of the catering company's loudspeaker and "its disturbing power." Referring to a 1736 English ruling, he wrote that "Long ago it was adjudged that one dwelling in a city, which with aid of a speaking trumpet made great noises in the nighttime to the disturbance of the neighborhood, must answer to the king"; in 1930, he continued, "The precedent is not one to be hastily renounced in days when trumpets have a power unknown to a simpler age." And what should one do, then, about airplanes circling the city in 1931 with loudspoken siren-calls for women's fashions? "The world, my friends, is sound," a preacher told his parishioners that year, but Sir Duncan Grey was not reassured, for in this "Age of Noise" it seemed to be the rule that "Lungs beat brains. The tub-thumper outshouts the thinker. The loud-speaker drowns the still small voice of logic." Wasn't it true that loudspeakers "Often Have Disastrous Effects Upon the Ear Dynamo That Sends Sounds to the Brain"?[123]

Like a mechanical parrot shouting "Progress" (wrote the authors of *England, Ugliness and Noise* in 1930), the loudspeaker problem was an audible instance of the everyhowness of noise after the Great War. Any and every means applied to control sound seemed to end up producing more interminable, and unfortunate, sound. The reversal was due not merely to unforeseen consequences, such as how the relative quiet of asphalt pavement, rubber tires, and internal combustion had prepared the route for "a motor car with a loud engine, a loud horn, loud brakes, and loud gear-changes." Or how advances in the cooling of well-insulated auditoriums had led to demands for ever-greater amplification of voices and music to overcome the thrum of powerful air conditioners. Or how there had arisen a tendency "to overcome local noises by making a loud speaker speak louder." (Civilization, grumbled *The New Statesman* in 1935, "has always a new noise up its sleeve; and it is usually a worse one.") Nor was the reversal due simply to the sensitivities of new equipment that could accommodate low and high frequencies, which in turn (as with loudspeakers and the first battery-powered hearing aids) amplified all source noise—a wheeze, a squelched cough, surface knurls on phonograph records, and "extraneous noises due to electrical induction which might be quite tolerable if the system did not respond efficiently to the high frequencies." The implications of such sensitivity had been anticipated by an English inventor of the microphone, David E. Hughes, who had been impressed in 1878 by "the pandemonium of sounds which . . . the microphone reveals where we thought complete silence prevailed."[124]

No: the everyhowness of noise was much more insidious, as it was in a 1938 dream reported by the German-Jewish thinker Walter Benjamin, in his fifth year of exile from Germany and visiting Bertolt Brecht in Denmark. Suffering from the noise infiltrating his room—as lifelong he suffered from "an extraordinary sensitivity to noise" (wrote his friend Gershom Scholem) "which he often referred to as his 'noise psychosis'"—Benjamin found himself that night in a "terrifyingly dreary and bare" landscape, studying a befuddling map. The landscape and the map were that of Hell and of the labyrinth of his auditory canal, an insidiousness that makes an auditor a captive of his ear. Analysts might skip past so obvious an oneiric inflection of Benjamin's waking hours. The noise historian cannot, given how neatly Benjamin's dream reproduced a theme he had spent a decade developing in works of literary history, radio drama, and cultural criticism: that (writes Benjamin's biographer, Gerhard Richter) "there can be no body of the self that is not disrupted by noise," nor any social body in which noise cannot be a political resource. For the Benjamin of the *Arcades Project*, it is the interruption of a Parisian idyll of the 1920s by the noise of the demolition of the old Passage de l'Opéra that cues historical consciousness. In order to be fully awake to modernity ("the time of hell") and to changes wrought by industrial capitalism's "colossal acceleration of the tempo of living," one has to attend as much to inner noises and feelings—"such as blood pressure, intestinal churn, heartbeat, and muscle sensation"—as to new forms of finance or architecture. Such inner noises anchor a person to the actuality, and the radical potential, of a present that fascists aestheticize and reactionaries rescind. For the Benjamin of *A Berlin Chronicle*, begun in Ibiza in 1932 in lieu of a planned suicide, the city of his childhood is recuperable best through "an intricate network of noise." So, writes Richter, Benjamin's memoir entails "an elaborate meditation on the corporeal transformation of the ear" by force of telephones, radios, brass bands, and loudspeakers. But as certainly as the ear could be constructed as a prison, it could be reconstructed as an organ of resistance to the monauralities of fascism, its labyrinth a preserve of remembered voices that maintain a self inaccessible, or inconformable, to tyrants. For the Benjamin of "Radau um Kasperl: eine Geschichte mit Lärm"—"a story with noise" that opens on those sirens and boat whistles that had upset Julia Rice—the microphones of Radioland intercept, confuse, and amplify; while the play achieves a child's hilarity of tumult, throughout there is a struggle (in 1932, on the eve of the Nazi takeover) to hear entirely on one's own and aright.[125]

Sound-effects, when most effective, were noise-effects: insubordinations of voice or memory or remembered voices which, because personal and tremulous, could be anti-nostalgic, anti-aesthetic, inconformable. For the Benjamin of "The Work of Art in the Age of Mechanical Reproduction," the close-ups of silent film and the multiple illusions of Talkies (by 1935) confirmed the episodic and counter-intuitive nature of film production. Because scenes were not only shot out of sequence but had their sound dubbed later, filmmaking had some revolutionary heft for challenging axioms of inevitability and the imbeddedness of property, for combatting the one-wayness of soundscapes dominated by rows of loudspeakers. This was the inverse of what Jean-Paul Sartre would call the "practico-inert," that "malignant destiny . . . which human beings create over against themselves by the investment and alienation of their labor in objects which return upon them unrecognizably, in the hostile form of a mechanical Necessity."[126]

Take for example the "'New Deal' in sound" at the Rochester General Hospital in New York, whose many "shrill-toned telephone bells" had been replaced by an announcer in a soundproofed room who sent messages through a voice-paging system to sixty "low-toned, low-volume" speakers in different wards. That was as good an idea, wasn't it, as had been the creation of an electric fire-alarm system in lieu of the ringing of churchbells? Well, hmmm. As the President of the American Hospital Association explained in 1938, "the modern (and desirable) insistence upon a fire-proof structure converts into a veritable 'loud speaker,' every major sound vibration being transmitted through the entire [hospital] by its solid ferro-concrete structure," its ventilation ducts, its plumbing, and chic *terrazzo* flooring, so the introduction of a public address system, however pleasant the tones of the voice or chimes, redoubled the problem of noise in wards beset by rattling carts and the clanging of dropped bedpans. Or take for example the fact that more sensitive studio microphones, stronger transmitters, more discerning antennas, and higher-wattage loudspeakers on home radios made it possible for crooners, their lips kissing the mike, to make quietly melodic love to millions at a time. That was as great an advance, wasn't it, as the recent development of resonator guitars and banjos which, absent their cast-aluminum bridges and interior cones, would not have been able to make themselves heard above the brass and drums of traveling bands and radio orchestras? Well, if you say so; and yet the same genre of acoustic advances made operatic arias feel emotionally "outrageous," wrote a German aesthetician, Rudolf Arnheim, and the entire sound-space of daily life seem "overloaded"; with every new radio station came "a spitting and a spluttering, then screams and squeaks and whistles" as the wireless was tuned across competing broadcasts. Or take for example the electrification of the recording and amplification of music, which made for smoother playback and easier control of volume. Wasn't that as kind to the ear and arm as electrically amplified jukeboxes, more than half a million of which were in place by 1937, protecting records from scratches and playing music at the flick of a nickel, with rubber mountings and "a 100% noiseless, smooth, even-running turntable"? Well, when a person no longer had to crank, pedal, bow, blow, puff, pluck, plink, or pound to get music, and when automatic record-changers (used first in jukeboxes) and continual radio programs overcame the four-minute limit of playback, a person might keep the music going through all the waking hours, a background sound-for-the-asking that only the wealthiest had commanded in the past. An English critic, Constant Lambert, knew for a fact that "Some business-men actually leave the wireless on all day so that the noise will be heard as they come up the garden path, and they will be spared the ghostly hiatus of silence that elapses between the slam of the front door" and their public exits and re-entrances. As much by reason of the easiness and loudness of electrically amplified music as by the "tonal debauch" of the tunes most replayed, Lambert was convinced that "The loud speaker is the street walker of music."[127]

Other critics would reserve their anathema for Muzak, a lovechild of the loudspeaker. Since the 1890s, industrialists had thought to use song and music as it had been used for ages in field labor and artisan shops, to promote (at little extra expense) a rhythm of work that inspired efficiency while lifting spirits. On behalf of manufacturers, or on their academic own, and later for war production boards, experimental psychologists began comparing the output of workers in different sonic environments: quiet; noise; spoken word; lively song; instrumental recorded music. Which sounds or silences dragged them down? Which happily sped them up? In 1916 the owners of a large laundry began playing ragtime on phonographs and their employees happily sped up: "The girls simply can't loaf 'in time' with ragtime... and the days hitherto humdrum and filled with the dry drudgery and routine that makes sullen subordinates, now speed happily on their syncopated way." The results of such experiments were odd and contradictory, some showing higher productivity, some lower, so the psychologist John J. B. Morgan of Columbia University in 1916 gave his subjects a tedious task to perform then interrupted them with unrelated sounds: a gong, hammer-struck bells, electric buzzers, or gramophone recordings of vocal solos, waltzes, comedic pieces. The results were still odd. When the first half of the session began with bells or music of any kind, work slowed; when the first half was silent, work hastened as soon as any kind of gong, bell, music, or buzzer was introduced. On follow-up, Morgan found that "irrelevant noise"—bells, buzzers, *or* music—slowed tasks that required thought and a retentive memory but doubled the pace of tasks that were mechanical, as if people at rote work *required* environmental variation to make a decent go of it. After the Great War, as experience with shock and battle fatigue informed research into industrial productivity by Myers and others, and as radios and phonographs acquired louder speakers, both management and workers welcomed music on assembly lines and such tedious tasks as mail sorting. William Green, the President of the American Federation of Labor, was unequivocal: "Music is a friend of labor for it lightens the task by refreshing the nerves and spirit of the worker." So arose Wired Radio, Inc. One of the inventors of multiplex telephony (1909/1910), Major General George O. Squier had been Chief Signal Officer of the U.S. Army during the First World War and founded its Fort Monmouth laboratory for radio research. In 1922 he had patented a system for using electric power lines to transmit news, music, advertising, and entertainment into private homes, a plan for piped-in sound with antecedents in Paris and Budapest. By 1934, when he renamed the service Muzak, the company's focus was on piping in "functional" music to restaurants, night clubs, and hotel lobbies, where loudspeakers would be mounted discretely overhead or behind potted palms. The music was thoughtfully varied, modulated, and inset with periods of silence, employing the first 33⅓ r.p.m. vinyl recordings for longer play and ease of programming. How thoughtful and well modulated would be a bristling question for piano-bar musicians put out of their jobs, music lovers distressed by the narrowed dynamic range ("otherwise," explained Muzak men, "loud passages are too loud and soft passages become lost"), and social theorists like Theodor Adorno, who in 1934 indicted "background music" for promoting a habit of passive listening that waxed the path toward dictatorship.[128]

Vain would be any wish for loudspeakers of themselves to collapse under their physical and cultural loads, as Dr. Edmund P. Fowler wished in *Hearing News*, organ of the American Society for the Hard of Hearing: "Many throats give way under a strain of constant loud speaking. What a blessing it would be if the mechanical loud speakers in our neighbours' radios and in the movies would do likewise." Vain was that wish not because loudspeakers in 1936 were so hardy; rather, because by then loudspeakers were already imbedded in thousands of cinemas, hospitals, churches, schools, assembly halls, factories, banks, department stores, and corporate offices. Like big-band guitars, buildings were being fitted with an electrically amplified sound-transmitting infrastructure. While the acoustic formulas worked out by Sabine at Harvard before the First World War and by Philip M. Morse and Frederick V. Hunt at MIT before the Second World War were being applied to the proportions of rooms and the textures of walls, floors, and ceilings, architects in America, Europe, and the Soviet Union were also beginning to plan for integral sonic instrumentation, like the telephone speakers built into the interior panels of limousines for communication with chauffeurs. Although Vern O. Knudsen, at the forefront of architectural acoustics in California during the 1920s, confronted daily "the superstition or belief, still held by many, that the acoustical qualities of an auditorium cannot be known until the building is completed," he had recognized by 1927 that any room with a volume of a million cubic feet or more, no matter how well designed, required a system of loudspeakers, and for those with frail voices, much less capacious rooms would also fare better with loudspeakers. But should one incline toward paranoia, such inconspicuous speakers could seem to be instruments in a vampirical conspiracy. One must remember, wrote Fowler, "that every sound wave is a push and a pull, a squeeze and a suck of every tissue and structure through which it passes or upon which it impinges." At worst, as it was to Enrique Muñoz of Colombia, writing in desperation in 1939 to Harvard's S. S. Stevens (junior author of *Hearing, Its Psychology and Physiology*), loudspeaker amplification might be used in tandem with nefarious Television Waves to drive a person to suicide by attacking him with "utterly disagreeable noises, single voices, dialogues, speeches, music, etc." whose volume could be increased at any distance and transmitted anywhere, an "utterly cruel work" that could cut short the sleep of its victims and "burn the organic resistance" with invisible baths of noise.[129]

Stevens (or his archivists) filed the missive from Muñoz under "Crank letters." As a psychologist drawn to the field by Edwin G. Boring's lectures on perception and by new research on auditory neurophysiology by Hallowell Davis (senior author of *Hearing*), Stevens might rather have suspected Muñoz's logic as a schizophrenic's attempt to explain inner voices in terms of the painful materialization of aetheric forces. This genre of psychosis had earlier fixated upon mesmeric fluids, as in the "mania" of a fifty-year-old laborer writing from the Royal Edinburgh Asylum to explain that "The Mesmeric Operator (REA) still continues from time to time to speak aloud transgrammar with the Medicines tongue. This is called the transgrammar nuisance." The vibration of aetheric fluids or (newly Hertzian) waves dominated the trance-grammatical madhouse memoirs of the German jurist Schreber analyzed by Freud and often reappeared in disturbed fantasies about radio-controlled bodies and infiltrated ears.

The discovery of "penetrating radiation" from beyond the Earth and their baptism as "cosmic rays" in 1925 had added another mysteriously acoustic power to the schizoscientific arsenal. Stevens himself could have taken a more personal interest in this tradition, for it was he who had revealed to the Acoustical Society of America in 1936 that "the human ear can act as a radio loud speaker, converting the impulses coming in on an ordinary radio set into music," given the ability of the basilar membrane's own electrical charge to respond mechanically to an alternating current. And it was he who noted that the ear might add overtones of its own to the sounds it received, like any imperfect microphone. And it was he (co-authoring with Davis) whose first paragraph of *Hearing* had begun: "The development of the thermionic vacuum tube has revitalized the science of acoustics. Whereas it was formerly necessary to discuss the production, transmission, and recording of sound-energy in terms of purely mechanical devices such as tuning forks, organ-pipes, strings, sirens, tubes, and reeds, we now treat these problems in terms of electrical and electromechanical systems such as microphones, amplifiers, and loud-speakers." And it was he (with Davis) whose final page of *Hearing* emphasized that a listener's attitude toward any sound "determines the character of the ultimate differentiation quite as much as does the stimulus itself." Stevens might therefore have acknowledged the link between the distortions of loudspeakers and the baths of noise that were sending Sr. Muñoz over the edge. After all, Stevens (with Leo Beranek) would establish Harvard's Psychoacoustic Laboratory in 1940, dedicated in part to researching the "effects of continuous din on the performance of men working in airplanes and tanks," and he would spend much of the rest of his distinguished career studying the exponential relationships between sensations and perceptions of loudness.[130]

Perhaps exponentiality was the secret to the loudspeaker's invasion of spaces public and private, cultaural and psychic. Ambient noise had grown so exponentially (hadn't it?) that only a loudspeaker could cut through it. "In some form or other noise permeates nearly everywhere, and has always existed," wrote Alfred H. Davis of Great Britain's National Physical Laboratory in 1937. "There is no doubt, however, that noise is one of the features of uncontrolled development in a mechanical age—an undesired by-product of the machines." Davis had the technical skills to devise a machine of his own, "an objective noise meter for the measurement of moderate and loud, steady and impulsive noises," so he should know. And in this diesel-powered, jackhammering, electromechanical age, wasn't the best way to deal with noise often to wall it off with more, loudspoken, sound? Some old-school architects, scorned by new-schoolers such as Floyd R. Watson (author of the 1923 study *Acoustics of Buildings, including Acoustics of Auditoriums and Soundproofing of Rooms*), were still stringing miles of thin wire across ceilings in order to distract or entangle unwanted vibrations, but ordinary folk, wrote Davis, had caught onto a more modern trick: "Some people, annoyed by intermittent disturbance from a neighbour, switch on soft wireless music at a strength sufficient to obliterate the irritating noise." Scarcely soft, of course, if *obliterating*: the paradox of a modern world in which one had increasing recourse to machines in order to minimize "the noise resulting from the ever increasing mechanization of all our activities." Some had been so deafened by the mechanisms of this "Age of Noise" (hadn't they?) that loudspeakers functioned as a kind of public hearing aid. "Out of every thousand industrial workers given hearing examinations today," the American Management Association would report in 1940, "less than fifty-two per cent will have normal hearing in both ears," a problem aggravated by the construction of dense industrial housing with thin party walls or semi/detached houses compressed onto small lots to allow (paradoxically) for park space and playgrounds. And wasn't something awful also happening to the ears of the elite, like those in the stands at the Yale–Harvard football game of 1937, when the rowdy crowd so drowned out the announcements from 200-watt loudspeakers that Frederick V. Hunt, another physicist in the postwar vanguard, had to advise that the wattage for the loudspeakers at Harvard Stadium be tripled or quadrupled?[131]

Insulated with six-inch layers of eel grass and a foot-thick inner lining of closely-packed cotton-wool hanging edgewise, the anechoic chamber of the National Physical Laboratory allowed Alfred Davis to establish a standard of silence by which to calibrate his noise meter. If there could be no perfect loudspeaker ("one which when receiving perfect broadcasting through a flawless set will create the illusion of reality" for radio audiences; one that did not buzz, rattle, and snap as the volume was turned up on a public address system), there could be a perfect silence in which one could assay the imperfections of a loudspeaker. The *Oxford English Dictionary* notes the first reference to a "dead room" in 1930 and the first to an "anechoic" chamber in 1948, but back in 1900 Sabine had already silenced the sub-basement at Harvard, in 1907 reporters caught an earful of the off-putting quiet of Prof. Hendrik Zwaardemaker's chamber at the University of Utrecht, and in 1911 a soundproofed room was central to Charles Myers's revamping of Cambridge University's psychology laboratory. Since then, soundproof rooms and echo-free chambers for acoustic and psychoacoustic studies had been refined, first by Sabine himself up to his death in 1919 and then by his distant cousin Paul Sabine (son of a Methodist preacher) at the Riverbank Laboratories in Geneva, Illinois. Financed by Col. George Fabyan, a Boston Brahmin with a fortune from textiles, Riverbank was devoted in part to the construction of an acoustic anti-gravity device: two sets of finely tuned strings, one set inside and another outside a rotating cylinder which, when spun rapidly, would create a force field of harmonic vibrations sufficient for permanent lift-off. This at least was the blueprint drawn from decoded pages of cipher in the hand of Francis Bacon (he of the *Novum Organum*, of the sound-houses of the New Atlantis and, thought Fabyan, of the plays of someone called Shakespeare). Neither Wallace nor Paul Sabine came close to levitating; they stuck instead to the risings of sound in acoustic isolation—with far better isolation in the "mystic atmosphere" of large chambers below the Illinois sod than in a cramped Cambridge basement. Their work resulted in lucrative patents for sound-absorbent wall and floor coverings and in contracts with U.S. Gypsum and movie studios requiring sound stages. So it was that echo-free or sound-deadened rooms spread from universities to manufacturing and entertainment; from the homes of Proust, Pulitzer, and Rice to the Compagnie Parisienne de Distribution d'Electricité; from the U.S. Bureau of Standards (which tested sound-absorptive materials up against deafening loudspeakers) and the Laboratoire des essais of the Conservatoire national des arts et métiers. By 1939, the idea of a "dead" room was so prevalent that Pauline Davis of Harvard Medical School, studying the "Effects of Acoustic Stimuli on the Waking Brain," felt compelled to explain to the sophisticated readers of the *Journal of Neurophysiology* why she did *not* use a soundproof room in her research —"because it had been previously observed that people coming into a sound-proof room elsewhere in our labs were profoundly affected by the unnatural quiet and the unusual sound of their own voices in such a room."[132]

"Utter stillness was a violent stimulus in itself," said one of her subjects. While Harris Trust and Savings around 1922 moved its steno pool into a "Noiseless Room That Speeds Production" with felt-covered walls and ceilings that reduced the echoes of rattling typewriters and created a calm "almost library-like," librarians themselves were slowly abandoning their conventional reclusiveness and role as silentiaries in order to reach out to the multitudes. A report from the U.S. Bureau of Education in 1876 had urged that Free Libraries supplant corner saloons as gathering places; the more vital their civilizing mission, the more crucial it was that each new public library be built on "the corner of important streets in the heart of town," accessible to all working people. "The natural desire to remove it from the bustle of business and place it in some quiet, secluded spot should be sacrificed to more practical considerations." In manufacturing areas, the library "must neither repel the masses by high standards and an atmosphere of dignified respectability which will give it the odor of aristocracy, nor lose its hold on people of culture and refinement by descending to low standards and becoming the meeting place of a disorderly rabble." Racks of blood-and-thunder dime novels might be a swell invite to nearby shelves of classics; public librarians must forsake an arthritic ideal of guardianship (of rare, expensive, or salacious books) and promote good order and decorum only insofar as congenial to a literate *demos*. No more snootiness or shushed defenses of a library's "mystique of quiet"; instead, words of welcome for a polyglot, hard-working population eager for stories spoken or written, fictive or historical.

> See the Children's gay Librarian! Oh, what boisterous joys are hers
> As she sits upon her whirl-stool, throned amid her worshippers,
> Guiding youngsters seeking wisdom through Thought's misty morning light;
> Separating Tom and Billy as they clinch in deadly fight;...

And if the bank stenographers of Pennsylvania found themselves speaking in hushed tones upon entering their silent sanctuary, "as though they were in church," ministers were finding themselves as hard-pressed as librarians to maintain a formal silence. Instead, they were installing loudspeaker systems in their naves to compete with the "Holy Ghost power" shooting through Pentecostalists like "electric needles" and with the acrobatic glossolalia of the Holiness Movement shaking congregations as if each member were "holding the positive and negative handles of the electric battery in the school laboratory." They might even fit out sound trucks to compete with street preachers using newfangled battery-powered megaphones: one more form of the "Magic of Communication."[133]

Mohawk Company advertisements announced a speaker with such "a depth and clarity of expression" that "to excel it will require necromantic powers," but that was not what the men at Western Electric had in mind when they produced the "Magic of Communication" as a "tell-you-how story" for the U.S. Signal Corps. Throughout this animated film of 1928, magic inhered in waves of dancing electrons pushing at the membranes of the ear—"Back and forth the air molecules dance, advancing, then retiring"—in the same way that molecules press upon the carbon granules of a microphone. For children, the implications were inescapable: the air is teeming with sound and each of us is a biological miniature of the sound-system of the aether. Silence (for adults) might be a spiritual ambition for the very reason that it was infeasible. Who could escape the dancing electrons, or the squeals of cheap loudspeakers inside radios or mounted on poles as public address systems? Perhaps monks in their cloisters? Painters in rural artists' colonies? Spelunkers at Mammoth Cave being studied for the "Effect of Darkness and Silence on the Optic and Auditory Nerve"? Music students in newly soundproofed school practice rooms? Sedentary gentlemen in city clubs, their stolid quiet so often spoofed? Cenobites, artists, cave crawlers, piano students, clubmen—those must have amounted to a thousandth of one per cent of humankind. Add another thousandth for technicians working at vibration-sensitive equipment in industrial or scientific laboratories, for guards locking up sound stages, and for the larger number of undergraduates required by courses in psychology to enter "dead rooms" for this or that experiment. Another hundreth, maybe, had stood alone in the motionless calm of great deserts, atop great mountains, on the ice of Antarctica as described by Carsten Borchgrevink in 1899, where "The silence roared in our ears, it was centuries of heaped-up solitude." Silence was where people were not, or were rarely, as within a bluish veil of light along the Swan River, where the Finnish geographer Johannes Granö found "Complete silence! We could not hear the cuckoo calling, nor the whistle of the golden oriole. We could not feel any breath of air. We were surrounded by a great sanctuary of flowers, cast in wax, as it were. It was like a fairy tale, a dream!"[134]

Most people, of course, never experienced such silent places, and to most, a thoroughgoing silence was less fairy tale than bleak omen. City people in particular, wrote Prof. Lawrence Cockaday of New York University, felt uneasy in the "almost totally quiet surroundings of the countryside. Everyone," he claimed, "has felt that odd feeling of almost throbbing silence of the first evening in the country," regardless of crickets, bullfrogs, mosquitoes, the creaking of rockers, barking of dogs, hooting of owls.... Still, though cityfolk needed a fillip of traffic noise just as they needed a morning cup of coffee or a couple of Martinis in the evening (noise, concluded a professor at Ohio Wesleyan University, "stimulates human beings to the degree resulting from a mild cocktail"), quiet was understood as the ground of well-being. "Genius," wrote the copywriters for the Mohawk Rug and Carpet Company, "puts a heavy premium on Quiet, and rightly so, for only Quiet can produce the things that genius gives the world." Rolls Royce of America cushioned life itself: "You step in and sink back in the cushions.... Let the miles ahead be a chaos of shrieking traffic—you ride in a well of silence. Let the road be rough as a mountain trail—Rolls Royce will sheathe it in velvet." On every side it seemed profitable to appeal to a basic human desire for quiet if not for absolute silence: soundproofed rooms on the transatlantic *Bremen*; "Hyatt Quiet" for the ball-bearings of trucks; absence of "A-C hum" with Majestic radio receivers.[135]

Radio audiences had become very discriminating, noted Sherman P. Lawton, author of the first college text on how to speak on the airwaves. Not long ago, radio audiences had "looked upon any noise which accidentally escaped the static as a miracle," and radio operators as early as the Titanic disaster had been battling the "mournful rumblings and grumblings of the static." Now, in 1932, with more radios than telephones in American and European homes, static had receded farther into the background. The Neutrodyne Perfected Radio et al. had "picked the air clean," and audiences expected long uninterrupted passages of music, smoothly dynamic sermons, play-by-play accounts of sports events and political conventions. Old-style orators had to learn to tone themselves down lest they shatter studio equipment. "What do you know about broadcasting?"

bellowed an irate William Jennings Bryan inside a radio broadcast booth, his silver tongue tarnished by age even before his last hurrah at the Scopes Monkey Trial of 1925, where WGN engineers would install four microphones. "What do you know about oratory?" retorted Graham McNamee, a trained baritone who was among the first professional radio announcers and had narrated the live broadcasts of the Republican and Democratic National Conventions in 1924. McNamee was right: inside a booth the art of speaking had to be reformulated, since one spoke not to forgiving milling crowds but through an unforgiving mike that amplified slurred syllables, stomach rumbles, the crinkle of a trouser crease. The mike on one end and loudspeakers on the other also amplified the awkwardness of silences. While radio broadcasting, like recording, relied increasingly on a broadcast architecture of relatively "dead rooms," its nemesis was "dead air," which reached listeners less often as silence than as static, a static that postwar audiences otherwise learned to "tune out" as the cost of doing sonic business in a universe of stations competing for frequencies.[136]

Electric refrigerators, electric vacuum cleaners, and other electric appliances in the house, apartment, or shop also affected radio reception, as did thunderstorms. Telegraphists had known the static in their headsets. Miners and sugar-processors knew the static that sparked explosions in coal and sugar dust. Cinematographers knew the static that clouded their negatives. Telephone companies and customers knew static on the line, and a telephone operator in 1927 would win a workers' compensation suit for injury to her ears and peace of mind from the "intense vibrations, static, loud noises, and ringing bells" circulating through her headset. Radio buffs knew all kinds of "natural electric perturbations"—"strays," "parasite signals," atmospherics—and the new audiences for radio broadcasts heard static as clicks, sizzles, hisses, squawks, rollers, and grinders. Static had become familiar enough to take on the energetics of metaphor, nowhere more so than in James Joyce's *Finnegans Wake* (1922–1939), where radio antennas captured "skybuddies, harbour craft emittences, key clickings, vaticum cleaners, due to woman formed mobile or man made static and bawling the whowle hamshack and wobble." At once mundane and cosmic (who really knew where in the aether it came from?), enduring and temperamental (who knew when it would show up out of the blue?), Static during the 1920s entered the same realm as Vibration but became its cultural nemesis. Where Vibration had originated within the *harmonia mundi*, the music of the spheres, and intimated at *concordia mundi*, a world of mutual understanding, Static obscured meaning, deranged harmony. Where Vibration enabled swift dialogue across oceans and stirring contact with the Other Side, Static signified the unreliability of all connections. Where Vibration moved with and through bodies and minds, Static the doppelgänger stood in the way, heckling.[137]

Round about 1924, Charles Fensky, peddling a hearing aid made of "high-grade vibratory sterling silver" charged up with "Hearium" (and later with Radium), understood that Vibration was natural, Static not really. Therefore his Radium Ear, not a "high-powered electrical device" but an elemental resource, allowed wearers "to hear voices and sounds in a natural tone and in a natural way" and neutralized the static that the hard-of-hearing had always to contend with. In 1924 an engineer for the Radio Corporation of America, E. F. W. Alexanderson, inventor of the duplex antenna, the phase converter, and selective tuning, hoped to persuade readers of the *General Electric Review* that Static and Vibration were both natural and that the "great enemy of radio communication, atmospheric disturbance, so-called 'static,' is by this time well understood and under control. It is really this fact that makes commercial radio communication at all possible." Copywriters would use the "fact" of this control to the advantage of radio manufacturers but never acceded to an electrical engineer's notion of what was natural, never alluded to Alexanderson's interest in designing antennae and circuits that could exploit an aether "imagined to be a disturbed ocean with waves of every length rolling in from all directions." Instead, as with "The New 6-Tube Super-Zenith (*Non-regenerative*)" radio of 1924, copywriters emphasized accuracy of tuning through that ocean and clarity of reception despite that ocean, as with an explorer who took a Super-Zenith with him "to eliminate the greatest hardship of the Arctic—*the solitude*" and radioed from near the North Pole that "Zenith has united the ends of the earth."[138]

Lay experience, however, was of a supervening Static, practico-inert and as much a product of interfering radio stations and inexplicable vacuum tubes as of an omnipresent force of nature. So a syndicated cartoon strip of 1924, Martin Banner's "Winnie Winkle," follows the man of the house struggling mightily to get something other than AWOOO SKRAW SQWEEK E-EE-YOW-RR SQWEEEEE WEEEEE OOMP out of a radio; just when the song "For me and my gal" is barely audible, the station signs off with a very clear GOOD NIGHT! When the Tucker family tuned in, according to the humorist Hilda Morris in 1943 but nostalgic for 1924 and the first loudspeakered radios, they had just learned the "onomatopoetic word... static," and that word "kept tumbling out from the radio corner. When Ab said it, it was with triumphant disgust. (Ab possessed no radio.) When George said it, his tone expressed hopeful apology." And when Cornelia heard the "Crack-crack! Whee-ee-ee, whish! Whoo-oo-oo!" she was thrilled, for "It was as if there were no space barriers, and all the banshees, all the eerie people in the universe could crowd into her snug little house." Eerily, radios all on their own would switch from static or music to news to "some crazy bit of dialogue, which brought [audiences] running, as to a rendezvous, either to listen or to silence it." And while early stations proudly aired heartbeats and snake rattles to demonstrate the power and finesse of their broadcasting equipment, they had to be cautious about introducing sound-effects, lest they be mistaken for more static. "That's another word I'm always hearing—'static,'" says Holly Sloan in a 1932 novel. "Just what is it?" "What isn't it?" replies her publicist. "In a general way it means interference. It is all the diseases of radio. All the world is radio and static is most of the trouble in it." In short, "It's our word for every kind of trouble, mental, moral, material, and spiritual." In 1942, when the radio personality H. Allen Smith received a letter from Moses Saint Matthew Ashley, the Negro inventor of the E-Attachment Adam Genesis Atmoswitch, asking "Who are willing to Angel along this line of radio research?" his interest was *pîqued*. What-all the Atmoswitch did Mr. Ashley was unsure—he was still studying *Understanding Radio*, the book (by Herbert M. Watson) that had inspired the construction of this "tone repressioner"—but his old shellacked suitcase with its wires, knobs, and a patentable secret locked inside was, as he explained when Smith visited Ashley's room in a Long Island boardinghouse, apparently a signal filter: "You have got to understand that the tubes in the radio have not been un-vacuumized or de-aired enough and this causes mechanical noises. The Atmoswitch gets rid of these." Improved as they had been with superheterodyne receivers, with Stewart-Warner's "Sensational New REPRODUCER" that overcame the "shrill 'blasting' sound called 'distortion,'" with Philo Farnsworth's "multipactor" tube with its "noiseless" amplification, radios had become at once more familiar (81.5 percent of American households owned a radio by 1941), more acoustically "faithful," and more occultly powerful, so that static invited interventions at once homely and fanciful. Meanwhile, however, the residuum of static had taken on a more figurative life.[139]

"Famous since 1847," the bearded Smith Brothers of Poughkeepsie had been selling their cough drops any way they knew how. With the acceptance of germ theory and the concomitant use of medical masks in surgery and the establishment of national campaigns against the spread of tuberculosis and diphtheria, coughing had assumed a more assaultive demeanor. Where the mid-nineteenth-century cough had been a private aggravation of the throat or painful wracking of the chest, symptoms of (at most) individual mortality, the twentieth-century cough could be heard as an attack on others, a violation of the decorum of public health. Claques and barristers had long used coughing as a weapon, and actors and judges had noted contagions of coughing among audiences and juries, but now a single cough might be a grenade launching "microscopic balloons" containing infectious bacilli. "It isn't fair to yourself or anybody else to go round coughing," declared the Smith Brothers during their "Drop that Cough" campaign just after the Great War. Drop that Cough: "That's what they used to whisper to us when a working party went creeping across No Man's Land, toward the German wire, and any old sound might start a machine gun going," confesses an anonymous veteran in a 1919 advertisement. "Believe me, any bird who had a cough never had to say 'stop your shoving.' We kept our interval and distance from him without being reminded of it." So "When will people learn what the soldier has to learn, that a cough gets you into all kinds of trouble—*unnecessary* trouble, because a cough itself is unnecessary." Coughing was as unnecessary and troubling as other noises, like those that betray one's position on a battlefield or make a golfer go crazy on a putting green; it was a public nuisance, like snoring in a sleeping car; it was a bad habit, like unwrapping hard candies in the middle of a concert. "A public cougher is . . . a public menace. He has no friends, and doesn't deserve any." Drop that Cough! By 1928 one of the Smith Brothers' competitors, Luden's, had thrust its Menthol Cough Drops into the same arena with a full-page advertisement in the *Saturday Evening Post*: in a livingroom sit two well-dressed couples listening to a fancy radio with a large, modern loudspeaker, but everyone is glaring at the younger of the two men, who is sneeze-coughing into a handkerchief. Punchline:

> STATIC! something ought to be done about this.

Just as top-of-the-line radios neutralized static, so Luden's Menthol Action ("cooling, soothing, refreshing") could neutralize interruptions to socio-acoustic occasions in top-of-the-line throats.

> COUGH and the Concert Stops

warned Luden's: a well-behaved audience sits quietly as a symphony conductor "weaves the golden thread of melody—then, from the shadowed audience, the crash of a cough—another—another—his hand raises sharply—the music stops in protest until the wave of vocal static has ceased."[140]

Still sonic, static was no longer electrical or radionic or atmospheric; it was psychic noise, social and personal, a generic, unregenerate interruption—and potent ("A COUGH will dim the hearing of a thousand ears": Luden's). So while the Smith Brothers stressed the anti-social nature of coughing in churches and theaters, with editorials by actors ("Please *don't* make TALKIES into COUGHIES": George Jessel) and commercials on the WJZ radio network, a newspaper cartoonist confidently extended the technical side of the metaphor, conflating the "coughing," sputtering, and backfiring of jalopies with static itself. Wilfully mistaking a news item that "Automobiles in the future will be powered by radio," he took this to mean that drivers would soon fill up their cars at WJZ, WEAF, or WOR pumps, so he sketched a roadster heading out of a gas station powered by the **XXX**¡ of STATIC. During the Second World War, in Hugo Gernsback's *Radiocracy for Xmas–New Year, 1944*, a patriotic cartoon of radio circuitry relied upon the capacity of static at its most generic to choke out reception and meaning—in this case, its capacity within a "Degeneration" system complete with an Amplifooee Tube, three Blooper Tubes, and bazooka relays to reduce swastikas and setting suns to garbage.[141]

"Bloopers" in radio lingo were basically Freudian slips, hilarious or revealing reversals of words or sounds; "Bloops" in movie-tech jargon were echoes of voices recorded in studios with bad acoustics; "Corduroy" was the wavering of voices due to irregular current or poor microphone placement; "Voice Grain" was a roughness of pitch or timbre that resulted from unmerciful amplification; "WowWows" were "noises that appear on the finished film that were not taken up on the set," glitches in the electrical transfer of recorded sound as strips of fluctuating light onto the margins of the filmstock itself. The movies became Talkies, after all, not by hiring actors to stand behind the screen and speak the words of celluloid figures, which was done early in the era of the Silents, nor by synchronizing a phonograph record of music and sound-effects with each reel, an occasional practice by 1910 and still used by Paramount in 1928 for its first "all-talking" film, "Interference"—at whose premiere in London the needle stuck on one of the grooves and kept repeating the words "another of those damned postcards, another of those damned postcards." Rather, movies became Talkies by reconceiving sound as light. The phonograph had transcribed sound-as-vibration onto scratchy media. The telephone had transduced sound-as-vibration into electrical impulses. The cathode ray tube and phonodeik had translated streams of electrical impulses into optical signals that could be graphically registered. All that remained was to make this last process reversible, which could be done with the new photoelectric cells developed during the First World War.[142]

Joseph T. Tykociner, a Polish-American electrical engineer, was among the first to grasp the technical implications. Sound-sensitive as a youth, by nineteen he had already imagined the photography of sound as a logical advance upon phonography. The next year, 1897, after viewing a Biograph silent film, he was further inspired, for "the absence of sound made the show unnatural and especially the mute dramatic scenes seemed to me unendurable." On his way home "in the rattling elevated railroad, I had ready in my mind a scheme of photographing sound at the same time with moving pictures on a common film and reproducing the sound simultaneously with projecting the pictures upon a screen." Twenty-three years later he realized his scheme, after working with advanced radio technology in Germany, England, and Russia, and on submarine detection and communications for the Allies during the First World War. At war's end he had the expertise to design a feasible system for recording and projecting synchronous film sound, but neither Eastman Kodak nor Westinghouse had the foresight to capitalize such a "double illusion." He took a position at the University of Illinois, where, after months of struggling with "humming and cracking noises" from poorly insulated conductors and fluctuating magnetic fields, he created in 1922 his first sound film, featuring "My wife, a bell, and a question," with which he impressed campus physicists, including the acoustician Floyd Watson. Meanwhile Lee De Forest, who had visited Tykociner's lab, was promoting his own "phonofilm" as totally unlike "the talking movies, the scratching, hop, skip and jump variety unsuccessfully attempted some years ago." Watching a phonofilm, he promised, one would hear, "when a fine picture is shown of a great waterfall, or the surf upon the shore, the impressive music of nature itself, reproduced with perfect fidelity—the lack of which now makes all such pictures strangely 'empty' or unsatisfactory." One would also hear, sadly, the noise of the celluloid film stock itself. The endemic problem of movie sound, as it turned out, was the imperfect coating of the film; our eyes are happy with images that are 98 percent okay, but photoelectric cells reading the sound strip on the edge of each frame responded to imperfections of one-tenth of one percent. It seemed then that there would always be *noise* in the Talkies, not just the "whispering, coughing, laughing, rattling of programs, etc.," uncontrollable by exhibitors, but on every Talkies reel itself—an audio-visual static that was a composite of fluttering sound, flickering light, and finicky emulsions, a stubborn static that would remain even after sound engineers learned to shield their microphones, temper their amplifiers, screen out interference, and cancel take-offs at local airports.[143]

"Now we know what causes static in the talkies," reported Cedric E. Hart in 1928. As editor of Hollywood's first trade magazine to cover Talkies, he encouraged letter-writing campaigns to nearby aerodromes in protest against low-flying planes, whose "vibrating noises can even penetrate a soundstage." (Static was in this sense an aerial bombardment reminiscent of the barrages of the Great War.) Each issue of *Sound Waves* included an advertisement for Hamlin Sound Proof Doors and counsel on how best to move from "Squawkies" to Talkies. You couldn't expect the public of 1929 to be satisfied with poor-quality synchronized sound in cinemas when people had good-quality phonographs and radios at home. Bad sound would play into the list of grievances of the Musicians Mutual Protective Association, whose members had been relying for two decades on gigs with cinema orchestras, and into the grievances of thespians convinced that "moving pictures interspersed with vocalization is against nature—like the marriage of the sphinx and the caw-caw bird." Was this the contrarian point of such a putative Talkie as "Noisy Neighbors" (1929), a full-length Pathé feature in which the Quillan vaudeville family played itself with just three minutes of audible dialogue and little else than celluloid static to be heard the other seventy-one minutes? Were they trying to prove that performers on the silver screen did not need synchronized dialogue in order to be "heard," did not need the support of "noise libraries" of howling gales and airplane motors? That film sound, ribbed as it was with corduroy, got in the way of the kinaesthetics of motion pictures?[144]

Soundmen, of course, had been hunting down all of the sources of noise they could, in telegraphy, phonography, telephony, radio, and now the Talkies. Some of the noise arose from cheap or eroded materials and broken connections, some (as laypeople knew) from storms or electric mixers. As Chen-Pang Yeang has shown with regard to early radio, engineers hoped to recast the proximate environment and shield devices from that environment so that soundsystems were free enough from external noise to operate according to specs, while technocrats wanted to dictate standards for equipment and control the spectrum so that broadcasting could be rationally administered and station cross-talk reduced to a minimum. These tweaks were "hum-bucking," compromises with bad-boy circuits, since the long-term pother of electrical noise stemmed neither from mismatched components nor from outside clamor but from the Brownian motion of a corpuscular world. One could define electrical noise as "the reproduction of any variation of voltage . . . not caused by the signal impressed," then deal with it by storing and reversing the phase of any "noise voltage" received and reapplying it to an amplifier "in such a way as to cause noise in the amplifier to be balanced or 'bucked out,'" but this did not resolve the underlying atomic randomness. One could repair terminals, amplifiers, and receivers by "signal tracing," using noise as both the signal and essence of problems with shorted coils or leaking condensers, but there were always limits to what one could do with agitated electrons, even if you had been installing, servicing, and repairing radio sets since 1921, like Lewis S. Simon of the Rexall Radio Stores of Brooklyn, whose manual on *Radio Service Trade Kinks* had numbered sections on how to handle Distortion, Hum, Static-like Noise, Intermittent Distortion after Playing a While, Weak Volume Together with Steady Frying and Crackly Sounds, Speaker Rattling, Cross Talk, and Choky, Distorted, and Weak Reception. "Even with absolutely ideal selective circuits," wrote John R. Carson of Bell Labs, "an irreducible minimum of [static] interference will be absorbed, and this minimum increases linearly with the frequency range necessary for signalling," so all compensatory mechanisms would be "quite useless."[145]

Rooting out the noise would therefore be next to impossible. Let's be blunt: impossible. Just as unassisted audition relied upon the smallest, most intimate bones of the body—the hammer, stirrup, and anvil—so the vibrations ultimately responsible for audible sound were bound up with the smallest of particles, and their randomness was (as it were) bred into the bone of the universe. Not that our instruments failed us in our efforts to uncover pattern and law: rather, our acts of observation require energy, and at extremely small scales that energy, no matter how spare, becomes a significant source of interference. If, as a physicist for RCA explained in 1932, "Noise is entirely random in nature; its wave shape, which cannot be predicted, never exactly repeats itself," so the arcane mathematics of quantum mechanics and the fleeting results of quantum experiments had revealed that Nature on (un)certain levels was probabilistic—and, statistically, noisy. Static, then, was the audible expression of an elemental condition of matter, and of all communication.[146]

Much of this had been suspected since 1918, when Walter Schottky, a German physicist, published his studies of two persistent sources of noise in radio amplifiers. Even apart from power line induction effects and internal problems with weak welds, faulty wiring, or unstable cathodes, powerful radios were fraught with noise. The first source of this noise Schottky termed the *Wärmeffekt*, now called thermal noise, which results from the randomness of the flow of heated ions through any input to any amplifier. The second and, in his thinking, more disruptive source of noise was the *Schroteffekt*, or shot effect, a spontaneous (unprompted) and irregular variation in the magnitude of currents produced by the amplifier itself. A decade earlier, Albert Einstein, Marian Smoluchowski, and Jean Baptiste Perrin had shown that Brownian motion was key to atomic (molecular-kinetic) theory, so the presence of predictably unpredictable currents could not be dismissed as a technical glitch. Something basic was going on, and Schottky could *hear* it: it was a hissing noise in the headset, "the effect of thermally emitted electrons pattering on the anodes of the amplifier tubes," each electron arriving independently of the next, in a series of "unrelated shocks." Yes, *shocks*. With the post-Great War development of high-gain amplifiers, J. B. Johnson, Harry Nyquist, and Schottky in 1925–1928 could demonstrate persuasively that fluctuating voltages were the result of agitation at the subatomic level as free electrons were randomly emitted from the heated cathode of a vacuum tube. It now appeared that within any active conductor, no matter how shielded or materially pure, a chaotic cloud of excited ions must arise as an unavoidable source of resistance, such that measured strengths of current would always diverge from calculated values. That cloud could be heard as clearly as the newly discovered "flicker effect" could be seen ("a chaotic variation of light intensity taking place over the surface of the cathode"). Investigating radio circuits and grids for RCA in 1930, G. L. Grundman took the shot effect, the flicker effect, and thermal agitation as facts of life, such that radio reception should not be described in terms of absolute clarity but in terms of a better or worse "signal-hiss ratio." By the 1930s it was "a fact, all too familiar" to physicists and electrical engineers "that a pair of telephones included in the last stage of a thermionic amplifier, of great magnification, emits a continous rushing sound which cannot be eliminated by attention to batteries, contacts, or screen cases." That staticky sound, wrote the physicist E. B. Moullin, was a function of the parameters of all circuits, each of which had a "noise pulse": it was the sound, almost literally, of rushing electrons; most literally, of the uncertainties of quanta.[147]

Quantum theory was flailing in 1920 when Werner Heisenberg entered university to study with Arnold Sommerfeld. Fifteen years before, Einstein had not only conceived of a universe that was relativistic and probabilistic; he had posed, against the reigning theory of light as waves, a theory of light as quanta, "tiny, discrete bundles of light energy, each behaving like an independent particle and carrying identical amounts of energy" (thus Heisenberg's biographer, David Cassidy, whom I follow here). These indivisible bundles moved on probabilistic paths that were incalculable using any prior system of equations. In 1913 Niels Bohr in Copenhagen and Sommerfeld in Munich had compounded the dilemma with their quantum theory of the atom. Thereafter neither the classical model nor the quantum model (abetted by Max Planck's equation for angular momentum) seemed to be able to resolve obstinate anomalies in electrodynamics and radiation spectroscopy.[148]

Sommerfeld in his *Atombau* of 1919 had exulted in the "musical beauty" of quantum numbers and the promise of "an atomic music of the spheres," but it was the dissonance that was remarkable. While a German politician was assassinated on average every four days between 1919 and 1922 and inflation in the Weimar Republic mounted to "astronomical proportions," Heisenberg (independently wealthy) busied himself with the turmoil at the other end of the spectrum, the orbital behavior of electrons, which leapt erratically from one quantum state to another. Einstein in 1916 had characterized the leaping in terms of "transition probabilities"; Heisenberg would summon up a little-known form of calculus to arrive at a matrix mechanics that, as it were, netted the electrons. Even so, the spectroscopic evidence of atomic structure seemed to be explicable only by invoking *both* classical wave theory *and* quantum theory, or what Heisenberg's friend Wolfgang Pauli called "two nonsenses." For this it was necessary to resort to "reciprocal dualities," a theoretical contortion if not also *ein Schwindel*, "a swindle," and then, in 1926, to Ralph Kronig's suggestion of electron spin—so long as each electron (putative and probabilistic) were rotating with an annoying *half* unit of angular momentum. If, as some mathematicians thought, matrix mechanics was "too ugly, too complicated, and too arbitrary to be a proper representation of atomic reality" or, as some physicists thought, too "ascetic" in its resistance to visualization, quantum conundra could also be resolved by a competing and visualizable theory of wave mechanics. As developed by Erwin Schrödinger in Zurich in 1926, wave mechanics merged Louis de Broglie's recent presentation of matter waves with Einstein's recent emphasis on vibratory processes in very dense ("degenerate") gases in order to arrive at a happier formulation: assuming that light quanta moved in vibratorily harmonic if indeterminate concert, like waves of matter, their collective behavior could be calculable through "wave functions" that required only the use of easily managed partial differentials.[149]

Bohr and Heisenberg, who resented de Broglie's and Schrödinger's reliance upon the comforting continuities of waves, would have to come up with something better, or bolder. For this, Heisenberg returned to noise. That is, to Brownian motion and the problem of "the behavior of fluctuation phenomena—digressions from equilibrium caused by statistical effects of large numbers of particles." He argued not that Schrödinger was wrong but that his wave mechanics did not deal "naturally" with those perturbations and discontinuities found in the many experiments inferentially tracking quantum particles; he refused to cede the awkward materiality of the world to a purely formal set of equations. (Informed by a graduate assistant that space was simply the "field of linear operations," Heisenberg replied, "Nonsense, space is blue and birds fly through it.") Hedwig Born meanwhile was showing mathematically that Schrödinger's wave function was successful not because quanta actually formed into waves but because his function accurately depicted "the propagation of probabilities to find particles at certain locations in space and time." In other words, the reality of quantum waves was statistical, and as physicists were learning from the work of mathematician David Hilbert, "one could have axioms for a logic that was different from classical logic and still was consistent. That," Heisenberg later said, was "*the* essential step" in quantum physics—learning to distinguish between inadmissible inconsistency and viable paradox. As Born announced in August 1926, "We free forces of their classical duty of determining directly the motion of particles, and allow them instead to determine the probability of states."[150]

Heisenberg was twenty-five when at last, in March 1927, he formulated his indeterminacy principle; Bohr was forty-one when later that year he framed the principle of complementarity. Neither would have said that they were giving noise epistemological standing and ontological footing, but that's how it would seem, later or sooner, to electrical engineers, information theorists, and radioastronomers. Heisenberg's subatomic realm was a young man's realm of exciting discontinuities, high speeds, and epistemological static. The very energies required for identifying small, fast entities intruded upon all experiments and made it impossible for anyone to know, at any instant, both the position and velocity of a particle. Since position and velocity were crucial to determining causal relationships among magnetically interacting or presumably colliding particles, the law of causality itself, as Heisenberg heard/saw, becomes inoperable in subatomic space. The predictions of a quantum mechanic (like the predictions of a sly garage mechanic) could amount to no more than calculations of "a range of possibilities." And since there was an inescapably reciprocal relationship between the space-time uncertainties of atomic measurement, there would always be a quantum cloud of unknowing, akin to the insurgent clouds of chaotic electrons that Schottky and Johnson had located in the most well-defended of circuits. Herein lay the theoretical, experimental, mathematical, and philosophical roots of the shot effect, thermal noise, and other hissing and hooing one heard through even the finest of headsets. A technoculturally tight relationship linked uncertainty at the quantum level to the ineradicable randomness of electrons rushing from cathode to anode and banging chaotically on human tympani.[151]

Complementarity established ontological noise. Whereas Heisenberg, in order to hold quanta conceptually distinct from waves, made a case for limits on what can be known about matter at any moment, Bohr made a case for duality: the physical, theoretical, and existential necessity of both quanta and waves. These alterities of state were neither hierarchical nor arbitrary; they were the alterities of a mature man ready to sacrifice some of the exactness of science so as to comprehend "wider fields of experience." Quanta and waves, he claimed, coexisted in the key equations for subatomic momentum and energy; in experimental situations they made their presence known simultaneously. In fact, an experimenter's *choice* of one or the other to describe what had occurred was itself responsible for introducing a degree of uncertainty; the decision, for example, to use wave functions to analyze the results was enough to plant the seed of indeterminacy. Existence, and not just the mutually exclusive suppositions of quanta or waves, required both to be always in play, a principle of complementarity that Bohr soon applied to biology, law, and ethics.[152]

Returning then to acoustics and static. George W. Stewart, a physicist at the University of Iowa and inventor of acoustic filters that altered the approach to improving voice transmissions, published a short paper in 1931 proposing that an uncertainty principle of sorts was already imbedded in the classical equations of acoustics and in the limitations of the human ear: both demonstrated a strongly complementary relationship between frequency and time that was not fully resolvable to one or the other, and the behavior of both suggested that there were limits to the precision through which any ear (and brain) could perceive changes in the duration, frequency, and intensity of sound (particularly vibrato). In the context of carbon microphones, telephone lines, and radio, the uncertainty and complementarities of quantum physics meant that the quality of effective reception and transmission was always conditioned—and limited—by probabilistic agitations at the most ungraspable of scales, where any intervention was also a disruption, where any wave could act like a particle, and no where had a definite when. Body—as Indian philosophers had argued for millennia, and as the Indian mathematical physicist Satyendra Nath Bose (in cahoots with Einstein) would show—was fundamentally a question of energy in different states, so communication between bodies was subject not only to entropy but to the random leaps of energetic quanta. To stretch the point on behalf of the radio listener: merely tuning "in" was willynilly an act of tuning "out" that—because leaping spinning electrons insist on the several privileges of duality—could never be perfectly clear but would always be, in acoustic terms, lively.[153]

Sartre understood this in terms of jazz. Between 1931 and 1934, he wrote a novel whose protagonist lives, as he did, in the port city of Le Havre (a.k.a. *Bouville*, Mud City, Dungtown). Antoine Roquentin is losing his moorings as objects around him lose their solidity, people their gravity. First entitled *Factum sur le contingence*, and in a second draft *Melancholia*, it was as *Nausea* that the book appeared in 1938, unfolding an existential condition that "grabs you from behind and then you drift in a tepid sea of time," lost to oneself. (As, at sea, the fluid in the semi-circular canals of the vestibular system of one's inner ear may be so shaken that one loses all orientation and experiences "nausea," from the Greek ναυς, "ship," which the linguist Leo Spitzer would bind etymologically to "noise" after the Nazis had swept up his native Austria.) With nothing to hold to but the absurd—in which being at last *becomes*—Roquentin is sustained by encounters with the shot effect. At a terminus café, "seized by a slow, colored mist, and a whirlpool of lights in the smoke, in the mirrors," he asks the waitress to put on a record: *Some of These Days*. That was the signature song of that red-hot (Jewish) mama, Sophie Tucker, written for her in 1910 by Shelton Brooks, a Native/African-American composer known best for his *Darktown Strutter's Ball*, but Roquentin thinks of it as "an old rag-time with a vocal refrain" composed by a Jew (George Gershwin?) and sung by a "Negress" (Tucker's own act and her "Coon Shouting" songs done in blackface until 1909, her deep voice and syncopated stylings on phonograph records often mistaken by Europeans for that of a black woman). He likes especially the vocals but first come the saxophone riffs, "a glorious little suffering" energizing the only "small happiness of Nausea" he will feel. In that music "there is no melody, only notes, a myriad of tiny jolts. They know no rest, an inflexible order gives birth to them and destroys them without even giving them time to recuperate and exist for themselves. They race, they press forward, they strike me a sharp blow in passing and are obliterated. I would like to hold them back, but I know if I succeeded in stopping one it would remain between my fingers only as a raffish languishing sound." Those are excited electrons, indeterminate, falling "from one present to the other, without a past, without a future," and that is jazz, whose improvisatory brashness enlivens Roquentin, giving him not so much a reason as a rage for going on. Now the "Negress" takes over, singing,

Some of these days
You'll miss me honey

and although there is a scratch at the refrain, and an odd noise "that clutches the heart," and although the steel needle leaps and grinds in grooves so worn that he wonders if the singer is long dead, still "It seems inevitable, so strong is the necessity of this music: nothing can interrupt it, nothing which comes from this time in which the world has fallen." And it and she go on:

And when you leave me you know it's gonna grieve me,
Gonna miss your big, fat mamma, your mamma some of these days.

The last chord dying away, he feels "strongly that there it is, that *something has happened*," feels his body harden, the Nausea dissipate. He is ready to embrace that situated alertness Heidegger had been advocating and Sartre pondering. By way of jazz and its improvisatory "noise," Roquentin feels himself to be dancing; across from him dip and spin his contingent partners, wave and quantum, unreachable. "It is beyond—always beyond something, a voice, a violin note.... and when you want to seize it, you find only existants, you butt against existants devoid of sense." This could be Heisenberg on uncertainty, Bohr on complementarity. "It does not exist because it has nothing superfluous: it is all the rest which in relation to it is superfluous. It *is*." Play it again, Madeleine: "So two of them are saved: the Jew and the Negress. Saved. Maybe they thought they were lost irrevocably, drowned in existence," but they "have washed themselves of the sin of existing. Not completely, of course, but as much as any man can." So Roquentin, who has shaken off his torpidity and is about to leave for Paris, understands *Some of These Days*. "This idea suddenly knocks me over, because I was not even hoping for that any more. I feel something brush against me lightly and I dare not move because I am afraid it will go away. Something I didn't know any more: a sort of joy." A going-away present.[154]

5. *Zones*

Jazz rhythms and universal laws were one thing. Local regulations and state laws were quite another. If noise everywhere and everywhen was everyhow affecting everyone, the best an Anti-Noisite might engineer would be acoustic reserves, like Julia Rice's Quiet Zones around hospitals and schools. No one, not even a Maxim, should expect to live in a silenced world, although there were certain cloisters—Cistercian monasteries, study halls, maximum security prisons—where abbots, monitors, and wardens enforced a thoroughgoing quiet. And there were certain anti-Anti-Noisite cartoonists who went to the satirical extravagance of de-barking dogs, muzzling parrots, muffling sparrows,[155] or disengaging the whine

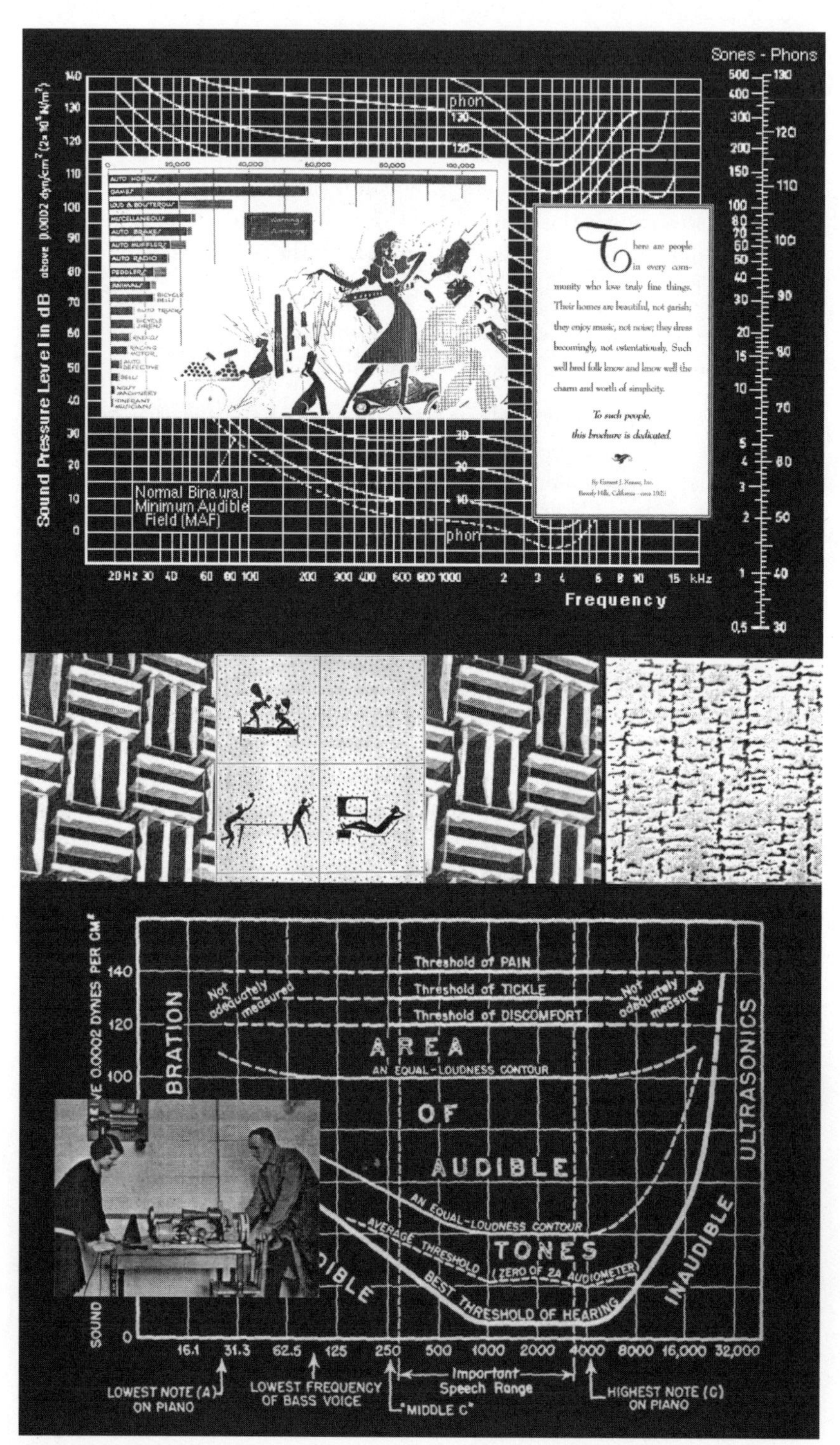

Sones - Phons
Sound Pressure Level in dB above 0.0002 dyn/cm² (2×10⁻⁵ N/m²)
phon
Normal Binaural
Minimum Audible
Field (MAF)
Frequency
kHz
AUTO HORNS
GAMES
LOUD & BOISTEROUS
MISCELLANEOUS
AUTO BRAKES
AUTO MUFFLERS
AUTO RADIO
PEDDLERS
ANIMALS
BICYCLE BELLS
AUTO TRUCKS
BICYCLE SIRENS
ENGINES
RACING MOTOR
AUTO DEFECTIVE
BELLS
NOISY MACHINERY
ITINERANT MUSICIANS
There are people in every community who love truly fine things. Their homes are beautiful, not garish; they enjoy music, not noise; they dress becomingly, not ostentatiously. Such well bred folk know and know well the charm and worth of simplicity.
To such people, this brochure is dedicated.
SOUND ... VE 0.0002 DYNES PER CM²
BRATION
Threshold of PAIN
Threshold of TICKLE
Threshold of DISCOMFORT
Not adequately measured
Not adequately measured
AREA
AN EQUAL-LOUDNESS CONTOUR
OF
AUDIBLE
AN EQUAL-LOUDNESS CONTOUR
AVERAGE THRESHOLD
TONES
(ZERO OF 2A AUDIOMETER)
BEST THRESHOLD OF HEARING
...DIBLE
INAUDIBLE
ULTRASONICS
16.1 31.3 62.5 125 250 500 1000 2000 4000 8000 16,000 32,000
LOWEST NOTE (A) ON PIANO
LOWEST FREQUENCY OF BASS VOICE
"MIDDLE C"
Important Speech Range
HIGHEST NOTE (C) ON PIANO

from the vocal cords of four-year-olds. Whether it was traffic and jackhammers in the city, or tractors and chainsaws on the farm, or the static that seemed to Chicago conventioneers and Blue Ridge radio audiences to be resident in loudspeakers and airwaves, or the spasms of electrons as posited by Schottky and Johnson, noisiness was evidently intrinsic to a world in motion—and to sentience itself. With modern technical skill, however, there just might be sound preserves almost as "quiet as the stars of night" at the Griffith Planetarium, lined with sugar cane bagasse (Acousti-Celotex) to keep noise out, or quiet as the Eastern High gymnasium, where bagasse kept noise in, all the "excited voices, clattering feet, and the whamming of balls... absorbed and confined to the room by the ceiling of Acousti-Celotex tiles"; quiet as the 300,000 American homes Celotexed by 1929, or as the factories and offices "Silent-Cealed" with multiple porous membranes that achieved (so it was said) 100 percent sound absorption.[156]

Strange, that the best design for sound absorption was proving to be porous and randomly perforated: as if the architectural remedy for noise had to reproduce the Brownian features of a quantum universe. Instead of fencing out sound as you fenced out light or wind or dirt by interposing taut surfaces, you had instead to construct a thicket. (With Burgess Acousti-Pads for Absorbing Duct Noises in Air-Conditioning Systems, balsam-wool was "covered with a facing of perforated metal" so that "sound waves are not deflected but trickle through the perforations to be dissipated in the pad by acoustic friction.")[157] Sound does not play by quite the same rules as light or wind, for it cannot be so blockaded; nor does it play by the same rules as dirt or smoke, for it travels through materials without impregnating them. Noise in particular was so much a function of an observer's referents and attention that it seemed paradigmatic of the principle of indeterminacy.

All along, the courts had had problems with noise precisely because it was so indeterminate. Other nuisances—smoke, sewage, obstructions, stench—leave lingering signatures. Although the sources of loud, persistent, obnoxious sounds could be located easily enough, their residues could only be tracked through cracks in roofs or the dislocation of walls, as from the sledgehammers of an armorer in 1377. Otherwise, the issue was the ongoingness of a noise: English law in the 1590s could find for the plaintiff against a blacksmith who pounded at all hours with an iron mallet whose blows made the rooms of Josue Bradley, gentleman, "uninhabitable," if not also structurally unsound. At all hours: for late medieval and early modern courts, time and timing established the seriousness of noise as a nuisance: courts were less reluctant to punish noisemakers of the night, when curfews were in force, or on the Sabbath, than during the common hurly-burly of a working day.[158]

Upon reports of public taunting, radical haranguing, or mobs a-gathering, local authorities could stifle potentially incendiary events as "disturbances of the peace" at any time in most any place; such events were surely noisy but their threat was rather political than acoustic. Noise *nuisances* were dealt with after the fact; one petitioned to reduce or (if need be) remove them, rarely to forestall them. Awards of damages were far more likely than restraint, particularly for noisy enterprises that had been licensed and properly situated. Even in an 1847 New York case in which the owner of a boardinghouse took exception to an adjacent blacksmith who had rented out part of his shop to a manufacturer of engine boilers whose work involved "a tremendous noise and pounding, commencing about six a.m. and until sundown," the lower court's decision favoring Mrs. Dodge (Justice Bronson: to prove a nuisance it is not necessary that a person first be driven out of her home) was reversed on appeal. The defendant himself had not been shown to be at fault, unless Mrs. Dodge could prove that the blacksmith knew aforehand how much noise would be entailed. It would take a change in both the sentiments and the presentiments of the law (how could a blacksmith *not* have known?) before noisemakers could be forestalled.[159]

Josue Bradley and (at trial) Catherine Dodge won their cases on the strength of property rights: noise had made their dwellings uninhabitable. The thrust of nuisance law was to restrict encroachments that were not explicitly territorial but otherwise interfered with the security and enjoyment of property rather than with physical or mental health or the moral order. Proceedings against nuisance were almost exclusively initiated by owners of property whose "quiet enjoyment" thereof could be framed in terms of habitability. Sound-encroachment being more intangible than harm from smoke (which registered on buildings and clothing and in throats and eyes), or from the fumes and unsightliness of sewage, or from obstructions that blocked the air and the sun, the case law on noise did not mount up so solidly that plaintiffs could parlay histories of sound into predictable mandates for relief. Despite some widely cited decisions and numerous local ordinances, courts handled noise cases as singular, with latitude for the oddities of each environment, each sort of sound, and each complainant. When David Owen in 1841 sued Abner Henman and the parishioners of the Brick Church Meeting House for making so much noise with their singing and loud scripture-reading that Owen could not hear the sermon of his own minister when he sat in his pew in a Presbyterian church close by, the Supreme Court of Pennsylvania warned that encouraging such suits would make a shambles of the law: "Could an action be brought by every person whose mind or feelings were disturbed in listening to a discourse, or any other mental exercise (and it must be the same whether in a church or elsewhere), by the noises, voluntary or involuntary, of others, the field of litigation would be extended beyond endurance. The injury, moreover, is not of a temporal but spiritual nature," with no earthly remedy. Upon the more earthly expansion of steam power and the earthy quickening of transport, the courts also tended to measure the extent of private inconvenience against the larger public good of, say, a new railway. And when it came to balancing industry against worship, Justice (not Learned) Hand of New York in 1848 was unconvinced by the Trustees of the First Baptist Church of Schenectady in their suit against the Utica & Schenectady Railroad Company for interfering with church services by "ringing or causing to be rung their bells attached to their locomotive engines, and by the puffing and whistling of said locomotive engines, and blowing off of steam thereof...and by the rumbling, jarring noise of said railroad cars and steam engines." None of this noisemaking was malicious, wrote Hand; besides, in nuisance suits "There must be some injury to the property, immediate or consequential," and he could find no precedent for "injury done to real property by mere noise." Further, although "unquestionably noise may amount to a nuisance," it was not actionable here because the state legislature had authorized the construction of the railroad, and "that which is authorized by an act of legislature cannot be a nuisance."[160]

As men and women of means began litigating against the smoke, smell, and noise of steam-powered industry, and as cities punched outposts into country lanes, old money was set against new, ancestral homes against generic real estate, and the "natural right" to absolute dominion against upstart theories of social utility. More fundamentally, a feudal sense of place as a stronghold of personhood was set against a bourgeois-modern sense of place as spaces of transit and of privacy as a portable right. The courts began to listen anew to complaints of excessive noise that reflected not only the shifting notion and experience of place but the shiftiness both of place and of time in a world where capital was indebted to markets that leaned toward speculative futures rather than reclining on static leases or steady rents. One after another, justices put aside the doctrine that individuals had no cause for complaint if they set up shop or house near a known source of foulness or noise; the world was now too much in flux to expect the character of a place to be other than provisional—the more so in North America, where any Huck with a chainsaw could light out for the territories. "However lawful the business may be in itself, and however suitable in the abstract the location may be, they cannot avail to authorize the conductor of the business to continue it in a way which directly, palpably and substantially damages the property of others," wrote Chief Justice Graves of Michigan with regard to a Detroit forge that relied on four steam hammers, one of which weighed 3,500 pounds. The same logic was implicit in decisions from London to California against stables close to residences, which for centuries had been accepted as a necessary evil, despite the dust, odor, and "the constant noise arising from the stamping and kicking of the horses, the rattling of the ropes and chains and blocks against the ring-bolts and mangers." Justice Finlayson of California was most grandiloquent in his judgment allowing the city of Sierra Madre to enforce its ordinance against the keeping of a stable for mules in a residential district: "We know of no heaven-sent Maxim to invent a silencer for this brute." Until nature evolved "a whispering burro or man invents some harmless but effective mule-muffler," it was right that the law remove to an acoustically safe distance "the loud discordant bray of this sociable but shrill-toned friend of man." The same logic undergirded rulings against the most stable of institutions, when plaintiffs could show that continual or early bell-ringing by churches "disturbs the quiet and comfort of our houses."[161]

Where before complainants had had scant likelihood of relief if they "came to an existing nuisance" or if a new nuisance unavoidably accompanied public improvements, now the encroachments of heavy machinery and the pressure of expanding cities worked from opposite ends to forge a rhetoric of rights that urged the granting of injunctions against future harm. Given that a nuisance by noise was "emphatically a question of degree," the courts began more seriously to entertain complaints against the makers of familiar noises that had come to be perceived as unreasonable or unseasonable, whether the perceptual shift stemmed from changes in habits of awareness, in demographics, or in the gendered and generational lines between domestic interiors and urban exteriors. (At home, women and men of different ages might inhabit different acoustic eras depending on their preferences in furnishings, cookware, and music, their access to various appliances, their acuity of hearing, and the density of traffic or trees just outside their doors.) As for new noises, these too had to be comprehended as matters of degree, whether in their tonal difference from earlier sounds, their volume, frequency or persistence. Thus Lord Selborne, presiding over what became a widely cited but ambiguous case for Anglo-American jurisprudence, had to determine in 1872 whether the recent introduction of a steam engine to a longstanding silk mill had brought in enough new noise adjacent to Gaunt's boarding house to warrant an injunction against the mill operating as usual, or whether the suit by two of the "unmarried ladies" housed by Gaunt resulted from an instance in which "a nervous, or anxious, or prepossessed listener hears sounds which would otherwise have passed unnoticed and magnifies and exaggerates into some new significance, originating within himself, sounds which at other times would have been passively heard and not regarded." Selborne sided against the ladies; in *Heather* v. *Pardon* (1877), another British decision turned the tables: although a steam-printing shop had operated for twenty-eight years in the same spot without upsetting its neighbors, refinements to its machinery in 1875 had produced sufficient new noise for the court to act on behalf of suddenly upset complainants. In a further English decision of 1879, a physician won an injunction against a confectioner, though it was the former who, by extending his consulting room into his back garden, had moved much closer to the kitchen wall of the latter, who for twenty years had been pounding pestles in large mortars to make his candies. By 1910 the Pennsylvania Supreme Court was willing to reconsider a noise complaint against an iron works, despite the evidence that (1) the complainant, Mrs. Collins, had built her house in 1899 in an area long used by manufacturers; (2) her house stood in close proximity not only to the Wayne Iron Works but to active rail tracks, an electrical plant, livery stables, a blacksmith's shop, and an automobile garage; (3) she had not brought her suit until December of 1906; (4) the value of her property during those seven years had risen by 27 percent, by her own estimate; and (5) Mrs. Collins had since moved away and rented out her house without problem to a tenant who had not joined the complaint. Refining but not reversing a lower court's blanket injunction against the Wayne Iron Works, Justice Moschzisker ruled that the works must keep its doors and windows shut during the day and cease all work between seven p.m. and seven a.m., so that its riveting machines, power chippers, steam hammers, and air drills not interfere—with what? with whom? Decisions against noise nuisances could now be expected where the scale or persistence was found to be intolerable by complainants with "simple tastes and unaffected notions," even in those neighborhoods that had grown up around mills, presses, quarries, rail-switching yards, or slaughterhouses.[162]

Unfair as this was to wheelwrights and blacksmiths who had been in the same quarters for decades, or to dairies that had grown to serve yet another generation of local customers with new bottling equipment and larger drays, the courts were taking fonder cognizance of citizens' senses and their vulnerabilities, a cognizance most blatant with regard to audition. As Horace Wood explained in *A Practical Treatise on the Law of Nuisances* (1875), "It is now well settled that noise alone, unaccompanied with smoke, noxious vapors or noisome smells, may create a nuisance and be the subject of an action at law in equity or an injunction." Of itself, this was unexceptionable, but Woods went on, citing recent cases in Britain and the United States (*Dennis* v. *Eckhardt*, *Crump* v. *Lambert*, *Walker* v. *Brewster*), to argue that "regard is not always to be had so much to the quantity, as to the quality of the noise." For example, a filing or rasping sound "may yet so affect the nervous system as to produce absolute physical pain in persons of ordinary sensibilities, while the heavy blows of a triphammer might produce no unpleasant sensations whatever." Furthermore—here Woods took a dramatic step—complainants need not prove that nuisances endanger health but only that they are "offensive to the senses."[163]

Context was critical: locality, trade, property, time, intensity, and motive. Accidental and transitory noise was never actionable, only that intrinsic to an operation (as of a rolling mill, a blast furnace, or the record-press of the Victor Talking Machine Company), precautionary (as of steam whistles and alarms), or intentional and inconsiderate (as of musicians). Animal noises were always tricky—how much barking was suitable to a dog on a respectable street, how much cackling to a hen in a village coop, how much crowing to a rooster in a Brooklyn backyard, how much scrawking to a parrot in a parlor—and so were cases of pots calling kettles black. A company of Philadelphia galvanizers, working in sheet-iron, sued a file-making company after it moved its equipment closer to the party wall between the two concerns, making it impossible, said the galvanizers, to give a sales pitch amid the "crash and din of ponderous steam hammers and the rattle of vibrating iron." But neither industry was known for its calm, and the shared building was located among other noisy factories, and the galvanizers had waited two years before filing against the file-makers, which delay implied a capacity to carry on business despite the nearer crash and din: no injunction resulted.[164]

If the law was loosening, it remained stinting, and skeptical of the petulant. Affected psyches had to be standard psyches to fit the bill in an 1888 Massachusetts ruling that the courts must "determine noise by the effect of noise upon people generally, and not upon those, on the one hand, who are peculiarly susceptible to it, or those, on the other, who by long experience have learned to endure it without inconvenience; not upon those whose strong nerves and robust health enable them to endure the greatest disturbances without suffering, nor upon those whose mental or physical condition makes them painfully sensitive to everything about them." Were this corsetry of conditions not tight enough, presiding judges would remind all within hearing that modern or urban life demanded daily compromises between the exquisiteness of civilized senses and the abrasive energies of commerce and industry by which civilization advanced. "The onward spirit of the age," the Kentucky Supreme Court had declared in *Lexington & Ohio Rail Road* v. *Applegate* (1839), "must, to a reasonable extent, have its way," however loud its whistling. Justice Hampton of the Pennsylvania Supreme Court had found in 1855 that the locomotives of the Ohio and Pennsylvania Railroad made "neither more nor less noise than would be expected of their business," and to enjoin their passage through city streets because of such noise would be in effect "to stop the passage through our streets of the hundreds of hacks, omnibuses, drays, and carts . . . to stop all machinery of every description, driven or propelled by steam—to stop all public markets which produce noise, and disturb the citizens residing adjacent thereto." Given "that we live in an age and a country of progress and improvement, in all the business departments of life," history was transforming what once may have been heard as noise into the music of progress. The next year Justice Lumpkin of the Georgia Supreme Court had forgiven a sash-and-windowblind factory "the continued noise made by letting off steam from the boilers, and by blowing the whistle attached to the engine, and by rolling, changing and working of the machinery." The world was switching over from horses to boilers, and "as well attempt to stop up the mouth of Vesuvius as to arrest the application of steam to machinery at this day." Get used to it, he told the plaintiffs, and let them be comforted by the historical record: "We know of no sound, however discordant, that may not, by habit, be converted into a lullaby, except the braying of an ass or the tongue of a scold."[165]

Similarly expansive, preemptive, and ostensibly exculpatory language was adopted by Justice Alvey of Maryland in 1878: "A party dwelling in the midst of a crowded commercial and maufacturing city cannot claim to have the same quiet and freedom from annoyance that he might rightfully claim if he were dwelling in the country." Yet he affirmed a lower court injunction against a brewery's new steam engine that caused "a loud, deafening and jarring noise of extraordinary force and volume" throughout an adjoining house. Again, Vice Chancellor Van Fleet declared in a New Jersey Equity case of 1881 that "Things merely disagreeable, which simply displease the eye, or offend the taste, or shock an over-sensitive or fastidious nature, no matter how irritating or unpleasant, are not nuisances." Only "stern necessity" would dictate an injunction against the printer Hardham, the vibration from whose steam presses so jarred the wall shared with Demarest's harness factory that his workmen became nauseous. Yet, although Demarest offered no proof of financial harm or production declines, Van Fleet ruled that Hardham must rearrange his presses so that their force was directed away from the party wall. Again, Justice Russell of New York in 1899 seemed to be siding with the Edison Electric Illuminating Company: "In the strict sense, the use of machinery producing noise or vibration injures neighboring property. But to some extent such results must come to all who live in a busy, prosperous city. The hum and throb of mechanical life cannot be wholly confined to the walls of any structure." Yet, though the complainant's own building was so derelict that it was more permeable than most to the noise, soot, and cinders from a power plant, Russell enjoined "the overloading of the machinery so as to produce unusual vibration or noise" and awarded Samuel Bowden $250 in damages and costs. If Justice Werner of New York in 1905 held stubbornly to the idea that "The rumble of trains, the clanging of bells, the shriek of whistles, the blowing off of steam, the discordant squeak of wheels in going around the curves" were incident to the operation of any rail system and unassailable as nuisances, other jurists were wobbling toward opinions that any louder-than-ordinary noise or heavier-than-ordinary vibration might be actionable, depending on how a machine were operated or a business conducted. "Discordant, startling, and terrific" whistles might be invaluable on locomotives to frighten off cattle that stray onto the tracks, but Chief Justice Butler of Connecticut saw no reason for any factory to install "a steam-whistle of such a character as to frighten horses of ordinary gentleness": decision for Elam C. Knight, M.D. and his poor nag, who had been startled into collision by "a shrill, sharp, piercing sound" from one of the whistles of Goodyear's India Rubber Glove plant, and by the "grum" sound of a second whistle. The ninety strokes of a 2,000-pound factory bell rung at five a.m., at six, and again at six-thirty were unnecessary "for any purpose of trade or manufacture," ruled the Supreme Court of Massachusetts. The principle of (un)necessity held regardless of whether physical injury had been proved—or the operations approved by elected officials. Thus in 1893 the Illinois Supreme Court had affirmed an award for damages against two railroads: "If the noise, confusion, and disturbance caused by the defendants' engines and cars is such as would, in the absence of legislative authority, have constituted an actionable nuisance, the existence of such authority in no way relieves them of their damaging effect or right to redress."[166]

Once disorientation and discomfort became persuasive of the damaging effects of loud, constant, piercing, and griding noises, courts began to give ear to complaints about injuries not only to property or person but to psyche. Since "comfortable enjoyment" of one's property implied a "mental quiet," the court (declared the Washington State bench in 1910) should take into consideration the dread created by noise: fears could still be real where their causes were imaginary. The sole reason that Justice Judd in 1896 had not enjoined the erection of another Chinese theater in Honolulu seemed to have been that the instruments to be played were left too much to the imagination, for "plaintiff has shown by the affidavits of a large number of credible persons living within the *acoustic sphere* of the noise of instruments used by the orchestra of a Chinese theatre that such noise would seriously interfere with their sleep and quiet." Those are my italics, calling attention to what was probably the first legal notice of an *acoustic sphere*, and to the fact that it was both sonic and psychic. Consider in these regards the 1914 suit of the Kestner family against the Homeopathic Hospital of Reading, Pennsylvania. Twenty-one of the windows of the large Kestner house faced the hospital and forty-eight of the hospital's windows faced the Kestners, making the family involuntary audience to the "moans, shrieks, and groans of persons receiving surgical aid," sounds "of such a character as to render wretched the lives of complainants, and of friends visiting them" and "such as to affect their nerves and impair their health." The trial judge had found no proof that the hospital mistreated its patients, or was slow to furnish painkillers, or used antiquated methods of analgesia, but its emergency operating room, its main surgery, and its maternity and delivery rooms were so close to the Kestners that "shrieks, groans, moans and yells of persons, and cries of children being operated upon, or in pain from other causes" could be heard nightly by the family at their dinner table thirteen feet from the operating room. A hospital is *prima facie* no nuisance, cautioned Justice Potter on appeal, nor was the Homeopathic Hospital inept in its care of patients, yet it did behoove the Hospital, as directed by the lower court, to reorganize its facilities so that its operating rooms lay elsewhere, out of the hearing of the Kestners. And he cited Justice Thayer's decision of 1888 in the Appeal of the Ladies' Decorative Art Club of Philadelphia, whose matrons had been enjoined from the merry beating and chasing of ornamental brass upon complaint by a nearby family that found itself unable to read, think, or converse during the Ladies' hours of artistic hammering. Thayer had no doubts that "The law upon the subject is well settled and very plain. Where a noisy nuisance is complained of, it is a question of degree and locality. If the noise is only slight, and the inconvenience merely fanciful, or such as would only be complained of by people of elegant and dainty modes of living, and inflicts no serious or substantial discomfort, a court of equity will not take cognizance of it. No one has a right to complain that his next-door neighbor plays upon a piano at reasonable hours, or of the cries of children in his neighbor's nursery. . . . On the other hand, if unusual and disturbing noises are made, and particularly if they are regularly and persistently made, and if they are of a character to affect the comfort of a man's household, or the peace and health of his family, and to destroy the comfortable enjoyment of his home, a court of equity will stretch out its strong arms to prevent the continuance of such injurious acts."[167]

Strong arms. Even a merry-go-round in West Virginia might have its rounds curtailed if it whirled to the music of a steam calliope too often or too late in the day. Dissenting from that puritanical 1895 decision were Justices Marmaduke Dent and Henry Brannon, who enjoyed merry-go-rounds and were unhappy with such strong-arming: they would not have the law protect spoil-sports, those who were "Always complaining, never satisfied, morbid, phlegmatic, continually annoyed, fault-finding, looking for slights, easily offended, hysterical, captious, haters of children and their enjoyments, a misery to themselves and a heaviness to their friends." Were the law to smile upon such grouches, courts would be committed to shutting up all theaters after sundown and penalizing churches for holding revival meetings late into the night, lest the invalid, the bitter, and the insomniac be discomfitted.[168]

Granted that legal systems are ever obliged to balance communal sensibilities against individual sensitivities, the notchings of nuisance law can be a revealing index of sensory shifts and biases in a culture. Due however to the intangibility of noise, courts tended to squirm when citing precedents for acoustic nuisance, as if obnoxious sounds had no theorizable matrix. Courts listened instead to witnesses who drew comparisons to more familiar sounds in order to make the horror of the new noise compelling—the 400-horsepower engine and huge cog wheel of an electric lighting company, for example, rumbled like an ice wagon; "continuous loud moaning" and combustions from a gas plant resembled altogether the rumbling of a train, firing of a shotgun, and exploding dynamite; steam-powered circular saws ripping through timber from six in the morning to six in the evening produced a "continuous buzzing, ending sometimes in a discordant shriek and sometimes in a melancholy wail." (The plaintiff in the latter case reminded the court that "the *shriek* of a saw was the technical term for that noise," while an engineer testified for the defense that the sound was like "the hum of a bee." Justice Chitty rejected such honeyed language and granted the injunction.) Before arriving at a decision against the Hotel St. George in 1887, Justice Brown auditioned the electric power plant in the cellar of the hotel, whose quartet of steam engines and dynamos was making the noise at issue. He returned to his chambers persuaded that no analogy to prior sound was feasible: "This is not one of the noises incident to city life, which residents of cities must submit to. It is peculiar in its sound and unlike any other that I have heard, except that produced by other machines generating electricity." And it was, to his way of hearing, a sound as unnecessary as it was unprecedented: injunction granted.[169]

Analogy furnished at best a general idea of the qualities of a noise. The more unusual a noise, the more irksome became the lack of objective measures for its loudness or acoustic insult. "Every good citizen is interested in knowing how much noise the law will compel him to endure at the hands of his neighbors," wrote a commentator in 1886, yet puritanical American courts listened less to amplitude or volume than to genre, in specific to the difference between "money-making noises" and "noises of pleasure," i.e., "useless noises." The Italians, he reported, "order this matter differently, and restrain blacksmiths, boiler makers, etc.," while allowing "musical merrymakers to make night hideous or beautiful, as the case may be, without any restraint whatever." Northern European law was rarely so Epicurean. In Germany, France, and the Netherlands, according to the historian Karin Bijsterveld, the "hammering trades" had long been shunted to the peripheries of cities and their hours abbreviated, but penal codes exempted from prosecution the noise of forges, factories, mills, and the workshops of gold, tin, silver, and copper smiths. If civil codes allowed suits for property (value) loss due to nuisances, not until 1914 did the Dutch Supreme Court accept noise (viz., "terrible din, heavy drones, and severe vibrations") as powerful enough to deprive one of (the enjoyment of) one's property, and German law protected noisy factories that stood shriek-by-howl with others in industrial sectors, so long as they were licensed and, in the case of drop forges, had double walls, double doors, and double windows. As of 1906, tenants in Germany could terminate their leases without prejudice if they could prove declining health as a result of nearby noise, but like other actions, such (de)terminations required time and money and a canniness in the staging of complaints so as not to appear peevish. French law was logically contradictory, favoring the freewheeling entrepreneur in some circumstances, the repose of citizens in other circumstances no more well-defined, for the politics of French tolerance was fraught with revolutionary and counter-revolutionary fervor when it came to all things voluble. Although English and American tort law was zealous in its defense of capitalism against the cavils of the senses, it was restless when confronting noise. With their gradual repudiation of the doctrine of "coming to a nuisance," the courts weakened the right of established businesses to stay in place while upgrading to more powerful—louder, shriller—or more reliable (night-and-day) equipment even as the U.S. Supreme Court, in *Richards* v. *Washington Terminal Co.* (1914), facilitated the rights of way of trolleys, railroads, and other noisy "public improvements" through dense communities. In addition, medical advocacy on behalf of neurasthenics and allied sufferers, whose hyperacuities appeared to be a sadly normal consequence of the speed, pressure, power, and stridency of the modern shift of gears, nudged the courts toward an embrace of rights of privacy that gave noise complainants greater leverage to sue for damages, and to ask for injunctions. Permanent injunctions were more prone to be issued once "the public interest" began to be understood in the context of continental welfare legislation and political theory as "a desirable and transcendent social goal" rather than "the combination of private interests of individuals." Guided by social reformist and Progressive ideas of the public interest and by the passage of public health acts, jurists could read decisions that enjoined a source of noise as other than additional insulation of the property rights of cantankerous plaintiffs; astutely issued, injunctions against nuisances might mute the principle of aggressive individuality in favor of responsible sociability with regard to shared, and sensual, space.[170]

Losers nonetheless outnumbered winners as the "public sphere" was contested along all radii, from the sensoria at the mindful center of personhood to sensationally loud technologies along the industrial spokes; from the nub of private reflection to a rowdy circumference of commerce; from the heart of empathy and quiet confidence to concentric political obligations and imperial-evangelical missions at every tangent. Noise in this geometry was ever ambiguous: if it drowned out a soft dialogue in the bedroom, it bound couples to the shouted conversations of the street and the public addresses of society; if it ground down workers in factories or at home, it was the off-gassing of a spirit of enterprise and the herald of a regular paycheck as well as of a coal-heated, steam-ironed, electrically vacuumed house linked by telephone or radio with the outer world. So, though dictating a respectful hush in their own halls of justice, architected for an echoic air of finality, the courts did not want to be heard restraining the world outside to such an extent that ordinary people could not accomplish their messy, smelly, noisy daily tasks.[171]

But the courts *did* pay increasing attention to how long and frequently a noise went on. Indeed, the recurrence and permanence of nuisances led American courts to recast such injuries as "takings," the annexation of others' property against their will, for which tort law commanded compensation. (Tort law elsewhere was national; in the United States it was federal, adjudged state by state, and consolidated only in law schools, where it was first taught as a separate subject during the 1870s, when Isaac Rice was going through Columbia.) As the legal historian Louise A. Halper has argued, this recasting of nuisance damages converted "into property the right to be free even of consequential injury"; perpetrators of nuisance could buy up those rights in the open market. G. Edward White has added a correlate: when American courts between 1870 and 1910 expanded the idea of negligence from a specific technical neglect to a "violation of a more general duty potentially owed to all the world," they transformed "guilt" and liability into a set of relations of risk in a capitalist arena. Damages for noise, and the noises themselves, were then not simply a cost of doing business but an effective means by which to assess the degree of risk taken, for in most techno-industrial and marketing arenas, noise was at the blade-edge of the intrepid and unheard-of, the newly ballyhooed and experimental—an avatar of, and advertisement for, the everyhowness of speculative venture.[172]

Who then should be held liable for such noise, and when? As the complement, in criminal law, of actions for nuisance at civil benches, liability law bore directly on problems of noise. Before American corporations gained legal status as persons, most liability suits against corporate entities like townships had been for nonfeasance (omission) of measures that could have prevented an ensuing nuisance, including noise. When corporate personhood was confirmed in the 1880s, the courts began to require proof of malfeasance in corporate liability—that is, of malicious or fraudulent breaking of express or implied warranties, or heedless interference with others' quiet enjoyment of their property. Knowing how difficult it could be to prove the mental state of any individual charged with committing a nuisance, civil courts had settled for testimony on the nature of the nuisance itself. It would prove much more difficult to demonstrate the mental state of a corporation (as Isaac Rice tried to do with regard to the manipulation of railroad bonds by holding companies), yet criminal courts on both sides of the Atlantic required findings of "fault," i.e., intent or active neglect, against companies selling unsafe products or polluting the air, the water, or the ears. Some economists and legal scholars at century's turn published briefs for strict liability—that is, for ignoring excuses of user error, intrinsic operation, or unintended consequence where the seriousness of risk and harm was uncontested—as already instituted in the first wave of workers' compensation laws in Europe. Findings of fault, however, remained indispensable well into the twentieth century in adjudications of corporate liability, particularly when it came to noise, which more than any other nuisance was presentable as an inherent hiccup of manufacture *and* marketing. For instance, who precisely was at fault in *Pawlowicz et al.* v. *American Locomotive Co.* (1915)? The railroad's drop-forge shop that lay across the street from Dennis Pawlowicz's saloon, with all the noise and jarring and smoke from the operation of the forge? Heavy trucks rumbling as usual through this manufacturing district and along the street in front of a new, ramshackle saloon? The freight trains thundering over the railroad tracks nearby? "The industry is one of the large industries of the city of Schenectady," wrote Justice Van Kirk, "on whose operation thousands of residents depend for their support. It is not shown that, in any respect, by a change of method or of construction [of the drop forge], any relief can be granted to the plaintiffs from such annoyance as they may suffer. Relief can be had only by restraining in full the operation of the shop. I find, therefore, that the use is a reasonable use, and a restraining order should not issue."[173]

Municipal aesthetics and mounting legal expenses, if nothing else, demanded a more coherent resolution to problems of industrial nuisance and corporate liability. "This is the age of cities," claimed a 1926 brief on "Municipal Aesthetics and the Law." Now that more than half of the United States lived in cities, now that home life was townhouse and apartment life, it was time to embrace the power of eminent domain for more than railroad rights of way. Municipalities had already been allowed to take land for playgrounds, parks, ornamental highway strips, and public beaches whose purposes were no less aesthetic than sanitative, and courts were no longer striking down laws against tall billboards and "garish advertisements in conspicuous places," so the time was ripe for asserting the police power of the state with regard to the size and siting of new buildings, as had been done for decades by Europeans. Wisconsin's Supreme Court in 1921 had held as unconstitutional a statute that prohibited the erection of buildings over ninety feet tall around Capitol Square in Madison, for this smacked of "an unreasonable exercise of the police power and an attempt to acquire rights that can only be acquired in the exercise of eminent domain," but Justice Crownhart in a passionate dissent declared that "If 'public welfare' has not done so already, it is high time it took on a meaning for the courts which it has for the rest of the world." Until then, "lords of the soil" would build more "Towers of Babel" that darkened the sky and loomed over the domes of democracy and justice. It was high time, in another words, for zoning.[174]

Zoning had been resisted by American businessmen, industrialists, real estate developers, and a compatible judiciary not only as a preemptive "taking" but as creeping socialism. That horror had been diffused (a double-edged word) by the successes of the City Beautiful movement advanced by the wives and daughters of bankers, industrialists, developers, and educators; diffused too by local frontage ordinances and national historic preservation programs; diffused finally by the city planning initiatives of Progressive economists, landscape architects, and sanitation engineers, whose designs for rational land use presumed the exercise of powers of eminent domain and whose "civic righteousness" implicitly invoked zoning as a way to ease the circulation of light, air, traffic, and sewage. Zoning was also implicit in such regulations as that passed unanimously by Chicago's aldermen in 1911, which forbade the operation between ten p.m. and four a.m. of all "pile drivers, steam shovels, pneumatic hammers, derricks, steam or electric hoists or other apparatus, the use of which is attended with loud or unusual noise, in any block in which more than half of the buildings on either side of the street are used exclusively for residence purposes." Zoning was explicit in a 1914 ordinance, to be administered by Chicago's Commissioner of Health, "Providing for Temporary Zones of Quiet Around Residences in Which Persons are Dangerously Ill," and had become explicitly permanent in Los Angeles in 1908–1909 when the city demarcated residential and industrial districts. By 1916 the retailers of the Fifth Avenue Association, using Düsseldorf as their model, had seen to the passing of the New York Zoning Law, whose division of the boroughs into residential, business, unrestricted/industrial, and indeterminate districts would be followed by four hundred towns over the next decade, as would its introduction of quieter "asphaltic concrete" paving. Whether by code, by custom, by campaign, or by fiat (as to what streets should be paved, which slums should be leveled to make way for rail lines, and in whose backyard sewage treatment plants should be built), municipalities were being effectively zoned years before the Supreme Court in 1926 sanctioned the practice—in large measure as a *cordon sanitaire* against noise.[175]

Citing experts who had testified that residential zoning would "decrease noise and other conditions which produce or intensify nervous disorders," the divided court in *Euclid* v. *Ambler Realty* validated residential zoning and approved the creation of "natural neighborhoods" to protect homeowners from invasions of "mere parasites," i.e., apartment buildings, which bring "the disturbing noises incident to increased traffic and business, and the occupation, by means of moving and parked automobiles, of larger portions of the streets, thus detracting from their safety and depriving children of the privilege of quiet and open spaces for play, enjoyed by those in more favored localities." Having long ago lost the legal umbrella of "coming to a nuisance" and newly uncertain of how acoustic analogy would compromise defenses couched in terms of the intangible risks and inevitable roar of progress, manufacturers, retailers, real estate developers, and their friends on the bench reconceived zoning as a way of preserving property values and ambits of action. Industries resting within their own zone remained thereby clear of fault in liability law and clear of nuisance suits regardless of the severity of the environmental transgressions of smoke from their chimneys, effluents from their chemical swamps, odors from their slag heaps, and the pounding and shrieking from presses, forges, and power saws. Zoning freed wholesalers and department store owners from the weight of laws that might otherwise hold them responsible for trespassing on the senses of citizens with their heavy trucks and delivery vans, loading dock whistles and annunciators. Zoning allowed police units to bypass the dilemma of finding quantitative measures by which to enforce ordinances against bad smells, bad air, and excessive noise, and obversely helped corporate attorneys defer the imposition of municipal limits on most forms of pollution. In return for agreeing to stay where they already were, and for not building over a certain height except in certain sectors they had already been ogling, business and industry were released of the obligation to be good neighbors.[176]

Cynical as that may sound, zoning was rather an act of segregation than of tolerance, nostalgic for a sound-past of acoustic privacies that had scarcely existed, deaf to a sound-future already in the air—ambulances racing down quiet avenues, their sirens blaring; ice cream trucks rolling past sedate cul-de-sacs with loudspoken melodies from old records; airplane overflights ("Noise has invaded the empyrean, once the realm of angels, and the end is not yet," wrote a consulting engineer in 1931).[177] Nor did zoning do much for acts of mutual defiance within a zone, like a London duel on Angell Road, Brixton, in 1892. Mrs. J. F. Holder Christie taught music and singing at a nearby school and also at home; her daughter, a medallist of the Royal Academy of Music, gave lessons at home on the piano and violin; her son, an amateur on the violoncello, practiced at home nightly after work, until eleven or twelve; and the family entire had musical soirées—most of which musicking made its way through a party wall to the ears of H. Fitzer Davey, his wife and son. Relations between the families had been "harmonious" for three years until the Christies took a boarder, a contralto who also practiced at home, whereupon Davey complained in writing about what he had heard first as the sunrise howlings of a dog and then knew to be "the frantic effort of someone trying to sing with piano accompaniment." The morning serenade, compounded by the "thump, thump, scrape, scrape, and shriek, shriek" of students throughout the day and the Christies' own practicing and performing well into the evening, might be "fine sport to you," wrote Davey, "but it is almost death to yours truly." The next day he "commenced a series of noises in his house whenever the playing of music was going on in the Plaintiffs' house—such as knocking on the party-wall, beating on trays, whistling, shrieking, and imitating what was being played in the Plaintiffs' house." The Christies (now the Plaintiffs) contacted a solicitor, who wrote to Davey that "we have yet to learn that it is an unreasonable use of a private house to play the pianoforte or sing," and intimating at action in Chancery against Davey's outrageous improvisations. Davey (about to be the Defendant) responded: "I have a perfect right to amuse myself on any musical instrument I may choose, and I am quite sure I should be the last person to do anything knowingly to annoy my neighbours. What I do is simply for recreation's sake and to perfect myself in my musical studies." Himself musically trained, he felt "perfectly qualified to distinguish the difference between music and noise," and every so often would host musical evenings of his own. The fevered practicing next door, however, was more than irksome. "Now, seriously, I put it to you, is it not most excruciating to have constant repetitions of the five-fingered exercises, and only receiving the higher notes of the vocal efforts?" As an engraver who worked at home, Davey had "the most difficult portraits to reproduce, requiring great thought and the most delicate treatment," and he had "yet to learn that I am compelled to be a martyr because my neighbour is musical." He wanted the Christies to be the first to know that he planned this winter to master the flute, concertina, horn, and piano. Now, seriously, the Christies went to court to restrain Davey from practicing with malice aforethought on any such instrument, or blowing whistles, banging on trays, or conducting "mock concerts." Davey in turn asked for an injunction against the Plaintiffs "making noises in their house by giving lessons, practising or playing upon pianos, violins, violoncellos, or other musical instruments, or singing in an unreasonable manner or at unreasonable times." His solicitor argued that "The Plaintiffs in fact seek to restrain the Defendant from making a noise, in order that they may be the better able to make a noise themselves." By 1900 such duels were being fought informally by angry neighbors with phonographs or player pianos, by 1930 with loudspeakered phonographs or radios. What most intrigues about the formality of *Christie* v. *Davey* (aside from the fact that Mr. Christie was, "perhaps fortunately for himself," almost deaf) is Davey's early claim that the constant and "atrocious hubbub" next door violated a tacit zoning, for the Christies' residence seemed no longer to be "a private house, but a public one, pupil after pupil coming and practising, and letting out their pianos for practising only, sometimes two pianos going at the same time." Davey's annoyance, and his decision to return like for like, was grounded less in a disagreement over musical taste or accomplishment than in the obnoxiousness of operating a "house of music" (which Davey described as if a house of ill repute) in a residential area. Although Justice North ruled against Davey, whose noises, he said, "were not of a legitimate kind," he did take notice of Davey's counterclaim that the Christies' music-making constituted a nuisance akin to the conducting of a ladies' school next door in a residential area. North decided that the making/teaching of music seventeen hours a week, and sometimes more, was proper to a semi-detached domestic residence, even one with so thin a party wall.[178]

Stray comments by the Christies' solicitors and by North made it seem that the true culprit was the builder who had erected such a flimsy partition; Davey himself deplored the fact that the Christies, unlike his prior—and musical—neighbors, had no carpeting. Had there been sufficient sound insulation, this feud/fugue might never have gone forward. A half-century later the need for interior sound-damping would be even more acute, according to the Health Organization of the League of Nations, whose *Report on Noise and Housing* addressed the growing "environmental problem" of noise from radios, motorized traffic, and the acoustic conductivity of cheap materials preferred by builders. One acoustical engineer went so far as to blame the decline of churchgoing on the poor acoustics of modern churches, whose designs baffled the high frequencies and distorted the low, necessitating "monotonous intoning, chanting (all at the expense of intelligibility) and generally low frequency sound" that was psychologically conducive to oppressiveness and sleep. Rarely were zoning laws accompanied by revisions of building codes toward the better insulation of floors, ceilings, and walls, which is why Anti-Noisites tended to propose dedicated sanctuaries of silence or suburban "oases of tranquility" as well as a familial "'Quiet Zone' in the Home" with rubber-rimmed utensils, bang-proof doors, and elimination of "the quite appalling noise made by the unsuitably planned or quite uninsulated lavatory." Nonetheless, zoning laws seemed to be a solution to problems of noise because of the association of the spatiality of zoning with those buffering distances and air pockets that effectively reduce the structural transmission of sound. Zoning also was of a piece with the widespread distribution during the 1920s and 1930s of other publicly private spaces—telephone booths, enclosed automobiles—whose walls and windows screened out traffic noise while securing intimate conversation in daylight and backseat intercourse at night. The protection afforded by zoning was stronger against obstructions, vibrations, and chemical outspills than against odor, smoke, and boom, but since zoning was early associated with sanctuary from noise as required by the sick, the studious, and the deliberative, its implementation appeared to signal an official endorsement of the value of acoustic privacy. As Julia Rice and Theodore Lessing had understood, hospitals, schools, and law courts enjoyed an advantage over private residences with respect to claims for protection from smoke, darkness, and noise because they were acknowledged to have medical, pedagogical, and jurisprudent needs for fresh air, light, and, above all, quiet. Rice and Lessing also knew that the posting of QUIET signs in the vicinity of each grammar school or hospital was but a gesture toward amenable quarantine, and that no city would have enough policemen to hunt down violators of these "islands of silence," which is why Lessing and other city planners proposed to create park-like surroundings for medical and educational complexes. Rice thought instead to muster a children's crusade, girded with pledges and badges, to give the signs a moral force that quiet zones had deserved since the time, say, of the revered Rabbi Akiva, who told his disciples in the second century that "A fence for wisdom is silence."[179]

6. Decibels Sones Phons

From Osaka in 1931 came a lament, translated in the *Literary Digest*, that Japan's cities had become "a virtual hell of noise" because the nation had no zoning regulations; shops and factories mingled indiscriminately with homes, and taxi drivers hit their klaxons at all hours in all districts. But wherever you were, Osaka, Bombay, Havana, London, Paris, New York, or Chicago, horns pursued you across sectors zoned or unzoned. By 1907, the year before Henry Ford started up his Model T assembly lines, Americans had registered less than 150,000 motor vehicles, which killed fewer people that year than did falling buildings; since cars trundled down the road at speeds under 25 m.p.h., advocates thought them safer than horse-drawn vehicles because they could more quickly come to a full stop. In 1931, 30,000,000 cars roamed the world at or above speed limits of 30–40 m.p.h., killing more Americans than measles, scarlet fever, and diphtheria combined, knocking off one in every 10,000 London pedestrians and one in every 25,000 Parisian *piétons*. Such statistics dictated a frequent use of the horn on the part of drivers encountering—according to one wry count of an hour's pedestrian traffic—"237 children on rollerskates, 116 on bikes, 98 playing ball, and 53 nursemaids wheeling baby carriages." Cities might rise to the defense of little old ladies by creating pedestrian "safety zones" down the middle of boulevards, but the onus was increasingly on the pedestrian to Stop Look *Listen*, a campaign rerouted from railroad crossings to sidewalks and street corners, where motorists in small towns were still as likely to be brought to court for failure to sound their horns as for frightening a mule with the "country tone" of a klaxon. More concerned that both drivers and pedestrians heard oncoming traffic than were protected from its clangor, the National Safety Council in 1922 printed up posters for trolleymen: "When the driver doesn't know what's behind him, WAKE HIM UP WITH YOUR GONG!"[180]

Did your aldermen or Ministers of Transport establish "silence zones" where honking was forbidden, or your Touring Club squelch horns in city centers, or your spiritual leaders "Stop that Noise!" in Jerusalem, or your local tyrant ban honking outright (e.g., Mussolini), you had yet to deal with the screech of brakes on faster cars, the grinding gears of heavier trucks, the loose bolts of aging jalopies, the gunning of motorcycles, and the sirens on every sort of emergency vehicle, none of which could be zoned out or even, it seemed, measured. "Is the general traffic noise level any higher than it was in the days of granite [paving] setts, steel tyres, and teams of horses?" asked a British engineer in 1933, pained by the absence of standards for quantifying and comparing sounds. And why, in the flurries of letters received by the *London Times*, was street noise more offensive than railway noise? Why did so few protest "as an express train passes through a station making a noise far in excess of any street traffic noise," or so many complain about the noise from airplanes when they were so scarce in contrast to locomotives? And how could one ever calibrate noises on the move, "the deep voice of the Packard and the shrill cry of the Ford"?[181]

Really, said Dr. G. W. C. Kaye of Britain's National Physics Lab and a member of the Ministry of Transport's Noise Committee, the challenge was that "Noise was very 'economical,' so to speak, and it took very little energy to make a lot of noise." For example, a thousandth part of the energy used in hammering went into the making of its noise. Aha. Might one compare noise levels across industries and geographies, if not across eras, by treating sound as energy, which physicists and engineers knew well enough how to quantify? If it was difficult to prove that noise caused anatomical damage, at least one could lay out correlations between levels of sound-energy and measurable levels of fatigue, hearing loss, or industrial accidents. "Almost I wish, at times, that we could adduce a mortality from noise, however small. It would be so much better propaganda," wrote the most prominent of British Anti-Noisites, Thomas Jeeves, First Baron Horder, M.D., in 1938. "But we cannot. Noise, like a number of other nuisances, does not kill." Charles Myers had revealed some subtler psychophysical harm from noise, and Horder himself had shown how noise could derange the digestion. (Loud noises of high pitch make you nervous and irritable, said fellow physicians; low notes upset the stomach.) As Chairman of the Noise Abatement League, Horder was certain that noise was a "dysgenic element" in society, but as a physician reviewing the data, he found the evidence against noise no better than the common opinion that an East Wind blows no good.[182]

Behind the curve as Americans had been with regard to zoning, they were ahead of the curve with regard to metrics, due in good measure to the efforts of the Bell Laboratories to assess the clarity of transmissions. More precisely, to determine how little clarity was needed for tolerable communication. The more that Ma Bell knew about the range of the human ear and the brain's compensatory skills at auditory processing amid noise, the less it might cost her for the construction and maintenance of telephones and telephone lines. ("Telephone engineers," read an article on "Noise, Nerves, and Business" in the *Readers' Digest*, "say that the current used in the ordinary telephone is 100 times stronger than it would have to be were it operated under conditions of silence."[183]) Research into the quantum limits of communication, the inherent noise in electrical circuits, the sound qualities of speech through microphones, and the psychoacoustics of listening was consequently of the highest calibre, though never disinterested. Nor was the promulgation by the Company in 1929 of an objective measure of the loudness of sound.

Victorian scientists, musicians, and physicians had tried to establish standards that would assure comparisons of intensities and tonalities of sound, such that all experiments in acoustics worked from a universal ground, all pianos were tuned to the same "international pitch," and all tests of hearing were referred to an invariant measure of audibility. This never happened (and the standard A of the international pitch kept rising). The closest they came to an instrument that could produce a somewhat steady frequency at a somewhat steady intensity was a set of Koenig tuning forks, and then the electric buzzer; the closest they came to an instrument for tracking frequency and loudness was the phonoautograph, which etched vibrations of sounds across a drum or roll of paper, and then the oscilloscope, on whose screen loudness might run vertically, as amplitude. None of these clarified what Charles Wead in the *American Journal of Science* in 1890 called the "perplexing subject of the measure of a sound-sensation," about which musicians seemed as unhelpful as physicists: Did "all parts of the scale *seem to the ear* to be of equal loudness?" This was not simply a question as to whether an acoustician should measure the amount of energy in the vibration of a tuning fork or the rate at which that tuning fork "gives up" energy to the air; it was a question of whether one had to correlate intensity and tonality in a systematic way in order to arrive at a reliable assessment of *heard* loudness. One could order from Rudolf Koenig's catalog a wagonload of acoustic devices—Cagniard sirens, Helmholtz double sirens, Helmholtz resonators, Seebeck Great Sirens, Scheible tonometers, Galton whistles, stethoscopes, organ pipes, phonoscopes, and Koenig's inimitable tuning forks—and never approach a whit closer to what loudness amounted to *in the ear*.[184]

Invalids, insomniacs, neurasthenics, and their friends could testify that no tone or intensity was felt always to be of the same harshness or loudness. Charles Myers could show that no two persons could be trusted to react identically to an identical tone or intensity. Vice versa, if idiosyncracy was built into our auditory apparatus, then that which *was* heard to be identical by diverse ears could be most acoustically persuasive, as with an alarm buzzer or the contralto voice of Carolina Lazzari as reproduced on an Edison record. During and after the war, Thomas A. Edison, Inc., sponsored theatrical "tests" in which a diva performed on an otherwise dark stage, at first singing and then mouthing a song as played by a NEW EDISON, at last putting a finger to her lips and bending an ear to the now-spotlit "Phonograph with a Soul." The seamlessness of the transition from live to recorded voice was a grand entertainment for "more than ten thousand music lovers and representative music critics"; for copywriters it proved "beyond all question" that "The Metropolitan Opera House may fade into memory. But the genius of Edison has perpetuated forever the real voices of the world's great artists. Not strident and mechanical travesties on their art—but literal Re-Creations, indistinguishable from their living voices." What's more, if more there could be, it was Edison's "method of scientific voice-analysis" that had led him to discover Lazzari and bring her before the music-loving public.

Sitting with his ears an inch from the horn of the phonograph, almost deaf and always disturbed by the extremes of soprano and bass, Edison sought out voices and melodies that lay within a temperate zone. No vibrato, no warbling, no Futurist antics that ended in tremor or static. A voice cautious in its range and effort, unmuddied in its lyricism, unseduced by syncopated rhythms, would meld best with his microphones and resins. The success of his "tests" of the recreative power of the NEW EDISON (he was seventy in 1917) relied upon voices that yielded audible matches between the unstrenuously alive and the soulfully mechanical.[185]

Bell engineers had also been investigating voices, but they wanted to know what happened to speech and hearing when the lines of transmission were not soulfully recreative. While Edison and his engineers sought to eliminate noise by technical improvements in manufacture and by redirecting audiences from the background scratchiness of a disc to foregrounded similarities of tone between voice and record, the Bell men anticipated and eventually assumed noise. Cable engineers had measured the strength of a signal by how many miles of cable it could travel without appreciable loss, and telephone companies had followed suit for telephone lines, but strength and loudness were not equivalent. Equivalences would be the crux of the problem for the U.S. Bureau of Mines in 1918 when it tried to establish rules of thumb for the distances at which a person using a geophone or a microphone could hear the underground blow of a pickaxe, the boring of an auger, the chipping of a chisel. Could the Bureau arrive at a "standardization of sound waves by character and by the blows or forces employed underground," it would help rescuers calculate the depth at which miners had been buried alive after an explosion. A chart of loudnesses could also help geologists distinguish among intervening media, for sounds travel "more readily through coal than shale than clay," and transmissions through what sounded like a coal seam could be music to company ears.[186]

Fixing upon "The Development of Modern Acoustics" as the theme for his presidential address to the Royal Society of Canada in 1920, the physicist Louis V. King of McGill University noted that "A few years ago, one would have pronounced the science of acoustics dead." Burial had proven premature even before the war, with Sabine emerging from his sub-basement tomb at Harvard, but the resurrection came full-blown with the wartime progress made in sound-ranging, fog-signalling, and submarine detection. King was most excited about advances in the visual display of sound. That same year a German engineer marveled at how much had been learned about the hyperbolic sound-curve of artillery and explained how the increasingly precise measurement of sound was the result of *des Stellungskrieges*, "wars of position." Shellshock or no such thing, Marinetti's *intonarumori* or Scriabin's *Prometheus* or nothing-of-the-sort-thank-you-very-much, sirens and klaxons or not-around-here-bub, sound had assumed a substantiality as scientifically estimable as it was culturally inestimable. So while one engineer in 1919 called for a better device by which to characterize the qualities of sounds AND NOISES, and another in 1920 called for wider use of "audible electric signals" that could be heard ABOVE THE NOISE of factories, other technicians sought more sensitive means for calibrating a broader range of tonal intensities, which now mattered as much to politicians and physicists as to composers and jazz quintets.[187]

They mattered, these intensities, because of what they had done, or were held to have done, to minds and ears during the war—and how they resonated afterward, as in the spiraling ovals of the ten "Noise" paintings by e. e. cummings, completed in 1921 after recuperating from months driving an ambulance on the Front and months more in a French concentration camp for putative spies and "undesirables." Because of the vacuum tube amplification of postwar loudspeakers in phonographs, radios, and stadiums. Because of the colonization of homes and offices by radios and other electrical devices that howled or squealed as often as they hummed. Because of the colonization of the streets by heavier trucks, larger buses, zippier sedans, and more guttural motorcycles, all with horns that seemed keyed to the few seconds between a red light and a green, making the new amber in the middle as much a noise flash as a caution. Because of the decline of fashionably long thick hair that had for decades shielded women's ears, now bobbed and thinly defended by new permanent waves against the standing waves of saxophones and vulnerable to the jangling of newly chic earrings of dangling metal or torpedo-shaped plastic that banged against the earpieces of telephones. Because of airplanes not so far above, whose "passengers and pilots must wear ear plugs and forego all conversation" amid a noise "of such stunning effect to some that they will be deafened for many hours after landing," and subway trains not far below. Because of the thinning of walls between apartments and the lowering of ceilings between floors. Because of the thousands of black-box cinemas for the Talkies, whose soundsystems had to overcome the whoosh of air conditioners and whirr of larger projectors. Because of the new symphonic music, which played out what the Dutch painter Piet Mondrian had called for in 1920–1922, "demolitions of melody and dry, unfamiliar strange noises that oppose rounded sound," a sonic "neoplasticity" that flexed its mightiest muscles in Baku in November 1922, where numerous conductors on special towers used colored flags and pistols to coordinate multiple choirs, the foghorns of the Soviet flotilla on the Caspian, two batteries of cannon, a machine-gun division, the rifles of several infantry regiments, a flock of hydroplanes, a steam whistle machine, bus and car horns, and all of the city's factory sirens in the *Simfoniya gudkov* composed by Arseny Avraamov; and even inside concert halls, with the klaxons, propellers, pianolas, and electric bells of George Antheil's *Ballet Mécanique* of 1925, which Antheil described as "streamlined, glistening, cold, often as 'musically silent' as interplanetary space, and also often as hot as an electric furnace," but which sounded to some ears like the "very noisy machine shop" across from which Antheil has been born and kept close company with other contemporary music, including jazz, all of which, wrote one critic, was "certainly shell-shocked, or, more properly, noise-shocked." Because of the secretion of public address systems, powered to outsound roadcrews jackhammering the avenues. Because, according to a New York physician who still made house-calls, "babies jerk and twitch in their sleep at the bark of a police dog or the continued tooting of an automobile horn, thus witnessing the birth, so to speak, of subsequent instability or neurosis." Because doctors knew that those with "A driving, restless, apprehensive, nervous personality" might be driven by noise to suicide, like Martha Bernard of Manhattan, whose life had been so upended by the loudness of the traffic at East 66th and Lexington that she leapt to her death seven stories down.[188]

How loud was too loud? How (in)constant did a sound have to be before it became a noise? And how much noise—now that it was as elemental as electrons, everywhere, and everywhen—was too much? If physicists before the First World War could count the scintillations produced on a fluorescent screen by disintegrating particles, what could be so difficult about enumerating and calibrating sounds with at least as much accuracy as a musician counts beats and calibrates the impact of a minor key? If physicists after the war could quantify such infinitesimals as radioactive rays passing through Geiger-Müller tubes, which converted radiation into electrical pulses and the pulses into an audible clicking (to become a soundmark of the Nuclear Age), what could be so difficult about quantifying already-audible noise? And if Hans Geiger, Walther Müller, and Niels Bohr delighted in giving demonstrations of the tube, its clicking amplified through auditorium loudspeakers to convey "the sensation of an immediate perception of radiation accessible hitherto only by the means of complex and lengthy experiments," what could be so difficult about reversing the process and building a meter that registered and displayed patterns of rays or waves of sound?[189]

Readers-aloud of the previous Rounds of this book may tick off some of the difficulties. Loudness and noisiness enjoy a temperamental marriage in which each is often unfaithful to the other—loudness may not seem noisy, noisiness may not be loud. Cochlear sensations of loudness are often independent of cultural relations of loudness. Rhythmicity or musicality can divorce intensity from loudness, dissonance from noisiness. What people say they hear, or swear they do not hear, may be a function neither of the biophysics of audition nor of the pathology of ear canals but of environmental stresses, personal trauma, psychosocial preferences. What is loud alone may not be loud in a crowd, and what is loud in one ear may not be loud in the other. Sound may be sonic but loudness is tonic: that is, the higher the tone, or frequency, the louder a sound may sound, although workers in heavy industry lose their higher-frequency hearing earlier than the general population.[190]

Asked the editor of a Michigan technical journal, "What is more universal than the presence of noise and of what significance is noise to the industrial world?" By 1925 these were rhetorical questions: "Noise means imperfection, which in turn implies wear, waste and inefficiency." His next questions, inset within an essay on the measurement of noises, were not rhetorical: "Yet how shall this criterion of perfection be evaluated in terms of positive and quantitative units? More plainly stated; how loud is loud and how much louder is too loud?" Those were questions posed by D. L. Rich, a professor of physics who had been hired in 1922 by the Timken Roller Bearing Company to devise an instrument for measuring the noise made by bearings. Was there any other field where the absence of measuring instruments and commonly understood units of magnitude was so galling as in practical acoustics? Elusive qualities such as texture and hue had recently been standardized but, asked Rich, suppose that a buyer rejects a carload of gears for being too loud or noisy? On what basis could you, the manufacturer, rebut or retool? Whose ear was the gauge of loudness? Of noisiness? Suppose that one of your own inspectors hears a day's production run as unexceptionable while another hears it as raucous. Suppose that the former, asked to reconsider, stays out late with his girl, goes on a bender, or gets a head cold, and the next day hears things differently? What you—and H. H. and William Timken of Canton, Ohio—needed for quality control was an impersonal ear, not an Edisonian audience of oohers and aahers. So, with the assistance of Floyd A. Firestone (who had worked at the Riverbank Labs and in 1929 became a founding member of the Acoustical Society of

America), Rich created a mechanical ear that gave "practically instantaneous," and reproducible, readings of the sound of Timken bearings doing their job of easing mechanical joints.[191]

Rich and Firestone had it easy. All they had to measure, it turned out, were numbers of vibrations. The more polished the bearing, the fewer vibrations it gave rise to when compressed between a pair of flat metal test plates. Their mechanical ear operated as a phonograph-in-reverse, "fundamentally a measurer of small alternating currents, generated by the vibration of the armature of the pick-up" placed under the bottom metal plate. Magnified by three or more stages of vacuum tube amplification, a technique developed during the war, the vibrations became the visual echoes of imperfections. Instead of listening for the whine, whistle, or "north-wind sound" of a defective bearing, a lovelorn or deaf inspector could watch as the needle of a galvanometer was deflected toward a red line of unbearable vibration. "It is evident from any angle," Stephen E. Slocum would write in *Noise and Vibration Engineering* (1931), that "noise is just so much sand in the bearings of business," and "it goes without saying that the same technical skill which has put efficiency into industry is capable of putting quiet into the environment."[192]

Originally Rich and Firestone had thought to transmit the test sounds of each compressed bearing through a condenser microphone patented by A. C. Wente of Western Electric Engineering Research, predecessor of Bell Laboratories. Wente's microphone, in use since 1916, translated soundwaves into electrical impulses that could then be transmitted by a vacuum tube amplifier and registered on a meter. This was fine so long as the bearing tests took place inside a chamber insulated from all other sounds, which was impractical for daily quality control in any factory. When the two men realized that test-plate vibration was most of what there was to the Timken north wind, and that the vibrations could be captured by a pick-up fitted to the plate housing, they were free from the vagaries of "aerial sounds" and free therefore to increase the sensitivity of the device, making their instrument, "in its most sensitive form," joked Firestone, something "not to be sneezed at—because the response may be so violent as to break the galvanometer."

Knowing the "spectrum" of a noise would be better than knowing only the scale of its vibrations, for then a technician might be able to locate the specific cause of a noise among the ten or twenty moving parts of a roller bearing or the 2,500 parts of a Model T sputtering down the road. Firestone in February of 1925 demonstrated his and Rich's "noise evaluator" to the Society of Automotive Engineers and announced that "work is now in progress on an apparatus which will confine the attention of the instrument to any given pitch or frequency." Soon it should be possible to assign "the proper noise values" to the various parts of a mechanism. The Western

Electric physicist H. Clyde Snook was more philosophical. If dirt be defined as matter out of place, then noise was "sound disagreeably out of place," and "although useful as a detector of mechanical imperfections of car operation," noise was now heard by the auto-buying public as "so extremely undesirable that elaborate methods for analysis with a view toward preventing or suppressing such noise are warranted." The real difficulty, as Western Electric/Bell engineers had determined, lay not with Ford assembly lines nor with new methods for the measurement of vibrations and analysis of frequencies but with the non-linearity of human hearing. The inner ear required only a thousandth of a dyne's difference in pressure against the tympanum to register a new tone but responded to complex sounds of mixed frequencies and amplitudes (as most sounds are) with an intimidating variability. Given the many who were partly deaf, and given that noise by engineering definition was a complex sound "to which no definite pitch can be assigned," it could be fruitless to conduct a frequency analysis for imperfect ears. In the ensuing discussion of Snook's paper, T. V. Buckwalter, chief engineer for Timken, added that the company had not known "what kind of noise would be acceptable" to users of its bearings, and had found meanwhile that in some vehicles "the noise set up by the sculpturing on certain kinds of tires" synchronized with, and compounded, the noise made by their bearings. In lieu of tightening all tolerances, maybe Timken should design its bearings to make noises at precisely the frequency that would be masked by the noise of associated gears.[193]

Dyne? What was that about a thousandth of a dyne? Who would give a hoot or holler to know that 1 dyne is the force required to cause a mass of 1 gram to accelerate at a rate of 1 centimeter per second squared? Why am I dropping the dyne on Buckwalter and Snook? It all goes back to the people of Jared, who came out from the confounding of tongues at Babel, and back to the Nephites, and to the Lamanites, ancestors of America's Indians, and to a certain "ancient record thus brought forth from the earth as the voice of a people speaking from the dust, and translated into modern speech by the gift and power of God as attested by Divine affirmation," through the glorified spirit-being Moroni. It goes back to Truman O. Angell and his brother-in-law Brigham Young, who in 1850 appointed Angell the official architect of the Church of Jesus Christ of the Latter-Day Saints and who in 1860, by reason of a vision, directed the Church's civil engineer, Henry Grow, to create a Tabernacle with a great elliptical dome that could hold eight thousand of the faithful. Built using a lattice-truss system more akin to the skeleton of an Ark than to the frame of a Victorian auditorium, the Tabernacle was finished in 1867 and its acoustics tweaked in 1870 by Angell, to the admiration of an unbelieving world and generations of architects. Let another scholar sift through the Book of Mormon's Book of Ether for a Latter-Day Saints theory of sound; let another track Angell's

1856 expedition overseas to survey the forms and acoustics of English music halls, theaters, and factories. It will suffice here to note that from the moment in 1828 that Joseph Smith, Jr., began dictating, hours each day, month after month, a first (lost) and then a second translation of the hieroglyphic golden tablets he had found in the hill Cumorah in Palmyra, New York, it had been manifest that the clairaudience of the voices of angelic spirits and the clear audition of Smith's inspired words were vital to the new revelation. "These days were never to be forgotten," wrote Oliver Cowdery, one of those who took down the words. They sat "under the sound of a voice dictated by the inspiration of heaven," often in view of the tablets themselves, which "seemed to be pliable like thick paper," wrote his wife Emma, another of the scribes, "and would rustle with a metallic sound when the edges were moved by the thumb." It suffices, I say, because it could yet have been happenstance that Harvey Fletcher of Provo, Utah (1884–1981) and his younger contemporary, Vern O. Knudsen of the same township (1893–1974), Mormons both and never apostate, should establish the modern scientific parameters for determing the volume and tenor of the spoken word, assessing how well that word is heard, and calibrating the loudness of an interfering, increasingly noisy world.[194]

Vern's paternal grandparents had converted to the new religion in Norway and emigrated in 1863 to the New World, where they followed the Mormon Trail through the Utah Territory to the State of Deseret. On the approach to Echo Canyon, their infant daughter died; once settled in Provo, Knudsen's father, Andrew, would return to Echo Canyon year after year in hopes of locating the body of his sister and giving her a decent burial. ("These things meant a great deal, as you know, to religious people, who believed in the literal resurrection.") Andrew married a Swedish woman whose parents had also converted. Vern, their seventh and youngest child, was thus on both sides a third-generation Latter-Day Saint, spending hours each day in family prayer and having his father and three older brothers serve for years as missionaries in Sçandinavia. He himself hoped for a mission nearer home and closer to a great university, though he held dear what Brigham Young had told the founder of his high school and college (Brigham Young Academy; in 1903, University): "You must teach mathematics, even mathematics, with the spirit of the gospel." A professorship in mathematics is what the University intended for Vern, but none foresaw the influence of Harvey Fletcher, who had returned to teach at Brigham Young after receiving a Ph.D. in physics at the University of Chicago under Robert A. Millikan. For Millikan, Fletcher had conceived and carried out the famous oil drop experiments by which Brownian motion could be further explicated and electron charge measured, which achievements would win the Nobel Prize for Millikan (and not Fletcher, who during Millikan's lifetime never made public the centrality of his role). For

Knudsen, who became his research assistant, and a lifelong friend, Fletcher shaped a career in physics and guideposts in ethics.[195]

Evangelism now summoned. Like every young Mormon, Knudsen had mission work to do, and in 1915 he was sent to Chicago, where he became a successful street-corner preacher with a partiality for 1 Corinthians 13, which begins in noise ("Though I speak with the tongues of men and of angels, and have not charity, I am become as sounding brass, or a tinkling cymbal") and ends in clarity ("For now we see through a glass, darkly, but then face to face: now I know in part; but then shall I know even as also I am known. And now abideth faith, hope, charity, these three; but the greatest of these is charity"). He became second-in-command of the Northern States Mission, exempt from the 1917 draft as a minister of the gospel. Fletcher meanwhile had thrown in his lot with Western Electric, and by January 1918, when Knudsen's obligatory twenty-seven months of proselytizing came to a close, he took a job there too. His assignment, denominated "essential war service," exempted him once again from the draft and enlisted him in the ranks of acousticians: he was to develop a speaker loud enough to "overcome the noise of the airplane and the battlefield," so that planes could issue orders to troops from the air (shades of Moroni!). Such celestial projection, when not angelic, was possible with the use of vacuum tube amplifiers, but the wrack-and-roll of biplanes generated so much noise in the tubes that Knudsen was next set upon the problem of protecting the tubes themselves from all the shaking, and thence to shielding transatlantic cables from the noise generated by cyclical currents produced primarily by the earth's variable magnetic field.[196]

After the war, Knudsen entered the doctoral program in physics at Fletcher's alma mater, where he was inspired by Albert Michelson's lectures on wave motion and where he learned that "electrons behaved like waves on Tuesdays, Thursdays, and Saturdays and like particles on Mondays, Wednesdays, and Fridays," with the Sabbatarian possibility that waves and particles were two aspects of the same thing. He pursued a dissertation under Millikan on "The Sensibility of the Ear to Small Differences of Intensity and Frequency," which banked on his wartime experience with electron (vacuum) tube oscillators that could be used to produce pure tones more evenly than tuning forks and to modulate intensity by small exact increments. His father a violinist, himself a clarinetist, Knudsen wanted to know how the ear picked up variations in loudness and pitch, and whether there were correlations among pitches and sensations of loudness. Helmholtz had shown that the ossicles of the ears amplify auditory vibrations, but experimental psychologists knew that for the ears to remark an increase in loudness, the intensity of a succeeding sound had to be on the order of 10 percent greater for each step up. At the threshold of audibility (that "extremely faint sound of which the ear is just dimly conscious"), the ears of Knudsen's graduate students required a 30 percent surge for the next step up to be heard. In other words: at feeble intensities, regardless of pitch, just-noticeable changes in loudness were a function of the degree of loudness itself, and not (as the standard Weber-Fechner Law had it) a constant ratio from step to step. As for the smallest perceptible change of pitch, the ear required a greater difference to remark a shift in lower frequencies than in higher. So long as tones came in quick succession, within 1.5 seconds, a listener with musical training did no better than the non-musical, since the cochlea was "a standard frequency analyzer, a physical instrument that is about the same for all hearing persons." Only with several-second delays between tones did musicians' memory for tonal relations stand them in better stead.[197]

Knudsen, his name pronounced with a silent initial "K" thanks to a tip from a Western Electric phoneticist after years of difficulty with operators who had found his K-nudsen noisily confusing over telephone lines, went on not to a career in otology but to researching the physics of sound and consulting on architectural acoustics. At Michelson's behest, he had investigated the acoustics of a high school auditorium in Chicago, and when he left to teach at (what became) UCLA, he sought out the newly published collected works of Wallace Sabine, a book so influential in his professional life that he "practically memorized" it in 1922 and fifty years later could still repeat passages verbatim. In Los Angeles he began a more concerted study of the hearing of speech, starting with "six of the world's worst auditoriums," all in local high schools, and in a campus men's room, abandoned because its sound insulation was "immodestly inadequate" for the three women working in the Purchasing office on the other side of the wall. He discovered that one had to supplement Sabine's formulas on reverberation and absorption with close analyses of room shape and room noise—"noise is very, very important, and we investigated that carefully"—and further consideration of the loudness of speech, now that it could be so powerfully amplified, and distorted, with loudspeakers. Among the first to acknowledge and factor in the noise resident or residual in a room and the effects of temperature and humidity on reverberation times, he was called upon to design Hollywood sound stages and much later the hall for the General Assembly of the United Nations. It was not that he lost interest in the inner ear; rather, as one whose religious scripture descended from an oral translation of hieroglyphic plates dug from a sacred mound, and whose scientific scripture had its origins in Sabine's isolation in a sub-basement at Harvard, the message was integral to its medium, truth to its tabernacles.[198]

Fletcher, who served for years as the President of the New York State Mission and at seventy-seven composed *The Good Life*, a Mormon Sunday School manual, devoted his professional life to the other side of the religio-acoustic equation, to physiological issues of articulation, the psychophysiology of auditory perception, and the technical problems of sound reproduction. While Knudsen shaped the milieux for that listening which was incumbent upon faith and charity, Fletcher (whose father became deaf at the age of fifty-five) mapped the contours of speech and the limits of hearing; he sired an artificial larynx and stereophonic recording, improved upon hearing aids and sonar. It was Fletcher who graphed the function defining the relationship of loudness and intelligibility and established the articulation index used by Knudsen to assess the acoustic merits of an auditorium. (Articulation was conditional upon the noise in the system, for it was "a joint property of the talker, the channel, and the listener," each channel being characterized by its frequency response *and* its noise spectrum. It was consistent with this concept that one form of articulation test involved a study of the percentage of disconnected, nonsense syllables that could be accurately identified.) And it was Fletcher who, with others in his Speech and Hearing Group at Western Electric, determined absolute limits for the hearing of sounds (from 1919 on), developed a practical portable audiometer for testing hearing across populations (from 1922 on), and began making "noise measurements" of subways (in 1923).[199]

Said measurements, to be exact, were not of noise but of sound-pressure as transduced electrically and amplified with the aid, once again, of the "miraculous" vacuum tube. Out of this would come, between 1925 and 1929, the decibel, or a tenth of a Bel, an honorific coined by Western Electric/Bell Labs to denominate not a force but a ratio. This can be confusing. During the 1920s physicists like Fletcher and fellow telephone engineers had begun using the term "sensation units" as a near-synonym for "transmission units" to describe the efficiency of circuits and apparatus across distances of cable. Once Fletcher, Knudsen, and W. A. Munson (Knudsen's research assistant, then Fletcher's colleague) began plotting the curve by which loudness along a scale of pure-tone frequencies was heard to increase by regular increments of intensity, what acousticians needed was a unit of measurement that referred back to the minimal sound intensity, or energy per area (in physiological terms, pressure on the tympanum) required to make a sound audible to a healthy adult ear. Since

1. that minimal intensity is about one-trillionth of a watt per square meter (or .0002 dyne of pressure), and
2. the ear's range extends beyond 1 watt (and 2000 dynes), and
3. for frequencies between 700 and 4000 Hz each just-noticeable increment in loudness requires about double the wattage of sound, and
4. within that frequency range, a doubling of the sensation of loudness (according to Knudsen, Fletcher, Munson, and B. A. Kingsbury, also of Bell Labs) requires ten times the wattage,

the most presentable form of a unit of measurement would be one that eliminated the troublesome trillionths while demonstrating the upward sweep of the curve of sensations of loudness. Hence the deciBel, or tenth of a Bel: the ratio of the wattage or sound pressure bearing upon the tympanum (or on the diaphragm of a condenser microphone in a noise meter) to the wattage or sound-pressure of silence, i.e., the energy just below that needed by a standard ear to hear a reference tone of 1000 Hz (round about the first note in "Somewhere over the Rainbow" as sung in 1939 by Judy Garland). Plotted on a common logarithmic scale, by powers of ten, decibels collapsed to presentable curves the ear's trillion-plus range of audition.[200]

Decibels indexed loudness; they bore no cogent relationship to noise. After all, loudness *per se* is not decisive in determining whether the buzzing of a fly is heard as noisier than a jazz quintet. A sound-measuring device calibrated in decibels is not, as commonly thought and written, a *noise* meter; it is a sound-pressure or "sound level" meter that is scarcely on the level. Since human ears hear an unevenly stepped pyramid of loudness for frequencies under 700 Hz and over 5000 Hz, maybe it would be best, decided the Bell engineers, for their sound meter to filter out lower and higher ranges that upset its physio-logic. Equipped with filters, such meters were half-deaf, compensating for their single-earedness (middle-toned, monophonic) by their handiness as audicons. Further, since people rarely encounter the pure, steady, vacuum-tube-generated tones by which auditory curves for sensations of loudness were first construed, and since each of our two ears registers amplitude somewhat differently, and since our brains sum up the pressures of environing sounds across a multitude of bandwidths using a neuromathematics idiosyncratic to each hearer, the half-deaf meters were half-*ss*ed. They failed to accommodate the hiss, rasp, susurrus, and gride that constitute much of what is heard as noise, and they dealt poorly with those impulsive releases or explosions of sound that are conducive to acoustic shock.

Fiddling with the condensers and circuitry, a savvy electrician could make meters that responded to the loudness of some frequencies more than others, approximating the responsiveness of a standard ear (*sans* mind/brain) to segments of the audible spectrum. Such "Subjective-Objective Sound Meters" never parsed sound, or noise, as do ears and brains together, which take into account not only sound power but the degree of powerlessness a person may feel with regard to the likelihood of quashing an irritating sound. The meters did not consider how taxing or soothing a sound might be; they merely expressed loudness, which Fletcher and Munson had determined was logarithmically additive in the same way that decibels are additive. "My pet peeve," said a Mr. Haynes at a convention of the Society of Automotive Engineers in 1938, fifteen years after the first vacuum-tube sound meters had been run through their paces and nine years after the decibel had been formally introduced by Bell Labs, "is the man with the trained ear. Isn't there any mechanical method for making uniform and consistent noise measurements?"—measurements that took into account "not just volume but the quality of a tone?" Wrote a physicist the next year, "We also need a general language for describing noise," for "most noise inspections are made by ear... and most attempts to quiet machinery are made on the basis of ear listening." And, as Dr. Edward Elway Free had explained to the readers of *Electronics*, "If an average person listens to two noises one after the other and decides that one of them is twice as intense as the other one, he is reasonably sure to be about 800 per cent wrong." This was because, "For some mysterious reason which no one yet understands, the human ear seems not to hear noises in proportion to their actual intensities, but in proportion to the cube roots of these intensities," a physiological datum uncovered by researchers at the Johns Mansville Corporation, manufacturer of sound-damping materials, and at New York University, where Free lectured. Evolutionary or at least megalopolitan advantage accrued to the ear's registration of loud sounds as less loud, for otherwise we might be deafened by our own prowess as *homo faber*; then again, what intensity or quality of sound should be used to calibrate a more "reliable" mechanical ear? What if the standard of comparison were tin cans dragged along the floor, imitating street noise? How was an engineer to choose among the eight competing "noise scales" of 1931, each with a different reference point, weighting, or range? And what of cross-cultural differences? As a Bavarian physician observed in a 1936 report to the industrial health division of the International Labor Office, the English, French, and Italians lumped the physical and physiological together as noise, *bruit*, *rumore*, while the Germans (and Dutch) distinguished *Geraüsch* or *ruis*, any sound with a vibratory non-periodic movement, from *Lärm* or *lawaai*, any unpleasant sound with a vibratory movement whether periodic or not. And what of the Rabelaisian shape-shifting of noise? Could any scale cope with the definition of noise given by the British physicist Kaye in 1931, "An acoustic disturbance which is unwelcome, whether because of its excessive loudness; its composition; its persistence or frequence of occurrence (or alternately, its intermittency); its unexpectedness, untimeliness, or unfamiliarity; its redundancy, inappropriateness, or unreasonableness; its suggestion of intimidation, arrogance, malice, or thoughtlessness." Competing units of sound measurement put forward during the late 1920s and early 1930s in Europe and America translated the testimony of half-deaf meters into numbers putatively more comprehensible or universal: the *phon*, which equilibrated loudness (aural sensation) across frequencies and was favored by the British; the Anglo-American *sone* and the German *wien*, which operated on scales meant to circumvent "the counter-intuitive character of the logarithmic scales of the phon and the decibel" (Karin

Bijsterveld) where each increase by 10 represents neither a tenfold increase in sensations of loudness nor a 10-percent increase but an exact doubling.[201]

Were it not for E. E. Free, the chemist and consulting engineer just quoted on cube roots, soil science would have no part in this account of phons and decibels, and we would be the poorer for it. Director of his own laboratory, editor of the *Scientific American* (1925), science editor of the *New York Herald Tribune* (1926–1928), and a contributor to *Popular Science Monthly*, Free had a broad reputation for brilliance in chemistry, physics, electrical engineering, and economic geology, all of which he had studied at university. His first job after graduating from Cornell in 1906 had been with the Agricultural Experiment stations of the University of Arizona, where he did field research on mine detritus that entered irrigation water. His professional interests henceforth focused on the detection and mapping of small particles, either as sources of pollution (soluble copper saturating the soil) or forces of nature (wind-blown sand and dust, as in the devastating Mediterranean siroccos of 1901–1903). His seminal work on *The Movement of Soil Material by the Wind*, published in 1911 in tandem with a ninety-page bibliography of "Eolian Geology," bespoke a fascination with the measurement of infinitesimals that, *en masse*, became consequential and (a Free earmark) often audible. These infinitesimals might be radioactive particles clicking up a hydrogen storm through Geiger tubes, or weevil larvae whose "minute munchings" of wheat kernels he amplified a millionfold; poisonous residues of lead in tetraethyl auto fuel (meant to reduce engine knock), or "noises you never hear"; invisible infra-red rays floating off fog-banks and (through Free's ingenuity) tripping a gong, duststorms louder than quakes, or almost imponderable sound-pressures drumming on the tympanum and raising Cain.[202]

Named science editor for *Forum* (the journal begun by the Rices), Free was asked in 1926 by the new editor-in-chief, Henry Goddard Leach, to conduct a noise survey of Manhattan. A scholar of Scandinavian life and literature and a fellow Anti-Noisite, Leach in effect was commissioning Free to resume Julia Rice's campaign against noise. With an engineer's passion for metrics, Free asked "How Noisy is New York?" and answered with a "noise map" of the city laid out in "sensation units" (soon-to-be decibels) as read off an audiometer lugged around Manhattan. Free assumed from the start—as the son of a coal-field physician and public defender of painstaking science—that "certainly it is easier to think straight, to reflect, to summon each tiniest resource of memory and logic, if one is free from the continual annoyance and distraction of noise." To his surprise it was not subways or elevated trains *per se* but the density of surface traffic that was responsible for the stridency of New York. Horse-drawn vehicles were "actually noisier, wagon for wagon," than autos, but 85 to 90 percent of the city's noise was due to poorly maintained elevated and trolley tracks and the rumbling of heavy trucks over uneven pavement. The point to be made, in his second report in 1928, was that "the relative noiselessness of cities is as possible and far less costly than the relative cleanliness which we have learned to demand." He who calculated to the metric ton the particulates of soil suspended in the atmosphere, and who knew how soil must be aerated lest once-fertile earth be whipped into duststorms, was optimistic that cities everywhere could as surely fend off noise as stench and dirt. The same capacities for measurement, management, and material design that were now eliminating open sewers and reducing coal smoke could be applied to reducing the noise of radios or rivet guns and minimizing the noise of city streets, where "people are already one-third to one-half deafened by the average noise."[203]

Numbers like that, pulled from whatever headset, coupled with a technotopian optimism, were "practical, dramatic, and spectacular" enough to prompt New York, St. Louis, Chicago, and Washington, D.C., to undertake official audiometric sweeps of, and campaigns against, noise. The resulting two volumes on New York's *City Noise* (1930–1932) would be quoted thereafter wherever an Anti-Noisite could not sleep, and by every historian of noise. I too shall do my share of quoting, but a question nags. Why was noise, of all things, taken up as a communal cause by so many cities in the first years of the Depression, when as much as a fourth of each national workforce in Europe and America was unemployed and food lines dragged on for blocks? Why did noise merit concern during a Depression so severe that the Dust Bowl—as anticipated by Free's work on soil and wind—could be described as a climatic example of what had gone suddenly wrong with the world entire? Why was noise still of moment in 1934 when, on May 12, a front of grey topsoil 1,800 miles wide and 2 miles high, stretching from the Great Plains to the North Atlantic, eclipsed the Chicago sun and choked Manhattan air with 617 particles of dust per square millimeter, which from the seventeenth floor of the Flatiron Building was equivalent (estimated Free) to forty tons of dust per cubic mile, the city suffocating for five hours under 1,320 tons of what had been the sandy loam of Texas, Oklahoma, Arkansas, Kansas, and Colorado?[204]

Noise, I would argue, was the most available, manageable metonym for the wider erosion of spirits and assets that was the Depression. Throughout the 1920s, industrial psychologists had been trying to show how noise wore workers down, made them less productive, less amenable. While Myers in England followed up on fatigue and shellshock, Donald Laird (now at Colgate University) followed up on John J. B. Morgan's wartime experiments at Columbia, trying to quantify "Just what effects the increasing noise of our modern machine civilization have on the workers in the normal course of the day's work." Putting gas masks on heroic typists, Laird quantified their exhalations in order to contrast the caloric expense of typing for six hours in two test chambers, one of bare brick walls, another paneled in Acousti-Celotex (440 holes to a rose-colored square foot), while subjected to a "noise machine" that reproduced the sounds of a busy office using the motor from a washing machine, a rotating metal drum filled with ball bearings, a gong, telephone bells, and honking horns. Results? The secretaries typed 4.3 percent more slowly and used 19 percent more of their own energy when surrounded by reverberant noise, and manifested "a general tenseness of the entire musculature." The cumulative intensity and general pitch of noise could induce an annoyance "similar to that experienced by rubbing one's finger over coarse sand-paper," and it was "undoubtedly [the act of] reducing this state of annoyance that brings the experience of blessed relief to the workers." Reviewing all experimental results on the effects of noise up to 1930, Laird reported that mechanical skills declined, reaction times lengthened, active memory weakened, cerebral functions lost their sharpness, breathing became irregular, and blood pressure rose in the presence of noise, be it the grind of pencil sharpeners, the sudden suction of pneumatic tubes, or sirens that triggered innate fear reactions. Consultant to Celotex, Dictaphone, the Master Bedding Makers of America, and the Order of Sleeping Car Conductors, Laird leapt easily from industrial to business psychology (*Psychology and Profits*; *What Makes People Buy*) and then to the writing of self-help manuals (*Increasing Personal Efficiency: The Psychology of Personal Progress*), for to his way of thinking, "Distractions are oftentimes due to your *inner attitude* rather than to the noises themselves." Whether mental or material—the clacking of a typewriter, an abrasive voice, or "inadequate personality adjustment"—noisiness was individually, and immediately, correctable.[205]

Factory floors might never be as quiet as the Hatchery in Aldous Huxley's *Brave New World* of 1932, where three hundred Fertilizers bent over their instruments "in the scarcely breathing silence, the absent-minded, soliloquizing hum or whistle, of absorbed concentration." Nor could one easily escape the Fordism of what Huxley called the "Physical noise, mental noise, and noise of desire" of this "Age of Noise" whose radio din "penetrates the mind, filling it with a babel of distractions—news items, mutually irrelevant bits of information, blasts of corybantic or sentimental music, continually repeated doses of drama that bring no catharsis."[206]

Yet, just as a keen plumber might resolve the problem of the sound of dripping water that plagued Huxley's own house, a sound "inconclusive, inconsequent, formless... asymptotic to sense, infinitely close to significance but never touching it," it seemed to Free, to the Bell men, and to many a mayor that the advent of reliable audiometry and a standard decibel made it far easier to deal with local cacophonies than national calamity or a global Age of Noise. Free himself had been agreeably surprised to find that noise itself was "astonishingly local" as decibel levels swiftly diminished in the short walk from a parlor fronting on a boulevard to a bedroom in the back of the house, or from hectic intersections to shaded parks nearby. There was something about the quiet neutrality of the meter and the calm breadth of a decibel scale that evoked what Huxley, on other pages, termed "the noiseless precision of conjurors." And there was something magical in the precision of physiological research at Cornell, at Harvard (Walter Cannon), at the University of Chicago (Edmund Jacobson), at McGill (Hans Selye), at Cambridge (Edgar Adrien), at Zurich (Walter Hess), and across Germany that was revealing how "stress" compromised not only the gut but nerve fibers, lungs, and heart tissue, so that deadening the noise of the world and learning to relax were part of the same, practicable, personal release from a "tense" Depression.[207]

Key to the promise of the decibel, and of noise and stress reduction, was the fixing of thresholds of exposure. Psychophysical thresholding had received its first impetus with nineteenth-century studies of "just-noticeable differences": how lengthy an exposure at what intensity was required before a healthy subject responded to a stimulus? Thresholding had expanded with bacteriology—how densely concentrated a microscopic colony would lead to infection?—and edged into common awareness with the imposition of public health and occupational safety standards for substances known to be easily contaminated (air, milk, water, meat) or highly toxic (mercury, lead, benzene). Since the 1890s, official reports had been issued and standards promulgated with "a series of values for a range of acute effects"; in the 1930s, authorities abandoned sliding scales for absolute limits to repeated exposure, such that an American researcher in 1945 could compile a list of published exposure limits for 136 different gases, vapors, soluble metals, and particulates. Audiometrists were attuned to this shift, testing one's capacity for discriminating intensities among the 300,000 different "pure sounds" a human ear was able to hear, then determining individual and median thresholds of audibility at different frequencies. These numbers were supplemented by studies of thresholds of audibility *in noise*. Before the Great War, a New York otologist had used a stopcocked apparatus to assault the ears with a variable rush of pressurized air while a subject was trying to make out spoken words. "The stopcock is opened until the patient fails to hear the voice, even when raised to an inordinate degree of loudness"; the extent of opening determined the subject's "noise index." After the war, Bell men further pursued thresholds of audibility in noise (seeking to determine the minimally acceptable levels of intelligibility over telephone lines) and found that 130 db at any frequency was more than could be borne even for a second by the hair cells of the cochlea. Thus was fixed an upper threshold of loudness as associated not only with pain but deafness. To this day, musicians, physiologists, acousticians, mining companies, labor unions, roadie sound techs, and air force physicians debate the harm done by exposure for *n* minutes each day to a sound of *o* db at *i* frequency with *s* degree of steadiness and *e* degree of explosiveness, but the idea of a threshold—of pain or permanent damage—was settled enough by 1930 that charts of decibel (i.e., *noise*) levels for everything from country cottages to skyscraper canyons could impress a listening public.[208]

Thresholds also made sociocultural sense during an era when the literate, the numerate, the illiterate, and the marginally numerate alike were being sensitized to the graphics of economic decline/recovery and to qualifications for welfare or Social Security. To the extent that the Depression was self-disclosing, so was noise, and to the extent that noise did not seem to be either an addiction or a mystery, it seemed feasible to calibrate and then dissipate it. Thresholding enabled acousticians and Anti-Noisites to rise above the fray of earlier debates over necessary noise, since auditory thresholds were set with relation not to the sources but to the receivers of sound. The exact measurement and active management of noise could be implicitly promoted as the advance offer to New Deals that worked across the board to calm an alarmed economy after a series of Crashes put most everyone at risk. Had not Laird and other industrial psychologists shown (and business executives learned[209]) that normal people became more sensitive to noise, and more disturbed by it, when stressed?

Political theorists might consider the many anti-noise commissions, surveys, and campaigns of the 1930s, from Melbourne to Milwaukee, New York to Havana, London to Lyons, Geneva and Vienna to Breda and The Hague as a *divertissement* during an era of bankruptcy and financial crisis, distracting people from a systemic economic insecurity with the canonical optimism of audiometric engineers. Of these campaigns, the most celebrated was also the most blatant of ploys, for it was endorsed by the jazz-loving, glad-gloving, songwriting, Deusenberg-driving mayor of New York, Gentleman James J. Walker ("Will you love me in December as you do in May?"). His citywide program began with the Health Commissioner's appointment of a Noise Abatement Commission in October 1929, the same month that Walker was featured in a new campaign movie, "Building with Walker," extolling steam shovels and high hopes for the Empire State and Chrysler Buildings. This was days before Black Friday, Black Monday, and Black Tuesday unhinged the stock market gong—not unlike the firestorm that engulfed a Reformatory run by the gong in "The Godless Girl" of 1928, a Cecil B. DeMille silent film ruled by clockwork, even in chapel, where inmates protest, "We ain't praying to God—we're prayin' to a Gong." It was "as if the primitive in us is roused to fight against marauding wolves every time we hear the cry of automobile horns in the street," said Dr. Shirley Wynne, the Health Commissioner—or every time we hear the gongs and grinding wheels of trolleys, the brass section of a "diabolical symphony" recorded in so many plaintive letters to the Health Department that noise abatement had become a paramount concern. Cancer was taking its highest toll ever among the elderly and cases of scarlet fever were suddenly doubling (most cases fatal, the rest ototoxic), yet it was still vital to report that "Hundreds of noises that would have struck terror in the hearts of all a century ago now daily assail the ears of residents of the City of New York," and Wynne guessed that "every hour of the working day the average New Yorker is stirred this way by as many alarming noises as the hardiest hunter in prehistoric times was in his entire lifetime." The steel-framed towers in which Walker took such pride had become "A veritable Babel piling itself to heaven in wave upon wave of buzzing, humming, pounding, shrieking din." After the publication in 1930 of the Commission's report on *City Noise* (ranking churchbells at 61 db, thunder at 70, steamship whistles at 94), Walker endorsed a second investigation, through to its several reports in 1932, at which time he and cronies were facing charges of venality pressed through the office of the Governor, Franklin D. Roosevelt. The political dissonance of the anti-noise campaign was obvious to the *Boston Evening Transcript*, commenting on "New York's Loud Noise about Noise," for the mayor and his coven of nightclubbing associates would have a devil of a time heeding some of the more strictly enforced noise laws. Speaking of the devil: "were his Satanic Majesty to roar like a lion on certain New York streets, he could make himself heard above a multitude of other sounds only for a distance of twenty or thirty feet." A Bengal tiger at the Bronx Zoo roared at 75.5 db, a Siberian tiger topped 79, but a good radio loudspeaker reached 81, a riveter 101, a tumbler drum 120.[210]

"Between beats the heart rests, and between breaths the lungs and diaphragm have an instant's respite," wrote the editors of the *New York Times* in 1928. "All the nerves and muscles of the body have some time when they are relieved of duty except those of the ear." Hospitals should offer citygoers soundproof chambers that would give ears some respite; otherwise "City noises going on day and night keep the ear drum and the tiny bones about it in a constant state of vibration." With its anvil, hammer, and stirrup, the ear was the dominant figure of a city never at rest; the ten thousand soundings taken in 1930 at 138 stations around the city by Harvey Fletcher and other Bell audiometricians were but a small obeisance to the larger noisescape. Although the soundings ramified E. E. Free's reconnoiterings of 1926 and gave weight to claims of a sonic "state of emergency," no corps of audiometricians could put in place the social baffles and legal bracings necessary to deal as wholeheartedly with that emergency as a bill being drafted by the Municipal Council of Budapest to banish noise "once and for all." The *New York Post* feared that if such a bill passed, so many Anti-Noisites would head for Budapest that "in their enthusiastic outbursts in praise of the quiet nobody [in Budapest] will be able to hear himself think."[211]

Exactly this notion of paradoxical reversal conditioned the public response to anti-noise campaigns of the 1930s, due perhaps to a stoicism about inevitable bull-and-bear economic cycles, or to an anxiety about inescapable reversals of relations between wo/man and machine, wo/man and metropolis, as reflected in Chaplin films, political and feminist tracts, and existential novels like *Nausea*. "There used to be a time when, by climbing up into a skyscraper, you were climbing into your own ivory tower of silence," R. V. Parsons of the Noise Abatement Commission told a WJZ radio audience on the day after the bleak New York Christmas of 1929. "That was when skyscrapers were few and far between. Now they huddle together and reflect the street sound back and forth," and with other bulky buildings "imprisoned the street noise." Asking, "Is There a Quiet Spot in New York?" the answer had to be, really, no. Towers of Silence became, too inevitably, towers of Babel. And as A. B. Crowley disclosed after a preliminary noise survey in San Francisco, "The exhaust of an automobile with a faulty motor, in a quiet residential street brings the noise level up to compare with that of an average factory." At the other end of the noise spectrum, the *Personnel Journal* taught business managers that the sound in an underground vault registered at no less than 20 db and one's own heart beat at 10 db. The League for Less Noise, centered in New York, had sent out scientific experts to all parts of the world and made "the unique discovery that in no place on earth is there absolute quiet. Even on the vast Sahara Desert in Africa, the sands, heat, and air create noise." Readers of the *New York Times* learned that biting their nails was a 3.3-decibel habit, scratching their heads a 2.1-decibel reflex if flaky and 1.5 if oily.[212]

Blame this acoustic pessimism on the decibel itself. In the cultural history of noise, the problem with the decibel was not only that it had too little to do with noise and too much to do with long-haired scientists "who waved a microphone, gazed at the meter, went into a trance, and came out with a lot of mysterious numbers." The chief problem was that the conversion of loudness into decibels made it seem more difficult than ever to reduce offensively loud sounds. At first hearing, a layperson would think that reducing the power of a sound by half would reduce the total sensible loudness by half. Hardly so. And, at second hearing, that reducing the power of a sound by half would be worth more than 3 decibels (abbreviated db). But, as readers of the *Baltimore Sun* learned in 1935, it took twenty-seven singers to sing twice as loud, in decibels, as a single voice, so any effort to cut by half the loudness of a choir would, in decibel essences, thin the choir to a soloist, a lay-off extreme even during the Depression. When it became clear that reducing the loudness of a sound from, say, an otologically harmful level of 120 db to a more acceptable level of 90 db required a 99.9 percent reduction in sound energy, the engineering needed for such quieting could seem pathological. While Anti-Noisites might wield decibels to show how greatly the volume of traffic burst in upon a parlor or schoolroom, industrialists could use decibels to show how much work and expense were entailed to reduce the loudness of a drop forge by a mere 3 db. As Paul H. Geiger wrote a generation later on the first page of his *Noise-Reduction Manual* of 1955, "Since the ear is an extremely sensitive device, vibration amplitudes of only a few millionths of an inch may produce a loud sound. For this reason, if an attempt is made to decrease manufacturing tolerances between moving parts by an amount sufficient to give quiet operation, such a high degree of precision of machining becomes necessary that the manufacturing process becomes very difficult or even impossible." The conversion of loudness into sound pressure and a scale of pressures into logarithmic decibels could make the noise problem seem as insurmountable as it was calculable. Spates of mass audiometry at hospitals and fairgrounds between 1929 and 1950 revealed that acuity declined with age, putting lower frequencies (women) and higher frequencies (men) out of range while making other sounds noisier, as sharpness of hearing declined toward a hearing of sharps. City sound-meter surveys suggested to Vern Knudsen that the twentieth century with its trucks and trip-hammers, roadsters and rivet guns, industrial chippers and high-speed grinding wheels (awarded the highest of phons in 1927) was getting noisier (i.e., louder) at an average rate of 1 decibel a year.[213]

Watch out, then: "The decibels will get you... if you don't watch out!" Such was the warning in 1938, by which time Laird had shown that when your ears were exposed to sounds over 80 db, digestion slowed by 37 percent and blood pressure rose, leading to "seriously premature old age"; over 90 db and you suffered a "continuous, progressive degeneration" of hearing, or of life itself. *Complimenti*, then, to New York's Finest in 1937 under a reforming mayor, Fiorello La Guardia, for issuing 300,000 citations to noisemakers, and to the municipal courts that convicted 20,000 noisemakers of civil infractions, upon the urging of a new League for Less Noise and in accordance with the recommendations of the Noise Abatement Commission's second report on *City Noise* of 1932. The report had drawn its epigraph from Jeremiah 10:22: "Behold, the noise of the bruit is come, and a great commotion out of the north country, to make the cities of Judah desolate, and a den of dragons," which was scarcely promising, for it would take an enormous collective effort to dislodge so many dragons. The Health Department should persuade the public of the miserable physiological effects of noise, the Transportation Board should buy quiet buses, the Sanitation Department should use rubber-rimmed trash cans, the Fire Department should relent on sirens near hospitals and schools, the

Post Office should silence its trucks, the Presidents of the Boroughs should maintain manhole covers (which if loose clunkclanked loudly under truck and jalopy traffic), the Federal Department of Commerce should require quieter steamwhistles of ocean liners in harbor, and the people, the suffering people should not expect the Commission "to come to their rescue like a magic prince." Whatever some correspondents seemed to believe, the Commission had no power to silence the crickets in a woman's hearth or a resident ghost dead set upon clanking its chains. Commissioners could support the updating and enforcement of noise ordinances, but they had to rely on citizens to flex their own muscles in defense of digestion, sleep, and sanity. "When people are made to realize deep down in the emotional springs of their collective mind that noise has a disastrous effect on their health, we will have silence," declared the Commissioners, and not a moment before. Reducing city noise was thus oxymoronically a question of publicity and persuasion: of radio lectures during the day and broadcasts at ten-thirty each night requesting listeners to lower the volume on their radios; of "hush days" when high school students were asked to keep their voices down; of motorists driving more intelligently so that they would not get themselves into jams and then try "to blow their way out." The blind could protest most feelingly about being unable to walk straight on the sidewalks of New York because city din obscured the sound of their own footsteps, obliterated the tap of canes, and blunted the soundmarks by which they were guided, but what was to be done when city fathers themselves acted at such cross-purposes as to hold a hearing on a noise ordinance while scheduling a performance of the Sanitation Department's brass band outside City Hall, its cymbals and saxophones celebrating the delivery of a book van to circulate through the streets (and honk its horn?) on behalf of the Queens Library Association? The band was so loud that a member of the Commission at the hearing could not hear his own voice, or so said Dr. Thomas Darlington, Tammany Sachem, former Commissioner of Health, and friend of the late Julia Rice. The hearing, on an ordinance to prohibit the operation of radio amplifiers within two hundred feet of a church, school, or courthouse, was adjourned.[214]

Ironies such as this, publicly remarked ("Trumpets and Drums Lend Point to Speeches"), might conduce to wisecrack or despair ("For relief from city noise we must direct our attention toward insulated construction rather than attempting to stop the source, a fairly hopeless prospect"). Or they might invoke the ironic physics of noise cancellation, an ancient dream recently recast by the Westinghouse Company, whose innovative electric ear, weighing sixty pounds, would tailor such a perfectly inverse soundwave that it negated the wave-shape of any offensive sound. Leon Theremin, who in 1928 had just patented an eerie electrical instrument by which one could "call music from the ether by moving the fingers rhythmically" between its antennae, spoke that year of the prospects for a pocket-size device that would "automatically shut off all offensive sounds" and allow a peace-loving man riding in a subway "to enjoy the profound silence of a forsaken cathedral." Or the ironies might persuade officials to listen to letters of complaint from correspondents. The Boston Noise Commission's newspaper questionnaire of 1930 netted 2,253 letters protesting, in order of frequency, the noise of trucks, auto horns, auto brakes, radios, muffler cut-outs, streetcars, buses, milkmen, dustmen, loudspeakers outside stores, factories, parties, fire sirens, newsboys. Such an ordering neatly reflected changes in media and mechanization (locomotive engines now ranked seventeenth and street musicians were nowhere to be heard), but it sprawled across the decibel scale: airplanes came in eighteenth despite a rating of 100 db, which was close to the sound of men hammering on steel plates at 110 db, the "threshold of feeling." And the complaints ranged unpredictably across tonalities, from the deep grind of trucks throttling between gears to the clink of milk bottles. Motorcycles, which "with their ear-splitting exhausts, clattering valve gears, and screaming horns" were the bane of the French, the English, and the people of Atlanta, Georgia, ranked twenty-eighth in Milwaukee, below "loud talking at night." New Yorkers bewailed the taxi whistles of doormen more than almost anything, while Milwaukeeans were most upset by streetcars—not just the screech of flattened wheels around corners but "street car motormen who keep up a constant clamor by stamping on their foot bells."[215]

Astute mayors or city planners could explain such rankings by reference to variations in traffic patterns, housing stock, law enforcement budgets, and zoning laws. But the run-on of complaints was as tumultuous as urban life and complex as the human ear, which was proving to be particularly sensitive to changes in frequency and masterful at teasing out simultaneous tones. The more that investigators refined the metrics of sound and metrology of noise (with "high brow" high-speed continuous level recorders and "low-brow" quick-and-dirty sound meters), the more ubiquitous, ineluctable, and dangerous noise appeared to be, and the more impossible to uproot. Although it could be edifying to realize that each city had its own tonality (New York a loud tenor, Chicago a bass, London a baritone), how frustrating it must have been to British Anti-Noisites to find that during the Omnibus Strike of 1937 the sound level on the streets in the absence of double-deckers declined by only 4 phons and most of London came in at 60+ db when a steady 50 db was thought by the more cautious of physicians to be injurious to nervous health. How exasperating it must have been to learn, from Dr. Foster Kennedy at Bellevue, that the noise from the bursting of a paper bag put "more pressure on the brain" than morphine or nitroglycerine. How French and American Anti-Noisites must have bristled upon learning, from the U.S. Bureau of Weights and

Measures and an article in *Le Figaro*, that they must contend with Paris fashions and the acoustic reflectivity of silk dresses that "send back the sounds which strike their fabric just as a hard wall makes a rubber ball rebound." Strange, joked the *New York Times* in 1930, "that no mention was made of the fact that skirts are now flaring, which must make them veritable loud-speakers."[216]

Countrysides, with more sows' ears than silk dresses, spoke no less loudly than cities, at least to sound meters that put "very quiet" radios at 40 db, Niagara Falls at 94, thunder at 95. The drumming of soldier termites, the ticking of the Grass Moth, the snapping of locusts, all of these now had decibel registrations, though lower than the scrawking of jays or squealing of pigs. Clouds of grasshoppers, which Darwin in Patagonia had compared to "the sound of chariots of many horses running to battle," as in the book of Revelation, came in at 86 db. One researcher even assigned a decibel signature to Longfellow's forest primeval: with a breeze of 12 m.p.h. but no prophetic Druids gathered beneath, a grove of Wisconsin pines murmured in the breeze at 12 db, which was 8 db less than a whisper. Out and about on a country lane, a Longfellow lass could not escape the intrusions of International Harvester's new Farmall tractor (1924, idling at 80 db, running at 90–100 db), Caterpillar's new Diesel Sixty (1931, ditto), Skilsaw's new handheld circular saw (1924–1928, about 100 db), diesel locomotives hauling longer lengths of rumbling freight cars (postwar, 80–110 db), and cropdusting biplanes (110–20 db). Farmers, cattlemen, and shepherds (and Monkey House zookeepers) had reason to be as concerned as any urbanite about the logarithmics of noise.[217]

"One-quarter of the population of England would have to talk simultaneously to generate the same sound power as [a] ship's siren," calculated a Scottish engineer, Norman McLachlan, for his *Noise: A Comprehensive Survey from Every Point of View*. The lesson to be learned from such arithmetic concerned not the acoustic inadequacy of our voices but the acoustic violence of our vocal proxies—electric alarms, klaxons, and loudspeakers (at 94–108 db). As these proxies were called into service, they raised the level of "background noise," an increasingly common phrase for the chorus that continued despite the two minutes' silence on Armistice Day and strained all listening when it rose above 40 db, equal to the hum of a 1930s electric refrigerator. Soundproofed rooms were therefore *de rigueur* for modern otologists. If the noise of street traffic was substantial in the Bell Labs ten flights up overlooking Manhattan, no doctor's office could be counted upon to be quiet enough for a proper audiometric exam—which, if conducted with regard to the narrow tolerances of the decibel, would likely show that 10 percent of Americans, young and old, had some degree of hearing loss. "There are some," wrote the New York otologist Walter A. Wells, "who maintain that since we habitually live in the midst of noise, it is all right to make your hearing test in a noisy environment," but this was a mistake (*pace* the Bell men), for in that case our ears and nerves would be put every year in greater jeopardy. Rather than tolerating the distraction, annoyance, headaches, tinnitus, fatigue, and confusion of background noise, wrote Hale Sabine of the Riverbank Laboratories, people should use perforated tiles, baffles, duct liners, rubber doorstops, and whatever else science might come up with to "push back" noise. In 1941, no longer New York City's Health Commissioner but still butting his head against noise at the age of fifty-eight, Shirley Wynne addressed the annual meeting of the Acoustical Society of America and made the same point he had been making for a decade, that from the wincing of newborns to the cringing of nonagenarians, we never overcome the devastations of noise, and "our unconscious reaction to noise continues in spite of our conscious indifference to it." Noise was Public Enemy No. 1, declared the president of the Acoustical Material Association, for "Noise is the true murderer of thought."[218]

7. Auroscope

"Gloomily futile," that's how Vern Knudsen himself, in 1970, would characterize generations of effort at noise reduction. By then the popular magazines had decibel readings for things that had not, or had barely, existed in the 1930s: home air conditioners (60 db), electric typewriters (64 db), garbage disposals (80 db), sonic bird-scarers (85 db), food blenders (88 db), power mowers (96 db), helicopters at one hundred feet (100 db), oxyacetylene torches (120 db), Saturn rockets (195 db), and H-bomb tests (250+ db).[219] I am skipping ahead, as I will henceforth skip about, through the Second World War, Cold War, and all posterior analytics, and back again to the 1930s, for this Round has reached a point where much of the noise will feel everyhow familiar.

An augury of what will be playing out on the coming pages may be helpful. From the broadest of soundpoints, the subatomic, anatomic, societal, and statistical irreducibility of noise, as theorized and then demonstrated during the 1920s and 1930s, will establish the *continuo* for an acoustemology in which noise is credited with an increasingly resonant, often advantageous, contribution to political culture, to economics and astronomy, to communications and philosophy, to the morphology of biological systems, to the dynamism of life itself. Were I to offer an auroscope with astrological/orchestral groupings, what lies ahead would sound like this:

Air/Winds: Out of studies of intelligibility across phone lines and through radio waves will arise information theory, which redefines noise as at once intrinsic and revealing.

Water/Voice: Out of more powerful, accurate means for the broadcasting and

SOUNDPLATE 25

detection of sound through water will arise a wholly different, surprising profile of the "silent deep" as oceans of noise braided with low-frequency channels, thousands of miles long, for the songs of whales.

Earth/Brass: Out of studies of language acquisition, fetal development, and child psychology will arise a new pedagogy that encourages children to listen without prejudice to the sounds of their world, in which noise is at once grounding and invigorating.

Fire/Percussion & Strings: Out of the crackling of mysterious cosmic rays will arise the radioastronomers' Big Bang and a half-figurative half-literal notion of noise at the origins of the universe.

Throughout, Anti-Noisites will become more sophisticated and occasionally more successful, but ghosts will also return, and a second Cage inspiriting silence with noise, and a new breed of geneticists, and a Merzbow investing noise with raw performative power. And there should be a finale, at the inauguration of a new American president to the words of a praise song by Elizabeth Alexander, where "All about us is noise and bramble, thorn and din," which the poet, like the historian, must work her way through.

8. Inhearancy

February 17, 1930. A woman lies prostrate on the floor, as if murdered. She rises slowly to her feet. A telephone rings. She answers, then drops the receiver like a stone. So begins Jean Cocteau's *La Voix humaine*, in which a telephone is at once lifeline and death knell to a romance heard only from one end of a conversation hexed by interruptions, static, cross-talk, wrong numbers. Scored for opera in 1959 by François Poulenc, the play was from the first ring as much about noise as about the human voice. Avital Ronell, a philosopher of the telephone, puts *La Voix humaine* into the context of opera as the staging of "the birth of language as a fall," a staging that each evening reenacts "the primordial shock in the splitting of sound into language and music." The couple in the Garden had sounded out the universe wholly and perfectly. After the Expulsion, language was one thing, the music of the spheres quite another; after Babel, languages were many, music something else again. Telephone diaphragms shiver still to this double shock, operating within the dialectics of long distance and close call, separation and intimacy, truth and consequences.[220]

Timbre, that quality of sound independent of pitch and volume but compounded of all fundamentals, overtones, subtones, and microtones in play, was a casualty of the mythic split of language from music after the disaster at Babel, and a casualty of actual telephonic research. Had corporate investments in the telephone taken their lead from Philipp Reis, Elisha Gray, or even the phonautographist Édouard-Léon Scott de Monville, all of whom were engaged with the transmission of music, the telephone might not have lost its attachment to timbre. Following as it did the vocal tractates of an oralist teacher of the deaf, the telephone was subjected to the pursuit rather of intelligibility than expressivity. By the 1920s, while Vern Knudsen was declaring an intelligibility level of 75 percent to be acceptable for auditoriums (which meant that 25 percent of syllables or notes might go safely unheard or confused), Harvey Fletcher was drawing up standards for acceptable intelligibility on the phone. In 1953, summarizing thirty-five years of research on *Speech and Hearing in Communication*, Fletcher would reproduce graph after graph on the "minimum perceptible changes in frequency and sound pressure level" and on "methods of measuring the recognition aspect of speech" with nary a reference to timbre, which is often vital to the recognition of a person's voice, mood, and urgency. On the basis

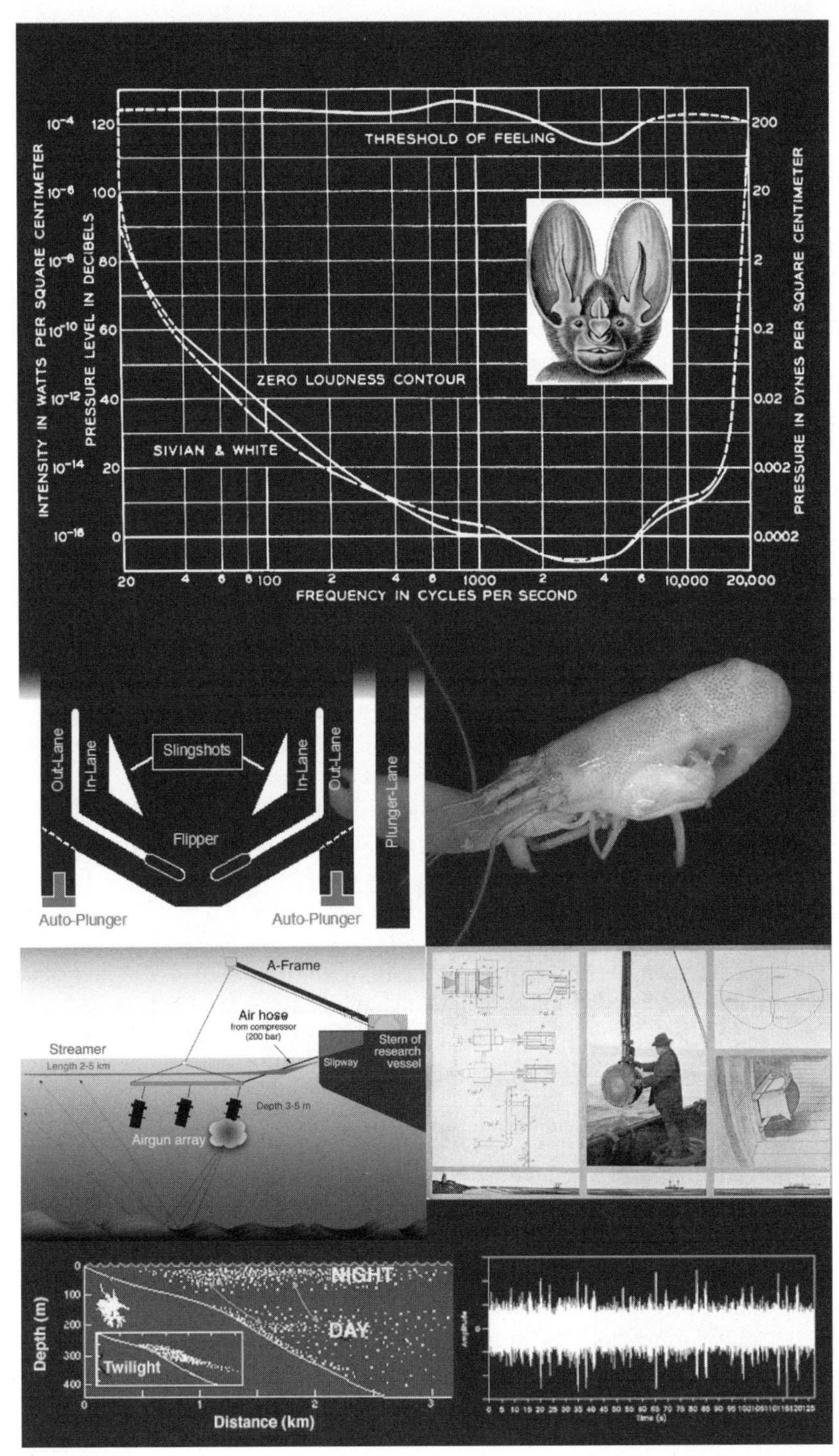

SOUNDPLATE 26

of Fletcher's decades of work, and his (Latter-Day Saint?) concern for on-point message, telephone systems were built to specifications that allowed for the loss of as much as a third of all sound and the scanting of timbre.[221]

Straining to hear through the crackle on the line, the cross-talk, the distortion, the "sea of noises in which a large number of people are daily submerged," what was one supposed to make of *la voix humaine* absent its defining timbres?[222]

Information. "Information" was a word in turmoil, about to lose much of its gravitas. Once grounded in moral instruction and religious inspiration, in social networks and judicial proceedings, "information" was becoming impersonal, neutral, ahistorical, and abstract. The turmoil, whipped up by the Leibnizian metaphysics of monads and Enlightenment ideals of impartial knowledge, had been given heavy top-spin in more recent quarters:

- At the headquarters of railroads and large industry: when management after 1880 began to insist on "scientific" systems of communication and information, the latter understood as timely analyses of operational reports (for "administration without records is like music without notes") but eventuating in the crankiness of the vertical file, the blankness of progress charts (one line, up or down), and the memo.[223]
- At Riverbank Labs, and in Room 40 of the British Royal Navy and kindred haunts during and after the Great War: when "intelligence" departments, faced with an efflorescence of coded and "overheard" messages from the blooming of the wireless, began to treat all communication as code.[224]
- At telephone switchboards: when Hello Central operators accustomed to furnishing local "Information," including news, gossip, and emergency assistance, were restricted to the dispensing of coded telephone numbers as mechanical direct-dialing systems were installed during the 1920s.[225]
- In police stations and bank personnel offices: when the first polygraph operators of the 1920s and 1930s began to scrutinize lines etched on rolls of blackened paper for changes in respiration, blood pressure, pulse, and galvanometric sensitivity—the body an involuntary snitch, giving up "information" under crafty questioning and physiological monitoring.[226]
- In psychological laboratories: when academic and industrial psychologists, turning their backs on introspectionist methods, quantified the effects of "distraction" (buzzers, telephone bells, falling hammers, gongs) on the efficiency of clerks, typists, and accountants with little regard for the nature of what they were asked to write, type, or calculate.[227]
- On the shores of Lake Como, at an International Congress of Telegraphy and Telephony in 1927: when Robert V. L. Hartley of Bell Labs presented a theory on, and a formula for, the "Transmission of Information."

A response to the spooling out of longer telephone lines with "increasingly severe standards of performance," Hartley's paper addressed longstanding problems of intrinsic noise while anticipating an era of electronic image transmission. He put forward, therefore, a quantitative measure "whereby the capacities of various systems to transmit information may be compared," whether telegraph, telephone, radio, fax, or television. This measure (log sn) was simply the logarithm of all possible symbol sequences, which amounted to the most general, inarguable formula of communication: conveyance. In such a logical (and explicitly non-psychological) universe, all symbols must be postulated as disturbances in a field; otherwise they would be indistinct from that field and go unconveyed. A spoken or written word, then, was nothing more than an "acoustic or electrical disturbance which may be expressed as a magnitude-time function" on a continuum with all other sound, and photons of light, and radio waves. So Hartley effectively neutralized noise by reducing all symbols to disturbances in a medium or, if you prefer, by elevating all disturbances to potential messages. Instead of being cowed by its intransigence or ubiquity, Hartley took noise to be a first-order phenomenon of conveyance and a second-order "intersymbol interference," a limiting condition on any conveyance. Leveling all media to modes of conveyance regardless of content, his "information theory" was predicated on the technical democracy of probability, where all symbols and sequences start off equally likely.[228]

Physicists could consider information theory another historic act of brilliant abstraction similar to those which had put vibration at the heart of acoustics, wave motions at the heart of vibration, and (one year earlier) the complementarity of waves and particles at the indeterminate heart of quantum mechanics. Like the ever-meddling observer of Heisenberg's Uncertainty Principle, noise in Hartley's formula was not just an antagonist but an agonist, a permanent player. When the English physicist James Jeans in 1937 published a small volume on *Science and Music* for the layperson, he commenced with a "most incredible" tale of the evolution of the ear from the "lateral-line" organ of primitive fish. Last of the human sense organs to be perfected, the labyrinthine ear was "sensitive to an almost inconceivable degree," which he demonstrated visually with a phonodeik illustration of "the sound-curve of a gramophone record of a baritone singing the word *rivers* to the accompaniment of an orchestra." Yet when Jeans came to discuss the transmission of sound, he had to restrain his enthusiasms, with the shake of an earmuff at Bohr and Heisenberg, for "Unhappily we live in an imperfect world in which perfect transmission is impossible." Consider how much happens, for example, when orchestral music is broadcast over the radio. "The sound-curve produced by the orchestra must then be handed on from the air of the concert room to the diaphragm of a microphone, from this to an electric current in a wire, from this to a shower of electrons jumping through a system of valves, from this to a current in an aerial, from this to waves of electric and magnetic force travelling through space, from these to another aerial and its connections in a receiving set, from these through more showers of electrons in valves to yet another current in another wire, from this to the diaphragm of a loudspeaker, from this to the air, and finally from this to our ear-drums." Each time the sound-curve was passed on from one carrier to the next, its shape was changed (call that "distortion")...and once the ear got hold of it, oh brother! It could add new notes to those played by the orchestra; if damaged, it might filter out low or high notes, "refusing to transmit them to the brain," or rebalance the middle ground, "so that the various sounds are heard in proportions quite different from those in which they were played by the orchestra." What a world. We were lucky to make out much of anything.[229]

Luck had nothing to do with it, according to the German social philosopher (and pianist, music critic, and radio disk jockey) Theodor Adorno. The problem with radio transmissions was less a problem of challenged or scattered hearing than of commandeered listening. Music-making, recording, distribution, and broadcasting were part of a sonic trap set by capitalism, which espoused easy listening as an adjuvant to easy consuming. Industrial society promoted an ethos of mechanical repetition and facile reproducibility that encouraged symphonic snippets and folksy tunes that relied on the comforts of familiarity for their appeal to the heart, the will, and the wallet. "Light" music was light on meaning, heavy on ideology—an ideology of immediate pleasure and quick spending; serious music, which required work of its audience, and cogitation, rewarded listeners with liberating alternatives of sound and sense. Recording technologies always threatened to turn music into a commodity by making it into a marketable material, and technical demands for a narrow mid-frequency tonal range and placid dynamics made serious music, whether twelve-tone, atonal, dissonant, or rhythmically venturesome, nearly prohibitive to record, so "record players" had become conveyances for compositional inanities and routinized performances, "pre-given and pre-accepted." Avant gardes could experiment with gramophone systems to create "legitimate, functional loudspeaker music" hitherto unfeasible, and audiences could benefit from repeated playings to achieve a keener interrogation of performances or pieces—so the technology was not altogether sinister. In the ordinary run of things, however, listeners were being trained to perceive "the complicated only as a parodistic distortion of the simple," that is, as noise; further, they were being trained to attend less to the musical qualities of a performance than to the celebrity of a well-advertised voice or widely photographed composer, which had more to do with commodity fetishism than music criticism. So popular (and recorded, and repeatedly broadcast) music had become a fixture, wrote Adorno in 1941, in a "system of response mechanisms wholly antagonistic to the ideal of individuality in a free, liberal society."[230]

Reframed in the language of Hartley's information theory as elaborated during and after the Second World War, Jeans's problem of hearing and Adorno's problem of listening were both problems in redundancy, at opposite ends of the spectrum of noise. Jeans the physicist made it seem that there was too much noise in all channels, biological and technological, to allow for trustworthy broadcast from a radio studio over the airwaves to the auditory cortex; Adorno the social philosopher made it seem that there was too little noise in most channels, cultural or political or jazz, to prompt healthy variations that would stimulate the intellectual labor necessary to a critical take on the world and, hence, a wider ambit of personal freedom. In the postwar terms of Bell Labs' Claude Shannon (juggler, unicyclist, and originator of digital circuit design) and MIT's Norbert Wiener ("ex-prodigy," and formulator of cybernetics), the greater the redundancy of symbols conveyed through any channel—with noise, *always* with noise—the greater the certainty of reception but the more impoverished the information. Call entropy, H, the freedom of choice in the symbols to be conveyed: not just the variety of symbols available but the rigidity of the rules for selecting one over another, one after another. The greater the randomness or freedom of choice, the greater the entropy and the greater the uncertainty about what has been conveyed—not, mind you, the meaning of the symbols, but the nature and number of the symbols transmitted. With H approaching 100 percent, what comes through can be no more than "white noise"—a statistical and (to Shannon) uncomfortably figurative term which, externalized in acoustics, sounds like an oceanic wash of random frequencies and amplitudes or (to the MIT psycho-acoustician J. R. Licklider) "like the noise of Brownian motion, like the noise inherent in a resistance... like ss-sh-hh-hh-hh." With H approaching zero, when there is scant choice of symbols to be transmitted and high redundancy, the conveyance can be close to certain but the information may be nil. With H around 50 percent, as in English messages (half of whose symbol-choices are determined by rules governing syntax and semantics), and with memory operative (so that what comes before affects what comes next), there can be greater certainty about what has been conveyed, since one can closely calculate the probabilities of sequences of symbols and syllables. Even so, if one considered, as did Wiener in 1949, the percentage of spoken words that reach the auditory cortex intact, then "the crude non-human measurements" of statistical mechanics "fail to give an adequate account of the tremendous losses of information inseparable from nervous reception and the transmission of language into the brain." Case in point: when leading thinkers had met in 1946 to discuss the significance of "Feedback Mechanisms and Circular Causal Systems in Biology and the Social Sciences," the psychologist and field theorist Kurt Lewin had spoken so quickly and deferentially that "from the noisy mechanical recording device we unfortunately have no record of the substance of any comments he made at the first meeting." A human receiver, self-noisy, must *always* contend with noise as "intersymbol interference" or "spurious" information, a function of all that comes through a channel of a specific bandwidth. The consequence, which Shannon's friend Warren Weaver found most "bizarre," was that "the two words *information* and *uncertainty* find themselves to be partners."[231]

Three hundred years before, the English physician Thomas Willis had written up second-hand accounts of "a certain kind of Deafness, in which those affected, seem wholly to want the Sense of Hearing, yet as soon as a great noise, as of great Guns, Bells, or Drums, is made near to the ears, they distinctly understand the speeches of the by-standers, but this great noise ceasing, they presently grow deaf again." Aurists came to accept such "paracusis" as a common paradox: the hard-of-hearing may hear better in loud environments. The physics of this phenomenon, in which word and noise were acoustically partnered, had been laid out for Knudsen by a genial otologist: "high frequency vibrations which are little ones *creep in* on the low frequencies which are big ones," making speech temporarily audible. This made no sense to someone of Knudsen's background, who found that noise *always* got in the way of accurate hearing. In 1929 he demonstrated to his own deep satisfaction that paracusis was paralogical: noise never directly abetted hearing; rather, in noisy situations people unconsciously and unself-consciously speak more loudly, enabling the hard-of-hearing to make out more of their words. Now, with information theory, paracusis seemed to be rearing its ugly ears again and (how could this be?) as no less than a paradigm of communication: no noise, no message.[232]

Correlative or cause? Did noise make communication possible? Was noise, like the bawling and babbling of infants, the nesting ground of all messages? Or, like the call of the parasitic cuckoo, did it lay eggs in the nests of other message-makers, confusing all species of sound?[233] Information theory was agnostic: it made no assumption about prime movers. Noise was axiomatic and affected all conveyance; this did not mean that noise effected conveyance. Cybernetics and the ensuing biophysics of stochastic (probabilistic) processes would be less reluctant to identify noise as omnipresence and effector—as a sort of creator. The momentum for such an acoustemology, where it did not harken back to Babylonian creation myths or Lucretian physics, stemmed from feedback systems as conceived for automata, steam engines, demographic curves, and thermodynamic processes.[234] More immediately, it stemmed from the experiences of electrical engineers who had suffered through feedback from microphones, radio receivers, and public address systems, and who by 1930 had begun to employ feedback in frequency modulation and noise-damping circuits, transforming what had been at worst an ear-splitting interference and at best a clue about crossed wires into a means for calming and even strengthening signals. The momentum came most directly from work done during the Second World War on statistically noisy problems—how to program a machine to decipher codes far faster than a cryptographer; how to incorporate binary electrical switches into that program; how to design anti-aircraft guns to achieve deadlier results against faster targets by linking them to radar; how to construct programs that computed new trajectories and re-aimed the guns more swiftly than the best artilleryman. As a mathematician and a theorist of Brownian motion, Wiener had been involved in those efforts at the proving ground in Aberdeen, Maryland, laboring over noisy calculating machines known as "crashers"; postwar, as the son of a world-famous Jewish philologist, he applied the principles of adjuvant feedback to the analysis of speech itself, which could only be acquired when an infant heard its own voice well enough to amend the sounds it was making and align them with the select, inflected sounds of its native tongue(s). Without error-correction and audible feedback all the days of our lives, our organs for speech production were too sloppy to prevent our slipping back into a vocal "deadness" and, as with severe adult-onset deafness, an inarticulate mumbling. In 1949 he and Jerome Wiesner of MIT were collaborating on "a method to replace hearing by tactile stimulation," so that the deaf, receiving patterns of stimulation through electromagnetic vibrations at their fingertips, got the appropriate feedback and could "participate in active speech."[235]

Deaf communities by 1950 were embracing Braille but in no mood to return to the brutal oralism of Alexander Graham Bell. In their Silent Clubs, their League of Elect Sourds, and their National Association of the Deaf, they mocked the oralist teachers who had not bothered, or had actively refused, to learn any form of sign language. They had no respect for educators who praised the dull stumbling grating noisiness of those who still struggled with spoken language as if it were their only recourse. How could it be "humane" to humor the deaf as half-articulate children rather than to honor them for an eloquent and adult fluency that happened to be, for the most part, silent? In their own way, deaf activists with an ever-expanding lexicon and syntax of signing would teach the hearing community that there was much to be learned from, and through, silence.[236]

Whether the quietness of a psychoanalyst who becomes a sounding board amplifying the words of a client while attending to his "own inner voice" (Theodore Reik, *Listening with the Third Ear*, 1948) or the postulate of a spiritual life for the Trappist monk Thomas Merton, whose *The Waters of Siloe* (1949) reintroduced the English-speaking world to Cistercian solitude, the embrace of silence after 1945 had more to it than attentiveness and insight. It had to be felt as an act of resistance, a form of orison, a tactic of erasure or detachment, for it had indelibly behind it the firestorms and shockwaves of Dresden, Auschwitz, and Hiroshima. "Silence is always close to history," wrote a Swiss physician who had turned to philosophy and then Catholicism for anchorage, or hermitage. "There was an example of this at the end of the last World War, the war that was a rebellion of noise against silence; when silence was powerfully present at least for a few days ... [and] more potent than all the horrors of the war. It could have been a healing influence ... if it had not been overrun and destroyed by the noise of the whole industrial machine getting down to work again." Cities these days were "enormous reservoirs of noise," and airwaves too, for "God, the eternally continuous, has been deposed, and continuous radio-noise ... installed in His place." Thus Max Picard, whose *The World of Silence* (1948) Merton held in high regard. If, wrote Picard, we feel like animals lying in wait upon our own extinction while "sinking ever deeper among the briars and bushes of the world of noise," silence could restore to us a hopeful life, for it was the "natural basis of forgiveness and of love." But silence was becoming rarer: "no longer an autonomous world of its own, it is simply the place into which noise has not yet penetrated ... the momentary breakdown of noise."[237]

Momentary as the *4′33″* of silence that became the soundmark of John Milton Cage, Jr., a performance piece he credited to his few minutes (4′33″?) in an anechoic chamber where no noise was supposed to penetrate and he himself was the carrier. Son of that California engineer who, as may be recalled, had patented a submarine diesel-propulsion system dismissed by the Navy as too noisy, and who later improved on hydrophone sound detection, Cage the younger seemed to pride himself on equal degrees of acoustic hypersensibility and musical insensitivity ("The whole pitch aspect of music eludes me. Whether a sound is high or low is a matter of little consequence to me"). In 1935, at twenty-three, while attending Arnold Schoenberg's summer lectures at UCLA and haunting Los Angeles movie theaters, he became intrigued by the animations of Oskar Fischinger, a German artist, engineer, and cinematographer whose special effects had helped put Fritz Lang's woman on the moon (*Frau im Mond*, 1929). Struck by similarities between the abstract designs in expressionist cinema and the graphic patterns visible on film soundtracks, Fischinger had worked on projecting an "optical poetry" whose "absolute" geometries vibrated in sync with the frequencies and rhythms of sound. His films, like the 1935 *Composition in Blue*, got Cage in the habit of "hitting and stretching and scraping and rubbing everything" at hand; a talk with Fischinger, invited to Hollywood in 1937, taught Cage to respect the "Buddhist belief that all things have a sound, even if we don't listen or hear it." By 1938, composing, accompanying, and lecturing in Seattle at Nellie Cornish's School of

Music (and art, and modern dance), Cage had arrived at a credo on the future of music: "Wherever we are, what we hear is mostly noise. When we ignore it, it disturbs us. When we listen to it, we find it fascinating. The sound of a truck at 50 m.p.h. Static between the stations. Rain. We want to capture and control these sounds, to use them, not as sound-effects, but as musical instruments."[238]

Cage's rhetoric recalled that of Italian and Russian Futurists, of Swiss and German Dadaists, and of Henry Cowell's *New Musical Resources* of 1919, not to mention Edison's phonographic interest in "capturing fugitive sounds." The music that Cage made, as in his "prepared piano" pieces, his *Imaginary Landscape No. 1* (1939, for muted piano, cymbal, two variable-speed turntables playing frequency-test records), and his *Living Room Music* (1940, for household objects), would recall, to afficionados, Cowell's *Banshee* (1925) and other pieces for plucked^! banged piano, or Henry Brant's *Music for a Five and Dime Store* (1932) with parts for violin, piano, and kitchenware. The streetwise might hear in Cage's performances the sounds of David Rockola's new pinball machines, which arrived alongside newly amplified jukeboxes, half a million in play by 1937. To Harry Partch, yet another Californian who had never been crazy about "pitch" but went bonkers over Hollywood movies, every sort of percussion, and the New Orleans riffs of "little Negro boys... who play washboards, tubs, tin cans, anything that intrigues their aural imagination," John Cage's "sound experiments" seemed "precious and vapid." What was so original in Cage's press release for a 1942 event: "Through organization, [city sounds] lose their nerve-racking character and become the materials for a highly dramatic and expressive art form"? Hadn't Chicago jazzmen been doing that for decades, notwithstanding Cage's dismissal of jazz licks as too tame? Hadn't Edgar Varèse, reacting to the soundscape of New York, already outpaced Cage in his studio at 188 Sullivan Street, "a cave of sounds, coming from bells, recordings, gongs; and the music seems composed of fragments of music, cut and repasted like a collage"? Thus Anaïs Nin, who heard Varèse's music as aptly deafening, with a power that "suits the scale of the modern world. He alone can play a music heard above the sound of traffic, machinery, factories... a universe of new vibrations, new tones, new effects, new ranges.... In his room one becomes another instrument, a container, a giant ear." Why not attend to Partch's new forty-three-tone octave that explored new intervals and tunings? "People may leave my concerts thinking they have heard 'noise,'" wrote Cage in 1943, "but will then hear unsuspected beauty in their everyday life." Precious. Or laughable: the forte of Spike Jones and his City Slickers, who had been using klaxons, gunshots, and "junk instruments" since the early 1940s, with chart-topping hits in 1942, *Der Fuehrer's Face*, each *heil* punctuated by a "birdaphone" or Bronx cheer, and in 1944, a *Cocktails for Two* interrupted by hiccups, gurgles, and honking. Later would come a *Duet for Violin and Garbage Disposal* and songs recorded with "barking dogs in hi-fido." "One night," wrote Nin, "after hearing a John Cage concert, we all gave an imitation of it, with pots, pans, water-filled glasses, paper and tin." Vapid. Zen.[239]

American Zen. A Zen informed by the journals of Thoreau, the lectures of John Dewey, and the interfaith ambitions of the Open Court Publishing House in La Salle, Illinois/e, to which in 1897 had come D. T. Suzuki, a high school teacher of English in Japan and a modest student of Zen. It was there, bicycling through the fields south of Chicago, that Suzuki found satori, and it was then, in the years before the Great War, that he began to shape a practice compatible, as Douglas Kahn has shown, with the universalizing mysticism of "perennial philosophy" as explicated by Aldous Huxley. A practice compatible too with the world-mythologizing of Jung, who wrote the foreword to the 1948 reissuance of Suzuki's *Introduction to Zen Buddhism*, which Cage heard re-revealed in Suzuki's lectures at Columbia. There it was (where? in the book? in the crowded hall where the master's words were broken up by the roar of planes leaving La Guardia? in the hothouse air of 116th Street?) that a person had to listen for the sound of one hand clapping, for "so long as the mind is not free to perceive a sound produced by one hand, it is limited and divided against itself."[240]

Maybe that was the meaning of those boxes on the ears that Zen masters were perenially administering to their disciples, that sound of the one hand boxing an ear: a blow to quiet the mind, an ear-violence to shake off all contraries of self and other, inside and out. And maybe that was the shock of that box-on-the-ear in 1951 when Cage entered an anechoic chamber, a box-around-the-ear built by psycho-acousticians at Harvard at the start of the Second World War, after the "Noise in Honolulu," when a torrent of intercepted but contradictory Japanese radio messages and a series of unavailing American alerts had been so mismanaged that the fleet at Pearl Harbor could be sunk of a quiet Sunday morning.[241]

"Father of Secrets, there were two noises that day when you went into the soundless chamber," wrote Lewis Hyde in his "Elegy for John Cage": "the low one was the sound of your blood and the high one the sound of your nervous system." That's not what Cage knew at first; that's what a technician told him after he found that in the deadest of sonic cages he was still hearing things. Things, said the technician, that Cage had brought in with him: a heart beating in a ribcage, blood pulsing through his veins, the hum of his nerves. Already in the 1930s, acoustic engineers had made clear that soundproofed or anechoic rooms could "eliminate all noise except that due to heart-beats, breathing, and the like," and that in decibel terms "the absolute zero of noise is impossible if one human being is present (the beating heart and circulating blood inside one human body make a noise of 10 to 15 db)." But *nerve* noises? Were these a technician's prank, a jest on the common misapperception of neural pathways as buzzing circuits and of headaches as jagged lines of electrical interference produced by blacksmith hammers pounding on the anvil of the brain? Or did the nerve noises hail from a theory of tinnitus as "an illusion of sound caused by an irritation of the auditory neural elements," and a concomitant epidemiology of tinnitus that claimed that "People entirely without tinnitus are extremely rare, if such cases exist at all"? Or was it the Zen residue of "no-mind": what remains whatever you do? Hearts had been telltale since Poe; it was not the manifest noise of blood "in circulation," I suspect, but the ostensible sonics of his nerves "in operation" that led Cage to be "constantly telling" his Zen parable of the anechoic chamber, how that which is the least audible is our lease on the self, and how in life there is no escaping noise, only a re-realizing of it.[242]

Re-realizing, for example, a polka by Richard Strauss performed in 1890 at Madison Square Garden, where Strauss raised his baton, his orchestra lifted up their instruments, but a dozen phonographs alone played the piece, "Le Phonographe." Re-realizing "In futurum," a movement from Erwin Schulhoff's 1919 piano concerto that consisted entirely of rests, echoing the Two Minutes' silence of Armistice Day. Re-realizing Yves Klein's 1949 *Monotone-Silence Symphony*, whose first movement consisted of a twenty-minute drone and last movement a twenty-minute silence. It matters not whether Cage was familiar with any of these precedents, or (as he would later say) was responding to the [   ] paintings of a friend, Robert Rauschenberg, or was recapping his own clever proposal that one of the Muzak Corporation's daily segments be 4 minutes, 30 seconds of recorded silence. Zen is not about precedents, it's about a quiet mind. ("Music's ancient purpose—to sober and quiet the mind, thus making it susceptible to divine influence.") Quietly, David Tudor comes on stage on August 29, 1952 and sits down at a piano, whose keyboard lid he shuts. Consulting a stopwatch [   ] he then opens the lid: end of first movement; shuts the lid, [   ] opens the lid: end of second; shuts the lid, [   ] opens the lid: end of third and final.[243]

Birdsong, "legend has it," could be heard through the open windows of the concert hall there in Woodstock, New York, but *4′33″* had no naturalist program. The singing birds were as inadvertent as they would be in August 1969 at the "Three Days of Peace and Music." As were the hitch-and-clunk of the keyboard lid and the coughing of the audience, its foot-shuffling, its scratching of semi-private parts. "Silence as the entirety of unintended sounds," says Cage. Question in *For the Birds*: "Would you say that what people continue to call 'silence,' by force of habit, belongs in reality to another domain? Or does silence indeed arise from this same domain: music?" Cage: "It's already sound, and it's sound all over again. Or noise." The Zen of background as foreground. "Not one sound fears the silence that extinguishes it. And no silence exists that is not pregnant with sound," he tells students at Juilliard in 1952. Write otologists the next year, "It appears that tinnitus is present constantly but is masked by the ambient noise which floods our environment." Dam that flood and the soundnoise that was "subaudible" becomes audible, sound all over again. Performance space as anechoic chamber, audience as tinnital roar, event exactly 273 seconds of indeterminacy. "The main trouble is *drift*," Cage Senior had written in 1951 with reference to polyphase rectifiers in his *Theory and Application of Industrial Electronics*; the more powerful a direct current, the more problematic the fluctuation of cathode temperatures. Was he composing a koan when he entitled a chapter "The Dynamics of Closed-Loop Systems"?[244]

Too many places to go from here. To the 1957 "Duet" by dancers Paul Taylor and Toby Glanternik, she sitting, he standing, for four minutes "looking calm in an exciting way," and reviewed in the *Dance Observer* with four inches of [ ]. To the enthusiasm of airplane and steel companies, school architects and air conditioner manufacturers for the progress being made by RCA toward the development of an "electronic sound absorber" that worked by feedback loops to suppress noise, which the physicist H. F. Olson's many correspondents took as a promise of soundlessness, as if one could clamp an anechoic chamber around any source of noise. To "hush records," 45 r.p.m.s of hazy silence first played on jukeboxes in Detroit in early 1960, apparently to intrude some downtime in bars and soda shops, amplified in 1961 with a Silent Record Concert and a 120-piece Hush Symphonic Band led by Soupy Sales, as if silence were no less marketable, and no less open to satire, than any single on the pop charts. To a six-figure suit won by the John Cage Trust in 2002 against Mike Batt of The Planets for his "A One-Minute Silence," a cut in *Classical Graffiti*, as if the very notion of a framed silence were as copyrightable as a sonata. Precious.[245]

Should one reflect for a minute, or four, on what might have prepared any audience other than Trappist monks, Zen Buddhists, or Lovers of Meher Baba for Cage's *4′33″* in the late summer of 1952, it could scarcely have been any tradition of "music appreciation," whose bias toward the symphonic classics had infected most schoolrooms since the 1920s. With the South Pacific atoll of Eniwetok evacuated that summer for the first test of an H-bomb, it could have been some tradition of silence among scientists and technicians in the aftermath of each nuclear test, but just now South Korean and American soldiers had been pushed back to Pusan after a Chinese onslaught that was the "damndest thing" an army lifer, George Zonge, had ever seen—"they came swarming over the hills making all kinds of noise, blowing bugles and rattling cans and shooting off flares and yelling their heads off in these high-pitched sing-song voices... like one of those old-time infantry charges you read about in the Civil War."

On quieter nights, wrote one Marine, the Red Army set up phonographs and loudspeakers on the front, playing jazz for the Americans and asking them (in polite English) to leave Korea to the Koreans, then broadcasting an "hysterically funny" version of "No Place Like Home" as performed by "a trio of women, accompanied by a deathophone, a senilaphone, and a b-flat fart horn." The real jeopardy, what woke soldiers up and set them on edge, was silence; when peepers and frogs stopped sounding off and they heard no attack whistles or musical interludes, something was up, and it meant physical danger, not the intellectual challenge of a piano solo where no finger touches a key. So the cultural traction of *4′33″* silence had to come from a different direction. Had Cage been politically outspoken, the silence might have been heard as a critique of those unwilling to face down Senator Joseph McCarthy, who heard Communist conspiracy wherever he turned. *Heard*: "This is a time when all the world is split into two vast, increasingly hostile armed camps—a time of a great armament race," said McCarthy at the start of his infamous "Enemies from Within" speech on February 9, 1950. "Today we can almost physically hear the mutterings and rumblings of an invigorated god of war. You can see it, feel it, and hear it all the way from the Indochina hills, from the shores of Formosa, right over into the very heart of Europe itself." In late July 1952, stung by criticism of his methods and manners, McCarthy had sent a letter to the advertisers in *Time Magazine*, asking them to withdraw from its pages lest they help "to pollute and poison the waterholes of information," and this Cage would have heard reported weeks prior to the Tudor performance.[246] Regardless, his *4′33″* was heard neither as a demurrer to McCarthyite intimidation nor as a claim to the right to remain silent.[247]

Instead, it could have been heard and, I propose, must have been heard, in terms of sonar. Such a proposition would not have sounded odd to someone who in in his youth had composed gong-music for the underwater ballet artists of UCLA, nor to physicists like Frederick A. Saunders, who in the 1930s was telling his students at Harvard that all we hear we hear underwater, through the waves of lymphatic fluid that fill the spaces of the inner ear.[248]

9. Oceans

Lazzaro Spallanzani in 1793–1794 wondered how bats avoided obstacles while flying through dark subterranean passages. He blinded some bats, coated others with flour paste, and asked friends in Italy and Switzerland to help him determine which sense bats relied upon for their swift, accurate flights. He arrived reluctantly at a suspicion that a "new sense" might be involved, since no matter the visual, tactile, or olfactory deprivation, or the loudness of environing noise, bats managed not to go bump in the night. Asked to revisit these experiments, the Swiss zoologist Louis Jurine found that when bats had their ears plugged up, they did start bumping into things, so somehow they *were* using their hearing to navigate. Contemporaries were contemptuous: What could bats be hearing in their caves except other bats? The Baron Cuvier had heard nothing in those caves, but he did observe that bats had peculiar nodules on their large ears. Citing Cuvier, and in ignorance of the work of Jurine, nineteenth-century naturalists credited bats with a delicate sense by which they brushed their way through silent nights. A few scientists did explore the pitch-limits of human ears and speculated that insects, dogs, cats, and perhaps bats responded to tones out of human auditory range. After the sinking of the *Titanic* in 1912, Hiram Maxim took bat-sensing as a model for the detection of obstacles in heavy fog, guessing that their wingbeats generated very-low-frequency (that is, infrasonic, long-wave) vibrations on the order of 15 Hz, whose felt reflections bats used to navigate. After the war, Hamilton Hartridge, an English physiologist, theorized that because bat feeding and survival depend on the pursuit of small flying insects a few inches away and the avoidance of walls in tight places, bats were instead using very-high-frequency (ultrasonic, short-wave) vibrations, which were best for close work. During the 1920s, when George W. Pierce at Harvard developed an electroacoustical meter that could register humanly inaudible sounds from 20 kHz to 100 kHz, and other researchers subsequently refined quartz magnetostrictors that produced reliably high-frequency tones, investigators in the 1930s could now confirm the ultrasonic capacities of bats. Even so, Donald R. Griffin, a Harvard undergraduate, naturalist, and bat fancier, with the help of Prof. Pierce (who had been using his meter to transcribe the "songs" of katydids, locusts, and crickets), failed to prove that high-pitched clicking-crying was of any value to bats on the wing. It remained for Griffin as a graduate student, in tandem with Robert Galambos (who had been studying the cochleas of guinea pigs under Hallowell Davis over at the Medical School), to borrow Pierce's instruments, string wires across a soundproof chamber, and set loose different species of bats. In 1941, they proved that flying bats attended to the echo of their own "supersonic" cries in order to bypass the wires, which when their ears were plugged with collodion they "touched lightly (T), hit but flew on through (H), or crashed (C) into." Each species had its own click-cry range between 30–70 kHz through which it continually corrected its flight path using what Griffin would call, in 1944, "echo-location." Since the limit of human hearing is around 20 kHz (and 38 kHz for dogs, 75 kHz for cats), all of this bio-acoustic feedback was happening in relative silence, though Griffin and Galambos did hook up an amplifier and speaker to the microphone of Pierce's meter to "hear" the cries, and for many years after, until his death in 2003, Griffin was still listening to bats. How did their auditory cortex process sounds? How did they tell their echoes from louder environmental noise? What should we make of their world?[249]

"Anyone who has spent some time in an enclosed space with an excited bat knows what it is to encounter a fundamentally *alien* form of life," the philosopher Thomas Nagel would write in a controversial essay of 1974. Although it was well-nigh indisputable that bats have what we call "experiences," argued Nagel, we could never arrive at an indisputable account of what it is like to be a bat. His was a thesis about the phenomenological strangeness of consciousness, which eludes analogy. Straining toward analogy, however, is a well-nigh indisputable part of what it is like to be human, and anyone who has spent time with an excited sonar operator in a submarine may not have far to stretch toward a few reasonable assumptions about what it is like to listen into the dark, on the nautical fly, and navigate by echoes.[250]

Long before the launching of the first submarines, and years after, the ocean deeps were spoken of as the most silent of realms, quieter than mountaintops with their winds and avalanches, quieter than the Arctic with its winds and calving glaciers (a noise "equal to that of thunder"), quieter than an ionosphere that crackled with auroras. Monsoons and hurricanes ("the most tremendous unearthly screech" one seaman had ever heard) might scour the surface and whirlpools shake the Japanese island of Awa "with great force and unchanging noise" audible to sailors far off. Several hundred feet below, the waters were indifferent,

> And no rude storm, how fierce so'ever it flyeth,
> Disturbs the Sabbath of that deeper sea.

So too the soul, wrote Harriet Beecher Stowe in "The Secret," evermore,

> And all the babble of life's angry voices
> Die in hushed stillness at its sacred door.

In 1818 an English sea captain had dredged up three live worms and a basket star, *Caput medusae*, from the bottom of Baffin Bay, which proved to him the presence of life six thousand feet below the surface, "notwithstanding the darkness, stillness, silence and immense pressure produced by more than a mile of superincumbent water," but most scientists and poets continued to believe that "the abyss" was *azoic*: barren, currentless, lifeless. "Quiet reigns in the depths of the sea," wrote Matthew Maury in 1858, having just published a bathymetric chart of the Atlantic based on 180 deep soundings. Aside from a constant snow of the microscopic dead, animalculae who upon passing "in vast multitudes, sink down and settle on the bottom," there could only be a "perfect repose."[251]

With the publication of Darwin's *Origin of Species* and the laying of the first transatlantic cables (using Maury's charts), shipboard scientists began to look for "living fossils" in the ooze from the sea bottom, hauling up from the blackness five miles down thousands of unknown species, some flashing, some glowing, others with gaping eyes. Alert to these discoveries, the editors of *Good Words* nonetheless told their middle-class readers in 1890 that "In the Abysmal areas all is dark, calm, cold, and silent; it is a region of dull uniformity. There are chemical processes at work, it is true, and deposits are being formed, but with geological deliberateness." In 1910 Agnes Giberne wrote in a popular *Romance of the Mighty Deep* that "It is not easy to picture to ourselves the changeless calm of those abysses—those five or six miles of underwater, with nothing from sea-level to sea-floor to break the dead monotony," this despite the persistent churning of the 28,000 ships of the world's merchant marine. "If there were ears wherewith to hear in those ocean waves, they would be even less useful than eyes," claimed John C. Van Dyke, the art historian and shill for Southwestern deserts who took up the subject of *The Opal Sea* in 1917. For effect, he quoted Rudyard Kipling's "The Deep-Sea Cables":

> There is no sound, no echo of sound, in the deserts of the deep,
> On the great grey level plains of ooze where the shell-burred cables creep.

Although fleets of steam- and then diesel-driven trawlers had been "plowing" the continental shelves with an offensive violence, traditional line-fishermen protested rather the devastation of habitats and depletion of fishing beds than the noise of power winches or 100-horsepower motors. How could there ever be enough submarines, trawlers, or ships laying cable to make an acoustic difference in ocean deeps that covered two-thirds of the Earth's surface? During the 1930s, when two explorers from the New York Zoological Society took their new bathyscaph deeper than man or woman had ever bodily gone, they saw in the dark waters of Bermuda, half a mile down, dozens of luminous fish and a "pyrotechnic network" of organic mesh. There was life below, extraordinary life, deeper down than any had imagined—and more ancient: in 1938 a steel-blue coelacanth, a paleozoic fish long held to be extinct, was trawled off the coast of South Africa. Not a one of these, however, was vocal, not a one soniferous, though a "marine biologist" (1893) might have expected creatures at such murky depths to have taken evolutionary advantage of the fact that sound travels four times more quickly through saltwater than through air.[252]

Saltwater itself may have damped expectations of sound. European physicians had long been issuing cautions against entering an ocean more abrasive than welcoming. Even after "sea air" began to beckon in the late 1700s, "sea-bathing" remained a dubious adventure, given how "excessively frequent" it was for ocean bathers and swimmers to "complain of noises and unpleasant sensations in the ears, amounting in some instances to temporary deafness," as if the seas must avenge themselves by rendering our bodies (90 percent saltwater at birth, 60–70 percent as adults) as immune to sound as the deeps were deemed to be. Should a woman sing at night along the shore of the Atlantic, as in "the Idea of Order at Key West" (1934),

> It may be that in all her phrases stirred
> The grinding water and the gasping wind;
> But it was she and not the sea we heard.

If the sea had a voice, it was "dark" and, as Wallace Stevens later wrote in "Somnambulisma" (1943), almost noiseless. On Key West, when the woman sang,

> The sea,
> Whatever self it had, became the self
> That was her song, for she was the maker,

with

> The maker's rage to order words of the sea,
> Words of the fragrant portals, dimly-starred,
> And of ourselves and of our origins,
> In ghostlier demarcations, keener sounds.

"Who has known the ocean?" asked Rachel Carson in her earliest nationally published essay, in *The Atlantic Monthly*. "Neither you nor I, with our earth-bound senses, know the foam and surge of the tide that beats over the crab hiding under the seaweed of his tide-pool home, or the lilt of the long, slow swells of mid-ocean, where shoals of wandering fish prey and are preyed upon, and the dolphin breaks the waves to breathe the upper atmosphere.... Even less is it given to man to descend those six incomprehensible miles into the recesses of the abyss, where reign utter silence and unvarying cold and eternal night." All in all, the ocean seemed to be a clearer, better bet for sonar and the pursuit of "ghostlier demarcations, keener sounds," than did the upper atmosphere for that other remote-sensing technology of the Second World War, radar.[253]

Methods for sounding the deeps had not changed much since Egyptians first lowered line and lead into the sea during the third millennium BCE, until Anglo-American ships, laying telegraph cable from the 1860s to the 1880s, using steam-powered hoists, lowered a thousand fathoms of weighted steel wire to map the Atlantic floor. Alexander Graham Bell, expanding upon the idea of a submarine foghorn, arrived in 1879 at what he called "an essentially new idea of very great practical value," viz., computing ocean depth by timing echoes off the bottom, but it was Bell's nemesis, Elisha Gray, working with the engineers of the newly created Submarine Signal Company (1901), who devised a practical system for underwater signalling and sounding using what Gray called a "hydrophone," a microphone adapted to receiving vibrations through water. Among the problems to solve was how best to shield the hydrophone from the considerable noise of the ship itself. This was accomplished by hanging a hydrophone within each of two water-filled tanks, port and starboard, so that distant vibrations from an electrical bell on the sea floor or a wave-driven bell-buoy would penetrate the hull and be registered by hydrophone diaphragms apart from most ship noise, allowing not only for accurate echo-sounding but for the stereo hearing requisite for echo-location. During the Great War, as I wrote a while back, submariners scoured the sea for prey with hydrophones connected to binaural tubes for more protected stereo listening, while destroyers defended convoys with the help of amplified hydrophones that could detect the wake of a U-boat's propellers. Near war's end, too late for deployment, French scientists had success with devices that generated ultrasonic waves using piezo-electrically deformed salt or quartz crystals that yielded high-frequency vibrations of a constant intensity, ideal for echo-sounding and, in the 1920s, for the echo-location of schools of fish by commercial fleets. Complemented by Pierce's acoustic interferometer, underwater sounding was further improved in the 1930s with the use of magnetostrictors for powerfully directional ultrasonic emissions, all of which demonstrated to the young American physicist R. Bruce Lindsay that the field of acoustics had not been put to rest by the work of Rayleigh fifty years before. There were wrinkles to be worked out with regard to the velocity of sound in the changing temperatures, pressures, and salinity of sea water at different depths, and also with regard to the variable density of material on the sea floor that affected the absorption/reflection of sound, but for the purposes of submarine warfare enough of the distance-sensing technology was in place that ASDIC, or "sonar" (Sound Navigation and Ranging), could be tested at sea by 1928. After the refinement of a bathythermograph that measured ocean temperature and pressure at any depth and the subsequent charting of underwater sound velocity by temperature, sonar was ready for general installation by 1939.[254]

Sonar operators, whether in sound-controlled chambers on surface ships or in the metallic quarters of submarines, heard more than admirals wanted them to hear, more than "oceanographers" (1886) expected them to hear. Trained to ignore the self-noise of their own vessels (screw-propeller noise, branging motors, whistles, bells, the banging of scrub buckets, gun practice); trained to listen through surface noise from stormy weather and the cavitation or bubbles formed in a vessel's wake ("a regular blanket of sound"); trained to recognize the frequency spectrum of the engines and propellers of ships in their own navy; trained to distinguish seamounts and forests of kelp from enemy subs—still, sonarmen went "ping-happy." Straining to identify threats within an underwater environment that behaved "very much like a large empty room with bad acoustic properties," they heard pings bouncing off what turned out to be whales and schools of fish, heard pips refracting off what turned out to be temperature gradients, heard roars from what turned out to be waves rushing at rocks on distant shores, and heard much better in mid-morning than in the late afternoon. Near harbors where coastline defenses were most sensitive, they heard so much that they could decipher nothing at all. The latter sounds did not proceed from ships, tugs, or sewage pumps; they were, it was presumed, biological.[255]

Fishermen for millennia had claimed to hear fish make sounds with their gills, fins, bellies, or air bladders in the shallows and sometimes out on the open sea, but such sounds had been considered by scientists—when considered at all—inadvertent. What earthly, or aqueous, purpose could there be to audible sounds when all fish, lacking a cochlea, were deaf? Although anglers in skiffs suspected that fish could hear disturbances on the water, and an English aurist, William Wright, had argued in 1858 that fish at least *feel* sounds, a more detailed accounting in 1874 by the British anatomist John C. Galton of fifty-two soundful species made no scientific splash, nor was there any cavitation to his claim in the *Popular Science Review* that "the majority of sounds produced by fishes are not casual utterances, but are truly voluntary," and that since fish become noisier during breeding season, their sounds were heard by other fish as "nuptial hymns." In 1905, appearing to confirm Helmholtz's theory that only the cochlea of the ear effectively perceives sound, the German otologist Otto Körner had ruled out all sound-sensitivity in fish after whistling at length to catfish, "with his mouth, through his fingers, on penny whistles, high or low," and asking a professional soprano to sing them arias, none of which got any visible response.[256]

Karl von Frisch was unconvinced. Son of a Viennese physician and nephew of two physicists and of the physiologist Sigmund Exner (known for his experiments on Brownian motion, sensory discrimination, and cerebral localization), von Frisch in 1910 had completed a thesis on the senses of fish. In specific, he had studied the response of minnows to colors, during which research he had no doubt come across Körner's monograph, *Können die Fische Hören?* (Can Fish Hear?). If, as von Frisch had shown (contrary to the scientific canon), fish with working eyes saw colors, what stopped fish with auditory labyrinths or ossicles from hearing sounds? His research on fish—and on whether insects could see colors—was interrupted by the Great War, from whose infantry he was rejected due to poor eyesight. Afterwards he resumed his zoological studies, first and Nobel-famously with bees, then, at Rostock University and Munich, with fish. Körner, a professor at Rostock in 1921–1923, would soon be invited by von Frisch to observe the behavior of another catfish, named Xaverl. Figuring that fish had no impetus to attend to aimless tootling or *coloratura*, von Frisch used Pavlovian methods to "give a biological significance to sounds": each time he fed his blinded *Amiurus nebulosus*, he whistled. After a time the sound of his whistling alone seduced the bottom-feeding fish from its refuge in an earthenware pipe on the aquarium floor, and it would snap in expectation of a meal. Xaverl in Rostock and then minnows in Munich were trained to distinguish a food tone from a cruel tone somewhat higher or lower; when a fish swam out in response to the cruel tone, he slapped it with a glass rod. It became undeniable that his fish, though lacking cochlear apparatus, were listening and, to an extent, recalling what they heard; they came snapping for food at the food tone and fled at the cruel tone. Dissecting some of his subjects, von Frisch determined that at low frequencies fish felt sonic vibration through their skins; at higher frequencies they had an identifiable cerebral structure, a sacculus, that enabled them to discriminate among quarter-tones. "It may be asked," he wrote in a summary of his work in *Nature* in 1938, "for what purpose fishes are able to hear so well in silent water?" Noting that there were many soundful fish ("croakers," to name one obvious species, and the Mexican "grunt," and dogfish "barking," skates "squeaking," catfish "humming," toadfish "gnashing their teeth," his excited minnows "piping"), von Frisch wondered if there might be a "language of fishes" as there was a syntax of dance among bees?[257]

Grammar and syntax were far from the minds and ears of sonarmen who had enough to do to get a handle on Doppler waves, distinguish "true" from false echoes, listen through the "flat thermal noise" of amplifiers infiltrating their headsets, and tell the "wow" of their instruments from the bow of an attack sub. Selected for training according to their "intrinsic aptitude for reporting small variations in the pitch of target echoes," sonarmen were prone to hear too much rather than too little, and the Psycho-Acoustic Laboratory (PAL) at Harvard struggled to establish the parameters for accurate hearing under the noise of battlefields, in long-range bombers, and on submarines—which at full speed could reach 124 db in the engine room, 105 db in the control room beneath the conning tower. In 1942, when an American submarine in the Macassar Strait between Borneo and Celebes picked up a "strange crackling noise," sonar operators guessed that it was a "newfangled gadget," analogous to flak, that the Japanese were dropping into the water to mask their naval movements. A psychoacoustician might have interpreted the crackling as an artefact of the strain of headset listening had not a similarly troublesome noise, akin to "static crashes or to coal rolling down a metal chute," been encountered in 1934 by a Coastal Survey team working off Oceanside, California, and had not equally loud, inexplicable cracklings upset wartime sonar at Beaufort Harbor, North Carolina.[258]

These sounded more than anything else like electrical resistance noise, which the Harvard Laboratory and the National Defense Research Council had found to be the best sound for jamming speech on radio frequencies, so the cracklings merited serious inquiry, especially since British and

American physicists during the war were seeking the most effective means both for producing and reducing noise in order to confound or shield electrical communications and remote-sensing devices. There was debate about the relative effectiveness of such distracting noises as bagpipe tones, county fair recordings, and the scratching of fingernails on glass as compared to thermal noises. Although the problems aboveground with radar and radio differed from problems underwater with sonar or submarine radio, the need to distinguish intrinsic electrical noise from external and possibly tactical noise was critical wherever (very bored) men were listening for hours through headsets. The problems were particularly grievous underwater because so much remained unknown about the acoustic impedance of fluids, sound attenuation at different frequencies, and the refraction indexes of saltwater. In retrospect, wrote J. W. Horton, technical director of the Navy's Underwater Sound Laboratory, it was "doubtful if in any other branch of physics has the need to know been more pressing." Like the first generations of stethoscopists who had had to construct a lexicon of sound-symptoms, diagnoses, and prognoses, sonar officers had to be able to identify the continuous waves of commercial radio, the scatter from ionized layers in the upper atmosphere, the patter from charged raindrops, sparking from dirty commutator contacts, squittering from oscillators, and howling from circuits used to jam nearby transmitters. Vice versa, like the sound-effects men and women active in early radio and Talkies, who turned gravel into "rain," acoustical engineers for the armed forces had to investigate every sort of interference, animal or atmospheric, intentional or inadvertent (as with the surprising capacity of German radar's 800 Hz note to disrupt Allied pilots' radio reception), for defensive and offensive prospects; every noise had to be considered for its promise as weapon or armor. Could you learn by studying the interference from ignition systems how to defend your own forces from the effects while deploying those noises against the enemy? Could bad transducers, scratchy earphones, and poorly matched engine valves tell you, first, what words or tones have the best chance of making it through the racket and, next, what refinements were crucial to headsets? Could the known susceptibilities of Allied radar be turned against German radar systems in the same way that echo-sounding devices for detecting distant objects had been turned into echo-repeating devices for masking and protection? Could frequency modulation (FM), gingerly introduced before the war to enable a new range of clearer radio signals, be used in wartime for offensive barrages? Could you get insight into enemy technology by considering jamming itself as a kind of information, or change radar displays so that they homed in on the jamming and arrived at directional bearings for the source? Could you adapt the wobble of vibrators and feedback of amplifiers to jamming equipment like Bell Labs' JOSTLE, used at the Battle of El Alamein? Could you create noise generators by which, in addition to playing recordings of battle noise and typewriter carriage returns, you could inject "scientific noise" into the training sessions of radio, radar, code, and sonar operators so that they became adept at distinguishing atmospheric/oceanic interference, weak reception, room scuffle, and ambient noise from the scrawking of malfunctioning equipment or intentional jamming? (Warrant Officer Masterson in 1943 knew of cases in which men had torn their radio sets apart "looking for a particular noise, not recognizing that it was jamming." Airborne radar had similar problems with weather echoes, with "sea return, [which] like the biblical poor, is almost always with us," and with jamming signals that could, when modulated by an inconstant frequency, look like "the normal grass on the screen.") It therefore behooved the Navy to find out as much as it could about such a mysterious offshore presence as the crackling.[259]

Besides, there were plans to lay acoustic mines and develop torpedoes that could be explosively sensitive to underwater noises. A hydrophonic survey of the approaches to New York harbor in March and April 1943 had revealed an underwater cacophony averaging 88 db at 300 Hz in the daytime and 73 db overnight, but this had been ascribed "to ship traffic and other mechanical sources." The underwater source of the noises that interfered with sonar equipment in the Chesapeake Bay in 1942 had been less decisively, and less prosaically, ascribed: likely it was a "chorus" of twelve-inch adult croakers inflating their internal air bladders and grinding their back teeth for two hours each evening from spring to fall, accompanied by cooing toadfish and "gobbling" porpoises. The interference, however, was more constant than any chorus of croakers, and neither croakers nor toadfish nor porpoises made the kind of "frying bacon" sound that most disturbed sonarmen, nor could this consort produce what resembled "a conglomeration... of a pneumatic drill tearing up asphalt, a steamboat whistle, and the cackling of geese." Such offshore noise held more than zoological interest, for 39 percent of ships sunk during the war would be sunk in shallow water. When representatives from the Army, Navy, and National Defense Research Council met in April 1943 to discuss underwater background noise, "One of the principal items discussed, and in which great interest was shown, was that of background noise of biological origin." After being briefed on fish noises by Charles Fish, who with his wife Marie Fish stood for decades at the helm of American research into the sounds of marine life, the conference participants put their faith in a friend of the Fishes to solve the "profound mystery" of the crackling.[260]

Born in a sod-roofed farmhouse on the South Dakota prairie about as far from saltwater as possible for a North American, Martin Johnson was shunted around the upper Midwest while his father tried his luck at homesteading, blacksmithing, cattle-herding, and hostelry. Martin worked as a cowhand, set pins at a bowling alley, and heard as much as felt his way from one new place to another. In his memoirs he remembered Minnesota for "little other than walking to school on a cold winter day and stopping to listen to the humming sound emitted by telephone poles along the road"; the prairie he remembered chiefly for its redwing blackbirds and the songs of meadowlarks, Saskatchewan for its "dazzling frozen brilliance stretching in all directions on a cold sunny day" and "the snow underfoot emitted a squeaky musical tone" while bobsleds drawn by teams of horses set up "a symphony of high-pitched music that could be heard from a great distance over the minus 30 to 40 degree F. air." When the family migrated to the Pacific Northwest, Martin got a job as a "whistle punk," relaying signals from rigging slingers to other loggers by jerking a long wire fastened to the shrill whistle of a donkey engine. Hired on at a coal mine, he was buried underground by an explosion: "Never before had I experienced such profound silence with only an occasional drip of water and the sound of breathing to indicate that my ears were still functioning." Rescued, he became the bass drummer and clarinetist for the Silvana town band and, back in school, for the band of the Pacific Lutheran Academy near Tacoma. In 1918, as a member of the Student Army Training Corps, he learned the physics (and sonics) of explosives, then was inducted as a musician. After the Armistice, he returned to pursue a B.S. and eventually a Ph.D. at the University of Washington and obtained a position as curator at the Puget Sound Biological Station, embracing the littoral that had been all along in his Danish blood.[261]

Martin's focus, when not on Lelia Truth Clutter, a high school teacher whom he married in 1924, was on copepods, the small, silent, shrimp-like crustaceans about which he began publishing in 1930 after a stint as a planktonologist. Though he may never have known that *Cyclops scutifer* and other copepods do tango duets to the fluid-mechanical motions of turbulent waters, it was his wont to sketch the cavortings of marine organisms and to poetize about "Life in the Sea," where "above all we hear the soft notes of Singing Fish as he plucks the strings of the Lyre Crab and croons a liquid melody while the Fiddler Crab fills the coastal air to charm the land-locked listeners." In 1932 he commenced a fifty-two-year association with the Scripps Institution of Oceanography in La Jolla, California. A decade later he was one of three co-authors of a classic volume, *The Oceans: Their Physics, Chemistry and General Biology*, which laid bare so many secrets of the deep that the Navy hoped to keep it from Axis eyes. By March 1943, Johnson was at the Point Loma Naval Radio and Sound Laboratory of the University of California Division of War Research (UCDWR), assigned to explore the reports of a "disabling high frequency underwater crackling noise" that resembled radio static.[262]

Synalpheus Gambarelloides shrimp are, like von Frisch's bees, "eusocial."

A scientism dating (of course?) to the 1960s, "eusocial" refers to "perfectly social" species that live in matrifilial colonies (or communes?) where male drones and sterile workers attend to the needs of the fertile (queens, or Big Mamas) and all daughters share in the care of the young. *Synalpheus* eusociality, discovered in the 1990s, had not gone unsuspected by Johnson, who knew these shrimp to be highly gregarious. Johnson's contribution in 1943–1946 was to prove that this and related species of snapping shrimp could, like angry beehives, make more noise than anyone thought, especially at the higher frequencies that upset sonar. A swarm of riled bees might rise in loudness above 65–70 db; as recently measured, the sound of snapping shrimp might be 189 db within a meter of the snapping; at a distance, according to Johnson's measurements, their noises rose 31–45 db above the ambient sound level in the water. Squillid or mantis shrimp and common brown shrimp, *Crangon*, in their own crowds could also be heard to crackle close up, sizzle from medium range, and hiss from long range with such volume that above 2000 Hz their noise overrode the "sea state" sounds of tides, wind, and waves. Crackle, sizzle, and hiss resulted from the eusocial, synchronic snapping of the ends of their front claws, with an audible striking of a chitinous ridge and a thin jet of water that frightens predators. To sonarmen this "bedlam of crackling," amplified by hydrophones and worst at night when shrimp went on the prowl, sounded like a "continuous fusillade."[263]

"Fusillade" had been the clue. After weeks of hydrophonic frustration, listening to the clapping shut of clam shells, the rheumatic clacking of the many-jointed legs of lobsters, the expansion-contraction of ship timbers, and the clicking of barnacles (*Balanus tintinnabulum*), Johnson happened to hear a group of marines at rifle practice, whose "overall sound produced in rapid gunfire by fifty or sixty marines closely simulated the mysterious underwater crackling." He realized then that "it would not take an overwhelming number of snapping shrimp within hearing range to produce the type of crackling that in nature had an overall sound pressure of 30 or more db above 'sea state 1' water noise." Leading long but reclusive lives (almost two full years of bottom-dwelling), snapping shrimp had hitherto been collected in underwhelming numbers. Agile, pugnacious, and burrowed into coral reefs, oyster reefs, sponges, or seagrass mesas, they were never represented in any laboratory in numbers large enough to replicate the total sound of a colony, but Johnson was no longer doubtful that such minute creatures could be tasked with having disrupted underwater communications in every tropical and temperate harbor and inlet across the globe. Surveys offshore of Miami, Seattle, and San Diego confirmed the oxymoronics of shrimp power, even in the presence of cancer crabs cracking the shells of small clams and mackerel smacking their pharyngeal teeth. None of these was as persistently noisy as the tiny eusocial *Synalpheus*. Their snapping was so "continuously uniform as a community," observed Johnson in a most eusocial analysis, "that it is possible to chart areas of known high noise levels by means of 'isocrackles' or contour lines of approximately equal noise spectrum levels." Tireless, they and fellow "pistol" shrimp, all twenty-seven genera, as well as *Crangon dentipes* whom Johnson renamed *clamator*, would only cease their snapping in the aftermath of such extreme underwater disturbances as the shockwaves from the mushroom cloud at Bikini in 1946.[264]

Johnson wrote up some of his research for the *Biological Bulletin* in 1947, once his findings on snapping shrimp had been (mostly) declassified: no longer was it tenable to regard the sea "as a realm of silence," for "great stretches of numerous coastal areas are exceedingly noisy, periodically or perpetually, due to marine animals." Reviewing all of the prior literature and her own wartime work, Marie Poland Fish in 1952 identified forty-two families of fish as potential noisemakers in the Pacific and twenty-six in the North Atlantic, up from a total of fourteen that had been calendared for the Navy in 1942. Yet evidence that fish made *meaningful* sounds was either anecdotal or (like Fish's recordings) still classified, so a young biologist, William Tavolga, stuffed a waterproofed microphone into a condom and lowered it into a tank of courting frill-fin goby to demonstrate that the "little grunting sound pulses synchronized perfectly with the male's [sexual] head shakes." To prove that many fish, some with such give-away names as schoolmaster and roughneck grunt, made meaningful sounds *and* could hear—even ultrasonically, or infrasonically—would take decades of further work by the Fishes, Tavolga, and others destined from baptism for marine biology, like Michael Salmon and James F. Fish.[265]

Another source of underwater noise was even less well known and all the more closely guarded. Not the first to discover it, Martin Johnson would again be the one best situated to explain it. In 1942–1943, the Reverberation Group of the UCDWR was finding that the limitations of sonar came more from problems of refraction than absorption, in particular from sound scattering at surprising depths, which created ghost seamounts and doppelgänger subs. Sound scattering on land and in buildings had been studied at MIT before the war, but the Deep Scattering Layer (DSL) in the ocean had an unusual diurnal pattern, which by mid-1945 Johnson had shown to be of biological origin: shoals of plankton (copepods, prawns) moving upward in the evening to feed on rich strata of phytoplankton near the ocean surface, retreating in the morning light to the sheltering darkness of 175–225 fathoms. Marine biologists, ocean chemists, and physicists were uncertain whether the sound scattering was caused solely by the sheer mass of DSL plankton, or also by schools of hungry squid and jellyfish that followed the plankton, or by air bubbles in the quintillions created by quadrillions of creatures rising and then descending at ten to twelve feet per minute. Whatever the cause, submariners and anti-submariners must now take into account the daily movements of this biological layer, the better to detect enemies or to hide beneath the false bottom that pinged back in encounters with a DSL.[266]

Aside from their cavitation, the plankton were quiet. It was their impact on sonar that was, not so metaphorically, loud, and it was this that had to be kept hush-hush, lest the Japanese, the Germans, and now the Kremlin use it to tactical advantage. For this reason, Vern Knudsen was refused clearance to attend a major conference on Subsurface Warfare in 1947. Although he had helped the Army with gun-ranging for field artillery and advised the Navy on underwater acoustics, he was also a designer of soundstages and had received notice in the pages of the *Hollywood Quarterly*, which the FBI in a hot flash of paranoia had identified as the organ of a Communist front organization. Noise was nearly as dimorphic as Communism itself, at once a countervailing voice and a subversive whisper, dagger and cloak. "Noise—friend and enemy—it can bring us into contact with a submarine we seek or it can betray our location to a hunter," began a Navy training film on "Submarine Noise Reduction" in 1955. Submariners, like solid citizens, had to become aware of all that might betray them and/or distract the sonar operators: rough-running diesel engines, rocking machinery, locker doors awry, orders shouted over a squawkbox, cans rattling in the galley. "The life and effectiveness of our subs depend on noise reduction to hear and keep from being heard." The political riptide of investigations by the House Un-American Activities Committee and invective from a former marine, Senator Joseph McCarthy, and a Navy man, Congressman Richard Nixon, would sequester much of what was known about what was now a "very, very noisy" ocean—so *fundamentally* noisy, in both statistical and acoustic senses, that Carl Eckart, a physicist studying the fluctuation of sound in the sea, would compare it to a radio station fading in and out.[267]

CBS-Radio listeners to a January 1947, University of California program on "Sounds in the Sea" heard some of what Johnson and colleagues had been hearing—toadfish, croakers, snapping shrimp—so all had not faded into official secrets. They also learned that scientists had been unable to identify such other noises in the sea as "a mewing which sounds like a lost cat" and, 200 miles offshore in water 2,000 fathoms deep, "an awesome moaning . . . that might be a good sound effect for a 'whodunit.'" Whodunit? was a question of cultural moment when atomic espionage was in the headlines and American and Soviet submarines were moving stealthily into polar waters loud with the low-frequency grinding, bumping, shearing, and pressure-ridging at ice floe boundaries. The Associated Press reported in 1947 that ocean noise was "Hindering Work on War Secrets," citing none other than Knudsen, who had told an audience at a meeting of the American Physical Society that "The ocean depths sometimes are a bedlam of noise like the busiest streets of a big city." So the Navy kept under wraps the Harvard Underwater Sound Laboratory's wartime work on an Electronic Aural Responder, or EAR, which overcame the difficulties experienced by flesh-and-blood sonarmen trying to discern softer signals through louder ambient ocean noise. And when John Dove Isaacs, III, a Scripps oceanographer, sought to put a "Brobdingnagophone" into the water in 1949–1950, he was only half-joking that it too ought to be classified. Not that a huge hydrophone would uncover a subversive race of Swiftian giants: its size would merely make it possible to pick up the very low frequencies of waves traveling below the surface, of pounding surf echoing through miles of ocean, and of "water drag" over rocks on the seabed. Collecting such data would help define the spectrum of ambient sound in the ocean and sharpen the calculation of acoustic "sea states" against whose background noise sonarmen had to listen for other, manmade, noises. Since the eighty-foot-tall Brobdingnags in *Gulliver's Travels* spoke in "very deep rumbles" that were "a natural result of the great size of their so-called 'vocal chords' and their head cavities," and since such beings, joked Isaacs, would require a "prohibitive amount of muscle" to stay erect on land, he and his friends supposed that Swift had to have been referring "in highly imaginative fashion" to a race that lived "deep in the ocean, where their weight was largely nullified by the water surrounding." Yet, given that the listening range of underwater ears had been twelve miles in the best of circumstances during the Second World War, there would surely be security issues should an EAR or a Brobdingnagophone or an array of hydrophones linked to (new) tape recorders reveal behemoths singing at low frequencies through unsuspected channels thousands of miles long and very clear.[268]

Killer whales, pilot whales, and other large dolphins (family *Delphinidae*) and porpoises (family *Phocoenidae*) had long been heard to make a variety of noises. Whaling captains told Herbert Aldrich in 1887 that they could hunt whales by their "singing," which Aldrich took for "a sophomoric joke, slyly intended for me to bite at"; he found that they believed it. Having chronicled the decline of the industry from the prospect of the widows' walks of New Bedford, Aldrich spent eight months in 1887–1888 with American whalers, half of whose small fleet of forty-one vessels came from that one Massachusetts seaport. Now that more than half a million sperm whales and right whales had been killed in pursuit of oil for lamps, candles, soap, and lubricants (and ambergris for perfume), crews could no longer expect a season's haul to match the 373,450 barrels of oil taken by 278 vessels at the peak of whaling in 1852. Perhaps the relative quiet and economic straits of a reduced fleet allowed for, or demanded, more intent listening than was possible in the time of Lt. Charles Scammon, whose voyages (1853–1863) involved the hunting of right whales, finbacks, bowheads, and California greys with "bomb-guns sounding like muskets, cries in Portuguese, Kanaka, English, whales breaching and splashing" in turmoil, and in a crowded fog the "blowing of horns, ringing of bells, firing of guns, pounding on empty caskets to indicate the ship's position and avoid collision." But other seamen knew that whales made "a fearful noise," and "you could hear them a long way off," as one oceangoer noted in a shipboard diary of 1876. In any case, captains were certain, wrote Aldrich, "that humpback whales, blackfish, devil-fish, and other species of whales sing, and that walruses and seals bark under water, and it is believed that all animals having lungs and living in the water, as these do, have their own peculiar cry, or as whalemen express it, 'song.'" The first cry that Aldrich himself heard was the "deep, heavy, agonizing groan" of a harpooned whale. Next he heard the cry of a bowhead, or Arctic Right whale, "something like the hoo-oo-oo of the hoot-owl, though longer drawn out, and more of a humming sound than a hoot," beginning in F "and rising to G A B even C." Humpback whales held to "the E of a violin."[269]

Brobdingnagophones, or suchlike, received the songs in a different key, as did *The Baby Whale, Sharp Ears*, who "could hear remarkably well, even though you" and your children, reading this book together in 1938, "would not have been able to see any ears if you had been there to look for them"—indeed, a baby whale could hear "some sounds that you could not hear." The preponderance of sounds made by larger whales were in the very-low (infrasonic) frequency range, below 20–25 Hz, where hydrophones and their cables created considerable flow-noise and "pseudo-noise," where slight shifts in temperature could have drastic effects on reception, and where distant shipping and offshore drilling could register magisterially. It was one thing to speculate (as did Arthur McBride in the late 1940s) about dolphin sonar, or to prove (as did Kellogg and Kohler in 1952) that dolphins could hear well into the ultrasonic range, or to suggest (as did William Schevill in 1956) that they use echo-location, or to consider the prospects for a dolphin language (as did Kenneth S. Norris in the 1960s). It was quite another to show that the infrasonics of whales had a private, unimpeded channel through the world's oceans.[270]

Maurice Ewing in 1944–1948 had been the first to identify and explain SOFAR (SOund Fixing And Ranging) channels in the ocean, not because he heard whales but because, like other geophysicists prospecting for oil or faultlines, he had become accustomed to assaying depths through the seismic analysis of the travel-time of shockwaves from explosions. At sea during the war, Navy hydrophones detected the "signals" of a four-pound bomb he had detonated 2,300 miles away at 660 fathoms; the detonation of a six-pound bomb was picked up 3,100 miles away. Extrapolating from this data, he and his student John Worzel hypothesized that "ranges of up to 10,000 miles are possible with this size charge" and that with sufficient sound power similar transmission distances were achievable due to the "natural sound channel which exists in the ocean," formed by temperature gradients in ocean layers and best exploitable by the long waves of low frequencies. Since

(a) sound travels more slowly in colder water but more quickly under greater pressure, and
(b) sound "rays" bend toward regions of slower speed, and
(c) ocean temperatures decrease to near-freezing at a depth of 700 fathoms in the Atlantic and 500 fathoms in the Pacific, slowing all sound, but
(d) farther below, sound regains speed under greater pressure,
(e) there should be in all oceans
(*f*) an audio channel at the low-speed nadir, or "sound axis," where refracted rays of sound "can travel to any distance without contact with the surface or bottom of the ocean."

Any distance. A small investment in sonar equipment and chronometers for a series of monitoring stations fixed at intervals along a great circle uninterrupted by islands, sea mounts, or continents would enable the Navy to do an "acoustical sweep...at a rate of several thousand square miles per hour." So while Jacques Cousteau was picturing in prose and film the elemental quietness of the deep sea, where fish may creak or croak but react to audible sounds far less than to water "trembling with emergency," the U.S. naval command was initiating Project Jezebel, plotting out a secret long-range submarine surveillance system whose listening arrays would be aligned according to the new research on ocean sound. And while Comrade Nikita Khruschev was announcing in 1955 that "we will abandon Communism when the shrimp learns to whistle," and Elvis was gyrating over waves of adolescent girls, and kids were now dreaming "of being 'frogmen' as they used to dream of being pilots," the U.S. Navy had moved on to Project Caesar, the secret installation of a Sound Surveillance System (SOSUS), a global network of multiple microphones anchored to the sea floor to listen through oceanic sound channels. (The Navy already knew that surface platforms for sonar would be "of little value because of local noises which such structures produce because of their attraction to fish and marine growth.") And while filmgoers followed the acoustic dramas of *The Enemy Below* (1957) and *Run Silent, Run Deep* (1958), Schevill, the Fishes, and others began to suspect that some of the sounds coming through these channels were the songs of whales.[271]

Pulses of unknown origin, at 20 Hz (below which humans feel rather than hear sound) had been tape-recorded off the coast of Bermuda in 1951 and classified as "Secret." In 1963 Schevill and two associates from the Woods Hole Oceanographic Institute identified these pulses, whose sound-power was almost fantastic (160–86 db), with one kind of whale. Thirty-two years before, as a curator of invertebrate paleontology at Harvard's

Museum of Comparative Zoology, Schevill had signed onto an expedition to the Australian desert, where he uncovered the bones of a large carnivorous swimming "lizard" of the early Cretacean. It took the Museum twenty-six years to tease out the kronosaurus from its bed of rock and mount its skeleton for display, by which time Schevill and his wife Barbara Lawrence, a mammologist, had moved on from silent fossils and (Barbara's) howler monkeys to groaning whales and squeaking porpoises. Despite wartime encounters with biological noises in the deep sea well beyond the shooting range of pistol shrimp, there were as yet no recordings of underwater noises made unmistakably by one or another kind of whales. The U.S. Navy Electronics Laboratory in 1948 did have recordings of whales of some sort, but even Martin Johnson was only guessing that the noises came from humpbacks, so the Schevills hied to the St. Lawrence River that year with a hydrophonic system to record pods of Beluga whales, known to sailors as "sea canaries" for the high-pitched chirps they made at the surface and, presumably, below. Due to shipping traffic and strong tides, the St. Lawrence was too noisy for recording; on Quebec's quieter Saguenay River, the Schevills made the first underwater recordings of noises assuredly cetacean and specifically Beluga. These they described as "high-pitched resonant whistles and squeals varied with ticking and clucking sounds slightly reminiscent of a string orchestra tuning up, as well as mewing and occasional chirps. Some of the sounds were bell-like, and a few rather resembled an echo-sounder." The whales could also be heard trilling at the surface; in fact, they were never seen without also being heard. "This loquaciousness contrasts markedly with most terrestrial herd mammals," wrote William and Barbara, "and compares with such chatterboxes as monkeys and man." By 1953 the Schevills had recorded the chattering of other toothed (Odontocete) whales and had tested the hearing of bottlenosed dolphins, who were sensitive to sounds below 20 Hz and all the way up to 120,000 Hz, far exceeding the capacity of human ears. They soon reported noises like the hammering of an engine and "a grating sort of groan, very low in pitch, which reminded some of a rusty hinge creaking," from another Odontocete, the sperm whale, and sets of sharp clicks so spectrographically distinct that William Schevill's electronics accomplice, William Watkins, could identify them as the sound signatures of individual whales. "Deep, loud, throaty sounds, noises reminiscent of cat fights and squeals" were recorded from right (baleen) whales in 1956, and the 20 Hz pulses identified by Schevill and Watkins in 1963 came from another baleen (Mysticete), the finback. Both suborders of cetaceans were now contributors to the soundfulness of ocean deeps that had for so long been thought silent.[272]

Heartbeats, couldn't that low doublet rhythm he and Watkins recorded be the resonant heartbeats of whales? No. Careful listening since the early 1950s had linked odontocete sounds, especially from beaked whales (ziphiids), with specific occasions of feeding, courtship, and distress. In the older literature, ziphiids were said "to have roared like a sea lion (1912), bawled like a calf (1898), bellowed like a bull (1870), or [emitted a] 'sound like the whistle of a steamer' (1926)," all of which were treated as kinds of noise; in the newer literature they were also heard to emit rapid clicks that sounded to sonarmen like barks and yelps, but these were treated as communications as yet indecipherable. It made as much scientific sense in 1962 to deny whales their many vocalizations as to believe the sea still silent, and as much cultural sense to deny the meaningfulness of whale sounds (even a strange "boing" or a sound called "the carpenter") as to gag free speech or block the publication of an unexpurgated *Lady Chatterley's Lover*. Still, to speak of "whale music," as Schevill began to do in 1962, was a radical move, far more radical than that of a biologist at the Navy Electronics Lab in San Diego who had told a reporter in 1958 that "We have a crazy collection of recordings, which sometimes are amusing, other times sad songs of the sea," like the California grey whale, which "sounds something like a child hammering an empty cup on a table," or the bull-horn groaning of a humpback, or the mewing of a pilot whale, which resembled a wood warbler. To speak of whale *music* implied that there was a purposeful rhythm and order to what other marine biologists continued to describe as "groans" and "moanings."[273]

Honored as "the most powerful sustained utterances" of any living being, these low-frequency sounds could also be heard as dirge. In 1961–1962 the international whaling industry took 66,000 whales in the Antarctic alone and by 1967 had reduced the population of blue whales in southern waters by 99 percent, from 250,000 to less than 2,500. Roger Payne and Scott McVay, marine biologists invested equally in the study of whales and in saving them from extinction, found in 1967 that the long sequences of 20–24 kHz harmonics produced by male humpbacks during breeding season were no less musical than birdsong, with repeated phrases, themes, and variations strung together without pause, "so that a long singing session is an exuberant, uninterrupted river of sound that can flow on for twenty-four hours or longer," with a pace that appeared to be "set by the slow rhythm of ocean swells—the rhythm of the sea." Listening through hydrophones at night to the swells offshore of Bermuda and the whales below, Payne likened the experience to "walking into a dark cave, dropping your flashlight, and hearing wave after wave of echoes cascading back from the darkness beyond.... The cave has spoken to you. That's what whales do; they give the ocean its voice, and the voice they give it is ethereal and unearthly."[274]

Unearthly, and intelligent. Payne's cave was the auricular obverse of Plato's, whose narrative of the unreliable optics of troglodytes moves upward from a society of shadows toward a psychically isolated image of the beautiful and true. In Payne's darkness, lamps no longer flicker but ears are quickened and the cave-dwellers engage with a new, remarkable voice from below, not the voice of a Jonah in the whale but of whales now our Jonahs. Payne and McVay published their first report in *Science* in 1971, just as the phrase "noise pollution" was entering common usage, as Congress was enacting a panoply of laws for environmental protection, and as the Citizens League Against the Sonic Boom was grounding Boeing's SuperSonic Transport. Floated into the audisphere during an era of ecological sensitivity that had been heralded ten years earlier by that "biographer of the sea," Rachel Carson, author of *Under the Sea-Wind* (1941), *The Sea Around Us* (1952), *The Edge of the Sea* (1955), and most famously, *Silent Spring* (1961), whale song so intimated a warmth of outreach that musicians, poets, artists, and filmmakers were drawn to affirm the breadth of cetacean intelligence as part of a larger circle of interdependency. The Friends of the Earth, inaugurating Project Jonah, campaigned for an international moratorium on whaling, arguing that "turning magnificent, intelligent, ecologically critical animals into shoe polish, car wax, margarine, and lubricating oil may be the ultimate nonsense of the modern world." The National Audubon Society aired a series of radio spots in 1974, one of which asked, "Have you ever heard a whale sing? Listen a moment. [Sounds of whales.] Is it a love song? Or a lament? It could be both. The whales' love life isn't going so well these days. Years of commercial hunting have left so few that it's hard even to find a mate, let alone win one. Whales have the biggest hearts in the world. Now it's going to take human love to keep them thumping." In this context, whale songs would be taken up across the cultural spectrum—by composers such as Alan Hovhaness (*And God Created Great Whales*, 1970) and George Crumb (*Vox Balaenae, Voice of the Whale,* 1971); by dancers and choreographers such as Merce Cunningham (*Ocean*, 1994) and Mary Seidman (*The Whales' Song*, 2002); by filmmakers (*Star Trek IV: The Voyage Home*, 1986), New Age mantra-makers (Don V. Lax and Melissa Proulx, *Ancient Ocean Harmonies*, 2005), and singers of children's songs, authors of children's books (Michellet and Périllat, *Le Bal des baleines*, 2008; Tony Johnston, *Whale Song*, 1987; Scott Schuch, *A Symphony of Whales*, 1999)—as splendid evidence of ancient, enduring, nonhuman wisdom.[275]

Ancient, enduring, and responsive: after years of listening, Payne and his wife Katharine discovered that whales change their songs, creatively, like musicians who work over traditional melodic material to produce their own variations. In the millennial year 2000, Michael J. Noad, et al. would go so far as to identify a "Cultural Revolution" in the songs of humpback whales off Australia's east coast, where the musical changes were deep and sudden. Once whales had been proven to sing in fomulaic stanzas, to spend years rearing their young, and to live as long as a middle-class

Westerner, they were attributed culture as well as intelligence, and by virtue of their millions of years in the fossil record, were granted a more intimate relationship with the earth than humankind. Changes in whale songs could then be interpreted as operatic commentaries on our mutual fate, getting louder, lower, and more insistent in response to acoustic and biochemical pollutions. If their "thousand-mile" songs were meant to sustain pod integrity on long migrations, as some research suggested, then any abbreviation of their songs might signal a broader disaggregation of the system some were calling Gaia, a homeostatically driven ecological community of imbedded mechanisms for sustaining life on the planet. John Cage was back on the scene here with his *Litany for the Whale, for 2 equal voices* (1980), both performers standing with their backs to the audience, chanting by chance any one of the fixed-pitch letters of the five-letter word "whale" in a blind exchange that might be heard as a call-and-response between two whales or between vanishing cetaceans and anxious *Homo sapiens*. If we did not yet comprehend whale lyrics, we could call out the letters of their generic name, garbled as that might be.[276]

Our fault, that garbling. During and immediately after the Second World War, the noisiness of the deeps had been approached as a problem of dumbly biological interference with national defense and maritime commerce. When, in pursuit of the origins and course of these noises, oceanographers discovered clear channels, Cold War navies were grateful, secretive, possessive, and not a little annoyed that other life forms might also be depending on them. Stories in the press about "mysterious loud noises in the ocean, which sound like a wrecking crew tearing down a building," alluded to anonymous but obstreperous organisms which, thousands of feet down, were indifferently hammering, grinding, bumping, wailing, gasping, whistling, sighing, and snoring. The director of the Tuna Research Foundation in 1953 hoped to use recordings rather than anchovies as a lure for larger catches, since scientists had learned that tuna were "happiest while making a noise." By the year of Cage's *Litany for the Whale*, however, scientific and cultural currents were reversing: now it was us, *Homo ridens*, Aristotle's laughing animal, who stood self-accused of messing with the lines of communication, and with the very lives, of whales. "In the deep places" (wrote the most anthropoetic of fantasists, Ursula K. LeGuin) and rising with a submerged continent, could be heard a "yearning music from far away in the darkness, calling not to us... the voices of the great souls, the great lives, the lonely ones, the voyagers." We were the noisemakers, they the caretakers and musicians.[277]

Nuclear submarines nestling ever-deeper inside Pacific trenches... cavitating icebreakers in the Arctic shifting the auditory thresholds of Beluga whales... active sonar and SOFAR pinging and thumping ever-more intensely: ecologists began to suspect that we ourselves could be the cause of mysterious beachings, cetacean "suicides." Such a *mea culpa* was as audiconic as it was ironic: marine biologists had been drawn to the study of whale life in part by the surface whistlings, squawks, "nervous plaintive cries," and "anguished, mouse-like squeaks" made by dazed, harpooned, wounded, or dying whales, and in part by the appeal of their underwater songs. Now, listening for individuated cries and ocean "dialects" and playing back recordings to pods of whales underwater, they were exerting the power of acoustic apparatus that just might be held responsible for pushing whale populations toward extinction. As navies barged in upon ocean sound channels to track the whales' closest unearthly kin, submarines, and as geologists deployed ever-more focused explosives in their prospecting for oil beneath seabeds, and as supertankers heaved through the waves, discharging ever-more toxic waste, we were poisoning the ears of our evolutionary elders, who fifty million years ago waded from coastal shallows back into the deeps of the Ocean Tethys and left off being what we humans were yet to become, amphibian. (And "If man did indeed come from the sea originally, as many claim," said the Sony Corporation of America around 1960, "perhaps that is where he must find solutions to the problems he has created for himself on land.") As fishing fleets used ever-more sophisticated echo-sounders that drove shoals of fish into vast nets, and "ringers," "squeakers," and "seal bombs" at 190–210 db to keep pinnipeds and porpoises from the fishing grounds, it had to occur to others than Greenpeace activists that the human proficiency at making noise, at car-alarming cities and snowmobiling through wilderness, could be deadliest to those who listened in the ocean darkness for the calls and songs of their conspecifics.[278]

Eschatology was all the talk at the approach of a long-anticipated millennial turn, and apocalyptic rhetoric inflamed a surprisingly vehement debate over ATOC, the Acoustic Thermometry of Ocean Climate, during the 1990s. On one side sailed the scientists of the Scripps Institution of Oceanography, who had been at the forefront of analyses of the mounting evidence for global warming and who hoped to study changes in ocean temperatures as key to an understanding of the nature and extent of that warming. Between 1981 and 1990 they had been conducting tests, sponsored by the Pentagon, on the feasibility of accurate basin-wide measurements using SOFAR channels and SOSUS transceivers to confirm the pace, sources, and degree of oceanic (and atmospheric) warming. On the other side sallied biologists and ecologists suspicious of the Institution's military ties and of the harm that might be done to marine mammals by the transmission of very strong low-frequency impulses across ocean basins. Scholars have framed the controversy in terms of the nature of proof demanded in public and scientific debates (when is enough enough? what can fallible humans mean by "incontrovertible"?) and contrary notions of ecological protection (defending us from our worst natures? or the "natural world" from us?). It was also, and poignantly, a culturally conflicted battle over the everyhowness of noise.[279]

ATOC would have used the SOSUS listening arrays as stations for measuring the time taken by powerful 60–90 Hz impulses to pass through the ocean at various depths. Since temperature, far more than salinity or pressure or crosscurrents, affects the speed of sound through ocean water, monitoring the speed of sound-impulses transmitted between fixed SOSUS arrays through SOFAR channels would establish baselines and shifts in the amount of heat being absorbed and released by the saltwaters that cover two-thirds of the earth and swirled the planet blue in the iconic earthrise photograph taken from Apollo 8. The Scripps scientists marveled at their own compassion and ingenuity: take what had been a covert defense network, convert it at little cost to global climate prognosis, and insteading of hunting for Tom Clancy's "Red October" (the ultraquiet, ultra-mean Russian submarine of a 1984 novel and 1990 film), save the world from itself—that is, from us. In this schema, loud sounds, at levels of 195–221 db (over 250 million watts in the air but just 250 watts of acoustic power in water), would be invaluably informative, for the faster they travelled, the warmer the ocean basins and the warmer the earth as a whole.[280]

Dolphins and whales lay in the way. Those who had been listening for them were now listening on their behalf, and they heard the ATOC transmissions as noise, as "sound blasts." These could disrupt dolphin-to-dolphin, whale-to-whale, and pod-to-pod communications at exactly the frequencies they appeared to prefer; the blasts might also endanger their survival, since cetologists were now able to demonstrate that whales as well as dolphins relied on echo-location for navigating and hunting. "Whales are acoustic animals," wrote Katy Payne in a *Los Angeles Times* opinion piece in 1994. "When their acoustic environment deteriorates, so do all the basic mechanisms of their sociality—their ability to meet one another; to establish, maintain and adjust their social relationships that support breeding; to locate food; to care for their young, and to navigate." If dolphins had sensitive ears, "coupled with a massive auditory central nervous systems," whales were known to have the widest auditory range of any mammal, from below 20 Hz to above 20 kHz. Found at the farthest reaches of all oceans, and at astonishing depths, with acoustic powers infrasonic and ultrasonic, they were, one could say, the ears of the earth. Ignorance of how they held dear each part of their auditory range was no excuse for blasting them with sound at some narrow band of frequencies, hoping that they would be forebearing or, like audiences upset with bursts of radio static, switch to another station. Given that studies were showing how high-intensity low-frequency vibrations could induce bubble growth in tissues and produce the bends in human deep-sea divers, there was the added risk that ATOC transmissions might be physiologically cruel to non-human divers. Sending twenty-minute-long blasts rumbling through the SOFAR channels used by whales and measuring the sound-speed six times a day for a decade or more, as Scripps scientists proposed, seemed a bad idea. "The ocean is a noisy place at best," wrote Payne, "but it is vastly noisier now than in all of the animals' previous evolutionary history. Windows of relative quiet in which communication can occur are becoming a scarce and perhaps a limiting resource."[281]

"Forget the notion of the silent deep," wrote a reporter on the science beat. "The ocean of today is a noisy place filled with the sound of human activity—an aquatic wilderness that is becoming urbanized," astir with 200,000-ton supertankers so loud (232 db) that they can be heard underwater "a full day before they appear on the horizon." Along continental shelves the 240 db sound-pressures of seismic air guns used in oil exploration could pound across 500 miles of ocean; louder yet was the dynamiting preliminary to the construction of piers (296 db, sufficient to crush a whale's earbones). What was a little added sound, in this "phenomenally noisy environment," when transmitted briefly at controlled frequencies at less than a thousandth of 1 percent of the sound power of ambient ocean noise? Why be so protective of behemoths who had weathered eons of underwater volcanic eruptions, or of dolphins, elephant seals, and sperm whales who often came up to nuzzle the SOSUS speakers? So far as Scripps Institution investigators were concerned, and they insisted in all venues that they were concerned, the ATOC signals would "sound more like natural seismic rumbles." At worst, those few curious creatures who stuck around within 500 feet of a transmitter during its twenty-minute set might experience a temporary auditory threshold shift similar to that of teens who cannot make out certain tones for a few hours after a rock concert.[282]

Imagining what was ruckus to a whale, or entertainment, or information, or companionship, was not much easier after listening to reel upon reel of recorded clicks and stanzaic phonations than imagining how a bat encountered the world. It was still, as Donald Griffin would entitle his autobiography, *Listening in the Dark*. Hints were to be had from anthropologists like Charles R. Adams, studying "aurality and consciousness" among the Basotho, for whom "the experience of sounding and hearing is... the principle mode for the production of significant interpretations of the circumstances of living," and philosopher-poets like David Abram, writing about *Perception and Language in a More-Than-Human World*, where sense organs ensure "that my body is a sort of open circuit that completes itself only in things, in others, in the encompassing earth," but Scripps biologists stuck to their hydrophones and computers. The divide between proponents and opponents of ATOC was rather a question of phenomenology than of political or moral philosophy; both groups knew themselves to be environmentalists, both were desperately trying to think ahead, both took advantage of finely engineered apparatus to produce their data, both hoped to make use of "acoustic daylight" (the ambient noise in the ocean) to map the darkest parts of the deeps, both had little use for national borders or ethnic rivalries, and both were engaged in an altruism of the noblest kind, working on behalf of beings who knew them not. ATOC proponents, however, pressed for a sense of noise as conveyance, its opponents for noise as disturbance and obstruction. Marine mammals, wrote five marine biologists in 1995, learn much about their surroundings from natural sounds such as surf and shoal noise, ice noise, "seltzer" noise (melting glaciers, which introduce tiny explosive air bubbles into the water), and the noises of predators large and small, including snapping shrimp. But what were they learning from supertankers, seismic guns, drilling platforms, low-flying rescue helicopters or, for that matter, from underwater Greenpeace broadcasts of amplified killer whale calls to scare off fur seals from the path of hunting ships? And what were whales *not* learning about their environment because of us?[283]

Twenty-five years before, in 1970, a National Research Council committee, musing on the "Present and Future Civil Uses of Underwater Sound," had predicted noisier deeps with more submarines, scuba divers, underwater oil fields, distress signals from research submersibles, and more "hazards," such as sonobuoys and trawlers. Whales had no part in their musings. In 1993, 1999, and again in 2004, the National Research Council charged committees of experts to listen as if marine mammals for ATOC transmissions as well as for other noises all-too-successfully anticipated in 1970. They found changes in migration paths and coastal approaches but no "obvious catastrophic short-term effects," no "mass strandings" due to specific transmissions. Their conclusions would have been all-too-predictable to any Anti-Noisite, for whom evidences of immediate catastrophic effects are always in short supply and mocked when adduced, and for whom proofs of the cumulative effects of noise are always hounded by the cumulative effects of aging and anxiety. Although the Marine Mammal Protection Act of 1972 had been amended in 1994 to include two levels of (non-catastrophic) harassment, with the broader Level B defined as "any act of pursuit, torment, or annoyance which has the potential to disturb a marine mammal or marine mammal stock in the wild by causing disruption of behavioral patterns including, but not limited to, migration, breathing, nursing, breeding, feeding, or sheltering," a NRC report of 2000 proposed that mammal-protecting authorities discriminate between the transient, biologically insignificant effects of sound-blasts, and their more significant, or long-lasting, effects. Given the difficulties of monitoring the individual health and generational robustness of whales, this was tantamount to a free pass for ATOC, but public and Congressional opinion had already been turned against the blasting by whale songs, whale tales, whale movies (*Free Willy*, 1993), and the first evidence of "whale falls"—the sinking of whale carcasses to the sea floor in a dying-off all the more dramatic for being silent.[284]

Extinction was a wrought topic at millennium's end. Newspapers, magazines, and scientific journals published graphs of the disappearance of species at a rate so accelerated that it raised the specter of a global silence: one species dying off every thirty minutes, or was it every twenty-five, or every twenty? This sense of urgency about saving whales, owls, wolves, and other endangered species, and preserving their ecosystems was transmuted by the U.S. Navy into an urgency about defending American grasslands (and their catacombed military communication centers) against sleek behemoths rising from the deeps to launch missiles with multiple warheads. If it was only a question of time before the biologically significant effects of noise on whales could be irrefutably demonstrated, it was time that the Navy felt it did not have when it began in 1996 to implement a LOw Frequency Active sonaR program that used intense low-frequency pulses to listen for echoes from devious Soviet submarines. So, while satellite altimetry, combined with acoustic tomography conducted by other nations, revealed that ocean temperature variability was too great and the contributing factors too complex to confirm a scientific consensus that the earth was warming at an alarming rate, the Navy was back in the business of subsurface noisemaking. In 2000, immediately after a sonar training exercise in the Caribbean, with LOFAR as intense as the blast of two thousand jet engines (claimed the National Resources Defense Council), thirteen whales of four species stranded themselves on the sands of the Bahamas, bleeding around the brain and ears. Stepped-up training and LOFAR exercises in the interest of national security, as the Navy itself projected in 2008, would "harm marine mammals more than ten million times during the next five years off the U.S. coast alone." We fearsome humans—not dolphins, not croakers, not snapping shrimp—were now and everyhow the ambient noise of the seven seas.[285]

10. The Fetal Sac

> Water is home
> As the body knows it:
> Without encumbrances.

So wrote Robert Gibb in his 1976 poem, "Cetacea." For that order of beings,

> The skin only
> Carries its belongings,
> And the ear begins
> In the dark aquatics
> To hear the sound of light.

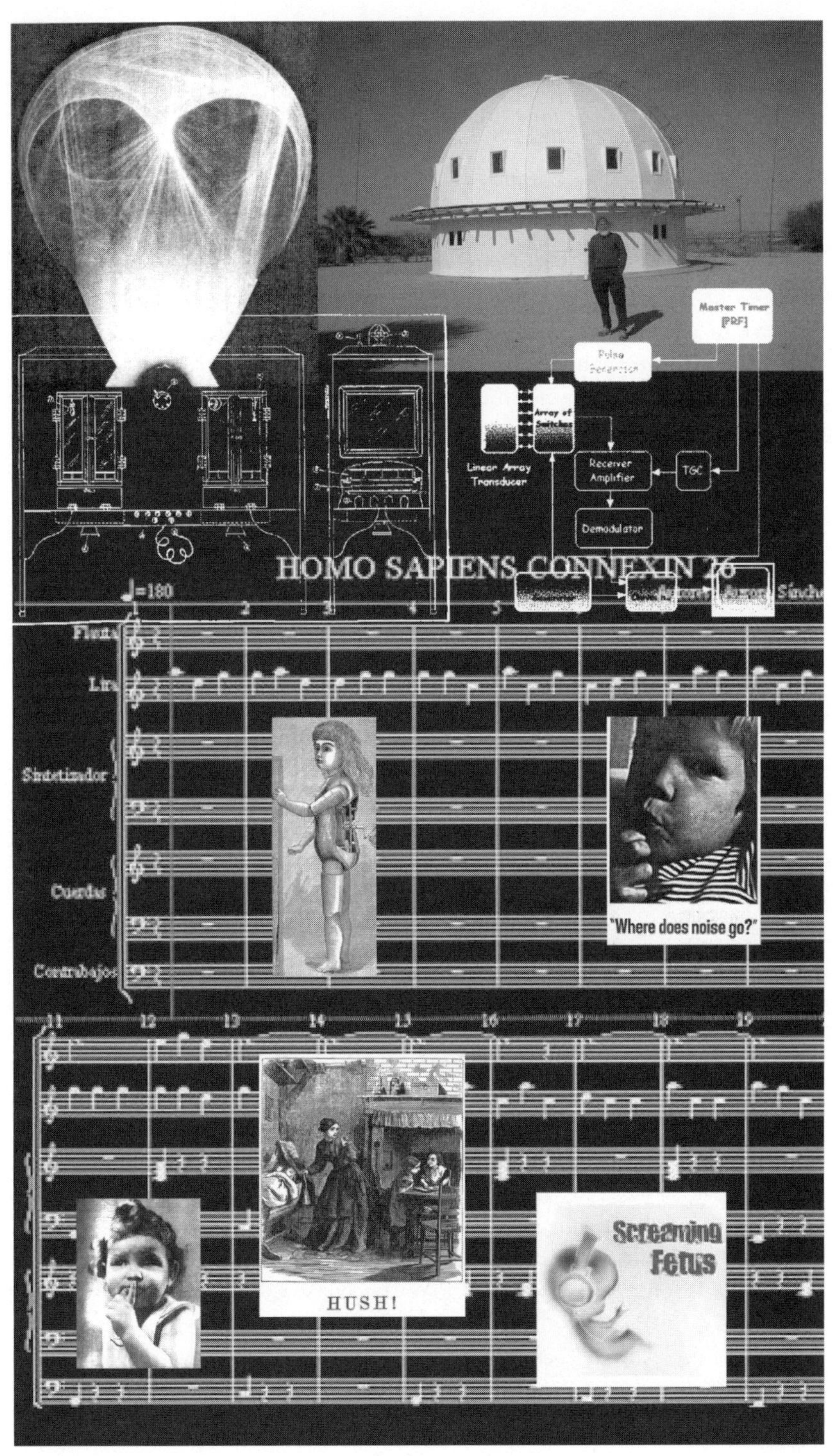

SOUNDPLATE 27

For them, and for us: as Gibb wrote in "Song for the Humpback Whale,"

> The years you hold are a
> History we come springing from
> To find this as our legacy:
> Your lives are our lines in sleep
> Swimming back to something deep.[286]

The homology of whale to fetus was mythopoetically tight, if inside out. Where once humans imagined being swallowed (and saved) by leviathans, now we in our billions were swallowing them, and being reminded of our obligations to keep them singing. Where once the deeps were dark and silent, home to nothing *not* monstrous, now with sonar and its offspring ultrasound, whale and fetus swam in saltwaters almost transparent and acoustically alive. Where once neither fish nor fetus had been supposed to have the wherewithal to listen and so glided in a world whose calm was an aspect of its soundlessness, now they heard us too well, pings and propellers, heartbeat and heartburn, talk and ATOC. Evolutionary harbinger and embryonic promise, whale and fetus were audiconically twinned: "Why Do Whales and Children Sing?" asked the sound artist David Dunn. In 1999 his question was rhetorical, a title not a riddle, and his album, *A Guide to Listening in Nature*, was issued by EarthEar, a company devoted to disseminating the sounds of deep ecology, which begins as it must in the "dark aquatics" of the womb.

Bathyscaphs—small, cramped, and tethered to a mother ship—had many of the qualia of amnionic sacs, and like humans in the womb, their occupants—one or two, at most three—had usually done better with ears than eyes. Since the first prototype of 1928, explorers in deep sea submersibles had learned more about their environs from hydrophones than from floodlights or mechanical tentacles. This could still be said of many of the later dives of marine biologists in the Woods Hole submersible *Alvin* (born 1964 and chronicled by Victoria Kaharl in/as *Water Baby*), although they would use high-resolution imaging sonar. When the anthropologist Stefan Helmreich joined a geochemist for a journey in *Alvin* in 2004, he found that this experience of immersion was, and could not but be, highly auditory. It was fated, he noted, by our lexicon: "sound" descending from the Old English *sund*, "sea" (if by sounding one means fathoming), or from *swinn*, "melody" (if one means vibrating). Sounds of "bleep-blooping, burbling and babbling... accompany and enable" *Alvin's* descent. The "Wa WA wawa WAWA wowowowo WOWO wawaWAWA" and "POP weewee woWOP ka POP weewee wo" that Kaharl described in 1990—a 9.5-kHz down-channel signal resonating each three seconds from *Atlantis*, the mother ship, and background pings from transponders deposited on the sea bottom—guide the pilot toward an undersea volcano. Internal to *Alvin* are the hum of onboard computers, classic rock from an MP3 player, small talk, and the banging of knees in tight quarters. A mile and more below the surface, among tube worms, bacterial mats, vent fish, and sea stars, *Alvin* maps a portion of the Mothra Hydrothermal Vent Field with an Imagenex system that transforms soundwaves, like those from volcanic eruption, into graphics. The three men inside are exiguous eyes, extrapolational ears.[287]

Beyond, as Jeffner Allen italicizes in her "Diving Planetarity," are "*multiple generations of resonance,*" listening as she is for a layered history of coral and the Caribbean, where the first bathyspheres descended and long-lived reefs are now dying off. In all diving, scuba or scaph, there is for us, for whales, and for other mammals, "*the necessity of sufficient air*" to contemplate

> reef: a continually shifting composition   the more components and
> greater the complexity   the less stable   intermediate disturbances long
> term oscillations   reefs without corals   corals without reefs
>   disequilibrium

but for the "skin" diver in particular, aqua-lunged and in

> suspension while
> stretched out in the water column   At the faintest movement feel bones
> in process of floating in a sort of sack   awareness given delicately by
> undulatory movement

what is heard below registers diffusely, from everywhere and nowhere. Helmreich explains: the human ear is too much of water to make a difference between self and surround, hence that term, and that feeling, "oceanic." Too much of water effectively to *transduce*, to translate motion in one medium, via a sensitive resisting membrane, to motion in another. In air, vibrations amplified by our tympanum and translated through cochlear fluid are transduced by TRPA1, a protein at the tips of the bending cilia of the inner ear, which opens ion channels in sync with the frequency and amplitude of the vibrations, creating an electrochemical voltage change that sends pulses along the VIIIth cranial nerve through the medial geniculate nucleus in the thalamus up to the auditory cortex. "Skin" divers, like early whales of the Eocene, receive underwater vibrations primarily through their bones, whose opacity and architecture obscures the source-direction of sound and most sonic qualities other than a damped rhythm. Unlike modern whales, human auditory physiology has not evolved from "air-adapted high frequency ears" into water-adapted ears with middle and inner capsules fused and separated from the skull by air-filled sinuses the better to accommodate and orient to a world in which light dissipates faster than sound.[288]

Heads hooded, bodies gently afloat in tepid water, John Lilly's subjects in 1955 were to see nothing, taste nothing, feel nothing, smell nothing, hear nothing but a slight lapping of water and perhaps, like Cage three years earlier, the pounding of a heart, the pulse of nerves. Trained in physics, medicine, and psychiatry, Lilly had been gathering accounts of sensory deprivation from polar explorers snowed-in during the long arctic night and people in life rafts surviving by dint of a near-perfect passivity. During their ordeals, he found, most experienced "an oceanic feeling, the being 'of the universe,' at one with it," on land or water. He had also been impressed by the work of Donald O. Hebb at McGill University, a psychologist celebrated for his 1948 book on *The Organization of Behavior*, which presaged the neuroscientific study of learning and emotion. Fascinated by a phenomenon discovered in 1929, that the brain manifests continuous electrical activity regardless of stimuli, and himself having run blinded rats through mazes, Hebb had a professional interest in "brainwashing" as brought to public notice during the Korean War. On covert commission from the Canadian Defence Research Board, Hebb and graduate assistants built a soundproofed room that would reduce all patterns of stimuli to a minimum, then watched what happened to undergraduates who volunteered, for pay, to remain as long as possible on a bed in an air-conditioned box with arms and legs restrained by cardboard sleeves, eyes covered by translucent goggles. They became confused, they thrashed, they lost track of the perimeter between wakefulness and sleep, and they became highly suggestible. If they made it past the first day, they hallucinated, which was a relief from the boredom, a result reported by the *Montreal Gazette* in January 1954 as "See, hear, feel nothing research shows bored brain acts queerly." By then Hebb's deprivation studies were done and their results with regard to boredom had been published; the results with regard to the effectiveness of sensory deprivation in changing attitudes remained confidential, though applied in more toxic and secretive trials by other Canadian and American psychiatrists subsidized by the CIA. In 1955, at a National Institute of Mental Health laboratory, a naked Lilly suspended himself (and in a second trial, a naked assistant) in a tank of slowly flowing water designed to reduce not just the patterns but "the *absolute intensity* of all physical stimuli to the lowest possible level," lights down, temperatures negligible. Within an hour each had relaxed, within two become antsy, within three begun fantasizing and then "projecting visual imagery"—Lilly's more positive take on hallucination. Released, each felt renewed, almost reborn, for "If one is alone, long enough, and at levels of physical and human stimulation low enough, the mind turns inward and projects outward its own contents and processes; the brain not only stays active despite the lowered levels of input and output, but accumulates surplus energy to extreme degrees." This could be useful data in dealing with the aftereffects of "involuntary indoctrination" of prisoners of war and with processes of "its opposite, psychotherapy."[289]

Lilly later preferred to describe the experience in the tank as "Peace in Physical Isolation" rather than "Sensory Deprivation." He would refine his tanks as avenues to *samadhi*, a state of intense absorption and focal consciousness, conducive to the insights of "The Deep Self" in a "private sea." Making a second career for himself during the 1960s and 1970s as an advocate of the possibility of deeply intelligent communication with dolphins, he helped design a computer interface called JANUS (Joint-Analog-Numerical-Understanding-System) to enable dolphins to converse with humans through an interspecies language. They would teach us, as implicitly wiser elders, to reach our "Quiet Center," that "void space" one can inhabit without encroachment from a dis-integrating planet.[290]

Amniotic life in three similitudes: bathyscaph traveler, "skin" diver, underwater monk. Each of these models of fetal experience had something going for it, a lesson scientific or sacerdotal, but in the postwar milieu they were all too quiet, too deaf. Ultrasound would do for the amnion and its denizens what hydrophones, sonar, and SOFAR did for the ocean deeps and their inhabitants: expose vulnerabilities to noise, appetites for music. In their fluid listening (part *sund*, part *swinn*) and their reliance upon vibration (part sound, part touch), fetuses were most like young whales or dolphins; the more patently they reacted to our lovetalk and quarrels, the more patiently would we bend an ear to them.

Stethoscopists had been eavesdropping on the uterus since the 1820s and all the more regularly since Adolphe Pinard's perfection of a fetal stethoscope in 1895, though using it could be "rather like listening to a clock ticking behind a waterfall." The possibility that the fetus was listening back had meanwhile faded alongside the "old wives' tale" that the shock of loud noises or shattering emotions experienced by a pregnant woman could be imprinted on her unborn child. Mothers knew full well the kicking and elbowing of an active fetus, but prior to a study in 1888 by the German obstetrician Friedrich Ahfeld, few mothers or doctors suspected that the fetus exercises its lungs in movements that rehearse breathing—and none knew that a fetus yawns. Rarely, a mother, midwife, or physician would report hearing clicks or clunks from within the womb; these turned out to be the sounds of a displaced ("clicky") hip, assessed in neonates at century's turn by Pierre Le Damany, a French orthopedic surgeon. Much more rarely, a child had been heard crying in utero, which doctors guessed was the result of a small rip in the uterine membrane through which air had been sucked into the amnion, giving the fetus a chance at airborne noisemaking. Crying in such events might be heard as evidence that the child was being strangled, its umbilical cord wrapped around its throat. An American doctor, S. Marx, therefore described the cry as "pitiful and whining... the weirdest call for help that can be imagined." Other obstetricians were not so alarmed; of the forty-four cases in print before

1914, most survived to bawl in the open air. Medical men who had never heard (of) such a thing were dubious, relegating it to saints' tales like those that swaddled John the Baptist, Muhammad, and Muhammad's daughter Fatima, so when the English obstetrician E. C. T. Clouston in 1932 reported a case of his own, he made a point of "the fact that the noise heard was loud and absolutely that of a child crying. It was not a pitiful whining noise as described by Marx, not a vague sound concerning which one had to ask oneself whence it came." Clouston explained the noise iatrogenically: he had thrust a forceps into the vaginal canal, inadvertently introducing air to the lungs of a child struggling to be born.[291]

Even more impressive than dysplastic clunks or perinatal cries would be the first ultrasonic images of a fetus in its amniotic deeps responding directly to the noises of our world, for this meant that the womb was no silent chamber and that prenatal life was not fully insulated from our hullaballoo. Mothers who had reported fetal kicks in reaction to loud sounds had been dismissed as overly attached or overtly hysterical. Besides, anatomical dissections of still-borns had revealed that their ears were "of inferior quality," with the three parts of the temporal bone yet to fuse, the axis of the tympanic membrane yet to shift upward and outward, the mastoid process yet to come into its own, and the labyrinthian capsule a structure independent of the rest, so fetal hearing must be at best rudimentary, with the Eustachian tubes flooded with amniotic fluid. Mothers of the 1920s would do well to consider buying a shielding "Audition Cap" for their infants, since "Babies, like puppies, are born with ears protected against sound" and highly susceptible to infection. As late as 2008, two Tunisian physicians would entitle their review article in a prominent European obstetrical journal, "Fetal Audition: Myth or Reality?" It was a reality, they wanted everyone to know. As ultrasound monitors had shown, a fetus hears and responds to sound at sixteen weeks, ten weeks before it opens its eyes.[292]

Ultrasound waves are by definition humanly inaudible, high-frequency (short-wavelength) vibrations which, on the verge of soundlessness, can be the most painful of penetrating noises, like the worst tinnitus, just before they wink out of range. That such a noiseful soundlessness should become the vehicle for revealing the rich acoustic worlds enjoyed and constructed by bats in caves, whales in deep seas, and the human fetus in utero was another manifestation of the everyhowness of noise. More than ubiquitous, more than elemental, more than thermodynamically axiomatic, noise could be instrumental. Acoustic, visual, or statistical, noise could be a tool intentionally wielded by physicists to comprehend subatomic uncertainties, by navies to defend against elusive submarines, and by physicians, now, to delineate soft living subdermal tissue to which x-rays are indifferent. Bouncing high-frequency pings off unseen or obscure objects of various degrees of solidity had manifold applications. Developed in France near the end of the Great War by the physicist Paul Langevin as a means for locating enemy submarines, ultrasound projectors and receivers were adapted by steel companies and structural engineers during the 1920s to detect cracks in metal; during the 1930s, the Submarine Signal Company promoted a Sonic Oscillator for the quick, safe, uniform production of a more digestible soft-curd milk. With ultrasonic waves (so short as to be "all acceleration and no motion," said one physicist), chemists could determine the viscous properties of liquids or gases and engineers could measure transient pressures. Such was the prospective power of ultrasonic "rays" produced by Robert W. Wood in his physics laboratory at Johns Hopkins that *Popular Mechanics* in 1927 vaunted ultrasound as the imaginal obverse of Maxim's silencers, the secret to a weapon of potentially immense destructiveness exactly because it could be so soundlessly disruptive. At 350–400 kHz, ultrasound was inaudible and deadly to almost all beings; it could split the membranes of blood corpuscles and coagulate egg albumen—a noise beyond noise. "All living matter, plant or animal, will perish when placed in the path of his ray."[293]

Detecting and destroying, ultrasound had been aggressively advanced by the military. In the postwar years, it was adapted to all kinds of healing, its narrow-beam heat apt to soldering and spot welding, its high-frequency vibrations apt to industrial degreasing and dental cleaning. Occultly modern, it was also adopted by medicine shows: who would know there were no innards to an "ultrasonotherapy" machine producing silent rays guaranteed to anneal frayed nerves, shake off paralysis, break up tumors? John J. Wild and John M. Reid of the University of Minnesota had hence some difficulty in the early 1950s convincing the medical establishment of the value of "echo-ranging techniques," borrowed from industrial methods for the inspection of metal seams, to produce readings of tissue in living bodies. During the Second World War, as a surgeon in England, Wild had been confronted with a number of often-fatal cases of bowel bloating, a condition attributed to shock from German buzz bomb blasts. (He himself had been buried for hours in the rubble of an explosion of a V-1 rocket.) After the war, in the States, he sought for a way to measure the thickness of bowel walls in ailing patients, at first by means of a device intended for the detection of stress fractures in tanks, then by adapting a navy-surplus radar simulator. Enlisting the aid of Reid, an electrical engineer, he then built the first clinical ultrasound scanner and, in 1953, an instrument that produced real-time images of cancer cells in a breast. They hoped, as did a couple of engineers at MIT and a Denver radiologist, that ultrasound could help identify tumors, invisible to x-rays, long before they presented palpably to a surgeon. The budding specialty of oncology, however, showed less interest in echo-ranging than did obstetrics, which had most in common with underwater naval pursuit.[294]

Inspired by his sister's war work with radar and by a Wild lecture, Ian Donald found in 1955 that an ultrasonic flaw detector used by a local firm for inspecting the welded seams of industrial and nuclear boilers could be used to tell a fibroid mass from an ovarian cyst. As Regius Professor of Midwifery at the University of Glasgow, Donald guided a team during the late 1950s and 1960s that developed a feasible diagnostic ultrasound system that would make visible the shape and position of a child in utero and enable the discrimination of abdominal tumors and cysts. When his Glasgow team—and others at work independently in Japan, Sweden, Australia, and the United States—added Doppler continuous-pulse scanning, nurses and physicians could follow by eye and ear in real time the motions of a fetus without compromise to its health or the health of the mother. Benefitting from a first wave of electronic miniaturization, medical technology companies would reduce the size of ultrasound equipment from that of a grand piano to that of an accordion with a television monitor. Benefitting also from oceanographic analyses of the acoustic properties of differently constituted homogenous liquids (vital to the effectiveness of sonar through layers of seawater), ultrasound technicians could assay the soft tissue of a growing fetus and the rate of blood flow from mother to child. "There is not much difference after all," said Donald, "between a fetus in utero and a submarine at sea." Accompanied by a Doppler soundtrack of heartbeats, the liveliness of fetal ultrasound had such lay appeal by the 1980s that ultrasound sessions were routinely requested by pregnant women in Europe, the Americas, and Japan.[295]

Extending the reach of physicians to more detailed and dynamic features of gestation at earlier stages, ultrasound technology could be held complicit in a further medicalization of childbirth. Yet, reduced again in size and cost, ultrasound devices in the form of over-the-counter handhelds (BellyBeats, Stork Radio, BabyFM) enabled expectant middle-class families to listen in on a fetus "anytime you like, anywhere you wish," this despite warnings of errors of interpretation (was that the mother's heart or the child's?). If the marketing of handhelds emphasized the autonomy of a fetal heartbeat from a mother's and so buoyed up the agendas of those on *all* sides of a passionate debate over abortion (all of which differently distinguished the being and consciousness of a fetus from that of its bearer), Stork Radio and BabyFM were semiotically perverse enough to be promoted also as harmonizers of parents and siblings with a "future child" and as pacifiers for anxious grandparents. With the Hi Bebe BT200, claimed the manufacturers, "Not only will the sound of your baby reassure you, but you can bond with your baby from the very beginning." Wrote Christie S. in a testimonial for Heartbeats At Home, an ultrasound rental company: "The doppler has also been the best bonding experience for my other two daughters. They enjoy listening to their baby sister's heartbeat, along with her kicks, rolls and hiccups."

Wrote Laura H: "I remember the early kicks and hiccups I couldn't feel, but could hear with your monitor. Thank you for such a precious gift."[296]

Hiccups a precious gift? What was going on? How could a noise that was in air a disturbance of breath-throat-sternum and in the amniotic sac a fetal stomach spasm be heard as a precious gift? Heartbeat, hiccup, click, clunk, or ostensible fart, the soundfulness of the fetus was completing the same cultural circuit that had just been completed by the songs of whales and would soon be completed by walruses, whose audible underwater belling and knocking would by 2003 be accorded the status of "diving vocalization songs" and "coda songs."[297] Passive and active sonar had proven that the ocean was a sonic medium occupied by hundreds of sound-making, sound-harvesting species; ultrasound scanning made it clear that the amnion was also a sonic medium whose denizens were highly auditory. Just as marine mammals listened for each other, called to each other, and were distressed (sometimes severely, perhaps mortally) by loud sounds from naval SOFAR, seismic testing, and supertankers, so fetuses listened for us, hiccuped toward us, and could be grieved by our noises.

Quoting "a good observer" (the French psychologist Bernard Pérez), William James in 1890 had written that "Children between three and ten months are less often alarmed by visual than by auditory impressions." A three-month-old might be delighted by flames licking at his cradle but frightened by firemen shouting through brass speaking trumpets. Not the flash of lightning but the rumble of thunder made infants startle and cry. James agreed with Pérez that "fear comes rather by the ears than by the eyes, to the child without experience." This had to be all the truer of a child in the dark of the womb, if it could hear—which was most dubious to a leading German authority on the biology and psychology of infancy, Wilhelm Preyer, who had written in 1885 that there was "a probability bordering on certainty" that the fetus had no auditory sensitivity. Forty years later, Albrecht Peiper at the Berlin University Children's Hospital had reservations about such a blanket claim after noticing that newborns moved to the tones of a toy trumpet; he then did some honking of a "loud and shrill" automobile horn in close proximity to frazzled late-term mothers, who reported fetal thumping and wriggling in response. Two American physiologists, having chanced to see a fetus "jump" in response to another sharp noise, sat with a near-term woman at a concert in 1926 to get an immediate account of her child's physical reactions to various pieces of music. In 1932, at New Haven Hospital, the child psychologist Wilbert S. Ray took more formal if "preliminary" experimental steps: first, he set a vibrating bell on an expectant woman's abdomen to observe the reaction from the womb; next, he loudly banged together a pair of spring-loaded boards two feet away from the woman (and child). Given that external noises could prompt fetal motion, he hoped to determine a set of sounds that could be used for "pre-natal education," heavy-handed as were his methods (and likely to engender lifelong claustrophobia, or so wrote a psychoanalyst in 1935). After the war, pediatricians and obstetricians still knew little about fetal audition and continued to believe that the fluid-filled amniotic sac, if not impregnable to sound, insulated the gestating child from most external noise, so that birth itself, wrote a psychoanalyst, must inevitably be a trauma of "onrushing multiple stimuli." Trauma there was, agreed the physiologist Sir Joseph Barcroft in his *Researches on Pre-Natal Life*, begun in 1932, written up during the War, and published weeks before he died in 1947, but with a Briton's stiff upper lip and a Quaker's resolve he argued that intrauterine development had physio/logically to include a hardening of the senses and a strengthening of the muscles of respiration to "withstand the shock of birth."[298]

Common sense dictated, in the form of Benjamin Spock's *Common Sense Book of Baby and Child Care* (1946), that affection and responsiveness, not rod and weal, be the keynotes to parenting. If one could not foretell a newborn's future from the shape of its ears, as had the Babylonians (sneered the acoustician Georg von Békésy in 1948), parents could help shape that future by listening empathically, from day one, to each child. The Children's Bureau during the Second World War had already revised its widely distributed *Infant Care* manual to promote a "gentle handling of baby urges, from thumb-sucking and masturbation to weaning and toilet training," and Dr. Spock was tagging along. A New York pediatrician with psychiatric leanings, Spock had been under analysis in the 1930s with Bertram Lewin, he who had explored the fetal origins of claustrophobia, and then with Sandor Rado, who in 1927 had framed the related problem of the "smothering mother." (Both Lewin and Rado departed from orthodox Freudianism in their psychoanalytic theorizing about women, Rado most vehemently after a disciple of Freud's, critical of Rado's work, claimed that "the birth of a baby was the most active, masculine thing a woman could do.") As a clinician, Spock was a sympathetic listener who learned from youngsters and from mothers in his care, among them Margaret Mead, who persuaded him of the virtues, urban professional and ethnographic, of flexible feeding. As an author, Spock was rather a talker than a writer, dictating chapters to his wife Jane, whose editing retained his conversational style, so that generations of mothers—a million copies on average sold each year for fifty years—would feel as if he were a "confidant." His oralist assurances that "you know more than you think you do" and that there could be no "letter-perfect" childrearing were attractive in an era anxious about democracy at home and authoritarianism on all fronts. Under cover of equilibrating shifts in gendered relations among mothers and fathers and approaches toward a more bivalent exchange between parents and children, the fetus got its hearing.[299]

Lester Sontag of the Fels Research Institute had been studying activity in the womb since the 1930s, correlating the patterns of fetal heartbeats (captured by a microphone taped to the mother's abdomen) with patterns of pure tone vibrations (amplified by a speaker set on a rubber baffle on the mother's abdomen). He and Jack Bernard in 1947 could demonstrate "fetal reactivity to tonal stimulation"; what they failed to prove was whether the reactions were due to vibrations felt or to tones heard. Using x-rays twice a day for ten days, David Spelt the next year was able to report that fetuses seven months along visibly jerked in response to the noise of a spring-loaded clapper, a *heard* noise physically apart from the bodies of mothers, in contrast to the vibrotactile stimuli of the electric doorbells that he had taped to their abdomens. Spelt even claimed to have used the clapper to effect some primitive conditioning of fetuses. He had thereby established the acoustics of fetal hearing, but almost no one else was willing to put mothers and fetuses at such risk of x-ray damage, so the degree of fetal auditory sensitivity remained primarily anecdotal until the arrival of ultrasound scanning.[300]

Mothers buying Dr. Margaret Liley's guide to *Modern Motherhood* in 1967 would nonetheless learn that "The unborn is exposed to a multiplicity of sounds that range from her mother's heartbeat and her voice to outside street noises. Especially if his mother has not gotten too plump, a great many outside noises come through to the unborn baby quite clearly: auto crashes, sonic booms, music. And the rumblings of the mother's bowels and her intestines are constantly with him. If she should drink a glass of champagne or a bottle of beer, the sounds to her unborn baby would be something akin to rockets being shot off all around." This simile was based not on ultrasonics but on superb microphonics, the product of a military-industrial/entertainment complex that during the Sixties amplified and expanded all means of acoustic surveillance, sound synthesis, sound transmission, and sound pick-up. Sensitive microphones registered uterine background noise at 30 db or more, a result of the placental pumping-in of nutrients and pumping-out of waste. That level rose when a mother's heart raced (at a rock concert or in mortal fear), when she had bouts of indigestion and borborygmus, or when carbonated bubbles from the first tab-top colas started popping in her gut. In 1966, Lester Sontag would argue that loud noise within or beyond an expectant mother could be a teratogen, a cause of monstrosity in the fetus or of personality disorder in the eventuating adult. Evidence for this came from reports of behavioral problems among children exposed in utero to bombardment during the Second World War, from social-psychological studies of crowding, and from his own success at using loud noises to induce seizures in pregnant rats, which had serious behavioral effects in rat offspring. Further evidence would come from studies of skeletal abnormalities in fetal rats whose mothers had been subjected to "noise stress." Sontag's argument was the converse of the supposition by the child psychologist Lee Salk (brother of Jonas) in 1960, of the physical anthropologist Ashley Montagu in 1962–1964, and of all artists who had painted Madonnas with an infant held against the heart-side breast: that a fetus, like a newborn, relies upon the steady beat of the maternal heart for its sense of well-being. Therefore any intrauterine noise that overwhelms the beat, or any external noise that causes the mother's heart to beat irregularly or be heard less clearly, represents a biological and psychological threat. Salk had sought to prove his hypothesis by extrapolation from an experiment in a postpartum environment; with the help of the Muzak Corporation, he serenaded newborns in a hospital nursery with night-and-day tapes of maternal heartbeats during the standard four days of their neonatal stay. Newborns blessed with the serenade gained more weight and cried significantly less often than control groups who heard no tapes. Most revealing was an unexpected datum: when the tapes became worn and began to hiss when played, babies who had been soothed by the heart-Muzak became restless and "fussy."[301]

Fussiness had a critical role in determining how parents and physicians would listen to, for, and with newborns and yearlings. By 1960 "fussy" had moved beyond euphemism into pediatric jargon: it described a well-fed infant who cried excessively and *less understandably* than one suffering, for example, from an ear infection, the most common infant malady aside from "colic," itself diagnosed rather by the extent of crying than the presence of pathogens. According to a defining study in *Pediatrics* in 1954, a fussy baby was one that had paroxysms of irritability or crying "lasting for a total of more than three hours a day and occurring on more than three days in any one week"; a "seriously fussy" baby had paroxysms (often and confusingly called colic) that recurred for more than three weeks or became so severe that narcotics were indicated. Supplanting traditional beliefs that certain infants had an innate character or constitution that fated them to gripe at the world from their first air-drawn breaths, fussiness also softened the sternness of Christian theological assumptions about the stubbornly selfish nature of all infants who, as carriers of the bad seed of original sin, required rigorous discipline. Fussiness was less a condition than a relationship between baby and mother in which ordinary acts of translation failed and crying became incomprehensible, an exasperating noise. In recasting baby and mother as a dyad, Spock and his American successors—and English and Canadian pediatricians informed by the developmental psychology of John Bowlby and Mary Ainsworth—recast crying as a problem not of temperament or discipline but of interpretation and mutual reassurance.[302]

Millennia ago, the earth-spirits caring for the infant Zeus had so clashed their weapons in a mock war dance that the infant's Olympic-length crying could not be heard by his father Cronus, who, once alerted to his hiding place, would have descended to slay him lest, in fulfilment of a primal omen of patricide, Zeus grow up to slay his father—which he did. Millennia later, Oedipal concerns were yet in play, refracted through a demotic Freudianism, demonic legends of sleepless fathers flinging bawling babies out of windows, and Mr. T. W. Orten's demonstrably anxious letter to the Children's Bureau in 1917, inquiring about the chances, "if any, of a young baby rupturing itself through crying." (Unlikely, the Bureau wrote back; if, however, the baby's muscles are weak, convulsive crying could increase the chances of a rupture of the navel.) But the focus of modern pediatrics was on mothers and, with regard to noise, on mothers' frustration with infants who could not be quieted. Crying itself was to be expected, of course, as an infant's sole medium for long-range contact and as physical exercise for immature lungs; however, when it persisted as a "crying jag" beyond all apparent reference (e.g., to teething), something was wrong. Whether the crying stemmed from poor sleep or poor diet, or from maternal indecision, pampering, or neuroticism, the remedy (said some experts and many a neighbor) was an unwavering regime of regular feeding and burping, changing and washing, sleep and darkness. Finding it hard to be so resolute, mothers would resort instead to the age-old pacifier, frequently sweetened or narcotized, recently rubberized. Now, though, resort to a pacifier might be an admission of failure. Deployed rather to appease a desperate mother (and forestall infanticidal fathers?), pacifiers delayed valuable early childhood lessons in separation and later lessons in identity and independence. "Dear Doctor," wrote a mother of Amityville in 1924, long before the Long Island horror films, about her screaming baby: "I was told not to let her cry very much, and thru 'neighborly advice' gave her a pacifier. Have tried several times to wean her of it but she cries so hard I get frightened and give it to her." This was her first child, a four-month-old with a ruptured navel who could no longer breastfeed; would the Children's Bureau please advise on "how to break her of the horrible pacifier—and if crying will hurt her?" Dr. June M. Hull of the Bureau duly replied: don't give her the pacifier but show your daughter that you'll always do what's best for her. This presumed a precocious capacity on the baby's part to appreciate what pediatricians had just discovered, that crying was nowhere near as hurtful as the "pacifier habit," which "forces the upper teeth forward, disfigures the mouth and jaw, interferes with chewing and distorts the air passages of the nose." Let your one-year-old "cry it out for a few times," the Bureau advised a Kansas mother in 1929. "It will take courage to do this," another Bureau physician advised another mother whose eight-week-old son some days would cry continuously from eight in the morning until ten at night.[303]

Dr. Spock himself had a dyslexic, and "fussy," first child, and "Though I knew I was wrong, I used to get so mad at my son when I thought he was a crybaby at three and four that he still remembered it 20 years later." Spock's published advice was more temperate: crying, and thumbsucking, were natural responses to an intimidating environment and would be outgrown when mother and child learned to trust one another. Mothers should realize that a baby "is born to be a reasonable, friendly human being," so "When he cries it's for a good reason," and "We aren't as scared, nowadays, of the danger of spoiling a baby as we used to be. If a baby is comforted when he is miserable, he usually doesn't go on to demand that comfort when he *isn't* miserable." Since crying is "seldom a sign of anything serious" other than hunger, wetness, or an undone safety pin, and since doctors still "don't know the basic cause of most colic or irritable crying," the act of comforting an insufferably crying baby would be helpful not only to the child but the mother, for "many mothers get worn out and frantic listening to a baby cry, especially when it's the first." Mothers too needed breathing spells, time away from baby, without grief or guilt.[304]

"Is it right for a small baby to have a temper?" Donald Laird the Anti-Noisite psychologist asked in *The Strategy of Handling Children*, co-authored in 1949 with his wife Eleanor, a research librarian. Answer: "children are born with healthy tempers," so be patient with them; yea, if the legends about Muhammad were true, children may cry before birth, and of course at birth, and when wet or hungry or gaseous or constipated or colicky, and when ill or croupy, and from night terrors, and while teething, and to stretch their lungs. If a toddler should launch into a tantrum, have him run up and down the stairs a few times yelling loudly; his tantrum will end in three minutes. The young mother in Richard Wright's powerful story of 1938, "Long Black Song," may feel desperately lonely and put-upon, but she quiets her crying baby by letting her beat on an old clock—"Bang! Bang! Bang! And with each bang the baby smiled and said, 'Ahhh!'"— an "ahhh!" of empowerment that the mother too is looking for in her own life. What a parent wants always to avoid is instilling in a child of any age a feeling of helplessness, wrote the Lairds, which leads to unmitigatable crying, which provokes parental feelings of helplessness and dread, which stokes the insecurities of a child, whose continued crying upsets the household sleep cycles, which makes parents exhausted and crabby, which cannot end well if it ends at all. Such circularity was noted by Lee Salk when he too, like Laird, made the move from experimental psychologist to pediatric expert. In *Ask Dr. Salk* (1980), he listed four reasons "Why Babies Cry": physical discomfort, tiredness, hunger, boredom. That last, new to pediatrics since the 1950s, implied an infant desire for engagement with the world. In any case, babies cry for a reason, though you "can't always tell what," wrote Salk. If ignored, they tune out and go to sleep (as if back in the womb), which may be a relief but etches a pattern of withdrawal that returns with a vengeance in later life. You should not, however, rush to pick up a crying ten-month-old who is otherwise dry, fed, and healthy; instead, pat its back, gently remind this veteran of the crib that he has not been abandoned, that she is in good company, and let the child lie, to cry as it will.[305]

Given then the cultural sense of the baby as a natural-born crier, fussy or not; given its need for making itself heard, from birth and possibly before ("pre-speech," one wit called it); given the popularity of remote-speaker monitors that came onto the market during the 1970s to alert parents and sitters to more episodes of crying than they would otherwise have heard; given a baby's responsiveness to the sound of a generic maternal heartbeat; given its delight in self-made sounds and noises; and given the companionate principles of parenting through which infants, mothers, and fathers were now bound within a circular acoustic field, it stood to reason that a fetus might share the soundscape of its larger environs. "Although it may be yet too soon to speak of the ecology of prenatal life," in 1969 it was obvious to the family therapist Antonio J. Ferreira (shortly before he left the profession he pioneered to become a scuba diver and marine biologist) that external sounds affected the fetus, whose cochlea was functional by the fifth gestational month. In his *Prenatal Environment*, Ferreira presumed that the sound-surround of a fetus had implications for life after birth—implications studied for decades by Ando and Hattori, beginning in 1969, at a time when Japan's Environmental Agency was receiving over 22,000 complaints annually about noise and vibration. Itami City, for example, had lowflying airplanes overhead every four to six minutes from seven a.m. to ten p.m. each day. It was there that Ando and Hattori observed the sleep patterns of newborns. Those who had not been habituated to airplane noise before the fifth month of gestation had more difficulty falling asleep and staying asleep through the overflights. Ultrasound would more directly confirm the intrauterine effects of extrauterine sound and, in particular, of extrauterine noise.[306]

Viewed on an ultrasound monitor and audited with Doppler, the response of a fetus to pleasing sounds would appear much subtler (a mild slowing of its typically fast heartrate) than its flailing or shuddering in startle reflex to sudden, shrill, or loud noise. Now that they could more closely and safely track prenatal growth and neural development, obstetricians could draw stronger causal connections between maternal or incidental noises and fetal motion, and between environmental noise, or overflights, and fetal viability. Yet, years before ultrasoundings became routine, and before real-time fetal electrocardiography was perfected, Anti-Noisites had picked up on the nexus between a blighted soundscape and fetal risk. In 1972 M. Barbara Scheibel wrote in *Noise: The Unseen Enemy*, "Thirty years ago, the home was a haven of peace and tranquility. The average household registered noise levels comparable to those of a whisper. Today it is a den of cacophonous bedlam, produced by dishwashers, mixers, blenders, refrigerator hum, ice crushers, electric can openers, knife sharpeners, exhaust fans, vacuum cleaners," in addition to clothes dryers, hair dryers, televisions, and stereos. Nowhere, she concluded, was anyone safe from the harmful effects of noise, not at work, not at home, not in the womb. "The womb," declared Theodore Berland of Chicago, warrior of *The Fight for Quiet*, "was once considered a quiet sanctum, apart from the world without. However, six months after conception, a fetus's hearing apparatus reaches normal development and the womb becomes a noisy place," penetrable by honking horns and the drone of garbage disposals. In 1971, the infant U.S. Environmental Protection Agency had held hearings on noise in eight cities and after reviewing "all aspects of current knowledge about the effects of noise," had issued a report that began: "Noise, commonly defined as unwanted sound, is an environmental phenomenon to which man is exposed before birth and throughout life."[307]

Kicking or wincing, its heart beating less smoothly, its blood pulsing faster, the fetus would be found by ultrasound to react in more ways to more noises than had been anticipated by anyone other than pregnant mothers. Physicians debated the degree to which external sounds were damped by the uterine wall or masked by the ambient noise level within the amniotic sac (85 db according to a microphonic study in 1971). Like ocean physicists calibrating the attenuation of soundwaves through various layers and depths, bio-acousticians measured the attenuation of different frequencies by maternal tissues and amniotic fluid. With every three or four years of research (on fetal lambs in particular), the intrauterine soundscape was revealed to be a decibel or two more open to the world. By 1975 some doctors were wondering whether outside noises could be penetrating and persistent enough to cause fetal distress. During the 1980s, layparents began to assume that the fetus was a captive audience to the talking, singing, and humming of the mother, the crooning of any father within lullaby distance, and the moaning and groaning of a loving couple's race toward orgasm. It was also assumed that a newborn could recognize the airborne voice of its mother after months of listening in the womb, as well as the voices, or cadences, of fathers and children who had been in the acoustic presence of the pregnant mother. How much of the rattlebang of daily life made it through to fetal ears *as noise* was still unclear, but the noise-startle effect was in operation at twenty-five weeks, and a committee of the American Academy of Pediatrics declared in a widely publicized report of 1997 that noise was "a hazard for the fetus and newborn." As toxicologists worked toward a statistically significant association between environmental noise and low birthweight, expectant parents were expected to treat the fetus to a pleasant bedtime story read aloud each night. And while otologists were developing "auditory-evoked response" systems to test the hearing, or incipient deafness, of third-trimester fetuses by the use of speakers projecting pure tones through the mother's abdomen, parents should play relaxing music for the benefit of the fetus, which had the taste to prefer, by head-nod or cardiac beat, easy listening to heavy metal.[308]

Mozart too? If the fetus made its share of sounds but, unlike walrus or whale, sang no audible songs, its progenitrix at least could sing to it, wrap it in music whose color-enhanced patterns on an Imaginex screen or computer monitor were as prettily rhythmic as the patterns of a fetal heartbeat on an ultrasound scanner. The notion that a fetus could be comforted by ear candy or enheartened by the late Baroque had grown alongside the microphonic and ultrasonic investigations that revealed the extent of fetal audition, the transmission qualities of amniotic fluid, and the levels of intrauterine noise. Fetal reactivity to sounds originating from within the mother, from the mother's environs, from rowdy siblings and rowdier experimenters, led to the idea that a musically thoughtful saturation of the proximate soundscape might be soothing, even pedagogical. The wide appeal of womb songs and "sound education" cassettes for the fetus complemented the pediatrics of the Spockian turn, which construed pregnancy and infancy as periods of intense reciprocity and mutual stimulation between mother and infant. Starting at the sixth gestational month, wrote Thomas Verny, M.D., in *The Secret Life of the Unborn Child* (1981), the fetus "leads an active emotional life." Not only was a six-month-old fetus listening "all the time" and learning speech rhythms in utero; it would grow agitated each time its mother "*thinks* of having a cigarette."[309]

Far-fetched as it might be to have the fetus acting as the mother's conscience, there was mounting evidence that if a newborn remembered its mother's voice from its time in utero, it could also have the equivalent of cigarette burns from much else it had heard/felt before birth. Veterinarians and medical researchers were finding that an adult's blood pressure corresponded to his or her mother's blood pressure during pregnancy, that daily blasts from an air horn increase the cortisol level in a pregnant rhesus monkey, which in turn decreases her offsprings' ability to respond to stressful social environments. Fetal lamb, fetal monkey, or fetal human being, wrote the veterinarian Peter Nathanielsz in *The Prenatal Prescription* (2001), "the baby is scanning and picking up sounds like a ham radio operator," unless hampered in utero by chemical or physical forces that disable the sensory transduction of sound by hair cells of the inner ear. ("Inner ear" was becoming a pun: it could refer now to fetal listening.) Janet DiPietro, a "fetal psychologist" at Johns Hopkins University, told *Psychology Today* in 1998 that as early as nine weeks a fetus reacts to loud noises, noises that could interrupt the sustained drowsing and periods of REM sleep vital to fetal development, causing psychic as well as somatic grief. DiPietro belonged to a new field that had been shaped in 1969 by Elizabeth Feher, whose Natal Therapy Institute worked to reduce the anxiety of the womb and trauma of birth. While more pacific birthing methods (under softer lights or in a warm bath, and welcomed by quiet voices and gentle hands rather than forceps) were contemplated in the wake of the French obstetrician Frédérick Leboyer's *Pour une naissance sans violence* (1974; *Birth Without Violence*, 1975), fetal psychologists encouraged a more affirmative ecology of the amnion. David B. Chamberlain in *Babies Remember Birth* (1988) speculated that "Remembering birth may be a special feature of life in the twentieth century," but Stanislav Grof in *The Adventure of Self-Discovery* (1988) was more sweepingly ahistorical: "we tend to live our lives as we lived our first nine months." As a result, mothers (and fathers) needed to bond early and deeply with the fetus, and to do what they could to help it cope with the turmoil of the uterus, "a place of considerable, constant noise in which the fetus spends much of his time in a dream state." Fetal dreams were weaker refuge from the blows of uterine plumbing and undertow of hormonal tides than a parent might suppose, if one trusted etymology as a clue to prenatal experience: the word "dream" could be traced back by the *Oxford English Dictionary* to noise ("OE. *dréam* = OS. *drôm* mirth, noise, minstrelsy—WGer. **draum-*. Kluge suggests that it is from the same root as Gr. θρϑλος, noise, shouting").[310]

Of the seven children born to Leopold and Maria Anna, five did not make it past their sixth month. The one male who did survive, to the age of thirty-five, had a recalcitrant afterbirth that had to be pulled out by an alert midwife. Whether or not the genius of Joannes Chrysostomus Wolfgangus Theophilus Mozart had been inculcated in utero by his father, who produced a textbook on violin-playing in the year of Wolfgang's birth, and whether or not Wolfgang had been reluctant to depart a womb whose exit had been so unlucky for so many, it appeared two hundred years later that his precocious music could quicken the formation of neural synapses and speed up cortical connectivity. His music could also reduce stress, as perhaps it did for Mozart himself, who began composing at five while growing up in the house of a Salzburg merchant whose many offspring were noisily in and out of the family's third-floor rooms. So Philips Records produced a profitable CD of *Mozart for Mothers-To-Be: Tender Lullabies for Mother and Child* (1996), then *More Mozart for Mothers-To-Be: Gentle Love Songs for Tender Ears* (2000). As promoted in 1997 by Don G. Campbell, who had previously written about the "musical brain," *The Mozart Effect* helped organize "the firing patterns of neurons in the cerebral cortex" and much else: under strains of Mozart, cows in Brittany gave more milk and drug dealers cleared out of downtown Edmonton. As originally adumbrated by Frances H. Rauscher, a cellist and experimental psychologist at the Center for Neurobiology of Learning and Memory at the University of California, Irvine, listening to Mozart facilitated in preschool children an ability to perform "spatial-temporal tasks involving mental imagery and temporal ordering." Rauscher herself never made any claims about fetal education, never supposed that Glenn Gould became a concert pianist because "his mother played the piano for him constantly while he was developing in utero." Like Lee Salk, whose concern for the comforting postpartum sounds of a mother's heartbeat had been transformed into an industry of records, tapes, and CDs for newborns *and* pregnant mothers (e.g., Dr. Jay Freed, *Sounds of the Womb: To Soothe and Comfort Your Baby*, 2000; Baby Sleep System, *Sound Sleep for Babies: Sounds In Silence*, 2004), Rauscher's inquiries into the positive effects of the musical training of children were translated back to the womb by "prenatal educators" who had a professional stake in fetal stimulation.[311]

Insofar as a "sound education" in the months before birth would mask noises intruding from beyond the womb, it was a bold extension of Anti-Noise campaigns that had begun with the establishment of Quiet Zones around hospitals, moved on to the quieting of hospital wards, and by the 1970s had reached Neonatal Intensive Care Units. On the model of chick incubators, Etienne-Stéphane Tarnier a century earlier had designed the first baby incubators in hopes of securing a higher rate of healthy births in a *fin de siècle* France anxious about depopulation. His "artificial wombs" were improved by Alexandre Lion of Nice, who introduced a forced-air ventilation system; successive editions of these were put on display by Dr. Martin A. Couney in "Incubator-Baby" sideshows at world's fairs, trade fairs, and Coney Island, where thousands of paying spectators would visit a row of "living babies" in glass enclosures attended by white-clad nurses; during forty-seven years of exhibition, Couney's shows provided free care for 8,000 premature infants (2–3 pounds at birth), 6,500 of whom flourished. Too scarce to reduce infant mortality rates as effectively as a diphtheria vaccine, these primitive incubators nonetheless demonstrated the power of medical technology to sustain the imperiled. With better systems for ventilation, aseptic handling, and electrically stabilized temperatures, the units by 1922 were considered reliable enough for Michael Reese Hospital to install twenty-eight. From Chicago their use spread across the Americas, Europe, and Japan, all operated according to the principle that, like the fetus, an incubator infant must be bathed in warmth and fed well but otherwise left alone to grow strong. Couney and his barkers (among them one Archibald Leach, later Cary Grant) told the crowds, "You may talk, ladies and gentlemen, you may cough. They will not hear you. They do not even know you are here." Incubators in the 1920s and 1930s were still temporary sanctuaries for beings thought so fragile that their senses had to be quiescent. But physiologists studying premature infants placed in incubators in the 1940s and 1950s would notice that their hearing was functional weeks before their other senses. Pediatricians and neurologists, inspired/horrified by Harry Harlow's maternal/sensory deprivation experiments on infant monkeys (1957–1963) began to advise that "preemies," like all infants, needed to be held, intrigued, and acoustically soothed—especially since they were incapable of making themselves heard. "A very premature infant," wrote the Yale child guidance expert, Arnold Gesell, "may exhibit all the facial contortions of weeping which culminate, however, in a perfectly soundless cry or a faint bleat." The revision of standards of care led paradoxically to noisier NICUs, what with the more frequent opening and closing of incubator doors (banging at 110–40 db) in order to caress or massage the infants, the new emphasis on "female vocals" to comfort at-risk infants, and standing instructions to thump on the glass of the incubator whenever apnea was suspected. Refined to cope with births considered viable at earlier stages of prematurity, incubators needed better pumps to aid respiration and blood flow, more sensitive monitors, and more precisely filtered air. In consequence, not only did NICU staff work more closely with the infants and with humming machinery that had warning bells or buzzers; the decibel levels within the incubators were so consistently high that they threatened permanent damage to the malleable auditory apparatus of the infants, whose cries were initially so weak that they might not be heard beyond the insulating glass.[312]

Otto Rank had broken with Freud in 1923 over *The Trauma of Birth*, which he heard in the "pre-Oedipal" cry of a newborn, "the cry which, by violently abolishing the difficulty of breathing, may presumably relieve a certain amount of anxiety." That cry was repeated in seeking the comfort of the mother, and again in infancy and childhood upon being left alone in a dark room, and once again in the neuroses of adulthood, the psyche still shaken by the violence of the primal expulsion. Other male spectators of parturition had presumed that the process must be so abusive for the child aborning—the pressure and pounding, the hubbub, the glare of light in the frigid air of the expulsion, the slicing of the umbilicus—that the "biological economy" of birth required insensibility until the fetus was terrorized into crying when slapped toward breath. As physiologists began to reveal a fetus long alert to worlds middle and outer, the compression down the birth canal with its attendant physical shocks warranted an even more anguished cry and a trauma that Rankian analysts would confront headfirst, for it was demonstrably more universal and palpable than any Oedipal complex. Ancient Romans at childbirth had invoked a god, Vaticanus, to oversee the opening of each infant's mouth for its first scream, as if the opening of the mouth had to be *vatic*, prophetic, and perhaps they were right. With further research belying the sensory "quasi-dormancy" of newborns and with microphonic and ultrasonic proofs of the "auditory competence" of the fetus and neonate, Rankian ideas were incorporated into a new therapeutic slant that attributed to an infant's gestational career a richer mental and emotional life than any adult could recall. Except, of course, those who undertook regression therapy or ritual rebirthing, or who reenacted, with the encouragement of Arthur Janov, a psychiatric social worker, the "primal scream." That scream had made itself known to him in the late 1960s as the full-throated screech of one of the men in a group-therapy session that he had been leading, a scream of "Mommy! Daddy!" with body writhing on the floor until the man "released a piercing, deathlike scream that rattled the walls of my office," after which he felt transformed and at peace. It was the scream, as Janov determined upon encountering another client who likewise screamed his way into a halcyon state, expressive of "central and universal pains which reside in all neurotics," pains arising from a primal neediness, and from primal needs inevitably unmet. It was a scream that did double duty, shattering and shriving. It was the fetal cry at its most explosive, on being born.[313]

It was the noise, the primal noise, of the everyhow: a noise that was at once a signifier of the inarticulate and an instrument of redemption; a noise as useful and telling as it was extravagant; a noise that could be heard bouncing off the angles of every (psychic) structure. "I can hear a deep human cry, a 'silent scream,'" wrote the psychologist Carl Rogers, "that lies buried and unknown far below the surface of the person." It was phons beyond the screams that had been echoic of Western modernity since Edvard Munch's paintings of 1893 or the operatic climaxes in Wagner and *Wozzeck*. It was more like the silent screaming horse in Picasso's *Guernica* or the silent stage scream, "harsh and terrifying," of Brecht's Mother Courage, after soldiers have brought her dead son before her. In the literal form of a newborn's first cry, as the composer and psychologist of music Robin Maconie would write, the primal scream is the loudest sound any of us ever hears, for it arrives at a pitch higher than any audible to us *in utero*, and is broadcast in the open air over "the shock of an expanded sound world, a world of harsh sibilants and brutal clatter," so at a volume hitherto (and just this once) incomparable. The cry at once announces our presence and our lostness, unleashing "a huge storm of high frequency information" that is more directional than anything heard in the womb and "uniquely adapted for learning about space, the final frontier." Declaring our selves but situating us in a vast world not of our making, primal scream and first cry reflected the amplified polyvalences of noise, which could ricochet with remarkable speed from the awful to the sublime, from the anarchic to the lawful. In 1984, for example, the noise of the primal scream emerged in two venues that could scarcely have been more different: an experimental rock band, Primal Scream, formed in Glasgow by Bobby Gillespie, the drummer for a gothic noise band, the Jesus and Mary Chain; *The Silent Scream*, an anti-abortion film, produced by Dr. Bernard Nathanson, an obstetrician who had previously served as director of CRASH (Center for Reproductive and Sexual Health in New York City) and been a vocal advocate for women's reproductive rights. An indie pop group, Primal Scream caught the public ear first with its *Sonic Flower Groove* album (1986/1987), then with *Primal Scream* (1989), most famously with *Screamadelica* (1991), a dance album fusing the sounds of "acid house, techno, and rave culture," and later with *XTRMNTR* (1997), "a nasty, fierce realization of a world that has lost the plot," defined by its chorus, "sick fuck fuck sick fuck fuck sick fuck." *The Silent Scream* was one of the final fruits of an increasingly public anguish expressed by Nathanson at the thousands of abortions he had performed or supervised, an anguish that had led to his "excommunication" in 1975 from NARAL (National Association for the Repeal of Abortion Laws, which he had co-founded) and would lead him from "Jewish Atheism" into Roman Catholicism while his film was being screened at the Reagan White House and in churches across America, its ultrasound images of the abortion of a midterm fetus so edited to make it look and sound as if it were screaming aloud.[314]

Projecting a scream into the mouth of a midterm fetus was compellingly audible only when part of a wider cultural introjection of noise. Its import and impact fortified by information theory, experimental music, guitar amps, and obstetric ultrasound, noise—as lossiness, gibberish, circuit hum, static, feedback, fetal cry or scream—was also being given neurobiological resonance in Ecological/Spiritual/Psychological impressions of a thoroughly sounding universe. Through such ESP could be appreciated the Screaming Trees, a mystical punk rock band formed in 1985 among the embattled rainforests of the Pacific Northwest, or successive variations on Peter Tompkins and Christopher Bird's *The Secret Life of Plants* (1972+), where crops cotton to universal harmonies, flowers brighten to lilting voices, weeds require a stiff talking to, and fruit torn from the vine may scream in agony. During the 1920s, the Indian botanist Sir Jagadis Chandra Bose had dedicated five thick volumes to the "life movements" and irritability of such plants as the *Mimosa pudica* and the "Praying Palm Tree" (*Phoenix elactylifera*) with nary a word about sound. By 1978, when a documentary on *The Secret Life of Plants* was finished, Dr. George Milstein had already distributed *Music to Grow Plants* (1970), J. G. Ballard had conceived a planet of singing flowers and Prima Belladonnas in his exquisitely speculative *Vermilion Sands* (1971), and Dorothy Retallack of Colorado Woman's College was recounting her horticultural research in a "biotronic control chamber" on the sonic preferences of squash, whose tendrils flung themselves into classical music, shrank like violets from radio static, and bent into grotesque shapes under the stress of rock. The Chancellor of the University of Denver wondered whether marijuana would flourish under a diet of acid rock, as claimed by audiophile pot-heads; an authority on *The Sound of Music and Plants* (1973), Retallack promised further experiments. Her new research, if done, was never published, nor was the *The Secret*

Life of Plants documentary commercially released, but Stevie Wonder's soundtrack to the film appeared in 1979, the first pop album to employ a digital sampling synthesizer on most cuts and later heard as an avatar of New Age music.[315]

New Agers were alive to communications from any part of the cosmos transmitted anyhow—empathically, electrobiologically, astrobiologically, astro-electro-acoustically. If some heard in *The Secret Power of Music* (1984) a New Age "Transformation of Self and Society through Musical Energy" from the good vibes of symphonic classics, L. George Lawrence had heard *Galactic Life Unveiled* since October 29, 1971 through the more radical humming of his three-ton Stellartron in the Hahamongna Watershed in Riverside County, California, and then in the Pisgah Crater of the Mojave Desert. Locked onto the Great Dipper in Ursa Major, this "large-scale astrobiological research instrument" picked up signals "possibly of a modulated anti-matter-type, that can be intercepted only by living organic (i.e., *biological*) transducer substrates." Capping forty years and dozens of accounts of UFO contacts around the Joshua Tree National Monument and its outlying boulders, Lawrence received (through a shielded cache of mustard seeds at the root of his Stellartron) "the first biograms of alien origin," direct from beings in Galaxy M81. His astrobiological instrument, "first of its kind," shared the skies with the nearby Integratron, a sonically pure hemisphere for time travel and rejuvenation of living cells, skillfully built during the 1950s and 1960s without nails, bolts, screws, or metal fittings by George Van Tassel. Using blueprints furnished by visiting Venusians, Van Tassel, an aeronautical engineer, wanted to eliminate all manmade electrical interference from his acousti-celestial system, as did Lawrence, who had been employed as an engineer by the aerospace industry to design jam-proof missile components. Their scientific and interplanetary diplomatic work was well-known to the flying saucer and "borderland science" communities that read their books and attended desert seminars, but they would need a Steven Spielberg to reframe their enterprises for a wider world. In 1977, a year before Van Tassel's sudden death, *Close Encounters of the Third Kind* opened in movie theaters across North America and Europe, with a climactic scene of direct audiovisual conversation between humankind and nearly incorporeal beings from the stars. That Spielberg's film, seen eventually by a paying audience of fifty million earthwide, featured as chief UFOlogist and communicator the French film director François Truffaut, was no mere *hommage*. Truffaut was fresh from his acting debut in his own *L'Enfant sauvage* (1970), where he had played Dr. Itard, the physician from the National Institute for Deaf-Mutes who had tried and failed to educate the "wild child" Victor into the words and world of Napoleonic Paris. In *Close Encounters*, Spielberg gave Truffaut/Itard a second chance, this time to help noisy inarticulate Earthlings rise to the challenge of a more technically and acoustically sophisticated galactic society, one of whose mother ships would return in the spring of 1982.[316]

E.T., with its lost, precocious (and extraterrestrial) child trying to "phone home" through suburban hominid static, rang the changes of soft science fiction on an unusually successful series of television commercials first aired in 1979 to increase long-distance revenues and forestall legislative action against the AT&T telephone monopoly. "Reach out and touch someone" did pretty much that, for years, while the astronomer Carl Sagan's thirteen-part series, *Cosmos* (1980), reached tens of millions of public television viewers and generations of schoolchildren. Sagan's pet project, SETI, the Search for Extraterrestrial Intelligence, soon subsumed Project OZMA, a radiotelescope search for signals from other starsystems (at 1420 mHz, the radio frequency of hydrogen) that had begun in 1960 with a chart recorder, a tape recorder, and "a loudspeaker as well, just in case." The problem here, as for Van Tassel and Lawrence, was one of filtering out the background noise or deepening its context, about which more in a parsec. In the meantime, on the punk side of the acoustic borderlands, No Wave musicians had begun to embrace noise *as* meaning, in tape loops and Noise Fests. A band like Mars may sound to the uninitiated, wrote one pseudonymous critic, Stella Doon, in language at once astronomical and down-to-earth, "like listening to a laundromat magnified. That's because every instrument is making a sound, but who is making which sound? Instead of one direct sound or beat the music travels in at least 3 different directions, speeds of rhythm making a totally orbital sound, one that never really enters the ear, instead spinning around the head. At times it sounds like tortured children singing in 7 different tongues." Another critic described the music of Mars as "Static transmissions from a distant dark star.... A cavernous musical universe riddled with eerie sound storms whose poisoned atmosphere seduced and threatened.... Mars unleashed a choking cacophony illustrating the body as machine in disrepair fueled by the impending repulsion and disintegration of a perverse romance with one's own demons." So the angular momentum gathered for the breakaway of "confessional pop-punk," or Emo, a heartfelt roar of amplified lyrics accelerated by the Rites of Spring in 1985, who departed from Industrial Noise and Noise Music on the strength of vocals that could more easily be made out and made something of, an Emotional Tenor (or E.T.) audible across staves of static, feedback, guitar licks, and suburban angst. After a decade of encounters with The Deftones, grind, and noisecore, Emo would morph into the "vocal-cord shredding" of Screamo and its abrupt shifts "from shredding metal grind to lulling acoustic interludes." Screamo arose from lumps in the throat and a twisted gut—from rage, loss, and love that could be articulated and authentic only when gut-wrenching, and in front of moshpits where the physical and laryngeal intensity of Screamo could be reclaimed at almost a cellular level.[317]

Okay, okay—I confess to chasing after the tortured arpeggios of music journalism because I want to make a moshpit transition to the research of Jim Gimzewski, a medal-winning UCLA biochemist and nanotechnologist who has likely never before been shoved into close confines with Screamo. But how better to come full circle from an expanding chorus of Primal Screams around again to the fetus except through the atomic-force microscope that Gimzewski held over yeast cells in order to listen in on their "screams" and give voice, during the millennial year 2001, to the discipline of sonocytology? Recording the vibratory sound-signatures of live cells, sonocytologists did not exactly drag dying cells "Screaming Under the Microscope," as the hip editors of *Nature* had it in their report on the new field in 2003. Sonocytologists did mean to use highly amplified soundings to assess cellular health, track cellular change, and understand apoptosis or spontaneous cell-death, and though their loudspeakers were also afterthoughts, the vibratory identities of cells had an acoustic flair that bioscientists would not easily let go of. The implication for intrauterine audition was that the earliest embryonic manifestations of life, from blastomeres on, might have something (barely) audible to tell us. Conversely, the implication for adult ears might be that the power of music "to reach the wellsprings of one's very being" (as proposed in 1986 by a philosopher, Jean G. Harrell) lay with the triggering of a "partial recall of very early auditory experience," one's deepest, vaguest memories of comforting sounds felt *in utero*, not just the E♭ and D♭ of systole and diastole but a "sound-rhythm complex" already in place in the thirty-two cells of the morula. On a similarly rhythmic note, in 2008 the biologist Peggy S. M. Hill would report on the preliminary findings of another twenty-first-century discipline, one lacking as yet a snappy moniker but of potentially enormous acoustic import: the study of vibrational communication in animals, including those humanly inaudible vibrations that travel through a local substrate such as soil, water, grass, or plant stems. Until the 1970s, writes Hill, "conventional wisdom held that substrate-born vibration could not transfer any biologically meaningful information among organisms," but just as physicians armed with ultrasound were able to show how sonically tremulous could be amniotic fluid and maternal tissues, so entomologists and animal biologists equipped with more sensitive microphones and seismic recorders and faster, cheaper, smaller computers for signal processing could begin to explore the soundlife of creatures long thought to be deaf or deaf-mute. Researchers had determined that vibrational communication "is a very ancient system, that it is perhaps the primary channel of communication in some animals, and that it is ubiquitous, at least in vertebrates and arthropods," with undeniable behavioral evidence that vibratory signals are understood within, and sometimes between, species. Since the 1980s, 150,000 insect species had been shown to communicate exclusively through vibrations, with another 45,000 using "vibratory signals together with other forms of mechanical signaling," as was also identifiable among a number of land mammals, including mole rats and elephants.[318]

Hubert Markl, a German neuroethologist, proposed in 1983 that the new field distinguish contact vibrations from near-field medium motions, and both from the boundary vibrations at interfaces between media (water/air, air/solid, solid/water). Those categories might help, agreed Hill, but she thought it no longer "a useful enterprise" to distinguish strictly between the soundmaking and vibrating of many creatures, since the same vibration-source may propagate simultaneously through pressure-waves in air and particle displacements in substrate solids, tensile filaments, fluids. French psychologists, psychiatrists, and obstetricians must have been onto something larger than they knew when they began to speak in the 1970s of *une enveloppe sonore* around the fetus and neonate—and now, it would seem, around all living things, from yeast cells to Uruguayan tarantulas to young hippopotami.[319]

L'enveloppe sonore du soi, "the sonorous sheath of the self," as described by Didier Anzieu in 1976, was an aspect of *le moi-peau*, "the skin-ego" whose contours he had begun to trace two years before in his psychoanalytic quest for the origins of narcissism and masochism. Twenty years before that, Anzieu had finished six years as an analysand of the French Freudian, Jacques Lacan, who twenty years before had written his thesis on the psychopathologies of Anzieu's mother, Marguerite, a postal worker who three years before (1931) had drawn a hunting knife from her purse and attacked the actress Huguette ex-Duflos, whom Marguerite did not know but believed to be conspiring to prevent the publication of her romance novels and also to kill her son. The year before (1930), ex-Duflos, née Hermance Joséphine Meurs ("I die"), had played a woman on the eve of her wedding "who narrowly escapes death at the hand of a mysterious assassin" in the first Talkies version of Gaston Leroux's locked-room detective drama, *La Mystère de la chambre jaune*, directed by Marcel L'Herbier, who some years before had narrowly escaped being shot and killed by a distraught lover. During the narrow-escape scene in the quite successful movie, ex-Duflos (making her first appearance in film) uttered a scream that reminded a *New York Times* critic of a scream in a nightmare, "for either the recording went wrong or Mlle. ex-Duflos lost her voice." It was, wrote Mordaunt Hall, "the most disappointing scream in the annals of audible pictures," but that almost-silent scream from within a locked yellow room is a fine soundmark from which to approach the theories of Anzieu. Like Lacan, Anzieu hoped to rescue Freud from himself. For Lacan, this meant an exploration of duplicities, including the doubleness of mirrors and mirror behavior, the reductiveness of self-analysis, and the psychic problem of rooms locked from within and without; for Anzieu, this meant investing analytic practice with a s/kinaesthetic. From his medical experience with dermatological disorders and his political experience with social psychology, Anzieu came to believe that once Freud had left off his benchwork as a neurologist, he had given up too completely on the intrinsic forces of the body. Not that Freud ever disregarded bodily functions or sexual energies: rather, he interpreted them as insignia of underlying complexes, epiphenomenal "conversions" or displacements of psychic conflicts. This misled Freud toward a life of the mind therapeutically distant from that embodiment which constitutes the self during the formative early years of being carried, birthed, and nursed. So Anzieu returned to the locked room of the fetus and the nursery of the newborn who must learn in the most loving of environs to tell its own skin apart from its mother's. As filter, as permeable membrane, as protective barrier, skin literally made space for a self separate from mother and world (beginning,wrote another Lacanian, Denis Vasse, with the rupture of the umbilicus and the navel involution so prominent in children's drawings). Early life had to be comprehended in terms of the vibrotactility and volumetrics of the body; however mythopoetic Anzieu's choice of "Marsyas" as the name for an illustrative patient who in infancy had neither been cuddled nor coddled by cooing, early life had to do with the toughening of layers of skin to protect and shape a body in motion. When Anzieu wrote of a "soundbath" of warm maternal syllables annealing an infant, that was no poetic excess; it was as close as he could come to describing a sensation for which there had been, literally, no words. Circumambient sounds necessarily shape an infant's first psychic space while it has little control over its own limbs or movements beyond that of its cry, which at its loudest entrains every muscle. Before an infant recognizes itself in a mirror, it recognizes its mother's voice and knows that voice apart from its own, a situation that (putting Lacan on auditory notice) Anzieu calls a "sonic mirror." Were the sounds discordant, brusque, impersonal, or absent, that mirror could mottle, become mercurial and poisonous, shriveling the ego. As for the Superego, which Freud had thought might congeal out of auditory fragments from a preconscious life, there was little room for it (cautioned Anzieu) if the newborn's sonorous sheath were riddled with dissonance, interruptions, noise.[320]

Porosity was the issue, to begin with and after all. "At that time I was coiled up like a shell," said A suddenly. "I was an ear, a growing ear, a body sculpted, already, by sound. From inside the cellar into which I had been flung, I could see nothing. The vivid scene hidden, the performance unfolding without me, behind the vibrant membrane, the quivering curtain hanging by my side. All I could discern was a sort of tremor, a flow of ripples, of waves, unfurling, striking—yes, beating, or was it stroking me: in my watery retreat, everything was deadened. Do you realize the pressure that sound can exert on you? Sound is tactile. Yes.... Sometime later, I was born, amid uproar and violence. Later, when my eyes opened, perhaps I became deaf." "To listen is to become porous," says B later in this radio dialogue scripted by René Farabet, founder of France's *Atelier de Création Radiophonique*. A responds, "Why don't we just let the world settle in us, re-sound in us . . . ?" To which Anzieu would reply: To be entirely porous is to be fetal or flayed; to be skinless is to be perilously self-less. The induration of ego requires a sonorous sheath, secure and comforting, and then, yes, a certain deafness, so that you and I may hear each other through the (tympanic) membranes of differentiated selves, selves that choose from all there is to be heard what shall be attended to. Otherwise you and I lose ourselves in a crowd of sounds, the "confused buzz of voices in the distance," says A, "welded together, indistinct, anonymous." B adds: with din comes inattention, and one must keep in mind that "an acoustic image has no circumference"—except within the sonorous sheath, that primordial harbor coeval (as Farabet would have it) with the amnion.[321]

Expelled from the womb at twenty-six weeks and just under three pounds, one newborn was so irredeemably premature (in 1919) that the midwife seized him by his right ear and tossed him into a trash basket. He was retrieved by his *grandmère*, who had borne twenty-four children and knew how to revive the weakest of infants, even one whose uterine walls had been compressed by the tightest of corsets cinched around a sixteen-year-old girl desperate to disguise her pregnancy. From that redemptive instant Alfred Tomatis was destined, as he wrote retrospectively in his autobiography, to investigate prenatal life, the months he had lost in utero, and wounded ears. A physician of the same age as Anzieu, Tomatis had preceded him as a theorist of sound and personhood but, attending rather to the mechanics of the ear, he had begun as a therapist of the strained voices of singers and the speech of stutterers, most of whom, he found, had hearing problems. Once he realized that the audiometric "hearing curves" he drew up for each client could also be maps to cerebral, emotional, and psychological traumas or losses, he became a counselor of the soul. All that too had been foreordained. Not only was Tomatis a sickly baby who cried all day and wept all night until his father took to stringing him up outside upside-down like wet linen on a laundry line; his father Umberto was a famous operatic bass. When he and his father were on holiday together, they went to hear the greatest male voices—Lorentz, Chaliapin, Melchior; when he studied, he read aloud; when, after a year of medical schooling, Tomatis was drafted to be trained in medicine and surgery, he was talked through each stage of doctoring; and when, after the Second World War, he took up his specialty as an otolaryngologist, he focused on the hearing loss of veterans, of workers in munitions factories, and of pilots in the air corps. Through careful audiometry he discovered that those who secretly wished to be certified as deaf (and receive workers' compensation) could unconsciously raise their auditory threshold levels, so that sounds at standard frequencies had to come at them at abnomally high decibel levels in order to be heard, while pilots and others who wished for certifiably good hearing could effectively reduce their thresholds. From this it appeared that hearing and listening were shaped as much by mind and memory as by neurophysiology, hence could be open to non-surgical interventions. Tomatis then applied his audiometric finesse to the treatment of professional singers who were losing their voices or their ability to hit the proper notes, for it turned out that they had been deafened by their own magnificent voices, which, issuing from larynx and diaphragm at 110–40 db, were a devastating self-noise.[322]

Rehabilitating the hearing, and thereby the voices, of famous performers in Paris would subsidize the pursuit of his "particular vocation," the study of acoustic phenomena within the womb. For years he experimented with detection systems that would give a better sense of the sounds in that locked room, which he compared at first to "that of the African bush at twilight (at least as we imagine it and as fictional films present it to us)" with its host of "distant calls, echoes, stealthy rustlings, and the lapping of waves." This led him to the study of what he would call "sonic birth," the adaptation of a newborn's ears from months of listening for those frequencies best transmitted in amniotic fluid to those frequencies most available in air, and to re-identifying the mother's voice, which provides a "vocal nourishment" that is "just as important to the child's development as her milk." And from this he came to sonic rebirthing, at the instigation of Françoise Dolto, a psychiatrist who suspected that schizophrenics in psychic actuality had yet to be born, so could not process or abide the sounds of atmospheric life. Using his patented Electric Ear, a device for ear-training through frequency-filtering, and weeks of prescribed sessions with therapeutically sound-edited tapes of Mozart, Tomatis had some success with the "acoustic integration" of schizophrenics, more success with the increasing numbers of children diagnosed with learning disorders, and most celebrity for his successes with the diva Maria Callas and the actor Gérard Depardieu. He had also spectacular if sporadic triumphs over the excruciating noises of the tinnital.[323]

Controversial as much for its infinite extensibility as for its slippery grasp of neuro-otology, the Tomatis technique deserves notice not because "What Tomatis has done with sound and hearing is virtually a Copernican revolution" (quoth the New Age impresario Jean Houston), but because Tomatis and his disciples in 250 centers around the globe made accessible and agreeable the emergent cultural logic about an acoustically active fetus. His notions about a listening embryo accommodated programs for prenatal auditory stimulation and the playing of tapes of mothers' heartbeats for infants in incubators. His notions of sonic birth and rebirth fit snuggly within the new schematics for a quiet childbirth in hospital or at home and coincided with Anzieu's more recondite schematics of the psychic space prepared and protected by *l'enveloppe sonore*. His retraining of ears by frequency-filtered music transmitted through headsets for regular listening practice at different decibel levels and rhythmic complexities could be easily accepted by younger generations to whom Ipod earbuds were almost second nature and who spent much of their lives within *une enveloppe musicale* (a phrase devised by Anzieu's disciple, Édith Lecourt) or an intense "circumspectral space" (Denis Smalley). His emphasis on sound energies was congenial to a swelling public awareness and swollen definition of autism, now a "spectrum" of disorders whose etiology, according to one recurrent theory, lay in congenital hyperacuities: quivering at the slightest dissonance or overwhelmed by quotidian peaks of sound, the autistic child retreats into a private sensory fortress to ward off a world too loud, distracting, anfractuous. Tomatis' curriculum for "acoustic integration" was in harmony with neurobiologists' and neuromancers' concerns with the brain's integrative functions and experiments confirming the plasticity of auditory cortical processing. While bioscientists were finding that hearing in utero could be damaged by prenatal cocaine exposure and that anything which jeopardized an infant's ability to listen for and imitate adult speech in turn jeopardized the normal development of communication skills, the Tomatis System promised therapeutic results for children with autism, dyslexia, and attention deficit disorder; for poor listeners, readers, and sleepers; for stutterers and sufferers from tinnitus and chronic fatigue; for pregnant mothers and their unborn; for college students and businessmen with "burn-out."[324]

11. Quiet Noisy Books

"Burn-out," as a noun brandishing a hyphen, entered common parlance from the jargon of engineers whose multi-stage liquid-fueled rockets lifted off during the first years of the Space Age, coincident with burn-out's hyphenated antidote, 2-Methyl-2-n-propyl-1,3-propanediol carbamate, a muscle relaxant and sedative familiarly known as Miltown. Burning out, in the sense of emotional and mental exhaustion, was an old figure, taken up by Shakespeare ("His rash fierce blaze of riot cannot last; / For violent fires soon burn out themselves") and reprised by Willa Cather in her assessment of the actresses of 1895: "Some women act with their senses, some with their soul. The soulful ones burn out the quickest. It is a bad thing for an actress to be too sensitive. Bernhardt's acting is a matter of physical excitement, Duse's of spiritual exaltation." Over the next decades, burning out lost some of that inner fire that Gaunt saw in Shakespeare's Richard II and Cather saw in Eleanora Duse, taking on a more impersonal, mechanical cast: "In running a car we know well enough that if we forget the oil we shall burn out the bearings," wrote James Truslow Adams in *The Tempo of Modern Life* (1931). "We Americans too often forget the oil in our own daily contacts everywhere, and we wear ourselves out without reason." During the 1960s, burning out still had something to do with personal dedication among, for example, counselors helping Vietnam War draft resisters, but by then it was as much about demands/assaults from without as compulsion/compassion from within. When the psychologist Herbert J. Freudenberger hyphenated and popularized the term in the 1970s after years of observing a syndrome of fatigue, irritability, depression, and ill health among volunteers in free clinics, burn-out was the result of multiple stressors—environmental, societal, and personal—which (emergency) physicians further compressed into burnout.[325]

Hyphenless, "burnout" was shadowed by "sensory overload," a newer concept often used to account for the cognitive and behavioral problems of autistic children. Coined in 1958 as a counterweight to sensory deprivation and "sensory distortion," the phrase could be applied with equal force to the shock of the neonate muscled into a world of atmospheric sound, light, odors, and breath; to the shock of a raw recruit thrust into battle; to the shock of a bumpkin set loose in a big city. Or to a wife trying to hold her own against a domestic tornado of screaming brats, jangling phones, blaring radios, wheezing vacuum cleaners, and imploring televisions, which were some of the "noises off" behind the glossy advertisements for Miltown (1955) and the benzodiazepines Librium (1960) and Valium (1963). By the mid-1960s, put-upon women and anxious men were being invited into a spider's web of "quiet pills"—pacifiers "for the frustrated and the frenetic," said *Time*; drugs for "transient situational disturbances," said the Food and Drug Administration; "Mother's Little Helper," sang the Rolling Stones in stereo. One exemplary mother received her first tranquilizer prescription when "absolutely going around the bend" with her second child, a "holy terror" who "screamed from the minute we brought her into the house and she never stopped screaming." A brand new baby, wrote the psychotherapist and social psychologist Fabian Rouke, was "in fact, a completely selfish individual," and on the way to juvenile delinquency if it could not accept, or did not trust, the inevitable delays and denials imposed by loving if frustrated parents. Asked in 1957 to organize a research group for New York's Committee for a Quiet City, Rouke used his Pathometer, "the most effective practical modern 'lie detector,'" to obtain graphic transcripts of New Yorkers' reactions to noise not always so obvious as the screaming of a holy terror. "One of the insidious aspects of excessive noise," he reported, "is the fact that an individual may be unconsciously building up nervous tension due to noise exposure, and be suddenly catapulted into an act of violence, or a mental collapse, by some seemingly minor sound which drives him beyond the point of endurance." Violence, *him*; mental collapse, more often *her*. Concerned with the psychodynamics of "negative self-concepts" among otherwise well-off women, Rouke told the *Journal-American*: "Many people who are using tranquilizers may be treating the symptoms rather than the disease. Exposure to noise may well be the cause of the nervous tension which requires 'tranquilizing.' What mother of a young family who experiences the daily 'horror hour' just before supper time is not acutely aware of the strain of noise when she is tired, when the children are hungry and restless? At this time of day or in this type of situation, even slight noises are magnified in their effect and contribute to severe tension and distress."[326]

Picture two French children, one blowing a horn, one banging a drum, the drummer think-speaking, *Maman dit de cesser ce bruit-là parce qu'il la rend nerveuse* ("Mommy says to stop making all this noise because it gets on her nerves"). It looks unlikely that he'll stop just because Mommy says so, and that's exactly what's captioned in black-and-white: "Children will never stop making noise." Worry not: mothers who take *Les Petites Pillules de Carter* will suffer little children more gladly. Eighteenth-century mothers under the strain of noise had taken to cheap gin, for baby first, then for mama: "Babies didn't cry so much when I was young," grumbled an old woman put out by teetotaling reformers in 1834, "and I never thought when I had a baby that I could do without a decanter of gin. There's nothing like it for the colic." Victorian (and Second and Third Empire French, Bismarckian German) mothers had administered opium syrups for the "grand cryings" of children, about which household medical books and physicians' advices seemed as much at odds as they were unhelpful; for their own nerves they had taken Carter's Little (Liver) Pills or Shedd's Little Mandrake Pills (heralded by images of a cat yowling through a tuba) or a solution of chloral hydrate, which packed a greater wallop. Early twentieth-century mothers were advised to use belladonna or chamomile for their children; for themselves they had, among other "time-tested sedatives," Nervine, a bromide (literally: sodium bromide). "Hello! Mary—What time will—Say, what's that noise? How can you stand it?" began an anecdote in Dr. Miles's *New Joke Book* of the 1920s. "Oh! That's the children playing—since I have been taking NERVINE nothing bothers me."[327]

Valium came next, and by 1969 it was the most widely prescribed drug in the United States, targeted at the anxieties of women but marketable also to professional men under high pressure. In 1973, with 15 percent of Americans on benzodiazepines, 17 percent in Belgium and France, perhaps 20 percent in Great Britain and 10 percent in lower-stress Spain, New York City's Environmental Protection Administration began a report on noise with what sounded like an advertisement for tranquilizers: "Living in a society where we are continually assailed by noise, it is as if we were scaring ourselves over and over.... Even when we become accustomed to a noise and think we're no longer bothered by it, we're actually expending a great deal of energy keeping it below the conscious level. As a result we get tense, irritable, and tired" and may burst into tears at the crying or whining of a child. Better a few pills than charges of infanticide for having smothered a screaming child, charges common in cases of Sudden Infant Death Syndrome, which was first identified as an epidemic in the 1960s but not well understood for at least another decade. A National Survey on Drug Abuse in 1982 found that 99 percent of "older women" (thirty-five and up) and 91 percent of "older men" had taken tranquilizers at some point in their lives; more regular usage was reported for 41 percent of mid-adult women ages twenty-six to thirty-four, 48.5 percent of older women, and 27 percent of older men. Prescribed to restore mental balance and emotional equilibrium, or to make it easier on physicians to manage "difficult" patients in hospitals and nursing homes, Valium and other benzodiazepines also helpfully buffered neurological responses to sharp sounds and the perturbations of ambient noise. In psychiatric institutions, where acoustic sensitivities were symptomatic of dozens of disorders and where auditory hallucination was diagnostic for schizophrenia, the drug of choice was the far more potent Thorazine, used since 1953 in French mental hospitals, to the relief of the staff: "straitjackets, psychohydraulic packs, and noise were things of the past!" The prescription of Thorazine for noise-management was inherent in its Greek derivation from *thorubos*, the confused noise of a crowd or an applauding assembly.[328]

Supposing sensory overload to be a compound of acoustic and figurative noise, Anti-Noisites could point to national sedation as a sociopolitical repercussion of noise. Another serious repercussion would be affective withdrawal, as societies became "autistic," unresponsive, uncaring, or obsessively focused on minutiae. The greater the environmental impact of noise, the deeper the presumed retreat from the public arena, whether through drugs or Harlequin romances or the earphones of portable radios and, later, video games and the earbuds of Walkmen and iPods. So sensory overload ran in mutual riot with "information overload," a phenomenon anticipated in 1961, well before media were digitized, up-linked, down-linked, and transmitted to vibrating mobile phones with personalized ring-tones. Too much data or unassimilable news could make people "tune out"—or "turn on" to the multisensorial circuses of hallucinogens. Once computers had emerged from engineering sub-basements, military caverns, and physics laboratories, notions of sensory and information overload merged in computer simulations of synaptic activity and the modeling of brain functions on computer processing, which in turn yielded theories of "parallel processing" cross-talk and "parasitic memory" as explanations for the auditory hallucinations of schizophrenics.[329]

Any which way, noise was understood to be a major contributor to—and chief signifier of—overload, for with overload more and more of the world collapsed into meaninglessness, horror story, or mental disorder. Sensory overload was Anzieu's sonorous sheath and its *simultanéité omniprésente* turned outside in. So Nicolas Jacob, a prominent legal analyst, told a gathering of French advocates and jurists in Royan in 1970 that the actions of moderns inevitably redound upon them as noise, and Pierre Léonard, a prominent acoustician, told the gathering that acoustics had originally been dedicated to the art and science of producing sounds but had been forced recently to devote itself to defending against the sounds so produced. Noise, said a member of the technical commission on noise of the French Ministry of Public Health, is an emission as dangerous and as devilish to eliminate as toxic gas or radioactive contamination. We could not hope and should not aim for a noiseless society, said Claude Leroy of the French League for Mental Hygiene, but surely we could reduce the chaotic stammering of all the information thrown at us. Otherwise we could end up with more cases like the ones Jacob cited, of men (always men) killing neighbors whose children made too much noise, or shooting down noisy youth leaving a dance, or dying of cardiac arrest from noise-shock. Noise, said Jean-Claude Oppeneau, a technical advisor to the Ministry of the Environment in 1984, was at the top of the list of things mentioned by citizens when asked about *perturbations* in their daily lives. As for fetuses, wrote Annie Moch in *La Sourde Oreille* (The Deaf Ear) (1985), they hear best the lower registers and vibrations of music and song, from which they draw comfort; pushed out into industrial and urban settings, younglings are inundated by noises of higher registers, which disturb their sleep and their digestion, make them fitful and "hyperactive." With all the *vacarme* or brouhaha of trucks, *motos*, sirens, radios, and *téléviseurs*, it became difficult to listen, even to learn to listen, and this was made no easier by the reverberant surfaces of slate, formica, and hard plastic in typical classrooms. No wonder that youth were taking to hard rock and claiming that their drug of choice was noise.[330]

Youth had need of an education of the senses, from early childhood on. Such a solution to the problems of inattentiveness, boredom, and overload harked back to the pedagogies of Rousseau and Johann Pestalozzi, then to Pestalozzi's student, Friedrich Fröbel, whose principles for the guidance of children-at-play had led to the creation of kindergartens. Progressive educators, some trained at Pestalozzi-Froebel Institutes, advocated a return to the playfulness and wonder that they claimed had been lost among the velours and bric-a-brac of Victorian stuffiness, though nineteenth-century childhoods, working-class or leisured, were rarely as subdued as twentieth-century reformers made them out. Aside, however, from Julia Rice's home-schooled half-dozen and others in equally fortunate households, children did not receive as extensive a sense-training as that proposed by John Dewey at the University of Chicago ("Without a constant and alert exercise of the senses, not even plays and games can go on") or by Rudolf Steiner in his anthroposophical Waldorf schools in Germany, with their curricular embrace of all Twelve Senses, including the Sense of Movement, Sense of Balance, and Sense of Speech, Word, or Tone. A parallel genealogy ran from Itard's work with deaf children through to his student Edward Séguin's *Idiocy and Its Treatment by the Physiological Method* (1866), whose American curriculum for "sensorial training" would influence Walter Fernald, Superintendent of the Massachusetts School for the Feeble-Minded, and many British minders of schools for the "dull." Maria Montessori, M.D., would use Séguin's teaching materials and theories for her own work with the "mentally deficient" and with slum children in her first Casa dei Bambini in Rome in 1907; she then translated her successes to "normal" children in a movement that spread across Europe and (back) to America, always with the presumption that education was about the carefully sequenced remediation of sensory and cognitive deficits in the young. A third genealogy for sense-education could be traced most acoustically from the work of Hermann Helmholtz and succeeding philosopher-physicists of musicianship, who argued that the ability to distinguish overtones (and play and compose music with the overtones in mind) demanded a special, well-trained, attentiveness.[331]

Invention and imagination, particularly the invention of sounds or noises and the imagination of sound- or noisescapes, were predictably neglected wherever teachers assumed that their primary role was straightening and steadying young children on their course toward the verbal, visual, tactile, and kinaesthetic languages of adulthood. Some, like the New Englander Horace Grant, did attempt to overcome this deaf-spot in sense-training; Grant's little book of *Exercises for the Improvement of the Senses for Young Children* (1887) included auditory puzzles and a game of guessing the origin of such noises as crumpling (of paper?) or shaking (of kerchiefs?). But it would be Lucy Sprague Mitchell, more than any other in America, who took seriously the promise of children's noise-hearing, noise-listening, and noisemaking. When she established the Bureau of Educational Experiments (BEE) in Manhattan in 1916, she came to Bank Street as a good friend of Dewey and "only daughter" of Alice Freeman Palmer, ex-President of Wellesley and now at the University of Chicago; as a friend of Jane Addams and volunteer at the Henry Street Settlement in San Francisco; as the first Dean of Women Students at the University of California, Berkeley; as a mother of four children and wife to Wesley Clair Mitchell, an economist who had agreed to leave Berkeley for Columbia, where he founded the National Bureau of Economic Research and studied "business cycles." She came to Bank Street, then, with strong leanings toward social and economic reform, believing that such reform could be best secured through early childhood education, as for example the Play School that Caroline Pratt had set up in New York City in 1912, unhappy with kindergartens where "You taught children to dance like butterflies, when you knew they would much rather roar like lions, because lions are hard to discipline, and butterflies aren't." The nursery school created by Mitchell, Pratt, and Harriet Johnson at BEE focused on assessing and assuring children's "readiness to react," which entailed a thorough program of sense-education for children *and* teachers and, in the long run, society entire.[332]

Brought up under the regime of a puritanical, haphazardly progressive father and a repressed, haphazardly musical mother, Lucy Sprague twitched "constantly and uncontrollably" whenever sent to schools public or private, so she passed her elementary years under the tutelage of a governess and in her father's large library. She called herself "the-thing-in-the-corner," a shy thing who kept secret notebooks and sobbed alone in an upstairs storeroom, yet as a girl she also regaled her younger brothers with sensational fictions and starred in a little theater of her own billing, with neighborhood kids as her claque. Contradictions between privacy and publicity were built into the Sprague's new house on the south side of Chicago, a mansion whose music room glowed with vases of yellow roses while her parents' dressing room was equipped with a telephone that had a button for "fire" and one for "mob," in case her father Otto (a director of the Pullman Palace Car Company and co-owner of the "world's largest wholesale grocery") were menaced by striking Pullman workers or anarchists. The brilliance of the nearby White City of the Columbian Exposition, and of the associated Parliament of World Religions (many of whose eminences passed through the Sprague dining room), was darkened by the death of Lucy's younger brothers Arnold (diphtheria) and Otho (typhoid fever), by the breakdown of her older sister Nan, by the subsequent deep depression of her mother Lucia and the gloom of her father, whose business was thrown into turmoil by the Panic of 1893. That fall, when Lucy was fifteen, the family departed Prairie Avenue for Southern California, where her father, now hemorrhaging from the tuberculosis he had acquired during the Civil War, hoped to fend off his consumption—which he did, in the company of a tubercular niece, for seventeen years. Nurse to a bleak family that recovered its fortunes sooner than its health (her mother with "recurrent spells of silence all the rest of her life"), Lucy escaped this private sanatorium in 1896, heading off to Radcliffe College, from which she graduated in philosophy (*magna cum laude*) in 1900, with an enduring love of language and prosody. In affect and effect, her experiments at Bank Street would be remediations of the oppressive silences, ominous coughing, and morphine sighs that had environed her youth.[333]

Rather than running children through classrooms like meat and potatoes through canning factories that pasted different labels on identical hashes, Mitchell proposed that schools be shaped according to the needs of individual children, much as she had shaped her own education. In order to substantiate such a curriculum, she had to prove that children had gifts and inclinations that could be reliable touchstones to a worthwhile schooling. She did this not by exulting in childhood innocence or exalting the new Intelligence Quotients but by trusting children's love of rhythm and listening to *how* they described their world. Listening hard. If "the old maxim that 'children should be seen and not heard' was losing its force," Mitchell still had to show that by listening closely to children an adult could learn things that would translate into better teaching, better students, better people—for the point of education was to fashion a more responsive, responsible, and inventive citizenry. She had the children create "songs of the city," their world as they heard it:

> Oh, to hear the din of the machines in the factories,
> Oh, to hear the everlasting call of the flower-men as they go through
> the streets of the city calling "Flowers, flowers."

Or, rawer:

> Oh, what fun it is to hear the great big machinery's songs of great big big heavy wheels.... And at home when mother sings to the baby, and the men are whistling at their work. And then there is an automobile with the engine going, and I suppose that the chugging is its song.

In 1921 she published her *Here and Now Story Book*, a set of "experimental stories" written for, and nearly by, children two to seven years old. For two-year-olds, Marni takes a ride in a wagon:

> Little Aa, he climbed into Sprague's wagon and Marni, she climbed in behind him. Then Mother took the handle and she began to pull the wagon with little Aa and Marni in it. And Mother she went:
>
> Jog, jog, jog, jog,
> Jog, jog, jog, jog,
> Jog, jog, jog, jog,
> Jog, jog, jog, jog,
> And Jog, jog, jog, jog,
> Jog, jog, jog, jog,
> Jog, jog, jog, jog,
> Jog!
>
> And the wheels, they went, (with motion of hands):
>
> Round, round, round, round,
> Round, round, round, round,
> Round, round, round, round,
> Round, round, round, round,
> And Round, round, round, round,
> Round, round, round, round,

Round, round, round, round,
Round!
And then Mother was tired. So she stopped.
And Marni said, "Whoa, horsie!"
Then Little Aa said, "Ugh, ugh!" for he wanted to go.
But Marni said, "Get up, horsie!" for she wanted to go too.
So Mother took hold of the handle and went:
Jog, jog, jog, jog,

and so forth, round and round, until Mother is exhausted. Then "Marni Gets Dressed in the Morning," which (as Mitchell wrote in a preface) "is told in the terms of her own experience, of her own environment, and of her own observations. It is nothing more or less than the living over in rhythmic form of the daily routine of her morning dressing," a celebration and a sanctuary. "I have found that any simple statement about a familiar object or act told (or sung) with a kind of ceremonious attention and with an obvious and simple rhythm, enthralls a two-year-old." For three-year-olds Mitchell retold a child's story about the "Many-Horse Stable" in a big city, where

> In the morning the men would come and harness up the g-r-e-a-t b-i-g horses with the g-r-e-a-t b-i-g harnesses to the g-r-e-a-t b-i-g wagons.... Then they would get up on the seats and gather up the reins and off down the street would go the g-r-e-a-t b-i-g horses. Clumpety-lumpety bump! thump! Clumpety-lumpety bump! thump!

The children listened, more happily than had Julia Rice's, to "The Fog Boat Story" that "grew out of questions asked before breakfast on foggy days, and was originally told to the sound of the distant fog horns," which went "Toot, toot, toot, too-oot, to-oo-oot!" (on many different keys) as sounded by tugs, steamers, barges, and more tugs. Best of all, according to reports from many in her audience, were the stories and poems about subways and Els:

> Rackety, clackety, klang, klong!
> and down the tunnel came a train of cars. "Yi-i-i-i—sh-sh-sh-sh!" screamed the cars and stopped right in front of Boris,

a nine-year-old recently arrived from the Russian steppes who was out exploring New York and already dizzied by the traffic:

> Kachunk, kachunk, kachunk went by an auto;
> Clopperty, clopperty, clopperty went by a horse;
> Thunk-a-ta, thunk-a-ta, bang, bang went by a truck.

And up above,

> Up on high against the sky
> The elevated train goes by.

Above it soars, above it roars
On level with the second floors
Of dirty houses, dirty stores
Who have to see, who have to hear
This noisy ugly monster near.
And as it passes hear it yell,
"I'm the deafening, deadening, thunderous, hideous,
competent, elegant el."[334]

Little of this may seem unusual to parents accustomed to reading aloud from twenty-first-century storybooks, although "competent" and "elegant" are odd adjectives for a "noisy ugly" elevated train, and few texts so directly incorporate the voices and word choices of young children. In 1921, Mitchell's trust in soundwords drawn from the collaborative storytelling of two- to seven-year-olds was about as radical as Russian Bolshevik concerts in factories by conductorless orchestras. Books for young children had scarcely broken ranks with their origins in rhymed primers, Sunday School lessons, and evangelical tracts, as in this doggerel from 1856:

I must not cry,
But try, try, try.
O guard my tongue
While I am young.

I must be mild:
A loving child.

The tone and tenor of such language echoed that of Susanna Wesley, the twenty-fifth of twenty-five children, when congratulating herself in 1732 on the upbringing of her own nineteen children, among them the founders of Methodism, John and Charles: "When turned a year old (and some before), they were taught to fear the rod and to cry softly, by which means they escaped abundance of correction which they might otherwise have had, and that most odious noise of the crying of children was rarely heard in the house, but the family usually lived in as much quietness as if there had not been a child among them." Such rigidity and recessiveness had been extreme even in her own era and mayhap her own house (where nine of her nineteen died as infants), yet schoolmasters and 'marms well into the next century had sought to enforce a like quiet that could stand as proxy for book-learning and moral probity. Mid-nineteenth-century literature still counterposed "Good Little Mary" to "Noisy Cecilia," who "if her maid would not let her out of the nursery, she would take up any thing she could get at, and drum upon the table till she awoke every creature in the house," so was bundled off to a far-away school and not allowed back home until she had ceased "her naughty noisy tricks."[335]

Admonitions to silence, though, had a more consistent presence in generations of etiquette books and primers for parents than could ever be mustered at home, where real children spoke out of turn, and wept or whined, and drummed or rapped on every resonant surface. According to more liberal, sentimental advisories arising at mid-century in tandem with the application of more loving "domestic economies," youthful noise was not only to be expected but welcomed everywhere except in pews and formal parlors. Addressing the topic of "Troublesome Children," Philadelphia's *Christian Recorder* in 1861 asked parents to consider, "When you get tired of their noise, just think what the change would be should it come to a total silence. Nature makes a provision for strengthening the children's lungs by exercise. Babies cannot laugh so as to get much exercise in this way, but we never heard of one that could not cry. Crying, shouting, screaming, are nature's lung exercise, and if you do not wish for it in the parlor, pray have a place devoted to it, and do not debar the girls from it, with the notion it is improper for them to laugh, jump, cry, scream, and run races in the open air." Even at Sunday School, advised the *Recorder*, it was better to accept noisy interruptions than to suppress the spirits of hymn-singing children. An essayist for the international *Chambers's Journal* in 1854 reserved "The Art of Being Quiet" to adults who had begun "to *think* about life, and how we should live." Quietness in children was ill-advised and unnatural: "So necessary is noise to the healthy development of children, that whenever we meet with a child who is remarkable for its quietness, we are apt to infer that it is in a morbid or diseased state." A cartoonist for *Harper's Monthly* in 1864 made more fun of a distractedly serious Uncle George "trying to prepare his Great Speech" than of the stomping children just above, in the nursery, "having a jolly party" of tag, tug, and tumble.[336]

Nurseries, children's rooms, playrooms, even "crying rooms" were the architectural correlates of the increasingly separate status of children, for whom much more than clothing was being tailored to their age, size, and energies. Between 1850 and 1920, middle- and upper-class children would have desks and chairs that fit them, and child-sized tools and musical instruments. They would stay in school longer and by 1900 be rounded up after school into playground teams, sports clubs, scout meetings. They would be tended by doctors of their own, "pediatricians" much concerned that children's ears be safely cleaned, so as to avoid the very common afflictions of ear ache and infection. They would read, or have read aloud to them, *Youth's Companion* (1827–1929, with 385,000 subscribers in 1885), *The Nursery: A Monthly Magazine for Youngest Readers* (1867–1881), *Our Young Folks* (1865–1873) which was absorbed into *The St. Nicholas Magazine for Boys and Girls* (1873–1941), the *Boy's Own Paper* (1879–1967), the *Girl's Own Paper* (1880–1956), two hundred other children's periodicals and scores of picture books and penny dreadfuls. Publishers would issue "juveniles" in natural history, whether *The Water-Babies* of Reverend (and Regius Professor of Modern History at Cambridge) Charles Kingsley, a stramash of fairy-tale, social criticism, and zoological fantasy on a journey to the Other-end-of-Nowhere across an ocean floor whose vents produced "a kissing, and a roaring, and a thumping, and a pumping, as of all the steam engines in the world at once," to *Charlie's Discoveries: or, a Good Use for Eyes and Ears* (a set of six books begun in 1839) that urged Charlie's little angels to listen for and compare noises: "but hark! What noise is that; so loud and harsh, just like the old watchman's rattle which I found in the lumber-room?" (A kingfisher?) What wounded bird could be making that "loud and piteous note," its affecting cry "very similar to that of a young child"? And while tenement children who would never hear a wounded ivory-billed woodpecker were out on the streets peddling and playing as they could, boys and girls in better-off households would be blessed with mass-produced mechanical toys sold as both noisy amusement and instruction—wind-up animals, miniature steam engines, air-guns, "scale model" electric trains; or as active domestic partners—dolls that crept, walked, talked, cried, called for "ma-ma" on the first push of a foot and "pa-pa" on the second, or laughed "at tight money, hard times and pessimists."[337]

Clacking, clunking, chugging, or bawling, the new mechanical playthings were noisier but ostensibly more intriguing than humming tops or bisque dolls, so they were promoted without a lisp of irony as child "quieters." Mothers, take your pick: an inveterately noisy child or an ingeniously noisy toy (like the latest novelty in 1908, a five-in-one marvel of horn, bell, ringer, rattlebox, and nipple) for a pacified child. That was more or less the choice, except for those children disposed by temperament to reading quietly or working for hours with paints. (Lucy Sprague Mitchell knew them all, and they knew themselves. She watched Eric and Philip load wagons with blocks and then announce that they had a truck. "Philip said, 'Yes, I have a noisy truck and he has a whisper truck.'") Most toys, observed the Wilkinson Company of Binghamton, "just make children smile. But when they throw a leg over their Choo Choo cars they *shout!*" Fine toys made children "Shouting happy!"[338]

"Hush!" said Mother to her older children in a vignette of 1874.

> If such a loud noise you keep making,
> The baby will surely be waking.
> I cannot permit such confusion,
> So let your talk have a conclusion.
> By and by, when he wakes, if the weather
> Is fine, we will all play together.

There were, as Aunt Laura wrote, reasons for children to be quiet for a while, so long as you understood that noise was elemental to their being, as it was when Uncle Harry came to watch over *Helen's Babies*: Budge, age five, and Toddie, three. He had been looking forward to a restful day in the countryside, with no urban rumbling of traffic, "nor any of the thousand noises that fill the city air with the spirit of unrest." What he had to contend with, as Aunt Laura might have warned him, were a pair of "Innocent, Crafty, Angelic, Impish, Witching, and Repulsive" children whose cavilling, bickering, slapping, and wailing had exhausted him by the time that breakfast was achieved, and Toddie's screaming, which could be subdued only by providing him with a loud whistle, a noise-prosthetic. Sleep, for children, could be defined as that biologically necessary interim between noises. Picture a boy beating a large toy drum for hours, and then, exhausted, napping:

Hush! The noise is over!
 Arthur is asleep,
Close in balmy slumber
 His tired eyelids keep....

And, for the sweet silence
 Which his slumber makes,
We'll forgive his noises
 When again he wakes.

A somewhat gendered forgiveness, this: boys would be boys, which rhymes with noise; girls would be Patience and Pearl. Nelly was surprised when Johnny Bixby, a newcomer to boarding school in 1860, turned out to be "such a very quiet boy." "He's the first *quiet* boy ever *I* heerd on, then," said her friend Comfort. In another of Josephine Franklin's Nelly books, Maurice and Joe whoop to wake up Charley, Bobbie, and Dick Jay, who really "weren't asleep. We were only keeping quiet. Uncle Dick said we were making the most something noise, and he can't read, and that if we would sit perfectly still for five minutes, he would give us ten cents apiece." Girls needed no bribes to play quietly; indeed, when one night Editha heard a "queer little sound coming from down-stairs—a sound like a stealthy filing of iron," and went down to find a burglar filing through shutter bars, she promised not to scream and instead bribed him with her own pocket watch to burgle as quietly as he could, so as not to disturb her sleeping mother. But girls would enjoy more sonic latitude once their mothers and Aunt Lauras awoke and took to the streets as suffragists lobbying for the vote, garment workers unionizing and picketing, miners' wives battling scabs, and Temperance advocates crashing saloons. Like boys, girls could then be shown setting off rockets on the Fourth of July, shouting proverbs in parlor games, making a racket among pushcarts, and acting like Palmer Cox's boisterous (and somewhat androgynous) Brownies who popped up wherever they wished and reappeared in thirteen books (1887–1907) in a million copies.[339]

Busy adults, however, rarely listened at length *to* boys at their sports or *to* girls at their tea parties or *to* either at hijinks, unless to afternoon headlines shouted by newsboys and newsgirls in the commuter rush. They listened *for* them, sounding out their presence rather than their games, their wares, or their storied inventions, for despite an improving infant mortality rate, the very presence of children was becoming more precious. The percentage of the American population between the ages of five and fourteen declined from 25 percent in 1850 to 22 percent in 1900 and would continue to decline to 21 percent in 1920 and 17 percent during the Great Depression as the average number of live births per (white) mother fell from seven in 1800 to five in 1850, four in 1900, three in 1940. Similar percentages were emerging on farms as in cities and (before the Great War) in northern Europe as in North America, so that rescuing "precious" children from the grasp of white slavers and itinerant musicians, from sweat shops and coal mines, from abusive parents or relatives, from bad air and busy streets, from toy cannon and firecrackers, and even from unorganized play ("usually noisy, often disorderly, and sometimes downright destructive"), would become the mission of visiting nurses, moral reformers, park directors, social workers, and other "professional altruists" on both sides of the Atlantic.[340]

Few of these were aware of the demographics. Rather, the rescue operations were driven by a reconception of childhood in which the child was less the "fruit" of a scriptural dictate or an economic asset than a sacred charge that must be given the time and space in which to mature; its work was not to earn but to learn, not to pray but to play. For this *Century of the Child*, as framed in 1909 by the Swedish feminist Ellen Key, not humility and obedience but self-mastery and social conscience must be the goals; instead of beating or sedating an obdurate cry-baby, parents must isolate the child and caution it that disturbing others will be profitless in the larger world. One communal statistic, however, was well-enough publicized to spur rescue efforts and a further replotting of childhood: traffic accidents, which in 1910 were the leading cause of death among American children five to fourteen, who were run over by trains, cars, trucks, trolleys, and breakaway horses. The acoustic consequence of such a statistic was an insistence upon louder and more frequent train whistles at street-level crossings, trolley gongs around corners, car horns at intersections, ambulance sirens down the middle of avenues, more bells on carriage horses, and a vociferous, well-funded Children's Safety Crusade. The economic consequence was a revaluation of the child. As Viviana Zelizer has shown, the more "useless" a child (at play or at school but otherwise unemployed) and the deeper the investment in its physical and psychic protection, the more its death might be mourned—and its life insured. Policies for children, primarily for a proper burial and funeral, had been widely marketed since the 1870s by private insurance companies in competition with fraternal societies and immigrant mutual aid circles. As (unnatural) accidents rather than (natural) illnesses claimed the lives or maimed the bodies of more children, insurance benefits from any source were cold comfort. Middling and well-off families sought emotional as well as financial recompense through suits for wrongful death, the legal outcomes of which were colder comfort. The less that a child contributed financially to a household, and the more costly its own upkeep, the less its worth in a court of law, a logical reckoning that outraged grieving parents—who, as members of juries, began to make larger, more socio-logical, awards.[341]

Pedagogically, the consequence of childhood reconceived and a child's worth revaluated was that children who were no longer expected to put bread on the table or tend to siblings would be best appreciated for their inner resources: what they heard, saw, and imagined that adults could not, or could no longer. If children were more than Romantic figures pointing back to a lost world of wonder, and if their stages of development were more than recapitulations of the prehistory through which humanity had struggled toward modern civilization, they were yet creatures of unusual sensitivity, at once enviable and endangered. Different enough from grown-ups not to be adults writ small, they were distant enough from blasé maturity to be easily fascinated, and so, perpetually fascinating. As the hope of a nation, one dared not stifle their voice: "We are told that it is natural for the very young to be noisy, and that it is unkind to suppress them," said Thomas Jeeves Horder, M.D., the first Baron Horder, in a radio address on Noise and Health in 1937, and as spokesperson for Great Britain's Anti-Noise League, he fully agreed: "I do not think any of us want to put a ban upon the fullest use of the lungs, or upon the loudest and shrillest toys that human ingenuity can invent, in the nursery or on the playing field." A National Park for the peace of mind of quiet adults might be in order, but as a columnist reminded his readers in *John O'London's Weekly* (reprinted in the League's *Quiet Magazine*), "in so far as we are natural we are all lovers of noise. If we had not been, the art of music would never have been invented. All those noble oratorios that exalt us into Paradise had their origin in a boy's yelling himself hoarse in the jungle." Children were to adults what wilderness was to civilization, the libido to the superego, an inarticulate interiority to a fraught self-awareness. "I am no advocate of the 'Silent School,'" said Sir Henry Richards at the League's Second Conference on Noise Abatement in 1935. "It is healthy and natural and proper for children and young people to make a noise.... To ask children to join a League of Quiet or a Society of Silence is to create a society of little prigs."[342]

Amnesia, however, was always dogging the heels of the young, blanking out their time in the womb, then the shock of birth, then their nursing at the teat and the indigestion and colic, then the learning to crawl, stand, fall. In this context, making noise could be psychologized, as had Theodore Lessing in Germany and Abraham Brill (Freud's designate in New York), as "one of the deepest and most fundamental tendencies of the human psyche." Noise arose from a basic, perhaps instinctual desire to be immersed in life and from a nearly simultaneous (and instinctual?) desire to numb consciousness in the face of life's traumas and setbacks. Or so supposed soured adults who spoke about music in much the same way they spoke about noise; both, wrote Brill, expose our tendency to avoid as much as possible "conscious reflection" and "the knowledge of our own empty existence." BEE children were not under psychoanalysis and had no such conflicted motives to their noisemaking. They did not scream, as did Brill's patients, with adult fantasies of revenge. The children made noise, and a plenitude of other sounds, because that was their job.[343]

Linguists would confirm this, and in so doing confirm the value of attending to odd sounds that most adults disregarded as mere noise or nonsense. While Lucy Sprague Mitchell, her colleagues, and Bank Street student-teachers worked up exercises to help young children articulate the world they heard, linguist-parents and psychologist-parents had taken up where Elizabeth Gaskell and Charles Darwin had left off in their diary jottings about their own first children. As a new mother, Gaskell had noted "the little triumphing noises" made by her infant daughter Marianne when she thought she was going to be picked up from her cot, and as a novelist Gaskell kept a running list in 1835–1836 of the sounds Marianne could make, all the while fearing that she might lose her, a fear that may also have heightened Darwin's acuity as a new father in 1840 as he observed the "various reflex actions, namely sneezing, hickuping, yawning, stretching, and of course sucking and screaming" of newborn William Erasmus, who impressed his father with his startle responses to loud and sudden sounds. (Marianne would live to be eighty-five, William seventy-four, but Gaskell and her husband would lose their first son, and Darwin and his wife would lose three of their ten children.) As for more creative soundmaking, William at forty-six days "first made little noises without any meaning to please himself, and these soon became varied," a babbling that did not seem inspired by the desire to imitate particular sounds or words. At eleven weeks he had a hunger cry audibly distinct from his cry of pain; at the age of a year, exactly, "he made the great step of inventing a word for food, namely mum," in which Darwin heard "a most strongly marked interrogatory sound at the end," an effort at intonation that he had remarked earlier when William would exclaim "ah" each time he was put in front of a mirror. "I did not then see that this fact bears on the view which I have elsewhere maintained that before man used articulate language, he uttered notes in a true musical scale as does the anthropoid ape Hylobates," wrote Darwin as he recounted his observations in response to a brief account in 1877 by the French philosopher and historian Hippolyte Taine on the "Acquisition of Language by Children."[344]

Taine had been impressed by the independence and audible inventiveness of an infant girl. He had listened as she took herself through "the same spontaneous apprenticeship for cries as for movements," groping after separate vowels and then consonants, arriving at a "very distinct twittering" whose sounds grew more precise, coming "almost to resemble a foreign language that we could not understand" and in which she took great pleasure. Whatever sounds she made she came up with on her own: "example and education were only of use in calling her attention to the sounds that she had already found out for herself." The same held for sundry intonations while twittering, such that "all the shades of emotion, wonder, joy, wilfulness, and sadness are expressed by differences of tone; in this she equals or even surpasses a grown-up person." By fifteen months, her "simple warbling" had resulted in "the beginnings of intentional and determinate language." So, as the Canadian Frederick Tracy summarized in *The Psychology of Childhood* (1894), "peculiarly charming infantile babble ... contains in rudimentary form nearly all the sounds which afterwards, by combination, yield the potent instrument of speech." The first cry, which the lyrical had exalted as "the triumphant song of everlasting life" and the cynical had cast as a cry of wrath or sorrow on entering a world of anguish and sin, was in any case more than the pain of "the rush of cold air upon the lungs"; varied by tonality and timing and then differentiated into vowels and consonants, it became language. As John Dewey argued that year, auditors of infant language must not underestimate the primitive, positive significance of sounds that grammarians would dismiss as mere interjections or exclamations. If one listened to what a baby's cries *did*, one found that most cries were verbs; from an infant's perspective, cries were meant to direct or solicit action. From nothing might come nothing, but from the possibility for making noise came the capacity for making meaning.[345]

Monkeys and apes also made different cries and seemed actively demanding. During the 1890s Richard Lynch Garner had gone on phonographic expeditions to Africa and to the monkey houses of zoos in an attempt to demonstrate a philological kinship between the chatter of other primates and the languages of Bushmen and Pygmies, but animal noises never achieved the phonetic or syntactic sophistication of the most "backward" of humans, or of three-year-olds. What the phonograph could do, while holding captive a soundscape and asserting the technological supremacy of one acoustic community over another, was to aid in isolating phonemes from the flow of babble and twitter (or from the sped up language of adulthood: the Association of Shorthand Reporters determined that Americans were speaking ten words a minute faster than they had at century's turn). After slowed, repeated playback, babytalk could be shown to contain invariable "chunks of sound" out of which speech took vocable shape, all phonemes universally accessible by infants at the height of babbling and every set of phonemes eventually bound up in a culturally-limited "semantic constellation." Two processes combined to nurture language: "a strict pattern of inner development," as the Viennese child psychologist Karl Bühler wrote in 1922, and an infant's oral-aural interactions with speakers up close.[346]

Teachers and student-teachers at the Bank Street College of Education had to learn to affirm in children their peculiarities and powers of expression, which meant that they had to listen as appreciatively to how young children wanted to be heard as to what they wanted to say. "The thunder we heard last night was as loud as...," prompts the teacher. "As loud as three horses fighting," says six-year-old Stevie. (By the age of six, a child has spent more than 20,000 hours listening to the world and can recognize 10,000 complex sequences of syllables; more if bilingual.) While linguists charted the sounds of infant speech in a new, supposedly neutral alphabet codified by the International Phonetic Association, Mitchell would arrange sounds for storytelling by rhythm first, honoring that "pattern of inner development" which led from babble and twitter to identifiable syllables and full-scale tales. (This could happen at anytime, even when BEE children were talking to themselves—as often they would, wrote the Swiss psychologist Jean Piaget in 1923.) And however much Mitchell drew upon an expanding archive of compatible observations by linguists and psychologists, her approach also benefitted from a cultural eagerness for listening that was apparent in every kitchen or parlor that had a radio. Audiences in 1922 might complain of a "terrific volume of noise battering at our ears...made by some far-off world as it flees shrieking in agony across the firmament," but they listened all the same through static that became Word in much the same way that babble became Story. As the custom of listening to favorite radio shows (and, soon, programs with folksongs and storytelling just for children) overtook the tradition of listening to a book being read aloud in the family circle, so the purposes of books for younger children, read aloud at bedtime or in primary grades, had to be transformed from the moralistic to the acoustic. A good teacher made thoughtful use of "noises that roar in the space between worlds," which could be static, or psychodynamic. "I can still remember standing solitary and alone on the corner of a busy street in the Village with my eyes tightly closed," wrote Edith Thacher Hurd, Bank Street '34, twenty years later, "for Mrs. Mitchell's assignment of that week had been for us to record *first hand* sounds and smells." In 1937 the editor of *Horn Book Magazine* (which had blown its horn since 1924 "for the best in children's literature") received a letter about Mitchell from Louise Seaman Bechtel, a graduate of Vassar and Yale, a teacher, an author, a radio personality, and director of the first children's book department at a major publishing house (Macmillan, 1919): "I think you *don't really* know how *dumb* most people are about *little* children. And how important her ideas are in relation to the beginnings of conscious use of language."[347]

Mitchell's *Here and Now Story Book* appeared a year before Carl Sandburg's *Rootabaga Stories*. Born in Illinois in 1878, just like Mitchell, Sandburg was dumb neither about little children (he had three daughters, the eldest with epilepsy) nor their use of language. The *Rootabaga Stories* began with Gimme the Ax ("And then one day I got a true look at the Poor, millions of the Poor, patient and toiling," as he had written in "Masses," 1916) heading out of town with his son Please Gimme and his daughter Ax Me No Questions because they have been living in a house "where everything is the same as it always was" and they want else ("Mary has a thingamjig clamped on her ears / And sits all day taking plugs out and sticking plugs in / Flashes and flashes, voices and voices / calling for ears to pour words in"—"Manual System," 1920). They sell their "pigs, pastures, pepper pickers, pitchforks, everything" except their ragbags ("I shall foot it / In the silence of the morning, / See the night slur into dawn, / Hear the slow great winds arise"—"Hobo Days," 1916) and purchase train tickets. The three go "chick chick-a-chick chick-a-chick" past balloon pickers to Rootabaga Country, where start their real adventures, which Sandburg intended "For People from 5 to 105 Years of Age," since the stories were "attempts to catch fantasy, accents, pulses, eye flashes, inconceivably rapid and perfect gestures, sudden pantomimic moments, drawls and drolleries, gazings and musings—authoritative poetic instants—knowing that if the whirr of them were caught quickly and simply enough the result would be a child lore interesting to child and grownup."[348]

"Do you remember when words and sounds and images had particularly curious meaning for you?" Leonard Weisgard would ask the audience at the Caldecott Awards in 1947. "When with a clothespin pinching your nose, mumbling noises would become a foreign language? Or a foreign language become music, making pictures or catching fantasy?" He had won the Caldecott Medal for illustrating Golden MacDonald's *The Little Island*, a book about a small outcropping off Vinalhaven, Maine, where Margaret Wise Brown (known also as Timothy Hay, Juniper Sage, Golden MacDonald) had a cottage. When she wasn't at a tiny nineteenth-century clapboard farmhouse hidden in Cobble Court between 71st and 72nd Street in (that's right) Manhattan, or at her editor's desk (1938–1941) at William R. Scott, Publishers, she was (by 1943) in The Only House, a quarrymaster's cottage on Long Cove, without electricity or indoor plumbing, watching "the sun rise and make a golden island for just five seconds in an early morning sea." Weisgard during his own childhood in England and America had been put off by the glum dullness of children's books and recalled distinctly a collection of nursery rhymes "that always made me angry. It was all so quiet and polite, all the children and all the world in it were so pale and washed out." Having studied art at the Pratt Institute and dance with Martha Graham, he much preferred the bold juxtapositions and contractions of modernist shapes and the subtle curves and shadings of neolithic cave paintings, which situated him, like "those original realists, the children," in primitive mysteries and wild places. "Isn't it strange," Brown had written from Vinalhaven to Lucy Sprague Mitchell, "what complete faith one can have in a place itself. I suppose wisdom is to know one's necessities and not to live without them. And this huge silence, with the woods and the ocean together, and the air full of kelp and the sound of fish hawks and seagulls and nothing else, seems to be something I cherish and get parched without." It was 1942, America had entered the war, and she was just out of hospital after "a minor breast operation," trying to "read and not write for awhile, only it is hard to sit still as long as it takes to read, sound by sound and image by image and word by word with an echo if I like it." At war's end she would finish *The Little Island*, where a fish tells a kitten "how all land is one land / under the sea," and the seasons pass in sounds, from the "howling, moaning, whistling wind" of summer storms to winter, when the snow falls softly "like a great quiet secret in the night." By 1947, at the age of thirty-seven, she had published forty children's books, including *Noisy Book*, *Color Noisy Book*, *Country Noisy Book*, *Seashore Noisy Book*, *Indoor Noisy Book*, and *Noisy Bird Book*, all of which Weisgard had also illustrated. "We experimented with color for sound and shapes for emotion, letting the child bring the magic of movement," he said. "So that a radiator would be placed in a shape suggested by the hissing noise it makes, and the round sound of a ticking clock would put it into a circle."[349]

Brownie as she called herself and sometimes Brownies plural as she was often in a whirl had been the first editor at a publishing house bold enough to ask Gertrude Stein for a children's book ("I think she is the first to write of many people without cutting them off from the grandparents who were once little children whose parents produced them and from the flow of life that goes on about them and went on before them and that will go on after them"). To everyone's surprise Stein in Paris cabled back Yes if on my terms. Stein had been contemplating the genre for a year or more in the form of an "autobiography of Rose," a girl whose sad perplexities arise from a world too confidently approaching war. In 1939 appeared *The World Is Round*, with whose "Once upon a time the world was round and you could go on it around and around" I began Round One, about how why and when it is we still read books aloud. Rose herself can put no faith in such a dizzying roundness as adults presume. It takes some doing and some listening, as Stein's prose circles vertiginously back upon itself, and a hard climb to the top of a mountain whereon cousin Willie shines a searchlight, before

Rose can find her footing. Brownie hoped that the sounds and rhythms of the prose would catch what Stein had caught in *The Making of Americans*, "a rhythm of American day to day existence and relationships that is as certain as the rhythm of the ocean and as binding as the relationship of the ocean to the little waves that crash on our shore." This children would like, for, as she wrote to Mitchell, "children's literature is a literature of the speaking voice, like the Bible." Even when children wrote about silence they wrote with oral cadences:

> Place your finger in the sand
> Burrow round and round
> All the twisting you will do
> Will not make a sound.

That quatrain came from Alice Siegal, twelve years old, and among the oldest of the children whose words were grist for the Bank Street Writers Laboratory of which Brownie became a charter member in 1937. She too had attended Bank Street, arriving after Edith Thacher. And it was Thacher's husband-to-be, Clement Hurd, just back from Fernand Léger's studio in Paris, whom Brown would ask to illustrate *The World Is Round.*[350]

She had pulled up to Hurd's house in Vermont in a convertible, hatless but arrayed in furs and so bursting with ideas talked out in an "odd, gentle, high-pitched voice" that he was exhausted once she left, but after this first visit he knew that she had a mind that ranged farther than Runaway Bunnies. Hurd would illustrate eight of her own children's books, including her most celebrated, *Goodnight, Moon* (1947), while she summoned the reserves to turn to opera, or maybe a study of Virginia Woolf, having been born in that character-changing year 1910. Of a younger, different generation than Mitchell, Brownie was more at ease with her wealth, which had come in her youth from her father's cordage company, with plants in Scotland, India, and Brooklyn. (Her father was one of nine children of Benjamin Gratz Brown, Missouri lawyer, abolitionist, senator, governor, and reformer.) She was born, she wrote, "on a wide street of cobblestones" that "ran down the hill / to the East River / where boats came in / from all over the world / and blew their whistles" past her father's docks where the rope was made, and when she was four they moved to Long Island where she had thirty-six rabbits, a collie, two Peruvian guinea hens, a Belgian hare, seven fish, "and a wild robin / who came back every spring." Educated in a private Swiss school, at a prep school in Wellesley, and at Hollins College in Virginia, she had then to cope with her parents' divorce and the repercussions of Black Friday, taking a series of odd jobs in Manhattan while her sister Roberta taught school and her brother Gratz worked for the A-C Spark Plug Company on the design of intake silencers for carburetors. The cost of her digs in Greenwich Village and her continuing education (in writing and at Bank Street) was borne by her father, who had not lost all, so whether a counter-girl, a rather oddly conventional Bohemian, or the granddaughter of a man who in 1872 had shared the presidential ticket of the Liberal Republicans with Horace Greeley, she was secure enough in her coming of age during the Depression to be at ease most everywhere, from the Village to Harlem, where she went in search of black writers who might merit a scholarship to the Laboratory. She found there, at the center of the Harlem Renaissance, a lifelong friend in Ellen Tarry, who would spend two years with the Laboratory and thereafter craft four picture books (most famously *The Runaway Elephant*) while co-founding Friendship House and working on issues of interracial justice. Brownie herself was rewriting some of the Uncle Remus tales that Joel Chandler Harris had "transcribed" from the Gullah dialect and whose 1880s language children Black, White, and Cherokee (whence Brer Rabbit) could no longer make heads or tails of. She kept the rhythm and figures of speech but did away with the little white boy that Harris had running interference between black wisdoms and white wisecracking. The African-American tricksters and innocents of the Uncle Remus tales must stand on their own, "out of slavery."[351]

Her plot for "How the Little City Boy Changed Places with the Little Country Boy for a Year" was the least original in her 1938 collection of stories, *The Fish with the Deep Sea Smile*, except that the two boys were "colored" boys who seemed to be much like any other boys. Or like other (city) boys who wanted to hear "birds that were not flying from garbage pails with crusts of bread, birds that sang, until it nearly busted your heart wide open to listen them, as they flew in the warm sunlight over endless fields and bushes," and other (country) boys who wanted to know, as Old Jack Lacy said, whether "the city sounded like a terrible storm." It was not that black children and white children led the same lives in the same places, but that, like all children, they used, *had* to use, their senses to secure the world. Or like "The Shy Little Horse" who, because he had "brand new eyes and brand new ears, and he heard and saw everything," was leery of men who chased him to put a halter over his head. But there was a tall man with a moustache who "knew a lot about shy little horses because he loved them," a man who knew how to whistle to tickle his ears, and to keep still and quiet as his mother the mare circled in toward lumps of sugar in his palms, until the colt would follow the mare would follow the man out of the field onto a road where "His brand new hooves made a clicking noise" and then onto a dirt road, where his hooves "made soft little thuds," to meet a shy little boy who would grow up alongside and then ride on a horse neither so shy nor so little any more, jumping fences and galloping "across the green grassy fields."[352]

Accustomed to turning heads with her strawberry-blond hair, green eyes, and insistently forward motion, Brownie herself was a fine equestrian but her forte was running with the hounds, chasing on foot after rabbits across Long Island marsh and meadowland. Her books, though, were usually quieter than the brass beagle horn she learned to blow—except for *SHHhhhh . . . Bang*, about a boy who teaches a city of whisperers the joys of noise (an oblique critique of sourpuss librarians: "Can't you see children in the hushed atmosphere of a library tiptoeing up to a librarian and asking for it—Sshhhhhhh B A N G!"). When Weisgard told Brownie how, as a child, he and his father had gone around London recording street cries and city sounds on a phonograph, the first of her ten *Noisy* books took shape. A little black dog gets cinders in his eyes. His eyes bandaged as in a Bank Street blindfold exercise, Muffin tries to identify all the sounds he hears, the grrrrrr grrrrrrr of his stomach rumbling, the sisss sisssssss of the radiator, the awuurra awuurra of car horns, sounds that Bank Street children had taught her. But Brownie knew from her work with Mitchell that full-on listening must entreat more than onomatopoeia; a listening ear *expects* to hear everything, even sounds that can only be imagined. Then the sun began to shine. Could Muffin hear that?" It began to snow: could Muffin hear that? And what of a

squeak
    squeak
        squeak

that Muffin could not quite decipher, a

squeak
    squeak
        squeak

that was not (of course) a big horse, nor an empty house. "What do YOU think it was?"[353]

"Listening all over, with eyes and ears and hands and feet," that was the point of Brownie's *On Christmas Eve* ("the quietest night in the world," 1938) and the point of all her *Noisy* books. Each had an elusive noise YOU had to guess at, pages YOU had to talk back to. Each entreated simile, poetry. "What is the quietest thing you know?" asks a Bank Street teacher. "Quiet," says one child, "as a thermometer goes up." Another: "Quiet as a splinter goes in." Another: "Quiet as mud." The *Noisy* books, Brownie would explain, "came right from the children themselves—from listening to them, watching them, and letting them into the story when they are much too young to sit without a word for the length of time it takes to read a full-length story to them. I mean three- and two-year-olds." What's more, "Dogs will also be interested in 'the Noisy Books' the first time they hear them read with any convincing suddenness and variety of whistles, squeaks, hisses, thuds, and sudden silences following an unexpected *BANG*." Writing for the very young was "not unlike writing for a puppy or a kitten . . . a world of pure language." So Muffin the little black dog is all ears in the summer for "a noise way up in the sky like chairs being thrown all around the room" which couldn't be, could it, "a Big Brass Band crashing their instruments up on a cloud"? And he was all ears in the winter for "a little noise, soft as the sound of people breathing." Could it be "A big balloon going up to the moon?" No. What do YOU think it was? And all ears in *The Quiet Noisy Book* in 1950 for something very quiet that nonetheless woke him up. "Was it a bee wondering?" No. "Was it a skyscraper scraping the sky?" "An elephant tiptoeing down the stairs?" No, no. "It was a very quiet noise. Such a quiet noise." ("Quiet as someone eating custard," she tried out in a draft. Quiet as "a toothbrush thinking?" As "A little house running in deep grass and a bay running away from it?") "Quiet as snow falling. Quiet as a chair. Quiet as air.... What do you think it was?"[354]

Leonard Weisgard knew: it was what every child needs to be assured of before going off to sleep after the last goodnights of *Goodnight, Moon*. It was (turn the page...) the sun coming up. It was, and Weisgard turned the lettering here from black or white to blue, "It was the day," once again. It was "a new leaf uncurling." Two years later, in 1952, traveling in Europe and about to be married, Brownie fell ill in Nice. Surgeons removed an ovarian cyst and, just to be safe, her appendix. Recuperating swiftly, as athletes do, she was so thrilled at the prospect of a quick release from hospital that she kicked one leg high over her head, dislodging an embolism that travelled to her brain and killed her. It was a very quiet thing, "a wheel turning halfway round."[355]

"Writer of Songs and Nonsense," read part of her epitaph. That was a playful understatement, though it was true that Burl Ives had recorded two of her songs and members of the New York Philharmonic had played an orchestral setting of *The Little Brass Band*. In sixteen years she had composed over five hundred works, many of them lyrical, few of them nonsense; two hundred books and stories had appeared under her several names, and another forty or so, like *Bunny's Noisy Book* (2002) would appear posthumously, along with translations into a score of languages and new or newly illustrated editions of her "classics." Ursula Nordstrom, her editor at Harper's for *Goodnight, Moon* (and "discoverer" of Maurice Sendak) boggled at a proposal for a Margaret Wise Brown (Memorial) Library for Children, for "I like to think of what Margaret's own idea for such a library would be. I'm sure she'd want the children to talk or sing wherever they wanted to, paint the furniture, draw with crayons on the walls, jump on the tables, kick any interfering adult on the ankle...when they weren't looking at books." Brownie had appreciated the noises and noisemaking of childhood. She had realized the sounds a child fears at different ages—at two, barking dogs and flush toilets, according to one authority, and at four, fire engines, at five, thunder, at six, doorbells and ugly voices. She had known, like Sendak, how children's fears of separation and darkness, of largeness and lostness, of contraction and abstraction, must be addressed out loud lest the "dolphins" (those pilings and moorings between groundedness and adventure) fall away, disappear. She had understood how important it is for a child to be assured, as Marsha Lippincott has written in her analysis of fear in children's books, that "you are safe; you are not alone; laughing helps; blankets and teddy bears help; magic words and magic tokens help. Most of all, stories help." Storied *sounds*: for sounds of all sorts are the ropelines from early childhood to adulthood, and Brownie had sensed how, from the noise of a fish breathing or butter melting, so much more can be made of the world than happenstance. "The important thing about the sky," she wrote in *The Important Book* (1949), "is that it is always there.... The important thing about you—is that you are you." Her hackles had risen when the "here and now" style was dismissed as the "beep beep crunch crunch" school, for she and Mitchell (and Harriet Johnson, and Edith Hurd, and Ellen Tarry) had gone well beyond onomatopoeia in their depth of metaphor, their range of language, their aurizons of cityscapes and darkened bedrooms. "There is a music I have heard," she wrote,

Sharper than the song of bird
Sweeter still while still unheard
There beyond the inner ear.[356]

"Juniper Sage" or "Golden MacDonald" or "Margaret Wise Brown" would never come to the tongue as trippingly as the "Dr. Seuss" of Theodore Geisel, who in 1937 began his move from cartoons advertising sugar and zippers and a bug spray called Flit into thin books for kids with rhymes anapestic and nonsense nonsensical. In the long run, however, Brownie's oeuvre had at least as great an influence on the dialogue between parents and children and, as a result, on the soundsystems of childrearing. It had been through Lucy Sprague Mitchell and BEE and then the Bank Street Writers Laboratory and particularly Margaret Wise Brown that children from the 1930s on would be encouraged to experiment with sound, their noisiness not put down, and children's rooms re-bounded as "rumpus rooms," and girls and boys alike found to be, if not prodigious musicians, then natural acousticians. Zabladowski, the plumber and helpmeet in Geisel's 1952 screenplay for *The 5,000 Fingers of Dr. T.*, may extend the principle of his Little Nemo Invisible Hearing Aid that "brings noises into my ear" so "it might also could bring noises into a bottle" full to bursting; Miss Brown the oenophile, who hid wine bottles in streams near The Only House, was everywhere uncorking phrases more poetic and noises metonymic. Hewing to a vocabulary of 250 words, Geisel/Seuss entertained his readers with nonlexical names, Ogden-Nashian suffixes, and chimeric figures in three or four basic shades; Brownie, Hurd, and Weisgard carved out a read-aloud territory for toddlers (and older) that deferred to no etiquette of sheltered lexicons and minimal palettes.[357]

Together, the Seuss cartoons, the Bank Street lessons, and the *Noisy* books of Margaret Wise Brown would prepare the soundstage for heroes and heroines of children's books who do not grow out of their noisemaking so much as they grow into the world through the making of noise. They will learn their numerals and their alphabets from jungles and farmyards of noisy animals, much as earlier children did, but now they will learn their acoustic manners from parents on soundwalks down city avenues past boom boxes and street vendors toward boisterous romps in city parks. They will master their fears of thunder, rural or urban, with the welcome sound of rain, as before, but also with the counter-insurgency of jam sessions with grandparents who welcome a little more noise in their lives. They will master their fears of bumps in the night not by dismissing ghosts and ghouls but by trading noises with the most rambunctious or terrified of monsters, like that eponymous little monster, Bump, imagined by Papa Hemingway's youngest grandson. Indeed, the *Noisy*-book children will befriend monsters anywhere, under the bed or in the closet, and like a young monster himself on the verge of dreaming in Michael Rex's *Goodnight, Goon*, they will first say their sleep-tights to all the other monsters in the room. They will wake to pages where girls and boys alike at last can be constitutionally loud and, in times of peril physical or ethical, commendable for their loudness. They will learn to stifle neither their laughter nor their music nor their tears—like Kate Klise's Little Rabbit, for whom crying is "not a sob story" but a perfectly human response. They will read Steve Parker's *Earsplitters! The World's Loudest Noises* with the same avidity as Jacqueline Woodson's *Hush*, the tale of a twelve-year-old African-American girl whose family must enter a witness protection program. They will go owling silently on a snowy evening in Jane Yolen's *Owl Moon*, the heat in their mouths "the heat of all those words / we had not spoken," then wonder at *The 39 Apartments of Ludwig van Beethoven*, each with five legless pianos that must somehow be moved each time neighbors complain about the increasingly deaf-and-piano-pounding composer. The generations who passed through childhood with *Noisy* books and "Silly Slammers, Toys That Make Noise" would give rise to protest songs and political chants, acid rock and heavy metal, industrial and machine music, emo, screamo, and Japanese noise bands like Merzbow, The Boredoms, Ruins, Ground Zero. And those who had sung in garbage-can chorus with Oscar the Grouch (whose favorite music was the "honk, bang, whistle and crash" of the "beautiful noise" of *Sesame Street*) would read to their children from the last page of David Elliott's *Finn Throws a Fit!* where even a tantrum ("Uh-oh! Thunder in the nursery! / Lightning in the kitchen!") has its moments: "May all that energy you have take you to the most amazing places!"[358]

12. Civil Defense

Left behind in Margaret Wise Brown's papers were scores of unpublished stories and nubs of ideas for stories. One nub, "The Bomb Proof Bunnies," had been inspired by artillery practice near Vinalhaven during the Second World War, if not also by the alarums of civil defense in a nuclear age. She had plans, too, for a *Ridiculous Noisy Book* and a *Squeaky Book* (was it the squeak of sugar dropped into boiling tea or the squeak of crying dolls when you make them sit up?), but the Bomb Proof Bunnies beg for exegesis.[359]

Bomb shelters are where we are heading: warren and amnion. First, however, I must resume the history of shellshock, which had resurfaced during the Second World War under other names, with new culprits to blame. Children's-book rhyme is apropos, for with "anxiety neuroses," "combat stress" and "convoy fatigue" arose again the question of a soldier's preexisting susceptibility to shock—and this time the blame was laid in the laps of overly anxious or "smothering" mothers who had not armored their sons with the confidence and independence to withstand life's blows.[360]

Blow by blow, the Second World War was louder than the First. The vibrations and reverberations felt and heard within the tight steel chambers of third-generation tanks, the curved bare-metal walls of long-distance bombers, and the cockpits of fighter planes could be as great as any experienced by a boilermaker; the larger field artillery, anti-tank guns, and anti-aircraft emplacements, as well as massive bombing runs, buzz bombs, and V-2 rockets, could make the earth rumble more furiously and the bones of the ear shake more sensibly than the munitions of the Great War. The sheer numbers of jeeps and trucks, motorcycles and tanks, escorted by whirroaring helicopters, made for louder, more mechanized battlefields. The home front itself had to cope with the clangor of round-the-clock assembly lines and accelerated production in heavy industry ("Every day," it was reported, "one East Coast airplane plant dispensary treats four or five women workers who are half hysterical from exposure to the intense racket of riveting"). Factory sirens competed with air raid sirens, military police patrol whistles with those of air raid wardens, the brankling of scrap metal collected on flatbeds with the clanging of fire brigades. And because the war was manufactured and fought with higher-speed engines than those of the Great War, loudness meant a distinctive shrillness. The shrillness of platoons of machine tools slashing metal. The shrillness of Hitler on the radio, "that high-pitched strident voice, rising sometimes to a hysterical shriek but never falling to a quiet, convincing level, keeping on without a break." A shrillness that was exploited by each nation's dive bombers—the American Dauntless, the English Skua, the Japanese Aichi, and above all the German Stuka, initially with a siren on the leg of its landing gear so that its engine roar and aerodynamic scream intensified as it dove steeply toward a target.[361]

Plant-Noise Measurement
CLICK GRIND ZONK ROAR
THUNK YELP PING KNOCK
CRASH FIZZ PUT-PUT BEEP
SWISH BONK SNORT THUNDER
HOWL CLAP
CLUCK DRIP SQUEAL
WANG PATTER KAPOW
TICK TOCK
CLANG
DONG
SLAM
NOISE METER ANALYZER
particularly convenient, lightweight, portable instruments, rugged, highly accurate, for laboratory and field use.
TYPE 410-C SOUND LEVEL METER $320.00
TYPE 420-A SOUND ANALYZER $619.00
Protect Your Workers' Hearing
avoid possible lawsuits by using
FLENTS
anti-noise EAR STOPPLES
The NEAR System INDOOR ATTACK WARNING
A little black box –
potentially the most valuable electrical appliance ever plugged into a wall outlet—holds the solution to a serious problem in national defense.
PROBLEM: How to sound an attack warning signal inside millions of American homes, schools, factories, and offices.
SOLUTION: The National Emergency Alarm Repeater system—NEAR.
FOR WAR NERVES
STOP NEEDLESS NOISE
Keep Fit - Keep Well - Keep Working
NOISE
IS OUR ENEMY TOO!
Noise wastes energy, dissipates manpower, slows war production
STOP NEEDLESS WAR NOISE
WARNING
FIRE·FLOOD·TORNADO
strike without warning!
The Atom Bomb is not the only, nor the greatest danger to American Communities. This year natural disasters have hit all parts of our country. Many of these areas felt immune to danger, but with nature there is no immunity — no four hour alert — she strikes now! Fast and furious!
All communities need disaster warning signals. Prepare now! Supplement your existing fire signal with modern sirens like the FEDERAL STL-10 and the FEDERAL Thunderbolt. Blanket your entire community with ample sound to prevent and reduce loss of life.
FEDERAL
SIGN and
Relieves Pain
Calms Nerves
Fights Depression
ANACIN
ACTS ALL 3 WAYS

Louder, yes—but *noisier*? more acoustically *terrifying*? The generals, avoiding trench warfare whenever possible, and military doctors, avoiding at all costs the word "shellshock," thought not. During the Blitz, England's Anti-Noise League never laid straw along the avenues "to soften the noise of falling masonry," as expected by *Punch*, but psychologists did work to damp the acoustic terrors of air raids. Noise there would always be, but tolerable, so long as soldiers and civilians were soundly educated in what crescendos to expect and how best to cover their ears. During training maneuvers, U.S. Army sound trucks broadcast recorded battle noises to give infantry a better sense of the cacophonics of war. Artillerymen and infantry were shown a short film of "Battlefield Sounds" as a first step in learning to distinguish the low cough of a 60 mm mortar from the higher-frequency crackle of an 81 mm mortar or the metallic, ringing discharge of the 105 mm howitzer. Officers in training were sent into underground chambers of horror whose "whole idea" was "the progressive and systematically increased harassing of all five senses so that individual officers may get to know themselves and their capabilities and limitations of action and reaction through stress." Loudspeakers amplified recordings of the shriek of shells, the roar of planes, the skid of crashing cars, violent laughter, a woman screaming, animals growling, a man sobbing in agony, a baby crying, and the crinkling of cellophane close to a microphone, "which is unbelievably nerve-shattering in its effect." For the war effort, the polyphonic Mel Blanc (a.k.a. Daffy Duck, Bugs Bunny, and Tweety Bird) took on the voice of Private Snafu, a character drawn by Geisel (a.k.a. Dr. Seuss) for a series of funny (a.k.a. instructional) animated features (a.k.a. cartoons) full of sound-effects, among them the dangerous "Flat-Hatting" of hot-shot young pilots who put a plane "into a screaming dive at some innocuous target." Sailors given rest cures for "convoy fatigue" were considered recovered if they could watch unfazed a movie "with lots of real battle noises." Meanwhile, physicians on both sides were testing new "ear defenders," plugs and muffs that infantry might find tolerable and use with conviction during a conflict that was expected to be loud and long. But not a single soldier in "Battlefield Sounds" was shown wearing ear protection, and Hallowell Davis, directing the Central Institute of the Deaf in St. Louis, would note as late as May 1944 that "Considerable doubt seems to exist in the minds of those with whom we have discussed the problem as to the desirability of providing the infantry with earplugs, as the risk of not hearing commands or approaching shells, etc., seems greater than the risk of blast deafness." It would take more than laboratory data to convince commanders on the front and soldiers in the field that the use of appropriate ear defenders could improve a man's acuity for significant words and sounds by masking much of the extraneous noise. Even so, the general staffs on all sides were concerned that the 31 millions mobilized by the Axis and the 62 millions mobilized by the Allies should lose neither their hearing nor their wits.[362]

Except that during training, soldiers at rifle ranges and artillery practice, with inadequate "ear wardens," had already begun to lose some of their hearing, as had their drill sergeants (why else did they insist on all that shouting?). And in combat it was the nature of noise and vibration *over time* that led to crack-ups. A soldier did not cringe at the mere rustle of a curtain because he had been in battle for a few days but because the noise of war went on for years, and the body's fight/flight mechanisms were "intended for short spurts," not for months under (sonic) assault. Nor were the nerves or the ears so constituted as to sustain hours of straining to hear a sussurus of leaves that could be an intruder with a bayonet; or to make out, through the wallop of mortars and whoosh of flamethrowers, the "cricket" squeeze toy used by Allied troops to identify each other as they slogged inland from the beaches of Normandy. In the midst of battle, whose ears could tell whether the unloading and banging of loose boards, the down-throttle of trucks, the laboring voices and intermittent hammering were elements of a recording used for "sonic deception" or a bridge being assembled to cross a strategic river? While German troops had fits of crying and panic on an Eastern Front bogged down in siege and trenches, American troops bivouacked on North African deserts stammered in near-constant fear and shambled across the sands listening to empty skies that echoed invisibly with low-altitude bombing and strafing runs that often enough *did* come out of nowhere, so that men would jump up in bed at night shouting "Dive bombers! Dive bombers!" when all was quiet. Field Marshall Montgomery as late as June 1944 had to disband an entire regiment, the 6th Duke of Wellington, because three-quarters of the men were "jumpy" under fire and audibly "hysterical." After four months under the arc of shrieking bombs a few yards from the screams of dying friends at Okinawa or Guadalcanal, men did not have to be suffering from what one colonel called "blast concussion" to be dazed and muted, but advances in field surgery made it possible for more men to survive mild brain injuries and go on from one bloody campaign to the next. In this total war, which killed at least 37 million civilians and 25 million in active service, no army could afford to limit its men to tours of duty of 200 days of combat, the limit (said the U.S. Army Surgeon General) of a soldier's endurance.[363]

Fight-or-flight became a pivotal analytic category for Wilfred Bion, a British psychiatrist who pioneereed group therapy for the shellshocked or whatever-you-called-them. Bion himself had been born in India, in the Raj of "Harsh sun and silence; black night and violent noise. Frogs croaking, birds hammering tin boxes, striking bells, shrieking, yelling, roaring, coughing, bawling, mocking," and he had loved it. As a nineteen-year-old Brigade Major of the Royal Tank Regiment in the First World War, he found the noise violent but unlovely, a groan from the mud and a distant cry as of bitterns mating, a "Dante's *Inferno*—but how much better we do these things now." He learned from the war, as much as from a postwar degree in history, a surgical M.D., and sessions at the British (Psychoanalytic) Institute, that he had to listen directly to the noises made by his patients, paranoid or schizophrenic or shellshocked: screaming, crying, howling, farting, stuttering, mumbling were to be understood as acts of articulation, ways of making sense in cases where other ways are, and must be, blocked. These were (not?) coincidentally the sounds of infancy, so U.S. and German psychiatrists (who hoped to clear all blockages with hormine, scopolamine, sodium amytal, sodium pentothal, and other "truth serums") tended to characterize "war neurotics" as immature men whose mothers had never given them the chance to cope on their own. Under pressure, "ninnies" would regress, go into a fetal position, literally or lexically. They were among the millions displaced by the war, though they were refugees across time, hearing themselves into the nightmare of a recurrent dive bombing, strafing runs, or thirty-six-hour bombardments, or into the sanctuary of a childhood from which they had never fully surfaced.[364]

Groups of such men might choose to work together toward useful projects or cling to the most paranoid among them, one whose fears were an apotheosis of their own. Group therapy was intended to help the shellshocked regroup, gather their thoughts individually and their resolve collectively, and restore to them a confidence in words spoken and shared. How and why group therapy was effective Bion would spend decades sorting out, he who himself, just before going into battle, had been gripped by a nightmare horror so grey and shapeless he could not cry out, could not put it into words, and who had awakened "to the relatively benign terrors of real war." Thousands of demobilized soldiers returned home unable to regroup, belying the assurances of a "Good War." (Virginia Woolf, writing her "Thoughts of Peace in an Air Raid" while at Monk's House in Sussex under the drone and howl of hundreds of planes rushing off to dogfights over the downs, urged women in 1940 not to plug their ears or cover their heads with pillows; they should go out and campaign to end the noise of air raids by ending war itself.) The United States Infantry Association in 1945 sponsored the publication of a paperback guide, *Psychology for the Returning Serviceman*, which took seriously the problems of soldiers returning to an America that bore no "outward scars of war," and where "Some civilians have the idea that all combat experience has some terrible effect on all men, and that practically all returning servicemen have had a bad case of combat nerves." Which was not so far from the truth: even the most unruffled veteran had to be prepared "for the shock of the feeling that you are stranger in your own home," and civilians had to realize that something so simple as "Carefree laughter can be extremely jarring to the ears of men who have forced themselves to get used to hearing men who are dying." The inner defenses erected by a man during wartime, whether on the front or in a POW camp, would be hard to let down in peacetime, and "Almost every man comes out of combat keyed up, restless and tense," so it would take time to settle down and settle in. Peggy Terry's husband, a paratrooper who had made twenty-six drops in France, North Africa, and Germany, came back from the war entirely changed, an alcoholic plagued by "the most awful nightmares. He'd get up in the middle of the night and start screaming.... We'd go to the movies, and if they'd have films with a lot of shooting in it, he'd just start to shake and have to get up and leave. He started slapping me around and slapped the kids around. He became a brute," hard-drinking but never "battle-hardened." Others simply came back hard of hearing.[365]

Millions of others. One otologist estimated that there were 150,000 hearing-loss casualties in the American army alone, their ears deranged by the sudden detonation of bombs or rockets, by 88-mm gunfire, by "blasting effects" from very strong signals during prolonged radio operation, or by pulling eight-hour shifts twenty-four days a month for fourteen months in the diesel-engine room of a patrol craft. Another estimated a million such casualties, based on his time in the First World War and the new noise of tanks and mechanized battle. The Veterans Administration stated that "the thunderous din of war damaged the ears of about 40,000 American servicemen," but that number, argued researchers, reflected only the most egregious cases. If infantrymen were like those of the Great War, or like run-of-the-mill young men anywhere, they would hesitate mightily before complaining of hearing loss, as they also resisted the use of earplugs. Instead, they would cunningly or half-wittingly compensate for both tinnitus and reduced acuity lest they be stigmatized by the visible prosthetics of a hearing aid or be temporarily deafened by its distortions at high frequencies. The Zenith Radio Company, which had learned from its wartime research how to maximize sound for deficient ears, how to minimize some of the distortion, and how to miniaturize the technology, distributed a pamphlet that claimed that 98 percent of people in postwar America had "some defect of hearing," that hundreds of thousands of soldiers had already "been put to the necessity of wearing ear-aids," and that "Today intelligence is winning over vanity," so that "Ear-aids are everywhere in evidence, even blossoming out of the ears of children." The blossoms were more likely the earplugs growing out of shirtpocket radios and (in 1954) out of the first transistorized portables, but in the first years after the war otologists had startling statistics from wartime audiometry: 30 percent of a run-of-the-mill passel of soldiers who made no fuss about their hearing were nonetheless found to have as much as a 30 db loss at one or more of the standard test frequencies. Such findings indicated that there must be "a very large group of armed personnel with varying degrees of undiscovered or unrecorded battle noise trauma."[366]

Airmen in particular returned home with their hearing awry. Not only had the vibrations of powerful engines damaged the cilia that respond to lower tones; not only were higher-frequency propeller noises louder inside a plane than outside, and ototoxic; the men also experienced great difficulty adjusting to "negative noise." Flight crews reported that "after lengthy trips, the sudden cessation of noise to which they had become accustomed—which may also be termed *negative noise*—is as painful as the *positive noise* and that, for hours after de-planing, they find it difficult not only to hear but also to quiet their nerves." Another kind of shellshock: unnerving silence. Aviation medicine, a field that got its wings in the 1920s, had been brutal in its approach to the otological, expecting pilots in due course to lose much of their hearing. Louis H. Bauer, Commandant of the U.S. Army's School of Aviation Medicine in 1926, asked only that flyers begin with healthy Eustachian tubes and inner ears, so that they could maintain their equilibrium while doing loop-the-loops and other evasive maneuvers. Cockpits should therefore be designed primarily for touch and vision; any auditory cues had to be loud and penetrating enough to be registered by a man with a third of the acuity of an average citizen and likely to be ear-fatigued after flying long hours each day, or subject to anoxia and vertigo if he had flown too high. All that Bauer could advise pilots contending with the "constant roar of a high powered motor" was to sew ladies' powder puffs into the ear-laps of their flight caps. By 1939, military aircraft had closed cockpits with earphones for pilots and, in larger planes, intercoms to radiomen and gunners, which made for a new set of noise problems. Due to the greater reverberation within closed cockpits and to "the clatter of aircraft static projected through tightly fitted earphones," pilots had to be screened for above-average hearing so that they would not mistake conning tower directions or radio commands and hear/head east for west, 80 degrees for 18. With excellent (initial) hearing, airmen would be that much more alert to noise, so they stuffed cotton in their ears under their headsets to damp extraneous noises but rarely used earplugs, whose shapes and textures they found to be a ----ing nuisance. As a result, they spoke loudly, almost shouting, into their onboard mikes, and their mates would suffer the auditory consequences of being shouted at through static while being shaken and shot at. When flight crews came home, after many more missions than they had signed up for, their hearing loss would have been significant but, in comparison to the norm, their acuity would have been just below average, and unremarkable.[367]

Unremarkable, because the hearing of the civilian population had also suffered during the war, especially in Europe, despite programs to reduce industrial and commercial noise for reasons as much of economy and civil defense as of otology. ("Quiet!" exclaimed one American war poster. "Noise Wastes Energy, Dissipates Manpower, Slows War Production. Stop

Needless Noise." Noise was the counterpart of Rumor, Careless Talk, and Loose Lips, which sank ships. Uncle Sam with a finger to his lips glared down at machinists and pedestrians.) The military funded research into earplugs and hearing loss, used conscientious objectors as subjects for hearing tests and hearing aids, and staffed centers for the "aural rehabilitation" of soldiers deafened in combat. Less was done for those at home, although in 1940 the American Management Association had already published a report that "of every thousand industrial workers given hearing examinations today, less than fifty-two per cent will have normal hearing in both ears." *Time Magazine* did feature advertisements in the form of "Facts from the Files in America's War on Sabotage by Noise," including "The Case of the 'War Nerves' Victims, or How an Alert Management Brought Blessed Relief to its War-Busy Office Workers." While otologists for the Army and Navy were discovering that "identical hearing losses are not necessarily most successfully compensated for by [hearing] aids of the same frequency characteristics" and that "There is no universal pattern of compensation," American companies remained reluctant to broach the question of hearing aids, lest they acknowledge that industrial noise could be a direct cause of debilitating hearing loss and so become liable for claims for workers' compensation, claims that had more standing in European welfare systems. While "Smitty" Stevens and his adjutants at PAL were seeking answers to such questions as those posed in 1943 by the psychologist and science writer Marjorie Van de Water—

1. What jobs in armed services for which hearing especially important? Communication, Radio reception, Submarine detection, or am I wrong?
2. Is pitch discrimination especially important anywhere?...
3. Can any simple statement be made about the pitches of common military sounds, like airplane motors, propeller sounds, rifles, guns, tanks, shells?
4. How can the ear be protected against loud sounds and from the shock of explosions? Does opening the mouth really help for explosion shocks?

—corporate lawyers and accountants were fretting over the financial consequences of recognizing the deafening effects of years of daily shifts on factory floors or airport tarmacs. While plastic surgeons during and immediately after the war had to deal with the "greatest number of facial injuries in the history of man," including ears disfigured, heavy industry put the best face it could on its efforts to lay the onus for ear protection on workers, who were rarely required to use earplugs or muffs in the proximity of riveting jigs, punch presses, drop forges, and power looms.[368]

Stevens himself gave an opening to industry with a widely cited article on "The Science of Noise" in the *Atlantic Monthly* in 1946, quickly reprinted in *Science Digest* as "Don't Be Afraid of Noise." It was "just the sort of article which short-sighted plane operators undoubtedly like," according to an angry reader in New Hampshire, "regardless of the discomfort of the much greater number of the non-air travelling public, who are driven crazy by the racket of planes overhead." Stevens wrote that people were happily productive in noisy milieus, and that "the walls of an ordinary room reflect sound more effectively than a good mirror reflects light," so that people were suited rather to noise than to quiet, anecdotal evidence for which he drew from the uneasiness of the many visitors he had escorted through the 72,000 cubic feet of Harvard's anechoic chamber, whose 20,000 wedges of muslin-covered fiberglass removed all but a thousandth of the sound. Given the finesse of our ears, which could pick up a soundwave whose amplitude was "less than the diameter of the smallest hydrogen molecule," and the power of our ear canals to reinforce wave amplitudes by a factor of five, it stood to reason that we were equipped rather for sound than silence. Ears (therefore?) were so built as to recuperate swiftly from transitory acoustic trauma, and despite any professed desire for quiet, "man is a complicated creature who nowhere exhibits the contrariness of his complexities more readily than in his reaction to noise." For example, "In World War II there were more and bigger clamors than ever before in history," the loudest of which were air raid sirens that achieved over 25,000 watts of "acoustic violence" and produced sustained shrieks of 165 db at the mouths of their horns—but no one complained. Where Dr. Stevens had gone wrong in his measurements, the reader from New Hampshire could only guess. "Perhaps your experimental personnel was made up of husky young brutes just off the football field, who would rebound quickly from the effects of such exposure as you describe. But if you are right, certainly banks, business houses and industrial concerns have been wasting a lot of money recently in deadening noise in their establishments with a view to increasing the general efficiency of their forces." American industry might underwrite the construction of sound partitions or acoustic ceilings but resisted any legal constructions of noise as an occupational hazard. While war-torn nations had to rebuild their manufacturing base from the ground up and so could more easily incorporate better acoustic designs and materials, America's infrastructure was intact, so manufacturers alone would have to bear the expense of costly refittings to avoid workers' claims for noise-induced hearing loss.[369]

Psychologizing noise, as Stevens did with his research into the unevenness of the relationship between physical stimulus and subjective response (to increasingly bright light, intense vibration, tactile pain, or loud sound), also gave industry room to maneuver. Despite the ambitions of the Harvard Psycho-Acoustic Laboratory (PAL) to graph the experience of noise in a generic—usually male—ear, companies could argue that complaints about noise were idiosyncratic and unsystematically related to the decibel levels of typical workplaces. A research physicist for the Johns-Manville Company, which fabricated many of the sound-control materials used by factories and offices, was "struck by a fact which I cannot explain," as he told PAL in 1954. "On the one hand, we cannot find any changes in metabolism, reaction, coordination or the like as the result of noise, but on the other hand, people seem to be willing to spend millions of dollars to provide quiet working conditions." Did Stevens have an explanation? Stevens did, returning again to that anechoic chamber where John Cage had stood three years before. "I am continually amazed, and amused, at the reactions of visitors who enter our anechoic chamber where for the first time in their lives they hear what absolute quiet sounds like. They feel a pressure on their ears, which they usually try to relieve by swallowing. They find it difficult to talk, singing is out of the question, and often they say, 'Wouldn't it be awful to live in here!' And they are usually happy to get out again into a normal room where only the ceiling is 'sound-treated' and where they again experience the reassuring reverberation of the normal environment." Bottom line: If people don't like much noise, they don't much like quiet either. "A man is like a circuit with a heavy dose of negative feedback—he is pretty stable" in an unquiet sort of way that Stevens had explored with subjects wearing staticky headphones or surrounded by loudspeakers fed by white-noise generators. "If I were to try to guess why it is that I dislike noise, and why I have spent a lot of Harvard's money for sound absorbent ceilings, I think I would name *masking* as the chief factor. It's frustrating—and frustration is annoying." Annoyance, however, had no demonstrable bearing on permanent hearing loss and, ergo, none on claims for compensation.[370]

Age was another story. Who was to say that workers might not have lost their hearing in any case, from the natural effects of aging (presbycusis) or from a general wear-and-tear (sociocusis) of years as a hunter, or stints at rifle ranges or shooting galleries or in the army, or from sitting each night with an ear to a radio whose volume has been turned way up to pierce the screech of traffic or the brass of a Big Band blaring from a neighbor's phonograph. Below 100 db, 110 db, or even 120 db, it was—and for a few industry apologists, still is—difficult to associate levels of sound-energy and median hours of exposure with consistent personal sensations of disorienting or painful noise, individual evidences of irreparable cochlear damage. Audiometric tests at fairgrounds and other mass venues had revealed characteristic, progressive hearing deficits among the general population at specific frequencies with increasing age, regardless of occupation, race, or gender. The war itself had revealed that audiograms based upon the ability to detect faint, pure tones were scarcely prognostic for the ability of a person in ordinary life or on the front to make out intelligible sounds through daily or military noise. Who could say, then, what percentage of any employee's hearing loss an employer should be held accountable for, given that almost no workers could prove that they had begun employment with excellent or even average acuity or that they had not over their years on the job injured their ears outside of work while operating power tools or drag-racing or hunting squirrels or playing the drums or rocking a screaming baby?[371]

Veterans back in the workforce made the issue of compensation trickier still. Eager for jobs, they disguised whatever hearing loss they knew they had and ignored all subtler signs of deficit in the interests of being hired on. Smaller, more effective hearing aids were increasingly available, using wartime advances: printed electronic circuits, invented for proximity fuses; walkie-talkie technology; acoustic filters that could selectively screen out disturbingly amplified high frequencies; miniaturized microphones; molded plastics. Some hearing aids were now tuneable, responding to the habit of "fine tuning" that had become ingrained in radio audiences listening for war news; the Acousticon Imperial A-90 hearing aid of 1947 had "not *one* but *three* positions for tuning out, *by fingertip control*, unwanted, irritating background noises as simply as you tune out stations on your radio." But returning servicemen, like most of the adult population, would not take up prosthetic ears unless their hearing loss was strong and sudden—or appreciated as a chief cause of the bungling of work or relationships, which was the thrust of postwar advertising for Western Electric hearing aids. (Directed to the Boss and Secretary: "*I don't think he likes me!* Wait a minute. Remember your boss doesn't hear well. That means he must strain to understand. That's hard work. It's bound to show in some of his actions and reactions." Directed to Mother: "*I'm Back in the Family Circle Again.* I never realized how much my family used to struggle to keep me in the conversation." Directed to Father and Son: "*You mean my dad can hear like other men? Mother won't even have to talk loud to him? And we can go places like other fellows?*" To Grandmother and Granddaughter: "*You're not a crosspatch any more!* Remember, sis, Grandma didn't always hear well, the way she does now [with a 'full color' hearing aid]. And we're glad you notice the difference.... Dispositions often suffer when someone doesn't hear well. Straining to understand can make a delightful person upset and nervous.") Western Electric knew well what it was up against, as did the Chairman of the Board of Dictograph Products, Inc., who explained in 1949: "The principle difficulty that every hearing aid manufacturer is faced with is the fundamental fact that the vast majority of the hard of hearing do not want to buy a hearing aid *and reject in advance their need for a hearing aid*.... Our advertising, therefore, is deliberately designed to induce these hard of hearing people to at least disclose to us who they are and where they live."[372]

Because invisible, easily hidden (and almost as easily feigned), hearing loss had little of the social or scriptural cushioning of blindness. For centuries it had also been enfolded within iconographies of imbecility, decrepitude, and senility. Although the wearers of thick glasses had been subjected to a humor no less cruel than the pasquinades of ear trumpets, modern eyeglasses were much easier to fit than hearing aids, and poor hearing, even when amplified, was more deeply implicated with isolation, unhappiness, or psychic imbalance. "Hearing loss, failing health and neurosis go hand in hand," wrote the Dictograph Chairperson. "The fundamental fact is that the hard of hearing are not well, physically or emotionally." Where there were no signs of neurosis or illness, there remained intimations of intellectual incapacity: Western Electric directed one of its advertisements to Son, Teachers, and Parents: "*They even thought at first he might be 'dull.'*" All told, as veterans returned to the assembly line or big rigs (displacing the able women who had taken their places during the war), they were not partial to disclosing otological infirmities. Years later, then, management could redirect the scrutiny of workers' compensation boards from the noise of metal stamping machines toward the lasting imprint of war.[373]

Paralyzing sound guns had been contemplated as tactical weapons in 1940, until investigators for the National Defense Research Committee proved that acoustic attacks could not be sufficiently disabling to merit further research and development. As late as V-E Days, German scientists were still toying with the idea of a lethal "sound bolt" that could stop men dead in their tracks, but even with the advent of jet engines that produced sound-energies greater than any other humanly contrived source except atomic bombs, the noise would harm only the ears, according to Hallowell Davis. As Executive Secretary of the BENOX (Biological Effects of Noise, Exploratory) team of nine physicists and physiologists assembled by the Air Force in 1952, Davis advocated the use of ear defenders, especially on Navy flight decks, but the team could find no extra-otological bodily evidence of the hurtfulness of very loud or persistent noise, despite common complaints of "excessive fatigue, occasional nausea, and loss of libido." The team did acknowledge that no laboratory was equipped to research the physiological effects of sound bursts over 140 db; as the thrust of jet and rocket engines increased, science would be unable to say whether such unprecedented sound-energies could trigger epileptic seizures or produce discharges in the brain's anterior reticular formation, activating the pituitary release of ACTH and creating a refractory state in the adrenals that led to chronic fatigue. What would happen to a launching crew routinely subjected to 130–40 db, given that earpain had been found not to be a reliably early indicator of injury to the more sensitive parts of the inner ear? Would the crew experience hazardous disequilibrium? Reduced reaction times? Permanent hearing loss? And how did the physical, emotional, and cognitive effects of exposure to intermittent but enormously loud noises compare to the effects of four or eight hours' constant exposure to quieter sounds around 100 db?[374]

Workers' compensation boards needed to know. Compensation laws had been in place in Prussia since 1884, in England since 1897, in France since 1898, in the Netherlands since 1903. In the United States, federal workers had coverage by 1908 but otherwise each state had to draft its own legislation, which began in 1910 with New York (invalidated the next year) and 1911 with Wisconsin, concluding thirty-seven years later with Mississippi. Until 1951, though, claims regarding work-related noise-induced hearing loss, pesky in Germany and the Netherlands from the start, had been fended off by American corporate lawyers on the grounds that the baselines for normal hearing were unreliable and the noisiness of industry a reliable measure of its own vitality. Carey P. McCord of the Institute of Industrial Health, who that year chaired the planning committee for a series of twenty-five lectures by twenty-five men on "Sound—Wanted and Unwanted" at the University of Michigan's School of Public Health, was willing to admit that "Noise is ubiquitous; the 'din of the tumult of things' is with us everywhere and all the time; the day when one who desires to 'go into the silence' may not be able to do so without ear defenders is already a close tomorrow." Yet man now had the power to reroute noise (his phrase) and use it as "a boon"—for instance, in medical ultrasound—so noise could not be all bad. Besides, en route to work or to play, travelers inside a Los Angeles bus, the Paris Metro (1st class car), the London Tube, or a DC-3 all had to put up with 100 db of noise for hours at a time, and at home their valve-type flush toilets (wrote Vern Knudsen) "roared to 88 db at each flushing," while factory floors averaged 80 db or below. Even so, the matter of noise deafness was "about to break over industry's head with even more dire potentials, in the form of claims, than those which the silicosis racket exploded twenty years ago." How many industries would be destroyed if noise-induced hearing loss became regularly compensable and "riotous programs of compensation payments appear"?[375]

Everyone also seemed to have deafened friends who compensated masterfully for their auditory deficit and would never want to be designated as handicapped. And since few became so deaf that they could no longer work, it was exceedingly difficult to correlate degrees of hearing loss with degrees of incapacitation, so any monetary reward would seem more arbitrary than for the loss of mobility or of vision, or of a limb, an eye, even the pinna of an ear. When Charles W. Ferst sued the Dictograph Products Corporation for compensation for hearing loss after weeks of assembling and testing the receivers and transmitters of Dictograph machines, New York's Industrial Commission in 1920 agreed that his ears had been "subjected to an extra hazard caused by the strong vibratory sound waves continuously striking his ear drum," but the Appeals Court reversed the decision, hearing nothing causal between his job and the reputed injury, which had left him with functional acuity in both ears, though slightly diminished in the right. A telephone operator in 1927 was able to convince a North Dakota Appeals Court that there could be no other reason for the pain, irritation, and inflammation of her ears than the fact that for seven years of daily work at the switchboard, she could not remove her headset and had involuntarily received in her ears "intense vibrations, static, loud noises, and ringing bells"; she won her case not because she could prove permanent hearing loss but because she had been driven so crazy by all the "low static noises" in her ears that she had had to quit for six months in order to recuperate. Only in 1951, when a sixty-year-old man who had been working for six years for the Green Bay Drop Forge Company sued for recompense due to hearing impairment as the result of exposure to (unavoidable) noise at work, did a state supreme court, reversing a circuit court decision, acknowledge noise-induced hearing loss as a disability to be compensated according to a schedule similar to partial blindness. This judicial turn of events had three pivots: a new section of a Wisconsin statute that made loss of hearing compensable whether by immediate trauma (as decided in 1916 in *Wagner* v. *American Bridge Co.*), or as an occupational disease; a 1948 New York ruling in *Slawinski* v. *J. H. Williams & Co.* that noise-induced hearing loss (with bad tinnitus) was a compensable occupational disease, and a 1949 New York ruling in *Rosati* v. *Despatch Shops* that the 53.7 percent hearing loss in Angelo Rosati's right ear and 46.2 percent in his left after twenty-seven years of riveting railroad cars were compensable as permanent partial injury, despite a lack of exacting proof that such hearing loss would be permanent; legal precedents in compensation for silicosis, awarded years after a miner left the mines, despite stipulations that the miners in question had not been ill before they left and so had not lost any working income. The analogy to silicosis was most significant, and not simply because hardness of hearing rarely forced early retirement; by analogy to the subterranean working environment in which silicosis developed, the Wisconsin court held the drop forge environment more directly responsible for the worker's hearing loss than his sixty years of age. As a seminal decision, the Green Bay Drop Forge case gave medico-legal standing to problems with hearing aggravated by, though not necessarily originating within, an industrial setting. The decision made hearing loss seem less inevitable and noise-control/hearing-conservation programs more urgent, though the standards for determing harm were wobbly, the range of state laws "a crazy quilt," and schedules of compensation continued to show "a striking disjunction between disability rating and functional impairment." Should workers' compensation be awarded only where there was certifiable deafness at the frequencies required for speech? Where job performance declined? Or where a worker's ears took a week or more to recover from a temporary threshold shift after lengthy exposure to certain frequencies at certain decibel levels?[376]

Temporary threshold shifts—increases in the sound-power required for a set of tones to be audible after acoustic trauma—could be taken contrariwise as evidence of resilience, since ears usually resumed their prior acuity. That is to say, an encounter with acoustic trauma, even a "blast concussion" or being bowled over by the thrust of jet engines, was not a doom. There might be some ringing or buzzing in the ears, an auditory dullness, a snippet of hair loss in the cochlea, but the ears were robust, and protected by the strongest part of the skull. This, at least, was one of the arguments against compensation for hearing loss: who could say what loss was permanent? Otologists even wondered whether a person who was about to make loud noises or knowingly enter a noisy environment could "'preset' a protective mechanism that might make the noise less hazardous." After all, not every worker who spent years beside a drop forge or a printing press complained of hearing loss that could be unimpeachably attributed to the forge or press, and among the 2,336 entries in the bibliography on noise compiled in 1955 by the Industrial Hygiene Foundation of America, there was no consensus on just what levels of loudness or noisiness an ear might tolerate for just how long. If, as Hallowell Davis wrote, "Noise continues to increase because it is intimately and probably inevitably associated with power," perhaps a knowing ear could physiologically accommodate the axiom that "Noise is the penalty of power, and cannot be engineered away."[377]

Cold War rhetoric made of resilience otological and psychological the riposte to penalties of power, atomic or acoustic. "Ever since the roar of a saber-tooth tiger sent our ancestors scurrying for their caves," declared the Civil Aeronautics Administration in 1957, "mankind has associated loud noise with danger." This was a deep-seated reaction, but reasoning citizens should understand that the opposite had to be true when it came to our *Upstairs Neighbors*, warplanes. "The loud roar proves that the engines are doing their job and that all is well. A silent airplane in the air would indeed be dangerous." So, from the standpoints of safety and national security, "there is nothing more reassuring than a good, deep engine growl." The same ilk of federal authorities in the United States hoped to persuade citizens that you could survive an atomic attack if you took a few precautions. For example, at the first scream of an air raid siren, make like Bert the Turtle and "duck and cover," steadying your nerves by "reciting jingles or rhymes or the multiplication table until the All Clear sounds." Kneel with your face to your knees in a fetal position, eyes closed, hands clasped behind the neck in the "atomic head clutch position" (which I recall rehearsing in the long brick-and-linoleum hall of my grammar school). If caught on a highway, crawl into a trench and pull tarpaper over your head.[378]

Better yet, enter the quiet confines of a bomb shelter, its pantry stocked with cans of food according to menus printed in *Shelter on a Quiet Street*: there were home economists of atomic aftermaths. The government's 1959 edition of *Facts about Fallout Protection* began with the reassurance that

"The Whole World Is Radioactive." A natural phenomenon, radiation in normal amounts was safe. Too much, right, could kill... unless you had three feet of earth above you or two feet of concrete between you and it, and a door as solid as that on Isaac Rice's chessroom, and one humdinger of an air filter. Yet, despite an "epistemology of the bunker" accommodating a national security state in a permanent condition of red-alert, and despite a cultural poetics that made of the yellow-and-black inverted-triangle insignia of fallout shelters an icon of the Fifties, few were ever bought or built. (Except those secretly destined to harbor cabinet members, senators, generals, and their families below Mt. Weather, Virginia, behind thirty-four-ton blast doors five feet thick, or under Raven Rock, Pennsylvania, or in bunkers beneath the Greenbrier Resort in White Sulphur Springs, West Virginia, or in the seven-story underground shelters of Sweden and the enormous complex that could house 300,000 beneath a Moscow suburb.) Years before, in a 1923 parable of "The Burrow," Kafka had drawn out the precariousness of subterranean refuges too well for any subsequent playwright, novelist, or screenwriter to be persuaded of their impregnability or even of their capacity to nurture—through enforced intimacy—a disintegrating "nuclear" family (that pun from civil defense historian Laura McEnaney). "The most beautiful thing about my burrow," wrote Kafka in the voice of the burrower, "is the stillness. Of course, that is deceptive. At any moment it may be shattered and then all will be over." Listening intently through his labyrinth for sounds of cave-in or invasion, the dweller hears his "own blood pounding all too loudly" in his ears and an indefinable whistling, "audible everywhere, night and day." The stillness is as shifty as the threshold is temporary, as it would be for the military unit holding out on *Level 7*, Mordecai Roshwald's novel of 1959 in which radiation seeps underground, level by level, after a 178-minute nuclear war that has eliminated all threats, including humanity, in a silence empty of recompense.[379]

Anyhow, a massive program of shelter-building would have been prohibitively expensive: the director of the Federal Civil Defense Administration estimated that the cost could amount to as much as the value of the Gross National Product for 1950, especially if one followed the recommendations of the RAND Corporation and built a catacomb of deep-rock shelters under Manhattan, with forty-one entrances leading to safety 800 feet below the surface. A massive public program would also have been dangerously provocative, intimating that America was planning on a first strike (as Godzilla the fire-breathing monster had been awakened in 1954 by the noise of nuclear tests, notes the historian Kenneth Rose). And after the Eniwetok tests of 1952 and the Russian H-bombs of 1955, most shelters would have been useless for much else than that modicum of quiet now scarce inside middle-class homes with their televisions, stereos, and electric blenders; unless burrowed inside a mountain, shelter occupants would be incinerated or suffocated within ten or twenty miles of the epicenter of the explosion of a single-megaton hydrogen bomb. If they were (un)fortunate enough to survive, they would experience sensory deprivation in the months before they could return to the surface. Money and mortality aside, a shelter program would be immoral and inhuman, said critics on the left, like Margaret Mead and the Physicians for Social Responsibility; it would encourage a garrison mentality and, if put into practice, an antidemocratic triage. For critics on the right, any proposal for communal shelters was further evidence of creeping socialism, forcing upon a democratic, self-helpful, minuteman society the dogmas of shared food, shared space, shared air; let those build them who had so prospered as to be able to afford them (at a cost, barebones, of half the annual income of an average family). In the American South, civil defense officers argued about segregating the shelters, and nationally the NAACP would only engage with civil defense if it came with guarantees of civil rights for blacks. As a "dead room," then, the bomb shelter was the cultural-historical converse of Cage's anechoic chamber music: a place where the beating of one's heart meant not an appreciation for the noisiness of being alive but the panic of people banging on barred doors, "the seismic quake of the ground all around, the loudness of the noise of the blast," and the fear, from within and without, of extinction. After a nuclear war, survivors would know true sonic deadness: "Most of the familiar sounds would be missing," predicted *U.S. News and World Report* in 1961. "Birds and insects would be dead, so you wouldn't hear them. Traffic would be dead, too—or very nearly so. And the factories, of course, would be out of operation. It would be a silent world to which you emerged."[380]

Fallout changed the terms of emergency and the sonic urgency of an alert. "The Atom Bomb is not the only, nor the greatest danger to American Communities," claimed the Federal Sign and Signal Corporation of Chicago, one of the nation's largest manufacturers of sirens in 1955. "This year natural disasters have hit all parts of our country. Many of these areas felt immune to danger, but with nature there is no immunity—no four hour alert—she strikes now!" Naturally, "Even a few seconds warning can save countless lives" and Federal's Thunderbolt siren could rally rescue teams from far and near. But hydrogen bombs had severe, prolonged aftermaths. Any rescue team braving the desolation, wrote the biologist Florence E. Moog, would find itself unable to quiet or comfort the millions burnt, mangled, and dying aboveground, with opiates in short supply and medics burning out. Where it might be reasonable to "Blanket your entire community with ample sound to prevent and reduce loss of life" due to tornado or flood, how could a person alerted by Thunderbolt and safely sheltered not hear in her mind's ear the screaming and groaning that would be the residual overtones of the explosion of bombs hundreds of times more lethal than those exploded over Hiroshima and Nagasaki? Would anyone in any shelter live to hear the sounding of an atomic All Clear?[381]

Evacuating prime targets might be a better bet, if early warnings could come early enough and if, at the first whirr of a prescient siren, 62,000,000 citizens scrambled calmly out of the forty-two American cities most likely to be targeted. And if workers left depots, oil refineries, and munitions factories with the choreographed orderliness of the evacuees in "Invasion of the Body Snatchers" (story, 1954; film, 1956). "When the siren sounds," said Glenn Ford, narrating a 1950s civil defense film, there would be no panic or societal collapse on "The Day Called X," since "each man knows his responsibility" should an enemy H-bomb obliterate Portland, Oregon. Cold War anxieties over evacuation paved the way for Eisenhower's request for the Congressional funding of an interstate highway system. The President himself was dubious of mass evacuations, knowing as he did that there would be less than an hour between the time a Russian missile was launched and its arrival over American targets. He and his general staff also knew how many hundreds of square miles, and hundreds of miles downwind, would be in the zone of lethality created by a single nuclear detonation as scientists developed more powerful payloads and longer-range rockets with multiple warheads. During a series of annual nationwide air raid drills called Operation Alert (1954–1961), trial evacuations were often fiascos: traffic jams and radio traffic jams upset "Operation Kids" in Mobile, Alabama, when parents tried to carpool 37,000 children out of town; the people of Peoria, Illinois, refused outright to participate; sirens were ignored not only by thousands at Rockefeller Center but by many of New York City's own civil defense officials. A traffic study of Milwaukee revealed that, under sunny skies and with all cars, trucks, trains, and buses in working order, it would take seven hours to evacuate the city's 640,00 inhabitants; by extrapolation, forty hours' advance notice would be needed to evacuate the 3,620,000 residents of Chicago, eighty hours to empty the boroughs of New York—if that could be done at all. The Fellowship of Reconciliation, the War Resisters League, SANE, and other peace organizations therefore argued that the posting of fallout shelter signs at the entrances to subways and the bolting of evacuation route maps to the exteriors of public buildings was a hoax on the public, the majority of whom would not survive the first hours of a nuclear war or the months of fallout thereafter. This accusation did not sit well with state legislators, who passed laws in New York and elsewhere that made the seeking of shelter mandatory during Operation Alert. When Dorothy Day of the Catholic Worker Movement and other like-minded New York women remained quietly aboveground in City Hall Park to protest the 1955 drill as "a military act in a cold war to instill fear, to prepare the collective mind for war," twenty-seven were arrested, found guilty, and given suspended sentences as murderers responsible for the deaths of three million New Yorkers mock-killed during the mock attack of Operation Alert. It was indecently pessimistic, and un-American, not to give citizens some sort of heads up about keeping one's head down at the approach of Armageddon.[382]

Whence the debatably redemptive wailing, warbling, and buzzing of new networks of alarms that would rise above the noise of traffic (sometimes literally, with loudspeakers in twin-engine airplanes), and at home above the crying of so many babies that one day they would be called "boomers." In England in 1947, air raid sirens were being sold to factories as surplus; across the Anglophone Atlantic, more powerful sirens in elaborate grids were gradually being installed around all major cities, 679 in New York City in 1954. Municipal officials resumed the making of sound maps that had been initiated during the war, charting the acoustic reach of factory whistles and mayoral sirens. Cities scheduled weekly siren tests in the open air and the federal government broadcast emergency tones on Conelrad, a new civil defense radio network. An exclusive reliance upon radio or television alerts would have been problematic during the 1950s, as people of all ages were not necessarily within hearing distance of electronic media; visuals (excepting the always-effective Batman searchlight) would be useless in dense cities, across hill country, through forest, and on fog-bound coastlines. Even with a network of sirens, civil defense administrators expected that farmers' first notice of a nuclear attack might be intense lights in the sky over the prairie. In urban settings, the acoustic design problems of civil defense devices were fivefold: they had to be easily powered, with reliable backup; they had to have different sounds for danger, imminent peril, and all clear; the sound spectra had to be recognizably distinct from those of sirens on emergency vehicles, police cars, and storm-warning towers; the sounds had to be audible above car horns on the street and metal lathes on assembly lines, above the new electric blenders and dishwashers at home, above copying machines and electric typewriters in offices; and they could not be so loud as to deafen nearby ears.[383]

"Automatic Wardens" of various makes had been tried out during the Second World War in various weather and wind conditions, with continuous wails, repeated blasts, or fluctuating tones. The latter, as Vern Knudsen had proven, drew and kept one's attention longer than any single pitch at any decibel level, so by 1959 the guidelines in Washington, D.C., were for intervals of a five-minute single-tone blast to signal that an attack was on the way, then three-minute bursts of a rising-and-falling wail to urge laggards to take cover instantly, and a single continual tone to indicate All Clear, which was not easily distinguished from the signal for an attack in progress—as if, once Cold War turned nuclear, All Clear meant All Over. People nonetheless had to learn the meaning of these signals, which led to regular sound-drills on every main street and in schools, offices, shops, factories, and armories. Since two-tone signals yielded many high-frequency harmonics, over long distances they would get lost in a general uproar, and since most sirens attained a nearly deafening 125 db, they had to be mounted high and away from on-listeners within a hundred feet, so these soundmarks of war-anxiety were at once physically remote and geographically local, often with different tonalities in each community.[384]

Earl Gosswiller, chief designer of the tones and motors for the Federal Signal Corporation, a major manufacturer of sirens, figured that the most identifiable signal would be a slow whoop, a sound that became possible with the electronics of the 1960s, but already in the late 1950s policemen and firemen had noticed that urban drivers failed to respond to standard sirens, so began (with Gosswiller's help) to put bars of flashing colored lights atop their vehicles, following the example of visual signals used to alert the deaf. He and they also sought more distinctive droning or growling sounds whose vibrations could penetrate the rolled-up safety glass of air-conditioned or well-heated drivers listening to car radios turned up high to make out the music behind the static. Like police, ambulance, and fire sirens, civil defense sirens had to be able to penetrate the acoustically secured mobile shelter of a Chevy or Chrysler and the earplugs of Japanese transistor radios rocking around the clock with Bill Haley or all shook up with Elvis. Which then was the noise to attend to? Clouds of music from car radios? The whistle, squeal, and high-volume distortion from portable radios? Sirens wailing overhead or behind in tandem with blinking lights? Or some putative incoming enemy missile whose explosion would sound no worse (said the Science Director of the U.S. Naval Radiological Defense Lab after witnessing the detonation of the twenty-kiloton Shot Diablo) than "a garbageman dropping a lid on a can in the morning"? When the *Boston Herald* wrote at the height of the Cold War of an "invisible agent... at work in the environment, causing men to become tired, women to snap at their children, dogs to gaze through narrowed eyes at a world their wolf ancestors never knew," it expected readers to assume that the agent was fallout, but that was a trick of the newspaper lead. The agent was NOISE! "It rumbles by on the new highways, it clatters over Boston's old cobble stones... it blasts up and out of the explosion that does away with the ledge in which a fallout shelter will be built." And, like radiation, you couldn't escape it in the suburbs. "Today, in the quietest of country lanes, the chopping beat of the power lawn-mower can be heard to shatter the silence of the stillest morning."[385]

Managers at the Office of Civil and Defense Mobilization objected strenuously in 1959 to a report from the Midwest Research Institute that "sirens are inadequate and so common they are virtually useless." Since 1955, however, the Atomic Energy Commission and civil defense directors had been investigating the feasibility of an in-home warning system called NEAR, the National Emergency Alarm Repeater, to compensate for research that showed that air raid sirens went unheard by most suburbanites, farmworkers, cattlemen, hikers, and people in well-cooled, well-heated, well-insulated buildings; when the sirens were heard, they were confused with those of police or emergency vehicles, or ignored because there had been so many local and nationwide practice alerts and siren tests. The Muzak Corporation volunteered its transmission lines to send unique civil defense signals into stores, restaurants, and offices: "Muzak, the piped music system that soothes your frazzled nerves with Montovani," reported the *New York Journal-American* in 1960, "is now ready to shrill a Civil Defense warning of impending disaster," but NEAR was nearer, as the "nuclear-age counterpart of Paul Revere." The system had been tested that year in 1,067 homes and 400 offices and businesses in Charlotte, Michigan, whose NEAR-wired residents were each to send up an anchored weather balloon when they heard the repeater go off, triggered from a radar base near the North Pole. Judging from the number of balloons that rose above the rooftops, NEAR seemed feasible, pending Congressional approval of millions of dollars for signals to be transmitted through local power lines (the 2nd and 4th harmonic voltages of the 60-cycle power-frequency, induced by exciting a direct current for up to three minutes) into dedicated household devices (manufactured by the AC Spark Plug division of General Motors) that would buzz raucously at 240 or 255 Hz and were "considerably louder than an electric alarm clock," loud enough to "wake up instantly the soundest sleeper."[386]

Basically, the civil defenders were seeking an uncivil tone, very loud but not deafening, disruptive enough to get citizens to stop what they were doing but instructive enough to get them to turn on their radios for emergency advisories. The problem with the 80 db buzzing of NEAR was that background noise everywhere had grown substantially since the rides of Paul Revere. As "a voice" said during the May 1962 proceedings of the NEAR Receiver Conference at the Pentagon, "I would question whether a housewife running a vacuum cleaner or a washing machine in another room would hear this." Mr. Drake of the Falcon Alarm Company: "I, too, question whether 80 dbs in a two- or three-bedroom house is going to be sufficient.... I think there is going to be a surprising degree of loss... under all conditions prevalent in the average home." Just outside, where Revere would have been galloping, "80 dbs would be about the same noise that you would get if you were standing on the street with heavy trucks going by," estimated A. P. Miller of the Federal Civil Defense Administration. As for the tone of the buzzing, NEAR should not be acoustically confusable with telephones, doorbells, or household motors, not should it be electrically confusable by other household devices, such as the rectifiers in light dimmer switches that could emit 4th harmonics of sufficient voltage to set off a NEAR receiver, creating false alarms and, god forbid, heart attacks.[387]

NEAR came no closer to implementation than the successful test of a NEAR inverter at a Michigan power plant in 1965, and despite the Cuban Missile Crisis of 1962, most home-shelter companies went belly-up by the mid-1960s. Nor did many buy into BANSHEE, "a radioactivity detector that fits into home radios [and] makes a clicking noise that rises to a wail when the radioactive level rises dangerously." What did go banshee was the soundtrack of the final seconds of the apocalyptically ironic film *Fail-Safe* (1964), where the clicks and static of a connection to New York City resolve to one prolonged screech, the sound of a telephone melting in the heat of a nuclear explosion. And so the Federal Signal Corporation redirected its outdoor alarms to fires, natural disasters, mining explosions, severe weather warnings, and nuclear power and military base alerts. Its SiraTone EOWS (Electronic Outdoor Warning System) of 1994, advertised for use in recreation parks, shopping centers, and industrial complexes, would be "capable of producing seven different signals plus voice—

- Wail (Attack)
- Alerting Wail
- Pulsed Wail
- Steady (Alert)
- Pulsed Steady
- Alternating Steady
- Westminster Chime

—plus any one of many other auxiliary plug-in modular signals, including the temporal slow whoop" that had been recommended by Earl Gosswiller. From a hundred feet distant, EOWS speakers would register at 119 db, enough to make themselves heard in the more lyrical intervals of concerts by The Who, Metallica, My Bloody Valentine, and Manowar, which in 1984 held the record, according to *Guinness Book of World Records*, for the loudest musical performance.[388]

13. Cosmotology

Why has NEAR, never implemented, occupied these pages more tenaciously than the trans-oceanic generation/s of *musique concrète*, bebop, rock and roll, punk (The Noise Boys), grind, and heavy metal, or the decades of military-industrial campaigns urging tolerance for sonic booms, longer runways, and larger airports, all also intended to afford ostensible noise and tangible vibration a more favorable hearing?[389] Because concern over the insufficiency of sirens revealed the extent to which background noise was so indelible a part of daily life that full-on civil defense demanded the bringing of noise into the home, into the very wiring of each house: noise

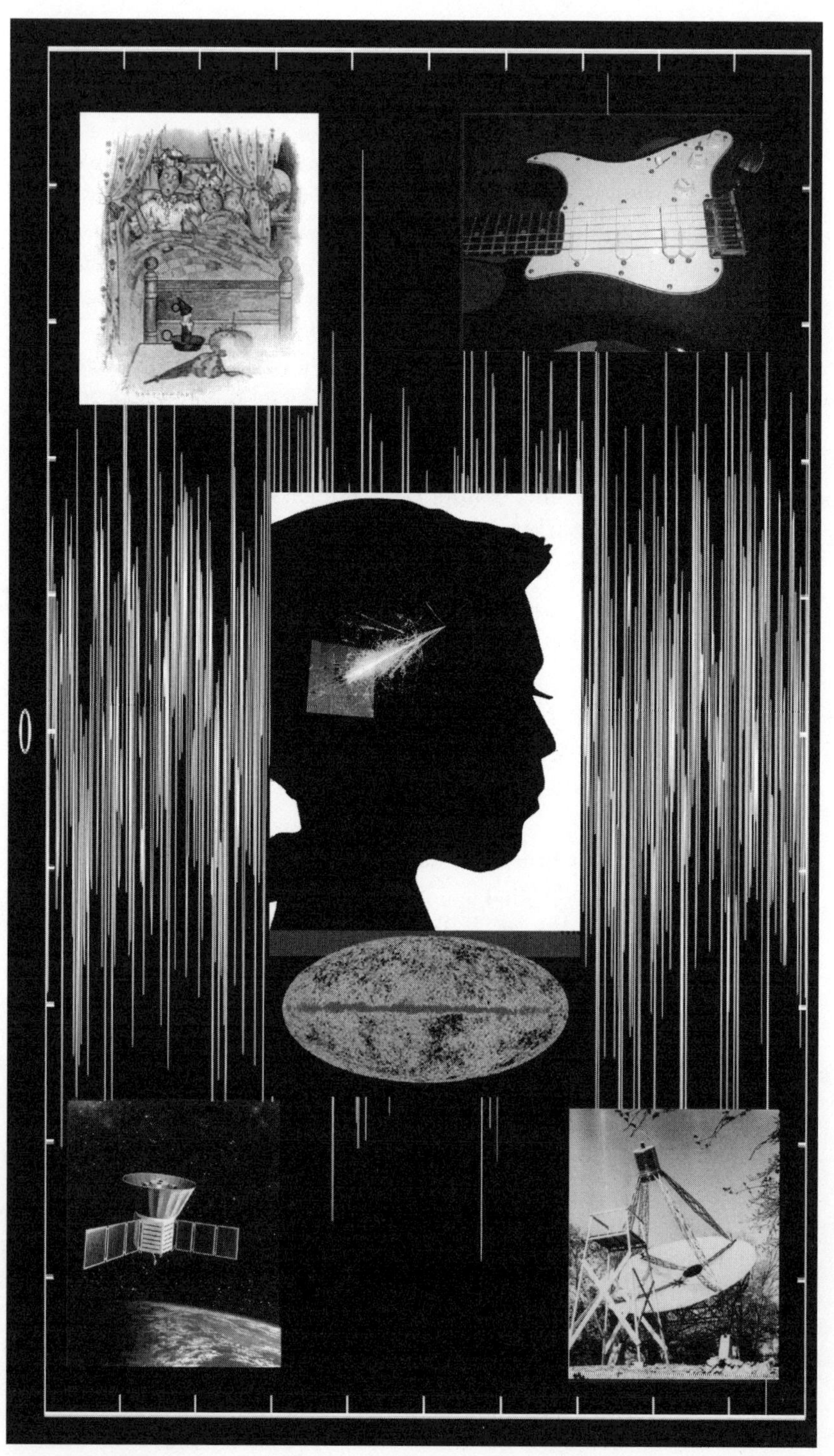

SOUNDPLATE 29

secured. Because, in technical terms, NEAR meant to restore the buzz of transmission lines that power companies had been struggling forever to tame but now were supposed to reconfigure, in the interests of a free and failsafe world: noise worth hearing, noise your life depended on. Because NEAR designers brought home, more willfully than jet fighters breaking the sound barrier and more intimately than guitarists shredding on a wired stage, the virtues of a stridency and sound-power that was built into the very dynamics of postwar technology, be it transformers, ramjets, or Fender amplifiers for Stratocaster guitars: noise enabled. Because NEAR was about the physical and cultural penetration of a noise that, to be effective, had to be as telling as it was intrusive and ubiquitous: noise enshrined.

Umpundulo surely knew what he was doing on Johannesburg afternoons when he swooped from the heavens to the earth, plumage flashing, wings thrumming, and Basil Schonland rather liked the image of a Zulu thunderbird as icon for his research into storms so loud across the corrugated iron roof of his laboratory that he and his wife Ismay could not hear each other talk as they recorded the background radiation passing through their ionization chamber. Noise itself, in terms atmospheric, atomic, visual, and acoustic, was at issue. Since the discovery of radioactivity in the 1890s, physicists had been tracking radioactive particles in and through cunning vessels. Early on, they had been troubled and intrigued by something that made its way indifferently through whatever shielding they used; Madame Curie called it "a penetrating radiation disseminated throughout the universe." After a series of balloon ascensions from a meadow in Vienna in 1911–1912, Victor Hess of the Institute of Radium Research found that this radiation rose as he did. He concluded that "there must exist a very penetrating radiation of extra-terrestrial origin which enters the atmosphere from above." Electrical engineers and radiomen during and after the Great War heard this radiation among the "atmospherics" with which they became all-too-familiar: clicks and scratches on telephone and telegraph lines; "sudden uncontrollable variations in the strength of reception"; hissing everywhere. In 1925 these obnoxious extraterrestrials were denominated "cosmic rays" by Robert Millikan, now at Cal Tech in Pasadena, who claimed in *Nature* that the "origin of the rays is everywhere in interstellar space" and vital to the health of the universe. According to the theoretical mechanics of William D. MacMillan, a former colleague at the University of Chicago, the universe maintained itself by creating atoms out of dissipated radiation; this steady-state universe little by little wound itself up as little by little it ran down, so that (wrote Harvey Lemon in 1936 in *Cosmic Rays Thus Far*), "now and then, here and there, in the great solitudes and silences of interstellar, intergalactic space, an atom, perhaps of hydrogen, is born of radiation." In this context, cosmic rays, which came in four different wavelengths, were none other than "the wail of newborn atoms," hydrogen giving rise to helium, silicon, oxygen, and iron. It appeared to Millikan in 1925 that the radiation consisted of high-energy gamma rays, hence the *rays*; later evidence suggested that it consisted of streams of charged particles, minute, diffuse, and as invisible as "a pig's squeal"—which led the English physicists James Jeans and Arthur Eddington to hear in cosmic rays (quoth Lemon) "the death rattle of dying atoms." Schonland wondered how these extraterrestrials, whether birth-cries or death-rattles, related to the ionized regions of the atmosphere that made possible long-distance radio transmissions around the planet but could also upset them, loudly and unpredictably. Some meteorologists had speculated that the ionosphere itself resulted from accumulations of free electrons ejected from sunspots with such velocity that they reached Earth and could, in times of high solar activity, disrupt radio waves; others speculated that electrons might have been freed by ultraviolet light from the sun, knocking away electrons from molecules of air. As Schonland and his wife would show in 1930, positively charged storms measurably reduced the amount of radiation coming through the ionosphere, a phenomenon best explained if the extraterrestrials consisted of other positively charged (antagonistic) particles. Elsewhere, physicists who had put cosmic-ray registers atop mountains in Panama, Peru, New Zealand, and Hawaii would confirm that cosmic rays could be cascades of positively charged particles. Where they came from remained a mystery.[390]

Sagittarius, maybe. Something as fractious, perhaps as energetic, was coming from the constellation of the Archer in the Hercules Cluster near the center of gravity of our galaxy, reported Karl G. Jansky in 1933. And anyone tuned to the WJZ radio network could hear what he had heard in Holmdel, New Jersey, through his 20.6-mHz receiver and an antenna bolted to a sort of merry-go-round that rotated automatically—and automotively, on four Model-T tires—to track "high-frequency atmospherics" hour by hour. It was a quiet noise: a hiss like steam escaping from a radiator. Son of an eminent electrical engineer, Cyril J. Jansky, and younger brother to Cyril J., Jr., who built some of the earliest commercial radio transmitters, Karl had joined the Bell Labs in 1928 upon earning a degree in physics. There he worked on radio telephony under C. R. Englund, who during the Great War had studied radio static, and Harald T. Friis, who after the war had measured field-strength and noise across a wide range of radio frequencies. Hams in the 1920s were finding that higher frequencies (>1.5 mHz) with shorter wavelengths (<200 meters) worked "surprisingly well" for long-distance conversations, though subject to interference locally from the electrical ignitions of cars and atmospherically from magnetic storms. Seeking to improve radio telephony for the conduct of transoceanic business, Friis and Englund set Jansky to profiling the atmospheric noises that disrupt short-wave communications, a serious concern when

AT&T was charging customers on one side of the Atlantic $75 to speak for three minutes to people on the other side. Aside from "first circuit" noise due to a receiver's own wiring, Jansky heard three major sources of noise: the static crash of nearby thunderstorms, like those studied by the Schonlands; the weaker, steadier static of distant storms reflecting off the Heaviside layer in the ionosphere; a persistent if variable "hiss type" static that he thought at first, in 1932, originated with activity in the sun. (Was it the noise of the sun coming up?) With more observation, and tutoring on astronomy, Jansky in July 1933 self- corrected: the maximum hiss came not from the direction of the sun but from a right ascension of 18 hours and a declination of -10 degrees on the celestial meridian. That is, from smack dab in the middle of the Milky Way.[391]

Jansky publicly denied any connection between cosmic rays and the hissing he had tracked back to stars in the Hercules Cluster; in private he was not sure whether the "hiss type" static came directly from a radio source 30,000–40,000 light years distant or was a secondary radiation caused by "primary rays of unknown character" (i.e., cosmic rays) colliding with molecules in Earth's atmosphere. Eager to publish his results, lest someone "steal the thunder from my own data," he was also eager to press forward on his discovery, which made international headlines. Sadly, once Bell Labs learned that the low-intensity hiss was untouchably galactic and too faint to upset short-wave conferences, he was assigned the more mundane task of clarifying the methodology of measuring electrical and radio static, while other Bell scientists chronicled the rumbles, splashes, surges, clicks, and crackling heard over submarine cable connections, and the hissing and frying heard on radio headsets and telegraph lines, along with whistles, tweets, hollow rustlings, chirpings, crashes, and a "swish" that sounded like "a flock of birds flying close to one's head." Some of these sounds could be explained by auroral excitements and ionospheric refraction; others were inexplicable, and Jansky in his spare time began to think that the "star static" he had heard came not from a single powerful source but from the thermal agitation of hot charged particles scattered throughout the Milky Way. He made the case to Friis that the permanence of this interference from "interstellar noise" put a practical limit on the minimum strength of radio signals that could be received by any device, which in turn should put limits on the time and money invested in improving radio receivers. Still, the interests of Bell Labs lay elsewhere, for example with the nature of electrical interference from motorboat engines. From the docks of marinas, Jansky was then shipped out during the Second World War to help the Navy locate U-boats by intercepting shortwave transmissions. He died of Bright's Disease in 1950.[392]

Astronomers had paid little notice to the work of a young man who was not one of their own and who at the time did not realize how lucky he was to have been listening in on the universe at the quietest point in the eleven-year cycle of sunspots, so that the extraterrestrial hiss had been clearly discernible and he could determine its orientation without having to compensate for disruptions by solar flares. *The New Yorker* commented wryly, "It has been demonstrated that a receiving set of great delicacy in New Jersey will get a new kind of static from the Milky Way. This is believed to be the longest distance anybody ever went to look for trouble." The *New York Times* doubted that there could be a single star or "transmitting station of such staggering dimensions and energy" to put out "the billions and billions of kilowatts" requisite to create even those feeble effects that Jansky had heard as "hiss-type" static. "And why," editorialized the *Times*, ventriloquizing for a number of orthodox astrophysicists, "should there be but one such transmitting station? All the stars radiate energy—heat, light, radio waves. It would be strange indeed if a unique type of radiation came from Hercules. But the poets will rejoice if the conclusions are verified. They will not fail to link with the Cluster that bears his name the hissing of the serpents that Hercules strangled in his cradle, or of the hydra that he slew in performing the second of his twelve labors." Only in Illinois did his (Karl Jansky's) labors find good reception, from a graduate of the Armour Institute of Technology, Grote Reber, a radio buff who used his own money, scrap lumber, car parts, and his mother's backyard to construct the world's first parabolic radiotelescope. His day-job was designing radio receivers for the Stewart Warner Company, which had begun with the manufacture of electric ignition systems, speedometers, and horns, but by the 1930s was producing radios of all kinds, including shortwave radios through which you could listen in "on the incessant alarms broadcast to cruising squad cars." At night, in a distant suburb of Chicago when and where his reception would not be muddled by ignition sparks or the shortwave reports of police cars, he listened in to the stars, and by 1939 heard the galactic radio noises that Jansky had heard seven years before. Reber did not stop there. Using his dish antenna with its high angular resolution, Reber compiled a radio-noise map of the northern sky, which had peaks of "cosmic static" in Sagittarius, the Crab Nebula, and Cassiopeia.[393]

Those peaks in Cassiopeia would turn out to be remnants of supernovae, and the extraterrestrial "noise" of cosmic rays would turn out to be of a different nature from the interstellar disturbances of "cosmic static," findings that had cosmological implications, though Reber's work, like Jansky's, was ignored until the late 1940s. Only then did the wartime radar experience of astronomers, engineers, physicists, and technicians conduce toward collaborations eventuating in new radiotelescopes, while Reber himself went on to put ever-larger parabolic dishes atop Mt. Haleakala in Maui, in Tasmania, and in Ottawa. "The cosmic shortwaves," wrote the German astrophysicist Albrecht Unsöld in 1946, "bring us neither the stockmarket nor jazz from distant worlds," as had shortwave radios. "With soft noises they rather tell the physicist of the endless love play between electrons and protons."[394]

But hssst! not so soft, as we shall hear, nor so endlessly loving. While Georges Lemaître was attending a Jesuit grammar school in Belgium, Edwin Hubble's seventh- and eighth-grade teacher in Wheaton, Illinois, was Harriet Grote, later Mrs. Schuyler Reber, mother of Grote, and in a *post hoc propter hoc* world one would enjoy this as a comedic set-up for a theory of the cosmos that was heaven-sent and hell-bent. The set-up would be braced when Lemaître soldiered through the Great War in an artillery unit while Hubble as a young astronomer squinted through the Yerkes telescope at fuzzy spiral-shaped nebulae that just might lie beyond the froth of the Milky Way. The set-up would be double-braced when Lemaître was ordained an *abbé* in 1923 while Hubble ascended to the larger telescope at Mount Wilson, through whose lenses he located a Cepheid variable star in the Andromeda Nebula, proving that star systems existed beyond our own. How far beyond depended on how one read the red shifts observed in spectrographic sightings of distant celestial objects: was this, or was this not, a Doppler shift in which color might be correlated with the distance and/or speed at which objects were moving away from Earth? Ordained a mathematical physicist in 1923–1924 at Cambridge University, ordained an astronomer at Harvard the next year, Lemaître crossed Hubble's path on the first day of 1925, when Hubble publicly announced his Cepheid. The Belgian would necessarily follow the American's work as, in 1929, Hubble formulated his namesake law, by which the (yes) Doppler shift in the spectra of stars could be used to show that "the bulk of the observable universe lies beyond our Galaxy" *and* was receding from us at speeds that increased with distance, strongly suggesting that the universe was expanding, at a (Hubble) constant rate.[395]

Expanding from what?

Cue Lemaître, who had been probing the equations of general relativity by which Einstein, with the preternatural help of a cosmological constant, had managed to keep the universe to an eternally steady state. Lemaître's calculations showed that the universe could hardly be so stable; it must be contracting or expanding. By 1927 he had opted for expansion; by 1931 he had taken the next, "repugnant," step of arguing as a good astro-mathematician and a better Jesuit that the universe had a starting point, rather more a time than a place, and more a temperature than a time. The universe had to have begun very very hot with all energy concentrated at a very dense zero-point, before time and space. *Ex nihilo*, or nearly, intense instantaneous pyrotechnics flung hydrogen nuclei and assorted particles into vast curving fields shaped by lines (or waves) of gravitational force. Over billions of years the universe had cooled, with smidgens of coalescence called galaxies and stars that were to this day speeding away from the primal event. "In the beginning of everything," said Lemaître, laying out his theory in the library of the Mount Wilson Observatory in 1933, "we had fireworks of unimaginable beauty. Then there was the explosion followed by the filling of the heavens with smoke." That smoke was an understandable metaphor from a thirty-nine-year-old who had been an artillery officer during the Great War. It made similar biographical sense for the "debris" of that explosion to be, as Lemaître proposed, cosmic rays. A fifty-four-year-old man in the audience stood up and applauded: "This is the most beautiful and satisfactory explanation of creation to which I have ever listened," said Einstein.[396]

Destruction, not Creation, and nuclear fission, not the fusion that powered the sun and must have energized "the first three minutes," damped further cosmological frictions in the run-up to the Second World War. Afterwards, the debate resumed hot and heavy as the Cambridge University astrophysicist Fred Hoyle sought to restore the credibility of a steady-state universe, which was, on a human scale, the *sine qua non* of a postwar Europe that would maintain displaced-persons camps for another twelve years. The theological dispute imbedded within the competing cosmologies—a "godless" and infinite steady-state, or an amorphously deistic Genesis with a furious Exodus of half-lives spinning out to a definite end—was also equally stark. Lemaître himself, in black robes, was not the best envoy for scientific neutrality on matters of faith, despite the even-handedness with which he admitted that his cosmology left "the materialist free to deny any transcendental Being" while perfectly "consonant with the wording of Isaiah speaking of the 'hidden God' even in the beginning." A better mediator might be George Gamow, an atheist but a grandson of the Metropolitan of the Russian Orthodox Cathedral in Odessa. Having escaped Joseph Stalin's Ukraine to teach at George Washington University, Gamow had a gift in English for making hard science accessible and an inclination, unusual for a theoretical physicist, toward organic metaphors, as in *The Birth and Death of the Sun* (1940) and his later work on the base codes of DNA. In his *One, Two, Three... Infinity* of 1947, Gamow took up first the topic of large numbers, as illustrated by ancient speculations about the age of the universe, which Brahmanic legend put at fifty-eight trillion years but which physicists confined to under twenty billion. In his last chapter he returned to "The Days of Creation" and the origin of showers of cosmic rays, which concerned him mightily as a theoretician of beginnings. The next year, with the assistance of a student, Ralph Alpher, and the connivance of Hans Bethe, an authority on cosmic rays who in this case had only to lend his last name, Gamow published on April Fools' Day the punniest of scientific letters on "The Origin of Chemical Elements," which essentially activated Lemaître's theory for observational confirmation. Steady-staters had long had the advantage of being able to point to new stars forming as old stars vanished and to embrace the ecological principles underlying homeostatic processes, on Earth or in the vicinity of Sagittarius. It had been much harder to know what *specifically* to point to, other than problematic cosmic rays, as proof of an explosive beginning, dissipative, self-annihilating, and inapproachably remote. Alpher, Bethe, and Gamow explained that the relative distribution of elements across the galaxies, at present, was a key to how the universe came to light; they also presumed the existence of a galactic background radiation as a remnant of that initially stupendous detonation of what Gamow called "ylem," matter-in-readiness, and what Lemaître called "*l'atome primitif*" (these days, a "singularity"). None of this moved Hoyle. In 1950, during one of his science programs on BBC radio, he dismissed the Lemaître-Gamow theory "that the Universe started its life a finite time ago in a single huge explosion." He found most unsatisfactory this "big bang idea."[397]

Aspersion or dramatization, the inflammatory phrase "Big Bang" smoldered a while before catching fire. It hung on not simply because it was cognate to the blare of the Big Bands of 1950. Or because the research behind the $\alpha \beta \gamma$ letter was financed by the U.S. Navy's Bureau of Ordnance. Or, to stretch the time-space continuum a bit, because J. Robert Oppenheimer had told Leo Szilard during the Manhattan Project that "The atomic bomb is shit. . . . It will make a big bang—a very big bang—but it is not a weapon which is useful in war," and had referred to that "big bang" on later occasions when he wanted to make forcibly clear that the bomb was primarily a spectacle of power, as it would be for many of the five hundred A- and H-Bomb tests conducted in the twenty years after Hiroshima.[398] The Big Bang hung on because of the cultural reach of noise.

Radioastronomy did help. In the 1950s it began to reveal how much larger the universe was than steady-staters had imagined, and how much louder, starting locally with our sun's "noise storms." Regarded as "a new index of solar activity and an important key to the vital problem of solar-terrestrial relationships," these bursts of radio noise were investigated in depth during the International Geophysical Year of 1957–1958, when the sun gave its most vivid performance in two centuries. Beyond our sun and beyond "the general noise from our own galaxy," radioastronomers also picked up peculiar noises from extragalactic clouds of gas and "screaming stars." Little of this, however, was documented or celebrated in the mass media. The Big Bang became such an audicon of astrophysics and so constant a cosmological figure in popular discourse because of the omnipresence and everyhowness of noise.[399]

Harken to the quotidian physics behind the decision in the Wisconsin case of *L. L. Olds Seed Co.* v. *Commercial Union Assurance Co., Ltd.* in 1950, the year that Hoyle first referred to the Big Bang. Aside from the firecrackers of Independence Days, Bastille Days, Boxing Days, and New Year's Eves and the weekly manhandling of garbage cans, the most familiar of bangs came from steam radiators, all of which from time to time experienced "water hammer" due to the abrupt interruption of flow when a valve somewhere in the system was shut. The question before the Seventh Circuit Court of Appeals was whether the intense water hammering that ruptured a pipe in the basement of the Olds Seed building constituted an explosion. If it did, under the extended coverage that Olds had paid for, Commercial Union must cover $25,249.01 in loss of inventory stored in the basement. The insurance company's lawyers argued that it was not an explosion but a mere (and uninsured) rupture, citing a 1942 Indiana case on the nature of explosions, which must have "a going away of material from the center" (the *ex*), a noise (the *plos*ive, as it were,), and a release of energy (the *ion*). Since only cold water had issued from the Olds Seed pipe, there could have been no release of energy; by definition cold water could not explode. The judge in the Wisconsin trial court, relying on the Indiana case, had defined an explosion as "a sudden accidental violent bursting, breaking, or expansion caused by an internal force or pressure which may be and is usually accompanied by some noise." In Indiana, however, workers had actually heard booming and rumbling before and during the explosion, so the noise had been obvious; here in Madison, the pipe had burst overnight, so the noise had to be imputed, which Commercial Union (like Fred Hoyle) would not do, lest noisiness itself be taken as proof of an event other than rupture. An hydraulics expert intimately acquainted with water hammer testified in Madison that for the basement pipe to have burst, the pressure had to have been far greater than normal and "undoubtedly noise would occur." The Appeals Court agreed: coldness was irrelevant; the pipe had given way with violence, and likely with a bang. Although "an explosion is an idea of degree," there could be no doubting, from the aftermath, that an explosion had occurred, if in the dark, unseen and unheard. Lawyers must not pretend to more abstruse mechanics. Pipes that banged with such impromptu gusto throughout ordinary days and nights under the vagaries of water hammer would surely give their acoustic· all under grievous pressure. The Big Bang was a natural.[400]

"Seal," wrote the copywriters for *britannica.com* beside a photograph of the soul singer Seal Henry Olusegun Olumide Adeola Samuel, "is thinking that whether it was the Big Bang or 'God Spoke,' the universe was definitely created by some kind of loud sound." That was in 1999, at the end of a century of internal combustion, elevated trains, and telephone crosstalk; of loudspeakers and anechoic chambers; of sonar, snapping shrimp, and whale song; of transistor radios, boom boxes, and information theory; of ultrasound monitors and mothers and fetuses listening to each other; of primal screams and quiet noisy books. It mattered little that the Big

Bang was a misnomer, that the birth of the universe could not have been audible and was rather a swift unfolding than an eruption or explosion. How else could the universe be heard to begin but with a bang? When *Sky & Telescope* sponsored a contest in 1994 to remedy the misnomer, the editors received entries from forty-one nations, including Bertha D. Universe, MOM (Mother of Matter), SOAP (Start of All Problems), BOOM (Basic Origin of Matter), You're Never Going To Get It All Back In There Again, and Topsy ("it just growed"). Among the 13,099 suggestions, ruled the judge, Cornell astronomer Carl Sagan, "There's nothing that even approaches the phrase 'Big Bang' in felicity." In 2000, when the soundtrack was composed for a "multisensory re-creation of the first moments of the universe" in the bottom half of the Hayden Sphere of Manhattan's Rose Center for Earth and Space, the Blast from the Past just had to burst out from the quiet electronic hum of ylem into cymbals, strings, and brass.[401]

Thirteen or fifteen billion years after the singularity, of course, the Big Bang is no more than a "fossil whisper," as Marcia Bartusiak has written, and the whisper would not be heard until 1963–1964, when Arno Penzias and Robert Wilson of Bell Labs in Holmdel confronted a faint, ineliminable noise in the "ultra-sensitive microwave receiving system" of their radiotelescope. First used to bounce signals off the ECHO satellite launched in 1960, their radiotelescope had been reconfigured to study microwave (7.3 cm) emissions from the Milky Way. The upsetting noise, as they determined, came neither from within their apparatus nor from atmospherics; it came from beyond the Milky Way and from all directions—unlike cosmic rays, which arrived unevenly on Earth and were deflected by its magnetic fields. At nearby Princeton, Robert H. Dicke had been on the alert for just such a noise as Penzias and Wilson had in front of them. Twenty years earlier, as a wartime physicist, Dicke had devised a radiometer that solved the problem of detecting astronomical microwave sources whose noise-power was less than that usual to the circuits of a manmade receiver; now he was listening hard for noises even quieter but persistent, uniform, and diffused through all quadrants of the heavens, for such could only be the residue of a "primordial fireball" that had unleashed the nuclear material for protogalaxies, instantaneously and in all directions, leaving behind a uniform Cosmic Microwave Background Radiation (CMBR) whose energy kept space from cooling farther than three or four degrees above absolute zero. Gamow had believed it impossible to detect this radiation from Earth, where solar storms, ionospheric turbulence, manmade radio transmitters, and millions of electric car ignitions would mask faint sounds left over from the start of time. He had not anticipated that the relic radiation would have unique spectral characteristics, as theorized by Russian astrophysicists, and he had not pursued radioastronomy techniques that might discriminate a primordial noise from all else. Nothing less than a Bang, reasoned Dicke, could account for the CMBR, even if that noise today was no great shakes along the decibel scale from the amps of a Stratocaster to the thrusters of F-104s. It was, regardless, the staticky remnant of white-hot hydrogen plasma that had suffused the neonatal universe during its first half-million years.[402]

Car alarms were much louder, and as evenly distributed on the streets as CMBR throughout the galaxies, and almost as constant (two to four hours at a go), until lawmakers put limits on their self-styled 120 db hysteria. They were so loud because they were designed in the 1970s to be heard above horns, sirens, and screeching brakes, although excited so easily and so commonly that the police came running only to shut the damn things off or tow offending cars away. Alarm manufacturers soon had to rely rather upon high-decibel otological deterrence: Hello Car Thieves. Welcome to the House of Pain. My point? The presumption of a perpetual background noise that was fixed cosmologically during the 1960s by the triumphant reach of radioastronomy had strong correlates down in/under the 'hood. The positive excitement of background noise would be fixed in 1960s poetry and theater by a gospel of spontaneity and a wide range of intoning voices; in music by amplifiers fed back upon themselves and Takehisa Kosugi scores for such "events" as breathing; in the visual arts by the Fluxdust sweepings of Robert Filliou and the post-Dada, Situationist International reinvention of "happenings" and installations; in neurology by the newly remarked *charivari* of incessantly firing synapses; and in social thought by a reconceptualizing of the crowd and mass media. Noise appeared at every interval to link macrocosm to microcosm, and not merely as "visual noise," as in the "rain" of vertical interference patterns on the picture tubes of the first postwar televisions, or the "snow" of more intemperate interference. RCA engineers had warned as early as 1956 that "Galactic noise" was "an important design consideration of VHF television tuners," and *Life Magazine* told readers, listeners, and viewers that "solar violence" could upset radio and television reception in the family room. Jocelyn Bell-Burnell of Cambridge University, sifting through four hundred feet of strip-charts from a radio-telescopic search for quasars in 1967, could not ignore "one little niggling bit" of visual noise on a quarter inch where "there was a little bit of what I call 'scruff,' which didn't look exactly like interference and didn't look exactly like [quasar] scintillation" and which proved to be the short regular radio bursts of the first identifiable pulsar (a neutron star, 200 light-years away, issuing radio bursts at 81.5 mHz). In suburban dens or the astronomic lens, background noise contributed to the sense of an expansive, expanding universe.[403]

"Practically everything unpleasant can be explained by the presence of static," claimed Roy F. Beebe in late August, 1945. With his Cosmic Ray collector he could keep "static pushed back out of the air," and in fifteen minutes he could have "every bit of static out of Long Beach," which would mean every bit of bigotry and paralysis, "but the way the city has been rawhiding me, I won't do it." To him, as to many a radio listener and electrical engineer, static was the universal background noise with which one had to contend, a limiting, prejudicial force, which Beebe conflated with the recent dreadfulness at Hiroshima and Nagasaki. By 1973, the German composer Karlheinz Stockhausen had in mind and ear Beebe's cosmic rays, but now *they* were the universal background; we, as "transistors in the literal sense," were responsible for the static. "A human being is always bombarded with cosmic rays which have a very specific rhythm and structure, and they transform his atomic structure and by that his whole system. And if we're too much involved with our personal ego, desires, interests, people, whatever it is, then these rays become focused on us as individuals with our tiny private problems. But when we more and more forget ourselves—I mean try to make ourselves pure in the state of reception and transmission—then this current passes through." This was the *ne plus ultra* of communication, and if you were in a pure listening state there were the mystical-musical *plus ultra* of shortwaves in the 9-mHz region, and "all the quasar waves." I do not mean to make light of Beebe, whom the medical establishment regarded as a charlatan but who drew several thousand Californians to his doorstep each day for boxes of cosmic-rayed wheat meal and good cosmic talk, or of Stockhausen, a richly commissioned avant-gardist who followed through on his cosmology with music astrological, astronomical, electronic, and aethereal. I mean rather to point up the cultural turn by which background noise was given mass, extension, and stability, then had its sign and spin reversed from negative to positive. Consider Rudolf Kippenhahn, director of the Max Planck Research Institute for Astrophysics in Munich, speaking around 1980: "At a lecture I gave around 1960, I asked the audience to imagine an instrument capable of transforming all the incoming radiation from space into audible sound. We would hear the constant rushing of the starlight and the radio eruptions of the sun, as well as the rushing of the radio waves." Now, twenty years later, "we would hear the heterodyne ticking of the pulsars—the low humming of the Cancer pulsar, for instance; sources of x-radiation would fire their salvoes, the source of MXB 1730–335, for instance, emitting pulses of very high energy from a spherical star cluster." All of this was fascinating in the extreme. "There is not only rushing to be heard in space, there is ticking and drumming, humming and crackling," and neutron stars were at the heart of the delightful ruckus. Like the deeps of the ocean and the amnion, the cosmos was no longer quiet.[404]

Yea, the crashing of ocean waves on the world's shores as well as cacophonies of storms atmospheric and solar and the thermal noise of starlight made it horribly difficult for physicists to detect the one cosmic phenomenon that *should* be audible: ever-elusive gravitational waves, "quakes in space-time" which, if found, "may even record the remnant rumble of the first nanosecond of creation." All of the rest of the galactic ticking, drumming, humming, and crackling is really a figment of complex instruments and how we choose to monitor, and talk about, the voids of space, which, in terms of earthly acoustics, can propagate no sound. Gravity waves would be different, supposed to come in somewhere between a low A and a high F sharp. As Bartusiak writes in *Einstein's Unfinished Symphony*, "Since their frequency happens to fall into the audio range, gravity waves will at last be adding sound to our cosmic senses, turning the silent universe into a 'talkie,' one in which we might 'hear' the thunder of colliding blackholes, or the whoosh of a collapsing star."[405]

Rarefied as the above might seem, at front porch or apartment level the density and diversity of complaints about noise in the 1950s and 1960s would perversely confirm the persistence and uniformity of a chthonic background noise. Working with the Office of Civil Defense and the League for Less Noise, New York's Committee for a Quiet City (1954–1958) was so cheered by a series of annual Noise Abatement Weeks, a clever $10,000 advertising campaign, and a "highly successful" drive to "eliminate the unnecessary sounding of automobile horns" (shades of Julia Rice) that it began a second phase directed against backfiring buses, unmuffled exhausts, and clanking manhole covers. The Committee also welcomed expressions of New Yorkers' specific noise concerns, which went far beyond the avenues, beyond "insidious harassment and slow torture by the ceaseless nightmare of noise caused mostly by trucks using the highway" or the whistling of ocean-going tugs on the East River (more shades of Rice). There were, wrote thousands of correspondents,

old-fashioned shutters that kept banging back and forth in the wind,

& "those horrible grinding noises" made by carting companies retrieving empty wooden crates,

& a drone from office buildings at 145 West 45th that was "driving us all crazy,"

& a power plant whose "grinding, clashing roar... goes on every minute of the day and night, sounding as if every demon in hell were let loose,"

& the "loud swearing" of postal workers on the night shift,

& a "shrieking soprano" who did her scales around noon every day to the barking of dogs and "men on the street yelling for her to 'shut up,'"

& burglar alarms that went off at five a.m. as owners opened their shops,

& the "terrific vibration" of air conditioning units,

& the blowing of taxi whistles by doormen outside theaters and hotels and the "ear-splitting whistle, which is equivalent to six auto horns," at Bonwit-Teller's,

& the "terrible thundering and shattering noise" of low-flying airplanes that made phone-calls impossible and caused Mrs. Richman's mother to wake up at three a.m. screaming, "We're being bombed!"

& "Radio and TV instruments blaring through walls and out of windows," degrading the home "to the status of a beer saloon where customers because of too liberal an intake have become hard of hearing,"

& the metal-stamping machines of the Angel Harp Corporation,

& the churchbells of St. Philip Neri and St. Simon in the Bronx.

"We have been asked," wrote Robert Watt of the Committee, "to stop the noises made by flying saucers—to prevent husbands from snoring—to poison the neighbor's dog—to have the tongues removed from small children—to place rubber wheels on roller skates—to outlaw air conditioners and to install fifty-foot-high loudspeakers on all of New York's skyscrapers so that sound waves could be broadcast that would dampen all other noise vibrations." But the Committee was only advisory. Though it might agree with Emil O. Boehnke that "The wages of din is death," it was at a loss to help some of the most desperate, like a woman of the Bronx, under a doctor's care for nervous tension, who suspected that her landlady was trying to get her to move out by keeping a fish tank with its "motor running and water bubbling right up against my side of the wall." And what was the Committee to do with a missive in soft blue pencil on adding machine paper from Clayton Hawks, who wrote that "noisyness is one of the prices for giving up the ways of our Forefathers who were silent, because Nature ruled that noise and impatience and clumsyness meant no food." Our Forefathers "had no automobiles or horns or shoes, or loud pavements except the native rock that was silent when their clever feet walked upon it" and were "not startled by loud and vulgar talking in unpleasant tones and tongues (because the offensive sound of foreign languages was unknown since there were no immigrants from other lands, however agreeable in their own countries)." Asked Hawks, "What would our ancestors have thought of the riveters and tools we use to construct our walls higher and higher, our skyscraper shells in which we quiver like soft snails when Nature's wind and rain, and their healthful properties, are all sealed outside?"[406]

Hawks had an eloquent, assured point, but his was the eloquence and assurance of romanticists for whom the natural world was either soft, sublime, or melodious. When Garry Moore on CBS-TV claimed that a "strange hush" had fallen on the streets of New York as a result of the Quieter City campaign and he could "hear sounds that haven't been heard in years," the example he gave was "the plaintive wail of a baby hummingbird calling to its mother... (Three-beat pause and then real loud) Hey MAAA-MAAAAA!!!" It was no longer reasonable to expect Nature to be silent or harmonic. The realm of "necessary" noise had so expanded since the era of Julia Rice that the best a person could hope for, with Edward L. Bernays, son-in-law of Freud and father of public relations, was the temporary transformation of New York "from a bedlam of noisy horns to an almost tootless city." Tootless, or toothless? "I can't imagine living in a New York City that has the serenity of a little village," wrote Jim Richards to the Committee, which filed his letter under "Screwball" and made no reply. "To me, the bleeps, honks and similar sounds are as much a part of New York City as are its world-famous buildings, theaters, restaurants, and yes, even its sidewalks. . . . They broke my heart when they did away with the El trains. Sure they were noisy but they had a bit of nostalgic romanticism to them, and their rushing Roarrrr was sweet music to my ears." Anti-Noisites in Manhattan hardly expected a return to the serenity of dirt roads, hardly expected Rice Krispies to cease its snapping, crackling, & popping or Kelloggs to introduce a new breakfast cereal called "Sh-h-h," as Johnny Carson said, whose grains "don't do anything, just lie there and stare at you." No, there would always be a residual noise in the air, as natural to humans as static was to the cosmos. "The frontiers of our universe are imbued with background noise," wrote the Hungarian (eventually Nobel) physicist Denis Gabor in 1958 in a preface to the work of two French scientists tracking *le bruit de fond* on scales microscopic, electronic, and astronomical; "this is one of the most fundamental discoveries of our century."[407]

Terrestrial background noise came to be characterized in almost cosmic terms. When a national panel on noise abatement reported in 1970 on *The Noise Around Us*, noise was described as "an environmental pollutant, a waste product, not a mass residual but an energy residual" with "a rapid decay time." This put noise in nearly the same frame of reference as the kaon and pion hadrons found in cosmic rays in 1947, the up, down, and strange quarks proposed by Gell-Mann and Zweig in 1964, and the charmed quarks added to the particle zoo in 1965 by Glashow and Bjorken. "The total amount of energy dissipated as sound throughout the earth is not really very large," noted the panel; "it is the extraordinary sensitivity of the human ear that allows this relatively small amount of energy to constitute such a problem." So, as physicists went a-hunting with exquisite instruments for short-lived particles (and new quarks called Truth and Beauty), Anti-Noisites had recourse to more sensitive and more complexly calibrated audiometers. "We don't know much about measuring noise," said Lewis Goodfriend of the Institute of Noise Control Engineers, "we're not positive of all the ramifications, but we know we've got to get rid of it, because it interferes with communication, hearing and sanity." The Panel had concluded as much: "allegations of complexity, inadequate technology, lack of conclusive proof of harm, or absence of suitable standards are not acceptable excuses for failing to implement an initial program of noise abatement and control."[408]

Every successive campaign taught Anti-Noisites a little more about the cultural physics of noise abatement. With so-bad-they're-good slogans ("Ear Today Gone Tomorrow"), nifty ambassadors (Quiet Quentin, a rabbit with earmuffs), and slam-bang weeks of publicity, quiet had a chance to catch on, as it did in Atlanta, Jacksonville, London, Paris, Philadelphia, Berlin, Stockholm, Lvov, Bombay, Buenos Aires, Manila. With concerted enforcement of ordinances, as in Memphis, Zurich, and Singapore, quiet could gather momentum. Once it had inertial direction, however, a brace of gravitational forces came into play, exerting heavy social drag and strong political friction; without an influx of new sources of energy, such as millions of dollars in workers' compensation awards for noise-induced hearing loss, quiet once more yielded to noise. The entropy, intrinsic to scientific definitions of noise, did not go unappreciated. Forced to rely on personal testimony after his sound meter was crunched by the backfire from a semi, a New York reporter in 1955 went up to the fourteenth floor of the Socony-Vacuum Building, where "15 gangs of riveters were tearing the air to little pieces." There he interviewed the foreman, who had been in construction since 1911 and believed that all the noise had been good for his nerves: after forty-four years on the job, he didn't jump at every little bang, and after thirty-eighty years of marriage, his wife had learned it would do no good to yell at him. Wives had their own problems: according to a psychiatrist and several environmental designers at the University of Wisconsin, the array of garbage disposals, high-speed blenders, vent fans, knife sharpeners, and wall exhausts in an American kitchen could make it as noisy as a boiler room. With vacuum cleaners, stereophonic record players, and a household average of two televisions (whose commercials "come on louder than the actual program," complained Edward C. Kranch), domestic din could cause "all the symptoms of combat fatigue." At the very least, said the psychiatrist, household noise so frustrates conversation that family members spend only twenty minutes a week actually talking to each other, much of this in bickering or yelling.[409] If abatement programs were to succeed, they had to compensate for a Second Law of Thermodynamics and the equations of information theory, both of which held that the universe tends toward noise and nonlinearity, at home, in the stars, and in our pitiful mimicries of stars, nuclear reactors.[410] We humans, it seemed, were less a sturdy local exception than an anxious demurrer.

14. Colors

"Woe is me!" wrote Mrs. Ruth Zoubek to the Committee for a Quiet City, about the extra-large window fans of her neighbors. "How does one go about trying to drown out the grind and groan of a neighbor's fan *all night?*" The answer was not to mask that noise with the grind and groan of all-night lovemaking. (Sexual activity declines, claimed environmental experts in Munich, when street noise rises beyond a certain decibel level.) Nor was the answer a counter-offensive, such as blasting music from speakers placed on her own window sill. ("Rest assured, our beloved

Franz Kafka would have jumped out the window" at such noise, wrote a complainant in 1972 to the Citizens for a *Quieter* City, New York City's next and most dramatic anti-noise effort, led by Robert A. Baron, a theatrical manager.) Edwin Newman, a critic for NBC News in 1970, doubted that there could be an answer for any Mrs. Zoubek: "Noise—in the cities it lies in wait, it follows you around, seeks you out, pounds at you, and reminds you that the expectation of quiet is no longer realistic." But the answer to the fans and much else was at hand, if only one embraced the shape and substance of background noise and made it one's own.[411]

Futurist, Dadaist, *concrète*, and rock composers had already done this for music, by foregrounding sounds that customarily went unnoticed or unappreciated; sound-effects and Foley artists had done this for radio and film, by locating prosaic correlates for exotic or complex sounds; acoustic ecologists and archivists of sound at the World Soundscape Project (Vancouver), the Folkways project of the Library of Congress, IRCAM (Paris), the British National Sound Library, and the Sound Space and Urban Environment Research Center (Grenoble), had done this for historians and policymakers by identifying, pursuing, preserving, and encouraging experiments with the sounds of the street, farm, factory, battlefield, bathroom, and surgical theater. We no longer accept sounds from outside our home or our control as information, argued the media analyst Tony Schwartz in 1973, so we treat that background as noise to be ignored, just as residents of Queens had ignored the screams of Kitty Genovese, "Oh my God, he stabbed me! Help me!" as she was knifed, raped, and left to die in the street on a cold night in March 1964. Schwartz himself had made his name during the 1950s by going out into the wilds of New York City with a tape recorder set to neutral, that is, without a foregrounding prejudice, to catch sound-profiles of neighborhoods. He also taught schoolchildren how to listen more skillfully, although, like Lucy Sprague Mitchell, he thought that teachers and administrators had much the greater need of such instruction.[412]

Tape recorders, which made the patching, editing, and over-dubbing of live performances faster and easier, would be exploited by John Cage, who spliced sounds and silences into flexible ribbons akin to the newly conceived double helix of genetic materials with their mutable pairings. In a review of some of the earliest electronic music compositions for tape recorder, created by Vladimir Ussachevsky and Otto Luening and played out at the Museum of Modern Art in 1952, the *New York Herald Tribune* heard "the music of fevered dreams, of sensations called back from a dim past. It is the sound of echo, the sound of tone heard through aural binoculars.... It is in the room, not yet a part of it." This uncanny feeling of an immanent detachment attracted an entirely different crowd, ghost hunters who made a fetish of the background noise on magnetic tapes, which they heard as messages from the Other Side. Inexplicable voices barely intelligible behind the scratches and static of a phonographic disk had, decades before, lured Thomas Edison toward the possibilities of contacting and recording the Absent. Similarly profuse noises on reel-to-reel tapes, whose magnetic inscription process was more sympathetically ethereal because less visibly analogous to writing or engraving, fascinated a Scandinavian portrait painter, opera singer, birdlover, archaeologist, and multilingual cosmopolitan, Friedrich Jürgenson, whose first tape-contact came in 1959 as he was outdoors recording the song of a chaffinch. On playback, which was encumbered by "a noise, vibrating like a storm," he heard a man's voice addressing him in Norwegian, which he suspected to be a message like those he had been receiving telepathically in 1957–1958 from a Central Investigation Station In Space, ever since the first Sputniks were launched. An inquiry in Swedish from his dear departed mother via a higher-end tape recorder persuaded him of a clear if polyglot channel to the Other Side, which he announced in press releases, interviews, and three books. His work was amplified by Konstantin Raudive, a Latvian disciple also multilingual, and by a growing number of researchers across the world. Listening to such tapes demanded the acoustic obsessiveness of a Harry Caul, the surveillance expert in Francis Ford Coppola's sound-saturated *The Conversation* (1974), whose plot hinged on what Caul could make of the background noise of his recordings. It took three months, wrote Raudive, for his ears to adjust to the "more rapid frequencies" and the polyglot sentences used by the (un)dead; slowing the playback speed and exerting his auditory skills as a psychiatrist, he amassed an audio library of 72,000 voices, among them a voluble Carl Jung. By the 1980s, national associations and international congresses had clustered around the study of "Transcommunication" and Electronic Voice Phenomena, which in their plentifulness seemed to prove that, as Tom and Lisa Butler wrote in 2003, *There Is No Death and There Are No Dead.* There were, instead, frequency modulation noises, drop-out noises, motor fluctuation noises, electromechanical flutters, and incomplete erasures, much of which could be silenced when a brand new tape was not run through the recording head but left to absorb the "atmospheric static noise" of the cosmos and, with it, the *Voices of Eternity.* If tape recorders were ideally neutral auditors and impartial cryers of News-from-Anywhere, then the words that came through on the playback of virgin bands of acetate or mylar had to be as disarmingly real as they were disembodied.[413]

Unsurprisingly, the plurality of voices chose to speak words of comfort, just as radiowaves from sources 180,000,000 light years distant were tra/ns/duced by Fiorella Terenzi, an Italian astrophysicist, into music that was compositionally calm when not downright meditative. Modern mediumship tended toward the ecumenical and anxiolytic, so séantists might well expect that the "noisy hissing cacophony" produced by "quick shifts in molecular and atomic energy levels in gases made hot by newly born stars" on the edge of the galaxy, and Jupiter's "[h]uge rapid sighs like an intense roaring of a distant surf... triggered by Jovian electrical activity" would be reconstituted as easy listening. While scientists became comfortable with the noise from supernovae, from solar coronas and flares, from lightning on Venus and whistlers on Jupiter, from the hissing of the magnetosphere and the telltale crackle of the ionosphere just before earthquakes, the Big Noise production company issued a "radio friendly" CD of pop folk music by Andy Kelli and the Big Bang.[414]

Elsewhere, background noise would be foregrounded by urban anthropologists, sociologists, linguists, and historians studying popular culture, political protest, everydayness ("a form of disquiet"), conversation, and race. None of this would help Ruth Zoubek; her answer came in an array of colors. Since the days of Charles Myers, experimental psychologists, engineers, otologists, and psychophysicists had been trying to understand how, and how well, people made sense of articulated sounds in the presence of noise; the increasing intensity of their research was itself a clue to the increasingly formidable presence of background noise. Psychologists had studied stimuli, attention, and fatigue, electrical engineers signal-to-noise ratios, otologists tinnitus and the cochlea, psychophysicists bandwidths and entropy. Out of their research during the Second World War came machines that did nothing but make noise, once the technical province of children, toy companies, siren designers, fireworks firms, shamans, and claques. The noise of the new machines had to meet newly rigorous specifications: it had to be noise at any decibel level, to any ear, through any volume of space, at any time. It would be noise because, by definition, it was random and carried no information; it could be heard only as itself. "White noise," this would be called, by inaccurate analogy to the optical white that is all spectra of light, and by improper descent from the "white voice" (*voce bianca*) of certain singers or passages, defined as a pure and colorless tone (that of castrati and choir boys) or an emotionless, depleted tone (in passages sung without vibrato, the tongue and jaw rigid, for depictions of despair, idiocy, or madness). White noise is patternless sound, "featureless and unpredictable at all frequencies," rather than an olla of all sounds. It is pure only in the sense that each succeeding sound is innocent of the influence of preceding sounds; it is depleted in the sense that it has no key.[415]

Black noise also has no key, but its sequence of sounds is so predictable that it is stultifying; a slave to time, it signals a loss of resilience, for resilience demands diversity.[416] Any two samples of black noise, in the statistical sense employed by physicists, economists, mathematicians, electrical engineers, and the occasional sound technician, will have a high correlation with each other; samples of white noise never correlate and are timeless because selfless: there is no self-similarity through time. Brown noise, far more correlative than white but less than black, can be galling, like grumpy window fans or the fiercely protective ventilating systems aboard space stations; at its acoustic best, brown noise yields a series of notes that "wanders up and down like a drunk weaving through an alley." The auto-correlations of red noise, as tracked by ecologists, drive a synchrony that is predictive for the extinction of a species. Orange noise comes off as citric, sour as the sound "generated by a roomfull of primary school students equipped with plastic soprano recorders," where you hear the *attempt* to play in concert. Blue and violet noise repudiate correlation—negative numbers, as it were, to the zero correlation of white noise. No color of noise can soothe the heart and mind as can low-intensity white, which is what Ruth Zoubek needed to mask the brown noise from outside and to soothe her nerves within. She needed a white-noise machine like those that, a generation later, became the mainstays of birthing chambers, nurseries, offices, and bedrooms, like the Sharper Image Heart and Sound Soother, "your own personal sound environment," with buttons for rain, ocean, brook, summer night, and white noise, "Real digital recordings. Not tape." Or maybe a CD for fussy babies, each track with "nine peaceful minutes" of "the soothing hum of a vacuum," "the constant whir of a hairdryer," "the continuous rumble of the lawnmower." Or truly Pure White Noise: "Soothe your crying, fussy baby by playing our comforting, soothing Pure White Noise® baby CD and MP3 baby audio files in baby's nursery, or wherever you hug or rock your baby. Our collection of baby soothing sounds includes Baby's First White Noise, a variety of nature sounds (ocean waves, surf, brook, rain), motor noises (vacuum cleaner, dishwasher, blow dryer, clothes dryer, washing machine, fan, car motor)." Motor noises? What an odd turn of events, that sounds once heard as obnoxious and disruptive should become a sonic backdrop marketed without qualm as "pure white noise."[417]

Qualms were more literary, as in Don DeLillo's *White Noise* of 1985, where Jack Gladney has reached "the age of unreliable menace." He must cope with the commonplace white noise of an expressway, "a remote and steady murmur around our sleep, as of dead souls babbling at the edge of a dream," then with its vaguely tactile, lingual, and darkly visual counterpart, an "airborne toxic event" that floats its emergency into Blacksmith, giving Middle American kids a taste of metal in their mouths and teachers the gift of tongues. Whiteness fades toward the generic as noise fuddles through dinnertime chaos, local static, national malaise, a universal babble, cosmic EVP. Toxicity and audicity: "What if death is nothing but sound?" his wife asks. "Electrical noise," says Jack. "You hear it forever. Sound all around. How awful," she says. "Uniform, white," Jack says. "Sometimes it sweeps over me," she says. "Sometimes it insinuates itself into my mind, little by little." Airborne and more, the toxicity may be "woven into the basic state of things," like white noise, like that "dull and unlocatable roar, as of some form of swarming life just outside the range of human apprehension." The billowing cloud causes "Heart palpitations and a sense of *déjà vu*," as do the air-raid sirens that haven't been tested for a decade and "make a territorial squawk from out of the Mesozoic," but the cloud is transient relative to the resident white of "waves and radiation," which are as preconscious and pervasive as radio and television, lust and rumor. DeLillo won the National Book Award for *White Noise*, in the same year that Bernard Malamud published a short, unassuming story, "Zora's Noise," hers too a residuum. Of the Big Bang? "They say the universe exploded and we still hear the roar and hiss of all that gas." Was that the noise she was hearing? No, it was more wretched than the music of the spheres, "an ugly thrumming sound shot through with a sickly whine" that seems to begin in the distance before entering the house through their bedroom. It is not Dworkin's irreparable snoring, not his beloved cello; it is "a vibrato hum touched with a complaining, drawn-out wail" that frightens her like a *déjà entendu* of something out of an unhappy childhood, maybe hers. Dworkin strains to hear it and the Milky Way crackles in his ears, a "great wash of cosmic static." When Zora wakes from a dizzying sleep noise-nauseous, he listens harder, hears the "whomp-whomp of machinery" from the direction of a paint factory. "No steady, prolonged, hateful, complaining noise?" No, but the newspaper is following a class-action suit against the D-R Paints Company for its "persistent harrowing noise," a "sneaky sound that goes up and down like a broken boat whistle," and soon Dworkin hears an "insistent, weird, thrumming whistle," an intrusive droning he cannot shake until D-R replaces a faulty ventilation system and all clears up. For him. For Zora the ghostly sound, "drawn out as though at the far end of someone crying," is still audible. "Suppose," she asks, "it's some distant civilization calling in, trying to get in touch with us, and for some reason I am the one person who can hear this signal, yet I can't translate the message?" "We all get signals we don't necessarily pick up," says Dworkin. The noise, he comes to hear, is the remnant noise of the house as it had been when enlivened by his departed first wife, of whom he catches a glimpse after a midnight session with his instrument under the constellation of The Cello. It is the sound of an originary universe, an earlier beginning whose residual message Zora can only hear as "the sound of my utter misery." If he and Zora are to stay together, he agrees, they must leave the house.[418]

Gallimaufry: a jumble, a confused medley. Any astrophysicist would rightly protest this gallimaufry of white noise and galactic hiss, CMBR and cumbrance. Regardless of the strictures of science, this was how cosmic rays, the Big Bang, white noise, Taos hums, gravity waves, and the songs of high-tension transmission lines were being culturally constellated in the years before and after 1984. *Le valeur moyenne d'un bruit blanc est nulle mais sa puissance moyenne est infinie*: "The mean value of white noise is nil, but its mean power is infinite." Thus Jacques Dupraz, a French theorist of communication, transforming a physicist's mathematics into a poet's calculus. Meaningfulness itself was limited by noise, everything depended on signal-to-noise ratios, and white noise was everywhere, everyhow. It was what the poet and historian of sound-motifs, John Hollander, called "this unsonorous / Reminiscence of winds that have passed among / Us." We ignore it at our peril, wrote poet Joe Basilone,

> the white noise
> that remains in black rooms
> as civilization presses on.

And the white noise *of* civilization pressing on: an unshakeable presence, sometimes acoustic, often allegorical—an indicium of loss and of "a new approach to infinite dimensional analysis." An elemental sound, and force, capable at high intensity of deafening, at low intensity of delivering a buzz.[419]

Alka-Seltzer's copywriters in the 1950s had already begun to rely as much on its audible effervescence as on the invisible action of aspirin, sodium bicarb, and citric acid that had calmed thumping headaches and grumbling stomachs for thirty years. With the Brownian motion of two white tablets dissolving in water to the score of "Plop plop fizz fizz, oh what a relief it is," white noise had its first popular audicon, but the idea of a welcome murmur went back at least to the first century, to Seneca's *Epistle* 56, written from lodgings over a Roman bathhouse. "Here am I with a babel of noise going on all about me," and yet "I swear I no more notice all this roar of noise than I do the sound of waves or falling water." He could hear the grunts of weight-lifters, their "hissings and strident gasps," the smack of hands pummeling clients on massage couches, the off-pitch singing of naked men indulging the resonance of tiled pools, the calls of professional hair removers advertising their craft and customers yowling as hairs were plucked from armpits. Nonetheless, his writing and thinking were unimpeded, which Seneca attributed in part to the fact that the noises blended together, in part to the Stoic axiom that "There can be absolute bedlam without so long as there is no commotion within." Technically, white noise *is* "absolute bedlam," sound compounded without internal logic, which frees one from attending to it, though rarely does it go without recognition: it reminds of waves, of rustling trees, of falling water. "Everything was still," wrote a reporter from the thronged pews of New York's St. Patrick's Cathedral at its dedication in May 1879, "save that irrepressible murmur that always comes from a great multitude, and resembles the rustling of forest trees or the far-off break of the waves on a sandy beach." It could be the sound of expectancy, of tremulous excitement, of the milling of crowds or subatomic particles. "The favorite noise in the experimental laboratory is wide-band continuous random noise," the MIT physicist J. C. R. Licklider told the Seventh Conference on Cybernetics in

1950. "It is like the noise of Brownian motion, like the noise inherent in a resistance, but in practice it is usually derived from the fluctuations of current in a gas tube," and it sounded like "ss-sh-h-hh-hh."[420]

Aha. White noise made its cultural way by operating ambisonically, as hush and indrawn breath, the EEG activity of someone going into deep non-REM sleep or the hum of traffic and anticipation. White noise was continuous and random, an exteriorization (just listen to those seven syllables) of noises in the brain and ears. "When the neon lights in my office go 'ngngngngngngng' all the time, I guess that is a white noise," wrote the satirist Roy Blount, Jr., in 1970. "You wouldn't think a machine would go about conquest so insidiously." No neutral entity, white noise was a player on both sides of the acoustic aisle, lullaby and lisp, "sound shield" and encryption technique. So Dan Flavin, a neon artist who clearly had no problem with the "ngngngngngngng" of his own cool-white fluorescent lights, would withdraw his work from a show at the Whitney Museum because of a sound like "the voice of a dying mouse" coming from a piece in the next room over, "Time Column—The Sound of Light." Architects who used double-paned windows, sound-absorbent panels, and acoustic ceilings to cope with the noise of typewriters and ringing phones in the open-plan offices that arose during the 1950s would by the 1970s use heating and ventilation ducts to add in white noise as "sound curtains" veiling desktop clicking and as "acoustic perfume" to give officeworkers a shared backdrop of productive, communal activity. Was the blow and hum of air conditioning units perfectly suited to the (white noise) job of masking unwanted sounds in apartments with low ceilings, weak walls, and thin windows, or was its "metallic threnody" an omen, as *Time* heard it, hanging "in the air above close-nestled, rich communities like the thrum of some giant insect infestation"? While psychiatrists and priests thought to purchase white-noise machines for their waiting rooms and nearby pews to mask the secrets revealed on the couch and in confessionals, a West German firm began test-marketing a tape-recorded system that combined the whoosh of air conditioning, the hum of heavy traffic, and the taptap of typewriters to restore an auditory vividness to air-tight office buildings. And when the secretary of the Golden Ring Club inquired of Robert A. Baron, head of New York's Citizens for a Quieter City, about a white noise machine specifically for the elderly, Baron would write back, "It is unfortunate indeed that our existence requires that we make noise to mask other noise."[421]

Jamming had been the way to think about this in wartime, as Mr. F. C. of Brooklyn had thought about it just after the war, in 1946, when he suspected that his wife had been made sick by shortwaves or radar streaming through the skies, for which he had fixed her an umbrella. "It has a piece of Christmas tinsel, like the tinsel our fliers used during the war to jam the German radar, hanging from each rib, and is grounded with a short chain hanging from the handle. My wife is not sick anymore now." During the 1950s, people began to think in terms of cancellation, nullifying an incoming waveform by creating its inverse, an anti-noise homological to science fictionists' anti-matter—and politically convenient to Cold War. As the readers of *Mechanix Illustrated* learned in 1958, "an offensive noise can be smashed to tiny, inaudible pieces by counter-waves of the equal and opposite pressure." The National Noise Abatement Council explained to supporters that same year, "When the two sound waves meet... they use up almost all of their energy fighting each other." If only Cold Warriors on all sides had taken this to heart. In any case, such "Active Noise Control" required instant analyses of incoming sounds that were formidably dynamic; this became practical only with the refinement of digital-signal processors in the late 1980s, as the Cold War was abating. By then, there was a third way to think about noise-on-noise—less confrontational, more at ease with paradox.[422]

> Query: if Cosmic Microwave Background Radiation, through the demonstrable constancy of its noise and evenness of its distribution across the heavens,[423] constrains astrophysicists to agree upon a Bang and a singular beginning;
>
> and if white noise, although resolutely random and axiomatically uninformative, has the power to calm infants and recall adults to paradises of beach and waterfall;
>
> and if feedback, hiss/screech and the wailing of hard rock, thrash, heavy metal, techno, and noise music can turn the young and restless into ticket-buying, time-scheduling, map-reading, fashion-coordinated audiences;
>
> then might there be an intrinsic role for noise other than as a lesson on the indeterminate and entropic? Might there be an affirmative, instrumental role for noise other than as alarm or expression of liveliness, collision, or disaster?

Think pink. Pink noise, or $1/f$ noise, or flicker noise, is "moderately correlated over *all* time scales and so, on the average, it should display 'interesting structure' over all time intervals," explained the astronomer John D. Barrow. That was an academic understatement. William H. Press, a Harvard astronomer, would explain with a bit more emphasis that "the main difference between the astronomer's outlook and that of people in other fields seems to be that astronomers refer to these stochastic fluctuations [e.g., flicker noise] as 'signal,' while in other fields the most common terminology is 'noise,'" so of course that signal would be "interesting." Pink noise, however, was much more than a latent or celestial signal; it was a permanent provocateur. While the power of the noise in white noise is distributed evenly over all frequencies, the power of noise in pink noise is distributed *logarithmically*. Like "natural" logarithms using the base *e* (Euler's irrational transcendental number, essential to the calculus for financial investment, biological growth, and population ecology), $1/f$ noise was suddenly found to be flickering almost everywhere that things or beings were in motion. It was in fact intrinsic to perception and judgment, to the firing and refreshing of neurons in the brain. Thinking pink is what we all do, all the time. Between 1980 and 2003, so many complex systems had been found to exhibit a $1/f$ power-spectrum that the French physicist and metrologist Michel Planat was tempted to regard $1/f$ noise "as a unifying principle in the study of time" itself, which lies "at the boundary between order and chaos"; if treated with mathematical respect, $1/f$ noise might reconcile the age-old (Western) divide between cyclic and linear time. No mere shenanigan or nickname for a shebang of shenanigans: pink noise made life feasible, gave us time to talk about it.[424]

Background noise had prepared the way, on the ground as in the galaxies. While civil defense planners and siren designers compensated for the masking and distractions of an industrialized, electrified, and widely alarmed world that rarely let up, otologists, psychologists, and sociologists had come to agree that background noise made people more alert and engaged. "Until a person loses some hearing," wrote Joseph Sataloff, M.D., and Paul Michael, Ph.D., "he can scarcely realize how important it is to hear the small background sounds around him, and to what extent these sounds help him to feel alive, and how their absence makes life seem rather dull." Someone hard of hearing need not be able to identify the source or nature of the sounds, or think them all pleasant, to benefit from them; it was the essential backgroundness that was advantageous, situating a person in place and time. (For the completely deaf, a sensitivity to substrate vibration could do much the same.) D. A. Ramsdell, an academic psychologist, had discovered the value of background noise during the war, working with deafened soldiers at the Aural Rehabilitation Center of Deshon Army Hospital in Pennsylvania. "We react to such sounds as the tick of a clock, the distant roar of traffic, vague echoes of people moving in other rooms in the house, without being aware that we do hear them," he wrote in 1947. "These incidental noises maintain our feeling of being part of a living world and contribute to our sense of being alive." A third level of hearing little explored, our unconscious attentiveness to background noise was "psychologically the most fundamental of the auditory functions," and in a sense the most primitive, for it guided us through darkness, fogs, tears, crowds, forests, and battlefields. What was more, and pink, was that "the most distinctive feature of these background sounds is that they are constantly changing," which quasi-randomness readies us to respond to our milieu, something that the designers of hip restaurants, swinging cafés, and social hotspots would take into consideration. Those who had become hard of hearing during the war were decoupled from their environment and depressed without knowing why; neither they nor their doctors had understood just how much the background noises meant, *as noise*. The cultural sociologist Richard Sennett, contrasting three centuries of public life in America, argued in 1976 that the practice of civility demanded respect for the background noise that was the turbulence of a robust, outspoken public, which must never be confused with the soundtracks of loud, brief, and spurious events as reproduced by electronic media under the thumb of late-capitalism. The latter reduced the person to a vacillating personality that pandered to the excitements of the hyperbolic instant and the twenty-hour sale; the former was a legitimating contest of voices, perhaps out of order but usually and fairly representative.[425]

Wild sound, filmmakers call it, an "atmosphere" recorded on location and dubbed back in under the dialogue of the actors, the noise of Foley artists, and the screen music. To the extent that it makes the scene seem "natural," wild sound is unnoticeable but for dramatic pauses or occasions when foreground fails to cohere with background. (Imagine Alfred Hitchcock shivering or coughing as he steps in to fit his profile to that of his motionless shadow on the wall: an issue of signal and noise, writes the poet Miriam Goodman.) Wild sound must seem as out of the hands of the recordist as it is recognizable and right, the *sonus loci* of the cinematic moment. Like a *genius loci*, and like Echo, it is pervasive yet detached. Wistfully indescribable ("you just had to be there"), it must nonetheless be graspable—a presence, like Echo, somewhat random in expression yet somehow predictable in effects.[426]

Echo returns here for good reason. Sonar operators and their friendly neighborhood physicists had noticed that the acoustic irregularities characterizing noise could arise simply from the steady transmission of echolocating pure tones. Small bubbles created in the water by these regular vibrations could generate a series of other frequencies and eventually a randomness of sound audible as white noise. The mathematics behind this, as the physicist Carl Eckart confessed in 1951, were daunting, due to the "Protean nature" of stochastic (probabilistic) phenomena, to which the "small stock of ordinary empirical equations [was] woefully inadequate." He seemed to be more stumped by the acoustic (and probabilistic) consequences of sonic cavitation than by underwater sound-scattering or cosmic radiation, other topics to which he made significant contributions. Beyond the "carefully controlled environment of the laboratory," he wrote, noise was hard to pin down, but once he had gotten a set of differential equations functioning, his successors would find that the equations could run in reverse: noise, usually pink or $1/f$ noise, could equilibrate changes of state and mitigate chaos.[427]

Itself an increasingly technical term, chaos would come to seem as "normal" to the observations of physicists, biologists, and ecologists as "a certain class of stochastic functions" whose "sufficiently common occurrence in nature," wrote Eckart, justified calling them "normal." Normal, *and* beneficial, for the Western sciences operate in the best of all plausible universes, where what is normal is presumed solely by its prevalence to be of measurable and predictable value, the abnormal by its rarity to be heuristic. If there was to be unavoidable background noise as a posterior condition of the Big Bang or an axiom of thermodynamics and information processing, then it must have some earthly good, and scientists seem genuinely happiest when they can, like alchemists, transmute the pathological into the normal. Nowhere more so than, these last decades, turning the trick of noise.[428]

Everyhow, and in every which way, by resonance. Stochastic resonance (SR): "Strictly speaking, stochastic resonance occurs in bistable systems, when a small periodic (sinusoidal) force is applied together with a large wide-band stochastic force (noise). The system response is driven by the combination of the two forces that compete/cooperate to make the system switch between the two stable states." I am quoting directly from *Wikipedia*, which ought to know, because it depends on SR to improve and grow. Non-negligible internet traffic—carping, harping, sniping, even mistyping—with regard to a new or existing entry, together with periodically added or edited text, should yield an increasingly reliable web-based global encyclopedia. When this traffic noise is minimal or maximal, entries are frozen in or out. For each entry there is "an optimal value of the noise that cooperatively concurs with the periodic forcing" in order to effect an insertion or worthwhile revision. Of course, those who launched *Wikipedia* want it to do more than oscillate; they want it to thrive, like an organism that spirals toward more complexly stable (trustworthy, comprehensive) states. That too can happen when speaking less strictly of SR, whose favorable conditions are "quantitatively determined by the matching of two time scales: the period of the Sinusoid (the deterministic time scale) and the Kramers Rate (i.e. the inverse of the average switch rate induced by the sole noise: the stochastic time scale). Thus the term 'stochastic resonance.'"[429]

Does that strike a bell? Remember paracusis, that claim of the hard of hearing to hear better in noise? Since the 1680s, aurists, audiologists, and acousticians had been debating the reality and circumstances of paracusis. Then, in 1925, a company man for Western Union had likened the phenomenon to "cases in which a more or less regular succession of vibrations is employed to jar a mechanism having a vibration of its own, to a state of higher sensitivity." That was a mechanic's version of stochastic resonance, which needed little more than a blurt of electricity to work its magic in telegraphy, microphonics, and the electromechanics of the human heart. Still controversial, paracusis may not explain how or whether the hard of hearing do hear better in noise; the point is that the plain notion of a resonant kick had been audible for some time and needed only some coloring to make itself known.[430]

Pink noise, or 1/*f* noise, seems to be the optimal noise for catalyzing phase transitions and rescuing systems out of whack. When added to a weak signal, pink noise can nudge it over a threshold crucial to awareness or stability; when introduced to a system in turmoil, pink noise can shepherd it back to homeostasis. That, I suspect, was the logic hidden behind Australian composer Percy Grainger's 1950s Cross-Grainger Double-decker Kangaroo-pouch Flying Disc Paper Graph Model for Synchronizing and Playing 8 Oscillators. From the acoustic situation of a seat in the audience, pink noise allows organisms to "hear" and respond more aptly to their environs; in physical and otological terms, it restores balance, as perhaps it does with Alan Lamb's recording of the humming of telegraph wires in the winds of the Outback. "Dithering" with pink noise can improve the intelligibility of electronic transmissions, strengthen the coherence of audio, and sometimes damp the out(r)ages of Tourette ticcing. Pink noise comes to the fore in the modeling of the fluid dynamics of sewers and crowds, of ecological catastrophe and protein interchange, of the pounding of earthquakes and hearts, of the songs of the Beatles and the riffs of a jazz clarinet. It may be the informing pattern behind a mass of serial historical data, as it was when SR was first put forward in 1981 to account for the periodicities of ice ages; it may be vital to the process driving sensory responses, as it is with the mechanoreceptors of crayfish, a phylum so ancient that biologists speculate that SR had a large role in earliest animal evolution. If not primordial, SR (through pink noise) was primal. According to neurologists and microbiologists, SR was "evolvable" for and integral to gene expression. In the long run it made possible the detection of "quiet" events or small changes in biological systems, which are "dominated by noise, or random neural firings"—as in the Brownian motion of the stereocilia of the cochlear hair cells or the fibrillations of the heart.[431]

"Fluctuations allow the different elements of the Universe to explore any state, irrespective of its degree of stability," wrote the physicists Jordi García-Ojalvo and José M. Sancho in 1999, putting the most intrepid of spins on what had been the bane of chemical, mechanical, electrical, and nuclear engineering and plasma physics. Forty years before, the Canadian physicist Donald K. C. MacDonald had allowed "that to many physicists the subject of fluctuations (or 'noise,' to put it bluntly) appears rather esoteric and perhaps pointless; spontaneous fluctuations seem nothing but an unwanted evil." Agreeing that "perhaps it is only a rather funny (funny-peculiar) type of mind that likes to deal with more or less random processes," he hoped nonetheless to persuade fellow scientists that fluctuations deserved more respect and were, if not basically good, then very useful. Forty years later, flux bore the multicolored emblem of noise as the imperial standard of adventure, invention, and manufacture, for "flux can produce new steady states that are completely absent in a deterministic scenario." Such newfound respect was anchored in a review of "noise-induced transitions" by Werner Horsthemke and René Lefever in 1984. The two physicists, situated on the opposite end of the spectrum from the continuing research on noise-induced physiological damage, had studied twenty years of experimental results in a dozen fields, starting with radio engineering, all of which demonstrated that external noise could induce "more structured behavior" in non-equilibrium systems. Through this "symbiotic relationship of order and randomness," they heard the answer to a conundrum common to twentieth-century science, philosophy, and literature: how can long-range macroscopic order "spontaneously appear and maintain itself in spite of molecular chaos and internal fluctuations"? With a nod to Darwin's suggestion that random mutations, leveraged by natural selection, were the triggers of change, Horsthemke and Lefever affirmed that noise was "omnipresent in natural systems," whose stable states were often the "creatures of noise." In this respect the Big Bang had been less a manic solo than a downbeat for galaxies continuously configured by noise, and we would do well to abandon point-to-point analyses of cosmic events in favor of stochastic "densities" akin to the density of auditory experience with its simultaneities of sounds. Noise acoustic, biological, electrical, statistical, thermodynamic, and sub-subatomic was the Eternal Gospel, the universe's way of perpetually revealing, renewing, creating.[432]

Partisans of complexity, like the chemical physicist Mark Millonas, began to appreciate "that in nonequilibrium systems noise plays an active, one might even say a creative, role in processes involving self-organization, pattern formation, and coherence, as well as in biological information processing, energy transduction, and functionality." Everywhere you turned, in physics and biophysics, chemistry and biochemistry, engineering and electronics, you happened upon noise-induced transitions and upon structures that, in the absence of noise, relax back to chaos. The passive ubiquity of noise was *passé*; now it actively sustained the qualities of complex systems. As García-Ojalvo and Sancho would write with metaphorical verve, "In convectively unstable regimes, the presence of noise seeds the system of small perturbations everywhere, and, as a consequence, spatial structures." The implications were grand: a universe seeded with, seething with noise must be one in which noise makes things solid.[433]

Atoms owed their very existence to noise, claimed the physicist Leon Cohen in 2005. Perorating in Manhattan on the centennial of the "glorious birth" of noise, Cohen explained that Einstein's analysis of Brownian motion in 1905 had established the physical reality of atoms, hypothetical entities since at least the time of Lucretius. And while vibrational noise proved atoms real, reality would prove to be substantially stochastic, such that "It would be a a dull gray world without noise." This would be beautifully illustrated in 1982 by the computer-generated color images in a book by the mathematician Benoît Mandelbrot, *The Fractal Geometry of Nature*, whose topology and morphology were generated by "scaling noises" or fractals whose self-similar "non-decreasing singular variation" accounted as thoroughly for the gustiness of winds as for the attractively jagged edges of mountain ranges. Though fond of Cantor dusts and other fine abstractions, Mandelbrot insisted that the scaling noises he had in mind were, and should be, sensible: "The term 'noise' has long become metaphorical, but every recorded fluctuation can be processed so that it is actually heard, and—even more often—actually seen on an oscilloscope or computer screen. In addition to quantitative criteria, it is essential to remind sober scientists not to despise the evidence of their senses." Sober scientists should be able to see, and hear, the $1/f$ noise underlying fluctuations in the annual flood levels of the Nile and the wobble of Earth's axis, see and hear the scaling noise of starclusters, snowflakes, coastlines. As summarized back in 1978 by Martin Gardner for the lay readership of *Scientific American*, "the changing landscape of the world (or to put it another way, the changing content of our total experience) seems to cluster around $1/f$ noise." Not only was the world literally shaped by noise; our brains required noise. Pink noise. Measured at the peripheries, the noise of the nervous system is white; in the brain, electrical fluctuations approach $1/f$. What better way to approach the "*f*ace of the deep" than as a continuing bodily engagement with the profoundness of noise, the very condition of possibility?[434]

Thus Catherine Keller's "theology of becoming," and thus Jacques Attali's more influential *Noise*. A theodicy of sound in the guise of a *Political Economy of Music*, Attali's 1977 *essai* was a delayed reflection on the failed French Revolution of 1968. What should one learn from the loud eloquence and transient alliances of factory workers, farmers, and students that had come to nought 120 years after the aborted revolutions of 1848? ("*Les murs ont des oreilles, vos oreilles ont des murs*," read a slogan scrawled across Paris in *mai 68*: "Walls have ears; your ears have walls.") Marx had drawn from the noise of '48 a lesson about the misrepresentation, and mishearing, of the masses; Attali drew from the mistakes of '68 a lesson about noise. Marx adopted the hyperbolics of sound to condemn those who claimed to act on behalf of the masses but did not understand, or feared too greatly, the power of capital, whose bluntest weapon is deafness and whose sharpest is an aura of inevitability: "Seldom had an action been announced with more noise than the impending campaign of the *Montagne*; seldom had an event been trumpeted with greater certainty or longer in advance than the inevitable victory of the democracy.... The blaring overture that announced the contest dies away in a pusillanimous snarl as soon as the struggle has to begin." Attali framed the collapse of '68 in terms of Muzak:

"Crisis is no longer a breakdown, a rupture. It is no longer dissonance in harmony, but excess in repetition, lowered efficiency in the process of the production of demand, and an explosion of violence in identity." In his political economy raw noise, residual, ineliminable, was the one force capable of twisting society in a more radical direction, provoking it into a more humane stage. For a feminist theologian (like Keller) and a would-be revolutionist (like Attali), as for an avant-garde musician, the colors of noise may be heard (thus the sound-theorist Virginia Madsen) as "the call of the wild," the call of the pure, untempered, or Dionysian that Richard Wagner had taken up as a composer-prophet in 1848: "I will destroy the order of things that turns millions into slaves of a few, and these few into slaves of their own might, own riches. I will destroy this order of things, that cuts enjoyment off from labor." But Wagner too, in defense of the tonalities he had dared, would betray the revolution while making his own and his riches; he spurned the next waves of political noise—the Communards, labor-socialists, anarchists—and fellow composers listening for new chords, untapped dissonances, machine hymns. So it would go: through noise, a new order, and with order, a remnant background that shifts restlessly from white noise to brown to pink, fulfilling—on no one's schedule, at no one's beck—prophecies of renewal.[435]

History *is* noise, noise history (Henri Atlan); without the colors of noise and their sacred violence (René Girard), there could be no *turn* of events. Attali had been reading Atlan, the French biophysicist, and Girard and Michel Serres, cultural/literary analysts, and Ilya Prigogine (the Russian-born Belgian chemist and Texan expositor of self-organizing systems); with them he heard in noise the fiercest guarantee of freedom. He was not thinking of the random composite noise that made Place Saint-Augustin in Paris louder than Niagara Falls but of expository noises, the disconcerting sounds of protest or half-articulated sounds of insurgent desire. Like rough music, which had always been an extra-legal assertion of the moral standards of rural communities. Like post-punk industrial noise bands, asserting the rights of a marginalized urban "unworking class." Or like glitch artists, who would exploit the tics of playback hardware and the errancies of recording software in the name of an ever-insolent humanity. The intentional making of noise was an ontological statement: I substantiate my historical being through the noise I can make. For the open system that is "man" (Atlan would cite rabbinic sources to this effect), as in any open system, there could be points of equilibrium but no final resting place that was not death, or the touch of Midas.[436]

15. Economies

Benoît Mandelbrot had been a market analyst in the 1960s, and Jacques Attali an economist of note who would later be appointed adviser, translator,

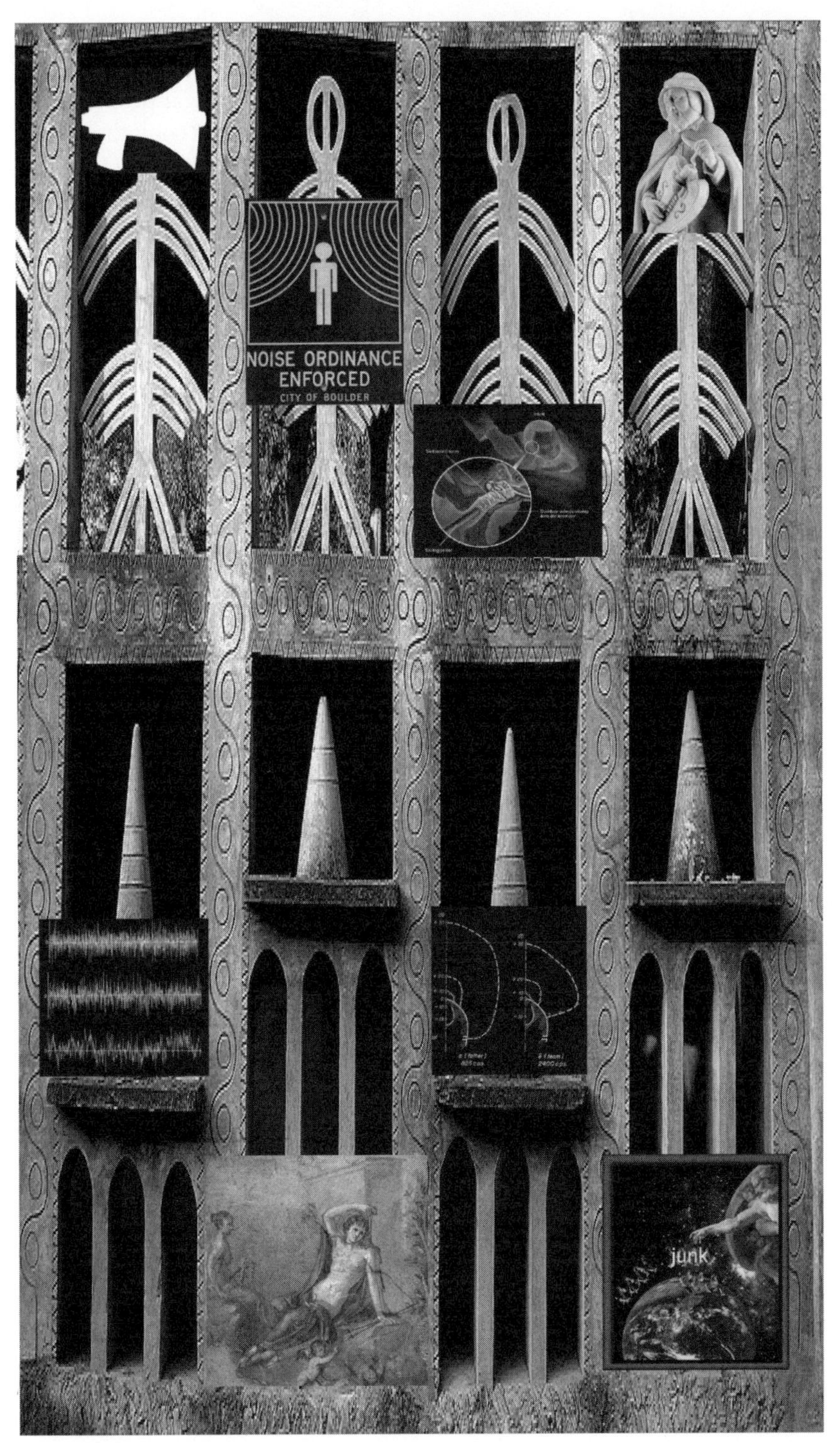

SOUNDPLATE 30

and Minister of Finance to France's socialist president, François Mitterand. After the student strikes and the massive general strike of *mai 68*, Attali had written *L'Anti-Économique* (1974), attacking the classical pessimism of an irreparable inequality in the distribution of wealth. His co-author, Marc Guillaume, would extend Attali's notions of an assertive, egalitarian noise in his own *Éloge du désordre* of 1978, which sought to tally the cost of "industrialization and its cortège of flux" in a world "devoted to excess: energy, desire, violence." Others outside of France had also begun to hear noisy currents in financial streams and shared Mandelbrot's early suspicions that economic activity in the long run did not follow a Gaussian (or Bell curve) but a Cauchy distribution that had fat tails on both ends and a peaked center ∩∧∩ that could suddenly *pîque* ∩∇∩ toward ruin. The graphics were tellingly phallic, for economic theory had long been phallocentric, disdaining the work of women as mere housewives or whores, dismissing the relevance of such "feminine" behaviors as gossip or intuitive risk-taking; economists had plotted the movements of industry and investment as if guided entirely by a rational, manly self-interest. Theorists of monetary policy and portfolio management, faced with threats of impotence that were explicit in Cauchy distributions, dashed to make noise safe for men. They figured that if they could fix noise within equations that clamped down on the ungovernable, they would not have to invoke the mathematics or the moral courage to face the pink noise of "infinite variation" and an economic system too loose at both ends to afford a reliable pair of drawstrings.[437]

Half-past one of an afternoon in 1845 and the Exchange Hall in Hamburg was at its fullest, everyone talking at once. Then, as the exchange rates were marked up, reported Ida Pfeiffer, a tourist, the "hubbub and noise of the thousands of eager voices" became "absolutely deafening." An English visitor to Chicago, reporting for the *Times* in 1887, wrote nothing about the noise of its square miles of stockyards; only when he entered the gallery of Chicago's Board of Trade did he comment on the noise—from the "swaying, squirming crowds" not of steers or hogs but "gladiators" below, fighting over futures in pork and corn and lard, who "with their calls and shouting make a deafening uproar, being a veritable Bedlam." But the noise of the trading pits, intrinsic as it was to the rush of gambling other people's money in 1887 or 1978 or 2007, masked another less audible noise increasingly constitutive of the finances of industrial, multinational capitalism. No, I'm not talking about noise-as-inefficiency, the engineer's rationale for sound-damping in factories and machine shops; or noise-as-waste, the accountant's way of pressing for quieter offices; or noise-as-fatigue, the physician's strategy for rescuing workers from their bosses; or noise-as-arousal, the modern architect's rationale for open office plans and exposed ventilation systems. Nor am I talking about economists' efforts to assign a value to quiet in proposals for pollution trade-offs, or their appeals for more government regulation because "[the absence of] noise is . . . a so-called public good," and "for noise, the market mechanism does not give people what they really want." (There is, declared a study by the National Academy of Sciences in 1988, "an inherent bias in the pricing arrangements that forces people to accept levels of noise higher than they themselves would accept.") I'm talking rather about the noise that drives economic systems with their ever-more opaque financial instruments, ever-greater trading volumes, and ever-larger institutions banking on secondary insurances, tertiary futures—a far cry from the economic history of medieval Europe through which Robert E. Lucas, Jr., had come to the mathematical analysis of business cycles and monetary policy.[438]

Son of the New Deal but a graduate student in Milton Friedman's conservative-libertarian Department of Economics at the University of Chicago, Lucas as a young professor at the Carnegie Institute produced in 1970 a paper which was as much a psychomachia as a Nobel Prize-worthy essay on "Expectations and the Neutrality of Money." The mental battle within the paper echoed the psychic and political battles without, the turmoil of the 1968 Democratic Convention in Chicago and a decade of protest against the Vietnam War, of inner cities in flames and assassinations out of the blue. It was a battle for stability in the midst of too-awful improbabilities and for confidence despite the lack, loss, or loopiness of data. What Lucas had in mind was a new model of equilibrium, based on mathematical "density functions," that could confirm "a systematic relation between the rate of change in nominal prices and the level of real output." For those to whom this may already be verging on gibberish, Lucas borrowed another economist's rhyme to frame the paradox he meant to resolve: "Money is a veil, but when the veil flutters, real output sputters." How could the abstractions of money—in the guise of bond values, currency exchanges, and interest rates whose ups and downs were keyed to independent trading momenta, hedging tactics, and liquidity—how could such abstractions directly impact the creation of goods and services? Most crucially, how might monetary policy be wielded to effect desirable changes in employment and production?[439]

Lucas's answer, in his "simple example," was to consider market equilibrium as an endlessly dynamic process rather than as a point-to-point series of plateaux, and then to solve each imbedded problem of vertigo recursively (one of Mandelbrot's favorite techniques, and favored by the "dynamic programming" models of mathematical optimization theory in the 1950s). In order to make all of this work, however, and "to keep the laws governing the transition of the economy from state to state as simple as possible," Lucas had to assume a helluva lot. Acknowledging that much was stochastic—the allocation of traders across markets, changes in the quantity of money—and that investors usually operated on incomplete, inaccurate, or untimely information, he assigned to all such fluctuations the perfect randomness of white noise; with that settled, and with a stipulated rationality on every side, he could marshall the aggregates of macroeconomics toward a triumphal arch, the usefulness of monetary policy in stabilizing a disordered or sluggish economy.

Except that one could not trust to "the rationality of each treader's behavior." The typo, which appeared in the last paragraph of Lucas's text, was the visible mistake of an invisible hand, and I take it for more than it's worth on the open market: as pink noise intruding a natural *e* that makes of traders treaders of economic waters, heads barely above the surface and unaware of—or hysterical about—riptides and storm surges. (Aye, the figures of speech grow wilder sentence by sentence, like the briefs of financial advisers during a panic.) Economists were starting to find that the recognized irrationalities of tr*e*aders, as well as the public and private errancies of information across increasingly electronic, swiftly fluctuating fronts, had a profoundly catalytic role in major markets. For that reason, the "common prior assumption" of lack of good information was not to be ignored as microeconomic pocket change, nor could one assume that ostensibly erratic trading behaviors were inevitably or preponderantly white noise. Nor could one "denoise" the market as one could now "denoise" digital images. A large, waxing contingent of "noise traders," uninformed or misinformed small investors more addicted to the act of trading than the turning of profits, could dominate and destabilize markets already cascaded by computerized brokerage algorithms and portfolio hedges. Although the Brownian activity from many investors acting with poor information may provide the liquidity essential to profitable securities trading, the greater the number of traders excited to act by nothing other than the "churning" of the market, the greater the chances that their noise will be "injected into prices" and the greater the consequent confusion about who has been trading according to reasonable extrapolation, who according to the extent of the noise. If the courts themselves were confused, for example, about what investors had a right to know, what stock-trading firms had legal cause not to tell them, and what companies could legitimately puff on behalf of their own securities, a few judicious economists with a grasp of stochastic equations might at least clarify the enduring significance of a hitherto unheralded noisiness at the heart of superheating financial worlds.[440]

"Disequilibrium" was the odd word by which economists in the 1970s began to allow for an apparent surfeit of noise. Referring ambivalently to market data that was unassimilable or incomprehensible or tardy and to market activity that was inexplicable or hard to graph, economic noise was sometimes an obstreperous hum, sometimes a soft nagging wheeze. Classical mechanisms predicted that such noise should remain in the background, a wallflower; once this noise became a player, once it moved onto the dance floor and "open outcry" of stock exchanges and up on the Big Board, it challenged all theories of "general equilibrium." Like the community of astrophysicists, economists were being compelled to incorporate within their cosmologies a forward, meaningful, and perhaps originary noise beyond the sadly traditional cycles of boom-and-bust. Disequilibrium might just be the base state of capitalist economies, a chaos kept within manageable bounds by "hidden attractors" (monopolistic practices) and stochastic processes.[441]

Call it what they would, disequilibrium, chaos, dissipative structures, Brownian flux (each with its own sorcerer's apprentice), underlying noise was the no-longer-incidental force that economists had to comprehend in order to account for changes of state. By the 1990s, changes of state provoked by $1/f$ noise were as much the province of economists as of chemists, nuclear physicists, and population ecologists, though rarely did institutional or individual treaders have the stamina or the glands to cope with the implications of noise of all colors. When Mandelbrot published *The (Mis)Behavior of Markets: A Fractal View of Risk, Ruin, and Reward* in 2004, he did not hear as clearly as he might have the imminent cataclysm, in part because the trends in minute-by-minute, day-by-day, month-by-month, and year-by-year stock and price fluctuations seemed too reassuringly similar, that is, manageable within stochastic margins defined by a self-affinity at all scales. But he did note that the variability of financial markets takes two forms, "almost-trends," or the Joseph effect, and "abrupt change," or the Noah effect, and that "Big changes often come together in rapid succession, like a fusillade of cannon fire." With Nassim Taleb, a derivatives trader and Professor of Uncertainty, Mandelbrot explained in the *Financial Times* in 2006 that sharp jumps or discontinuities were better plotted along a Cauchy distribution than a Bell curve, and that economic systems were subject to "wild randomness," where "a single observation or a particular number can impact the total in a disproportionate way," an idea previously popularized in film and fiction as "the Butterfly Effect." Wild randomness would be exaggerated in economic milieux of "extreme concentration" where "1 per cent of the US population earns close to 90 times the bottom 20 per cent [and] half the capitalisation of the market... is concentrated in less than 100 corporations." Where tails were so fat and peaks so steep, the academic Bell curve yielded nothing like "an acceptable approximation," and the managers of investment portfolios could be at much greater risk than they knew, or let on they knew. In fact, radically humanistic economists were now diagnosing the classical *Homo economicus* as autistic and calling for a "post-autistic economics" that would incorporate sociopolitical values and ethical alerts. "We live in a world primarily driven by random jumps," wrote Mandelbrot and Taleb, and some of that noise, absent the discipline of fractal self-affinity, was "truly bad news": it could jump one from a desk to a precipice.[442]

16. Memory

Lemmings do not follow each other over cliffs, although they appeared to do just that courtesy of a hidden turntable onto which a few captives had been crowded during the fabrication of a scene in *White Wilderness*, an Academy Award-winning documentary of 1958. Lemmings do not make a darkly funny splat when falling atop one another from great heights, as they do in the Amiga video game, "Lemmings." Blind mass suicide is something that has been projected onto them since the Cold War, as proxies of suppositious mass evacuations and the anxieties of genocide or nuclear Armageddon. From birth, lemmings do make ultrasonic noises we cannot hear; Amiga's sound designer pitched them cartoonishly high and shouting "Yippee!" when escaping with their lives. And lemmings do reproduce so swiftly that their populations experience chaotic cycles of boom-and-bust, declining to near-extinction after devouring all the moss in sight, then multiplying furiously once again in a four-year pattern best explained by a statistical model of "noise with memory." Memory? In quasi-technical terms, the greater the similarity in samples of a noise over time, the greater its imputed "memory" for earlier intervals. White noise has no memory of itself, which makes it remorse-less, therefore guiltlessly restful; pink noise has memory unpredictable enough to make it "interesting," like an amnesiac re-collecting her history; black noise has far too much, as if shellshocked.[443]

Remembering who we are and were, this too was a noisy process. It stood to reason. Where our cosmic origin, subatomic vibrations, and capacity to know and to grow were bound up with noise at its most elemental, how could noise not be woven into the helixes by which we are constituted, cell and bone, through time? While the music historian David L. Burrows was claiming that "sound has played a massively liberating role in human evolution, the key role in the development of what is most distinctive about humans as a species," molecular biologists were centrifuging toward a stochastic teleology of genetics, indebted to information theory and the related issues of coding and redundancy. During the 1970s, they had discovered that 90 percent or more of human DNA was "selfish" or "noncoding," an astonishing amount of "nonsense" or background noise for a reproductive system that needed redundancy but had energy limits on the carrying of freeloaders. "Junk" DNA did occasionally mix with that system, stealthily contributing to various cancers and hearing loss, but its disproportionate presence so riled biologists and popularizers that they repeatedly sounded the junk for hidden codes, topological, fractal, maybe alien, with much the same fervor as SETI investigators, like Dr. Jill

C. Tartar (model for Jodi Foster's character in the feature film *Contact*), sought *audible* signals from interstellar intelligences hidden within the noise of cosmic background radiation.[444]

Junky as was DNA, so noisy and gnarly were the processes of gene transcription, translation, and regulation. Transcription and translation were bollixed by the constant wriggling and "random birth and death" of molecules of messenger RNA. Regulation was messed up by fluctuations in biochemical reaction times within individual cells. Gene expression itself was "inherently stochastic," as demonstrated by the large variability "among identical cells in the number of protein molecules for a given gene." Still, humans and fruit flies survived from generation to regeneration, despite mutations that turn the courtship song of the male *Drosophilia melanogaster* into a "cacophony" or cause deafness in humans. Behind such resilience lay pink noise, working its magic, whose name was "homeorrhesis," a term coined in 1957 to describe a homeostatic process sustained in and by the presence of perturbations. Reformulated by Marcello Buiatti and his collaborators in animal biology and genetics at the University of Florence in 2004, homeorrhesis was "the capacity of living systems to maintain their harmonic dynamical networks through continuous change in response to external and internal noise." That response need not be resistance: organisms found ways to use some intrinsic noise to counterbalance the remaining noise, or to reorient noise as an anchor against genetic drift.[445]

Evolutionary or intracellular, robustness was definable in terms of degrees of "adaptation to environmental change and noise." I should have italicized that: *and noise*. Noise now had to be understood acoustically, neurologically, biologically, and ecologically as a force more commanding than insurgent. The body, from its amino acids to its acoustic nerves, fairly reverberated with noise, and with a "DNA music" as uncovered by composers, microbiologists, and performance artists using algorithms that transformed sequences of nucleotide base-pairings into audible notes whose patterns of adenine, thymine, guanine, and cytosine resembled pieces for baroque clavichord or an avant-garde electronic suite. The ear itself, responding to sound with (surprise, surprise) a complex nonlinearity, made noises: otologists in the 1970s proved that a healthy cochlea issues a small echo upon receipt of each acoustic stimulus of the outer ear, an "otoacoustic emission" helpful to tests for hearing loss in newborns and others equally inarticulate. Neuropathologically amplified, that cochlear echo may account for severe tinnitus, which isolates a person from the world (Ovid's Echo and Narcissus); ethically or "ecosophically" amplified, that echo partners us with the world beyond (Longus's Echo and Pan). In short, the "experience of everyday listening" could never be so meagre as a passive auditory cortex quietly registering or calmly fending off intrusions of noise. Even in the absence of provocation, electromotile proteins in the cochlea's outer-hair cells propagated mechanical energy through the middle ear, beating the eardrum from the inside out, as it were, and producing spontaneous otoacoustic emissions. We were sounding beings, noisy as all get out, ribosomes to brainstem, muscle fibers and mitral valves to basilar membranes and auditory canals. With the nanosecond reaction-times of our ears, the fastest reaction-times in the body, we heard ourselves into the cosmos.[446]

Ears reacted so quickly because their cilia had a passion to forget, as soon as humanly possible. Otherwise the never-silent world would become too much for us to manage, like the worst of concert halls, new sounds perpetually at cross-blows with old. The simultaneities of acoustic experience were rich enough without such lingering after-effects or the "jitter" of cilia at high frequencies. Since 1927, the Hungarian biophysicist Georg von Békésy had been investigating how the mind and the cochlea conspired to recruit, delay, or disregard sound; for this and related work, he won a Nobel Prize in 1961. In the years following, it would be shown that what Békésy called "sensory inhibition" was as critical to the viability of a sensory system as was acuity of reception; organisms would otherwise be flooded with sensory data. From its evolutionary origins, as Békésy noted, the human ear had had to contend with the noise of heartbeats. Its organ of Corti so sensitive that it "can almost hear the Brownian movement of molecules," the ear would have been totally prepossessed had it registered each loud pulse supplying blood to its tissues, a constant streaming of blood cells involving "a completely irregular series of impacts with the capillary walls" as well as periodic pressure variations due to the rhythms of the pumping heart. Since circulatory pulsations are very low-frequency sounds, the ear had evolved so that its sensitivity to such low frequencies from any source was thousands of times weaker than its sensitivity to higher frequencies in the range of the human voice or infant cry. One could learn from this example, and from Békésy's much earlier work on disruptions in telephone lines, an opposite and equally valuable lesson for the psychophysiology of hearing: through masking, delay, and distortion, intrasystematic noise could contribute to damping the initial reception of loud or penetrating sounds. As both a warden and contractor of thresholds, noise could "get rid of information" that had no physical imperative.[447]

Freedom of information was a shibboleth of the twentieth century, so there was a conflict, macrosocial and microbiological, otological and oh-so-legal, between the democracy of immediate audition and the meritocracy of memory. For example, did the acknowledgment of noise as a potentially invaluable and universal catalytic force entail the subordination of an implicit "right not to listen" to an explicit "right to be heard"? In 1948, in *Saia* v. *New York*, an estimable Supreme Court had dealt with this question and divided almost down the middle. Writing for the narrow majority, Justice Douglas argued that loudspeakers were "today indispensable instruments of effective public speech." The police could not arbitrarily forbid the use of sound-amplification devices in public; this would be a prior restraint of free speech. "To allow the police to bar the use of loudspeakers because their use can be abused is like burning radio receivers because they too make a noise." Justices Frankfurter, Reed, and Burton dissented: "The native power of human speech can interfere little with the self-protection of those who do not wish to listen. They may easily move beyond earshot." Mechanical or electrical amplifying devices, however, "afford easy, too easy, opportunities for aural aggression," and too easily invaded our privacy, like the loudspeakers of the Jehovah's Witnesses in question, which could be heard 600 feet away. "Surely there is not a constitutional right to force unwilling people to listen," wrote Frankfurter et al. Let none forget that "The men whose labors brought forth the Constitution of the United States had the street outside Independence Hall covered with earth so that their deliberations might not be disturbed by passing traffic." So "To the Founding Fathers it would hardly seem a proof of progress in the development of our democracy, that the blare of sound trucks must be treated as a necessary medium in the deliberative process."[448]

Privacy and publicity were, in 1948, an antique frame for debates over the necessariness of noise. Douglas, more contemporary, was worried that "annoyance at ideas" could be "cloaked in annoyance at sound," an information-theoretical position: wilfully confusing the messenger with the message. By the 1990s, the problem of noise, so far as there was a "problem," had to be dealt with inside-out. By this I mean, first, that across the board the signs of noise, positive and negative, were reversing or at least reversible. I mean also that the usefulness or usualness of noise could no longer be judged according to whether it was intrinsic or extrinsic, since noise as a recombinant, stochastic entity appeared to have shifted from the intransitive to the transitive and was carrying meanings or purposes foreign to prior generations. Only now that noise had so changed its tune could the director of a university writing program claim to be "channeling Jimi Hendrix" on behalf of better feedback while teaching English composition. Only now could ATTIK, a design and advertising collective, claim that

> Noise. exists as an all-encompassing idea.
> Noise. filters our thoughts and inspirations simultaneously through each individual to create a single, holistic experience.
> Noise. Is our process, direction, and experimentation, our creative journey—our creativity and passion. A position. Now and the future.

Ecologists were as responsible for this shift as astrophysicists or geneticists: reframing ecosystems in terms of energy flow and balance, they had to reframe noise itself as a form of energy ("In our stampede for bigger and better we have literally set the environment vibrating with misspent energy"). Anti-Noisite engineers had earlier tried something similar, damning friction noise and vibrations from poorly adjusted engines that led to metal (and mental) fatigue: "We have tried to make a virtue of necessity," wrote C. Fenno Faulkner in *Factory, The Magazine of Management* in 1927, "by adopting the insensate clamor and roar of industry as a symbol of its vigorous activity. We have used it to weave an atmosphere of adventure and romance around modern industry; whereas, in truth, noise often is merely a sign of wasted energy, of poor design, of hurried ignorance... a symbol of all that is worst in modern mass production." The difference now was that noise within large systems seemed more autonomous, as if meant to be there from the start. Occasional architects and urban planners had the perspective to contemplate "noise control" as an overarching issue in the design of spaces or places, and Herbert Marcuse would theorize environmentalism of all sorts as a struggle against capitalism to open up a nonviolent "living space" for the People. In practice, it was a matter of creating and defending private preserves of sound through local insurgency (white-noise machines, the background babble of radio or television), ambush (noise-cancellation headphones), counter-terrorism (boom boxes, boom cars, Mike Battle's Echoplex tape echo delay machine), or armored defiance (the earbuds of Walkmen or Ipods, volume at the max).[449]

Sonic preserves afforded a greater latitude of action than anechoic chambers. They were portable... and exportable. Guiding a Protestant missionary through Irayan villages in the Philippines, Modesto referred "to rain booming on the metal roof of a small bamboo chapel as 'White man's noise.'" Thinking he meant "white noise," Tom Montgomery-Fate nodded, for "The hammering rain on the metal buried any sounds from the jungle. The buzzing insects, whistling birds, and screeching monkeys were all drowned out by the white noise." Maybe that was all to the good, thought Montgomery-Fate, compelling worshippers to turn inward? But Modesto had been exact. The metal roof was not only noisier than traditional bamboo but harder to fix, and hotter, and blocked out those voices from the mountains that the Irayans listened for and remembered in all worship. A gift from a previous white missionary, the roof was the sonic enclosure that the Irayans associated with whites, who heard so little and had such a limited notion of what was worth the hearing. Montgomery-Fate composed *Beyond the White Noise* in 1997 in protest against the unthinking restrictiveness of such sonic enclosures in a multicultural world. Back home in the United States, however, where he moved on from theology to Thoreau—and where, according to the first director of the Environmental Protection Agency, William D. Ruckelshaus, environmental problems were "hitting us from every conceivable direction"—sonic enclosures were all around, at each intersection, on each interstate, courtesy of Delco Batteries. Asked by General Motors and Delco to create a soundsystem for cars, the electrical engineer Amar Bose of MIT had found to his shock that the perfect space in which to listen to music was the interior of an automobile. There alone could he be certain of listeners' physical positions vis-à-vis the speakers, the distance of ears from walls, the volume of air to be filled with sound, the frequencies of the air conditioning fans that allowed for windows to be rolled up in hot weather, and the sound-absorbent qualities of the upholstery. Ambulance sirens might not get through to listeners coaxing the best music out of Bose soundystems in 1984, but that was to be expected. The quality of the sound depended upon the insulation of passengers from wind noise, tire noise, weather, and any *sense* of emergency. And the drivers and passengers, many of them, already had higher expectations of quality because, since the debut of *Star Wars* and *Superman*, they had sat in yet other sonic enclosures, movie theaters, where the physicist Ray Dolby had taken his expertise in defuzzing the noise from long-wavelength x-ray microscopy and adapted it for a noise-reduction system that "washed" noise out of analog soundtracks by selectively boosting the volume to overcome inbred hiss and whoosh. And speaking of inbred: according to several post-modern spiritual guides we ourselves are sonic enclosures, and "The crowd within you and me that we call 'myself' is a noisy bunch."[450]

Quietude was scarcely an option, though Dolby had come to his algorithms one Sabbath in Chandigarh, quietest of India's cities, a sequestered capital of white concrete and artificial lakes where he had been sent by UNESCO to set up a scientific instrument institute. ("On the one hand the mass of people look for a decent dwelling," wrote Le Corbusier in 1923, thirty years before he designed Chandigarh, "and this question is of burning importance. On the other hand the man of initiative, of action, of thought, the LEADER, demands a shelter for his meditations in a quiet and sure spot," which would be this new capital of Punjab, far from madding crowds. Dolby found a sure way, at least, to punch up the quieter spots of recordings, without distortion, so that they did not get lost in the scuffle.)[451] Quietude had become, like punctuation on electronic screens, a desperate ambition, a gesture in the direction of an imaginary realm where a pause asks an audible breath, where waiting is never impatient, listening never forced. Gordon Hempton, a "sound tracker" and silence activist, hoped in 2009 to preserve at least "one square inch of silence": in the Hoh Rain Forest of (where else?) the Olympic Peninsula (Washington). This is not to say that Anti-Noisites abandoned the field to leaf-blowers on suburban streets or snowmobiles in Yellowstone National Park or the 90,000 annual flights of helicopters and light planes over the Grand Canyon. Hempton himself wanted a law prohibiting all aircraft from flying over the most pristine national parks.[452]

Cases could still be made for quiet, but not because it was incontestably "natural." Quiet had to be a right, a civil claim—as it was for the Right to Quiet Society in Vancouver, Canada, and the Right to Peace and Quiet Campaign (1991–1996) in the United Kingdom. Like claims for a range of other human (and animal) rights, it was asserted on behalf of what was recognizably on the verge of being lost: between 1993 and 1994, the constabulary of England and Wales fielded 131,153 complaints about noise, out of which came 372 convictions; "No peace in our time," wrote an English columnist. "Silence, in its way, is fundamental to life, the emotional equivalent of carbon," wrote the novelist Mark Slouka in *Harper's Magazine*, but "everywhere I turn I see a culture willing to deny that essential truth." It was 1999 and we all were living in "a discordant, infinite-channeled auditory landscape.... Everywhere a new song begins before the last one ends, as though to guard us against even the potential of silence." Like recurrent demands for civility, the right to quiet was presented prophetically and holistically, in the name of a grand polity and with allusion to global torture, though commonly taken to be motivated by festering personal wounds or ongoing trauma, with suicide looming. If quiet was a right, it was reactive: a right not to be bothered *any more.*[453]

17. Exit Lines

Timeless and untimely, noise is the noisiest of concepts, abundantly self-contradictory. Profligate. It compels us at every stage to reorganize, take our lives up a notch; or it does us in, deafening us to our relations, obligations. Noise must be what we were waiting for all along, an encounter with the chaotic that loosens the lug-nuts of routine. Or, grating and incessant, it sends us over the edge. Sound and unsound, something or *other.*

Bound up with bone and tissue, with solids, liquids, gases, and plasmas, with the tactile and cortical, with the chthonic and the cosmic, all those vibrations that are soundmusicnoise have been historically re-cognized, from era to era, within a cultural logic as nonlinear as the coils of the hairs of our inner ears. Distinctions between sound and noise, or noise and music, or music and sound, can only be provisional—not because they are matters of taste but because they are matters of history and histrionics: of what becomes audible through time and how the acoustics are staged, in auditoria or bedrooms, in laboratories or courtrooms. "A thousand departments of nature and of art, many a deed and passion, the scintillations of the midnight torch, and the lightnings of battle, own themselves tributary to sound," wrote a reviewer with regard to Benjamin Pierce's *Elementary Treatise on Sound* (1836). "Consider its meanderings in the walks of philosophy, in the hum of the throng, from the cathedra to the forum, from the roar of Niagara to the mystic melody of Memnon's statue, from the warbling of the nightingale to the thunder-peal. See it mingling with history, arousing the echoes of the past, and accompanying the inventor whose voice is lost in futurity."[454] Noise, as I have argued throughout, is no tributary of sound but a fellow traveler in the walks of philosophy and the hum of the throng, in chambers legislative and anechoic, in hospitals, laboratories, and prisons.

Bugbear and Bugaboo are tricksters and second-story thieves. They climb trees to get into children's rooms and scare the nightlights out of them. Parents say

> There's nothing there—a moth, a breeze,
> a house's creak, a mouse's sneeze,

but children hear the rattle of windowpanes and understand that they themselves must catch Bugbear and Bugaboo before all can be calm. For the noises creeping in at night through Rhonda Gowler Greene's *Eek! Creak! Snicker Sneak*, are a residue of the noises they have been making all day, like Henry, who when he plays is loud, when he walks is loud, and when he talks is loud. Henry is so noisy, say his parents, he could wake the dead. Which he does, according to Monica Harris. They rise grumpy from their graves, out to lay their bony hands on whoever is making all the racket. Henry meanwhile is teaching a dog to roll over and play dead. When the Dead find him, they do not roll over, and Henry's new-found, hard-pressed silence is not enough to persuade them back to eternity. In desperation he announces a sleep-over in the graveyard, at which he reads aloud to the retiring Dead a bedtime story, *Goodnight, Goon*. There we have it: hearing as first and last of the senses, first to come into its own in utero, last to vanish in extremis.[455]

Bugaboo and Bugbear will be back for us, because we are always there for them. Without noise, we would not be in the world. Without us, Bugaboo and Bugbear would have neither space nor time to prepare for the sun coming up, once again, ever so softly, with a noise "as quiet as air."

SOUNDPLATE 31

Endnotes

Due to the length of this book, the publisher has decided to make the endnotes available digitally. They can be downloaded from our web site at: *www.zonebooks.org/makingnoise* and on the web site of our distributor the MIT Press at: *www.mitpress.mit.edu/makingnoise*. A digital file may also be requested via e-mail by writing to *makingnoise@zonebooks.org*.

We apologize for any inconvenience.

Soundplates, in Fine

When a child wails in the dead of night or an alarm rousts us from deepest sleep, we may hear a sound in isolation. Ordinarily we hear congeries, sounds clumped or clustered, fragmented or undertoned. The illustrations (better: earshots) to this book are therefore collages of images that have been stretched, compressed, collided, overlain—audiconic haunts of the themes and people appearing on nearby pages. I have made two exceptions: the risibly arrowed set of earshots for BANG, which is after all an introduction to "what follows"; Soundplate 18, a test chamber for measuring the sound-absorbency of materials in near-perfect isolation.

I have drawn many of the earshots from Wikimedia Commons (wc). Color images have been variously traduced.

FRONTISPIECE (P. 2) *rightside-up alongside upside-down*
Michael Barton Miller, "aroundsound #2 (elpasoyodel)" (2000), sculpture in foamcore, wood, and steel.

SOUNDPLATE 1 (P. 16) *follow the arrows*
Marduk battling Tiâmat, Babylonian cylinder seal, wc by Ben Pirard. Harpocrates figurine in terracotta, from Myrina, ca. 100–50 BCE, in Louvre. Photo by Marie-Lan Nguyen, wc. Cicada species, from color lithograph (Wiesbaden, 1923), wc by Teacoolish. Fulton Aermore Exhaust Horn ad, *Saturday Evening Post* (May 22, 1926), 166. TDK 60-minute Audio Cassette, wc by grahamuk. George Cruikshank, "Pit, Boxes, and Gallery," etching (1836). Bain News Service, Enrico Caruso at Victrola phonograph, publicity photo (ca. 1910), restored by Michel Vuijlsteke, wc. *USS Turner Joy* (1964) and *USS Maddox* (DD731, early 1960s), with sonar graphic inset. *Turner Joy* official US Navy photo by PH1 Moen of *USS Kitty Hawk*, at www.navsource.org/archives/05/951.htm; *Maddox*, photo NH97900, www.navsource.org/archives/05/731.htm. Cover by Leonard Weisgard to Margaret Wise Brown's *The Quiet Noisy Book* (1950) by permission of Abigail Weisgard. CIA, Map of Sudan (2000) at Library of Congress, wc by Spangineer. San tribesman with audiometric headphones, by permission of Reuning Archive, Rock Art Research Institute, University of the Witwatersrand, South Africa, www.sarada.co.za.

SOUNDPLATE 2, LEFT (P. 38) *boustrophedon—right to left, then left to right*
Joseph Wright of Derby, "An Experiment on a Bird in the Air Pump," oil on canvas, (1767–68), National Gallery, London, wc by Johnbod. W. Wylde, "Manchester, from Kersal Moor," oil painting (1857), often reproduced as an engraving by Edward Goodall, "Cottonopolis," wc by Bwwm. Brownian motion, plotted. Ernst F. F. Chladni, vibration figures, *Die Akustik* (1802), wc by Basavaraj Talwar. ["Soundplates" alludes to these figures.] Calico hood worn by prisoners at Old Melbourne Gaol when outside their cells. Photo by Ciell, wc. Athanasius Kircher, plate of earbones of different species, *Musurgia Universalis* (1650), wc by Mattes. Jean-Baptiste Greuze, "Young Girl Weeping for Her Dead Bird," oil on canvas (1759), wc by Makthorpe. Domenico Bagutti (architect), Echo relief, Parc del Labirint d'Horta, Barcelona, wc by Till F. Teenck. Leonardo da Vinci, detail of "John the Baptist," oil on panel (1513–16), wc by Amandajm. John Bulwer, hand signs, from *Chirologia; Or the Natural Language of the Hand* (1644), wc by 99fresve. Gerard ter Borch the Younger, "The Reader," oil on canvas (ca. 1660), wc by Yorck Project. Lucas Cranach, "The Papal Belvedere," woodcut of peasants farting at the Pope, in Martin Luther, *Depiction of the Papacy* (1545), wc by Epiphyllum.

SOUNDPLATE 2, RIGHT (P. 39) *boustrophedon—right to left, then left to right*
Pieter Bruegel the Elder, "Tower of Babel," oil on wood (1563), wc by Gakmo.Marcantonio Raimondi, etching for Pietro Aretino's *I Modi* (1524), wc by Shakko. Athanasius Kircher, spiral listening device, *Musurgia Universalis* (1650) II, 02–03. [oval] Marcellus Laroon, "The Merry Milk Maid," *The Cryes of London* (ca. 1688), wc by Yomangani. Sandro Botticelli, "Chart of Dante's Hell," pen and ink on parchment (1480–95), wc by Maksim. "Noises for All Pockets," satirical ad, Boddery Handbill, Branch, Newton (1920s), reproduced in *This Britain* (NY, 1951). Victor the Wild Boy, frontispiece to *Der Wilde von Aveyron* (1801), wc by CopperKettle. Michelangelo, sculpture of Giuliano di Lorenzo, Medici Chapel, Sagrestia Nuova, Basilica of San Lorenzo, Florence (1524–50), online at http://it.wikipedia.org/wiki/Sagrestia_Nuova. Gottfried Hensel, language map in *Europa Polyglotta* (1730), wc by Nikola Smolenski. [oval] John Lilburne, frontispiece portrait ae 23 in his *An Answer to Nine Arguments* (1645), wc by Bkwillwm. Frances Hauksbee's "Influence Machine" or electrical generator, from his *Physico-Mechanical Experiments* (1719), wc by Gerben49. Robert Boyle's Air Pump, from *New Experiments . . . Touching the Spring of the Air* (1661), wc by Astrochemist.

SOUNDPLATE 3, LEFT (P. 88) *down left side, then up right side*
"In the Nursery Above," cartoon in *Harper's Monthly* 28 (March 1864), 574. Francisco Goya, "Qué alboroto es este? [What is this hubbub?]," Plate 65, aquatint print, from *Los Desastres de la Guerra* (1810s), wc by Sparkit. John Leech, cartoon of antagonism toward organ grinders, in his *Pictures of Life and Character* (1869). Unknown artist, "Compassion," *Our Boys and Girls* (October 1871), 692–93. Richard Wagner, signature, final bars of manuscript of "Tristan und Isolde" (1857) in Frankfurter Stadt- und Universitätsbibliothek, wc by Tjako van Schie. R. T. H. Laennec, his first stethoscope, *De l'auscultation médiate* (1819), wc by McLeod. D.-M. Lévi, drawing of Brunton otoscope, *Manuel pratique des maladies de*

l'oreille (Paris, 1885). George C. Frye Co., Cammann's Binaural Stethoscope and Knight's Binaural Stethoscope, from *Illustrated Catalog and Price List of Surgical Instruments* (Portland, ME, 1881), [oval] Charles Babbage, portrait, *Illustrated London News* (Nov 4, 1871), from photo taken at 4th Intl Statistical Congress, London, 1860, wc by Mschlindwein.

SOUNDPLATE 3, RIGHT (P. 89) *down left side, then up right side*
E. Hader, portrait of Thomas Carlyle, after photo by Sophus Williams (1870–80), Library of Congress, digital ID cph.3c34797. Floorplan of Thomas Carlyle's "soundproofed" garret study, from Reginald Blunt, *The Carlyles' Chelsea Home* (1895) p. 60. Nineteenth-century sketch of the garret. Samuel Lawrence, portrait of Jane Baillie Carlyle (before 1870), wc by Dcoetzee. Walden Pond, at site of Thoreau's cabin (1908), Library of Congress, digital ID cph.3a40169. Thoreau postage stamp, US (1967). Fire Sirens, *Harper's New Monthly Mag* 270 (Nov 1872). Screech owl, image contributed by Pearson Scott Foresman to Wikimedia, wc by PatriciaR. Pieter Huys, "A Woman Enraged," oil on panel (ca. 1560?), reproduced by permission of Cliff Schorer.

SOUNDPLATE 4 (P. 249) *as you will*
G.-F. Tournachon, known as Nadar, photo self-portrait in his hot-air balloon, wc by G.dallorto. Honoré Daumier, "Free Day at the Salon," from series, "Le Public du Salon," *Le Charivari* 10 (May 17, 1852). [overlay] Early steam whistles.

SOUNDPLATE 5 (P. 251) *as you will*
William-Adolphe Bouguereau, "La Nuit" oil on canvas of Nyx, Goddess of the Night (1883), wc by Thebrid. Hoosac Tunnel Guide cover, online at www.catskillarchive.com/rrextra/htpage.jpg. Matthew Brady, Wounded soldiers in Civil War hospital, US National Archives, 111-B-286, http://arcweb.archives.gov/arc/action/ExternalIdSearch?id=524705. [oval] Goodman of Derby, photoportrait of Florence Nightingale (1858), wc by Scewing.

SOUNDPLATE 6 (P. 276) *as you will*
TCY, photo of Greenwood Cemetery, Brooklyn (estab. 1838), wc. Mrs. Winslow's Soothing Syrup ad (1885), online at http://quackdoctor.wordpress.com/2009/01/09/mrs-winslows-soothing-syrup. New Haven Clock Company, "The Junior Tattoo" alarm clock ad, *Literary Digest* 42 (April 15, 1911), 750.

SOUNDPLATE 7 (P. 301) *as you will*
Edwin Howland Blashfield, "Christmas Bells," painted before 1907, online at http://publicdomainclip-art.blogspot.com/2009/10/three-angels-ringing-bells.html. [overlay] Tuning forks.

SOUNDPLATE 8 (P. 315) *as you will*
Alexander Graham Bell speaking into a telephone (1876), photo from Official Catalogue of the Centennial Exhibition (Philadelphia, 1876). Alexander Melville Bell, chart of Visible

Speech, fig. 1 in his *Visible Speech* (1867), wc by Primordial. [overlay] Michelson-Morley experimental apparatus, fig. 3 of their paper, "On the Relative Motion of the Earth and the Luminiferous Ether," *American Journal of Science* (Nov 1887), 337.

SOUNDPLATE 9 (P. 340) *as you will*
Chart of conditioning sequence for Ivan Pavlov's dogs, from Sarah Mae Sincero (2011), "Classical Conditioning-The Most Basic Type of Associative Learning," retrieved 6/23/2011 from Experiment Resources: www.experiment-resources.com/classical-conditioning.html. Henry Gray, location of superior temporal gyrus, center of functional hearing, fig. 726, *Anatomy of the Human Body* (1918 edition; [originally 1858]). Schematic of a Rayleigh surface acoustic wave, wc by Woudloper.

SOUNDPLATE 10 (P. 353) *as you will*
Alphonse Bertillon, schema for measuring the ear, in *Signaletic Instructions* (Chicago, 1896). Best & Company ad for "Claxton Patent Ear Cap," *Babyhood: The Mother's Nursery Guide* (March 1895). [overlay] H.E. Waite, Ear Tube, US Patent 312,308 (February 17, 1885).

SOUNDPLATE 11 (P. 363) *as you will*
Quinine bottle (mid-1800s) online at http://antiquescientifica.com/catalog8.htm. Blacksmith shop (1800s), online at www.as.wvu.edu/ihtia. Rail collision in Lagerlunda, Sweden (1875), wc by Thuresson. Scarlet Fever quarantine sign, Connecticut (late 1800s).

SOUNDPLATE 12 (P. 375) *as you will*
Commonwealth of Pennsylvania Municipal Police Officers' Education and Training Commission, Instruction form for testing hearing, SP 8-300C (April 2008), online at www.temple.edu/cjtp/pdfs/MPO_Physical.pdf. Photograph of 254 mm/50 naval gun, Imperial Russian cruiser Ryurik (July 19, 1911), wc by BezPRUzyn. [overlay, left] Waterbury Watch Co., catalog page of pocket watches (NY, 1880–90), wc by Grossus. [overlay, right] Terence Davidson, schematic of Rinné hearing test, CMO fig 1.02, by permission.

SOUNDPLATE 13 (P. 384) *as you will*
"Perils of Pauline" silent movie poster (1914), wc by lhcoyc. Titanic radio room, at www.astrosurf.com/luxorion/qsl-ham-history-titanic.htm. El Toro wooden roller coaster. Photo by Dusso Janladde, wc. Edvard Munch, "The Scream" (1893), reduction. [overlay] J.A. Miller, "Safety device for pleasure railways," US Patent 979,984 (Dec 27,1910).

SOUNDPLATE 14 (P. 392) *as you will*
Mahlon Loomis, drawing of atmospheric static (ca. 1870), in his papers, Library of Congress. Andrew Crosse, drawing of *Acarus electricus* (1837), online at http://antonyhall.wordpress.com/2007/07/20/electrical-life-andrew-crosse. Mahlon Loomis, Electrical Thermostat or Alarm Thermometer, US Patent 338,090 (March 19, 1886). Mahlon Loomis, Convertible Valise, US Patent 241,387 (May 10, 1881).

SOUNDPLATE 15 (P. 409) *counterclockwise from top right*
Annie Besant and C. W. Leadbeater, painting of a thought-form, from *The Music of Wagner—a Thought Form from Thought-Forms* (1901), wc by Haovo. [circular overlay] John Worrell Keely, illustration of vibratory energy of the ether (1890s), reproduced by Donald Simanek for his website, "The Keely Motor Company," www.lhup.edu/~dsimanek/museum/keely/keely.htm. Double-exposure publicity photo of Nikola Tesla in his Colorado Springs Laboratory amidst an apparent electrical storm (ca. 1900), wc by Yelm. [overlay, lower left] H. D. Ruhmkorff's induction coil, engraving from *Bibliothek allgemeinen und praktischen Wissens für Militäranwärter*, Bd III (Berlin, 1905), wc by Kogo. Augustin Privat Deschanel, images of Geissler Tubes, in *Elementary Treatise on Natural Philosophy, Part* 3, 13th ed. (NY, 1896), Plate 2, wc by Chetvorno. [overlay, lower right] Heinrich Hertz, wave sending/detecting apparatus, fig. 25 in his *Untersuchungen über die Ausbreitung der elektrischen Kraft* (Lepizig, 1892), wc by Joergens.mi.

INSET WITH TEXT (PP. 424–25)
Rudolf Koenig, "Manometric flame," *Acoustic Catalogue* (1865), online at www.phys.cwru.edu/ccpi/Flame_manometer.html. Wilhelm Busch, "The Virtuoso, or a New Year's Concert" (1865). Wilhelm Busch, "Fourth Trick," *Max und Moritz* (1865, scanned from a 1925 Munich edition by Robert Godwin-Jones, 1994–99), online at www.fln.vcu.edu/mm/mmeng4.html.

SOUNDPLATE 16 (P. 426) *as you will*
Bromo Seltzer Billboard in countryside, 1918–19, and New Jersey State Dept of Health, Venereal Diseases billboard (1919), both by R. C. Maxwell Outdoor Advertising Company, in R.C. Maxwell Company Collection, Database #M0047 and #M0064, Emergence of Advertising On-Line Project, John W. Hartman Center for Sales, Advertising & Marketing History, Duke University Rare Book, Manuscript, and Special Collections Library, http://library.duke.edu/digitalcollections/eaa/. [overlay] Henry Collins Brown, drawing of overhead telephone and telegraph wires on Broadway in New York City ca. 1890, *Book of Old New York* (1913), wc by JJMesserly.

SOUNDPLATE 17 (P. 445) *as you will*
[underlay] floorplan of Joseph Pulitzer house (1901–03), inlaid with facade, both from McKim, Mead & White, *A Monograph of the Work of McKim, Mead & White, 1879–1915* (Architectural Book Pub. Co., 1915–1920) plates 180 and 181. Sargent door-closer ad, *Saturday Evening Post* (Oct 24, 1925). "Escaped by Screaming," *National Police Gazette* (Dec 6, 1890). Dutch patent for revolving door (1920), based on a series of US patents by Theophilus Van Kannel, 1888–1912.

SOUNDPLATE 18 (P. 457)
W. C. Sabine conducting an absorption test in his experimental acoustics chamber, by permission of the Riverbank Laboratories.

SOUNDPLATE 19 (P. 471) *as you will*
[underlay] Western Electric, page from Klaxon brochure (ca. 1915). Klaxon ad, "All in a Day's Driving," *Saturday Evening Post* (March 17, 1923), 57. Jericho Horn ad, "Warns without offense," *McClure's Magazine* (June 1911). Physicist Arnold Sommerfeld at blackboard (ca. 1920), online at www.astro.physik.uni-potsdam.de/~afeld/einstein/einstein.html. Nicola Perscheid, photoportrait of Theodore Lessing, 1925, wc by Paulae.

SOUNDPLATE 20 (P. 485) *as you will*
[underlay] Robert D. Heinl, photo of Julia Barnett Rice at piano, "Mrs. Isaac Rice," *The American Magazine* 75 (February 1913), 32. Hamilton Williams, "The Noises of New York," *Harper's Weekly* (Nov 9, 1907), 1659. Isaac Rice playing chess, online at www.chesshistory.com/winter/extra/rice.html. Edward Sylvester Morse, photograph courtesy of the Maine Historical Society. "Quiet Hospital Zone" sign.

SOUNDPLATE 21 (P. 539) *counterclockwise from top left*
F. Moras, lith., "A Musical Party on the Fourth of July," postcard (1880s–90s), Warshaw Collection, Archives, National Museum of American History. Three-in-One Oil ad, "Silence," *American Magazine* 86 (October 1918), 130. Fourth of July Postcard (postmarked July 4, 1911); written on verso: "They are not selling fireworks in this town," Post Card Collection, Miscellany 11, E-M, by permission of Rare Book, Manuscript and Special Collections Library, Duke University, Durham, NC. John W. Grove, toy and fireworks dealer, Pittsburgh, "Wishing You a Glorious 4th of July" (postmarked June 1911), Warshaw Collection, Fireworks file.

SOUNDPLATE 22 (P. 552)
[above] Map of Torres Strait, from David R. Moore, *Arts and Crafts of Torres Strait* (Aylesbury: Shire, 1989) by permission. [inset] Wyndham Lewis, cover of *Blast* 2 (July 1915). [inset as if another island] Sigmund Freud, schema, *The Ego and the Id* (1923), wc 3/27/07 by Chaos. [below] Soldiers (one shellshocked) at an Australian Advanced Dressing Station near Ypres (1917), taken by Official War Photographer, wc 5/8/08. Group Portrait of the Torres Strait 7 and native informants, N.22900.ACH2, Museum of Anthropology, Cambridge University, by permission. Captmondo, photo of pair of steatite shabtis of the pharaoh Senkamanisken, from Nuri, pyramid No. 3 (ca. 643–623 BCE), in British Museum, wc 8/19/08.

SOUNDPLATE 23 (P. 605) *top to bottom, left to right*
Luden's cough drop ad, "Static!" *Saturday Evening Post* (March 9, 1929). J.A. Glassman, Loud Speaker as parrot, US Patent 1,662,742 (March 13, 1928). [inset] Meher Baba, digitally enhanced photo (1925), wc by Cott12. "Fill it up" cartoon, *World Telegram* (May 2, 1938), from Clark Radioana Collection, Archives, National Museum of American History. Photo of "Stop Look Listen" warning near railroad tracks, in "Automobile Scarecrow Makes Motorists Think of Safety," *National Safety News* 3 (May 1921), 8. Hiram Maxim, Gun Silencer, US Patent 9,126,885 (March 30, 1909). John B. Johnston cartoon, "El Rivetor,"

in response to Antheil's "Ballet Mécanique" at Carnegie Hall, *The World* (April 12, 1927). RCA ad for Police Radio, "Silent as the Sphinx," *Municipal Signal Engineer* (January–February 1941), 36. Loudspeakers photo, "New Discoveries," *American Weekly* (February 23, 1936). Hugo Gernsback, Radio Diagram of the Month, in *Radiocracy for Xmas–New Year, 1944* (1943), 15, in Clark Radioana Collection, as above. Smith Brothers ad, "Drop That Cough," *Literary Digest* (March 29, 1919).

SOUNDPLATE 24 (P. 654)
[top] International Labor Office, "Sound Levels in Decibels, Sones, and Phons," *Industrial Environment* (1936). Ernest J. Krause, Inc., pamphlet cover, Beverly Hills, California (1923). Janeen Hunt, graphical representation of Heisenberg's Uncertainty Principle, wc. [middle] Baffle designs for anechoic chamber (1957/58). Nu-Wood ad, "Keep it quiet with beautiful Nu-Wood," perforated ceiling tiles, *Better Homes and Gardens* (Aug 1956), 114. Simpson Forestone Acoustical Tiles ad, "Suddenly it's quiet," *Business Week* (May 7, 1960). [bottom] D. E. Wheeler and Aram Glorig, acoustic field, fig. 1 of "The Industrial Hygienist and Ear Protection," *American Industrial Hygienist Assoc Q* 16 (1955), 44. Arthur Vincent, photo of psychologist Donald Laird and assistant (his wife?) moving his noise-generating machine into place, in "Taking Noise Off the Pay Roll," *Office Economist* (1930). Meyer Berger, graphic from "It's still bedlam on the subway," *NY Times Mag* (Sept 29, 1940), 21 .

SOUNDPLATE 25 (P. 699) *clockwise from top of zodiac*
Across several sectors of zodiac: "Humpback whale, *Megaptera novaeangliae*, jumping," plate in *Brehms Tierleben* (small edition, 1927), wc by Petwoe. Cosmic ray shower (see Soundplate 29). George Wiora, sonar principles schematic (Oct 3, 2005), wc by Xorx. Roman Feiler, photo of Torún Radiotelescope (7/17/07), wc by EAJoe. ChaV Stabber Bad Fucking Noise Absolute Audio Destruction poster, wc. US Navy Integrated Undersea Surveillance (enlisted) Insignia, wc. B. Reeves (dir.), title frame from film serial, "Undersea Kingdom," with Ray "Crash" Corrigan (Republic Pictures, 1936), wc by AdamBMorgan. Yngve's extension of the "model of information transmission" presented in Claude E. Shannon and Warren Weaver, *The Mathematical Theory of Communication* (1949), rendered illegible; legible version at wc by J Steinke. NASA WMAP science team, cosmic microwave temperature fluctuations from the 5-year WMAP data seen over the full sky, from WMAP's ILC data reduction; ecliptic coordinates (March 2008); wc by Mike Peel. NOAA, spectrogram of humpback whale song, online at www.pmel.noaa.gov/vents/acoustics/whales/sounds/sounds_akhump.html. Nevit Dilmen, ultrasound scan of womb with remnants of spontaneously aborted fetus after pregnancy of fifteen weeks (2011), wc.

SOUNDPLATE 26 (P. 701) *top to bottom, left to right*
Harvey Fletcher, "Loudness Spectrum chart," *Speech and Hearing in Communication* (NY: Van Nostrand, 1953) 134. [inset] Ernst Haeckel, Lesser False Vampire Bat, Plate 67 of *Kunstformen der Natur* (1904), wc by Pengo. Joe vom Titan, pinball machine schematic: in-lane, out-lane, plunger-lane, slingshot, wc. NOAA, photo of snapping shrimp (*Synalpheus fritzmuelleri*) online at http://oceanexplorer.noaa.gov/explorations/04etta/

background/decapods/media/synalpheus.html. Hannes Grobe, Alfred Wegener Institute, air-gun array for underwater prospecting, wc. Reginald Fessenden, electric oscillator used for submarine signalling, in *Scientific American Suppl.* 2071 (Sept 11, 1915). Deep Scattering Layer graphic, wc. NOAA, Spectrogram of song of Alaska humpback whale, online at www.pmel.noaa.gov/vents/acoustics/whales/sounds/spectrograms/akhumps_128_016_0_500c.gif.

SOUNDPLATE 27 (P. 739) *top to bottom, left to right*
Walter Hodge, Picture of an Echo, soundwave reflections creating image of a skull, in "Sound control and noise elimination," *Personnel Journal* 15 (May 1936). June A. Sekera, photo of George Van Tassel's Integratron, Landers, CA (2008). [overlay] Edwin B. Cragin, MD, "The Sloane Hospital Incubator," *JAMA* 63 (Sept 12, 1914), 947. [overlay] Kieran Maher, simplified block diagram of a B-mode ultrasound scanner using a linear array transducer, wc. [underlay] Aurora Sánchez Souza and Richard Krull, "Homo Sapiens Connexin 26," first bars of a composition based on DNA-music algorithm: on a base of bass and string is the genome with different bell sounds (2nd score); on it a flute melody, at www.reviberoammicol.com/2005-22/242248.pdf. Albert A. Hopkins, ed., drawing of phonographic doll (1880–90), in *Magic: Stage Illusions and Scientific Diversions* (NY, 1897), 402. Bestwall/Certain-Teed ad, "Where does noise go?" *House Beautiful* (April 1963), 184. Mead Johnson ad, Hush image, from "Better Babies supervised by physicians," *Hygeia* 19 (1941), 161. Aunt Laura, "Hush," *The Nursery: A Monthly Magazine for Youngest Readers* 15 (1874), 83. Screaming Fetus, album cover of music group performing Comedy/Electronica/Thrash (Matt West, Daniel R., Daniel C.; New South Wales, 2007?) at www.myspace.com/screamingfetus.

SOUNDPLATE 28 (P. 795) *top to bottom, left to right*
US Civil Defense bomb shelter poster (1950s). General Radio Co., cover of *Primer of Plant-Noise Measurement* (West Concord, MA, 1969). Herman Hosmer Scott, Inc., ad for "Noise Meter Analyzer," *AIHA Quarterly* 17 (December 1956), 467. Flents' ear stopples ad, "Protect Your Workers' Hearing" (1955). [background] US Office of Civilian Defense, sound-map, "Location [and range] of present plant [factory] whistles, City of Toledo [Ohio], 1942," in National Archives, RG 171, Box 5, folder 1-31, Oct 1942, Chart II. US Army Poster, "For War Nerves, Stop Needless Noise" (ca. 1943). US War Department Poster, "Noise is our enemy too" (ca. 1943). Federal Sign and Signal Corporation ad, "Warning," *US Civil Defense Council Bulletin* 28 (June 1955), 3. Federal Sign and Signal Corporation, cover of Catalog 36, *Civil Defense and Disaster Warning Systems* (1994). Anacin ad, illustration within "What do doctors do to relieve tense nervous headaches," *Saturday Evening Post* (Nov 8, 1958).

SOUNDPLATE 29 (P. 816) *top to bottom, left to right*
[backdrop] White noise graphic. Harrison Cady, illustration to Frances H. Burnett, *The Racketty-Packetty House* (1906), "'Do you hear a noise?' said Meg." wc by FloNight. Fender Stratocaster, wc 4/14/2008 by Nettrom. Unknown artist, silhouette of Hillel Schwartz (mid-1950s). [inset within silhouette] Dinoj Surendran, Mark SubbaRao, and Randy

Landsberg of the COSMUS group, University of Chicago, with the help of physicists at the Kavli Institute for Cosmological Physics and the Pierre Auger Observatory, a simulation image of a Cosmic Ray shower—using the AIRES package by physicist Sergio Sciutto of the Universidad Nacional de La Plata, Argentina, online at http://astro.uchicago.edu/cosmus/projects/aires. NASA, bottom third of "A Brief History of Background Radiation Discovery" (2003), online at http://map.gsfc.nasa.gov/media/081031/index.html. NASA artist rendering, COBE satellite exploring background radiation (1992) online at http://www.nasa.gov/topics/universe/features/cobe_20th.html. Grote Reber's first radioastronomy dish, Wheaton, IL (1937), wc by Gothic2.

SOUNDPLATE 30 (P. 847) *left to right, top to bottom*
[underlay] Giridhar Appaji Nag Y, photo of decorated wall, Rock Garden, Chandigarh, wc. Nicolas Bouliane, vector image of a megaphone, wc. Georg von Békésy, sound field around the head during the vocalization of "father" and "team," fig. 156 from *Sensory Inhibition* (Princeton, 1967), 193. Noise Ordinance sign from Boulder, Colorado, from *Noise/News* (1972), 127. Figurine in ivory representing hearing, Dieppe (Normandy), 18th/19th century, at Museum of Fine Arts, Lyon. Photo by Marie-Lan Nguyen, wc. Diagram of use of otoacoustic emissions to assess infant hearing (2010), online at http://www.familydoctor.co.uk/info/assessing-hearing-loss, by permission of Mark Thornton. Benoît Mandelbrot and Richard L. Hudson, "Spot the trends" figure of models of price increments, *The (mis)Behavior of Markets: A Fractal View of Risk, Ruin, and Reward* (NY: Basic, 2004), 188 by permission of Aliette Mandelbrot. Dolby Laboratories, Inc., Stereo Cinema Processor schematic, *Surround Sound, Past, Present, and Future* (SF, 1994). Echo and Narcissus, fresco at Pompeii, wc by Bolo77. Junk DNA collage, online at www. galaria.blogspot.com/2004/10/extraterrestrial-junk.html.

SOUNDPLATE 31 (P. 860)
Marcellus Laroon, "Crier of windmill toys," *The Cryes of London* (ca. 1688). [overlay] Futures Wheel.

Reverb/Index

The reverb here is constrained. Figures who entered these pages for historical context, with little to say about noise, have been jettisoned. Place names and geographical features appear only when noise nodes. Titles of books, plays, paintings, operas, films, and law cases rarely appear except as identifiers. (A separate bibliography of children's books may be found online at www.zonebooks.org/makingnoise.) Contemporary scholars appear under "authorities." References to illustrations go by Soundplate number, e.g., SP1, SP3R. Roman numerals refer to Rounds One, Two, Three and appear only with regard to substantive entries in endnotes, as in II,n31. I include a few sounds that recur in the text, as sound studies in future will need to track individual acoustic phenomena (buzzing, ticking) from era to era.

Typesetting by Meighan Gale
Soundplate creation by the author
Soundplate production by Julie Fry
Printed and bound by Thompson-Shore